D1738681

MINERAL PROCESSING

Distributors

For: *Gt. Britain and the Commonwealth*

Elsevier Publishing Co. Ltd.
Barking, Essex, England

For: *U.S.A. and Canada*

American Elsevier Publishing Company Inc.
52 Vanderbilt Avenue, New York, N.Y.

For: *all other areas*

Elsevier Publishing Co.
335 Jan Van Galenstraat
P.O. Box 211, Amsterdam

MINERAL PROCESSING

by

E. J. PRYOR
A.R.S.M., D.SC., D.I.C., M.I.M.M.

formerly Reader in Mineral Dressing
University of London

THIRD EDITION

ELSEVIER PUBLISHING CO LTD
AMSTERDAM - LONDON - NEW YORK

1965

Library of Congress Catalog No. 65-28772

Printed in Great Britain
by Galliard Printers Ltd., Queen Anne's Road, Gt. Yarmouth

CONTENTS

PREFACE TO THIRD EDITION

The formidable progress in our understanding and technical development of mineral treatment in the past few years has required considerable addition to the contents of this book since its 1960 edition. Re-grouping of its material and changes in chapter planning have been used to aid consecutive development of discussion. The industrial treatment of most common ores is now concentrated in the last chapter, separate consideration of gold and coal being placed there, with the bibliography appropriately expanded. The mounting importance of chemical extraction required an extensive increase in its discussion (Chapter 15). Consideration of the fundamental forces underlying froth flotation has led to the replacement of the older three chapters by two, one dealing with flotation principles and the other with practice. This is in line with the original intention of this book—to instruct the student and mill worker as simply as is possible with this complex and wide-ranging science, and to illustrate teaching by selected samples of good practice.

In the Apocrypha (Ecclesiasticus 38; 32–34) the vital role in society of the anonymous but highly skilled technical worker may be worth considering today:

32. Without these cannot a city be inhabited: and they shall not dwell where they will, nor go up and down:

33. They shall not be sought for in publick counsel, nor sit high in the congregation: they shall not sit on the judges' seat, nor understand the sentence of judgement: they cannot declare justice and judgement; and they shall not be found where parables are spoken.

34. But they will maintain the state of the world, and [all] their desire is in the work of their craft.

E. J. PRYOR

DEVON,
September, 1965.

PREFACE TO FIRST EDITION

The purpose of this book is to set out the basic principles which underlie sound milling practice and to show their relation to standard commercial operations. The sources named in the bibliography are but a few among the number which have been studied in its preparation. Detailed acknowledgement would be as impossible as would adequate recognition of private correspondence and discussion with the teachers, research workers, consulting engineers, and milling men who have so generously shared their experience. The first draft of this book was made some twenty years ago. Revision kept step with growing field experience and accelerated from the year 1944, from which time the author enjoyed the research and testing facilities of the mineral-dressing section of the Bessemer Laboratory of the Royal School of Mines. Generous grants from the Nuffield Foundation and the Imperial College, and practical aid from industrial sources, made it possible for him to visit a wide range of laboratories and plants in America, Canada, Europe, and Africa and to study both the principles of mineral dressing and their commercial application.

The author wishes to express his indebtedness to the following publishers for permission to reproduce from the textbooks mentioned the various illustrations indicated:

American Institute of Mining and Metallurgical Engineers. From *Mining Engineering* Figs. 218 and 224. From *Mining and Metallurgy* Figs. 101 and 103. From *Coal Preparation* Figs. 110, 201, 202, 203, 207, 208 and 209.

Australasian Institute of Mining and Metallurgical Engineers. From Wark's *Principles of Flotation*, Figs. 174, 175 and 177.

Messrs. Charles Griffin & Co., Ltd. From Rose and Newman's *Metallurgy of Gold* Figs. 148, 149, 155, 157, 160 and 163.

Institution of Mining and Metallurgy. From *Transactions* 1938–39 Figs. 96 and 97. From *Transactions* 1945–46 Figs. 51 and 52, and from *Recent Developments in Mineral Dressing* Figs. 45, 57, 165, 179, 229 and 230.

McGraw-Hill Book Co., Inc. From Gaudin's *Principles of Mineral Dressing* Figs. 64, 75, 212 and 231. From Dorr and Bosqui's *Cyanidation and Concentration of Gold and Silver Ores* Figs. 74, 134, and 158. From Richards and Locke's *Text Book of Ore Dressing* Figs. 4, 36, 62, 73, 126, 129, 132, 137, 200, 204, 205, 206, 214 and 228.

Mining Publications, Ltd. From Miller's *Crushers for Stone and Ore* Figs. 25 and 27. From Rabone's *Flotation Plant Practice* Figs. 5, 22, 34, 43, 63, 72, 182, 187, 190, 191, 215 and 223.

Reinhold Publishing Corporation. From Riegel's *Chemical Machinery* Figs. 6, 50, and 213.

Messrs. John Wiley & Sons, Inc. From Taggart's *Handbook of Mineral Dressing* Figs. 7, 15, 38, 54, 61, 113, 114, 138, 186 and 233.

Messrs. Macmillan & Co., Ltd. From Truscott's *Text-Book of Ore Dressing* Figs. 8, 9, 10, 11, 12, 14, 16, 20, 28, 29, 30, 31, 32, 33, 35, 37, 59, 60, 65, 66, 67, 68, 69, 100, 109, 111, 112, 115, 119, 133, 136, 225, 226, 227 and 232.

Acknowledgement is also gratefully made to the following manufacturers in respect of illustrations of machinery taken from their catalogues and other sources:

Allis-Chalmers Manufacturing Co., American Cyanamid Co., British-Geco Engineering Co., Ltd., Denver Equipment Co., Ltd., Dings Magnetic Separator Co., Dorr-Oliver Co., Ltd., Fraser and Chalmers Engineering Works, Hadfields, Ltd., Head Wrightson & Co., Ltd., Humphreys Investment Co., International Combustion, Ltd., Kehoe-Berge Coal Co., Link-Belt Co., Mine and Smelter Supply Co., Nordberg Manufacturing Co., Rheolaveur General Construction, Ltd., Ross Engineers, Ltd., Simon-Carves, Ltd., R. O. Stokes & Co., Ltd., and Wilfley Mining Machinery Co., Ltd.

Special tribute must be paid to the memory of the author's first teacher, the late Professor S. J. Truscott, whose practical encouragement now bears fruit. Despite the precept and example of that lover of accurately used English, many ambiguities of phrase and misuses of words were brought to light when the final draft was scrutinised by various colleagues, whose critical aid has been invaluable.

EDMUND JAMES PRYOR

LONDON,
July, 1955.

CHAPTER 1

A GENERAL INTRODUCTION

This book is designed to describe the basic principles used in the concentration of specific minerals from their ores. It further deals with the development of these principles in industry and the requirements for effective process control. The earlier editions were written in connexion with the teaching of mineral technology at the Royal School of Mines in London, and were designed to be used together with the laboratory study and plant experience essential for a sound grasp of this science and its applications. This correlation of principles and practice is continued here.

Physical science is operative in the whole field of human activity. In the study of mineral dressing, which calls for special application of the laws of physics, chemistry, and electricity, the term "physical science" is here used to embrace these three categories. Its technology derives from fundamental physical laws, and operates at two levels—basic and applied. The word "technology" is here used in its sense of the combination of technical skill and economic justification, since successful field operation is mainly judged by financial criteria. "Basic technology" considers the general principles applicable to a standard type of operation, while "applied technology" connotes specially imposed variations designed to improve the treatment of one specific ore body.

A mineral deposit normally contains at least two distinct types, or species, of mineral. In the ore mined and sent to the mill these are closely interlocked. Mineral processing starts by freeing (liberating) the desired mineral or minerals, by comminution. It then separates them.

Scientific application of the principles of physical chemistry to industrial processing has led to rapid expansions in the chemical engineering industry, with which mineral dressing has much in common. Indeed, the cyanide process, used for the solution of gold from its ores, was among the earliest continuously applied chemical processes, and was developed before chemical engineering had become a profession.

The engineering crafts are called for in handling, transporting, segregating, and disposing of large tonnages of rock with smoothness and precision. Here, as with the chemical controls used in the flotation process, automation is widely used in ensuring consistent quality of material at each stage of treatment. Surface chemistry and physics, magnetic force and electrostatics are important factors in modern milling. Leaching is reinforced by the use of pressure and heat, and ion-exchange methods have been developed to deal with the dissolved products, notably in uranium recovery.

It would be impossible to cover this wide field of scientific application in one textbook. Basic description is attempted, and the bibliography has been specifically chosen to guide the reader toward a fuller treatment of his specialised interests.

No fully satisfactory term has yet emerged to describe the processing of minerals, which is also called "ore dressing", "mineral dressing", "mineral engineering" and, in the University of London degree course "mineral technology". The dressing of ores was an excellent description of the older processes which aimed to break down rock to appropriate sizes, grade it, and separate the heavy fraction from the light one in each grade or size by gravity methods. The work done in the mill today goes far beyond these simple operations, and requires some knowledge of physical chemistry, particularly the branches which deal with the physics and chemistry of surfaces and of the interphase between solid particle and the surrounding liquid. At the same time, the engineer must not become so absorbed in the study of fundamental and applied technology as a physico-chemical science that he overlooks the mechanical, economic, and humanistic aspects of his work. He is an engineer, a chemist, a physicist, and an administrator and, as such, should have a sound scientific and cultural education. Technically, his work is to extract the valuable minerals from the ore sent to his mill; economically, it is to balance all the financial costs and returns in such a way as to ensure the maximum profit from the operation.

As far as possible, details of machine sizes, tonnages, power requirements and the like are omitted. Information of this kind is specific in every case to the ore being treated, and the current technical press is the proper medium for its dissemination. Reputable manufacturers maintain highly specialised testing services to ensure that their machines are suitable for the work given them, and in addition there is the *magnum opus* of Professor Taggart[1] which deals compendiously with such matters.

The approach to a specific problem in mineral processing is governed by fundamental laws, the scientific effect of which is reasonably predictable when applied to pure mineral specimens. Such purity of chemical and crystal structure is rare in Nature, so statistical mathematics are in wide use in the elucidation and industrial control of random variations in the composition of the ore constituents. At any moment uncounted millions of tiny particles are being processed in the average plant and their mass behaviour must be sufficiently understood to be brought under control. An individual particle in this mass may have unpredictable characteristics of its own. In ore treatment there is therefore an element of empiricism. The fundamental laws, concerned with such matters as thermodynamics, adsorption isotherms, particle dynamics, reaction equilibria and the like are applied in research, while statistical appraisal of routine sampling information is used in day-by-day process control and adjustment of treatment to variations in ore texture.

Mineral processing combines a series of distinct unit operations. These include the transport of material and water through all stages of the process as continuously and smoothly as is needed; liberation by comminution of the desired mineral particles from their associated gangue-stuff; separation of the desired from the undesired particles; collection and disposal of products.

After this division, which must be carried down to the resolution called for by the problem, the possible techniques which can be used in developing the desired effects can be considered and tested out in the laboratory, the pilot plant, and the mill—in that order. These techniques embody methods of applying fundamental scientific principles in such wise as to influence the behaviour of the material under treatment. Preparation, including liberation, is the first step in mineral engineering and is dealt with in the early chapters. Separation, or concentration, comes next, and disposal last.

For easy reference some definitions of words or phrases which have a special meaning in mineral dressing have been grouped in alphabetical order in Appendix A (Glossary) at the end of the book. The definitions are sometimes given in the text when they first arise.

Objectives

These are of two kinds, technical and economic. To bring the marketable (or shipping, selling) product or "concentrate" into the technical condition required by the customer, unwanted constituents in the original ore must be removed or reduced below some specified percentage. The product may have to conform to requirements as to particle size, assay grade, moisture content. If more than one valuable mineral is present, the mineral dresser may be obliged to separate them so that each can be marketed separately, or so that the purchaser can handle them economically. The smelter or other purchaser protects himself from financial loss by imposing a penalty on all concentrate failing to reach the agreed grade, and it becomes the duty of the mill manager to ensure that this grade is reached or exceeded. Broadly, the requirements for a concentrate include its assay grade (with respect to the mineral or minerals contained); its freedom from associated minerals; its moisture; its sulphur content; and its grain size. Where the same element occurs in two different combinations requiring different forms of subsequent treatment, appropriate separation may be economically justified. An example of this is ore in which copper occurs as sulphide and as carbonate. The former concentrate is smelted and the latter leached at N'Changa, in Zambia.

Some of the valuable product is inevitably lost in the tailings finally rejected. A second objective is to keep this loss as low as is economically justifiable. The cost of treatment must therefore be balanced against the revenue obtained by sale of concentrate.

A further requirement is that the plant shall handle an adequate tonnage from the mine. Failure to do so might raise the overall operating cost or otherwise complicate the orderly conduct of the enterprise.

Scope

The mill receives its raw material from the working places whence ore is being mined or quarried. After concentration, the products pass from the mill to a smelter or other consumer. These stages may be depicted in flow-sheet form

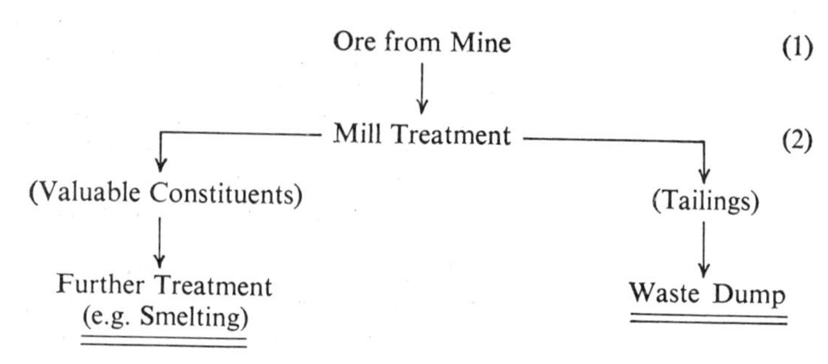

Fig. 1. Flow-sheet. General Milling Treatment

(Note. *Double underline signifies end-product for flow-sheet concerned throughout this book.*)

In terms of function, stage (1) is concerned with severing ore from its parent body and transporting it to (2). Most mills work continuously, and most mines deliver their ore periodically. The successive stages of ore treatment can be listed thus:

(*a*) Reception of a periodic feed from the mine.

(*b*) Liberation of values by comminution.

(*c*) Steady delivery ("feed") of properly liberated ore into the concentrating section of the mill.

(*d*) Shipment of values ("concentrates") in a suitable condition.

(*e*) Disposal of rejected products ("gangue" or "tailings").

Usually, the mill receives ore from the mine during part of the 24-hour day, five or six days weekly (except during holiday periods). Part of this is temporarily accumulated in ore bins, in order to turn from irregular reception to a controlled rate of feed to the plant. In addition, part of the arriving ore may be diverted to a stockpile, whence it is reclaimed in due course. If the various sections of the mine ("stopes") vary in the nature of the ore and its consequent response to treatment, special binning and blending arrangements may be used at the reception stage. The feed, after liberation by grinding, is treated by methods of concentration which separate the desired minerals (called "*values*")from the gangue. The values are then roughly dried and shipped, while the tailings are sent to a dump. Save in leaching or roasting processes performed in the course of milling, no chemical alteration of the mineral species occurs in typical ore-dressing operations. The cost of the work must be kept low as compared with that for an equivalent tonnage treated by any alternative method. This economic necessity provides a broad distinguishing feature between mineral processing and other extractive methods such as pyrometallurgy, where the original structure of the mineral species is modified chemically. The distinction, once watertight, has been shaded by the increasing application of hydrometallurgical processes in the mill. These may be used on the whole of the feed, as is common in the recovery of gold and uranium, or on selected fractions. "Mill" and "milling" in this book embraces comminution, separation and disposal during the linked processes of mineral treatment. So far as the grinding mill is concerned, what is aimed at is a continuous stream of "pulp" (the name given to the finely ground ore when it has been mixed with water and flows like a stream). This pulp should move smoothly, without checks or surges, from point to point along the flow-line. As it progresses, the required treatments are given. Chemicals may be added; grain size may be reduced; the solid-liquid ratio may be varied; one or more of the minerals may be removed.

In some ways this is like the technology of mass production, save that in mineral processing we remove substances from the flow-line more often than we add them. A quiet and steady flow is needed if high efficiency is to be maintained, since the whole of the work is planned on a time-and-change basis and irregular running would give too little or too much time from point to point, thereby upsetting the rate at which the process is being developed and applied at each stage. Later it will be seen that when the flow of ore pulp is interrupted, the contained solids begin to settle out. Since this would very soon cause machines and communicating channels to become choked, the design of the mill must be such as to minimise stoppages from such a cause. To a lesser degree, the sand trying to settle out is liable to set up irregular running of the pulp through the conducting pipes and launders. The more successful the layout has been in preventing this, the smoother will be the flow of pulp, and the more efficiently the mill will work.

Mineral Characteristics

The minerals associated in a piece of ore are given generic names according to their composition—*e.g.*, galena (lead sulphide), apatite (calcium phosphate), quartz (silica). Samples of a given mineral from different sources usually react similarly, but traces of impurity are sometimes present in quantities too small to be discerned by ordinary assay, yet sufficient to modify the response of the mineral to treatment. In theory chromite consists of $FeO.Cr_2O_3$, but it frequently contains aluminium or magnesium and has wide variations in both Fe:Cr ratio and the Cr content. Again, malachite ($CuCO_3.Cu(OH)_2$) occurs in nature with a copper content far below that indicated by its formula owing to intergrowth with silicates, dolomite, etc. Cassiterite rarely approaches the 78·8% Sn indicated by the formula SnO_2, its crystal lattice being intimately penetrated by other elements. It is therefore useful to remember that the generic name of a mineral does not guarantee that specific behaviour of samples from two different deposits will be identical. This will be further discussed when the physics of froth-flotation are considered.

Hardness, together with toughness, are the main determining factors in crushing and grinding. Its hardness affects the brittleness and friability of a mineral. Toughness determines its resilience and elasticity. The interplay of these two qualities in a homogeneous piece of rock determines its response to the compressive forces (crushing) and abrasive ones (grinding) comprehended in the word "comminution". A lump of ore is not homogeneous, however, since it contains more than one mineral species. Assessment of its response to comminuting methods is therefore empirical. Values must be liberated from gangue as an essential preliminary to their separation into concentrates and tailings. Comminution for this purpose is normally the most expensive stage in treatment, and in addition, can lead to difficulties and loss in the later sections if performed inefficiently. The hardness and toughness of an ore body are of critical importance, and are assessed by grindability test methods applied to representative samples. Installations, maintenance, wear replacement, and power requirements must be able to handle the rated tonnage sent from the mine. Hence the more resistant the rock, the more costly is its comminution.

One definition of hardness[2] is: "Hardness is that property possessed by solid bodies in a variable degree (which) defends the integrity of their form against forces of permanent deformation, and the integrity of their substance against cause of division."

Another is that of the mineralogist, who uses either Moh's or Knoop's scale of relative hardness (Table 1). This series of minerals is arranged in ascending hardness or resistance to abrasion, each compound being scratched by those above it in number. It gives no guidance to elastic yield, however, and is therefore of little help in assessing grindability.

TABLE 1

HARDNESS SCALES OF MOH AND OF KNOOP

Mineral	*Moh's Value*	*Knoop's Value*			*Knoop Average*
Talc ..	1	(softest) 16	to	25	21
Gypsum	2	42		67	54
Calcite	3	144		177	132
Fluorite	4	171		199	188
Apatite	5	427		514	476
Orthoclase	6	547		729	682
Quartz	7	899		985	958
Topaz	8	1391		1561	1435
Corundum	9	1802		2250	2004
Diamond	10	(hardest) 5500		8500	7000

Toughness is influenced by the manner in which the crystals of the ore are interlocked, and by their grain or size. A very fine (crystallite) structure is usually more resistant than a coarsely crystallised rock. Many ores consist mainly of valuable metallic sulphides and unwanted silica minerals, the former often being friable and the latter tough. There is a tendency for the smaller particles to be richer than the bigger lumps, and this can lead to losses in treatment if it is ignored. It can also lead to sampling errors when a heap of ore is being reduced to a small sample for assay purposes, if all the rock is not systematically blended with respect to particle size.

Mined ore which is treated by mineral processing consists of two or more minerals which vary in their resistance to applied crushing forces. The more friable mineral or minerals are most likely to yield under crushing stresses and strains. The weakening of the structure of a given particle by these stresses depends partly on the percentage of each constituent mineral it contains. Broadly, a micro-crystalline structure with the values disseminated evenly through the gangue minerals is far more resistant to crushing than one which is coarsely crystalline. If in addition the values occur as occasional coarse crystals instead of being minute and disseminated, these crystals may crush even more readily. Grindability must be assessed specifically for each ore complex. It cannot be predicted from knowledge of the behaviour of the individual minerals of which a particular ore is composed. Various assessments, such as the Hardgrove rating[1], based on the attrition resulting from dry grinding of material under controlled conditions, have been published. This extremely important subject, which affects both treatment cost and efficiency of concentration, is developed in later chapters. One of the qualities affecting grindability is the abrasiveness of the ore. This bears directly on the cost of replacement of machinery worn during grinding contact and on the cost of the power used in this unavoidable wear.

Methods of Treatment

When the mineral species in a given ore have been sufficiently freed from one another by comminution, the process of concentration becomes possible. The liberated particles must differ sufficiently in their physical, electrical, or chemical properties to respond to an appropriate differentiating force. Physical differentiation uses the shape, size, surface area, specific gravity, porosity, colour, and gliding friction to sort out the mixed species. Electrical treatment works on magnetic susceptibility (natural or induced), conductivity, and radio-activity. Chemical attack depends on the reactivity of minerals in a given bonded or lattice structure. All three types of attack shade into one another and may be mixed in a given process. A list of the most important exploitable characteristics, together with the mode of application, is given in Table 2. These characteristics are developed by selected methods of grinding and further treatment, which should always be based on preliminary study of the specific ore.

Treatments in which chemical solutions attack mineral suspended in an aqueous pulp are in growing use. The cyanidation of gold and silver ores is one of the oldest continuous processes. The first stage of uranium recovery is a chemical attack by acids or alkalis. Other elements treated by leaching include copper, aluminium, manganese, nickel, tungsten, and zinc. Mineral processing normally confines its operation to separation of ore constituents without changing their physical state. In hydrometallurgy the chemical processes are conveniently applied to ore in transit through the mill, and this leads in practice to the entry of the metallurgist into the plant. These hydrometallurgical methods are discussed briefly in connexion with their use of mill machinery in Chapter 16.

Pyrometallurgy is sometimes combined with mineral dressing. In one process tin ore is chloridised and the metal chloride is then volatilised. In another a mixture of smelted metal sulphides is separated by froth-flotation. Other processes include the selective coating of one mineral in a mixture with magnetic paint, the electro-chemical separation of antimony and zinc from their ores, and other treatments, such as the extraction of magnesium from

TABLE 2

PRINCIPAL EXPLOITABLE CHARACTERISTICS

Selective Mineral Characteristic	*Type of Separating Force*	*Operation*
Colour, Lustre.	Visual, manual, automated.	Hand-sorting of graded ore, to remove detritus, waste rock, special constituents. May use fluorescent lighting, or impulses triggered by reflected light.
Specific Gravity.	Differential movement due to mass effects, usually in hydraulic currents.	"Gravity" separation of sands and gravels by D.M.S., jig, sluice, shaking table, spiral.
Surface Reactivity.	Differential Surface Tension in Water.	Removal of relatively aerophilic mineral as froth from aerated pulp by froth-flotation. Widely used process.
Chemical Reactivity	Solvation by appropriate chemicals.	Hydrometallurgy. Ore exposed to solvating chemicals, perhaps with heat and pressure, then filtered. Dissolved element/s recovered from filtrate, chemically, electrolytically or by ion exchange.
Ferro-Magnetism.	Magnetic.	Magnetic devices remove the preferred mineral. Also used to remove "tramp" iron.
Conductivity.	Electrostatic charge.	Particles pass through high-voltage zone. Rate of dissipation of induced charge influences subsequent deflection. Differential conductivity.
Radio-activity.	α or β rays.	Emissions are signalled by G.M. Valve which also activates a separating or "picking" device.
Shape.	Frictional.	Sliding force is opposed by "cling" of particle, resultant movement depending on cross-section and area, hence on shape.
Texture.	Crushing. Screening. Classifying.	Characteristic shapes and surfaces are developed during comminution.

sea water. In this book, which is concerned with the basic principles used in standard methods of treatment, these specialised techniques are not discussed.

Choice of Method

A method of ore treatment must be chosen which is appropriate to a specific ore as well as to a set of general principles. There are three main types of ore complex—the massive, the intergrown, and the disseminated. Coal, and some bedded deposits such as iron seams, are examples of the first-named type, in which dressing operations are simple or even unnecessary. Most lead-zinc ores are intergrown and need controlled grinding. This must be taken to a point where galena is adequately freed from sphalerite, but must stop short of the production of particles too minute to be treated by the concentration process used. Disseminated values are characteristic of an enormous tonnage milled today, with the wanted mineral distributed sparsely through a valueless rock matrix, the whole of which must be reduced by grinding to a fine sand before the desired contents can be separated. Choice of method, then, is bound up with grinding cost and efficiency of liberation. Right at the beginning of our study, in this introductory chapter, the essential factor in all ore-dressing problems can be observed—correct liberation. This is the key to success. The valuable mineral must be freed, but only just freed. A process which over-grinds the ore in this unlocking is wasteful, since it consumes grinding power without justification and makes efficient recovery more difficult to attain.

Sequence of Operations

The flow-sheet is sometimes presented diagrammatically or in an embellished form. Since it is functionally a chart showing the sequence of operations, it should be simple. On the flow-sheet are marked any details required for the operator's guidance. The author uses it in a form which can be typed and cyclostyled, so as to make it a simple method of record, change and instruction. Thus considered, it can be presented (Fig. 2) as a block flow-sheet in which all operations of one character are grouped. In this case "Preparation" deals with all crushing, grinding, and preliminary rejection. The next block, "Separation", groups the various treatments incident to production of concentrate and tailing. The third, "Product Handling", covers the disposal of the products.

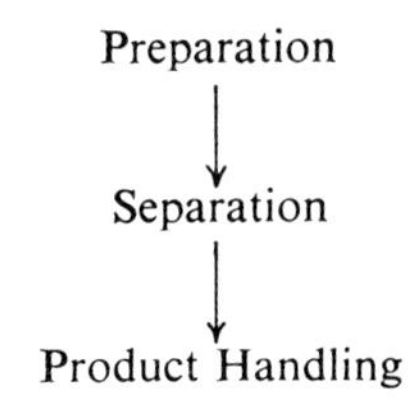

Fig. 2. A Simple Block Flow-Sheet

The main purpose of this flow-sheet is to analyse the general scheme, which in a large mill can be very complicated, and to fix attention on the need for "quality control" of the condition of the ore or pulp before it is transferred from one stage of treatment to the next.

The second, and for many purposes sufficient, type is the simple line flow-sheet (Fig. 3). On it can be shown details of machines, settings, rates, etc., or a figure may be added to each descriptive label for use in a separate reference file. This makes possible the recording of detail, without loss of the essential simplicity and compactness which is so desirable. A large sheet of paper can be handled in an office, but is a nuisance in the mill.

Run-of-mine ore is long-ranged in its sizes, and the biggest lumps may be difficult to transport. It may be mixed with small quantities of waste rock and debris. The primary purpose of all comminution is the unlocking of values in the ore, but crushing may be combined with sorting, and performed in stages. Screens are used to prevent oversized lumps of ore from entering machines not built to handle them. Screens can reduce the size range of ore fed to a concentrating appliance, if this improves its efficiency.

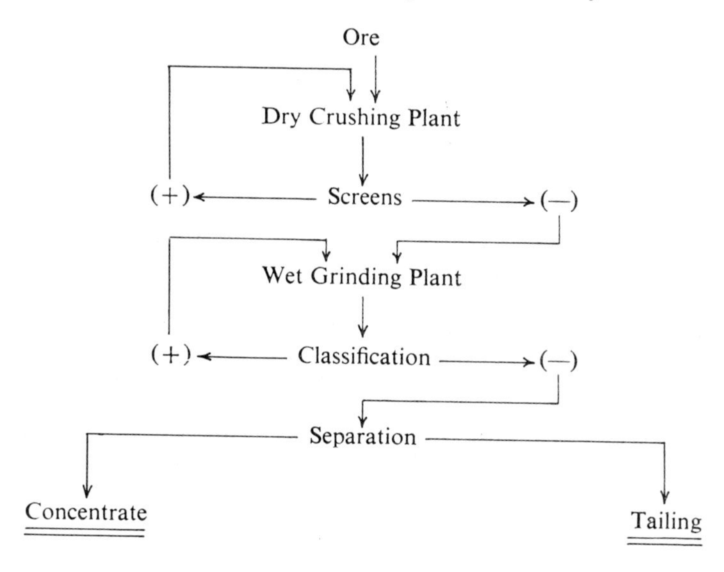

Fig. 3. A More Elaborate Flow-Sheet
(+) indicates oversized material returned for further treatment and (–) undersize allowed to proceed

When dry crushing has produced pieces of rock so small that they can no longer be gripped readily between crushing faces, in the way a nutcracker grips a nut, wet grinding normally follows, with random blows rained upon a passing stream of ore suspended in water. (For some purposes dry grinding followed by dry treatment is practised. Minerals thus treated include talc, cement, clinker, asbestos, and those treated in waterless country.) As the wet-ground mixture, now called ore pulp or pulp, leaves the grinding device it may be sent direct to the concentrating section. This form of transfer is called open-circuit grinding. More usually, the pulp is passed through a device which returns oversize for further attrition and sends on ore which has been sufficiently ground. This arrangement is called closed-circuit grinding, and is widely practised. Sometimes screens close the wet circuit (see Fig. 3), but usually the solid part of the pulp is now too finely ground for their effective use, in which case it can be sent through devices called classifiers. In these the ore particles are separated into coarse and fine fractions relatively to their rate of fall through water. Grinding can be done in one or more stages of size reduction. The larger the tonnage treated the greater is the justification for breaking down every operation into several stages, each arranged to be applied within restricted limits. Mills which handle 50,000 tons of ore daily have not the limitations of the small 50–ton outfit, where one machine must perform a wide range of duties at each stage.

The grinding section is controlled so as to deliver material in an appropriate condition to the next stage of the treatment. Some easily won concentrate or some undesirable ore-constituents may have been "scalped out" in the grinding circuit, but the main work of concentration is done in a section of its own. If more than one mineral is to be concentrated selectively, two or more such inter-related yet independent sections are provided in series. Here the work of separating a sufficiently high-grade concentrate is performed, together with the production of a tailing so low in value that further treatment would not be profitable.

Each product leaves the machines either as moist gravel or sand, or as a fluid pulp. In the latter case it can be run to waste if a tailing or de-watered and disposed of (shipped). The various processes are controlled by sampling and assay. Tailings must be disposed of either by dumping or running them into a retaining dam, and the water used in the milling treatment may either be reused or run to waste. Precautions are sometimes needed in the latter case to ensure that no contamination of the district water supply results.

Products

Separation of the ore's constituent minerals is not carried to completion in mineral processing. It is stopped when an optimum economic or technical stage of concentration (or a desired physical condition of product) has been reached. In an ore body which carries one value together with gangue minerals it may be decided that 91% recovery of concentrate assaying 40% of the desired mineral or metal will yield a higher profit than would 93% assaying 39%. For another similar deposit the figure could be quite different since the economic factors which determine the most profitable rate and degree of exploitation depend in part on local conditions.

Where two or more valuable products are recovered, either as a bulk concentrate or successively, further considerations enter, which are developed in later chapters. The trend today is for industry to impose conditions as to the purity and physical state of the materials it purchases, particularly where highly stressed metals and alloys are to be manufactured and even minute traces of impurity must be removed. It is often possible to separate deleterious minerals during concentration and thus obviate or simplify the more costly processes which otherwise must be applied in further treatment of the values. Concentration is never pushed to its limit, since the entailed cost of regrinding, re-treatment and loss of value in tailings must not be taken beyond economic limits. Processing must therefore be controlled so as to conform to optimum schedules of cost, percentage recovery and assay grade.

With some minerals, notably rare earths, spars and silicates, the physical condition may be important to the purchaser. A very finely powdered dry concentrate could be treated in a reverberatory furnace but would be blown out and lost from a blast or shaft furnace. Colour, size and crystal form can affect selling (vend). Barite used by the pharmaceutical and paper industries must be white, but for drilling muds its colour does not matter. Asbestos is priced on length and strength of fibre, together with freedom from entrained dust. Mica is valued for transparency, size and cleavage. Where-ever specialised requirements are laid down by the customer, mineral processing is looked to in the expectation that it can handle the problem. The best and cheapest stage at which an unwanted constituent can be removed is usually in the concentrating section.

Economics

In this chapter the economic factors which influence the choice of a process and the efficiency with which it is to be operated can only be indicated in general terms. They include reduction of transported bulk and weight, standardisation of selling grade, and the nice balancing of production cost against the market values of saleable products. Ore as it is mined can rarely be smelted direct. Since pyrometallurgy is far more expensive than ore dressing it must be reserved for material rich enough to bear that expense. Mineral processing is the first stage of extraction metallurgy, applied to run-of-mine ore because it provides the cheapest way to discard unwanted material. Smelting is a later stage. The point at which processing hands over to smelting is determined by relative treatment cost, partly by the technical aid which can be given to the smelter by evening the grade supplied it, and partly by the question of transport. Freight to a distant smelter should only be paid on what ought to be sent there, and if impurities and moisture are shipped, it should be because that is cheaper than removing them in the concentrating plant and because the smelting process can tolerate their presence. Loss in the smelter varies, among other things, with the amount of slag, and the more impurity the smelter must remove the more slag it is forced to produce. Hence pyrometallurgy benefits technically and economically by receiving ore concentrates which have been processed to an optimum grade of metal content and a permitted tolerance of undesired associated minerals. The relative costs of mill and smelter treatment can be assessed with reasonable precision, and a decision can then be made in each case as to the final flow-sheet.

A large producer usually operates its own smelter, if labour, fuel, and flux are available. The value to the mine of its ore is the selling value of the product it ships, *minus* all the costs incurred in getting it to market. In addition to mining these include concentration, transport and smelting costs. Whether the mine runs its own smelter or ships concentrate to market, the cost of smelting is determined by grade and tonnage of the mineral treated. Although smelter schedules seem complicated, they are really technical documents relating (*a*) to treatment cost and (*b*) to transport and storage. (*a*) varies with the assay of the mill concentrate; deductions covering slag and volatilisation losses; bonuses for additional desirable constituents; and penalties for unwanted impurities which add to cost of smelter treatment. (*b*) affects the concentrating plant since valueless tonnage shipped increases the freight charge. As will be seen later, the higher the grade of concentrate the higher will be the tailing loss, and the greater the treatment cost. These considerations can be balanced out when deciding what should be the optimum plant performance.

Milling costs include a proportion of capital cost and depreciation which are affected by the ore reserves (a wasting asset); the nature and complexity of the selected treatment for which the mill must be equipped; and the working capacity, which limits the tonnage treated, and hence the rate of mining, and the allocation of fixed overhead expenditure per ton mined or milled. This last item, in practice, is apt to lead to quiet but persistent pressure on the mill to try to exceed its designed and rated capacity. The equation of all these considerations is fixed in the first instance by the consulting engineers, but once the mill is "run in" and has trained its personnel, it is common to find that alert management can adapt and modify the treatment if the mine should wish to increase its output of profitable tonnage. The domestic accountancy of the mill is analysed on some such basis as cost of each part of the work per ton treated. A textbook is too static to go into details on such matters, which are richly provided in the current technical press. Economically, the plant is not isolated, but is intimately bound to the prosperity of the mine on one side, and of the smelter or other market on the other. It is a unit in a mass-production process that starts underground and ends in some purchaser's factory. In its inter-company relationships the mill is the "customer" of the mine and in turn "sells" to the smelter. If this relationship is co-operatively understood, smoother running will result and all concerned will benefit. Nothing so helps assembly-line techniques (and modern ore treatment is work of that nature) as steady throughput, with clear directives and objectives governing all mutually interdependent sections of the plant.

Reference

1. Taggart, A. F. (1945). *Handbook of Mineral Dressing*. Chapman and Hall.
2. Anon. (1957). *Mining Magazine*. October.

CHAPTER 2

ACCEPTANCE INTO THE MILL

Preliminary

Run-of-mine ore contains not only rock which has been deliberately severed for treatment but also a variety of valueless or even embarrasing adulterants. Past generations of miners used personal craftsmanship to select ore suitable for milling, so that the laborious handling methods of those days characteristically produced a small but fairly clean tonnage for treatment. The mining of large low-grade deposits by modern methods relies on mechanisation and bulk handling, in which careful discrimination between valuable lode-material and waste country rock is not usually feasible. With the severed ore may be carried pieces of iron or steel lost from the machines. These would damage the mill's crushing plant if fed in with the ore. Wood, unexploded gelignite, lubricating oil, grease, and other undesirable materials also find their way into the run-of-mine ore and should be removed before they can become run-of-mill feed. Some deposits contain clays and primary slimes of little or no economic value. These tend to cling to the working parts of crushing appliances and to impede crisp action and smooth passage. Slimes may give trouble in the concentrating and disposal sections if allowed to pass that far.

When the magnitude of the operation justifies it, preliminary rejection of waste rock and ore too low in grade to justify treatment may be undertaken. This results in a reduction of the tonnage sent to the mill without the rejection of an appreciable part of the values mined, and therefore is a concentrating treatment. The special form of this dealt with in Chapter 12 (dense-media separation) is an extremely important development of recent years which has made it possible to up-grade many uneconomic deposits to a point where they will repay more costly treatment.

"Acceptance" Operations

If the mine keeps records of tonnage dispatched to the mill, and the mill records the tonnage accepted for treatment, any difference between these should be brought into account as waste. Discrepancy between the two sets of figures may arise from accounting error, but could be the result of planned rejection *en route* of rock which on inspection in daylight proved too poor to be treated. It is common to remove such material by methods described later in this chapter. The result is to raise the head value of the ore accepted for treatment as compared with that of the ore mined. This is therefore a concentrating process, and can be expressed by the basic formula for weight balance:

$$F = C + T \tag{2·1}$$

where F is the weight in specified units of *feed* (the milling term for material sent to be treated), C the weight of *concentrate* (the valuable product), and T the weight of *tailing* (the valueless rejected fraction). Suppose 1000 short tons to have been mined and 100 tons of this to have been rejected as waste. Then

$$1000(F) = 900(C) + 100(T) \tag{2·2}$$

and the tonnage accepted for treatment is 900. Useful as this figure is in showing what quantity of the Company's money is spent in milling costs, equation 2.1 is insufficient for control purposes. An "ingredient balance" is essential to check the efficiency and justification of the process of rejection, or *sorting*, to use the milling term. This, in equation form, is:

$$F_f = C_c + T_t \tag{2.3}$$

where f, c, and t are the assay values of feed, concentrate, and tailing respectively, in appropriate units. In the case under consideration let the assays be $f = 2{\cdot}0\%$ $c = 2{\cdot}2\%$ and $t = 0{\cdot}2\%$. Then

$$1000^* \times \frac{2^*}{100} = \frac{900^* \times 2{\cdot}2}{100} + \frac{100 \times 0{\cdot}2^*}{100}. \tag{2.4}$$

The value in each constituent is therefore

$$20\ (F_f) = 19{\cdot}8\ (C_c) + 0{\cdot}2\ (T_t). \tag{2.5}$$

The items which comprise this equation are not all measured in ordinary working. Those which are usually checked are marked with an asterisk(*). Although this preliminary operation has done but little to raise the grade of the ore, it has been checked by an equation of metallurgical balance (2.3) and the information yielded could be summarised as a metallurgical balance sheet.

TABLE 3

A METALLURGICAL BALANCE SHEET

Product	(1) *Weight*	(2) *Assay*	(3) (1) x (2)	*Distribution of value*	*% Dist. of value*
Concentrate	900	2·2%	1980	19·8	99
Tailing	100	0·2%	20	0·2	1
Feed	1000	2·0%	2000	20·0	100

In the foregoing case, the facts thus displayed would be considered in relation to the cost of milling the *T* fraction, as against that of sorting and dumping it, and in consequence it would be rejected. The first stage of acceptance is to know what is received, both as regards tonnage and value per ton. This involves weighing the incoming ore and taking control samples of adequate accuracy. Sampling must check the assay grade of the waste product rejected as well as the ore accepted for treatment, to ensure the efficient working of this section of the operation.

If rejection involves removal by hand (called picking or sorting), it requires display of the ore in a reasonably clean state, in sizes which can easily be moved by hand. Primary crushing and screening are used in order to reduce lumps too big for lifting by hand, and to bypass ore too small to be picked. Washing aids this work by removing adherent coatings of dust and slime, so that the true surface of each piece of ore is displayed.

Sampling

Ore sampling falls into two categories, exploratory and control. When prospecting for, proving, and developing a mine, the former type is used. In routine exploitation of the ore and control of the mill operation, samples are taken in order to check the efficiency of the work. This last type of sampling is the main concern in mineral dressing. In a given sampling system the quality of the material at the point under scrutiny is checked with sufficient accuracy for the immediate purpose. This may include check of lump size, moisture content, assay grade, purity, chemical state and other matters dealt with later. A sampling nomogram has been constructed by P. Gy[1] which indicates the best reduction treatment under correctly stated conditions. These take account of the liberation mesh of the valuable ore constituents, the assay grades and the densities of the minerals present.

Despite the obvious importance of good sampling as a means of technical and economic control, it is on the whole one of the least satisfactory operations in milling. The physical difficulties involved in taking a truly representative sample and reducing it to the few pounds weight needed for assay are formidable, and the chances that such a final sample will accurately represent the thousands of tons from which it was drawn are best when finely disseminated ore of fairly even grade is being smoothly carried through the transporting system. When, as in many gold ores, much of the value exists as tiny scattered particles, or in irregularly distributed rich patches of ore, some experienced workers consider the chances of accuracy to be so remote that close daily check of recovery based on the entering tonnage is not possible. In this section ideal practice will be considered.

The control sample cut from the original feed is called the head sample. It may be taken by hand or mechanically, according to the way in which ore comes to the crushing plant. An elementary method is to make a "grab" sample, in which a small portion is taken from the contents of each passing car-load of ore. Since the valuable constituents are frequently more friable than the gangue, there is a risk in such cases that the richer ore has worked down out of reach, so that the sample thus taken from the top will not be representative. With even distribution of values in a disseminated type of ore this should not occur to a serious extent, but there are other objections to hand sampling. If it is to provide reliable data for technical control it should be foolproof. To take an adequate sample by hand and to reduce it accurately to assay dimensions is work calling for conscientiousness, hard labour, and monotonous repetition. Workmen combining such virtues are rare, and would be better employed on other duties than routine hand sampling. Methods of hand sampling are described later.

With mechanical sampling, the uncertainty of human partiality can be avoided. In the ideal layout a point is chosen or contrived where the whole ore-stream is falling freely under gravity. The sampling device cuts through this falling stream at regular intervals and removes a percentage of ore commensurate with (*a*) the importance of accuracy and (*b*) the facilities for reducing a large sample to manageable size. The whole stream should be cut for part of the time so as to remove a representative fraction from its

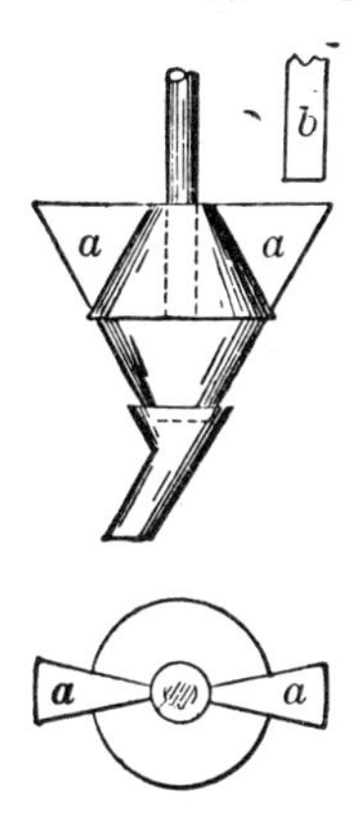

Fig. 4. *The Vezin Sampler*

full cross-section. Observation of moving streams of ore, dry or wet, shows segregation to occur, and it is therefore important so to arrange the cut that no part of the stream is disproportionately represented. Cuts should be taken at intervals which minimise the chance of missing the passage of an unusual delivery of ore, either rich or poor. Where storage ahead of the sampling device has given the ore opportunity to segregate, sampling intervals should be closer than where run-of-mine ore comes through without check, since this segregation leads to alternate deliveries of coarse and fine material. Excess of either in the sample would imperil its accuracy. The cutting aperture of the sampling device must be big enough to give ample passage to the largest piece of ore presented to it—usually not less than three times the width than the largest piece of rock in the passing stream.

Mechanical samplers mostly favoured are of two types—arc path and straight-line path. One of the oldest types is the Vezin (Fig. 4). Two hollow truncated cones joined at their bases are mounted vertically and rotate so that at each revolution an attached scoop (*a*) which projects from the upper cone cuts through the falling stream of ore (*b*) and withdraws a sample proportional to the opening in the sector of that cone. The scoop edges must be truly radial and its sides vertical in order that an accurate cross-section of the ore stream may be taken, and the sampler must run smoothly. A jerky motion would result in an unrepresentative cut. For a 5% cut the scoop must subtend 18° of arc; for 10% 36°, and so on. Alternatively, two diametrically opposite scoops may be used. The former arrangement is better since it allows a larger aperture and reduces the risk of a piece of rock "bridging" the

Fig. 5. Stokes Mechanical Sampler

scoop and thus deflecting into the main stream ore which should be included in the sample. Several other stream-cutting samplers are standardised manufacturing products. These samplers can only be inserted where there is an appreciable fall of rock. They cannot work satisfactorily on feed containing large lumps.

Straight-line cutters take several forms and can deal with a wide size-range in the feed, beside taking but little headroom. One form consists of a sampling conveyor set under and across the delivery from the main conveyor. Two sprocket chains carry the sampling bucket, and normally the ore falls between them. Periodically the bucket traverses the stream and cuts a sample, which is tipped as the bucket turns round at the end of its journey and proceeds back, bottom up, through the stream of ore. The sample taken is a proportion of the tonnage, being the total bucket area on the sprockets divided by the area between the leading edges of successive buckets. At Midvale, Utah, a 2% cut of the lead-zinc ore is thus taken, and dumped to a hopper. This discharges its contents in such a manner that a further cut is made by the buckets, thus reducing the sample cut to 1:2500 in one mechanised operation. Typical of straight-line samplers are the Geco and Geary-Jennings, which use chain-drive and screw-drive respectively to reciprocate a sample cutting device across the falling stream of ore. One mine uses a pusher mechanism to "wipe" a sample sideways off the travelling conveyor-belt at intervals.

Usually the mined ore contains lumps too large to be sampled before they have passed through the crushing plant. This work is then done in the conveyor system ahead of the fine-ore bins. If sorting has been practised before this point, the fraction thus removed should be weighed and sampled in order to check the efficiency of the operation and reconcile the mined tonnage with that accepted for milling. This practice would give no protection against "*high-grading*"* on the picking belts in the case of a spottily rich gold ore. The closer the size range of the feed at the time the sample is cut, the more representative it will be, since a big disproportion between the largest and smallest lump of mineral is liable to upset accuracy of cut.

Reducing the Sample

The size of the original sample cut from the ore stream should be as large as can be conveniently handled. Even with a daily throughput of only 100 tons, a 1% cut means that one ton of initial sample must be reduced to a few pounds in weight for assay. This bulk reduction must be made in the simplest possible manner, subject to working accuracy. The "grain" and distribution of the minerals in the ore affects both accuracy of reduction and the decision as to a safe routine procedure in cutting down the bulk. Where all or part of the values consist of friable sulphide in a tough gangue, there will almost certainly be a high concentration of value in the "fines". In such a case the ratio between the weight of the largest lump of ore and the total weight of the sample must be kept much lower than where a fairly homogeneous rock is being handled. A 1000:1 ratio would mean that in a one-ton lot† no piece should weigh more than about two pounds. In Chapter 21 a typical procedure for heap-sample reduction is described.

For routine work, a few empirical tests should be made to establish the maximum ratio it is wise to tolerate. At the same time the amount which can be rejected after each size reduction can be experimented with in order to obtain adequate sampling accuracy with minimum effort. If the daily original sample is of the order of ten tons, a simple crushing operation followed by a 10% cut would bring it down to one ton, which might be

* Where a term is italicised, its definition may be found in the Glossary, Appendix 1.

† Throughout this book "ton" refers to the short ton (2,000 lb. avoir.)

delivered to a small roll-crushing unit before a further similar reduction of the product to 200 lb. With some ores a more drastic cut might be possible in each of a series of stages, while with others a cut of one-in-ten would be too drastic. It is not wise to lay down hard-and-fast rules. The aim is to get reasonable accuracy with minimum effort, and the work should be as fully mechanised as possible. A few simple tests on a trial heap of representative mine ore will facilitate the selection of the best procedure. When a sample has been obtained by one procedure, the heap is remixed and resampled by another method, the various samples obtained being compared by assay. The human element is a far more serious source of error than the method chosen. For this reason it is important to shift all the drudgery of sample reduction to suitable machines. (See Chapter 21).

Objectives in Sampling

The plant can be thought of as if it were a trading concern, buying its raw material from the mine, processing it, and selling it to a smelter. Seen thus, operating control is a form of technical accountancy, even though by its nature it cannot achieve the complete precision of an arithmetical system. To arrive at a metallurgical balance, the mill must be debited with the total units of valuable mineral (or minerals) entering as feed accepted for treatment. It is also credited with the total units of value in its products (concentrates *and* tailings), in terms of equation 2.3. Although an exact balance is unattainable, the closest possible agreement should be attempted. In the case of gold ores, the high specific gravity of the metal causes it to accumulate in launders, classifier beds, scoop-boxes and the crevices of such places as mill linings. Retention of values in the circuit cannot be avoided and will nullify any attempt to strike a short-term metallurgical balance. Over a long period, if there is no theft, the head assay should be reasonably well accounted for from the known recovery of gold and from the tailing assay, together with gold tied up in cyanide solution in the plant. Although it is not always possible to relate the head assay to the full working control of the mill circuit over a short period, the information it gives is vital to long-term check on operations. The mining engineer uses stope assay plans to calculate the value and tonnage of ore mined and sent to the mill and the mill manager may be asked to account for the difference between promise and performance. No stope assay sample can compare in accuracy with an actual proportional cut made through the moving stream of severed and hoisted ore entering the processing section. Serious lack of reconciliation between two such sets of figures is *prima facie* more likely to be due to technical difficulties in stope sampling than to returns based on accurate sampling of the mill-head ore.

In the case of the base metal and non-metallic ores, theft is not an element in the ordinary problem of striking a metallurgical balance. It is important, however, to have a satisfactory check on what comes from the mine. A head sample is used for this purpose. Provided it is properly taken, carefully reduced to assay proportions, and accurately assayed, and that the tonnage from which it was cut is known, the first essential in operational check-up (units of value accepted for treatment) is established. Subsequent failure of the plant to show a consistent metallurgical balance sheet will give warning that something is wrong. The modern plant is scientifically designed and should be controlled at key points by impartial sampling. The keen engineer takes care that these sampling points are properly maintained and reasonably tamper-proof, and he uses the reports they yield to maintain an acceptable standard of work.

The original sample gives information on three important points—assay value, size of passing material, and moisture. Assay value has been considered. Size of passing material is important since it shows what further work must be done in the crushing and grinding sections. If the plant is supposed to receive its ore from the mine *via* an underground crushing station set to deliver rock broken below a certain maximum size, this becomes a condition of acceptance, and the mill is entitled to insist that the terms are kept. This checking test must be made on part of the original sample, and in the earlier stages at present under discussion, visual examination of the passing ore stream or of the cut sample should show whether all is in order. The third piece of information given by the sample is its moisture. The plant is "buying" dry ore from the mine. If it receives over its weighing devices 1237 tons of material at 4% moisture, it is only getting 1187 tons of value-containing rock, and that dry tonnage is what it should accept. In the dry season in the tropics, damp and sticky ore from underground will register overweight. If the sample is left lying in the sun, it will lose much of its moisture. Conversely, ore first exposed to streaming rain after leaving the weightometer, but before being sampled, would lead to under-estimation of the tonnage received.

Weighing the Input

When one company runs both mill and smelter, an approximate figure for tonnage received is usually sufficient, provided always that it permits the plant to establish a head check on its metallurgy. In the case of a customs mill, greater accuracy is called for. Track scales and platform scales can be used, either with a scale-clerk to do the weighing, or with automatic recording. The tare of the tub, truck, or ore car must be known and if the container does not empty itself completely, a correction factor must be established. This will vary seasonally and according to the amount of clayey ore coming from the mine. Snow, ice, and such live loads as vehicle drivers vary the tare, and must be kept out of the reckoning. Automatic hoppers worked in pairs are used for finely broken ore in Germany, filling and tipping in alternation. These give an accurate record provided they empty completely at each tipping.

Weighing devices which register and record the weight passing over a section of a belt conveyor are widely used. They are compact and accurate to $\frac{1}{2}$% provided the belt can be kept clean and the weighing mechanism regularly checked and serviced. If sticky material builds up on the conveying surface and fails to discharge, it registers as received ore unless a correction is applied. In the Merrick Weightometer (Fig. 6) a short length of the moving belt is picked up by a system of rods hanging from weighing levers, which bear upon a beam held in balance by an iron weight floating in mercury. Movements of this float are recorded on an integrating device which presents the speed of the belt and the load per unit length in terms of total weight passing over the weighbridge. Sticky material on the belt can be compensated by balancing an equal weight of return belt against loaded belt. The weightometer can be calibrated either by passing known weights over it, or

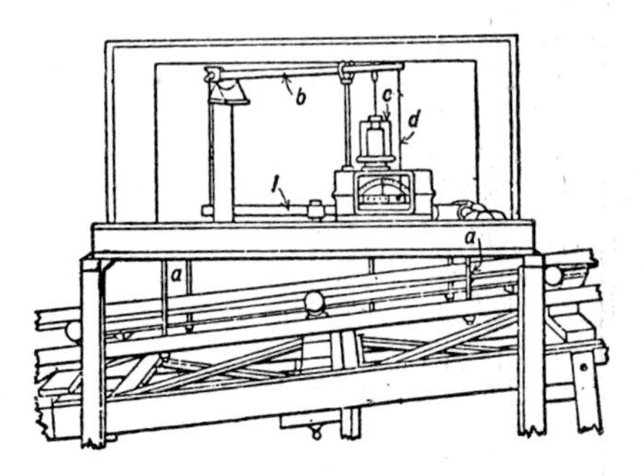

Fig. 6. The Merrick Weightometer

a—a *Suspended length of conveyor belt*
b, c, d *Continuous weighing and recording arrangement*
l *Sliding adjustment of weighing device*

by causing a length of heavy roller chain to trail from an anchorage over the suspended section while the empty belt is running.

Approximate measurement of tonnage can be made by using some form of revolution counter (such as could be improvised from a cyclometer or car speedometer) to register the footage travelled by the conveyor, and combining this with knowledge of the weight of ore resting on the conveyor per foot run.

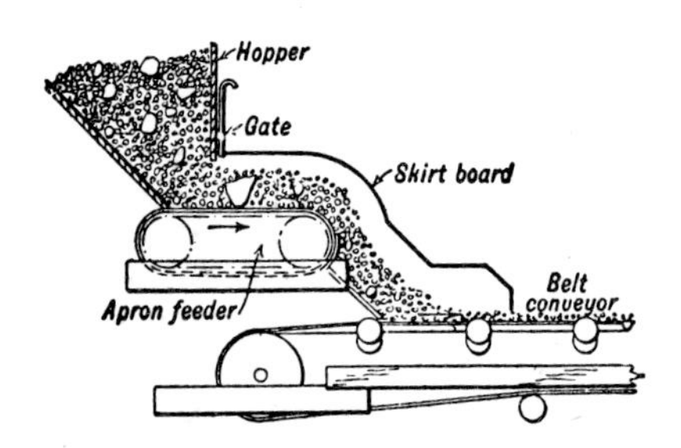

Fig. 7. Controlling Rate of Feed

This weight is found by stopping the belt, removing the ore resting on a measured length, and weighing it. This routine check must be made frequently. If ore is being drawn on to the belt from a guillotine form of gate (Fig. 7), the clearance of the gate above the belt determines the height of the "ribbon" of ore. If the gate is periodically adjusted in order to regulate the rate of feed, the weight per foot run of belt must be ascertained for each gate setting, and must be corrected periodically as the lip of the gate wears away. The shiftsman in charge must book the time and cyclometer reading whenever he alters the height of the gate. Another method of recording the footage travelled by fixed-speed belt is to mount on the framework a portable recorder of the type used to register the travelling time of a lorry. With this there must also be an alarm system which gives warning when the belt is running unloaded. This can be improvised from a trailing arm which is held off the belt by the ribbon of ore. When the feed fails and the belt runs empty, the arm falls and an alarm-circuit is closed.

Washing and Scrubbing the Ore

Washing of run-of-mine ore removes obscuring dust and casual dirt from the surface of the pieces and facilitates recognition when hand picking is to be practised. A second effect may be the removal of primary slimes of no value. Clays, colloidal fines, and similar valueless material can easily be taken out at the earlier stages, but may be difficult to remove once fine crushing has liberated a substantial quantity of finely shattered mineral. Such primary slime may at times be a nuisance in the plant. It clings in dry-crushing machines and interferes with their work. It chokes the apertures in screens, consumes valuable reagents in flotation, and gives trouble in the thickening and filtration of finished products because of its reluctance to settle and its tendency to choke filter-cloths.

Some ores can receive a measure of beneficiation by the gentle disintegration at scrubbing or washing action. Loosely bound gravels, agglomerates, and clayey ores containing limonite or manganese nodules are examples. In the dredging of cassiterite in Malaya the richest tin-bearing gravel often lies immediately above a softened kaolinised clay, which can embed and carry to waste good tin if it is not dealt with early in the concentration. The *angle of nip* below which a crushing machine can seize and fracture a piece of rock is increased when the system becomes slippery, as can happen with slimy ore in the feed. Where fine material has been weathered down from the true ore body, the valuable minerals therein may have oxidised to form more or less soluble salts. It is sometimes desirable to wash such salts away in order to prevent their entry into chemical reactions in the flotation or cyanide section of the plant. This can be done while scrubbing the ore.

The disintegrating forces used are of a gentle nature, not intended to fracture the ore but simply to tumble particle against particle until the desired amount of cleansing has been accomplished. A water jet, or simple stirring, may suffice to detach slime from the lump material so that is flows out of the cleaning system as a dirty effluent while the cleaned ore particles leave by another channel.

In placer-mining the hydraulic monitor, gravel pump, hydraulic nozzle elevator, and dredge trommel help to disperse the clays they handle. The interior of the trommel, a cylindrical heavy screen through which the pay-gravel tumbles and is washed by jets of water, may be fitted with disintegrating blades rotating about a central axis, which chop up the clay.

In ordinary hard-rock mining, a jet of water from a nozzle may play on ore at any suitable point in transit for simple washing. More thorough in its action is the drum washer (Fig. 8). Run-of-mine ore is fed in at one end of a cylindrical or cylindro-conical shell as it rotates horizontally, and clean water enters at the opposite end. Lifting plates may be used in the shell to help raise the ore, which is thoroughly tumbled as it progresses toward the discharge. There it is elevated by perforated scoops, or raised by a spiral and discharged. This *counter-current* arrangement ensures that the entering ore is first washed by the dirtiest water as the latter runs out. After travelling through the machine, it receives its final washing from the cleanest newly-entered wash-water. In various forms, this counter-current principle is used in many mineral-dressing operations. Gentle agitation together with washing is sometimes performed on washing screens, which are specially adapted from the standard screens discussed later and are fitted with spraying arrangements or do their work in water. Some slime can be removed by washing ore on a rising conveyor-belt. The belt must be troughed and have provision for discharging the resulting pulp clear of the pulley mechanism. Removal is incomplete, but the cleansing of the rock surfaces aids recognition and hand picking further along the belt. A development from the relatively gentle action of the drum washer is that of the Telsmith super-scrubber. This machine rotates faster and gives a vigorous tumbling and rubbing action. It accepts (in the 96″ drum diameter size) material up to 8″ in size and can handle up to 180 tons hourly with a water consumption of 1100 gallons, using a 125 H.P. drive.

For heavy work, where more strenuous tumbling action is required, the log washer is widely employed. It is a shearing disintegrator, a tumbling device and a washer, and can give a measure of classifying action. It consists of an inclined tank (Fig. 9) in which one or two box girders rotate. These are armed with stirring blades (about four per foot run), set at a pitch angle of about 65°, so as to form part of two interrupted spirals. The blades rise inward and force the ore up-slope against a down-current of water. The raw ore enters from above, falling into a pool of water in the tank of the log washer, and pulp laden with slime escapes over a weir at the lower end. The volume of washing water must be so adjusted as to allow all the wanted ore to be worked upward against the stream running down to the discharge end.

Hand Sorting

Hand sorting or hand picking may be defined as removal from the run-of-mine ore (as hoisted with its associated country rock and detritus) of material which is not to be sent to the crushing and/or concentrating plant for treatment. The substances thus removed may be valueless gangue or waste rock not worth milling; unusually rich ore which does not need up-grading by

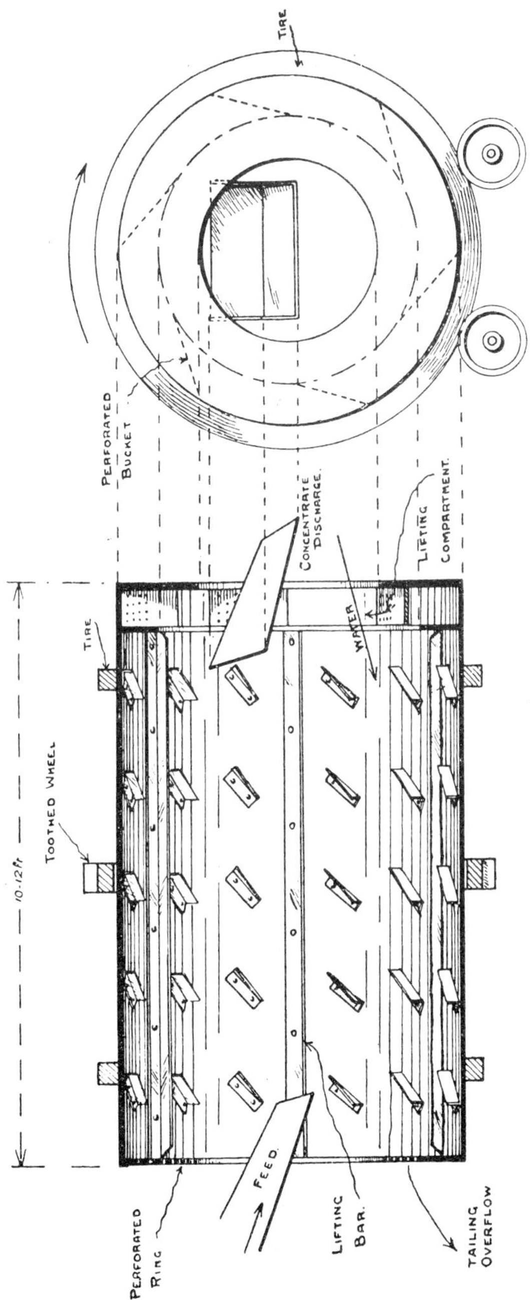

Fig. 8. The Drum Washer (after Truscott)

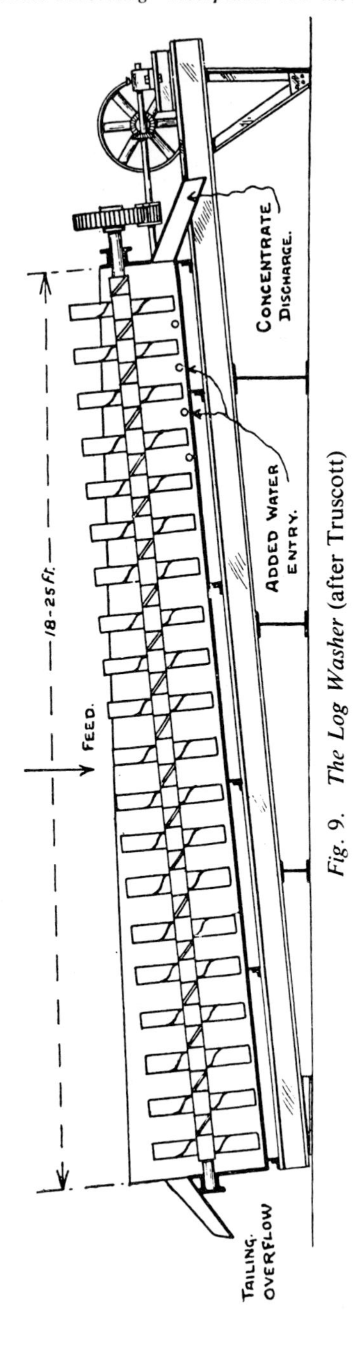

Fig. 9. The Log Washer (after Truscott)

ore-dressing methods; a specific ingredient of the ore which, for some reason, is not provided for in the concentrating section of the plant; or such mining detritus as timber, iron, and unexploded dynamite which might damage the plant or interfere with the concentrating processes. In the sorting of coal this hand picking was once the main treatment used to grade, to size, and to remove pyrite, slate, and shale. Sorting alters the assay grade by producing a richer and a poorer product and is to that extent a concentrating operation. Since it is usually a minor process which takes advantage of easily contrived display conditions in the earlier crushing stages, it is discussed at this point. On some Rand mines up to 30% of the mined rock is hand sorted. Removal of such a high percentage is called "hard sorting". With the development of mechanised methods of dealing with large tonnages, hand sorting has declined in importance, but in a small plant anxious to produce something saleable by cheaply contrived, though rather primitive, methods it can form a useful part of the flow-sheet. With cheap labour or primitive conditions all hand methods of beneficiation are important, as witness the large proportion of the world's tin which is concentrated by panning and hand sluicing. Sorting of some kind is done at most mines, if only ahead of the underground jaw-crusher whenever its operator removes broken mine timber from the feed.

The mineral to be sorted out must have a sufficiently distinctive colour, lustre, shape, or general appearance, so that it can be readily recognised and removed. Since the task is monotonous working conditions should be made comfortable, with good space and proper lighting. If the ore comes past the worker in a steady stream, and is displayed in a single layer three feet or so from the ground to minimise stooping, fatigue is lessened. Too large a lump of rock is difficult to deal with, and too small pieces take a lot of finding and removing, thereby slowing down the work. The size limits between which hand sorting is considered most efficient are a maximum particle diameter of 12″ and a minimum of about $2\frac{1}{2}$″. Shadowless "daylight" lighting from a diffusion lamp source is a good alternative to natural daylight. The minor constituent should be handled by the sorter, and the major one left to be transported mechanically.

In British and North American practice little or no hand breaking of the ore is attempted during the sorting operation, so there is no need for the supporting surface to act as an anvil. In Latin America and some European and Asiatic practice the ore may be broken, "cobbed", or "spalled" with hammers to remove a desired portion. The most elementary arrangement is a level floor where dumped ore can be picked and cobbed, the separated fractions then being shovelled or swept away. Moving surfaces which bring the work to the picker make for higher efficiency and bigger throughput. A pan conveyor withstands shock loading and can handle large rocks. For all-round convenience, the belt conveyor is the most adaptable arrangement and is in wide use. Operators can stand or sit on one or both sides, from three to six feet apart, and push or pull rock off the passing belt into appropriate chutes. A wide and flat belt is best, fed so as to be clear of material for a few inches from the sides. It can run up to a speed of 60 feet/minute. If the feed is moist from washing, the belt should slope to aid drainage. An advantage of belt sorting is that it can be combined with transport and elevation of the ore, thus giving this operating service at little beyond the cost of the labour required to remove and dispose of the products. The revolving sorting table has fallen into disuse, as its sole advantage over belt sorting is centralisation of supervision when an unusually valuable material is being handled. Youths, women, and convalescent labour are commonly employed in sorting.

Consideration must be given to the relative cost of milling run-of-mine and sorted ore, and the variation in total recovery with and without such sorting. From time to time the rejected fraction should be assayed, to ensure that no milling-grade ore is being lost, and the weight removed should be recorded and brought into account when calculating the accepted tonnage. Hand methods of sorting are sometimes assisted by the use of fluorescent light, which cause some minerals to glow and thus reveal their presence. Geiger-Muller tubes are also used, suspended above the picking belt so as to signal the arrival of a piece of radio-active ore.

When the tonnage to be handled justifies the cost and the difference in density between payable ore and that too poor to mill is sufficient, the dense-media process (DMS) has replaced hand sorting. It can work cheaply on a wide range of sizes.

Mechanized Sorting

Low-grade ore deposits can only be economically exploited on a large scale. With the increase in this class of mining visual recognition tends to be replaced by mechanical sensing devices. These typically assess a specific quality in a passing particle of ore and send a signal to a mechanical or electrical device which removes that particle from the traffic. Such devices may be solenoid-operated or worked by an air blast, deflecting device or trap. The requirements for this rather young technology are:

(*a*) A quality in the selected mineral which differentiates it sharply from the main ore minerals
(*b*) Sufficient cleansing of all passing particles to prevent blurring of the signal by superficial contamination
(*c*) Sufficiently close sizing of the fed particles
(*d*) Display of the particles one by one to the sorting device
(*e*) If helpful, special excitation of an otherwise dull signal
(*f*) Adequate robustness, simplicity and cheapness of operation

One quality, the relative densities of sub-marginal and pay-grade ore, is widely used in the dense-media process, dealt with in Chapter 12. With current advances in applied physics it is possible that further useful methods will emerge. Waste rock can be used underground as roof support, but is a loss-item all the way from the stope to the final disposal dump. The economic factors in mechanised sorting, as in hand sorting, must consider not only process cost, but also the saving on transport, reduced mill head accepted for treatment, and disposal of rejects.

Characteristics of the particle which may activate an appropriate sensing device include its:

(*g*) conductivity
(*h*) specific gravity
(*i*) reflection and refraction of a light beam
(*j*) light sorption
(*k*) radio-activity, natural and induced
(*l*) ferro-magnetism

These characteristics must be considered conjointly with those listed above (*a–f*). For example, the quantity of radiation emanating from a particle weighing, say, 10 grammes might be insufficient to activate an ejecting mechanism which would act on a 20 gramme piece of the same mineral, with twice the signal strength. Conductivity measurement can be made by allowing suspended wires to brush the surface of passing lumps of rock. Density differences are exploited in the DMS process (Chapter 12). A beam of light can detect specific colour, lustre or reflecting angle, the last being used to sort diamonds. Ferro-magnetism is considered in Chapter 22, and in the next section of this chapter. Short-life radio-activity can be induced in some minerals.

Although mechanized sorting—apart from the DMS process—is a newcomer to mineral processing, the following industrial applications are now well established. The La Pointe picker was developed in connexion with the sorting of moderately coarse radio-active concentrates, prepared by jigging, during which they are washed and sized, so that there is no random activation by adherent radio-active dust. A single line of particles moves on a narrow belt, and passes below a Geiger-Muller tube. If radio-active emission exceeds a set level a sorting device is operated, which removes the signalling particle from the moving line.

Another arrangement triggered by the particle's own radio-activity is the K. & H. equipment. Run-of-mine ore of between 2″ and 8″ lump size is washed and delivered piece by piece, so as to fall past three successive scanning devices. The first is a neon light working on A.C. at high frequency, so arranged that its beam falls on a photo-electric cell save when interrupted by the falling particle. This interruption, by measuring the transit time, notes the size of the lump. Next, the particle passes a *scintillometer*, which determines whether it is above or below the pre-set level of radio-activity for this size of particle. Finally, it falls past a series of air jets which are electronically controlled in accordance with the signal received from the previous scanners. Pay ore continues to fall unhindered, but a piece which is below millhead grade is blown sideways as an air blast operates, and falls to a waste conveyor. One of these units treated between 25 and 50 tons of Rand gold ore hourly in a demonstration run. In many of the banket ores there is a fairly consistent ratio of gold to uranium, and this can be used to signal low-uranium (and therefore low-grade gold-ore and waste rock) out of the feed. At Eldorado in Canada, ore larger than 3″ is similarly sorted.

Removing Tramp Iron

Although iron and steel which have inadvertently found their way into the ore stream may be removed by hand during such operations as have just been described, more positive protection against their entry is usually made. If a piece of steel enters a crushing machine damage may result and production may be held up while repairs are made. "Tramp iron" is normally magnetic, and can be removed by hanging an electro-magnet over the belt conveyor or other suitable point. From time to time the magnet is swung to one side and unloaded. This arrangement may fail to catch small items such as nails and bolts which are held down by pieces of ore, so belt conveyors are sometimes furnished with an electro-magnetic head pulley supplied with current from a D.C. source. Iron thus held next to the belt is not discharged at the delivery end of the conveyor but is carried round and dropped well clear of the ore stream, after it has left the zone of magnetic attraction. These devices are discussed in the chapter on magnetism.

Another protective device is the search coil. This is a sensitive device which can detect a slight concentration of magnetic flux such as arises when a feebly magnetic material passes through its coil. Some alloy steels used in the mine are non-magnetic, and there is always the further danger that a piece of metal may be pinned down or be too heavy to move. One mine uses three search coils on its feed conveyors to deal with this possibility. When a piece of iron or other metal passes the first of these coils, a current is generated which can be amplified and caused to trip the motor of the feeder supplying ore to the conveyor belt. The second search coil detects the same piece of metal and stops the conveyor, the distance between these coils being such that the crusher has worked itself empty before the conveyor stops. Thus the inconvenience of a build-up of ore on a stationary conveyor is avoided. The third search coil is an extra, which would operate in the event of either of the first two failing, or if the shiftsman failed to remove all the iron. Normally, he locates the tramp iron, picks it off, and then restarts the belt. In another type of induced-current detector, flour is dropped on the belt as an alarm sounds and the driving motor trips out, so that if the appliance continues a short distance before stopping the trouble can be quickly located. The use of guard magnets and hand picking is sometimes combined.

Reference

Gy, P. *Minerais et Metaux.* Paris

CHAPTER 3

PRIMARY CRUSHING

Preliminary

Crushing, grinding and other words or phrases associated with the size-reduction of ore and other rock are all comprehended in the word "comminution". This (Truscott) is "the whole operation of reducing the crude ore to the fineness necessary for mechanical separation, or for metallurgical treatment..."[1] It is usual to make an arbitrary division of comminution into convenient stages. Primary crushing brings run-of-mine ore down to a maximum size of the order 4″ to 6″ in average diameter; secondary crushing receives feed at —6″ and reduces it to below ¾″. "Dry" crushing includes work on ore as mined, which may be somewhat moist when delivered. It is succeeded by comminution in water, arbitrarily called "grinding". Although a considerable amount of fine grinding is done by dry methods, this book follows usage by reserving the word "crushing" for an operation predominantly dry and "grinding" for work on a suspension of ore particles in water. One important difference between dry and wet comminution lies in the mode of seizure of the particle. In the former case the particle is large enough to be gripped between two solid steel members as they are pressed together by mechanical forces. One or both of these members moves to and from in a fixed-path cycle. The rock gravitating through the rapidly expanding and contracting gap thus produced is nipped and crushed. In wet grinding the bulk of the ore is already too finely divided for a particle to be seized in this manner. It is therefore exposed while more or less free to move, to random blows. There are exceptions to this generalisation.

Machines used in dry crushing must work in dusty conditions, even when the main cause (escape of fine particles at transfer points) is dealt with. They are usually worked intermittently, to fit in with the hoisting and delivery programme of the mine. On completion of the dry crushing their product is delivered to bulk storage (in the mill's fine ore bins). From these it is delivered at a controlled rate to the more continuous grinding and concentrating processes.

The following classification of basic reduction steps is proposed by Hukki.[2]

Explosive shattering	From infinite size to minus 1 metre					
Primary crushing	,,	minus	1 m	to	minus	100 mm
Secondary crushing	,,	,,	100 mm	,,	,,	10 mm
Coarse grinding	,,	,,	10 mm	,,	,,	1 mm
Fine grinding	,,	,,	1 mm	,,	,,	100 μ
Very fine grinding	,,	,,	100 μ	,,	,,	10 μ
Superfine grinding	,,	,,	10 μ	,,	,,	1 μ

The nett energy consumed during equal reduction ratios in comminution increases with increased fineness of the ore being treated. From experimental evidence Hukki suggests an apportionment of this power consumption of 0·35 kWh/ton in primary crushing of a brittle solid, rising to 0·6 in the secondary crushing, 1·6 in coarse grinding and 10 kW/h in fine grinding in the ratios stated above. These, in his view, suggest an important change in the use made of the applied power through these stages. At the primary crushing level, the results correspond statistically with the requirements of Kick's Law. This phase is followed by approximate agreement with Rittinger's Law in the intermediate stage, while Bond's formulation becomes increasingly good at the finer end of comminution. These laws are discussed below. The enormous rise in power consumption when a product well below a micron (μ) in size is aimed at more or less rules out the use of standard grinding techniques on the score of cost.

Crushing

The main purposes are:

(*a*) Convenience in transport
(*b*) Production, for use without further treatment beyond screening, of graded sizes and shapes
(*c*) *Liberation* of specific mineral/s as a step in separate recovery from the ore
(*d*) Exposure of contained values to chemical attack
(*e*) Production of granular material suitable for treatment by gravity methods
(*f*) Development of particles suitable for feed to froth flotation

The methods of treatment are discussed later in this book, but must be recognised from the start as depending for efficiency on correct comminution. The proposed end use should dictate the stages and methods employed. A granite or limestone ballast for railroad beds or roadmaking would have no problem of specific liberation, but would be concerned with particle shapes and sizes as these affected packing, drainage through voids, and structural strength. Preparation for the processing of an ore would call for much finer crushing and grinding, in which the cost—which rises sharply when fine sands must be prepared as feed to the concentrating plant—would be an important factor.

The earliest stage of rock breaking is in connexion with the severance of the ore from its lode. This is performed by the use of suitable explosives applied in such a manner as to produce lumps of rock of a size convenient for handling in the mine's transport system. The interest of the alert ore-dresser can well commence at this early point since the manner in which the abrupt stresses of explosion are applied to the rock *in situ* is one determinant of its *size range*, crushability, and total surface per unit volume from that point on to the next crushing stage. The lavish use of high explosive underground is to be deprecated, first on grounds of cost, second because the finely shattered ore thus produced is difficult to gather and transport, and third because the delicate reactions used in treatment are jeopardised by random exposure of small particles between stope and process control in the mill.

Ore, as broken, may range in size from lumps weighing several tons downward, but delivery passages, chutes, gates, trucks, and skips work best when they are not exposed to the shock-loading of large pieces of dropped material, and "pack" best when the sizes transported are reasonably *close-ranged*. Severance may be followed by the use of explosive on large pieces of ore lying in the stope, by sledging, breaking down lump ore on a grating protecting an ore chute underground, or at an underground crushing station, where a jaw crusher of the Blake type is frequently installed. This machine should be set in the updraft ventilating zone so that the dust produced in crushing does not contaminate the mine air. Broken mine timber coming to the crusher with the ore must be removed. At all stages of ore treatment it is good practice to remove *tramp iron* and other mining detritus as early as possible. The feed to the underground crusher must be adequately displayed and illuminated, to facilitate hand picking by the crusher attendant.

Big tonnages of rock are crushed for use as graded sizes of stone or homogeneous rock for road-metal, ballasting, etc. Such "dressing" has nothing to do with mineral processing, in which crushing is a stage in the liberation of values. Washing and sorting may be applied to the ore in transit and simple concentrating treatments may be used at the same time, but the main purpose in crushing is to reduce the size of the rock particles by suitable stages so that the most efficient use of force is made and unnecessary comminution is avoided. For some purposes dry crushing must be used throughout the work—e.g., in the grinding of cement, dressing of mica, talc and some other minerals. Generally comminution commences by dry crushing the ore to below a size established by tests, and finishes by wet grinding to the required liberation size or *mesh-of-grind* (m.o.g.). The changeover from dry crushing to wet grinding lies between $\frac{3}{4}''$ and $\frac{1}{4}''$ and tends toward still lower sizes with the introduction of tougher alloys, improvement in design, and better methods of lubrication. These have led to the development of crushing machinery which can withstand the severe working stresses involved when large tonnages are crushed to gravel size.

Crushing Theory

The forces used to produce fracture of a perfect crystal are of two main types. The structure is bound together by its inter-atomic forces of attraction. Stress as considerable as 10^6 p.s.i. (pounds per square inch) is required to disrupt this bonding or "theoretical strength", which can be calculated. If tension is applied the crystal stretches elastically until it reaches its yield point, and recovers if the stress is removed before this point is reached. Once the elastic limit is exceeded a flaw is produced, usually in the form of a minute crack, which becomes a focus for incipient fracture. In his classic paper Griffiths[3] noted that the stress which the crystal can thenceforward withstand is inversely proportional to the square root of the length of the crack. This means that it will now fracture at a much diminished stress loading. During the stressing which created the imperfection, work was done to overcome the mutual bonding of the inter-atomic forces. This work was stored as elastic energy, and was released as the atoms returned to their normal positions. Since a crack existed, the atoms in its vicinity were able to shed elastic energy while the crystal was being stressed. If this released energy was sufficient to overcome the weakened inter-atomic bonds at the tips of the crack, it grew rapidly (at a speed of about 15,000 ft/sec). The fact that stress of the atomic bonding is focussed at crack-tips has been proved by polarised light studies of plastic materials.

Such a crack in solid material may start as a scratch or a surface blemish (a superficial discontinuity), as a minute fissure in a crystallite structure, or as a defect in the atomic lattice of a crystal grain which, by yielding, permits the start of plastic flow. Most rocks and man-made materials contain such foci of weakness, so that practical strength falls far short of theoretical. Ideal glass should withstand stresses up to 10^6 p.s.i., but in practice failure occurs at about one ton p.s.i.

Where dislocations exist, crystals which would otherwise deform plastically as one plane slides freely over another, have this plasticity blocked at each such dislocation. Stress builds up at this point and the surrounding atomic bonds are ruptured. The result is a minute crack which, given a little relatively light further stressing, becomes a complete fracture. Once crack propagation begins it proceeds at nearly the speed of sound. A corollary is that once a crack is running at this speed, no further stress applied to the main structure can catch up with the advancing fracture.

It is thus clear that two kinds of stress can operate—the reversible, which is removed if the crystal is not loaded to its elastic limit, and the irreversible in which surface discontinuities are formed or plastic deformation occurs. This last is the result of the slipping of one plane of atoms over another. The slip is local and so produces a deformed area, with one side in compression and the other in tension. Energy is thus stored ready for use in further deformation. Similar storage builds up in zones surrounding dislocations which prevent the free sliding of one atomic plane over another. Naturally occurring rocks contain random focal points where stresses due to defects are stored till an initial crack is produced by a relatively small further stress. Further weakening factors are the explosive shock during blasting of ore and its chemical oxidation between severance and milling.

In a piece of ore at least two mineral species are inter-crystallised in various patterns. The situation is far more irregular than that in the perfect and completely pure crystal with which this discussion commenced. The practical application of crushing theory must therefore add statistical methods of testing and an empirical approach to the considerations outlined above. No two lumps of ore are precisely similar. It is common to think in terms of such varieties of disruptive force as compression, tension, shear, torsion, abrasion and shatter. The last four are compound forms of the first two. When a beam suspended at its ends is centrally loaded till it bends, the lower part is in tension while the upper part is in compression. When seizure of a piece of rock occurs in such manner that the seizing forces move in opposing directions while the rock is prevented from rolling, tensile shear predominates, with local compression where its high points are gripped. Most crushing force is applied compressively. A piece of rock can be thought of as either a column or a beam loaded beyond bursting point. A piece of ore sufficiently large to be gripped forms a short pillar between approaching faces and is loaded to failure. If the ore bridges the crusher faces, beam-loading results. With the exception of the dynamic class of crushing device, in which the particle becomes a projectile in a fast-moving gaseous stream and acquires sufficient velocity to cause it to shatter itself on an impact plate (special cases which will not receive further attention in this book), application of stress by the crushing device is very slow in relation to the rate of propagation of a running crack. This is true even for such impact crushers as stamps and hammer mills.

Many attempts have been made to establish crushing principles on an unassailable basis of fundamental law. Kick's "law"—which Gaudin[4] rightly says should have been regarded as a postulate—states that "the energy required for producing analogous changes of configuration in geometrically similar bodies of equal technological state varies as the volumes or weights of these bodies" (Stadler). The Rittinger law, which should be regarded also as a hypothesis, states that the energy necessary for reduction of particle size is directly proportional to the increase of surface. If the sole force operating to produce crushing was used to disrupt molecular bonds along planes, and thus to produce completely severed new particles, Rittinger's law would be in line with current physical concepts, but this is an over-simplification.

Kick's law can be stated in equation form thus:

$$E_r = a \tag{3.1}$$

where E_r is the energy required to achieve a specified reduction ratio. The equation for Rittinger's law can be written:

$$E_r = \frac{b}{d_o} \tag{3.2}$$

where a and b are constants and d_o the original particle size. If d_p is the product size, the Kick and Rittinger hypotheses can be reconciled by Dobie's[5] equation:

$$E_r = a + a_1/d_p \tag{3.3}$$

Something of the difficulty facing the research worker in the field of comminution may be appreciated when it is recalled that four types of binding force are recognised in fundamental physics.[6] Taking nuclear force (that binding a proton to a neutron) as unity, then electrostatic force (proton to electron) is 10^{-2}, nucleon decay force (emission of β particle) 10^{-14}, and gravitational attraction indefinitely weak. Nothing is at present known of the balance between these forces which must be upset for cleavage to occur. Crystal study shows, for a homogeneous material, three types of imperfection which affect resistance to shear. Micro-defects are lattice imperfections due to irregular ion-distribution. They probably have a sub-threshold effect on comminution, though as will be seen when the physical chemistry of mineral surfaces is studied, they modify flotative reaction. Macro-defects are incipient strain areas, flows or discontinuities in an otherwise regularly repeated lattice structure which in the perfect crystal would be an orderly multiplication of its unit cell. "Mosaic defects" are typical in crystals in which orderly blocks (type 2) are constituents of the overall imperfect particle. Fortunately for the mineral engineer, these considerations are of less importance than the empirical approach.

Basic research on the crushing characteristics of multi-phased rock such as a piece of ore is hampered by the fact that no two pieces are alike. Repetition tests cannot therefore offer reliable evidence. Since comminution is the most expensive part of treatment and also exerts a critical influence on both concentration procedure and percentage recovery of values, empirical test procedures have been developed for use in each specific problem and for each specific ore body. The main practical interest in a given case centres on two factors. The first is the manner in which the association of different mineral particles in a piece of rock can best be disrupted. The second is the effect on subsequent treatment of any new development in grinding methods.

A new approach to the theoretical consideration of crushing has been made by F. C. Bond.[7] He commences by challenging three assumptions which weaken the classical theories. These neglect the work previously done on feed particles under examination, although such effort is part of the total input of crushing work. Secondly, the old theories were arrived at from study of breakage of cubes, not of the irregularly shaped particles handled in milling. Thirdly, these theories equate useful work input against energy increase on breakage and neglect the energy released as heat as "wasted" or external to the problem. Bond proposes a theory intended to give consistent results over all size reduction ranges for all materials and machines.

"It appears that neither the Rittinger theory, which is concerned only with surface, nor the Kick theory, which is concerned only with volume, can be completely correct. Crushing and grinding are concerned both with surface and volume; the absorption of evenly applied stresses is proportional to the volume concerned, but breakage starts with a crack tip, usually on the surface, and the concentration of stresses on the surface motivates the formation of the crack tips."

This argument follows the widely accepted hypothesis of Griffiths, discussed above. His theory postulates, from experimental and mathematical development, that the bulk of the input work is used to deform particles and is released as heat through internal friction. If local deformation exceeds critical strain a crack tip forms, the surrounding stress energy flows to it, and a breakage follows. Although the new surface may finally represent conversion of input power into new surface energy, the main purpose is to form crack tips. Bond suggests that the total crack length multiplied by the net energy used to form a crack of unit length gives the theoretical net energy input for a given amount of comminution and, expressed as a fraction of the total energy input, gives the mechanical efficiency of the work.

He states his theory thus:

"The total work useful in breakage which has been applied to a stated weight of homogeneous broken material is invariably proportional to the square root of the diameter of the product particles."

This is formulated thus:

"When F is the diameter that 80% of the feed passes, P is the diameter that 80% of the product passes, and W is the work input in kw.-hr. per ton; then Wt, the work input required to reduce from an infinite feed size, is

$$Wt = W\left[\frac{\sqrt{F}}{\sqrt{F}-\sqrt{P}}\right] \tag{3.4}$$

When F and P are in microns and Wi is the work index, or total kw.-hr. per ton required to reduce from infinite size to 80% passing 100 microns, or to approximately 65% passing 200 mesh,

$$Wi = W\left[\frac{\sqrt{\mathrm{F}}}{\sqrt{\mathrm{F}}-\sqrt{\mathrm{P}}}\right]\sqrt{\frac{\mathrm{P}}{100}}\text{."} \tag{3.5}$$

The Rittinger theory deals with measurement of surface areas, the Kick theory deals with volumes of product particles, and Bond's "Third Theory" with crack lengths formed. None of these theories can be regarded as adequate, though to some extent they are complementary. A present weakness of the "third theory" is that it considers only the linear dimension of a surface rupture, whereas this is part of a new area. Against this must be set the fact that a comprehensive set of Work Index figures has been prepared as a result of extended tests on a variety of ores, and as its author says:

"The Third Theory and its work index permit closer predictions and more accurate comparisons of all crushing and grinding installations than have heretofore been possible and should result in more efficient operations." A step towards reconciling these three theories has been taken by Holmes.[8] He points out that the statement for new surfaces

$$\text{Energy} \propto \text{surface area}$$

fails to consider the energy absorbed during the process of elastic deformation. The initial size of the particle undergoing deformation and fracture must be a dominant factor. Kick's law, though applicable to homogeneous and annealed material, is not valid for non-homogeneous rock including various weakness zones. Bond's theory, being developed from much testwork, provides an excellent basis for calculating the relative performances of crushing systems. The "work index" values, however, are not constant for fixed conditions of reduction, and the theory has the limits of empiricism. Holmes propounds a modified form of Kick's law, suited to ore, in the equation

$$B = kD^{1-r} \tag{3.6}$$

where B is the theoretical elastic energy absorbed in deformation to produce unit area of fracture in a cube of side D, r and k being constants. For a reduction ratio R and product size P this can be written

$$W = Wi\left[1-\left[\frac{1}{R}\right]^{r}\right]\left[\frac{100}{P}\right]^{r} \tag{3.7}$$

where Wi is the work index and r the "Kick's Law Deviation Exponent" which expresses the degree of variation of particle strength with variation in size, specific to the material and the mode of stress application.

In a discussion of these hypotheses Schumann[9] summarises his own conclusions thus: ". . . readers may find it helpful to think of comminution from the following point of view: Any actual comminution product may be regarded as a mixture of "finished" comminuted material and coarse material not yet completely comminuted. Any comminution process simply converts the coarse material into finished material but does not change the size distribution or average size of the finished material itself. Only the relative proportions of coarse material and finished material are changed, and it is these changes in relative amounts of coarse and fine material that account for the continually decreasing average particle size of the overall product during comminution". This would seem as good a generalisation as can be made at our present level of understanding of this complex subject.

Physical Aspects of Comminution

An impression exists that one effect of crushing and grinding is to wear down the projections of a particle so that it becomes rounder. This can occur through prolonged chemical exposure, climatic erosion or corrosion, e.g. tidal action or the stirring of sand by wind. Although macroscopic particles are very irregular, they tend to conform to a prototype, in the shape of a truncated tetrahedron having a ratio between its three axial dimensions of 1:1·2 to 1·5: and 1·6–1·8. The apparent rounding of microscopic particles is due to optical refraction, and is seen as the wavelength of light is approached but actually the smaller the particle the nearer it approaches the true crystalline shape of the mineral. Such a shape, however, is unlikely to be developed during comminution.

When a piece of homogeneous rock is fractured by a single blow it breaks down into particles of all discernible sizes. Gaudin and Hukki[10] have investigated this shattering action, and find that when the resultant products are separated by screening into a series of products, the screens being in a geometric series such as the Tyler, the total surface produced will be similar for each close-ranged screen product. Ore is not homogeneous, and experimental difficulties in tests of this kind are formidable. Nevertheless, these authors' conclusion throws light on milling procedures designed to minimise *over-grinding*.

In the laboratory sieves are used to measure the size of particles down to some 37 microns (μ) in diameter. Below this point *sorting* methods are used.

These include *classification, elutriation, sedimentation* and other *sub-sieve sizing* techniques. Measurement with the microscope is used to estimate dimensions down to 0·25μ (250Mμ). The unit crystal is thought to have a maximum length of between 0·5 and 1·0 Mμ.

The energy expended in comminution is almost all converted to heat and sound. In terms of absolute grinding efficiency the effectively used fraction of the energy is measurable by the total amount of new surface created. This new surface increases the total amount of surface tension, a form of potential energy, and in theory the amount of power usefully employed in creating new surface should express the efficiency of the grinding operation, in the form:

$$\text{Efficiency} = \frac{\text{Power converted into new surface tension}}{\text{Total power applied}} \tag{3.8}$$

Unfortunately, no reliable method of measuring the surface tension of minerals has yet been found, helpful as it would be in many ways.

Only a fraction of the power used to energise grinding and crushing machinery appears as new surface (i.e. broken ore). Power loss in the mechanical system can be kept low by good engineering maintenance. Only a small percentage of the input power is consumed in the creation of new surface. In later chapters the methods used to avoid its waste and to limit *over-grinding* are considered. Good technical control leads to lowered cost and better recovery. In absolute terms the efficiency of comminution is estimated by various authorities at figures varying between 0·03% and "a few" %. The inter-atomic cohesive energy which must be overcome in order to create new surface—to crush the ore—is calculable. Hence, the low yield of used energy in ratio to that fed into the system is a challenge to research. Known machines and methods are unlikely to produce substantial new improvements. For practical purposes the best working measure of efficiency is the ratio between the weight of ore ground to "finished size" and the energy input.

The Crushing Sequence

If the ore sent from the mine is reasonably coarse, dry and free from slippery and clayey matter, it can be gripped by fixed-path crushing devices and the resultant fragments (or progeny) can gravitate rapidly through the system. In most operations of any size there are two crushing stages or more, on the general plan outlined in Fig. 16. In the first, or primary, section the run-of-mine ore (which has been reduced to an agreed maximum size before dispatch) is crushed to an agreed transfer size, at which it must pass through a screen to the secondary system. This is usually between 4″ and 6″ maximum particle size. It is determined by the *set* of the primary crusher. There is a working relationship between the size a machine can accept (its *gape*) and the maximum size which will pass through the discharge end—its *set*. This relationship is called the reduction ratio. It is governed by the stresses and strains imposed on the crusher during passage of feed through its crushing throat. The quantity of rock being crushed at a given moment, the progeny created at each nip, the percentage of voids between the particles, and the toughness of the ore are main determinants of a safe reduction ratio for a given type of machine.

To understand this clearly, consider a theoretical 24″ cube of rock which is to be crushed neatly into either 4″ or 6″ cubes (measured as length of one side). The original surface is 24 sq. inches on each of the six sides, or 3,456 sq. inches. On reduction to 64 cubes 6″ square the new area created is 10,368″². On reduction of a similar cube to 216 cubes each 4″ square the new area is 17,280″². The total crushing work in the two cases is in the ratio 1.7/1. Since the driving motor can only deliver a fixed maximum thrust and only part of this force is able to create new surface, the actual crushing ratio would be much higher than 1.7 to 1 and the machine would either stall on overload, or something would give way. Though reduction ratio is conveniently stated as ratio of gape to set, the basic consideration is the development of new surface. No known method can evaluate this, as part of the surface (e.g. internal fissuring) is invisible and, in any case, original lumps and their progeny are of all shapes and sizes. The percentage of input force ending up as new surface is probably much higher in the primary section than later, but at present an empiric approach based on practical experience governs manfacture and use of equipment.

Primary crushing machines are of two main types—jaw and gyratory. The largest can accept rock up to 84″ by 60″. At the discharge end the setting or "set" is adjusted in ratio to the biggest piece the crusher is to receive, usually from 4 to 1 up to 7 to 1. The latter would limit a machine with a 6″ set to feed smaller than 42″ tri-dimensionally. The bulk of the ore leaving a typical underground *stope* is well below such an awkward size and weight, which would be unwieldy in most mechanized handling systems. A 36″ cube of rock weighs over two tons. If through inadvertence *tramp oversize* (lumps too large to be seized by the crusher) arrives it obstructs the feed arrangements. Such material can be arrested on a grid, made of old mine rails flange up and suitably spaced, and be periodically broken down with a sledge hammer. Primary hands over to secondary crushing somewhere between *minus* 6″ and 4″. In opencast mining, where the problem of handling huge lumps of rock is not complicated by the need to get it through a mine tunnel and shaft, larger maximum sizes may have to be dealt with.

PRIMARY CRUSHERS

Jaw Crushers Gyratory Crushers	} Feed range from 84″ x 60″ down. Set, down to 4″.

SECONDARY CRUSHERS

Gyratory Crushers Jaw Crushers Cone Crushers	} Feed from 6″ down. Set, down to ½″.
Hammer Mills	
Rolls	

FINE CRUSHERS

Rolls Dry Ball Mills Hammer Mills Disc Mills Buhr Mills	Feed from about 1″ down.
Ring Mills	Feed from about $\frac{1}{4}$″ down.
Disruption by high pressure gases.	Feed already fine
Explosive shatter Fire-setting	From the solid rock.

At suitable stages, hand-operated hammering, pestle-and-mortar, etc.

Crushers which handle ore intermediate in size between some 4″ and $\frac{1}{10}$″, using added water in their work, include:

Stamps. Receive wide size range (up to —2″), deliver down to —60 mesh size.

Rod Mills. Receive —$\frac{3}{4}$″, deliver between —14 mesh and —18 mesh.

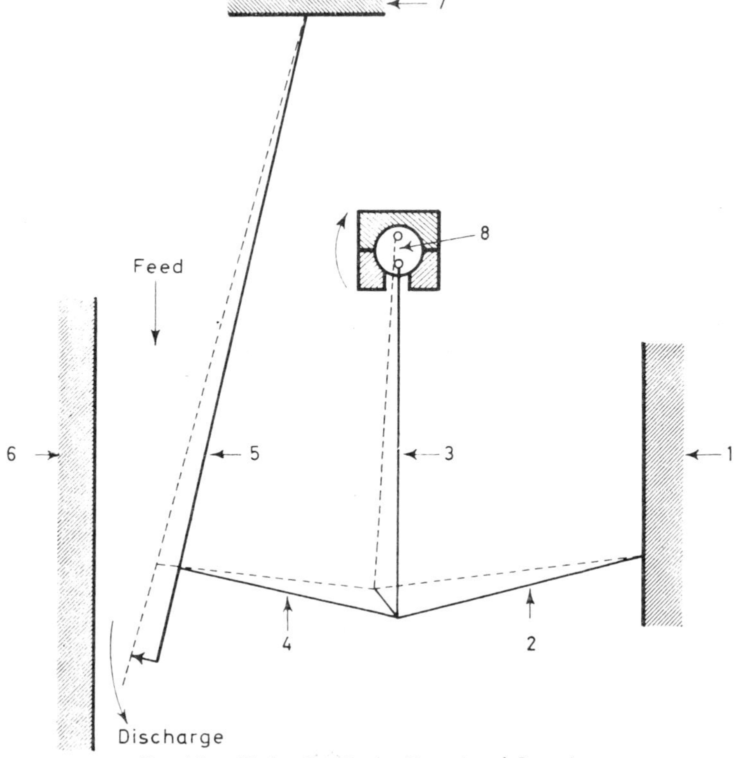

Fig. 10. *Blake Crusher. Functional Drawing*

Jaw Crushers

The Blake Crusher was patented by E.W. Blake in 1858. It was soon improved to the final form in which the entering feed received the least and the departing product the greatest crushing movement. Variations in detail on this basic form are embodied in the bulk of the jaw crushers offered by manufacturers today.

The functional drawing (Fig. 10) shows the essential details and Fig. 11 gives a typical cross-section. Parts (2), (3), (4) and (5) form a loose linkage which oscillates with a compound eccentric movement in the fixed framework (1), (6), (7) and (8). The size of the largest escaping particle is governed by the set, which is the horizontal distance between (6) and the tip of (5) when at the widest opening. The back *toggle* plate (2) pivots loosely from a bearing in (1). It is oscillated radially by the pitman (3), which is driven by the eccentric (8). As this toggle rises it presses the lower end of the pitman forward, a movement transmitted via the front toggle (4) to the swing jaw (5). The horizontal displacement is greatest at the bottom of the stroke and diminishes steadily through the rising half of the pitman's cycle. Thus, though the driving force applied through the eccentric does not vary, the horizontal travel of the swing jaw diminishes rapidly. Crushing force is least at the start of the rising half-cycle when the angle between the toggles is most acute, and is strongest at the top, when full power is being delivered over a reduced travel of the jaw. Since the jaw (5) is pivoted from above, it moves a minimum distance at the point where a large lump of ore has newly entered and a maximum distance at the discharge end.

Consider a large piece of ore falling into the feed end or "gape" of the jaw

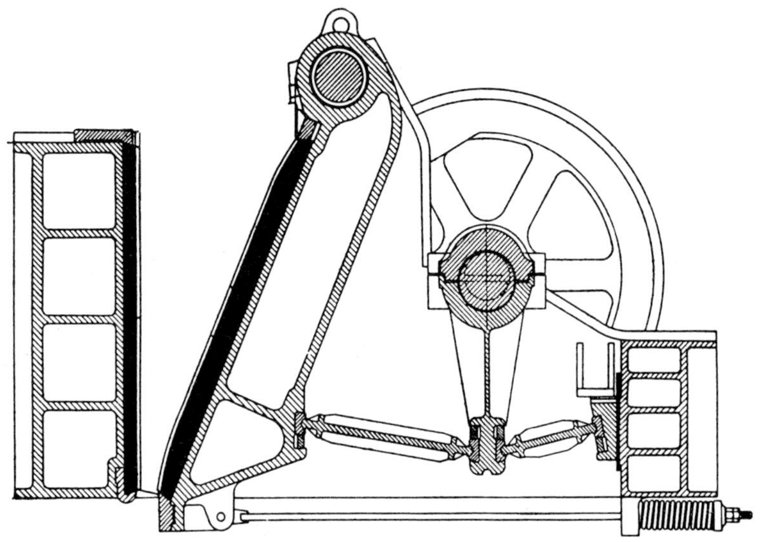

Fig. 11. *Section through Double-Toggle Jaw Crusher* (Allis-Chalmers)

crusher. The swing jaw is moving to and fro at a rate depending on the size of the machine and of the material it must crush (see Table 4). The running speed should not be so high as to strain the moving parts, which must withstand reciprocal action, severe loading on the compression stroke and sudden release on the return. It must also give time for rock broken at each "bite" of the jaw to fall to a new position in the constricting space of the crusher throat. The piece of ore falls till it is arrested, either above other ore, or by

TABLE 4

DATA ON BLAKE CRUSHERS

Gape	*Set*	*Reduction Ratio*	*Tons/hr.*	*R.P.M.*	*H.P.*	*Loss of Head*
7″ x 10″	$\frac{3}{4}$″	9·3	2	300	8	2′ 6″
	$1\frac{1}{2}$″	4·7	4	300		
12″ x 24″	$1\frac{1}{2}$″	8	20	300	25	2′ 6″
	3″	4	30	300		
18″ x 30″	3″	6	40	280	55	3′ 6″
	7″	2·6	80			
24″ x 36″	3″	8	55	250	80	4′ 3″
	5″	4·8	90			
30″ x 42″	3″	10	65	250	115	5′ 0″
	6″	5	150			
36″ x 48″	4″	9	100	225	130	6′ 3″
	8″	4·5	260			
42″ x 60″	6″	7	220	225	160	7′ 3″
	9″	4·7	320			
48″ x 72″	7″	6·9	310	150	210	8′ 0″
	10″	4·8	400			
60″ x 84″	9″	6·7	430	100	300	9′ 6″
	10″	6	450			

being nipped between the fixed and swinging jaw. Within a fraction of a second the moving jaw again closes on it, fast at first and then more slowly but with increasing power to the end of the stroke. Though the jaw only squeezes the ore for a short distance of its movement, this suffices to break the big lump. The fragments now fall to a new arrest point where they find themselves somewhat crowded, since the total cross-section is now less, while the overall volume has been swollen by newly created *voids*. At this arrest point another squeeze is delivered, this time with greater amplitude, since the radius of the swing jaw from its centre has increased. Crushing continues stroke after stroke until the crushed particles reach the lower end and fall clear. At each arrested fall the crowding together of the fragments would increase, owing to the combined effect of the increase in voids and the decrease in cross-section, were it not for the steady increase in amplitude of swing.

This accelerates the discharge of finished material, which works down and out at a rate sufficient to leave space for material arriving from above. Crushing under these conditions, in which particles are relatively free to fall between successive squeezes, is termed "arrested" in contradistinction to "choked" crushing, in which the volume of material arriving at a given cross-section would be greater than that leaving it if the rate of feeding was unrestricted. In arrested crushing the main force exerted upon a particle is directly applied by the jaws of the machine. In choked crushing a substantial amount of comminution results from impact of particle upon particle. The character of the product is different. Since in arrested crushing any particle small enough can escape at the discharge area, much of the crushed rock is finally delivered at a fairly coarse size. In choked crushing comminution continues even when particles are smaller than the "set". The difference is analogous to that of the orderly departure of a theatre crowd and of a panic in which people are crushed by other bodies arriving at the exits at too great a rate. The "set" of the Blake is the maximum opening between the jaws at the bottom, measured with the "V" of the toggles at the steepest point, with the eccentric full down. It is adjusted by using toggle plates of the desired length. Wear is taken up when required by adjusting the back pillow on which the end of the toggle bears. Since the toggles are loose in their sockets, a tension rod is used to hold the system together and to aid the return stroke of the swing jaw. A vertical spring may also be used to preserve smooth contact of the eccentric by acting upon the bottom of the pitman.

Arrested crushing can only take place if the rock broken during each nipping stroke falls with reasonable freedom during the return half of the jaw's swing. Since the overall volume swells with each stroke (owing to newly created voids), yet must drop into a decreasing horizontal cross-section, there would be congestion if the rate of fall was not steadily accelerated on the journey downward. This is made possible by a proper inter-relation of the following factors:

(*a*) Gape to set—the reduction ratio
(*b*) Rate of change of vertical cross-section of crusher throat in respect of fall of ore between strokes
(*c*) Speed and amplitude of swing-jaw strokes
(*d*) Sizing analysis of entering ore
(*e*) Crushing characteristics of ore

To minimise damage when uncrushable material enters with the feed, a weak point is built into the crusher. This can break and be quickly and cheaply replaced. In some crushers this is a weak belt-fastener on the drive; in others the driving pulley is weakly bolted to the very heavy flywheel of the Blake; the eccentric may be held down by weak bolts which break and allow the whole pitman to rise; or one toggle-plate may be scarf-jointed by a line of weak rivets. These protective devices should not be called upon. *Tramp iron* should be dealt with ahead of the crusher.

The heavy flywheel stores energy on the idling half of the stroke and delivers it on the crushing half, thus saving on driving power and smoothing the inevitable vibration of the machine. Since it works on half-cycle only, the reciprocating jaw crusher is somewhat limited in capacity for its weight and size. Owing to the alternate loading and release it must be very rugged, and requires strong foundations designed to avoid the transmission of vibration. The tough work the crusher must do calls for rugged mechanical details, good bearing lubrication, and generous cooling of the eccentric, possibly by circulating water. Forced-feed oiling is usual, with protection against the entry of abrasive dust from the ore into working parts. The lubricants must not be allowed to leak in such a way that they can contaminate the ore. Maintenance must be systematic but as the crusher plant usually works in rhythm with the underground hoisting programme and not continuously as does the concentrating plant, this can easily be organised. The dust generated during dry crushing should be trapped, drawn off, and safely disposed of. The "throw" or moving-jaw displacement varies from a minimum of $\frac{3}{8}''$ in small crushers to a minimum of an inch in big ones, the maximum possible being about three times the minimum in any case. With brittle material a minimum throw may be best. When the rock has pronounced elasticity so that it cracks, or deforms locally, as is the case with slabby and decomposed ore, far more movement is needed. Some adjustment is possible by varying the toggle-V, but the usual method of varying throw is to change the eccentric. The greater the throw, the better the evacuation of crushed material from the discharge end, and the less the danger of choking. Clay or "sticky" ore is liable to cling to the jaws. If it builds up in so doing, the set may be reduced to the point where arrested crushing is no longer certain, and increasing strain is thrown on the toggles and eccentric. Capacity is reduced and the quality of work suffers. If such conditions are serious enough to call for remedial action, the ore should be washed before crushing. Packing of the crushing throat by clay or fine material could lead to breakage.

The crushing jaws, which take heavy punishment in addition to abrasive wear from the passing rock, are protected by replaceable plates. This is standard in all machines which come in contact with ore, and with most conveying systems, gates and pulp launders. Since local crushing stress is severe, the wearing part must fit snugly on its supporting steel structure so that the load can be to some extent distributed. Heavy-duty machines frequently have an intermediary filling medium which is poured while liquid into the space between wearing part and support, so that it can set solidly and provide continuous backing. For lighter machines plastics or wood are used (e.g., behind ball mill liners). Jaw crushers lend themselves to close fitting by careful machining, but where jaw plates are reversible end for end, to take up wear, a backing medium may be used. The usual metals are zinc or babbitt alloy. One difficulty with molten metal, particularly when used between support and crushing head of a gyratory machine, is premature solidification, despite super-heating, so that non-continuous fingers or blobs form and the cavity is incompletely filled. Premature breakage of the wearing plate or cone may then occur, owing to "working" and deformation. Another arises when gold ore is crushed, if by mischance breakage leads to the entry of zinc into the cyanide process which follows. An organic chemical mixture[11] can be used, which sets soon after its two constituent compounds have been mixed.

Variations on the Blake

Variations from the simple Blake crusher are numerous. They fall into two main divisions:

(*a*) Variations in application of toggle motion
(*b*) Variations in slope of crusher throat

In the Telsmith crusher the movement is transmitted directly from an eccentric to the moving jaw. There are no toggles and the need for imparting reciprocating movement to a heavy pitman is avoided.

In the single-toggle crusher the swing jaw is hung on an eccentric and the whole of its surface is in lateral and vertical motion. In another crusher both plates move.

The effect of using curved jaws instead of straight ones is shown in Fig. 12.

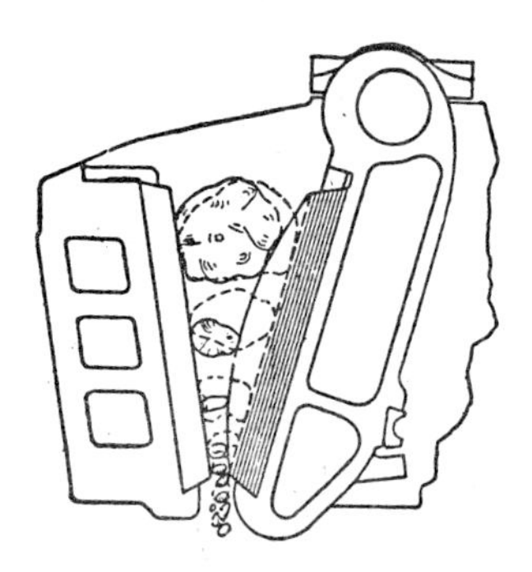

Fig. 12. *Crusher jaws, curved and straight*

The wearing plates which line the crusher throat are frequently cast with vertical wedge-shaped corrugations, so as to impart a measure of beam-loading to the applied force. Of the alloys used for these plates, manganese steel is most favoured.

The Dodge Crusher

This crusher (Fig. 13) reverses the jaw action of the Blake, in that it applies the maximum movement to the largest piece and the minimum to the smallest. The fulcrum is below, and only slight variation of set occurs as the moving jaw advances and recedes. Hence, for laboratory purposes where throughput is less important than close control of rock sizes, such a crusher, if lightly fed, can be made to do arrested-crushing work. If it is choke-fed, the choking becomes serious as the cross-section decreases. Rock crushes rock and there is not enough dilation as the moving jaw recedes to expedite departure of the finished material. This results in over-crushing, in which the crusher does work that would be better handled by grinding mills, and itself suffers

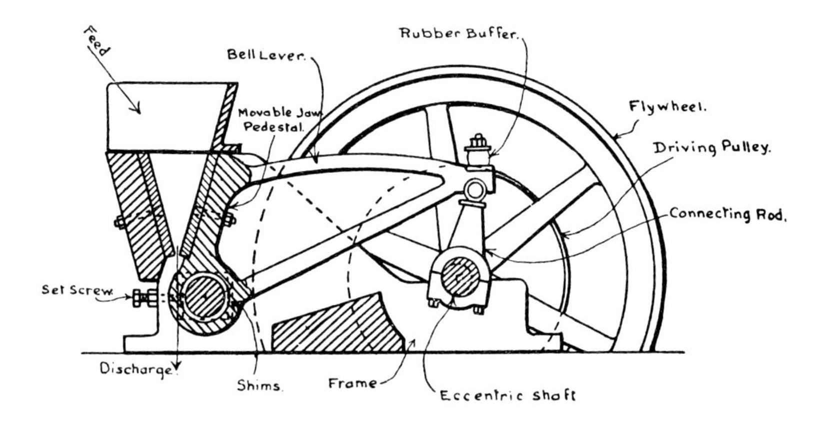

Fig. 13. *Dodge Crusher* (after Truscott)

severe strains in doing it. The practical result is that the Dodge cannot be built for heavy duty, and is rarely, if ever, incorporated in a flow-sheet.

Gyratory Crushers

The essential features of the gyratory crusher can be reduced to three elements (Fig. 14). In the diagram the spindle (1) is shown suspended from above. It is free to turn axially. The bottom end (1) is seated in an eccentric sleeve, and when this sleeve revolves the spindle sweeps out a conic path. Attached to the spindle is the crushing head (2), which gyrates with it. The combined system (1 and 2) is held in low-friction bearings at top and bottom and can either revolve as it gyrates or turn on its axis so as to roll on to rock lying between it and the fixed crushing throat (3). This is an inverted conic frustrum, called the concave or concaves. While the gyratory crusher (Fig. 15) runs unloaded, the head usually turns, rather than slides, in the eccentric sleeve. When ore is fed in, it is seized between head and concave. As the head rolls round, the gap between fixed and moving wall becomes steadily restricted and later relieved, pressure being brought to bear on the material in the gap. The maximum movement occurs at the bottom of the head, since it is at the longest radius from the fixed point. This differential dilation of cross-section favours relief of choking due to swell and downward contriction, and makes the machine a good arrested crusher.

The commonest type has a suspended spindle. It is marketed in two styles—long-shaft and short-shaft. In the latter the gyrating eccentric is set over the bevel drive, thus allowing a sturdier construction. Bearing and eccentric are sealed to prevent the entry of dust. Gears run in an oil bath and special attention is given to lubrication. The problem of the entry of an uncrushable body, such as tramp iron, is solved in one type by carrying the

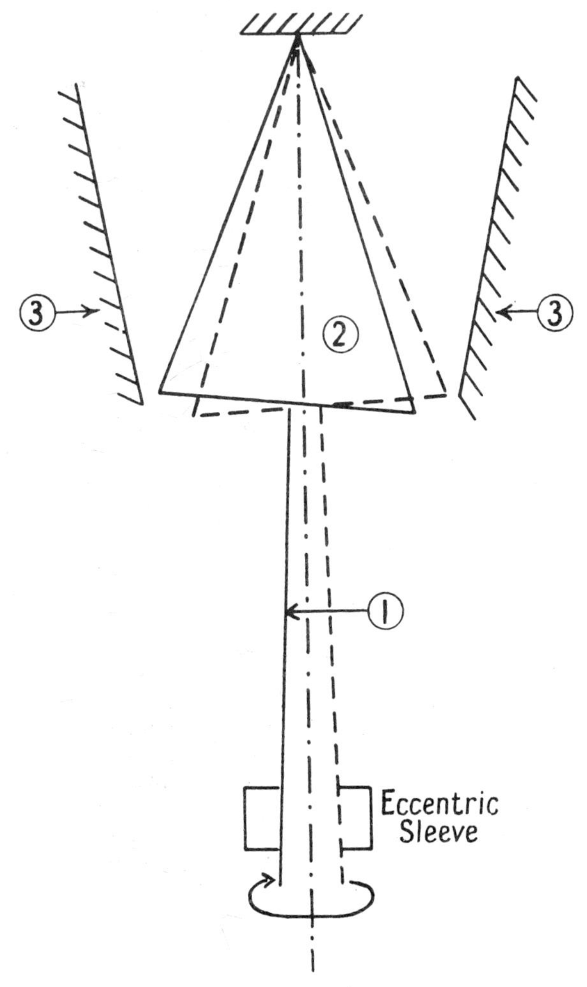

Fig. 14. *Gyratory Crusher. Diagram*

spindle on a hydraulically held mounting. If the machine becomes overloaded a valve opens, and the hydraulic fluid is released. The spindle then drops and the obstruction falls clear. With this machine, as with the Blake, an overload device can be used to trip the driving motor. The set is determined by the maximum opening at the discharge end, and this is either fixed by vertical adjustment of the spindle on its suspending nut or by hydraulic lift. Unlike the reciprocating jaw crusher, the gyratory works whole-circle, and thus runs more smoothly and with greater capacity. The smallest machine takes a 2″ feed down to $\frac{1}{2}$″ or even less, at 700 r.p.m., with a vertical

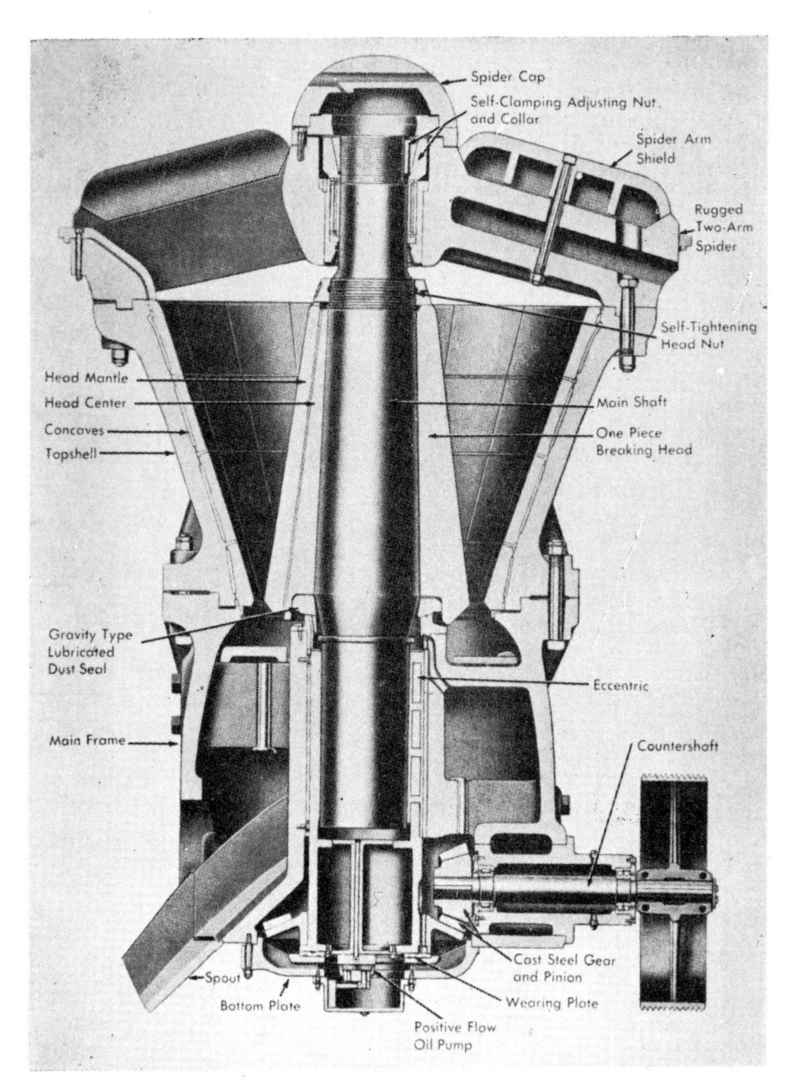

Fig. 15. *McCully Gyratory Crusher* (Allis-Chalmers)

drop of under two feet and an output of half a ton hourly for 4 h.p., while the largest model can receive 5-foot rock and reduce it to 12″ or less. These giants weigh up to 700 tons, need 500 h.p. and can drop over 3000 tons hourly through their 32 feet of machine height when running at 175 r.p.m. The throw, determined by the eccentric design, is less with the gyratory than with the jaw crusher, and is made greater for soft, tough rock than for brittle ore. Somewhat different in its application of the breaking force is the Telsmith crusher. Here, instead of a spindle which sweeps out an acute cone, the whole spindle remains vertical in its eccentric housing, thus applying the same amount of lateral movement all the way down the breaking head.

Comparison of Jaw and Gyratory Crushers

First cost, installation, and ease of feeding favour the gyratory crusher, which has not the problem of reciprocation of heavy and irregular loading stresses to meet in such an acute form as with the Blake. In jaw crushers half of each revolution is heavily stressed and half relaxed. This stress alternation demands rugged design, fatigue-resistant metallurgy and construction, and sturdy mill foundations. A crushing bowl can receive its feed from any direction and work with the top buried in waiting rock. Power consumption, whether idling, lightly loaded, or under full load, also favours the gyratory. Maintenance is heavier than with the Blake, and where transport is difficult, the machine is less easy to sectionalise. With clayey material the greater amplitude of motion of the jaw crusher is advantageous and its set can also be adapted more readily when there is a question of adjusting the amount of size reduction done between the primary and secondary crushing system. The most telling point in favour of the jaw crusher where large tonnages are not in question is that its gape allows it to handle more awkward oversize than the gyratory, the biggest machine, with an opening 84″ by 120″ taking a 6-foot rock. A gyratory capable of handling this would have a far higher tonnage capacity. With both types reduction ratio varies in practice between 4:1 and 7:1 and tends to be raised as improved engineering skill, metallurgy, and lubrication permit higher working stresses.

Mobile Crushing Units

These are increasingly used in connexion with the exploitation of small mineral deposits, in quarrying and where engineering projects call for a temporary production of local crushed and graded stone. Trains of machines are marketed which incorporate Diesel power units and various rock-crushing machines, together with screens and transfer bins. In opencast (surface) mining a mobile crusher can be fed direct by a digging shovel, and deliver its product *via* a bridge conveyor to a more permanent belt conveying system. Preliminary crushing is necessary, since large rocks should not be carried on belts. The competition between truck haulage and such a system is mainly one of cost, tonnage handled, ease of rock fracture and distance to be hauled. For long-distance work heavy trucks directly loaded from the digging machines are cheapest, but when a substantial tonnage must be elevated from the quarry floor and then moved a mile or so to the processing plant, on-the-spot crushing followed by belt conveying is substantially cheaper. In one installation a portable impact crusher reduces 30″ rock to *minus* 3″. The machine is mounted on rubber tyres and manoeuvred by the mechanical shovel it serves. For easily broken rock such as is quarried in the cement industry, equipment able to handle over 500 tons hourly is available. Heavy-duty machinery, in self-contained trains which deal with up to 70 tons per hour and use jaw or gyratory crushers, is made in Great Britain.

By-passing the Undersize

The run-of-mine ore delivered to the crushing plant contains rock of all sizes. A substantial proportion is already well below the set of the crusher, and if fed to it may produce certain disadvantages. The first is that the space in the crusher which could be occupied by ore in need of treatment is taken up by material already undersized. If this is by-passed (Fig. 16, line (5) to (7) from (—)), wear is reduced and capacity increased. The second is that it is cheaper to keep undersize outside the machine than to pass it through, as this would cause wear and extra dustiness. The third is that undersize tends to pack the voids between the big rocks and thus gives rise to choked crushing. The fourth is that any natural stickiness in the ore tending to cause hold-up in the crusher will be still more of a nuisance in the presence of undersize. Against these considerations is one which is always in the mind of the alert mill manager. Every manipulation introduced into the flow-sheet is an added liability, and every one which can sensibly be avoided tends to simplify operations. In a mill where the crushing plant is more than adequate for the tonnage treated, it will probably be better to send all the milling ore straight through the primary crushing section. If, however, the operation is so small that all the crushing is done in one machine, it is almost certain that the undersize should be removed first so that the available force will be applied to the coarser material.

The larger the tonnage treated, the more important it becomes to break down the work of size reduction into definite stages and to use machines set in series so that each deals with one limited range of size reduction. Thus a jaw crusher might be handling all the primary work, and substantial relief would be obtained by by-passing undersize. If a jaw crusher were used chiefly to deal with the very big material and to provide more efficient feed to a following gyratory, the by-passing might be done between the two crushing machines. Layout with respect to subsequent handling problems would be an important factor. A typical layout is shown in Fig. 16.

The surge bin (2) receives dumped loads from skips, lorries, or mine cars, and has enough storage capacity to maintain a steady feed through (3) to (4). Some hand-picking may be possible on (4). The grizzly at (5) is spaced in accordance with the ratio of reduction required. If no individual piece of mined ore exceeds 24″ in average diameter (or two-thirds of a ton in weight) and a four-to-one reduction ratio is used, then the primary crusher (6) will be set at 6″ to give this ratio. If ore is delivered at —20″ and a five-to-one ratio is satisfactory, then (5) and (6) can be set at 4″. The screen (7) will similarly be chosen to correspond in aperture with the set of (11), having regard to the reduction ratio. Item (8) is required to establish the *metallurgical balance*

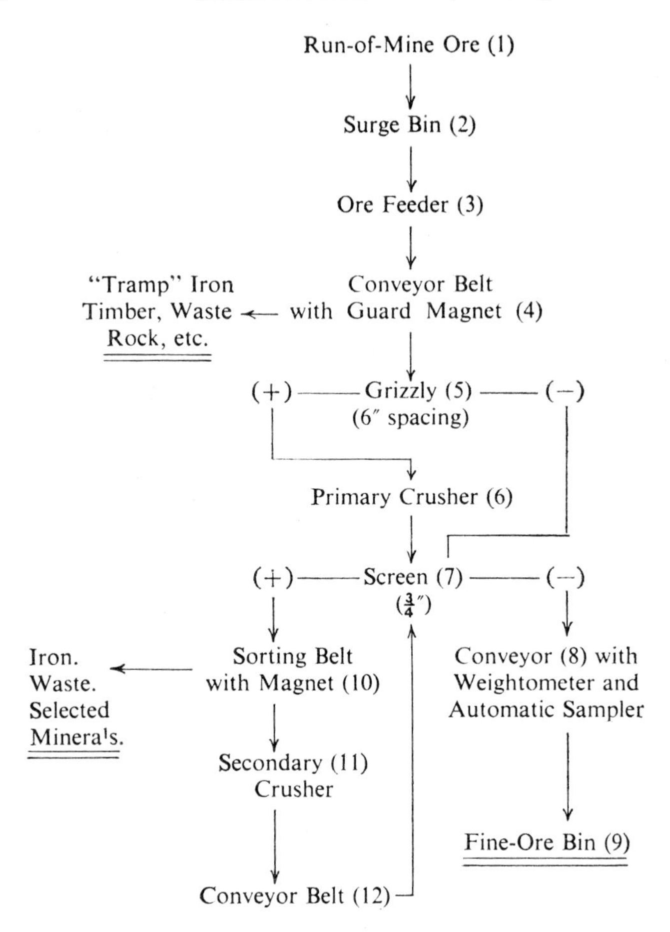

Fig. 16. *A General Crushing Flow-sheet, including Hand Sorting*

sheet (total units of value accepted into the fine ore bin). Waste and selected minerals removed at (10) and possibly at (4) are checked separately, to reconcile the mill's tally of ore with that sent from the mine.

Many operations involve "loss of head" where the feed gravitates through a machine. At one time this was a serious matter, and mills were sited on hillsides or provided with considerable height to ensure adequate fall through the process. Handling by modern methods has so improved in simplicity and efficiency that if the machines used to move ore or pulp are correctly chosen, properly installed and looked after it becomes a minor item, both for cost and maintenance. Loss of head need not, therefore, be feared when considering by-passing of undersize. At the same time, it must always be avoided where possible.

Two main types of appliance are used to sort the stream of ore into "oversize" which is to be sent through the crusher and "undersize" which by-passes it. Both are screens. The grizzly (Fig. 17) consists of an arrangement of steel bars having a tapered cross-section. These are held apart by distance-pieces and set with the thick end of the taper on the upper side of the sloping deck they help to form. Thus any piece of ore small enough to pass between two bars falls clear.

Old mine-rails set with the bulb of the rail down can be used. The grizzly must be sturdily made, as it suffers shock from falling rock. If stationary it must be set at a slope sufficient to ensure that ore slides freely. Slopes vary between 25° and 50° according to the stickiness of the ore, a usual grade being 40°. With a cantilever construction ore sets up vibration as it hits the grizzly, sliding is helped and loss of head can be reduced. Sometimes an arresting grid is set above the grizzly or between its discharge end and the crusher. The purpose is to prevent the passage of an occasional piece of ore so large that it might give trouble. This grid should be flat and stout enough to serve as an anvil on which oversize can be sledged by hand. Moving grizzlys have been developed in which alternate bars slide or otherwise move.

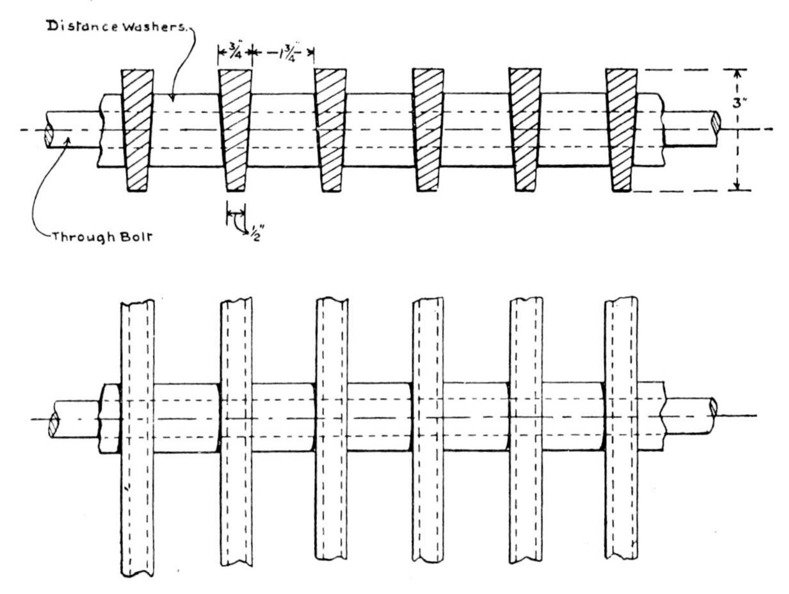

Fig. 17. *Section and Plan of Grizzly* (after Truscott)

One increasingly popular arrangement is the ring-roll grizzly, in which the passing ore rotates the rollers at slight cost to its kinetic energy, the rock thus being screened with but slight loss of working head. An advantage of a system with the lowest slope consistent with smooth passage of ore is that it is easy to arrest the passing ore stream and thus stop the feed.

Simplicity of control is worth planning for. If passage and delivery are gentle, there is less wear and tear. If a machine "stalls" or a transfer point becomes blocked by ore, or there is hang-up of material anywhere along the line, valuable time is lost and the smoothness of the general operation is upset. The subsequent feed reaching the blocked point must be cleared, usually by hand. One obvious precaution is to plan things so that as little rock as possible is in uncontrollable motion at any time.

Screens are not customarily used in the primary crushing circuit, where great robustness is wanted to withstand the shock of falling rocks which may weigh a ton or more. Screens are important in the secondary crushing flow-line.

Feeding Arrangements

Practice varies, but the broad problem of feed to the primary crusher must deal with the following points. (*a*) The mine or the quarry does its hoisting, transport of ore, etc., during part only of the twenty-four-hour day and probably only on a five-day week, perhaps with an occasional three-day holiday shut-down. The grinding section in many plants is feeding a process which it is uneconomic to interrupt, and is usually arranged for continuous operation. At some point, or points, therefore, a stock of ore must be held which can be built up during mine working hours and drawn upon continuously. The best place for this has been found to be storage immediately ahead of the grinding section of the plant. Ore bins having the necessary capacity are therefore planned between the crushing and grinding operations. (*b*) Most mines buy their power, or generate it, with an eye to peak-loading (maximum demand). The crusher house makes a substantial draught on power supply periodically, while the grinding section and following operations take their power continuously at a level demand. Hoisting is the other periodic power consumer which can be balanced against crushing. The hoisting rhythm is dictated by the needs of the underground management, and it may therefore be economic to hold the hoisted ore temporarily in bins until finished, when the crushing house can take over the power. If hoisting in its turn is being balanced against some other power-consumer such as air-compression for underground mechanisation, this pushes crushing toward the night shift. Ordinarily, ore is crushed as soon as possible after its arrival at surface. Expensive binning is not desired in two places for one job, and heavy loading near the shaft is sometimes undesirable. There is the further consideration that the ore coming to the crusher house is *long-ranged* and therefore awkward to restart after it has packed down, whereas if it is kept moving the big lumps will go along fairly easily. (*c*) Skips, dumper lorries and trucks, and other handling vehicles are intermittent in arrival, whereas the crushing section, once started up, calls for steady feed. To provide this, surge bins can be used. A surge bin is simply a convenient holding arrangement able to receive all the intermittent loads normally to be expected, and to feed them steadily through gates at controllable rates to the receiving machine.

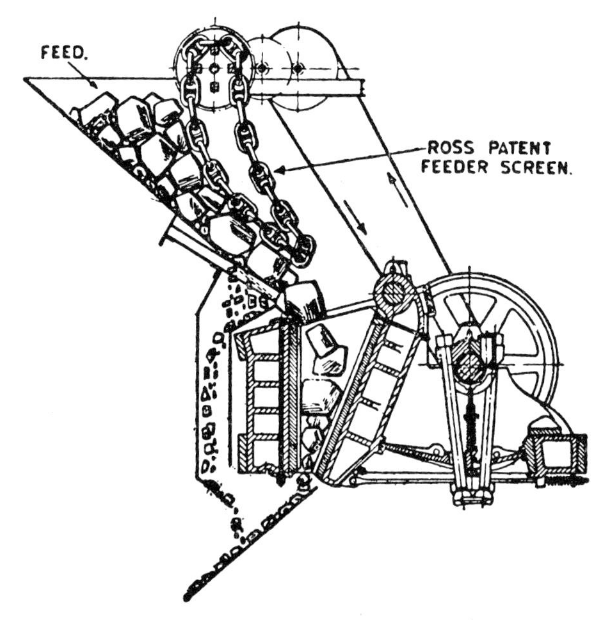

Fig. 18. *Ross Feeder and Jaw Crusher*

In the primary crushing stage large lumps of ore are handled and transfer points must be tough and readily accessible in the event of blockage. A machine widely used for smooth control is the Ross chain feeder (Fig. 18). This consists of a curtain of heavy loops of chain, lying on the ore at the outfall of the bin at approximately the angle of repose. It is controlled for rate of feed as desired, and can either be started or stopped by the crusher attendant or worked automatically. When the loops of chain move down, the ore on which they rest starts to slide. The whole arrangement is easily accessible for service.

Other feeding arrangements for coarse ore include the apron and a variety of reciprocating mechanisms, which push forward the ore lying at the bottom of the bin with strokes controllable for rate and amplitude. One heavy-duty machine, the jar-bar grizzly (Fig. 19), is particularly effective in handling clayey or sticky ore. It screens out undersize in the primary crushing circuit and is built to withstand shock loading.

Protective Devices

A Forum on protective devices for jaw crushers was reported in the *Canadian Mining & Metallurgical Bulletin* for October 1957. Several interesting points were brought out during discussion. In addition to the obvious main cause of overload (tramp iron or steel) which frequently leads to

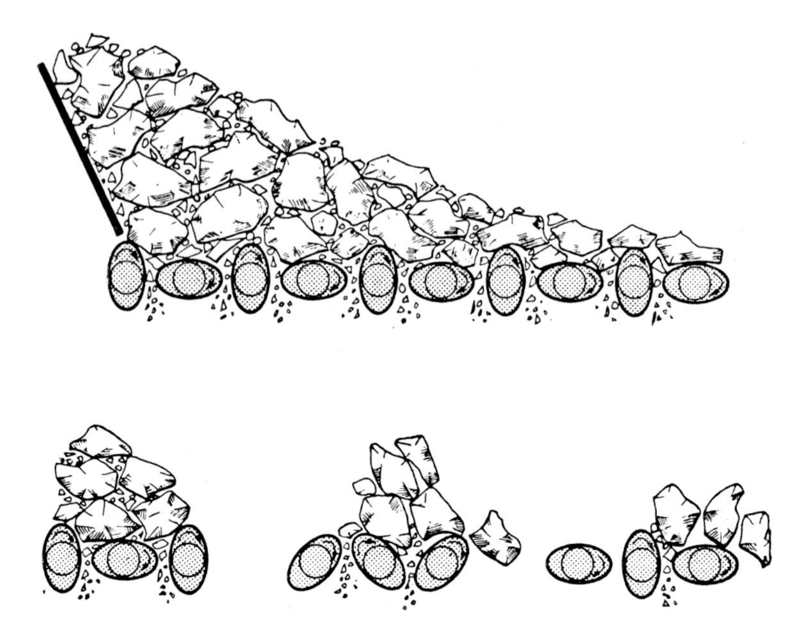

Fig. 19. Illustrating the Operation of the Jar-Bar Feeder. The lumps in contact with the bars are alternately lifted and dropped and at the same time moved forwards as the tips of the elliptical sections rise and turn over.

breakage unless the weak link in the crusher fails, there is an important secondary cause. The jaw crusher depends for normal operation on the fact that broken rock contains some 40% of voids. Near the discharge end, however, there is a reduction in volume between the open and closed side settings. If in this area feed conditions permit a substantially lower percentage of void space, the conditions of strain increase. If all the feed goes to the jaw crusher without a preliminary removal of fines, there can be danger when there has been segregation of coarse and fine material in the bin or other preceding part of the operation. Such fines could pass through the upper zones of the crusher and drop into the final (sizing) zone so as to fill what should be the normal voids. Should the bulk arriving at any level exceed that departing, it is as though an attempt was made to compress solid rock. This condition is commonly known as "packing of the crushing chamber". It is just as serious as arrival of uncrushable material such as tramp iron and can cause major breakage. Several speakers stressed the point that any protective device designed to fail on overload should act behind the heavy flywheels and not ahead of them. Such breaking devices as were originally used to give a weak anchorage of the driving pulley to the flywheel therefore fail, as do weak fuses in the drive motor, since they leave the stored energy in the flywheel to continue its work on the jammed crusher and thus lead to breakage. The usual breaking point is provided by a toggle plate. This is the least expensive part of the crusher, and toggle failure stops all further crushing movement. Various types of toggle for this purpose have been designed, the general preference being for a shearing type with the scarfed joint held together by tapered fitted bolts reamed home. Importance is attached to rigidity of this fit. Some development has been made with alternative methods which are not open to the objection of a failing toggle, that the crusher is shut down during replacement. Hydraulic protection is making some headway. In one form this is energised by the oil pump and accumulator system and arranges for a release of the drive linkage at a given maximum stress. This appears still to be on the drawing board. Fluid couplings, although useless ahead of the flywheel, are helpful for start-up, according to one speaker. Attention should be given to the toggle seats, which are at their best when made of manganese bronze and well lubricated. Special lubrication systems have been devised and are said to pay for their installation. A newly installed toggle seating should be run in carefully. Restraining ropes fastened to the toggles can prevent danger due to pieces being thrown into the crusher at the moment of failure. A fluid drive together with a pre-stressed friction clutch has been devised for use in the flywheel hub on the eccentric shaft. This combination protects against overload and also allows the jaw to stop immediately under excess loading. One speaker recommended development of an air-operated clutch for this purpose. The advantage would be that once the cause of overload had been cleared the crusher could immediately be re-started and thus avoid the present delay while replacing a broken or sheared toggle. Another recommended taking more care ahead of the crusher so as to reduce the danger of undesirable material being fed to it. A picking conveyor ahead of the crusher was suggested. This arrangement has since gained ground. Among other suggestions were those of using a molybdenum-based pre-coating compound followed by high-pressure oil for lubrication of the toggle seats, together with careful grinding and polishing of them to a high finish. Reduction ratios also were commented on and preference expressed for a series of low-ratio crushers rather than a dangerously high single-ratio reduction.

References

1. Truscott, S. J. (1923). *Text Book of Ore Dressing*. MacMillan.
2. Hukki, R. T. (1962). *Trans. A.I.M.E.* 223.
3. Griffiths, A. A. (1920). *Phil. Trans. Royal Soc.*
4. Gaudin, A. M. (1939). *Principles of Mineral Dressing*. McGraw Hill.
5. Dobie, W. B. (1953). *Recent Developments in Mineral Dressing*. I.M.M.
6. Dyson, F. J. (1958). *Scientific American*. Sept.
7. Bond, F. C. (1952). *Trans. A.I.M.E.* 193.
8. Holmes, J. A. (1956). *Trans. Inst. Chem. Eng.* 35.
9. Schuhmann, R. (Jr.) (1960). *Trans. A.I.M.E.* 217.
10. Gaudin, A. M. and Hukki, R. T. (1946). *Trans. A.I.M.M.E.* 169.
11. Nordberg Mfg. Co.

CHAPTER 4

SECONDARY CRUSHING

The Duty of the Section

In primary crushing the largest lumps of ore mined must be dealt with. In secondary crushing the maximum sized piece is unlikely to exceed 6″ in average diameter and some of the unwanted material coming from underground has probably been removed. The feed is therefore easier to handle. The crushing machines need not have so wide a gape nor so sturdy a construction. The transporting arrangements can be less robust, since the large pieces of rock have now been reduced to more manageable fragments.

Washing and sorting, if practised, may be done in the primary section, but is more usually combined with secondary crushing, where the rock is smaller and more easy to handle. Secondary crushers are usually arranged in series with the primaries, so they must be able to handle similar loads. Their main task is to reduce the ore to a size suitable for wet grinding. It then goes to the fine-ore bins, which must have sufficient storage capacity to receive all the ore accepted for treatment and to keep the plant running continuously, although the mine only delivers its ore periodically.

Lay-out and Equipment

A generalised secondary crushing scheme in which *minus* 6″ rock is sorted and reduced to *minus* 1″ in one operation or "pass" is shown in Fig. 20. The numbers in this flow-sheet refer to:

(1) Transport and feed regulation from primary crushers to screen 2. The feeder stops, starts or modifies rate of delivery and the guard magnet removes magnetic iron ahead of 4. It could be used at 3 instead of 1, or be dispensed with if hand-picking were used on 3 to remove iron. This might be done where the danger of damage to 4 lay in the passage of manganese steel or other non-magnetic and uncrushable material.

(2) Separation of finished *undersize* from *plus* 1″ rock which is to be crushed. A robust screen system is used.

(3) Sorting, picking or transporting belt (belt conveyor). This delivers to 4, and may elevate the ore in transit so that the crushed rock leaving 4 falls by gravity to 1. The — 6″ + 1″ ore is of a convenient size for handling, so removal of waste and detritus is possible. The side arrow shows that provision for collection and disposal of finished waste is then needed.

(4) Secondary crusher, set to 7/8″ By making the set a little below the 1″ screen aperture re-circulation of *near-sized* waste is kept down

(5) Conveyor belt or chute returning crushed ore to 2.
(6) Conveyor belt, perhaps equipped with weight-recording equipment (weightometer) and automatic sampler, which delivers screen undersize to storage.
(7) Fine ore bin/s, where ore accepted for treatment is received and fed at a controlled rate to the plant.

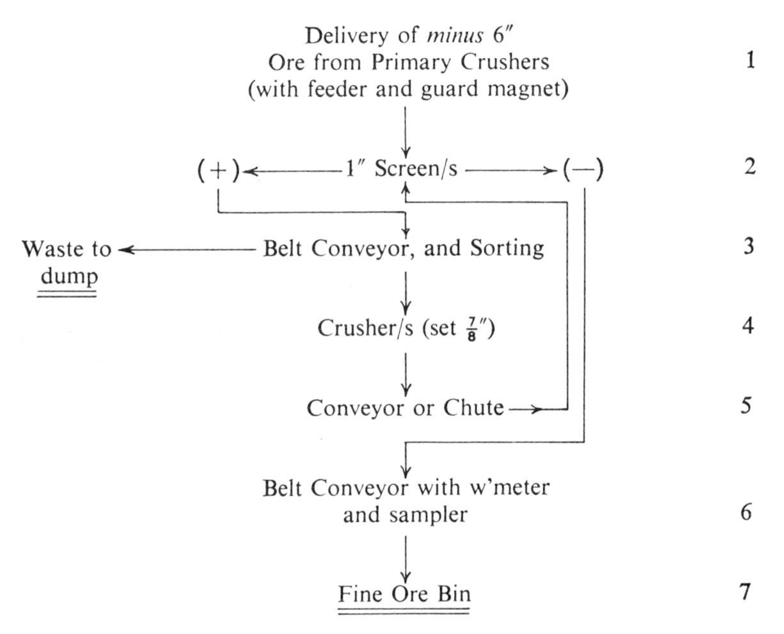

Fig. 20. *Flow-Sheet. Closed Circuit Secondary Crushing*

The purpose of the weightometer (6) and sampler is to record the tonnage accepted for treatment and to sample it for assay grade, moisture and particle size. Process control requires knowledge of dry tonnage treated and the values contained in that tonnage.

If sorting or dense-media separation (Chapter 12) is also practised, it will probably be introduced in this section. In the case of material requiring fine crushing by dry methods, special equipment of kinds not considered in this book may follow. The fixed-path crushing machines discussed in this chapter normally deliver to the bins ore crushed to below between 1″ and ½″. Closer settings are possible, particularly if the secondary crushing is performed in two stages so as to reduce mechanical strain by keeping each machine's reduction ratio below 7 to 1. The considerations which influence lay-out and installed capacity include the crushability of the rock itself, the question of waste elimination and the maximum rate of delivery of ore for immediate handling. During the mine's transport period the fine ore bins are filled, though throughout the mill's working day (usually 24-hour) they deliver steadily to the next stage of treatment.

Secondary crushing is today characteristically performed "dry". This word is used relatively, as the ore may be moist, or may be wetted during washing operations or when water is run in through a crusher to prevent the build-up of clay. (This last is bad practice, and may lead to mechanical trouble.) The machines mainly favoured are modified forms of gyratory crusher, though other appliances, such as crushing rolls, are in limited use. Beater mills, with their variants (hammer, whizzer, pin-disintegrating and ring-roll types) deal with large tonnages of coal, asbestos, limestone and easily broken mineral. The gravity stamp, though worked with added water, is briefly described in this chapter, since it is a fixed-path machine. It is obsolescent, its place being taken by the rod mill described later, but is still in use in a few older plants.

The Symons Cone Crusher

As with the gyratory, crushing results from interaction between three essential parts (Fig. 21). The important difference is that in this case the spindle (1) is not hung from its upper end but is supported in a universal bearing below the gyrating head or "cone" (2). Normals to the arc through the universal bearing carrying the breaking head intersect at O. This breaking head gyrates inside an inverted truncated cone (3), called the bowl, which flares outward and thus allows for the "swell" of the broken ore by providing an increasing space for it to enter after each "nip" has released crushed ore for a further drop. Two types are made, the standard (Fig. 22) and the short-head (Fig. 23). These differ chiefly in the shape of the crushing cavities. Standard crushers deliver a crushed product varying from $\frac{1}{4}''$ up to $2\frac{1}{2}''$ usually in open circuit and can be fitted with fine, medium, coarse, or extra-coarse crushing cavities. Short-head crushers have a steeper head angle, a longer parallel section between cone and bowl, and a narrower feed opening. They deliver a crushed product ranging from $\frac{1}{8}''$ up to $\frac{3}{4}''$, and usually work in closed circuit as in Fig. 20.

The clearance between cone and bowl helps the ore to spread as it works its way down. Choked crushing is avoided since the rock gravitates into an increasing cross-sectional area. The "throw" of the gyrating cone is greater than would be practicable in primary crushers, which must withstand heavier working stresses. Higher speeds are also used (from 435 r.p.m. up to 700 r.p.m.), and particles appear to flow through the machine, partly because of their freedom of movement and partly because, at each revolution, the gyrating head drops from under the recently nipped material. Unlike that for primary crushers, the set is the minimum discharge opening, since during its last two or three nips, the particle of ore must pass through a parallel-sided crushing zone. Tramp oversize can escape, when uncrushed oversize particles succeed in forcing the bowl up, or if clayey feed accumulates on the crushing faces. In the former event the bowl, normally held down to its job by nests of springs, yields sufficiently to allow the object to pass through.

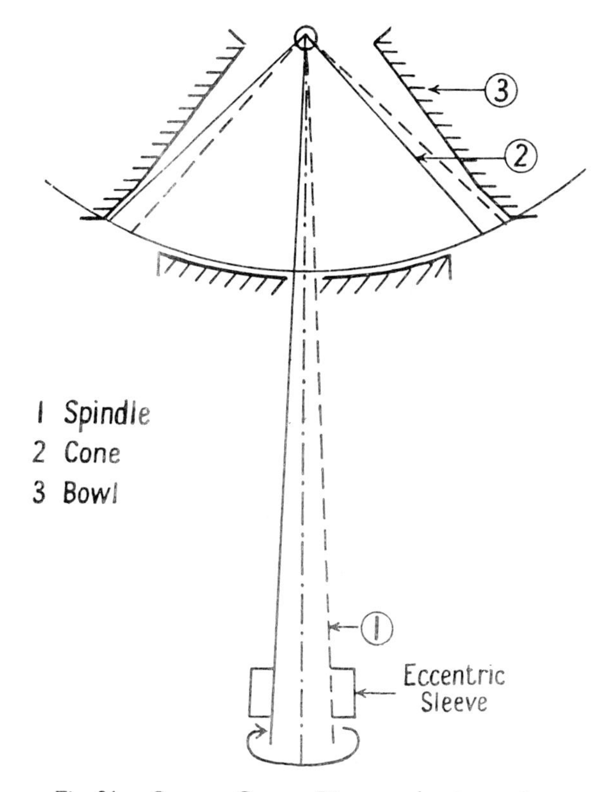

Fig. 21. *Symons Cone. Diagram showing action*

The springs then return the bowl to its correct clearance. While this happens or when choked with clay, the Symons crusher is apt to let oversize escape. Such clay is sometimes dealt with by introducing water with the feed. Better practice is to remove it by washing (see Chapter 2). It is dangerous with any dry-crushing machine to risk the entry of abrasives into its bearings or bevel gears.

The springs which hold down the bowl yield when the load is too severe. It is therefore usual to run the cone crusher in closed circuit with a screen, thus ensuring that "tramp oversize" is returned for further treatment. This recirculated material, which may include fragments of non-magnetic steel not removed by the guard magnet, may call for removal by special methods.

With some ores there is a tendency for extra tough particles to "spring" the crusher at a slight oversize to the set. This leads to accumulation in the closed circuit of such particles, and if neglected can build up sufficient pressure in the crushing throat to present a problem not met with in the types of crusher considered in Chapter 3. (In passing, it should be noted that in all closed circuits there is selective retention of one fraction of the ore stream

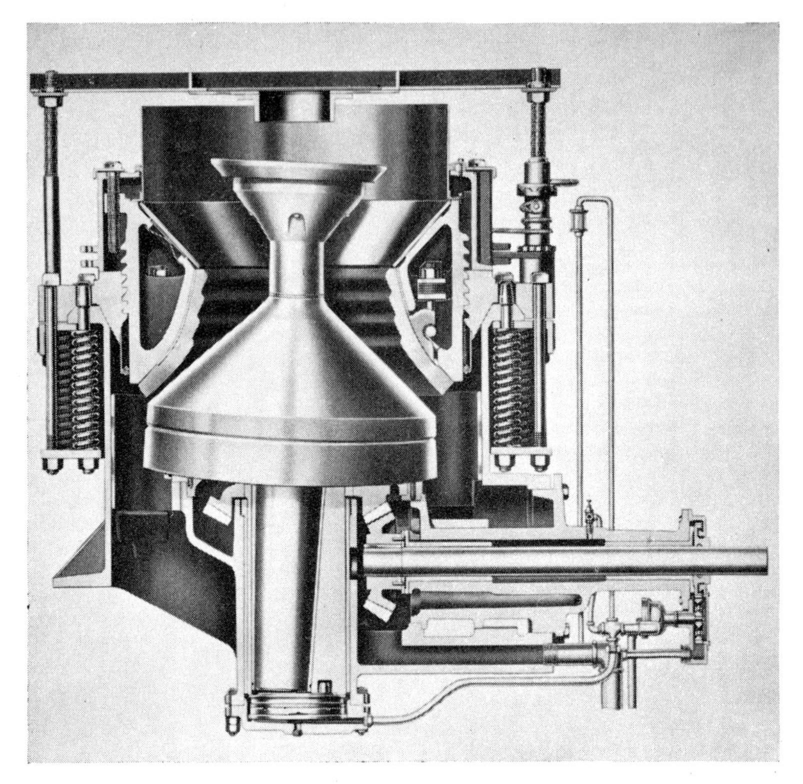

Fig. 22. *Symons Standard Cone Crusher* (Nordberg Manufacturing Co.)

which may call for special measures.) A simple solution is to use a screen slightly oversized to the set of the cone crusher, and thus increase the severity of the crushing action upon the most prominent (largest) particles.

Ratio of reduction is controlled by screwing the bowl up or down by means of its capstan and chain. It is usually held between 3:1 and 7:1, though higher ratios can be worked with some ores.

A 2-ft. standard crusher receiving —$2\frac{3}{4}''$ rock is rated to deliver 15 tons/hour —$\frac{1}{4}''$ in open circuit, or 60 tons/hour when reducing —4″ feed to $1\frac{1}{2}''$. A 2-ft. short-head crusher receiving $1\frac{3}{8}''$ feed and delivering —$\frac{1}{8}''$ in closed circuit has a capacity of 6 tons/hour. When reducing —2″ feed to —$\frac{1}{2}''$ product the capacity is 20 tons/hour. The coarser 7-ft. standard crusher has a capacity of 900 tons/hour when reducing —18″ rock to $1\frac{1}{2}''$. Such general figures, taken from manufacturer's literature, would require checking for performance on samples of the specific ore if it was proposed to work a crushing plant at full capacity.

In the Hydrocone crusher the cone is held up to the correct setting by a hydraulic jack instead of by springs. An automatic method of reset (Fig. 24)

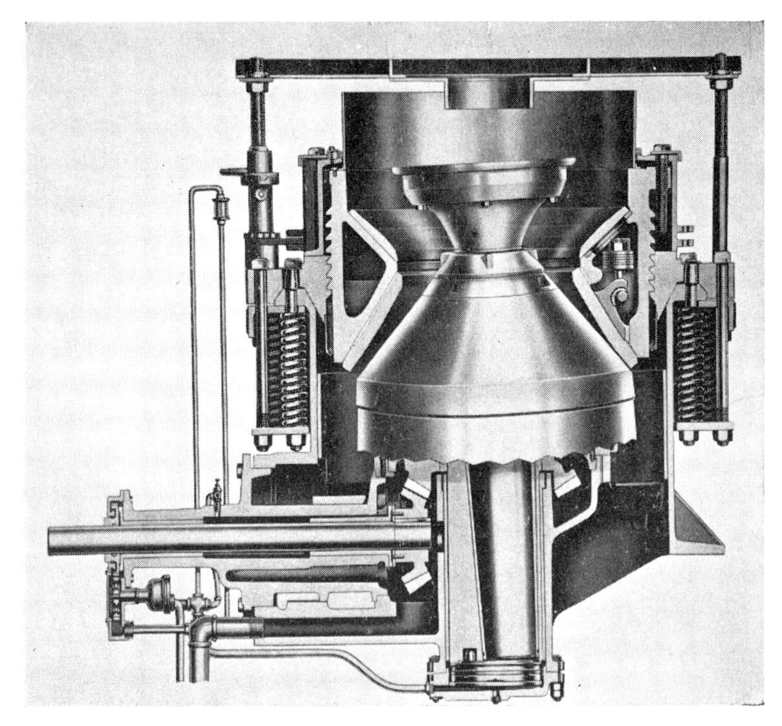

Fig. 23. Symons Short-head Cone Crusher (Nordberg Manufacturing Co.)

allows the cone to yield when an uncrushable object enters, and to return to the correct setting after this has passed through.

Gearless Gyratories

Several crushers are marketed which avoid the complication of bevel-gearing below the rock-treating section, by making the drive a direct extension of the rotor of the electric motor. The Newhouse, which is hung from cables to absorb the vibration of its running, has the motor above the grit zone. It is run at 500 to 600 r.p.m. with an eccentric throw of about $\frac{1}{4}''$. An advantage claimed for this design is that, by avoiding the use of gears, little power is consumed when the machine is "idling"

ROLLS

Although much work once done by rolls has been taken over by cone crushers, these machines still handle a considerable tonnage. Standard

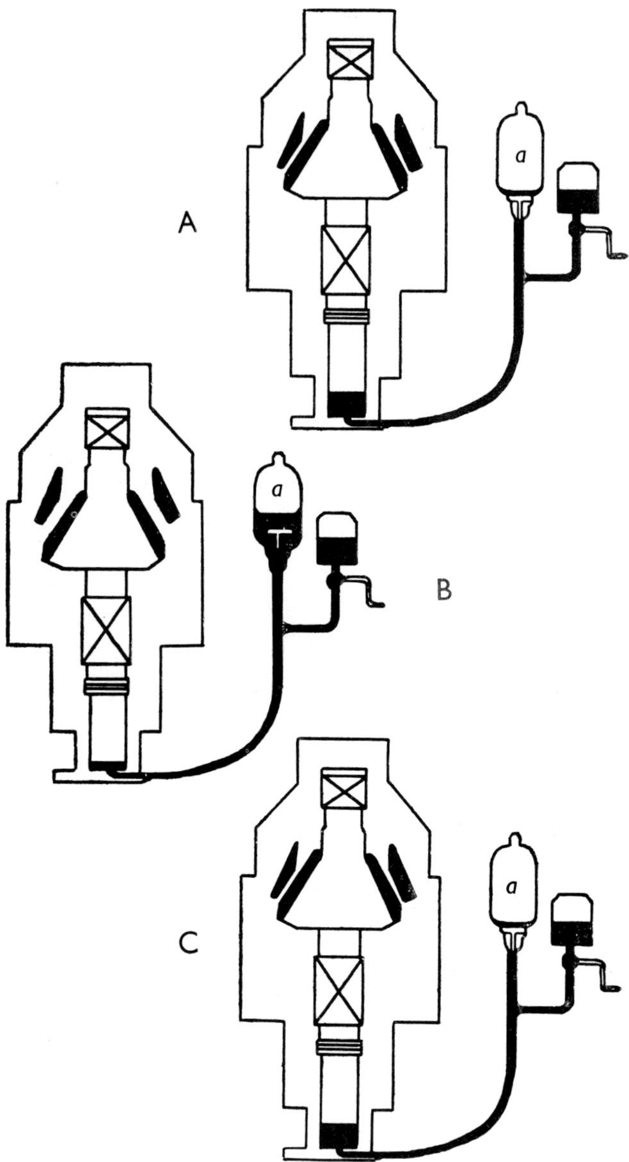

Fig. 24. *The Hydrocone's Automatic Reset* (Allis-Chalmers). A. *Normal running. Cone held up to work by gas pressure in accumulator* (a). B. *Uncrushable body forces cone down, pressure rising in accumulator as oil is forced up.* C. *Uncrushable body has passed. Gas in* (a) *forces cone spindle back to correct set.*

spring rolls (Fig. 25) have two horizontally mounted cylinders. The set is determined by spacing pieces (shims) which cause the spring-loaded roll to be held back on its sliding mounting from the solidly mounted roll. Modern rolls have both cylinders positively driven by separate motors, so that they rotate inward and downward. Rolls crush by nipping the feed between the approaching roller faces and it is essential that the entering material is seized and drawn down.

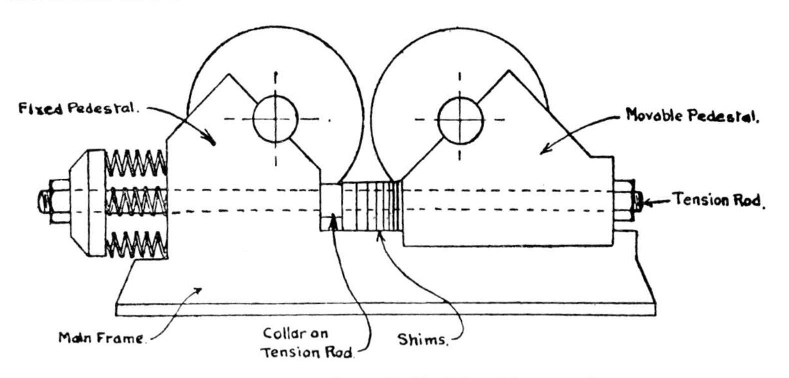

Fig. 25. *Crushing Rolls* (after Truscott)

One method used to calculate maximum size of feed is based on the angular relations shown in Fig. 26. Tan B is the coefficient of friction (usually taken as 0·3); A is the angle of nip, or wedge angle below which a particle is seized, and above which it skids; R is the radius of each roll. $2X$ is the thickness of the particle which can be gripped at zero setting, and S is the set, or distance apart, of the roll faces at their nearest point of approach.

Then,

$$\cos B = \frac{R-X}{R} \tag{4.1}$$

$$R \cos B = R-X \tag{4.2}$$

$$X = R\,(1-\cos B) \tag{4.3}$$

$$\text{Size of Particle} = 2R(1-\cos B) + S. \tag{4.4}$$

For a spherical particle this could be written

$$\cos B = \frac{2R+S}{2R+d}, \tag{4.5}$$

where d is the diameter of the particle. Maximum sizes of rock in relation to roll diameter are as follows, the calculation being made for a coefficient of friction of 0·3, with faces of rolls touching.

Roll dia.	*Max. size of rock gripped*
9 inches	0·36 inches
12	0·48
18	0·72
24	0·96
30	1·20
36	1·44
42	1·68
48	1·92
54	2·16

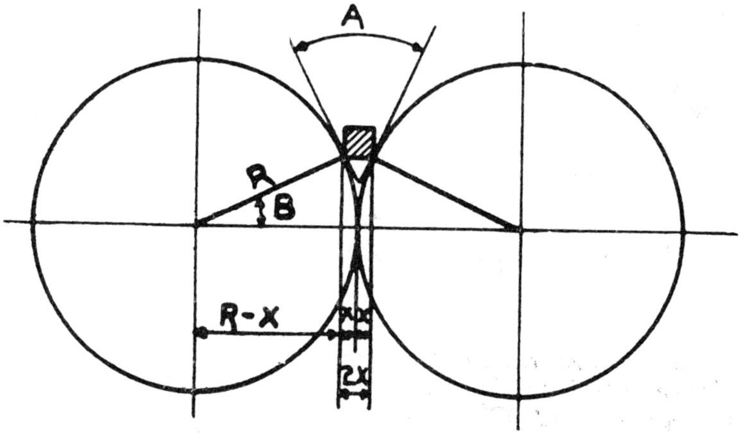

Fig. 26. *Rolls. Diagram for Calculating Maximum Size of Feed*

The method of feeding is important. Unless the entering ore is spread evenly over the whole width of the rolls, partial wear occurs, causing the surfaces to become grooved or flanged. The Traylor heavy-duty crushing rolls incorporate a "fleeting" mechanism which causes one cylinder to move to and fro on its axis, thus reducing this type of wear. A good practical rule is to arrange the feed so that some ore falls outside the crushing area at each end. This helps to even wear over the full width of each roll. Another is to raise the feeding device so that ore arrives on the rotating surfaces at their peripheral speed. This gives the best conditions for seizure. Rolls can only work as "arrested" crushers if lightly fed, because the breaking ore swells in volume as voids are produced, at the same time as the particles fall into a more restricted space. If not "starvation" fed, rolls are choke-crushers, ore grinding on ore. Unless rolls of very large diameter are used, the angle of nip limits reduction ratio, so that a flow-line may require coarse-crushing rolls to be followed by fine-crushing rolls. Although the floating roll is only supposed to yield to an uncrushable body, the choked packing of ore in the crushing throat sets up so much pressure that the springs are usually "on the work" during crushing, and a moderate proportion of unfinished material is let through. For this reason, rolls should be worked in closed circuit with screens wherever control of maximum particle size leaving the crusher is important.

Hammer Mills

In these machines, which can be used either as primary or secondary crushers, the breaking force is mainly due to a sharp blow applied at high speed to free-falling rock. The moving parts are beaters (hammers, rectangular plates, hanging bars or heavy metal rings). They move in a more or less fixed circle of swift rotation, though they are loosely suspended from pins on discs mounted on a driving shaft, inside a robust stationary casing with a grid through which broken undersize leaves the mill. The beaters weigh from a few up to 250 lb., and the larger machines can work on feed as coarse as 8″ cube. Fracture is chiefly produced by the flailing action as the beaters

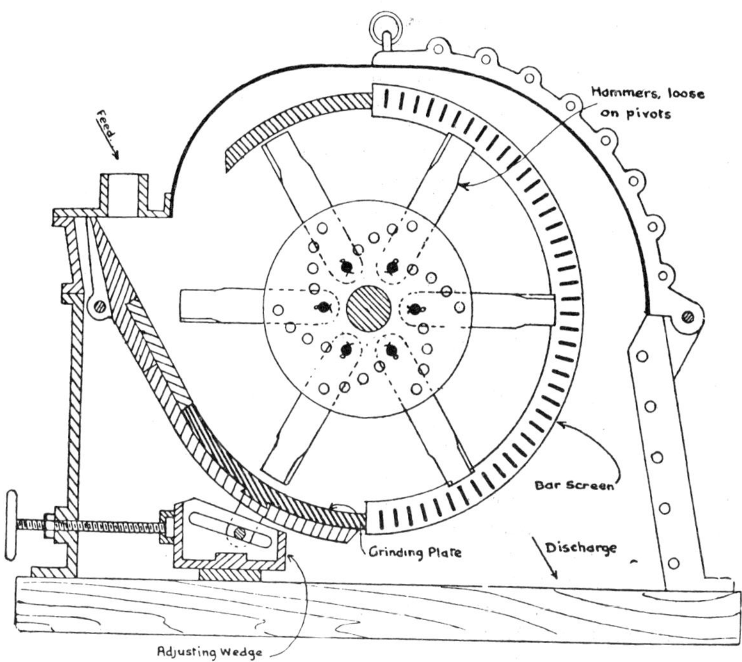

Fig. 27. *The Hammer Mill* (after Truscott)

hit the ore as they spin at from 500 to 3,000 r.p.m., though part of the comminution results from shatter of particle against particle or casing plate. The hammers shown in Fig. 27 can be moved to pins nearer the disc's periphery as they wear, and are readily reversible and removable. Wearing parts are of high-carbon steel or tough wearing alloys with manganese, chromium or molybdenum.

Both uni-directional and reversible hammer mills are manufactured. The former can be used over a wide size range of soft or friable material, in both primary and secondary crushing and the latter in secondary crushing. Reversal of the rotor direction obviates the need for turning the hammers round. A hammer mill with its breaker grid arranged like a miniature belt conveyor has been developed for sticky or wet feed liable to clog a fixed grating. Material which fails to fall through climbs against the down-running new feed.

Another type of hammer mill, the "Impactor" (Fig. 28) is designed to obviate stoppage for hammer adjustment. The rotor is reversible, and end-wear can also be taken up while running, by movement of the anvil blocks which regulate the set of the mill. Elimination of a retaining grid makes the machine able to cope with frozen or sticky feed and aids quick passage. The velocity with which the blow is struck in impact crushing is the main determinant of the severity of shock loading. The vertical distance of free fall as the feed enters is therefore a control factor. Care in arranging a suitable dropping height influences product shape, size and production of fines.

Among the minerals broken by this type of milling are limestone, spars, gypsum, shale, clays, coal, asbestos, gravels and rock required for ballasting or concrete aggregates. The products are characteristically sharply fractured. Moisture affects efficiency adversely, and wear on the beaters is heavy with abrasive material. Product size is controlled to some extent by varying the escape grid apertures, clearance and speed. Unlike the dry crushers hitherto considered, these machines are impact breakers.

Gravity Stamps

These machines, which crush by impact and work wet, are briefly considered here, though obsolescent.

The stamp (Fig. 29) is an impact crusher. From the bottom up, the parts composing it are the mortar box in which five stamps work; the die of each stamp set in the mortar box; the stamp shoe which falls upon ore spread over the die; the head into which the shoe is socketed; the stem which holds the head and is part of the falling weight in this impact crushing; and the tappet, which is keyed to the stem. Below the tappet is the cam, one of five revolving on a camshaft. These cams lift a battery of stamps so that they fall in some such sequence as 1, 3, 5, 2, 4. They are driven by a belt with a jockey pulley. A lever system worked from a tappet on the middle stamp actuates a horizontal rotating plate at the back of the stamps (the Challenge feeder) whenever the ore-level drops below a set height above the central die. This feeds new ore into the mortar box, thus maintaining a bed of rock undergoing attrition. Water is piped in, and at each fall of a stamp water and ore splash outwards from the die. A screen is placed along the front of the mortar box and when a crushed particle rises high enough, is small enough, and also is carried truly to an aperture in the screen, it escapes. Thus, through choice of a suitable height of screen and the size of its aperture, and by regulating the volume of transporting water available, the material stamped can be controlled as regards its maximum escaping size, but not as regards its minimum size. The stamp has a very big reduction ratio, being able to accept 2″ material and reduce it to fine sand.

It will be realised when modern grinding methods have been considered that this big reduction ratio is one of the reasons for the stamp's obsolescence. In theory, nothing could be more efficient as a comminuting system than the

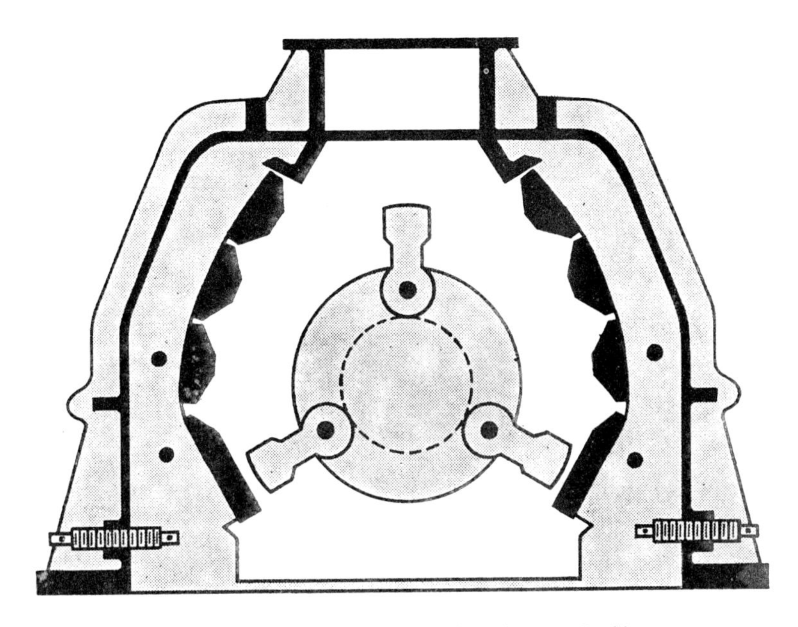

Fig. 28. *The "Impactor"* (G.E.C. Eng. Ltd.)

fall of an unobstructed weight upon rock awaiting crushing. The inefficiency of the operation arises because "finished-size" material has no positive way of escaping from the crushing zone, and may be the recipient of many more blows, each of which is doing unnecessary or even undesirable further work at the expense of capacity, power, wear, and attendance. It will be seen later that the modern replacement of the stamp—the rod mill—can operate without these disadvantages.

The stamp battery has done yeoman service in its day. It is transportable, simple, and an excellent "polisher" ahead of the amalgamation process which was for so long a period the most important means of recovering gold. The fact that stamps overground the ore did not matter much in the happy days when cost-accountancy and efficiency engineering were newfangled fads, not to be taken too seriously.

The Nissen stamp was developed from the five-stamp battery. It has an individual box for each stamp, with a peripheral screen. This improves efficiency of escape of finished sand. Pneumatic and steam stamps have had a vogue, and power stamps are still in use in one field.

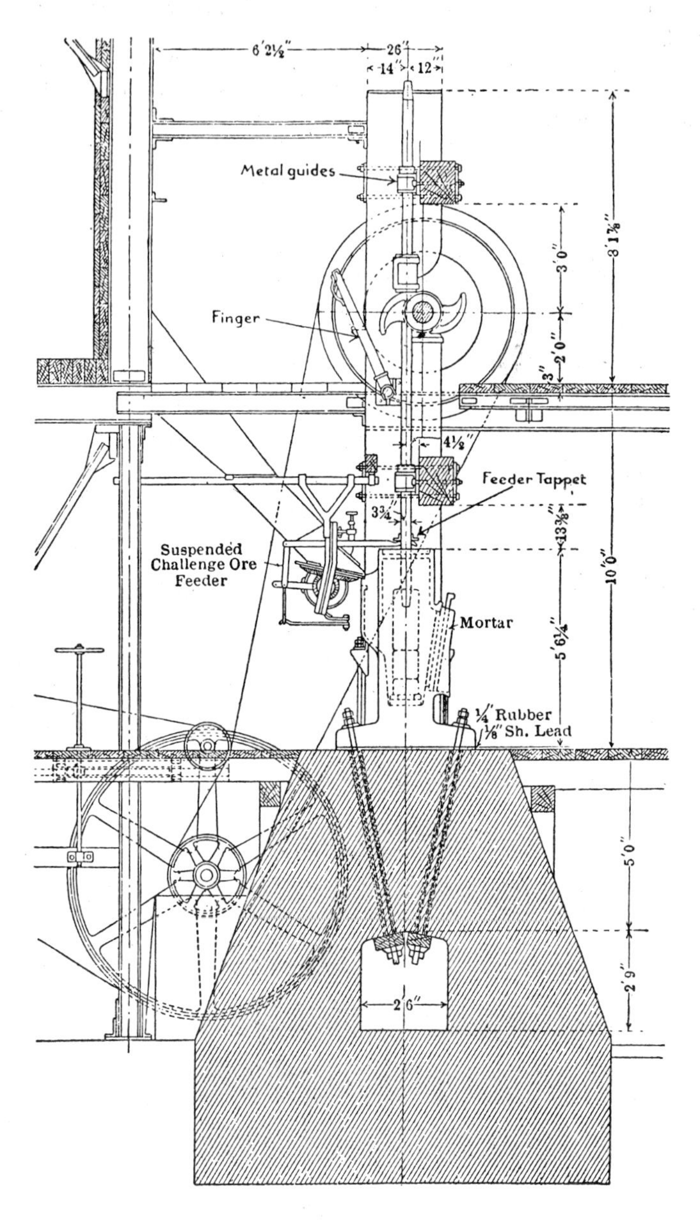

Fig. 29. Section through Stamp Battery (after Truscott)

Dry Crushers, Summarised

The gyratory crusher works continuously with half its surface, but the jaw crusher works half the time with all its crushing surface. For the same maximum-sized piece, the gyratory has considerably greater tonnage capacity than the jaw crusher, but this is not necessarily an advantage. Where large throughput is needed, and the need to avoid vibration is important, the gyratory is better than the jaw crusher, but if transport to the mine is difficult it is less easily delivered.

The running speed of a crusher is limited by the time needed for partly crushed material to fall between squeezes. On slabs or flaky material the gyratory is more efficient than the jaw crusher because of its more pronounced beam action. In secondary work, the Symons cone has a higher ratio of reduction than its nearest competitor, rolls. Hammer mills are rarely used in hard-rock crushing.

The usual sequence of crushing machines is a jaw crusher, feeding to a gyratory which sends its product *via* by-passing screens to cones or rolls, to yield a final product anywhere from $\frac{1}{4}''$ to $\frac{3}{4}''$ maximum size. The trend is to take secondary crushing still smaller so as to ease the work of the grinding circuit. Below this size, any further reduction is effected by tumbling mills in the grinding section. When properly fed, jaw crushers, gyratories, and rolls are all arrested crushers, but if the discharge arrangements become obstructed or choke feeding is indiscriminately practised, rock grinds upon rock, and waste of power, reduced throughput, and perhaps trouble in later sections of the flow-line will result.

With wear, the capacity of the Blake crusher rises somewhat, since the effect is to lower the pitman and increase the reciprocating stroke. Wear of the working parts of the gyratory has an opposite tendency. It is easier to alter the set of the Blake than that of the gyratory crusher, save with the type having a hydraulically held spindle. Clearance of the crushing throat after a stoppage under full load is also much easier with the Blake. Gyratories have better protection against the entry of dust to bearings, and they are more economical of power than reciprocating machines. Ore passing through loses more height than with the jaw crusher.

CHAPTER 5

WET-GRINDING MILLS

Preliminary

In this chapter the mechanisms of wet-grinding mills are considered. The nature of the forces at work, together with a detailed consideration of their interplay and means of control, are discussed in Chapter 6. The general term "tumbling mill" includes the rod mill, pebble mill, and ball mill. It is of cylindrical or cylindro-conical shape, and rotates about a horizontal axis. A load of crushing bodies called the grinding media forms part of the *crop load.* They bear upon the piece of ore in the tumbling mixture with abrasive and/or impacting force sufficient to reduce the mineral to particles of the desired size. As in the case of dry crushing, grinding may be divided into two stages, primary and secondary, if the scale of operation justifies such elaboration. Milling speeds (r.p.m. and rate of feed), types of liner, and size and shape of crushing bodies are chosen to develop shattering or impact milling in the primary mill and a more gentle abrasive action in the secondary one. The object is to bring the ore to the *mesh-of-grind* called for by the concentrating section of the mill, and this is best done in separately controlled stages. Often the primary grinding mills must work vigorously on a quickly passing stream of ore, and the secondaries more gently and with longer retention of the pulp.

Milling Action

The tumbling mill is a horizontal cylinder which is nearly half filled with crushing bodies, usually cast iron or steel balls. Ore can be fed in at one end and discharged at the other, continuously, and water added at the feed end helps to flush it through the mill. When the cylinder is rotated, this mixture of balls, ore, and water (the "crop load") is churned with, or battered by, flying balls, according to the speed at which the mill is run. The kinetic energy of this tumbling load is dissipated as heat, noise and comminution of ore, grinding media and mill linings. Its useful energy is expended during impact, abrasion, shear and compression of mineral particles, but it bears on all moving surfaces in contact. The conditions are quite unlike those in dry crushing. The inside of the mill is slippery with water and slime, and the particles undergoing comminution are usually too small to be firmly gripped. Further, the only fixed surface in movement is provided by wearing plates (liners) which line the inside of the cylindrical shell of the mill. These are buried under the crop load during part of each revolution and then rise clear. Chance contacts for each individual particle in a churning mixture of water,

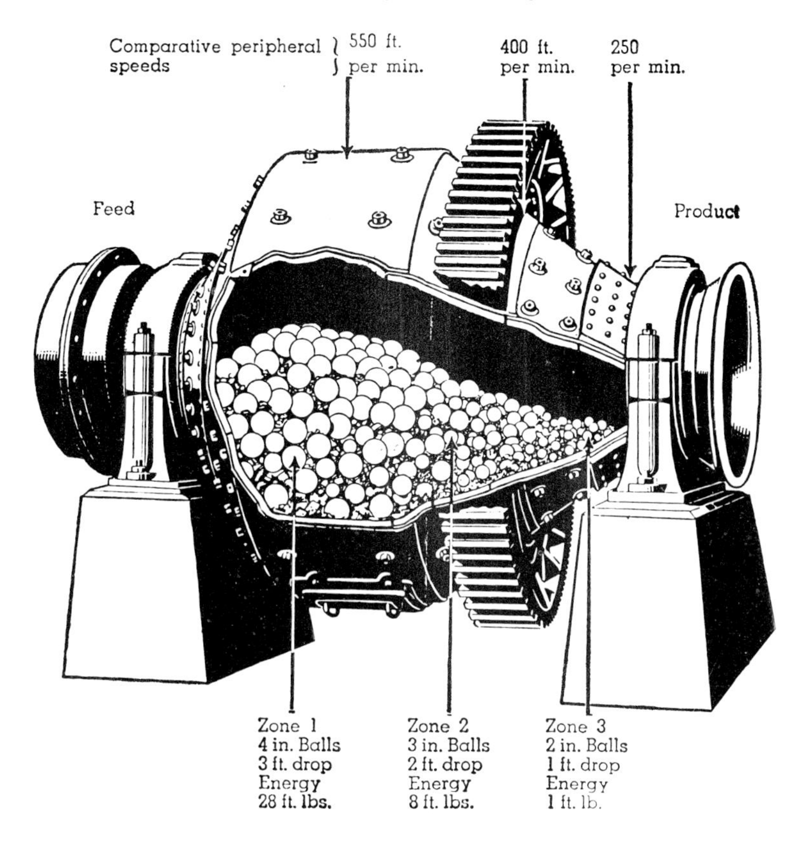

Fig. 30. *Hardinge Conical Mill*

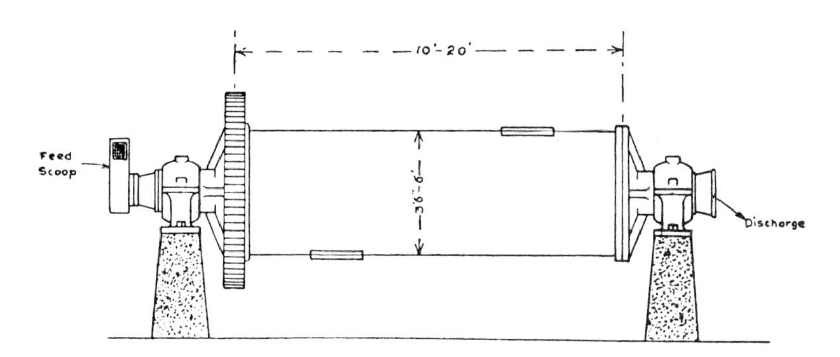

Fig. 31. *The Tube Mill* (*High Discharge*) (after Truscott)

steel, and rock characterise the condition in which grinding is achieved. By adjusting the balance between the various ingredients and the manner of their agitation, it is possible to maintain some control of the mixture and to influence the average size of the solids issuing from the mill. Two kinds of grinding are practised—batch and continuous. In batch work, a load is put into the mill and ground for a determined period or to a determined condition. This is common practice in re-treatment of minerals for such special purposes as laboratory testing, preparation of paint mixtures and pharmaceutical products, but is exceptional in ore treatment. In standard mineral processing continuous grinding is used, in which ore is fed in at a controlled rate and leaves the mill after a suitable dwelling time as part of the crop load.

Types of Mill

Tumbling mills may be classified according to shape into two types, cylindrical and cylindro-conical. In the latter (Fig. 30), two cones are joined by a cylindrical section. At the feed end is a flat cone. After passing the zone of maximum diameter, sometimes called the "drum", the pulp climbs a steep cone to the discharge. This shape has been adopted in the Hardinge mill to develop specialised grinding forces at each stage of the passage of the feed, suited to its changing condition as it progressively disintegrates. Another and perhaps preferable classification of mills is into two types—high discharge and low discharge. From Figs. 31 and 32 it will be seen that although the feed enters through a hollow trunnion at the centre of the feed end, the discharge arrangements are very different in the two types. In the high-discharge mill the only possible down-gradient is a larger trunnion at the discharge than at the feed end, so that in effect pulp only leaves the mill because it has been displaced by entering feed. In the low-discharge mill the support of a hollow trunnion at the discharge end is either avoided in design, or lifting scoops are used, so that a gradient through the mill is created from feed to discharge. In addition to displacement of discharge by entering feed, this adds the effect of gradient to accelerate passage of pulp through the mill at a more or less controlled speed. Since grinding force is applied at an even rate, the faster the ore passes through the mill the coarser is the discharged product.

In the type illustrated in Fig. 32 the feed, introduced through the scoop at *A*, is retained in the crop by the grating *C*. Material sufficiently fine to pass through this grating is elevated by the radial lifters to the overflow trunnion *B*. In another, and heavier, form of construction, the discharge end is not hung on a trunnion, but supported on rollers by means of a steel tyre which encircles the mill shell. The mill discharge can then flow straight out.

It is usual to say that the Hardinge is high-discharge only, but this statement fails to take account of the effect of the steep cone. The effluent of a mill consists of ore and water flowing as pulp. While it is true that this pulp must climb to the discharge trunnion of the Hardinge, it is also the fact that the cross-section of the flowing stream is continuously shrinking, so that the pulp stream must run faster and faster as it flows toward the discharge

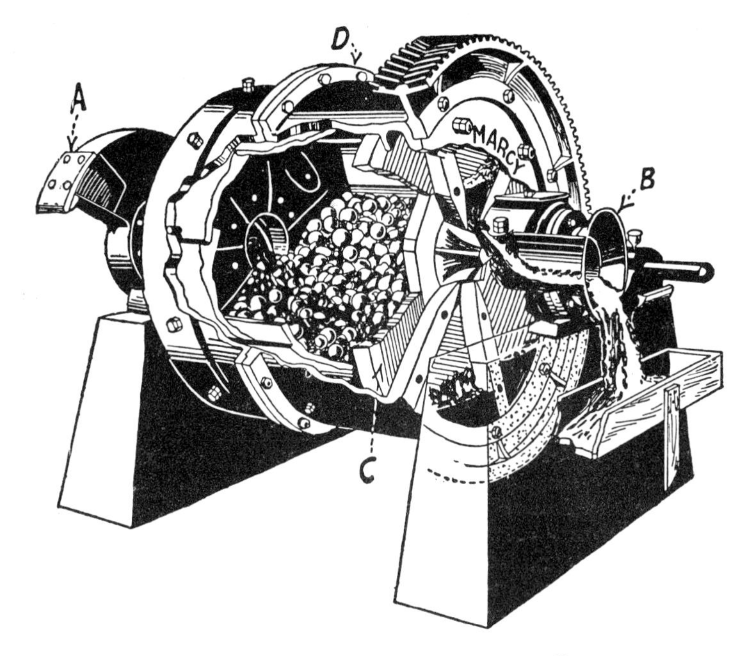

Fig. 32. *The Low-Discharge Mill*

trunnion. Descriptively, the Hardinge is a high-discharge mill, but the effect of this acceleration is to give it qualities intermediate between high- and low-discharge. Another point concerning which assumptions have repeatedly been made without experimental verification arises here. It is usual to say that a low-discharge mill permits rapid passage of the ore, with dwelling times for the average particle of as little as a minute, while high-discharge mills (particularly the 5′ x 18′ pebble mills of older Rand practice) hold their feed up to twenty minutes. This is correct, with certain significant qualifications regarding selection by the mill of what it will retain. (The importance of dwelling time is discussed later.) What is not necessarily correct, however, is the further statement that high-discharge mills set up a pool of pulp on the down-running side into which balls drop after breaking away, thus dissipating much of their kinetic energy uselessly. Rotation causes the mobile pulp to mould itself under the influence of centrifugal force as a crescent-shaped body along the rising side of the drum. The solid portion of the crop load is being shaped into a roughly pear-shaped cross-section somewhat loose at its centre. There exists some evidence that in a properly run mill, whether high- or low-discharge, such a cushioning pool can be avoided by correctly balancing pulp consistency, mill lifting-power, and loading.

One type of mill little used today is the Krupp screen-faced cylindrical mill with peripheral discharge. In this machine (Fig. 33) the feed is introduced through a trunnion and discharged when it has been ground sufficiently to pass through the fine screens lining the cylindrical part of the mill. Heavy perforated plates protect these screens from injury, and a coarse screen is mounted concentrically inside each fine one to give further protection. External sprays provide water and the external casing can be flooded so that the mill dips into water.

The Hardinge Mill

The shell of this mill consists of a flat cone followed by a drum, with a steep cone at the discharge end. The shell is carried in two hollow trunnion bearings, which permit feed and discharge. Drive is by a crown wheel bolted round the steep cone and driven by a pinion. The gear may be straight, single helix, or double helix. Straight or double helix are to be preferred, since no end-thrust is set up in operation. Feed may be introduced direct into the feed-end trunnion or, more usually, through a feed scoop which gathers ore from the bottom of a feed launder and elevates it to the entrance level. The latter arrangement is particularly useful when the mill is in closed circuit with a "mechanical" classifier. This is a power-driven appliance which sorts ore particles discharged from the mill, returning the coarse ones (oversize) for further grinding and allowing the finer ones to overflow as a pulp to the next section. To effect this return (close-circuit the oversize), the returning sands must gain height. They are raked up-slope and discharged so as to gravitate to the gathering scoop, which lifts them, together with new feed, to the entry trunnion. This trunnion may have a wearing plate with a conveying spiral to force the feed forward into the body of the mill.

A retaining grid can be used at the discharge end, permitting the mill to be loaded nearly to the centre line without risk of discharging balls. The interior of the shell is protected from direct contact with ore by cast-iron or steel lining plates, called "liners". For the cones, segmental sections are used, and for the drum portion curving rectangular liners. They may be backed with plastic material such as rubber sheet or old belting, or fitted directly on to the shell. The liners are held in place by liner bolts kept tight by external nuts and rendered leak-proof by plastic washers.

These mills are listed according to the diameter and width of the drum. A small Hardinge has a drum 2′ in diameter by 8″ wide, weighs over half a ton and when loaded to half volume carries over $\frac{1}{4}$ ton of steel balls. A large mill is 12′ by 72″, weighs 62 tons, including 27 tons for its set of unworn liners, and carries a ball load of over 50 tons. A 2 h.p. motor would drive the small mill, but between 700 and 800 h.p. are needed for the large one. The power needed per ton of ball load rises with drum diameter, and is somewhere between 9 and 15 h.p. per ton of balls at 75% *critical speed.*

The critical speed of a mill is $\dfrac{76{\cdot}6}{\sqrt{d}}$ in r.p.m. when d is the mill diameter in feet less the diameter of the largest ball in the crop load.

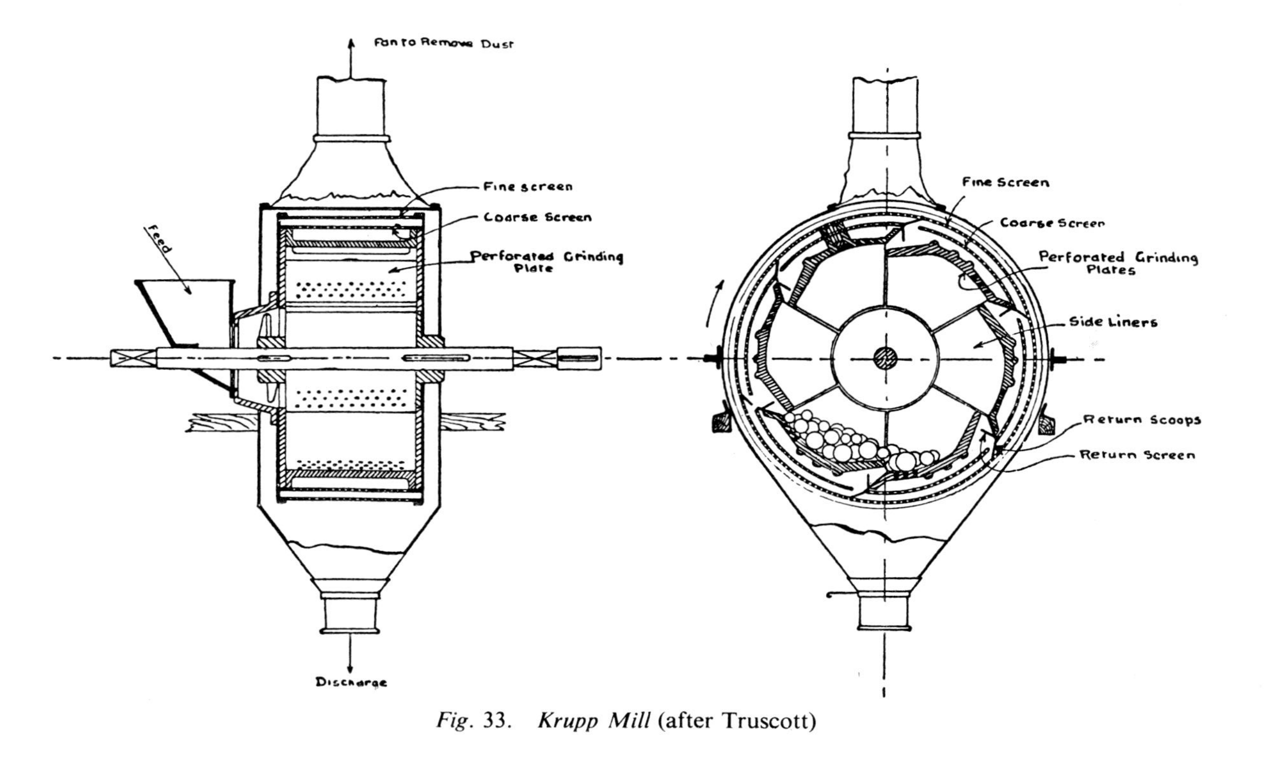

Fig. 33. *Krupp Mill* (after Truscott)

Critical speed being a function of peripheral speed, the rate at which these mills can be run depends on the maximum diameter inside fully worn liners. Centrifugal effect is strongest in the drum section, where the ability of the mill to lift its crop load up the rising side of the mill is therefore highest. At the same time the load inside the steep cone tends to work back down-slope, so the resultant is a pronounced heaping-up of load in the drum and a tapering off toward the discharge. Another differentiating action is also at work, as in all ball mills. The largest balls tend to work to the top of the crop load with the largest lumps of ore. They are then freest to roll or slide down-slope on top of the turning load, where they then are most liable to be caught between the bottom of the sliding mass and the downward-moving liners. There is thus a tendency for the biggest balls and rock to work to the periphery of the drum-section. The makers of this mill claim that the balls segregate themselves somewhat, in such wise that the biggest are in the drum and so disposed that when they rise in the turning load they fly or tumble the furthest distance, while the smallest balls work up toward the discharge. Any such tendency is useful, since it causes the biggest balls to work upon the newly-entered and therefore biggest pieces of ore, while the smaller balls, with less interstitial space, handle the partly finished material as it works its way out. At low speeds the pressure of the heap piled up in the drum has crushing value, and at high speeds balls break clear and rain hammering blows on the mass of churning metal and rock several feet below, giving impact crushing. It is obviously of value to be able, so to speak, to control one's punches, and to use a heavy weight against a tough piece of ore with some selectivity of target while keeping the light hits and tight jostling for the small stuff which does not need such strong treatment. The effect of the conical shape of the Hardinge mill on peripheral speed and kinetic energy on the crop load at various cross-sections is illustrated in Fig. 30.

In the attempt to increase the tendency toward segregation of the largest balls at the extreme radius, the Hardinge Co. also market the Tricone mill in which the drum, instead of being truly cylindrical, has a slight back-slope toward the feed end.

The Low-discharge Cylindrical Mill

Two types of low-discharge mill are available. In both, the feed end is served by a hollow trunnion which supports part of the weight, and through which ore and water can either be fed by gravity or "wormed in" by scoop. At the far end the problem of permitting pulp to discharge at nearly the full diameter of the mill has been solved by two alternative types of construction. In one, the weight of the discharge end is taken by a tyre mounted on rollers, thus leaving the whole end of the mill free for a bolted-on retaining grate or a loosely fitting stationary door, which when closed leaves an annular gap through which pulp can escape once its solid fraction is small enough. In the other type a high-discharge trunnion is the supporting member, an internal grating with lifting scoops behind it serving to evacuate everything passing through that grating. The grate aperture is chosen at the size at which it is desired to pass worn balls out of the mill, and varies from $\frac{1}{4}''$ to $1''$. Grates are cast with a slight flare outward to avoid "blinding" and are procurable in high-grade alloy steel. The size of the mill is limited at present by manufacturing possibility. 10′ by 10′ mills carrying a 45-ton ball charge and using 800 h.p. are in use. A characteristic feature of the low-discharge mill is that the diameter is made as great as possible while length is either about the same as the diameter, or at most twice as great. This follows logically when it is appreciated that the purpose of low-discharge milling is to cut down the dwelling time in the mill. A mill in which diameter and length are approximately equal is called a "square mill" in the United States.

Tube, or High-discharge Mills

Any low discharge mill can be converted wholly or partly to high-discharge by suitably plugging the outlet end, but a better appreciation of the difference between the types is suggested by Taggart, who states that modern practice tends to apply and confine the name (*tube mill*) to cylindrical mills with a length-diameter ratio greater than 2. They were developed, according to Truscott, from the Cornish "barrel pulveriser" which was used to liberate cassiterite middlings, and from cement-grinding mills, and were adapted to the needs of cyanidation practice on the Rand. Before the use, now universal, of closed-circuit grinding (in which circulation of the ore through the mill and classifier is continued until the latter permits it to leave the circuit) it was of paramount importance that all the auriferous pyrite should be so finely ground as to expose its burden of gold to the chemicals used for its dissolution. The tube mill was therefore made long. In the modern form these machines are from $5\frac{1}{2}'$ to $6\frac{1}{2}'$ in diameter, up to 22′ long, and are loaded with such crushing bodies as pebbles, mine rock, steel balls, or steel scrap.

The Cascade Mill

This recent addition to the range of wet-grinding mills (Fig. 34) departs radically in shape from those thus far considered. Like the *autogenous* dry-grinding Aerofall mill it has a high diameter/length ratio—in this case 3 to 1. Slightly concave liners give maximum diameter at the centre of the drum. This directs the crop load away from the vertical sides, a process aided by deflectors. The mill is made with diameters ranging from 6′ to 36′, the latter needing up to 6,000 h.p. Feed and discharge are through two support trunnions, finished pulp being lifted to the latter after being ground small enough to pass through a retaining grate.

The vibrating ball mill (described in Chapter 7) is finding increasing use in industry, particularly where an inert grinding atmosphere must be maintained.

Mill Liners

Mill liners perform two major duties. They protect the mill's cylindrical

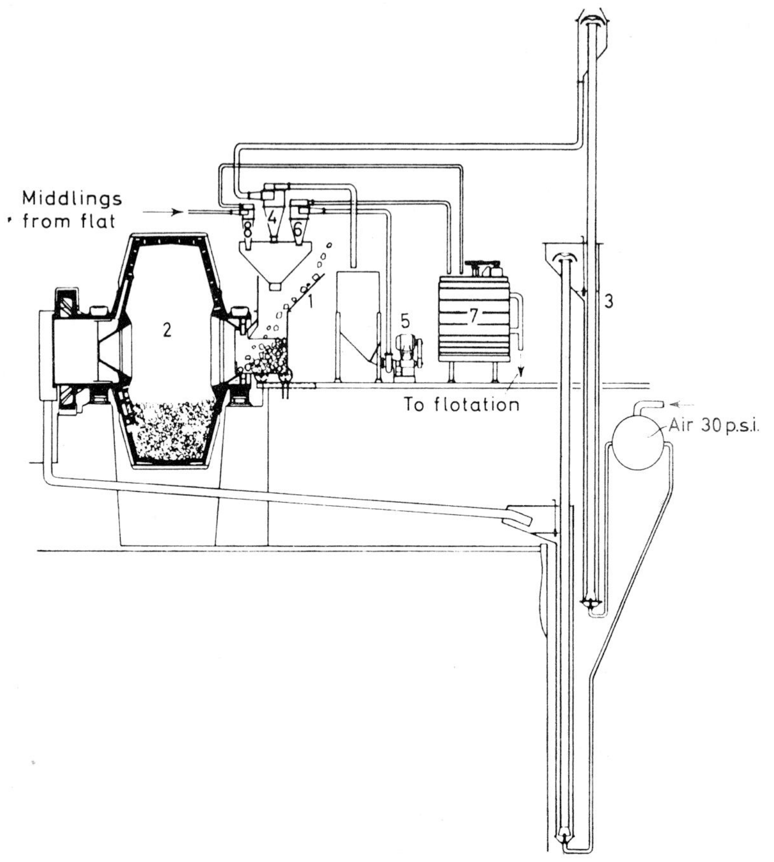

Fig. 34. *The Cascade Mill* (Hardinge Co.)

shell against abrasion by the tumbling mixture of ore, water, and crushing bodies, and thus preserve it from wear. The shell not only provides the frame of the mill, but must also be leak-proof. The liners also help to tumble and rotate the crop load. The shell, geared to the driving mechanism, revolves at a fixed peripheral speed. The liners, being bolted or otherwise fixed, revolve without slipping. Resting on the liners is the "crop load", a mixture of crushing bodies, ore at various stages of comminution, and water. Ignoring transmission losses, the kinetic energy consumed in grinding is that used to unbalance the crop load. This load is displaced by frictional drag against the rising side of the rotating mill. Being somewhat fluid it resists the displacing force by tumbling downward. The further the load is lifted out of balance the greater is the counter-input of the kinetic energy consumed during comminution. The grip of the liners transfers this grinding force into the churning crop load. As the mill revolves, the crop load is carried upward by the rising body-liners (the liners which cover the horizontal inner surface of the cylinder, as distinct from the end liners which cover the vertical circular ends). The binding forces causing the load to rise consist of (1) the centrifugal action of the revolving mass; (2) its collective *pseudo-viscosity* and (3) interlock produced by mutual adjustment of balls, liners, and ore. If the liners are smooth, these three factors play the chief part in determining how much of the mechanical energy being used to turn the mill shall penetrate to the crop. If the liner surfaces have projecting ribs, positive lifting force can also be applied to the load. Instead of losing contact soon after it starts to rise and churning round quietly, the outside of the crop load can thus be helped to climb higher, so that it falls back with sufficient violence to add shattering action to the abrasion characteristic of churning motion.

Body liners receive heavier punishment than end liners, and wear down about twice as fast. A contributory cause of wear is chemical corrosion, either due to mine water or acid from decomposition of sulphide minerals. Fortunately, it is frequently necessary to add lime in the mill circuit for reasons connected with the concentration processes, and this gives protection to the ironwork of the mill, neutralising any acids present. Although the purpose of grinding is the comminution of ore only, the forces employed also act on lining and contents of the mill, and all are subjected to the grinding action. Liners and crushing bodies are therefore selected for their hardness, toughness, and resistance to such wear. When made of alloy steel, they must not introduce any element into the circuit which can cause trouble in the concentrating section, since the steel abrades during grinding and is carried forward to that section. Consumption of steel and cast-iron liners varies, and is usually of the order of 0·03 to 0·3 lb. per ton of new feed.

Observations made by Bond[1] in which an experimental glass-ended mill loaded with $\frac{1}{2}''$ steel balls but no ore, showed that slipping in the layer of balls next the smooth shell increased with the milling speed, in the 60% to 80% critical range, whether the mill was run wet or dry. This effect in a 12″ diameter mill not containing mineral would not necessarily be repeated under normal operating conditions. Abrasive grinding is proportional to slipping of the charge. Bond's observations showed 15% slip between the outermost layer of balls and the smooth liners, and a further amount varying from 5% to 10% with each next layer, radially inward to the fifth or sixth layer where observation was not possible.

Rise in speed (and therefore in centrifugal pressure) increases the energy of outward travel of the heavier components of the crop load. This should cause a tendency towards strong keying of the solids near the shell and toward mobile liquid conditions at the centre of the churning load where centrifugal force is lowest. The keying effect on the solids of increased speed is, however, modified by any change in volume or weight of the crop load. The behaviour of the upper portion changes with acceleration. Part of the charge breaks loose and reduces the overall out-of-balance weight.

In a cylindro-conical mill the difference between the peripheral speed of the drum liners and of the segmental ones which line the steep cone leads to a change-over from pronounced impact grinding at the greatest diameter to pronounced abrasive grinding toward the discharge.

Mill liners are made in many shapes, but they fall into three main groups:

(1) Smooth, having simple frictional contact with the crop load.
(2) Grate (pocket, grid, El Oro, self-renewing) liners, in which the outermost layer of ore and/or balls tends to wedge in the openings (Fig. 34) and form the wearing surface.
(3) Lifting liners, with longitudinal waves, ribs, lifters or wedge bars which to some degree grip the crop load on the rising side.

Coarse-grinding mills, which draw their feed from the fine ore bins and reduce it mainly to coarse sand, usually have longitudinal projections on their body liners which provide a straight-line key along which the rods can lie as they are being lifted. Slip in zone AA–BB is thus prevented and impact crushing dominates. Primary ball mills also have ribs, steps, bars, "shiplap", or other projections, but these are not necessarily continuous horizontally. Secondary ball mills usually have wave liners or smooth ones. The profiles of some of these are shown in Figs. 36–37. Ribs and bars are shown in *a, b, c, i, j* and stepped liners and shiplaps in *f, g, h,* and *k*. Some types of wave liners are shown in *d, e, l,* and *m*. The coarse ore selected from the crop load by Group 2 liners reduces consumption of steel. They are designed

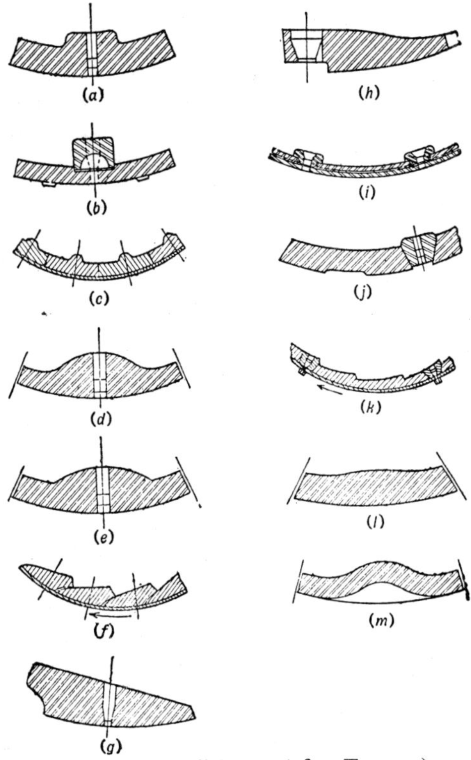

Fig. 35. *Mill Liners* (after Taggart)

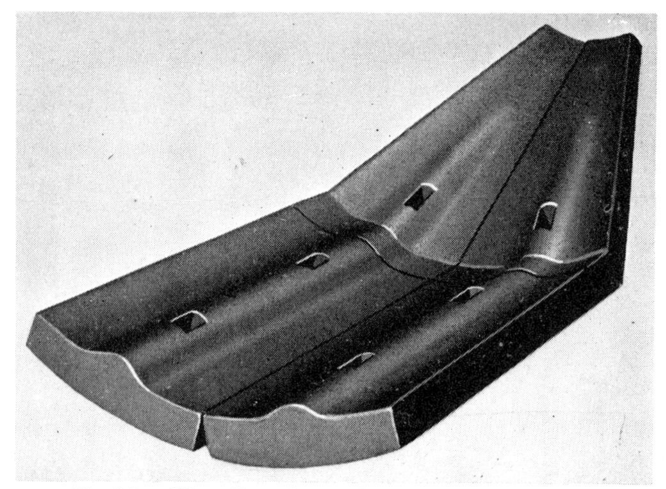

Fig. 36. *Wave Type Lining Sections* (Hardinge Co.)

Fig. 37. *Wedge Bar Type Lining Section* (Hardinge Co.)

so as to grip larger pieces of ore, which take the brunt of the wear and thus act as a self-renewing lining. This reduces the amount of steel consumed (a major cost item), and lengthens the period between shutdowns of the mill for relining. Against this, heavy particles of mineral find lodgement in the protected crevices thus created, and this sometimes ties up valuable mineral.

Liners must not be too heavy or big to be handled through the manhole in the shell of the mill, or to be manoeuvred into place. The Britannia liner consists of 3″ to 4″ lengths of old rails set on end and grouted in cement mortar.

Fig. 38 shows a cross-section through a cylindrical mill rotating anti-clockwise, as seen from the discharge end. The toe of the charge is at its characteristic "8-o'clock" position. If the direction of rotation were reversed the toe would be at the "4-o'clock" position at normal speeds. The descending body liner under-runs the toe of the load. From then on until it emerges clear of the crop on the ascending side, it transmits grinding force from the power supply to the crop load. The kinetic energy thus acquired by the crop load is the only source of grinding force in the mill. Only a small portion of this force is specifically applied to the desired work. The liner provides the

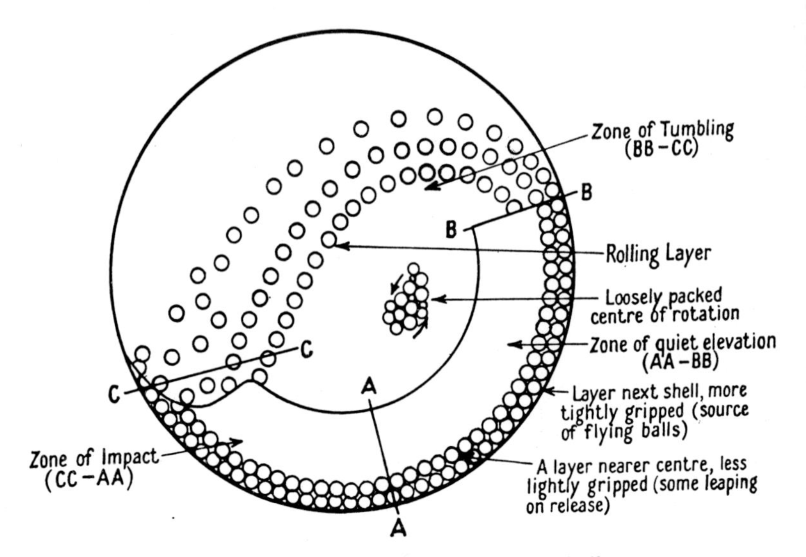

Fig. 38. *Cross-section through Mill, showing ball movement*

connecting link between shell and crop. If the mill runs fast enough (at "cataracting" speed) the balls nearest the liner are thrown clear of the crop load as they emerge on the rising side and then fall freely to the toe. If the liner projection exceeds one-third of the ball diameter, efficiency is reduced, the throw being such as to send the cataracting balls too far horizontally, so that they strike the down-running breast of the mill above the toe, thus wasting their blow. Pitch of the leading edge of the ribbing varies in practice. Wave liners are preferred to other types in most primary mills.

Abrasive comminution (characterised by the fineness and slime content of its product) is proportional to the slippage in zone AA–BB (Fig. 38). If the peripheral grip on the crop load is strong it is lifted quietly and the impact effect is increased, giving a coarser end-product and reduced liner wear. When the charge slips as it is dragged up abrasive grinding predominates, as shown by finer grinding, increased slimes, and greater wear.

In all technical operations a number of factors must be reconciled. Considerations affecting choice of liners should include:

(1) Capital cost per ton ground between fitting and scrapping of liner.
(2) Power used per ton ground with various types of liner.
(3) Grinding time lost periodically in liner renewal and stoppages for running overhaul.
(4) Effect of grind (coarse, fine or slimed) on recovery and grade of concentrate.
(5) General process economics affected by liner action.

Item 4 of this list directs attention to the need for consistent delivery to the processing section of an evenly ground pulp at a steady rate and pulp consistency. As liners wear the diameter and capacity of the mill increase, but this wear is concentrated on liner projections and boltheads. One result is a slow change of product fineness and increase of very finely ground material. Another is stoppage to deal with loose liners due to wear of bolts, since pulp must never be allowed to leak through to the shell. Leakage round the bolts gives useful warning of approaching liner trouble.

Where the milling ore is particularly abrasive there has in recent years been a significant return to pebble milling, sometimes combined with Group 2 liners. If the mine ore can provide suitable crushing media the increased pebble consumption which accompanies the use of self-renewing liners is all to the good, provided a proper balance is maintained between large and small lumps of ore in the crop load, a subject discussed later in this chapter. Some operators consider that a softish ore is best milled with an element of slip and attrition grinding. The specific problem should always be solved by study of the particular ore involved, which will show how the basic principles can best be applied. In later chapters it is shown that grinding is more than a matter of liberation of the various species in the ore, and that it is also an energising factor in the physical chemistry of a processing treatment.

The profile relationship between projection, width of projecting rib, steepness of sides, and distance between sides when new, has been studied in its relation to maximum ball size by Howes and others.[2] The radius of the ball should exceed the height of the projection. If the valley width exceeds three ball diameters, wear is concentrated between the ribs, and there is also lateral wear. If the width is less than two ball diameters, the reverse condition obtains, wear concentrating on the crests. The leading wall of the projection takes the load on the climbing wall of the ball mill. Hence, it receives the bulk of the wear between ridges. The nearer this wall's projection angle is to radial, the higher the balls will be carried and the further they will be thrown when they leave the side of the mill in free flight down to the toe. Shiplap liners are designed to equalise the wear in valley and along the crest, so as to preserve uniformity of impact grinding throughout the life of the liner, since loss of liner grip on the crop load is accompanied by increase in the proportion of over-grinding.

End liners are usually smooth. When they also form retaining grids, "blinding" of the apertures can be largely avoided by casting them with a slight outward flare, so that material small enough to escape into the aperture is readily discharged. Wear on end liners is relatively light.

Cylindrical mills are usually run with between 45% and 50% of the volume occupied by the crop load. After relining, the capacity of a mill is therefore at its lowest and it increases with wear to a maximum. This is reached when the liners have worn so thin that there is a danger of damage to the shell if they are kept in service any longer. As the mill diameter increases, the total weight of crushing bodies is correspondingly increased, together with the weight of ore in the crop. Power consumption and throughput, including finished output, rise in accordance. Further, a definite increase in peripheral speed occurs, which increases the centrifugal force available to lift the crop load on the rising breast of the mill. If rib wear exceeds valley wear, the extra lift is partly compensated by increased slipping of the load. This can be adjusted a very little by raising the pulp viscosity, effected by a slight reduction of the liquid-solid ratio in the crop. Since the mill speed is usually fixed, no running variation of this factor is possible.

Trunnion liners can be fitted from outside with only slight delay. They are either smooth or have a helical spiral. At the feed end this spiral, if used, helps ore in. At the discharge end it retards departure by working against the outflow. Liner bolts have an oval or square taper head. Liner backing is often used in heavy-duty grinding to mitigate shock and danger of liner fracture. Such backing may be 1″ planking, rubber sheet, old belts, or the liners may be set in zinc. Circumferential grooves may be worn in the liners by the frictional drag of the down-slipping crop load, the balls bearing on the liners without spinning. As soon as the grooves are deep enough for the balls to be in good contact with the liners this form of wear ceases.

The cost of liners is a relatively small item in grinding, so it is customary to buy those best for the work, even if expensive. Time lost in renewal of quickly worn out liners would more than offset the extra cost of a suitable steel. Alloy steels are preferred for heavy work in primary mills using balls larger than 2″ in diameter. For secondary grinding, where impact is lower, cast-iron or self-renewing liners are favoured.

Feeding

The grinding mill receives new ore from the fine-ore bins, returned classifier sands if in closed circuit, new supplies of grinding balls as required, and water. Usually, the feeding arrangement must elevate part of the load to trunnion height in order to close the return circuit without the need for installing a special device. The most widely used feeder is a scoop working in a feed box that slopes downward to the gathering zone. The scoop (Fig. 39) turns as part of the mill and its tip digs into the downward sliding ore. At the same time it picks up mill-head water which is being fed to the box at a controlled rate. The clearance between tip and box must exceed the thickness of the largest piece of arriving ore to avoid risk of damage. The balls fed in periodically to "top up" the mill charge enter through the discharge trunnion. Only the scoop's tip is armoured with abrasion-resisting alloy, and no other part of the outside of the scoop should be in contact with ore. Single scoops are most used, as they have the greatest capacity. The rate of feed (new ore *plus* classifier returns) is limited by the handling capacity of scoop and input trunnion. Ample clearance in the inside elevating spiral is important, in order to avoid obstruction or bridging by large pieces of ore. The spiral must be developed sufficiently slowly to ensure that feed is trapped and sent to the trunnion with no risk of falling out.

In open-circuit work, or when the return circuit is closed by a hydrocyclone instead of a mechanical classifier, the scoop feeder (Fig. 39) is not essential. Its place may be taken by a drum feeder (Fig. 40). When the crushing bodies are pebbles, or big pieces of rock, new supplies must be added frequently. These may be delivered *via* a feed hopper in light contact with the feed trunnion, which must be suitably large.

Most modern plants are planned for continuous ore treatment on lines which call for a certain amount of chemical and physical control of the pulp stream. Efficiency requires equable and smooth flow of a steady tonnage of properly prepared pulp and the first step toward achieving this state of things

Fig. 39. *Combination Drum Scoop Feeder* (Hardinge Co.)

is taken in the grinding circuit. The primary mill receives (*a*) somewhat long-ranged feed from the fine-ore bins, limited to perhaps $\frac{3}{4}''$ maximum size, and (*b*) returned classifier sands. From its closed-circuit should be released a steady tonnage of pulp of specified limiting mesh and water/solid ratio. Classifier action is discussed later, but it is desirable at this stage to consider the governing influence of the closed circuit used in wet grinding.

If a five-to-one circulating load is maintained in the primary circuit, this means that of six units of ore entering the mill one consists of new ore and five of material returned from the classifier after partial grinding during its previous passage through the mill. If the maximum entering size is $\frac{1}{2}''$, and the maximum discharge 60 mesh (12,700 microns and 211 microns respectively), this represents a reduction ratio of about 60:1. Since this reduction is not achieved in one pass through the mill the true figure is not 60 to 1. For the average particle it is 10 to 1, since it is the progeny of rock which has passed six times through the grinding zone before obtaining release from the circuit. Not all the particles are recirculated. Some are already undersize before entering the mill, and these should be passed on when they reach the classifier. Some, either through weight, toughness, or excessive size, may make more than six journeys round, being broken down progressively to the finishing sand size.

Fig. 40. Hardinge Conical Feeder

The largest and heaviest particles tend to be retained in the mill longer than the average pieces of new feed. This would lead to accumulation of oversize and loss of capacity were it not countered by methods to be described in a later paragraph.

The effect on the total feed (new feed *plus* returns) of its heavy dilution by partly ground ore is to reduce the average size of the particles undergoing comminution. The predominating size in the return circuit is of ore which has progressively been reduced toward a condition which will permit of its release from the closed circuit. As the return feed/new feed ratio is 5:1, this "near-release" particle tends to dominate the behaviour of the ore in the mill crop. The equalising action of these recirculating sands is a steadying factor in the grinding work. So long as five-sixths of the total feed is steady, a little "surging" from the average, either of tonnage or size of new feed, is smoothed out. If, however, the character of the new feed becomes steadily different, adjustment of its rate of admission is necessary.

The circulating load in a closed circuit can be calculated from the formula

$$R = \frac{F(b-a)}{a-c}, \tag{5.1}$$

where R is the weight of returned oversize/hour; F that of the feed to the ball

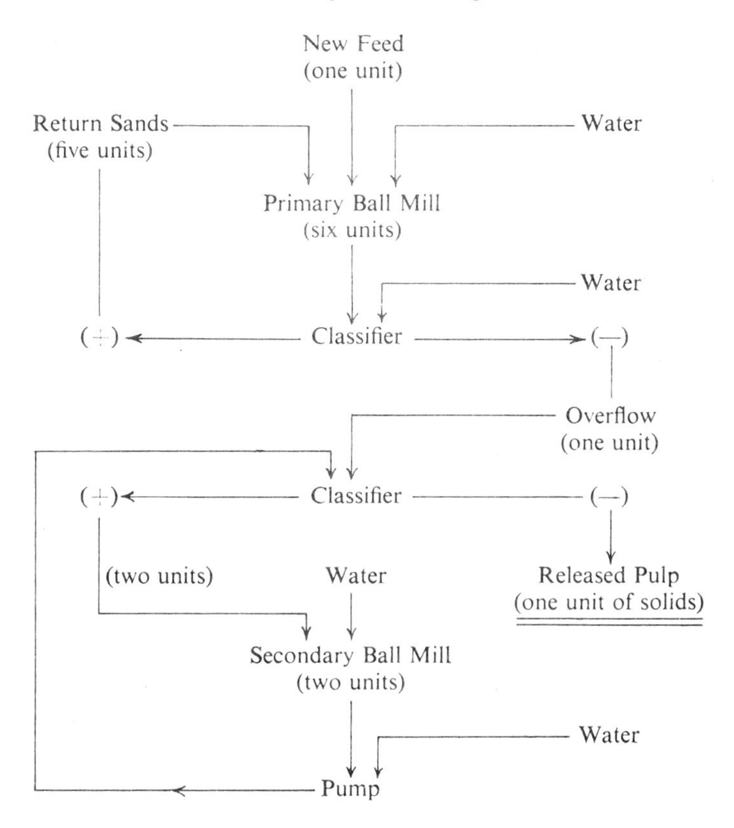

Fig. 41. *Flow-sheet. Units of solids distributed between closed circuits (primary and secondary grind)*

mill, and a, b, and c are the percentages passing a specified screen (say 100#). The a percentage is for $R+F$, b for the classifier overflow and c for the returned sand R.

The mill's charge of crushing media is so composed as to cope with new feed which is reasonably consistent in its proportion of various particle sizes, held below a specified maximum dimension. There must be good working liaison between crushing and grinding. A difficulty often encountered is segregation of sizes in the fine-ore bin, which leads to alternating deliveries of fine and coarse feed, even when no individual piece of ore exceeds the limiting mesh. Obviously, the less work the grinding plant has to do on the entering material, the greater is its throughput. Conversely, when the feed is "lumpy" its rate of entry must be slowed down since the available grinding power cannot be increased mechanically, and most mills have a fixed running speed. Farrant[3] (Fig. 42) has worked out an optimum change-point from crushing to wet grinding. 3″ is a rarely used maximum feed size, the usual figure being

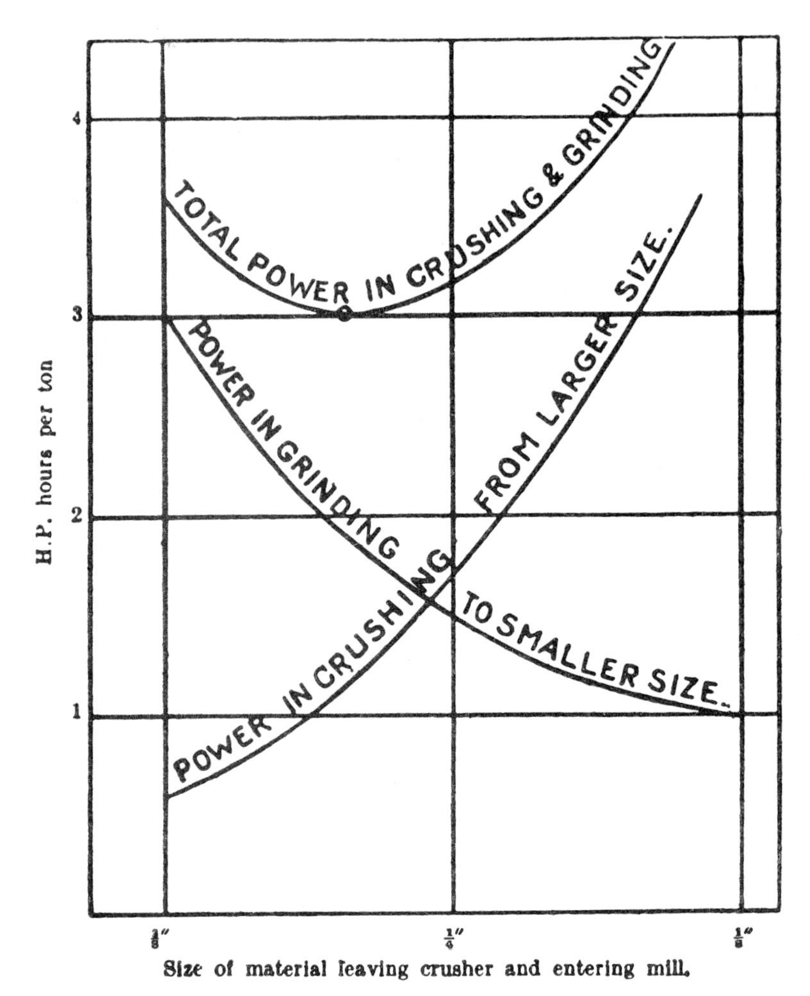

Fig. 42. *Optimum Change-Point* (after Farrant)

less than 1″ and tending toward ½″ in recent practice. If a rod mill is used between the secondary crushers and the ball mills, it is probably in open circuit and in one pass brings down the crushed rock to *minus* 16 mesh ($\frac{1}{32}$″ or 800 microns). Since a proportion of the new feed is already undersize with respect to the primary grinding circuit, it is sometimes by-passed direct to the secondary grinders *via* the primary classifier. In practice, some undersize is returned. It aids the passage of feed through scoop and trunnion, thus helping the mill to handle a big circulating load.

When the secondary circuit receives all the ore released from the primary circuit, the feeding problem must deal with the fact that only the entry of new ore to the primary circuit can be regulated. It is, however, simple to vary the amount of ore circulating in the primary closed circuit. If the secondary mill becomes overloaded, the primary circuit is set to grind finer. This throws more grinding work on the primary mill and enables the skilled operator to keep both loads in balance.

Two features of the flow-sheet (Fig. 41) are returned to later, but may now be noted. In the primary circuit a higher circulating load is possible, because the return sand is coarse enough to settle quickly in the classifier. The pulp escaping from the primary circuit must not be sent direct to the secondary ball mill, for two reasons. First, and most important, the solid-liquid ratio must first be adjusted by removing part of the overflow water. Second, any of the solids now sufficiently ground should by-pass the secondary mill. If a rod mill is used ahead of the primary ball mill circuit, its discharge must enter the primary classifier for de-watering.

In a simple operation where the ore does not vary much in quality, it may be drawn direct from the fine-ore bin through one or more gates. The use of multiple draw-off points helps to reduce size segregation. If there are no big pieces, the gates may deliver down a steep slope on to a belt conveyor. The height of the ribbon of ore is regulated by a vertical plate attached to the gate. From time to time the conveyor is stopped and the weight resting on a measured length is taken to provide a tonnage check.

A more robust arrangement is the apron feeder (Fig. 43). It has a variable speed, and the height of the ribbon of ore moving along it is adjusted to give the desired rate of feed by means of a vertical gate. The roll feeder is a large cylinder revolving below the withdrawal opening of the ore bin. It is simple, and readily accessible when obstruction occurs at the ore-bin discharge.

Fig. 43. *Apron Feeder* (G.E.C.)

A cut-off gate controls the height of the ribbon of discharging ore, and the rate of turn can be controlled if an adjustable driving mechanism is used.

Another method of checking the tonnage is provided by the Humboldt feeder. Two hoppers alternately receive ore from the bin, one filling while the other automatically tips its load. A further method, used for unusually sticky ores, is to push ore forward into a chute by means of a reciprocating ram. The length and speed of the stroke are adjustable, and the height of the moving ribbon is regulated by a cut-off gate.

Where the ore differs in various parts of the mine, it is sometimes necessary to keep the varying types separate. They may then be treated independently or may be blended to produce a consistent grade of milling ore. This calls for separate bins, each of which can deliver on to a blending conveyor. As the proportion of each type of ore is an important factor in this treatment, the draw-off feeder may be of a self-regulating type which delivers and registers its quota of tonnage to the main conveyor belt. One of these is the Hardinge constant weight feeder in which a short conveyor belt is pivoted below the ore bin. When the ribbon of ore upon it varies from the desired weight, the whole belt swings up or down under the influence of its adjustable counter-balance, thus altering the area of the delivery aperture a little and correcting the rate of feed.

Another type of feeder uses electrical vibration. It works best on dry ore. preferably fairly free from fine and clogging particles. The power unit (Fig. 44) throws the feeder forward and upward. The ore flows in imperceptibly short hops with little abrasive sliding against the conveying chute. Feed rate is controlled by varying the amplitude of vibration, by means of a rheostat. Any device operated by A.C. current to induce shaking must be carefully protected against even a slight change in frequency, which has a magnified effect upon the rate of feed.

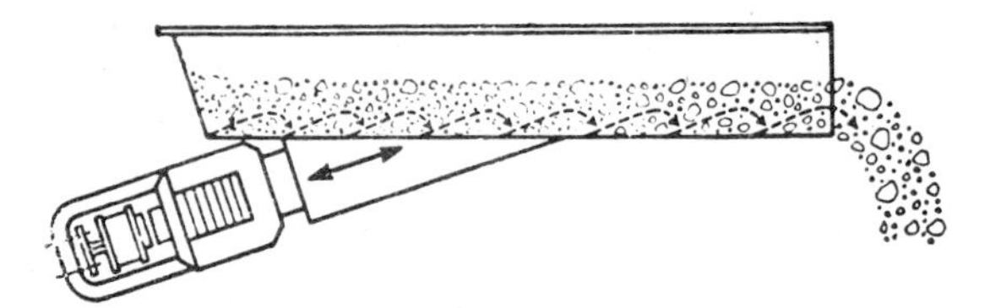

Fig. 44. *Electrically Vibrated Feeder* (Locker)

Crushing Bodies

As the power transferred *via* the liners is dissipated in collision and interfacial friction, impact crushing and abrasive grinding occur. For impact crushing to be effective there must be sufficient kinetic energy to disrupt the particle on which the blow falls. It is applied either direct or by transfer from the area of impact to the point where a particle is being gripped. The relation between the particle and the crushing body must be such that the energy applied exceeds the total resistance—or, in other words, the bigger or tougher the rock, the larger the crushing force needed. It follows that larger or heavier grinding media (balls or pebbles), capable of delivering more powerful blows, are needed in the primary than in the secondary grinding circuit. Each particle in movement in the crop load can interact with any others and may at any moment be striker, anvil, or both simultaneously. The liners, besides transmitting the driving force needed to generate grinding energy, act as fixed anvils at parts of the revolution. An interesting point is that so long as a free crop load exists the mill can absorb grinding power, regardless of the efficiency with which that power is used. If only balls were present, they would tumble and wear, and grind one another and the liners, producing abraded iron and heat. In practice, this would be worse than useless. If only ore and water were in the crop, some grinding would result, but the kinetic energy of the colliding and abrading components would be insufficient to produce efficient grinding, unless an adequate percentage of sufficiently large pebbles or pieces of ore were present.

The composition of the crop load influences the quality of grind. The proportion of water to ore, and of ore to the grinding media, can be controlled. The limiting mesh of ore is also controllable, but except for some hand picking it is not possible to regulate the nature of the feed. The density, size, shape, hardness, and toughness of the crushing bodies can be regulated when new, but unless the mill is periodically stopped and its contents are dumped out and sorted over, the shapes into which balls wear during their working life is uncontrolled. Their shape is a factor in keying together the tumbling load. There must be enough crushing bodies of appropriate weight to deal with the variously sized particles being ground. Steel balls worn small cannot grind by impact. Further, they reduce abrasive grinding by filling interstices with metal instead of ore.

Impact crushing can best be applied by the unobstructed drop of a heavy piece of steel, yet the stamp battery, which applies its energy in this way, is obsolescent. The shattered products cannot be cleared away sufficiently between blows. The rod mill applies part of its force as a dropping weight, and can both shatter and abrade the ore the rods fall on or roll over. It also discharges the crushed particles expeditiously and with a minimum of further comminution.

Rod mills can be run at high speeds[4] in specially designed mills. The lifting projections of the liners must assist the rods to lie axially and unentangled, and the rods should resist bending. The crushing load must not become matted so that it falls irregularly and increases the already severe shock-loading produced when each weighty steel rod crashes down from its break-away point. The mill components must be stronger than in ball-milling, in order to stand up to the punishment shell, liners, and bearings receive. Most mills are not, however, "cataracted", as running under free-falling conditions is termed, but "cascaded", the emerging rods rolling down the slope of the crop load to the toe of the mill. In this action the rods act as multiple rolls, the mill speed being well below two-thirds of critical. An important point in this cascading roll-action is that the whole weight of the rod (reinforced by that of all the rods pressing on it from above) bears upon a few particles of ore as they hold the rods apart. Since rod mills are nearly always run in open circuit, the biggest bridging particles will be near the feed end of the mill, and the smallest near the discharge. This results in a fanning out of the rods, visible as a tendency to work down and attempt to leave the mill by reason of a definite slope toward the discharge end.

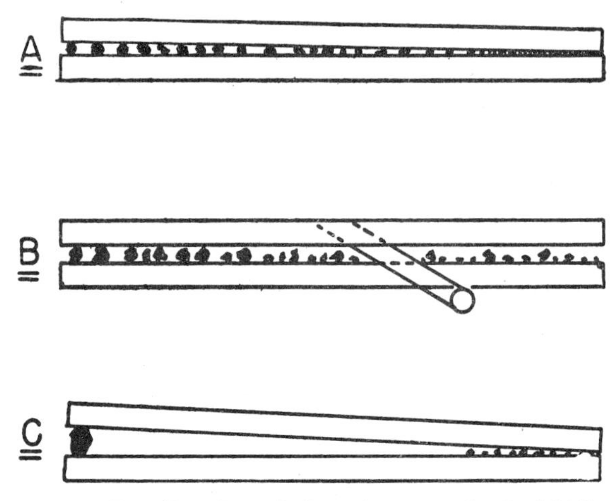

Fig. 45. *Particle Distribution in the Rod Mill*

In Fig. 45 *A* the aperture between two rods formed by the bridging effect of the largest newly entered particles (left) and the smallest (right) is seen to form a tapering slot. Collectively, the rods thus arrange themselves to act as a screen through which the smallest particles can flush clear of the crop load and then leave the rod mill if not again entrained by rods falling to the toe of the charge as the particles are carried toward the discharge end. The effect of a broken length of rod in upsetting this action is shown at *B* and the exaggerated aperture produced by too large a piece of rock at *C*. The wedge-shaped spacing presses oversized particles back toward the feed end and helps to distribute the load. The particles at the discharge end determine the "set" of the rod mill. All particles able to escape near this point leave the mill.

It has already been observed that every shattering blow produces a complete range of sizes. The rod mill cannot give a finished product all of which is just below a certain desired size. What it can do, given the necessary mobility of the particles in the crop load, is to direct the crushing force specifically toward the largest particles in the mill, and on the whole to avoid acting upon the smaller ones which are being protected by the bridging and screening action. This can happen so long as the voids between the rods do not become choked with fine sands to the extent that they are sprung apart, or choke fed. The analogy with crushing rolls under similar conditions of loading would be satisfied if those rolls had so weak springs that they yielded before sufficient crushing force could be directed upon the passing ore.

Rods, being lightly loaded, would yield if choke fed, grinding efficiency would suffer, and the nature of the product would change. It has been found that rod mills work best in open circuit, and suffer if a closed circuit is used in which quantities of fine sand are returned. Loaded below choke conditions with reasonably coarse material flushed by plenty of water, the rod mill can do its own close-circuiting by reason of the wedge-screening action above discussed. In one set of tests the water-solid ratio at feed and discharge ends was 30:70 but it was found that in the mill itself a 70:30 ratio was operating. The rest of the water was acting as a sluice, flushing out finished-mesh ore.

Rods wear down in use, and this affects the angle of nip. Tests have shown that, regardless of diameter, rods circulate through the crop load without marked segregation with respect to the turning centre of the load or of the mill axis. High-carbon steel, hot-rolled and straightened, is the preferred material for rods. This steel breaks when worn small, whereas mild steel bends and tangles, matting the load undesirably. Efficiency is highest when worn rods are removed periodically. Mills with big discharge openings facilitate this. Rods can be charged in by crane in bundles fairly quickly and worn ones removed. Trunnion-ended mills are also used, rods being charged in through the discharge trunnion individually and the worn ones "grinding themselves out".

Rod mills are not used for fine grinding, but, increasingly, to bridge the gap between secondary crushing (to —$\frac{3}{4}$″) and secondary grinding, the rod mill discharging at —16 mesh. Because of their value in this transition range they are said to give a "controlling grind". This means that their discharge is of a regulated fineness, thus simplifying the loading of the ball mill with appropriately sized grinding media. Rod mills are run at about 40% of their volumetric capacity.

Many shapes have been tried in the search for the most efficient steel crushing media, but the approximately round ball is the only one in world-wide use. It is cheap and efficient. A mill, when loaded to its full capacity, has nearly half of its volume filled. In this load, the ratio between the weight of crushing media available to exert pressure on the bottom part of the crop load, and the ore packed into the voids between those media, depends on the space made available by the shapes comprising the packing system. Rods give the highest proportion of metal, consequently the highest static pressure. Balls, for reasons discussed below, are random sized, but they tend to stay truly spherical till worn small enough to become trapped in the voids between larger balls, when they wear into tetrahedra with concave faces, which pack the voids and reduce reservoir capacity and mill efficiency. Some primary mills are stopped regularly, for removal of such material and broken balls. The load varies in practice between 50% and 40% of mill volume, according to the percentage of the grinding capacity being used. The percentage void in a ball charge is about 38%, and the weight per cubic foot for an average mixture of balls is about 280 lb., as against 390 lb. for rods (see Table 5).

A sphere has minimum surface area for a given volume or weight. It has also equal mobility in all directions under crowded or uniformly restricted conditions. It is better able to spin on any axis than is a non-spherical body, since the latter can only turn freely on a limited number of axes, and hence only when preferentially aligned. The ball thus has good spinning quality until worn out of its spherical shape. Wear is caused by abrasion against other balls, liners, and ore; breakage (of small balls) when hammered by large balls; damage when arrested after flight, due to brittleness without adequate toughness; and chemical corrosion by acids derived from ore or milling water, which can be lessened with an alkali such as lime.

Rate of wear has been the subject of considerable research and discussion. It is said to vary as the cube of the diameter D of the ball by Davis[5], as $D^{2\cdot 3}$ by Bond[6] and D^2 by Prentice[7]. Wear is proportional to abrasive rubbing contact when a ball spins and grinds while cascading, or while slipping down in a rising crop load. Under such conditions wear should be proportional to surface, or to D^2. When the ball is cataracting a greater amount of impact crushing is introduced. Impact is proportional to the kinetic energy of the flying object, or to its mass multiplied by the square of its velocity.

TABLE 5

DATA ON SIZES, WEIGHTS, AND SURFACE AREAS OF RODS, STEEL BALLS AND FLINT PEBBLES. MILL DIAMETER (FT.) INSIDE SHELL LINING

	Area				
	per cu ft.	*per ton*			
RODS (10′ long)	sq. in.	sq. ft.			
5″ dia.	1100	3·28			
4″ ,,	1384	4·09			
3″ ,,	1680	4·96	Approximate weight in rod mill 390 lb./cu. ft.		
2″ ,,	2760	8·15			
1½″ ,,	3680	10·88			
			Wt. of one ball	*No. per long ton*	
BALLS			lb.		
5″ dia.	1188	4·91	18·5	108	
4″ ,,	1487	6·14	9·5	211	In mill
3″ ,,	1980	8·19	4·0	505	about
2″ ,,	2980	12·31	1·2	1670	280 lb./cu. ft.
1½″ ,,	3950	16·32	0·5	4010	
QUARTZ CUBE (S.G. 2·7)			*Wt. of one cube*	*No. per long ton*	*No. per cu. ft.*
12″			169 lb.	13	1
10″			98	24	1·7
8″			50	45	4
6″			21	107	8
4″			6·2	358	27
2″			0·9	2490	217
1″			0·1	22400	1728

Since mass is proportional to D^3, the rate of wear varies with volume. This seems due to greater abrasive power in the heavier ball rather than to impact, which would tend to reduce wear as it work-hardened the metal surface. The faster the mill is run the higher the charge rises and the greater is the loosening of its upper portion. This increases slippage and wear.

Tests are made in order to evaluate the wear rates of mill liners and balls at Mount Isa have been reported.[14] They were made in an 8½′ by 12′ trunnion overflow mill charged with 2″ balls, to which test balls of various compositions were added in batches of fifty. These were of 3½″ diameter, making them readily identifiable, and were removed at intervals of 448, 920, 1592 and 2504 hours and examined for wear. The results are given in a series of Tables, starting with the analysis, hardness and specific gravity of each of the ten types used. Some of the findings have been condensed in Table 6. The wear rate factors were calculated from the formula

$$\text{Wear Factor} = \frac{W_O - W_N}{D_O - D_N} \qquad (5.2)$$

where W_O and D_O were initial weight and diameter and W_N and D_N the final. This factor represents the loss of weight sustained per unit surface area. Comparison was with Ni-hard as 100 (column 3 of Table 6). A wear rate factor for time taken to change in size from 3·45″ diameter to 2·8″ is given in column 4.

TABLE 6

COMPOSITION	(1) *Final diam. ″*	(2) *Diam. loss* 100 *hrs.*	(3) *Column wear rate factors*	(4)
1. Cast steel	*	*	—	183
2. White cast iron	2·57	0·0399	153	154
3. Austenitic Mn Steel	2·52	0·0391	153	154
4. Pearlitic Cr-Mo steel	2·75	0·0327	122	121
5. Heat-treated forged steel	2·63	0·0327	127	129
6. Climax 6–1 steel (Mn 6·4 Mo 0·9)	Not recovered	—	—	131
7. Martensitic Cr-Mo steel	2·80	0·0287	110	100
8. Ni-Hard	2·85	0·0268	100†	100
9. 28% Chrome steel	2·85	0·026	97	95
10. 16% Cr 3% Mo steel	2·95	0·024	91	90

* Too small to be identifiable
† Ni-Hard was chosen as reference standard

The authors found the reduction in diameter of the test balls to occur at a decreasing rate. Under the local conditions Ni-hard was shown to be the most wear-resistant and economical liner material readily available, though alloys 9 and 10 (Table) would wear longer. Martensitic chrome-moly steel would be a favourable alternative to Ni-hard.

Thus far mill speed has been considered mainly as a percentage of critical speed. It must now be considered more closely in relation to its effect on the crop load. The speed of rotation of this load determines the quantity and quality of grinding, in which the amount of slip between liners and load is one controlling factor. Others are discussed in the next chapter. At critical speed, the contents of the mill are not seized to the shell by centrifugal force. Indeed, this figure could be greatly exceeded as is shown later. A mill run with light slipping could handle coarser feed than a similarly loaded mill with heavy slip, because the kinetic energy available in free-falling balls would be higher.

Fig. 46 shows in graph form the percentage critical speed for mills of diameter up to 12′ at various speeds.

The ability to crush a coarse particle of ore depends on the size relation between the hammering ball and hammered particle. If no "controlling" rod mill grind is used, feed varies over a wide size-range. This necessitates a correct proportion of balls of various sizes corresponding to the percentage of each size of particle in the feed entering the mill. The ore is changing in character continuously as it proceeds from feed end to discharge, a consideration which helps to account for the wide popularity of the cylindro-conical mill, which tends to keep its largest balls at the feed end. Sizing tests are made as part of the routine of mill control, and they show the percentage of each size of ore in feed and discharge. If the discharge is too coarse, more work is needed upon the coarsest particles. If too fine, energy input is excessive for the rate of feed. One grinding control is ball-rationing, the use in the crop load of a ratio between balls of various sizes, proportioned to the feed and discharge products. Two methods of attaining and maintaining this proportion are used. Wear rate of the balls is ascertained, partly by observation and partly by trial-and-error variations of ratio combined with sizing analyses. The balance is then maintained by charging into the mill daily a suitable tonnage of new balls of various sizes in a proper proportion. No good rule-of-thumb method can be given, since ores and mills vary widely.

For its effective comminution a particle must be entrained and seized between the crushing bodies at the moment of impact or abrading stress. The firmness with which this grip is produced depends largely on the solid-liquid ratio for a given size of particle. Thus a pulp density can only be correct for one size fraction of the particles in the crop load.

A study has been made by Davis[5] of the size distribution of balls when equilibrium has been reached. If the rate of wear varies as the weight of the ball then

$$\%W = \frac{d_a^3 - d_b^3}{d_f - d_r}, \tag{5.3}$$

where $\%W$ is the percentage (weight) of balls in the size range d_a—d_b, and d_f, and d_r are the sizes of balls fed to and rejected from the mill. Though the d_s relationship for wear is not generally agreed, this formula provides a good working guide for replacement in the crop load.

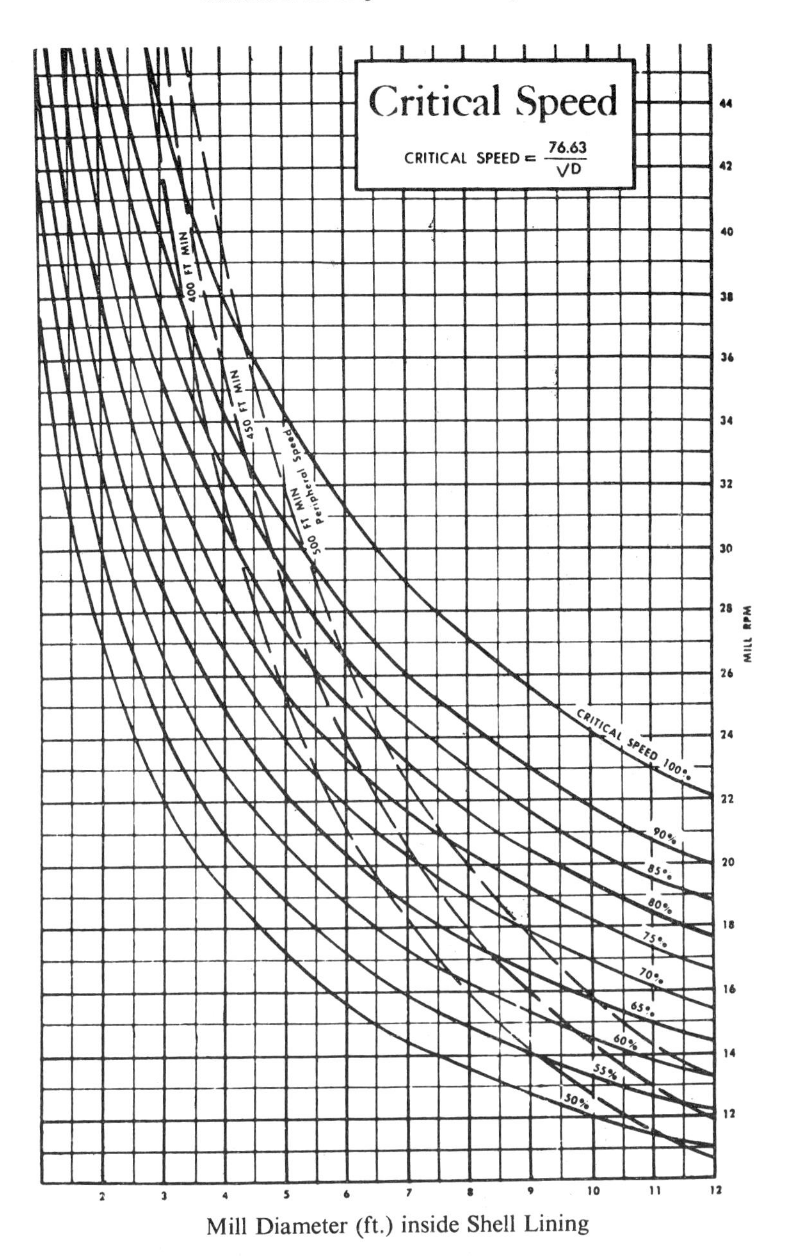

Mill Diameter (ft.) inside Shell Lining

Fig. 46. Relation between R.P.M., Diameter, Peripheral and Critical Speed (Allis-Chalmers Mfg. Co.)

Unless pebble grinding media form part of the crop load and are correctly allowed for in their relation to grinding capacity tramp oversize in the entering ore feed is bad. A properly loaded mill contains no balls big enough to shatter such material by free-falling methods. It remains in the crop, cascading and wearing slowly down till it is either small enough to be nipped and abraded efficiently, or has suffered random attrition sufficient to reduce it to shatter-size. An occasional piece of oversize cannot be avoided, but an appreciable quantity would build up in the crop and interfere with throughput. Where an open-circuit rod mill is used between the secondary crushers and the ball mills, the question of ball ratio is unimportant, since its discharge is delivered at —14 mesh and size disparity between ore particles is then a minor factor in grinding. An expression for the ratio between ball and particle size had been worked out by Coghill and de Vaney[8]. If D is ball diameter in inches, d— particle diameter, and K is a constant for rock grindability varying from 55 for tough chert down to 35 for softer dolomite, then D^2 is Kd.

As the balls spin, ore is drawn between their surfaces, where it receives abrasive grinding and shattering from any transmitted impact force generated by the blows being rained down upon the crop load by flying balls. Thus the ball needs to have surface hardness, expressed as a high "Brinell number", in order to withstand abrasion. It also needs toughness in its core in order to give resilient resistance. It is anvil to a descending ball, hammer when itself a descending ball, and a transmitter of shock through the crop system. When balls are worn out of shape they no longer spin freely and are less able to coat themselves continuously with particles drawn from the interstitial reservoir between the balls. Then one of two things happens—either the same coating of rock receives too much attention and is over-ground because it is not constantly changed and renewed, or ball rests against ball and metal grinds metal. Either condition is inefficient.

Balls are made from chilled cast iron or forged steel, which may be alloyed. If a delicate chemical concentration follows the grinding, care must be taken that no endangering alloy is used, since ball wear can be as great as 4½ lb. per ton of ore in extreme cases. Ball hardness decreases wear rate, while ball toughness decreases breakage. Wear is proportional to the mill diameter and speed, the liner roughness and hardness, the ball diameter, the solid-liquid ratio of the pulp, which affects its coating of the ball, and the acidity derived from the ore and mill water. Wear is normally between 1 and 3 lb. per ton milled, something between 1½ and 2 being usual. Coghill and de Vaney find ball wear to be proportional to the input of useful power, and to be of the order of 0·15 lb./h.p. hour for steel balls.

In pebble mills sea-rounded flints or selected pieces of rock, between 2½″ and 4″ in size, have long been used in grinding gold ore and in cases where iron contamination must be avoided. At one time finely divided iron was considered to be harmful to cyanidation of gold ore, and it was considered that the use of large pieces of ore to complete the comminution of very small pieces achieved several purposes—provision of crushing media, comminution of the large pieces, and avoidance of iron in the cyanide section. When, as with auriferous sulphides, the main purpose in grinding is to bring the included gold to the surface of the sulphide particles, the metallurgical considerations are quite different from those which precede efficient flotation, and in many cases, plant economics favour their use. For many years the Rand gold mines have screened 4″ to 6″ material (not necessarily of milling grade) from the *banket* ore for this purpose. Where iron has been found inadmissible, as in the final stage of grinding at Climax Molybdenum, pebbles are used for the selective grinding of a relatively small percentage of the ore after everything which could not benefit by this final grinding has been rejected from the feed. This is a special case dictated by the metallurgy of the concentrating section. Another kind of special case is the preparation of iron-free ceramics. The use of tungsten-carbide balls is reported, their high initial cost being compensated by their high density, toughness and long life.

Several factors have influenced the increasing use of *autogenous* milling in current practice. The term, which is loosely used to include extraneous rock as well as "pebbles" sorted from the mined ore, describes a crop load in which selected larger pieces of ore are used to grind the more finely crushed run-of-mine material. These are preferably of milling grade since they wear down quickly and thus form part of the tonnage ground and sent for treatment. Thus, they are part of the rated capacity and cost only the work of selection (usually by screening), storage and controlled handling into the mill. The general requirements for steel balls apply also to their efficient use. Pebble size must be appropriate to the force required, in terms of the size analysis of the mill feed. Since pebbles wear quickly, topping up of the charge must be performed much more frequently than with steel, and handling arrangements into the mill must be suitably designed. In one plant this is today automated by a linkage between the meter which measures power drawn by the crop load and the supply hopper gate where a reservoir of pebbles is maintained.

The mill shape suited to rods and balls must be modified for the best use of pebbles, and current developments are moving dramatically away from the old concept of a long narrow mill to the opposite shape, as seen in the Cascade or the Aerofall. Capacity and lift must take into account the larger bulk occupied by pebbles of S.G. 2·7 than of steel (S.G. 7·9) and also the logistics involved in moving a pebble of equivalent mass and much larger size with the required force.

If the pulp is to receive chemical treatment the presence of abraded iron may be undesirable. As much as 2 lb. or more of worn steel ball can join the process feed with each ton of ore. Liner wear also can be considerably reduced by the use of self-renewing liners which capture pebbles. These then take the peripheral wear and the great bulk of the grinding is done by ore on ore. Liner wear and lost time for renewal is also reduced. In grinding with steel the cost of liners and balls is a major item.[9,10]

Although the use of autogenous mills has not yet spread widely the savings reported by industry show that where the ore is suitable for their somewhat gentler comminution, they must be seriously considered in new developments.

Among recent studies of autogenous grinding are two Papers presented at the 6th I.M.P.C.[11,12] The relatively gentle liberating action, lower power consumption and wear, and elimination of preliminary crushing of a *minus* 1″

feed favoured autogenous grinding in comparison with the use of steel crushing bodies.

Capacity

Grinding is a means to an end—optimum profitable recovery. Mill capacity is governed by the requirement that the arriving pulp shall conform to certain specifications as regards the size of solid particles it contains. Capacity must therefore be provided in terms of the tonnage to be milled and the *mesh-of-grind* required. Comminution is carried out in stages, each stage doing part of the work better than it can be done elsewhere. If the plant is being worked to its full capacity, these stages must be mutually adjusted so as to keep the final one running at full load. As this final stage must bring everything in its feed down to the optimum *release mesh*, increase of capacity is achieved by adjusting the duty along the line till every machine is well loaded and the last stage of feed has been brought to such a condition that the final grinding circuit can handle the required volume efficiently.

This loading, while taking care that no oversize leaves the final grinding circuit, does not give maximum throughput unless it is supplemented by the satisfaction of a further condition. It would be possible to grind the ore in such a way that the largest particle was just small enough to escape from the closed circuit at the same time as the bulk of the particles were ground to a far smaller size than was necessary, or even desirable. In addition to preventing particles above a specified maximum particle size leaving the grinding circuit, the operator may also need to liberate every particle at as large a size as possible, provided it is under that maximum. He therefore tries to apply grinding force only to the oversized particles. For the purpose of most concentrating operations, this also improves the recovery of values. This policy is not followed where extra fine grinding is used to expose minute specks of gold to the chemical attack of cyanide solution.

If a ball mill were run dry and empty, save for its load of balls, it would consume practically as much power as when it grinds ore. In this it differs from nearly all other comminuting appliances which use less power when idling than when working (stamp batteries are an exception). To obtain maximum capacity it is necessary to balance several variables. These include:

Group A. Mill speed; liner contour; percentage of mill volume charged with a crop load; ratio of mill length to diameter.
Group B. Solid-liquid ratio; ball size ratio; new feed sizes.
Group C. Feed rate; circulating load; dwelling time in mill.

They are here grouped in this manner in order to simplify the problem of bringing nine main variables into good control. All those in Group A can be regarded as fixed for the purpose of an ordinary test not lasting more than a few hours, though the two latter items will change slowly through wear. Those of Group B can be maintained fairly constant during test. Any change of feed rate (Group C) immediately affects the other variables in this group.

This interdependence will be dealt with in detail when classifier action has been studied (Chapter 10). Several other variables exist, but they need not now be discussed. The interplay of these variables requires specific study in the case of each ore milled.

The main cost items in grinding are for power and wear of steel. Power consumption in a charged mill varies slightly between its fully loaded and underfed condition, good and bad ratio of grinding media to ore and water, good operation and bad. The economic concept of optimum capacity is therefore associated with rate of wear of steel (balls and liners). From the technical viewpoint capacity is qualified by product size (mesh of liberation desired), which in turn is dictated by the requirements of the concentrating section of the plant. The capacity is directly affected by the grindability of the ore. Unfortunately, no fully acceptable definition of grindability has yet been agreed on. It includes the qualities of hardness (or brittleness) and toughness (resilient resistance) of the ore and the characteristic resulting particle shape of the finished product (influenced by any tendency of rock gangue to break along grain boundaries and of valuable metal-sulphides to shear across them). To add to the difficulty, the purpose in grinding varies considerably according to the process by which the ground product is to be treated. This leads to diverse criteria of its suitability as a feed to the concentrating process. In a given mine the association of the minerals in the ore body and the grain of the ore may vary from stope to stope, with corresponding variations in the grindability. The effect in the mill is that some ore "goes through" faster than other, thus calling for varying feed rate so as to maintain equal loading. Bond's "Work Index" (see Chapter 3) gives a useful empiric guide to grindability for an ore of consistent character.

Even if the mill is handled correctly it is still limited by design to a definite order of flow capacity. "Flow" refers to the sum of new feed, return feed, and water passing through the mill in given time. For good flow capacity, the scoop must be generously proportioned and produce a pressure head when revolving, so as to force material into the mill along the feed trunnion. The latter should either be flared or given a helix at least half the projecting height of the maximum particle fed in, in order to keep the feed moving briskly into the grinding zone. At the discharge end, arrangements for evacuation of products should be yet more generous so that a downgrade through the mill is assured at all times. The combined area of the grate openings must exceed the cross-section of the feed trunnion. The pick-up scoops must be able to draw the pulp-level well down toward the bottom of the grate openings. In a high-discharge mill the discharge trunnion should be of greater diameter than that at the feed end. It can be restricted when desired. It should also flare outward to promote quick discharge.

There is a definite relation between mill capacity and speed. A speed of just under 60% critical is favoured in many mills. Capacity falls off at high speeds, markedly so in an observed case when 80% of the critical speed was exceeded.

Much remains to be clarified in our understanding of the balance of the forces at work. A Cascade mill at Vassbo[10] which is equipped with variable-speed drive has demonstrated the relationship between change in circulating load and charge level. With increased toughness of ore the discharged sands are coarser and the circulating load consequently increases, since the classifying system sends them back. This increase is detected by a metering device, and the speed of the mill is decreased. The charge level then rises and the power drawn into the crop load increases somewhat. This produces finer grinding from the extra kinetic energy available, and the mill discharge product is correspondingly finer. The circulating load now falls and the automatic control system speeds the mill up once more.

A limited amount of specialised wet grinding is still performed in pendulum mills, such as the Huntington and Griffin, in roller mills like the Chilean (developed from the primitive arrastre), and in grinding pans such as the Cobbe. Descriptions of these obsolescent machines are given by Truscott.[13]

References

1. Bond, F. C. (1959). *Bull. Can. I.M.M.* Aug.
2. Howes, W. L. *Trans. A.I.M.M.E.* 169.
3. Farrant, J. C. *Trans. Inst. Chem. Eng.* 118.
4. Myers, J. F. (1953). *Recent Developments in Mineral Dressing.* I.M.M.
5. Davis, E. W. *Trans. A.I.M.M.E.* 61.
6. Bond, F. C. *Trans. A.I.M.M.E.* 153.
7. Prentice, T. K. *J. Chem. Soc. S. Africa.* 43.
8. Coghill, W. H. and de Vaney, F. D. (1938). *Bull. Mo. Sch. Min. Tech. Ser.* Sept.
9. Crocker, B. S. (1959). *A.I.M.E.* May.
10. Fahlstrom, P. H. (1962). *World Mining.* Sept. & Oct.
11. Jacobs, B. W. and Feik, J. (1963). *I.M.P.C. Cannes.* Pergamon.
12. Pasquet, M. and Joco, G. (1963). *I.M.P.C. Cannes.* Pergamon.
13. Truscott, S. J. (1923). *Text Book of Ore Dressing.* MacMillan.
14. Gilbert, I. and Wingham, D. W. (1963). *Australian Bull. I.M.M. Proc.* 207.

CHAPTER 6

FORCES IN WET GRINDING

Optimum Grind

The previous chapter was concerned mainly with the mechanisms used in obtaining grinding force. Discussion now turns to its controlled application, and to balance of the grinding components at work.

Optimum size of release from the grinding circuit into the concentrating section is determined by technical and economic considerations. The finer an ore is ground the more this grinding costs. Up to a point, finer grinding usually results in higher recovery of values, but beyond this over-grinding leads to poorer recovery. Optimum grind defines the mesh-of-grind at which a maximum profit is made on sales, when both the working costs and the effect of grinding on the recovery of values have been brought into consideration. Such an optimum point is determined in the first place by test-work in the laboratory when the flow-sheet for a specific ore is being prepared. It then becomes the operator's aim to achieve maximum throughput at this optimum grind. With care and attention it is frequently possible to improve on the figures obtained in preliminary testing.

"Optimum grind" or "*release mesh*" refers to the sizing analysis of the ore particles finally leaving the grinding section. In its simplest form it can be specified as, say "a 100-mesh grind"—meaning that substantially all the particles in a carefully taken sample pass through a 100-mesh screen. A more exact optimum might call for, say "*minus* 5% on 100 mesh and *plus* 85% *minus* 200 mesh" indicating that a little coarse and rather light gangue-stuff may safely be allowed to leave the circuit, and that the needs of the concentrating section of the plant will be best met if overgrinding is avoided by stopping the process when 85% of the ore, by weight, passes a 200-mesh screen. Such a specification calls for care in order to bring all the material, undersize as well as oversize, between certain size limits so far as skilful control permits. It results from the scientific application of forces now to be described. Careful grinding preparation develops the latent characteristics of each mineral species it unlocks from the ore in the form of individual particles. It thus simplifies the work in the concentrating section and is of great practical importance.

When ore characteristics vary from section to section of the mine it may be found economically desirable to avoid mixing, so that each type can receive specialised treatment. Separate bins are provided in such a case. Ores may also be bedded so as to be drawn in a uniform grade through one grinding circuit. In a small operation the dispatch from the mine of markedly different types of ore should be so regulated as to provide a steady run-of-mill head feed.

Applied Power

Of the electrical power fed in, a loss of the order of 10% occurs in the motor, and between 10% and 15% in the gears and mechanical friction of the mill. The balance is available as "useful power"—as kinetic energy in the tumbling crop load—but the fact that it is available does not by itself lead to its efficient use. If the crop load is not properly constituted, part or all of this kinetic energy will be wasted by conversion to avoidable heat, sound, and ground-up metal.

When the mill is at rest the crop load lies more or less horizontally across the lower half. If it slopes a little, the out-of-balance effect due to this slope is being held by bearing friction. When the mill starts to rotate, much extra power is needed to get the system in motion. The crop load must be displaced in the direction of the rising side and its components must accelerate from rest and build up their kinetic energy. As soon as normal running speed has been reached and the system has settled down, the intake of useful power falls to a steady level. The liners grip the load and carry it upward on the rising side. Toward the limit of this rise, the upper part of the load breaks

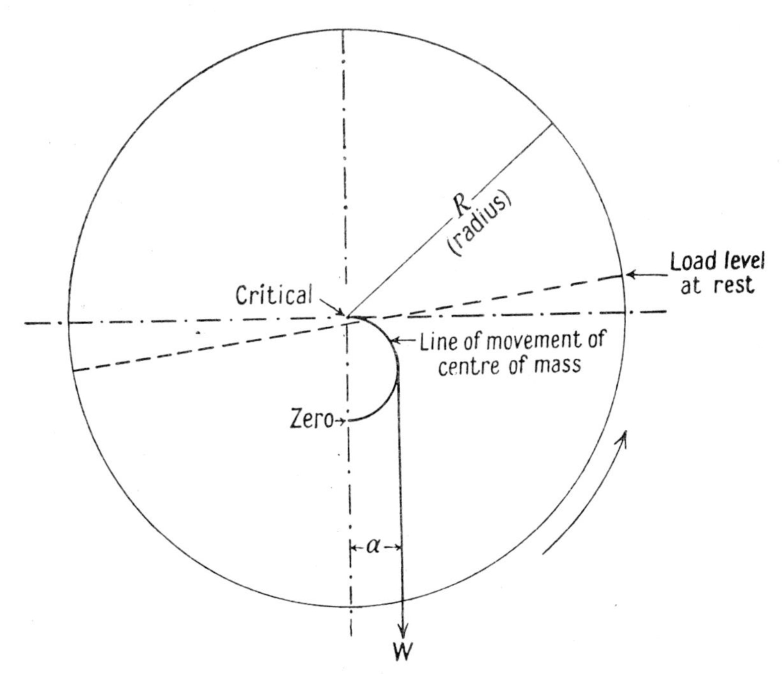

Fig. 47. *Movement of Centre of Mass of Crop Load*

away and cascades or cataracts down to the toe of the load. This turning mass is rotating about a centre *W* (Fig. 47), located somewhere along the path shown at a point where it balances the useful power at work. The intake of useful power is balanced by the displacement of the crop load. This power is continuously converted to kinetic energy and from this into heat, sound, and newly developed surface (of ore, balls, and liners). If R is the mill radius, W the weight of the crop load, and a the horizontal displacement of the centre of mass, then

$$\text{Torque} = WaR. \tag{6.1}$$

Fig. 47 shows torque (in the sense of useful power) to be nil at zero and also at critical speed. At critical speed (*plus* the extra speed necessary to compensate slipping of the crop load and hold it by centrifugal force) grinding stops.

Such a position is largely theoretical, since it could not be reached in a normally charged tumbling mill. The relationship between centrifugal force and its radial tangential thrust at various depths along the mill's radius passing through the crop load, precludes such seizure. The many interacting factors at work in the churning charge tend to confuse the picture of its dynamics. To clarify discussion, consider the case (admittedly over-simplified) of a variable-speed mill with smooth liners, no ore or water, and a load of steel balls of one size. As it starts very slowly from rest the load surface tilts (Fig. 47) until the slope is reached where load stability fails and the top layers of balls slide down. Neglecting slight irregularities in breaking away, the power draft is steady for a given speed when the stable displacement is at its maximum. This draft (ignoring mechanical loss outside the mill shell) corresponds with the displacement of the centre of the turning load from the vertical diameter by a distance a along the theoretical path (zero to critical) in Fig. 47. As the mill speed is increased this displacement reaches its maximum. With further speed increase the distance a begins to recede and power input falls. The concept of freeze-up at critical speed is not valid, in view of three main forces at work, of which centrifugal fixation is only one. First, before this stage is reached any peripheral balls rising clear of the down-slipping load after it has passed the plane of its horizontal diameter fail to maintain tangential direction because they are now acted on by gravity. Losing contact with the shell they take a falling trajectory to the down-running side of the mill. In the course of this they collide either with other balls, loosening the upper part of the charge, or with the shell itself, thus transferring part of their kinetic energy back to the shell from which they had received it. This acts against the input of new energy once a balancing peak of flight has been passed. *Second,* the packing structure of the charge changes steadily as mill speed increases. At rest, ball rested on ball and voids between these spheres was at its minimum. With rising speed the core of the charge, and also the upper layers, are loosened so that the volume increases. For a given mill speed there is a critical volume of charge at which the centre of mass a is at its maximum displacement from the vertical diameter. If the volume of the charge had been subcritical at this speed it would have been possible to increase the out-of-balance loading by adding more balls, and the power draft would have risen. Similarly, it would have been possible to increase speed without increasing charge to obtain the same effect. If, on the other hand more balls were added, or speed were increased to the super-critical point, the combined effect of the reduced out-of-balance dead loading and kinetic impact of falling balls on the down side of the shell and the toe of the load would be to reduce the useful input of kinetic energy to the system, and the current or wattage drawn by the driving motor of the mill would fall. The *third* force, the consolidation of the charge by centrifugal force, is modified by the first two.

The practical crop load is, of course, a mixture of grinding media of various sizes and shapes, perhaps even of varied density, since steel and large pieces of ore may form part of the crushing bodies. Next there is the ore, partly a new feed from dry crushing and partly a return load of partly finished sands from the closed circuit. Further, there is the make-up water added at mill head, which modifies the plasticity of the tumbling load in accordance with its *specific surface* and the percentage of water used. Finally, there are the liners, of various types, shapes and degrees of wear, with grip modified by changes in the amount of slime or slippery sulphide mineral anointing their surfaces. Again, over-simplifying somewhat, the crop load can be pictured as a loosely plastic body being continuously moulded into shape by tumbling action and influenced in its mass cohesiveness by the frictional hysteresis of its components. In most operating circuits the mill speed is fixed, but in all the crop load varies slightly in volume with the grindability of the ore. Since only "finished-grind" pulp is allowed to leave the grinding section increased resistance of newly entering ore results in increased retention in the closed circuit, part of which takes place in the crop load. If the load then goes super-critical (in terms of the above discussion) less grinding power is available, less grinding is done and the overcharge increases. If more easily ground ore comes to the mill the charge is diminished and too much energy seeks too little ore in the sub-critical loading which follows. The significance of the Vassbo experiment, referred to in Chapter 5, in providing automatic change of grinding speed geared to mill loading, can now be seen.

An equation for "best operating speed" (n) in terms of internal stability of crop-load and frictional grip from the shell has been proposed by Davis.[8]

$$n = \frac{0.8158}{\sqrt{r.}\ \sqrt{1 + K^2}} \quad \text{revs/sec} \qquad (6.2)$$

where $K = r_c/r$, (r_c being inner radius of charge and r the mill's radius). A Paper by Guerrero and Arbiter[4] modifies this to include the effect of slippage.

In a ball mill the rate at which power is converted into kinetic energy is fairly steady, but in a rod mill it varies somewhat abruptly owing to entanglements, hold-ups, and momentary seizures of the rods as they turn.

It can be shown (the mathematical proof is developed by Rose and Sullivan)[1] that the power input to a grinding mill (P) is

$$P = KD^{2 \cdot 5} \qquad (6.3)$$

where D is the mill diameter. Gow *et al*[2] find that in practice the exponent to

D should be 2·6. Bond[5] points out that in a commercial installation an exponent of 2·4 is approximately correct, or that "the power required per ton of . . . grinding media varies as the mill diameter to the 0·4 power, while theoretically it should vary as the diameter to the 0·5 power. The difference can be ascribed to the energy returned to the mill shell by the falling balls impacting against it."

Hukki[6] defines the "base mill" as a horizontal cylinder 1 metre in inside diameter and in length. The power index P_i for this mill is

$$P_i = \frac{P_n}{D^{2.5}.L} \text{ kW} \tag{6.4}$$

P_n being the nett power drawn and D, L the internaldia meter and length in metres.

If the kinetic energy is correctly applied, a maximum of properly finished ore results, but whether correctly applied or not, the power continues to enter the mill so long as the crop load is being dynamically held out of balance. Unlike nearly all other comminuting appliances, the grinding mill uses about the same amount of power all the time it is running. Substantially all the kinetic energy is finally dissipated as heat, which warms the transient pulp. In the endeavour to find a fundamental expression for grinding efficiency it has been suggested that, since new surface is proportional to grinding energy (Rittinger's law), and since this involves the creation of new surface tension or surface energy on the newly sheared particle surface, the efficiency of grinding is measurable as the proportion of useful power to new surface. Calculations along these lines have produced several sets of figures, all agreeing in the conclusion that grinding efficiency in the ball mill is very low—well under 0·3%. For practical use, industry needs something more concrete to measure the efficiency of its daily operation. Performance in the plant can most conveniently be judged by its relation to some selected standard of throughput. There is no good direct way of measuring the surface energy of solids. In any case the increase in surface energy of the particle which is due to the transfer of some input (grinding) energy to the newly created area is probably only part of the total rise in energy. No method exists for assessing the internal changes in energy level (physical, chemical, and electrical) which accompany comminution. Efficient or not, the tumbling mill is the best machine at present available for the work of grinding, which is the most expensive cost-per-ton item in the flow-sheet.

Rittinger's law has been the subject of much research, and may be regarded as a good approximation. The operator will work on sound lines if he thinks of grinding force as resulting in new surface (in an efficient operation). The next step is for him to ensure that:

(*a*) As much of this new surface as possible is created on particles neither too big nor too extravagantly small for treatment in the concentrating section;
(*b*) As few as possible of those particles shall consist of steel abraded from the balls and liners;
(*c*) As many as possible shall consist of value-containing ore.

This mental approach enables him to see the problem of grinding realistically, and to translate his vision into effective operational control.

"Useful" or Net Power

The kinetic energy generated in the crop load by transfer of driving power through the mill liners should be maintained at its peak value, in order to obtain the maximum amount of grinding from the system. This maximum draft is achieved by means of a correct balance between four main factors:

(*a*) Speed of mill rotation, expressed as % *critical speed.*
(*b*) Liner grip, notably of body (horizontal) liners.
(*c*) Constitution of crop load or charge C (media, ore and water).
(*d*) Volume of C under running conditions.

First consider item *d, a* and *b* being fixed and *c* varying only as to ratio of grinding media (m) to feed (consisting of new ore, classifier returns and mill-head water). Take as the starting point peak power draft, with displacement of C to the rising side at its maximum unbalance. If C now increases part of this increased volume is re-balancing the load by overspill to the down-running side. It thus reduces the unbalance of C and also feeds more driving energy to the shell liners. This transferred energy does not, of course, increase the total input. The reduced unbalance does, in fact, result in a power drop, registered on the ammeter of the driving motor or motors. However deployed, the net input power is proportional to an exponential value of new surface produced, and any reduction of this input leads to a corresponding diminution of useful grinding. This is why the mill must be run with a maximum input of useful power in order to maintain peak efficiency.

Some qualification is desirable at this point. The purpose of grinding is not solely technical. It must contribute to the maximum overall profit, which depends on a balance between all processing costs and the grade and percentage recovery of concentrate in the best condition for further use. The size analysis of the solids in the mill discharge has an important bearing on subsequent treatment, and grinding is a major cost item. This frequently raises the operating question of shattering *versus* abrasion in the grinding section. The fierceness of shatter at constant speed will obviously be reduced as C increases in weight and volume beyond peak displacement of its centre of mass. Slippage will then increase, balls be more blanketed and impeded in falling, and the toe of the charge will be more abraded and less hammered. The question always arises: "What kind of grind is best for a specific ore, treated by a specific method?" The reader should re-read this section after the chapters which deal with various methods of concentration have been studied.

A change in loading volume due to variation in retention of ore upsets the balance between m and C, and also alters the frictional characteristics of C. If C increases, shatter is reduced and the coarser particles of ore are less adequately reduced to sizes which m can seize and abrade. At the same time the specific surface of C decreases, though no change has been made in the carefully controlled solid-liquid ratio which is being maintained in relation to optimum specific surface. From this point there is cumulative deterioration in the efficiency of comminution. Slowly the mill discharge size analysis increases its percentage of coarse material. This discharge is returned by the closed-circuit classifying system in increased volume though there has been no increase in new feed of ore. Thus, both C's texture and volume are changing from their optimum balance at an increasing rate. This readily observed connexion between reduced power consumption and loss of grinding efficiency (which may have adverse effects right down the flow-line if it leads to overloading with wrongly ground material) underlines the vital importance of a well-controlled grinding section.

Items *a* and *b* may now be related to this discussion. When mill speed is increased without any other alteration C is reduced, and *vice versa*. This fact was brought out clearly in the Vassbo experimental work referred to earlier. There, the mill's best performance in overall terms was found to be 60% critical speed. Taking this 60% as the index of 100% efficiency, each increase of 1% of speed was accompanied by a drop of 1% or so in efficiency. At 80% critical it was necessary to use 25% more power to do the same amount of grinding. This extra power was obtained by increased unbalance of the crop load. The point is significant, since it shows that maximum unbalance (accompanied by maximum draft of useful power) cannot be taken in isolation as the criterion of efficiency. In this case it showed that the mill was too big for its job and in consequence its diameter has since been reduced. Maximum power draft must be related to the required finished grind, and achieved by the correct composition and volume of C. In the Vassbo operation autogenous grinding and a variable-speed mill were used, with automatic linkage between speed and change in the volume of returned circulating load. With ore as m change in size composition of m-components is far faster than with steel balls, and variation is easier to arrange and observe. The work was done on a full working scale and tied in with the subsequent treatment, so that effects on recovery could also be seen. The inter-acting factors thus revealed should affect grinding research and development.

A few further points may now be noted as accessory to the above discussion. When the volume is steady, the net power is highest with the interstices between balls full of ore and lowest with them full of water, which has a much lower density and therefore reduces the total crop weight. Power used is higher with plenty of "sharp" sand in the crop than with slimy sand only, since the extra friction helps the liners to grip the load more firmly and raise it higher. It is higher with a low-discharge mill, because pulp rises centrifugally on the rising side but can escape near the periphery of the grate, whereas in the high-discharge mill it can only overflow from the trunnion, so that a larger volume of pulp must be retained on the falling side of the mill. A drop in the ammeter reading of power input to a low-discharge mill might show that the discharge grates were partly clogged.

The greater the diameter of a constant-speed mill, the higher is the centrifugal force and the stronger is the liner grip. More power is then used, so available grinding capacity increases with wear of liners. The greater the lifting grip of the liners, the more the power used. If the wave-contours of the liners decrease with wear, the grip is reduced and this effect tends to neutralise the one mentioned in the previous sentence. The higher the density of the ore, the greater the power draft. With too little water in the crop load a paste may be produced, lubricating the liners and reducing power consumption because of increased slip. Power cannot be "pumped" into the crop load. It can only be drawn from the energising system, and then only to the extent that the crop load is dynamically out of balance.

Grinding and the Particle

When rock is worn down very gently by such agencies as wave action or the blowing of desert sand operating over long periods of time, the edges of the particles are more or less rounded. In ore grinding such gentleness would fail to deal with the tonnage. Rounded particles in the mill discharge are a symptom of wrong loading, in which balls too light for their job are rubbing instead of shattering the ore. Particles properly broken by shattering impact or severe abrasion are sharp edged. They are random in shape but tend to have the dimensions 1, 1·2 to 1·5, 1·6 to 1·8[7]. Rock of a soft character may, through compressive squeezing in the dry-crushing section by such machines as rolls, acquire a schistose or laminated appearance. Many minerals are naturally flaky.

The absorption of grinding power during size reduction can best be understood by the consideration of a theoretical case. Suppose the length (*l*) of each side of a simple cube of rock (Fig. 48) to be 4 millimetres. If it is now broken down by stages into smaller cubes, *l* being halved at each stage, the following relationships result:

Length of Side (l)	*Area of each Face*	*No. of Faces*	*Total Surface*
4 mm.	16 mm.^{2}	6	96 mm.^{2}
2	4	48	192
1	1	384	384

and so on. The figures in circles (Fig. 48) show the total number of cubes at each stage of size reduction. For each halving of size, or mesh, the total surface is doubled. With each reduction to cubes of half the previous linear dimensions or mesh an amount of new surface equal to that already in existence is produced. (It would be impossible, of course, to produce an exact geometrical crushing of this kind by any milling procedure.) It will be seen that when particles are reduced in even steps of mesh size then at each size there is a geometric increase in the total surface area. Using the accepted generalisation of Rittinger's law—that useful power consumed in grinding is proportional to the amount of new surface created—there is therefore a corresponding geometrical increase in the power needed from stage to stage to give evenly spaced reduction of mesh size. In relating comminution to the effect it has produced, change in specific surface gives valuable information. *Specific surface* is the total surface per unit weight of dry ore, or for special purposes connected with transport and treatment, per unit volume of a pulp.

In practice, the effect is that the power used to crush big rocks down to gravel size is far less than that required to grind the same weight of gravel

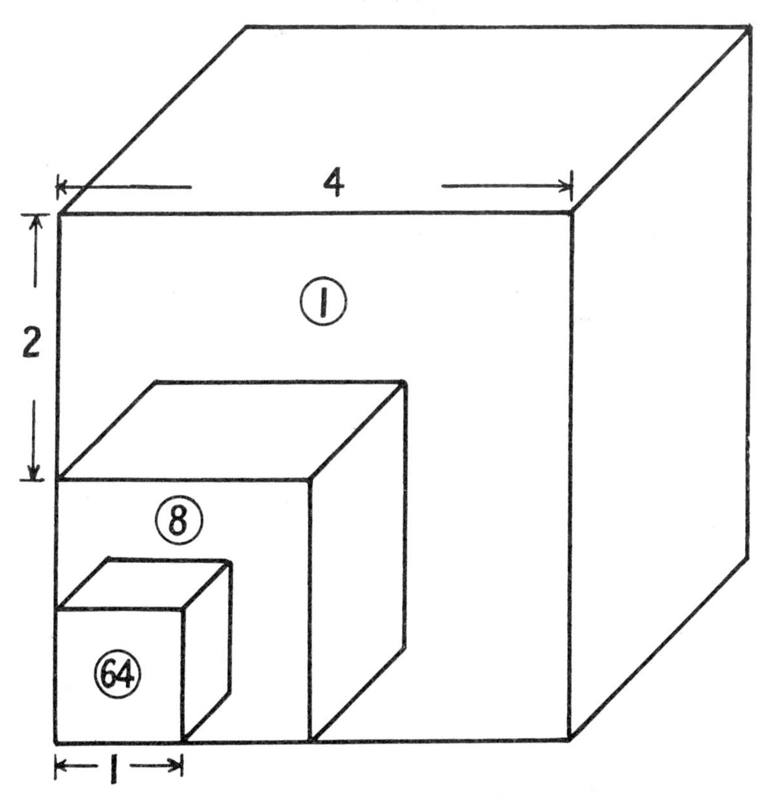

Fig. 48. *Effect on Total Surface of Size Reduction*

sizes to fine sand. The finer the finishing mesh required, the more rapidly the cost rises and the greater must be the provision of fine-grinding equipment.

Homogeneous rock all drawn from one source usually behaves consistently, but ores subjected to mineral processing are never homogeneous, as they carry at least two different minerals. Occasionally these constituents are loosely associated and can be separated by differential grinding, which frees lightly bonded grains of one mineral from a semi-cemented adherence to the other or others. Usually the minerals of the ore are interlocked. Each has a toughness and hardness (or grindability) of its own, so that a solid lump of ore is weaker at some points than at others. The percentage of each mineral varies in different parts of the mine, causing a corresponding variation in the grindability of the ore.

The work of concentration would be far easier and simpler if minerals in a piece of ore could be ground in such a way that they parted from one another along their crystal boundaries, but grinding does not usually work in this way, though some such tendency is mildly developed in *autogenous* grinding. For *liberation*, it is necessary to reduce the size of the particles until individual fragments are sufficiently rich or poor in desired value to be worth separating, and until they have developed characteristics which ensure that they will respond to the concentrating treatment. For this reason, release from the grinding circuit is determined by particle size or mesh.

Research has shown that when a 4-mesh to 8-mesh mixture of quartz and limestone is batch-ground in a laboratory ball mill, the initial rate of production of fine particles is proportional to the grinding time, but that the coarsest material is attacked selectively since it shields the finer particles. This initial rate is linear, but with continued batch grinding the more obdurate quartz becomes protective, so that the rate of comminution of the limestone slows down.

Grinding Objectives

Methods of concentration vary in what they demand as suitable preparation of the feed. From one point of view, grinding spends money to produce new surface, and from another, it produces specified sizes. As little new surface should be generated as possible, consistent with the production of the correct mesh of grind.

Broadly, ore concentration requires one of three types of grinding preparation. For gravity concentration the desired constituent of the ore must be liberated at the coarsest practicable mesh, and overgrinding or "sliming" must be avoided. In chemical extraction the constituent to be dissolved must be adequately exposed at the surface of each particle. Here over-grinding does not hinder recovery, but usually improves it. For froth-flotation, a particle not larger than 200μ to 300μ or smaller than 5μ to 10μ is usually required, with the desired mineral exposed at part of its surface. Hence, the adopted grinding practice depends on the method of treatment which is to follow. Where mixed methods are used, grinding may be done by stages, linked with stage withdrawal of concentrate or gangue. In such a case a series of grinding objectives may be pursued, each suited to a stage of treatment.

Comminution of Particles

The kinetic energy in the crop load can be applied to a given particle by:

(1) Collision between pieces of ore.
(2) Pressure loading of a particle pinned between balls, or between a ball and a liner.

(3) Shear and abrasion, the particle being dragged between spinning balls.
(4) Impact of falling ball.
(5) Shock-wave transmitted through crop load by falling balls.

The particle may be seized and dragged between balls if the *angle of nip* permits, in which circumstances it is subjected to pressure, shearing tension, and abrasion. The particle may receive impact either through a direct blow when it is resting suitably on a ball or liner, or when a shock-wave is propagated at the toe of the charge by a descending ball and is transmitted deep into the load.

In quiet conditions particles tend to settle on the most stable base. In a fluid current there is a tendency to turn so that the minimum cross-section is opposed to the stream. If a particle is hit when lying flat, any natural tendency to become flaky is accentuated. Should it be hit edgewise, so that the maximum impact is applied to the minimum area, a higher shattering effect results.

The mobility of the particle in the crop load is a determinant of its behaviour in the grinding zone. When the mill is run with too dilute a pulp the ore does not coat the metal surfaces properly, so grinding force is wasted, ball hitting ball. At the right consistency the balls are properly coated by a clinging layer of particles and a maximum amount of kinetic energy is directed against the ore. If too little water is used, the crop load becomes sluggish and pasty, kinetic energy is wasted in overcoming viscosity instead of being available for comminution, and material remains sandwiched between two grinding bodies instead of continuously making way for a fresh supply.

Too much fine material in the crop load, such as primary slime, clay, overground friable sulphide, or gangue, is sometimes a nuisance, since the metal surfaces become coated with a film which acts as a lubricant, increasing the slip and reducing impact grinding.

The hardness of the crushing media is important, for on it depends the crispness of delivery of the crushing blow. When pebbles or lumps of ore are used instead of steel balls they both grind and are ground. Their shape tends to become spherical while they remain large, but it is unlikely that sands are entrained between them by spinning, as occurs with balls. At what is called their critical size, pebbles become too light and small to act as crushing media, and for grinding purposes become *tramp oversize* until further reduced. An autogenous charge is more angular than one of steel balls, and tends to lock and break away crisply rather than to slide under comparable conditions.

Effect of Peripheral Speed

One element in the conversion of torque into kinetic energy is the peripheral speed of the mill. The rate at which the mill revolves is fixed for any given gear ratio in the drive, but can be altered. The other determinant of peripheral speed ($2\pi rn$ where n = r.p.m.) is the internal diameter, which increases slightly as the liners wear thin. Rotation of the mill causes the crop load to climb to a slope exceeding its angle of repose. This angle of repose is higher than it would be for a heap of static material, because centrifugal force is pressing the whole load outward. The higher the peripheral speed, the higher the load is carried before its outer layers free themselves and fall, and the greater is their kinetic energy on arrival at the toe of the charge.

The three types of tumbling action that can be produced at successive increases of speed are called *cascading*, *cataracting*, and freeze-up. At cascading speed the top of the load turns quietly over as it emerges, and rolls down to the toe of the charge. At a higher speed (cataracting) some rising balls are thrown clear of the charge and then fall parabolically toward the toe, where they apply shock-loading at the point they hit. Between the cascade and cataract speeds is an intermediate one, sometimes referred to as avalanche speed, at which grinding is effected mainly by abrasion, but to some extent by free fall of the crushing bodies. The final, but for all practical purposes unattainable and theoretical, stage is freeze-up, which occurs when the mill is turned at more than its *critical speed* ($v^2 = gr$), r being the radius from the axis of the mill to the centre of a ball lying on the shell in the trough of a liner depression. Slip prevents freeze-up from becoming complete under working conditions. The concept of peripheral speed only applies to material in contact with the body liners. Hence, only the outermost layer of the charge is in a good position to attain full cataracting action. Some differential action occurs in the Hardinge mill, since peripheral speed varies along the cone (Fig. 30).

Not only is the freedom with which the ball descends important in determining the kind of grinding being done, but also the point to which it descends. For efficient impact the flying balls must fall well inside the trough of the crop load in the 8 o'clock area shown in Fig. 38. If the balance between crop load, diameter, and mill speed is wrong, the balls may be thrown too far horizontally and batter wastefully against the down-running breast of the mill. Then metal grinds metal. Shatter produces more new surface than does abrasion for the same input of energy since there is less frictional loss. In Fig. 38 the action in the various parts of the crop load is shown. In the segment A—A to B—B, starting roughly at the 6 o'clock position and rising to a little before breakaway point, the load rises quietly, with a certain amount of ball-spinning due to slip, but with comparatively little grinding. Next, from B—B to C—C (the latter is the area undergoing bombardment at the toe) comes free fall during which the ball converts its potential into kinetic energy. If the fall is cascading, much of this energy is used in abrasive grinding during the descent. At the toe is a churning mass where the bulk of the grinding is done. Not only is there direct contact between falling balls and pieces of ore lying above balls and liners, but ore is packed between balls and receives the shock transmitted through the C—C to A—A segment by the continuous battering. At the extreme toe of the load the descending side of the mill continuously folds the churning mass at the toe into the crop, and carries it up to the breakaway point. The material next the liners is held with more firmness than is any more central part of the crop load. The larger the dropping ball or lump of ore, the less able it is to penetrate the charge near the toe, and the more likely it is to be worked out to a point where it is seized and under-run by the liners. Thus there is a tendency for balls and particles to align themselves concentrically with respect to the centre of rotation of the crop load, with the largest outermost and most stable during rise and the smallest most loose at the charge centre. This alignment causes the largest balls to be lifted at full peripheral speed and to be thrown outward from the breakaway point with the greatest force. The dynamic conditions in the crop load can therefore be pictured at a cross-section through a cylindrical mill. Outside, and travelling at a little less than the peripheral speed, are most of the biggest balls and pieces of ore. They are last to leave the shell at breakaway point because they are held to it by the inside layers and also by the extra centrifugal force working on them in this position of maximum mill radius and minimum load slip. The cataracting effect required must be applied in terms of these largest balls, which have the maximum individual mass, velocity, and inertia, to ensure that they do not miss the proper target (line C—C at the toe) and carry on to hit the liners. Next, by stages inward through to the more or less oval-shaped part of the cross-section, come smaller and smaller balls and pieces of ore. These have less superincumbent weight resting on them and therefore slip more easily. They are circling more slowly and this lessening of peripheral speed is accompanied by increasing slip. Thus there is less centrifugal force available to maintain the inside of the crop load in its climb, so it sags away from the rising breast of the mill earlier than the outer part of the load. Between A—A and B—B the whole load is in its most compact condition and brings most of its weight and centrifugal force to bear on any ore being abraded during slip or spin of the balls in this section. Above B—B and also above the centre of turn the texture of the crop load opens outward and upward, becoming progressively less dense as it gets away from the centre. This again favours flight of the largest balls. In all this the effect on the peripheral speed of the crop load produced by liner contour is obvious.

In plant practice despite some increase in slip and abrasion a relatively high speed low-discharge mill is usual in primary grinding, followed by a lower speed for secondary grinding. Mills can be geared for two speeds where a variable tonnage input must be handled, though only one of these speeds can be fully efficient. In the Hardinge cylindro-conical mill (Fig. 30) the diminishing peripheral speed in the cone leading to the discharge trunnion is responsible for a change-over from shatter to abrasive grinding. Partly "finished" material (*i.e.* ore almost fine enough to be released from the closed circuit) tends to get its final reduction here, an effect helped by segregation of most of the smaller balls in this cone. This segregation has the further effect that the large balls are more free to develop shatter in the drum, where they can act at their best peripheral speed on the newly entering (and therefore coarsest) feed.

The foregoing discussion applies to mills run at below critical speed and with a heavy crop load. Hukki[9] examines the effect produced when a lightly loaded mill is run at super-critical speed. In Fig. 49 this is taken to 240% critical with a 15% ball load, and the maximum speed of the outermost layer of balls remains constant. This means that the increased shell speed is compensated by increased slip in the crop load, made possible by the use of smooth liners, low discharge, and light loading. In the original research

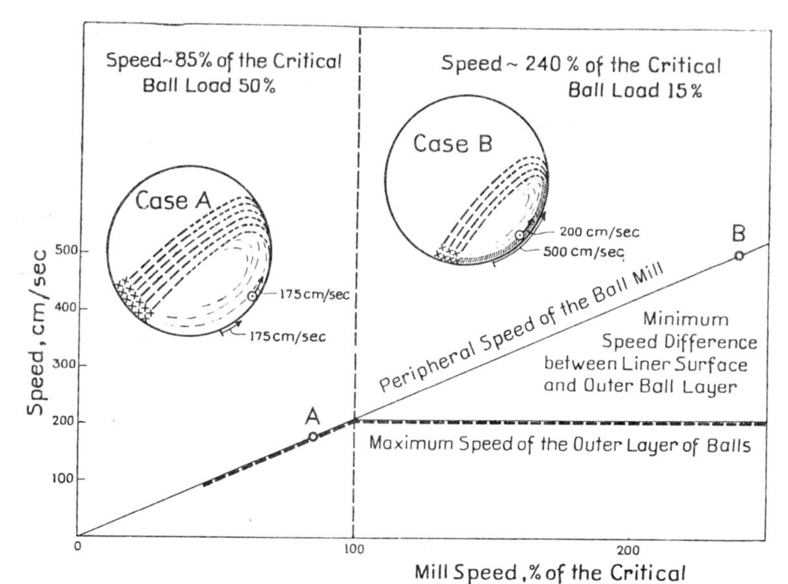

Fig. 49. Grinding Action at (A) *Sub-Critical and* (B) *Super-Critical Speed* (after Hukki)

work this, while increasing abrasive grinding, had the drawback of heavier wear on liners and balls. Despite the low crop load output of finished material rose steadily in the experiments quoted (Table 7). More recent tests with a large pilot mill have shown that when ore is used as grinding media and mill lining, grinding at super-critical speed indicates a capacity considerably above that of any sub-critical mill. Test work is still at the experimental stage but it has been proved that no centrifuging takes place during wet grinding at 150% critical speed.

The Return Load

The usual grinding arrangement provides that a mill works in closed circuit either with a mechanical classifier or a hydrocyclone. Discussion of these appliances must wait until the laws governing movement of solids in fluids have been studied. The needs of this section are met by the understanding that a classifier is a sorting device into which the discharged pulp flows from the mill. In the classifier it is separated into pulp of "finished mesh", the solid portion of which consists of particles ground to or below optimum liberation point, and oversize sands which have not yet been sufficiently ground. The pulp flows out of the circuit, and the sands are returned to the feed launder of the mill.

In an ideal grinding appliance, a particle of ore would be discharged as soon as it had been reduced to the required size. This, however, is not

TABLE 7 (*after Hukki*)

NUMERICAL RESULTS OF GRINDING TESTS IN PILOT PLANT ROD AND BALL MILLS

Test No.	1			2			3			4			5				
Type of mill used	Rod mill			Rod mill			Rod mill			Rod mill			Rod mill		Ball mill		
Mill speed, % of the critical	48			72			96			130			130		200		
Tumbling load, % of mill vol.	22·5			22·5			22·5			22·5			20		14		
Dry feed, kg/h	2030			2980			3920			5050			2000		b		
Motor, kW—rpm	11—1450			11—1450			11—1450			25—2940			11—1450		25—2940		
Energy used, kWh/t	2·74			2·83			2·97			3·15			6·3		6·16[a]		
Screen analysis Cumulative % undersize	Feed	Product	Diff.	Feed	Product	Diff.	Feed	Product	Diff.	Feed	Product	Diff.	Feed	Product[b]	Feed[b]	Product	Overflow
6 mesh	33·2	99·5	66·3	40·6	98·6	58·0	38·2	97·5	59·3	42·2	98·8	56·6	34·4	99·7	99·8		
8	28·8	97·8	69·0	35·5	93·5	59·8	32·8	92·4	59·6	36·8	96·1	59·3	29·7	99·2	99·7		
10	25·3	93·7	68·4	31·0	87·9	56·9	28·3	83·6	55·3	32·3	90·2	57·9	25·5	97·9	98·9		
14	21·7	84·6	62·9	26·7	77·2	50·5	24·2	72·2	48·0	28·0	81·2	53·2	21·5	94·5	96·8		
20	18·7	72·9	54·2	22·8	66·2	43·4	22·2	61·8	39·6	24·3	70·3	46·0	18·2	87·7	92·1		
28	15·8	61·5	45·7	19·3	55·9	36·6	17·0	52·8	35·8	20·7	60·5	39·8	15·0	78·0	84·8		
35	13·2	51·8	38·6	16·4	47·6	31·2	14·5	45·2	30·7	18·0	52·2	34·2	12·8	67·6	74·1	93·0	100·0
48	10·7	41·8	31·1	13·2	39·0	25·8	11·8	37·4	25·6	14·8	43·7	28·9	10·3	55·8	56·4	81·3	99·5
65	8·4	33·7	25·3	10·4	31·6	21·2	9·2	30·6	21·4	11·8	36·2	24·4	8·0	44·8	33·2	61·0	95·7
100	6·5	26·2	19·7	7·7	24·8	17·1	7·1	24·3	17·4	9·4	29·2	19·8	6·0	34·2	13·7	38·5	81·9
150	4·9	21·0	16·1	5·8	19·8	14·0	5·2	19·4	14·2	7·5	23·8	16·3	4·5	26·2	5·7	26·8	65·4
200	4·0	17·8	13·8	4·5	16·6	12·1	4·0	16·5	12·5	6·0	20·5	14·5	3·6	19·1	3·7	19·0	51·7

[a] Figure is based on original feed of 2000 kg/h.

[b] Classifier sands including circulating load.

practicable and any attempt to reduce all the feed to finishing mesh in one operation would result in wasteful over-grinding. The *circulating load* built up by running the mill in *closed circuit* with a classifier solves the difficulty.

The basic requirement of this arrangement is that the ore shall move fairly rapidly through the mill, dwelling there for as short a time as a minute if mill design permits. The shorter the passage (or dwelling time), the less grinding work is done on it. Since each fall of a ball should produce some finished material, the sooner this is removed the less it will be exposed to over-grinding. Hence, low-discharge mills with generous feed and discharge arrangements, and big diameters in proportion to their lengths, have supplanted the earlier high-discharge long mills through which it was not possible to push feed at a sufficiently fast rate.

Experimental work has shown that the faster the ore is rushed through the mill, and the greater the classifier capacity, the more effective is the grinding work, both in terms of output and in avoidance of over-grinding. There is a limit to the practical application of this, since it costs money to transport a heavy circulating load. This limit is usually found to be between 5 : 1 and 6 : 1 as a ratio between return sands and new feed in primary grinding. It is lower in secondary grinding, being limited by the sorting speed of the classifier.

When working at a 5 : 1 ratio, the mill receives one ton of new feed and five tons which have already passed more than once. If the feed enters at —$\frac{3}{4}''$ and is released from circuit at —60 mesh (= 1/120″), the ratio of reduction for new feed is 90. Since the average particle passes through six times (once as new feed and five times as return sands), the true ratio of reduction is 6 : 90, or 15. Owing to the tendency of grates to hold back large pieces, and for small sand to work out to the discharge end more easily than coarse ore, the actual condition in the mill is not accurately given by the above figure. Much, if not most, of the return feed has been ground nearly to finishing mesh. (It is characteristic of grinding in closed circuits that what is called "*near-mesh*" builds up in the circulating load.) Only a small proportion of the ore in the crop load consists of maximum-sized pieces. The average size of feed being thus kept steady, grinding itself becomes steadier and more controllable.

The effects of closing the grinding circuit with a large return load include:

(i) Reduction in the mean size of entering feed.
(ii) Marked increase in the circulation of nearly finished material.
(iii) Decrease in the retention and over-grinding of finished material.
(iv) Shorter dwelling time for finished material.
(v) Less need for ball ratio extremes.
(vi) Better interstitial loading of the ball charge.
(vii) Closer adjustment of the solid-liquid ratio.

The Solid-Liquid Ratio

In order to use the kinetic tumbling load to grind ore, the rock particles must form a coating on liners and balls. The pulp must also be sufficiently fluid to flow steadily through the mill. At a suitable ratio of particle surface to water, the smallest particles settle slowy and behave as if they were part of a heavy fluid, the proportions of water and ore determining its specific gravity. This, combined with the effect of *pseudo-viscosity* retards the settling rate of ore particles in the mill, although with the intense agitation in the tumbling load there is very little settlement.

With too thin a pulp (too low a solid-liquid ratio) the solids tend to settle and centrifuge outward, and coating of the balls by solid particles becomes patchy to non-existent, while the voids between the outer layers become overpacked. Even pulp distribution through the crop, and a clinging layer of particles on the metal surfaces everywhere in the mill, are essential to good grinding and reasonably high capacity.

Finely ground dry rock flows through a ball mill as though it were fluid, and has good transporting power. When the moisture content rises beyond 8% a stiff mud is produced, clogging the mill. This condition prevails with moisture up to 15%, beyond which fluidity begins to return, the material behaving in a sticky, treacly fashion. From 20% onward efficient movement begins to show. When a ratio of about 40% water to 60% ore is reached the pulp is in danger of being too watery to coat the grinding media properly.

This is the character of the pulp in general terms, but it is modified by two other factors—average particle size and density of ore. Solid-liquid ratio is usually measured by the mill operator in terms of weight of a known volume of pulp. The greater the density of the ore the smaller will be the volume of solids needed to maintain a given percentage. The fluidity of pulp depends on the amount of surface friction between the particles in a unit volume and is therefore governed by its specific surface. Consider a pulp of solid-liquid ratio 75 : 25 by weight, the solid fraction being ore of S.G.3.0. 75 grams of this ore will occupy 25 c.c., and so will the accompanying 25 grams of water. The volumetric ratio of this pulp is therefore 1 : 1. If the ore consists of a stone weighing 75 grams, it will have very little surface and will sink swiftly to the bottom of a vessel containing the 25 c.c. of water. If this lump were cubic it would be nearly 3 cm. along each side and would have a total surface of about 50 sq. cm. to rub against the water as it settled down. If the ore were now pulverised into fine sand. the total surface of the resulting particles would be measureable in tens or hundreds of square metres, and when they were dropped into the same 25 c.c. of water the friction between the faces and the cling between particle and water would slow down the settling rate very markedly. Later in this book the word "pseudo-viscosity" is used to described slow settlement due to such friction, which is a function of the total surface per unit volume of pulp or the *specific surface*. True viscosity refers to the internal friction in a fluid arising from its molecular cohesion. When finely ground ore is moistened it coheres to form a sticky mud. As more water is added the specific surface is reduced and particle mobility increased. Thus, *pseudo-viscosity* varies with the area rather than with the weight of solids involved. The flow characteristics of ore pulps are factors in most processing treatments. "Pulp ratio" defines the percentage of solid by weight, as does "solid-liquid ratio" Finely ground particles in movement respond far more to specific surface effects than to those produced by their collective or individual mass.

A coarse pulp might be very fluid at 70 : 30 ratio, but if the ends of the mill containing it were closed and a prolonged grinding were given so that the ore was reduced to extremely fine particles, it could on opening up be too viscous to flow until it had received further dilution. Thus, as the average mesh of the mineral decreases the solid-liquid ratio should be reduced in order to maintain optimum fluidity. This effect is produced by an increase in specific surface during grinding.

The author has made practical use of this change, which in the Hardinge mill is continuous along the discharge-end cone. Extra make-up water was piped in through the discharge trunnion and delivered halfway down the cone. Thus two solid-liquid ratios were maintained, the higher being in the drum where specific surface was lower than near the discharge end.

In ordinary continuous ore treatment, an optimum ratio is established at which conditions in the crop load are best satisfied. In working out the correct ratio, care is taken to suit it to the desired coarseness of grind, so that the pulp gives enough frictional grip to the crop load to help it to the desired amount of lift during rotation, and to ensure that the grinding media shall be properly coated. The thicker the pulp, within limits imposed by these considerations, the less wear of steel will take place. The higher the pulp density, the more ore passes through the mill in a given time provided the through rate is kept constant, and the greater is the lifting friction.

A rod mill is usually employed in open circuit and grinds at a mesh coarse enough to make its grinding action very different from that in the ball mill. Rods are kept apart by relatively large pieces of ore, and are not usually coated. The ore is not pulped, but is coarse enough to permit of rapid settlement of the solids till an appropriate solid-liquid ratio is reached. Supernatant water serves to transport ore through the mill. Thus, while most ball mills are worked at a solid-liquid ratio of between 70 : 30 and 80 : 20, a rod mill may be run at some such figure as 30 : 70 as regards ratio at discharge, though if it is stopped and the conditions in the crop load are measured, the apparent density there may be much higher. An important difference from ball-mill loading is that small particles are either excluded from the feed or flushed through, so that the coarse ones on which the rods are working cannot coat the metal at the working pulp density.

Control

Control of the grinding circuit cannot be fully understood before classification has been studied. Consideration of operational control is therefore deferred. This section deals with the correlation of the matters thus far discussed with the principles underlying control of the action of the crop load. The load consists of:

(*a*) A charge of grinding media, adjusted as regards weight, composition and percentage of mill volume occupied.

(*b*) Ore in ratio to the charge and limited as to the maximum size of new feed and the circulating load.
(*c*) Water in ratio to the ore, in such a proportion as will maintain both fluidity and good coating of metal.

The first control is concerned with sampling ahead of the mill, to prevent entry of an undue amount of *tramp oversize.* (This, like "critical size" worn autogenous media, is ore too large to be expeditiously broken down by the crushing bodies, but itself too small to act as such a body.) In a small plant this can be done visually, but methodical cutting of a head sample is better, and at the same time yields a sample for assay of entering feed. Next comes the question of ball ratio. Here the problem is to impress upon the crop load the right proportions of abrasion and shatter, as judged by the end product. Usually it is not convenient to alter mill speed or liner contour, except when periodically replacing liners, but it is possible to vary the dwelling time in the mill, the solid-liquid ratio, and the kind of blow struck. Dwelling time is a function of speed of feed, since feed displaces a similar volume at the discharge end. Solid liquid ratio is changed by altering the setting of a water cock in the feed launder. The blow struck depends, *inter alia*, on ball size (or weight). With other conditions held steady, if the percentage of heavy balls in the charge is increased, the impact grinding is also increased. The space available between balls is greater, and there are less points in the crop load where ball bears upon ball through the coating of ore. Hence abrasive grinding is reduced and impact grinding increased. The ball size required, or the blending of sizes, depends on the finished mesh-of-grind. The finer this is to be, the more abrasive grinding must be developed.

Bond[10] has produced a formula which relates ball size and work index and which amends an earlier one in important particulars. It reads

$$B = \left[\frac{F}{K}\right]^{\frac{1}{2}} \left[\frac{S \, . \, W_i}{C_s \sqrt{D}}\right]^{\frac{1}{4}}, \text{ where} \qquad (6.5)$$

B = Diam. (inches) of ball, rod, or pebble.
F = Size in microns passed by 80% of new feed.
K = Proportionality constant tabulated for various media and circuits.
S = Specific gravity of material being ground.
W_i = Work index at feed size F.
C_s = % critical speed of mill.
D = Mill diameter (feet) inside liners.

Unfortunately, much as this discussion would be simplified if only one variable were thus altered, the whole character of the crop load immediately begins to change with each single variation. With a coarser average particle the solid-liquid ratio hitherto suitable is now too thin, and this calls for cutting down the water if coating is to remain efficient. The discharge to the classifier bcomes coarser and the recirculated sands in turn are also coarser. This is accentuated by the fact that the coarser texture of the crop load is reducing the slip against the liners and thus increasing the cataract action.

It is thus obvious that the actual adjustment of a circuit must take simultaneous account of all these factors.

Returning to the question of proportionment of ball size to ore size, this is best worked out by trial and error, with study of any recorded experience of other operators working with similar ore. There is so much difference between ores in grindability, that no safe generalisation as to ratio can be made. The $D^2 = Kd$ formula already given is an excellent starting point. The mill discharge should be sampled and the sizes of its solid content ascertained by screening (methods are given in a later chapter). If more impact is indicated, the daily addition of new balls necessary to make up for wear should be adjusted so as to include a greater proportion of bigger ones. If, for instance, a particular mill is "topped up" daily with one ton of 2″ balls, $\frac{1}{3}$ ton of 3″ balls, and ten 4″ balls, the effect of changing this to $\frac{3}{4}$ ton of 2″, $\frac{3}{4}$ ton of 3″, and 12 4″ balls might be tried, the effect on the solid-liquid ratio being watched. If the next week or so showed that too much coarse grinding was now being done and not enough fine grinding, some adjustment in the reverse direction would be needed. Since the purpose of grinding is to facilitate efficient concentration, the mill manager would watch the effect of this change on recovery in the concentrating section, and would be careful not to make any other kind of alteration in the plant while this one was under observation, so that its effect could be properly judged.

The grinding mill does not respond rapidly to change. The high circulating load has a steadying effect, smoothing out surges. Time must always be given before the full effect of a change is visible, and this time may be anything up to a week. Each change of one factor necessitates adjustment of the other factors affected, and since this is to some extent hit-or-miss, it takes a certain amount of patience and skill to rebalance the crop load after each change. The golden rules in this kind of operational research are: "Go slow; only alter one thing at a time, if possible; wait after each alteration for the plant to settle down; only make each alteration little by little." It is far better to come slowly up to an optimum grind by a series of small adjustments lasting several weeks than to "over-modulate" so that the whole flow-line is upset. Grinding and concentration must be watched simultaneously during this very important business of gradual correction.

Since total grinding effect becomes visible in the form of new surface produced, and new surface can be found by measurement of the diameters of the particles in a sample (sizing analysis), the laboratory control of milling efficiency is carried out by measurement of the sizes obtained. For this purpose, samples are collected at regular intervals and sent for test. The solid-liquid ratio is also maintained at the required setting by testing the pulp density. If an ore with a specific gravity of 3·0 is ground at 75/25 solid-liquid ratio, then:

75 grams of ore in pulp occupy	25 c.c. space
25 grams of water in pulp occupy	25 c.c. space
100 grams of pulp occupy	50 c.c. space

In other words, the pulp density required is 2·0. If a can of known volume is filled with pulp from the mill discharge until it brims over, and is then hung on a spring balance or steelyard it should register a certain weight. If lighter, then the pulp is too dilute and the volume of water being piped to the feed launder must be reduced a little. If too heavy, then insufficient water is being used. The optimum grind can be established, in part, by careful variation of this solid-liquid ratio and in part by observation of its effect on finishing mesh and mill capacity.

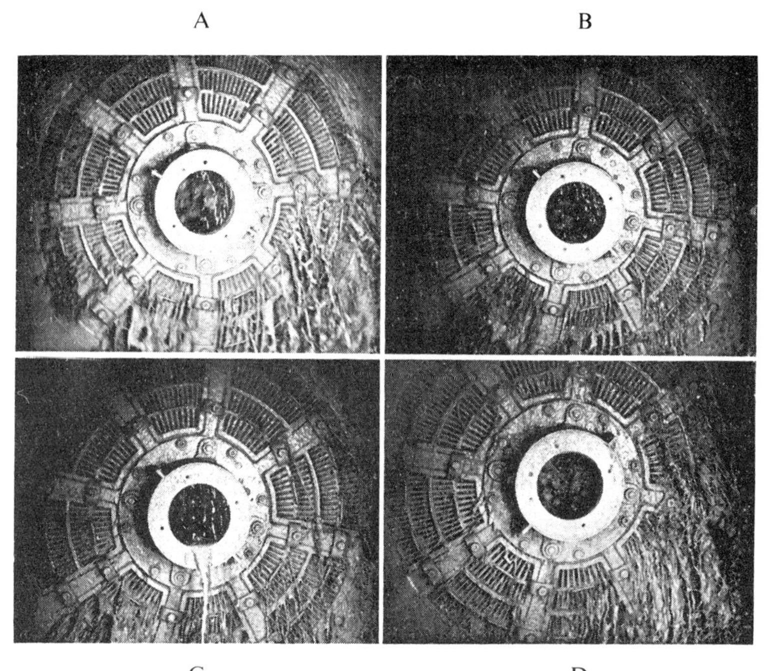

Fig. 50 *Effect of Overloading a Low Discharge Mill*

	% *Solids*	*Amps.*	*State of Grind*
A	80·0	300	Mill overloaded
B	80·4	300	Overload increasing
C	79·2	295	Badly overloaded
D	79·7	320	Normal operation

Given good all-round mechanical condition and a correctly constituted crop load, maximum production corresponds with a maximum intake of power. This is sufficiently marked in the case of primary grinding low-discharge mills for the ammeter to be used as a means of control by the operator. In the series of photographs (Fig. 50) made during a large-scale experiment at Hollinger, where control was exercised by maintaining maximum ammeter reading with the mill running quietly, the effect of overloading the mill with ore is reflected in drop of power consumption and consequent reduction of capacity. Case C, where power intake has dropped to 295 amps shows the pulp running from the centre discharge, clear evidence of overload approaching a height sufficient to upset the desired out-of-balance. If the mill were run with no ore or water, even higher readings would result, but the mill would not then be running quietly. In the case illustrated, the mill had a ratio of balls to ore of 6 to 1, was loaded rather more than half full, and average dwelling time was about a minute. 0·9 ton of ore was finished per horse-power hour.

In a large plant one operator can control from twelve to twenty primary mills, or from twenty to thirty secondary ones, a helper dealing with the daily addition of new balls. Repair work is the concern of the maintenance gang.

A form of control sometimes practised is to vary the rate of new feed in order to compensate changes in the grindability of the ore. For this purpose the Hardinge "electric ear" can be used. This is a microphone, set to respond to the noise made by the mill. The microphone controls a feeder belt, and holds the feed rate steady at the point where the volume of noise emanating from the crop load remains constant. If the noise increases, feed rate is quickened, and *vice versa*. The microphone setting is varied by trial and error until optimum setting is achieved. Useful improvement in capacity is reported by users of this device.

References

1. Rose, H. E., and Sullivan, R. M. E. (1958). *Ball, Tube and Rod Mills*, Constable.
2a. Gow, A. M. (1930). *Trans. A.I.M.M.E.*, 87.
2b. Campbell, A. B., and Coghill, W. H. (1934). *Trans. A.I.M.M.E.*, 112.
3. Davis, E. W. (1919). *Trans. A.I.M.E.*, 61.
4. Guerrero, P. K. and Arbiter, N. (1960). *Min. Eng.*, May.
5. Bond, F. C. (1959). *Bull. Can. I.M.M.*, Aug.
6. Hukki, R. T. (1960). *I.M.P.C.*)*Lond.*), I.M.M.
7. Pryor, E. J., and Heywood, H. (1946). Trans. I.M.M. (London), 55.
8. Fuerstenay, D. W., and Somasundaran, P. (1963). *I.M.P.C.* (*France*), Pergamon.
9. Hukki, R. T. (1958). *I.M.P.C.* (*Stockholm*), Almqvist & Wiksell.
10. Bond, F. C. (1958). *Trans. A.I.M.M.E.*, May.

CHAPTER 7

DRY GRINDING

Preliminary

Though most ores are reduced by wet grinding before being processed, some can better be ground and treated dry. Many minerals and synthetic substances require size reduction only. Other grinding problems arise in which chemical instability, contamination, corrosion or risk of explosion call for special precautions, such as milling in an inert atmosphere or one where moisture is undesirable or must be removed. In an arid country the chronic shortage of water may dictate the use of dry grinding methods. Where a dry end-product is called for and can be processed up to the required state without the use of water, dry grinding is to be preferred. Among the raw materials thus treated are asbestos rock and "crudy", coal for powdered fuel, cement clinker, talc, metal powders, drugs, and chemical salts. In addition to open and closed-circuit grinding, batch treatment is frequently used. In this method, grinding media and feed are loaded into the grinding mill and worked dry until the desired state of attrition has been achieved. The product is then discharged.

In the treatment of ores by chemical methods, such as the cyanide process, experimental dry grinding has shown promise. When comminution is followed by froth flotation it is usual to protect the newly developed mineral surfaces, and this is best done by grinding under water to which any required protecting chemicals can be added. The technical applications of dry grinding in mineral dressing are at present limited by this consideration.

Fixed-path Mills

Taggart[1] classifies dry mills into two groups: "in which the comminuting elements are relatively few and follow definite paths (fixed-path mills)"; and those in which "the elements are multifarious, and not constrained as to individual paths (tumbling mills)". The latter do the bulk of industrial dry grinding, but the former, of which there are several types, handle an important tonnage.

Burr mills range from the old-fashioned grindstone, originally used in grinding cereals, to vertical types. Two discs of stone, either horizontal or upright, are rotated in opposite directions, or worked with the lower one fixed and the upper revolving. Feed is central and finds its way along grooves in the stone faces, maintained by stone-dressing, to a peripheral discharge. The material is ground by attrition during its journey, being dragged between the stone faces. Soft rocks such as clays, barytes, talc, lime and limestone and gypsum, are treated in these mills. Feed is minus $\frac{1}{4}''$ and discharge can be as fine as minus 200 mesh. The mills are used for grinding material not likely to be injured by frictional heat, and also where staining by iron must be avoided. Care must be used to keep hard or uncrushable material out of the feed. Developments of this principle include vertical-disc mills with steel grinding-faces. The laboratory disc-grinder is widely used. Its discs can be parallel or slightly offset to one another, the latter arrangement reducing choke and improving throughput. Capacity is low.

The colloid mill has a vertical spindle on which is mounted a conical grinding unit, grooved vertically. This rotates at high speed in a fixed, close-fitting bowl. The grinding elements can be of ceramic or metal. This mill is used for grinding foods and soft minerals.

Hammer mills and rolls were discussed with intermediate crushers. They can also be used for fine grinding. A special application of the hammer mill is in the crushing of asbestos. The requirements are unusual, in that the material, as mined, carries the valuable fibre sandwiched between layers of shale. Hammer mills with heavy manganese steel plates are used to "fiberize" the blocky fibrous rock. The beating action opens the fibres and loosens attached shale. This is thus reduced to a fine grit which can be screened away.

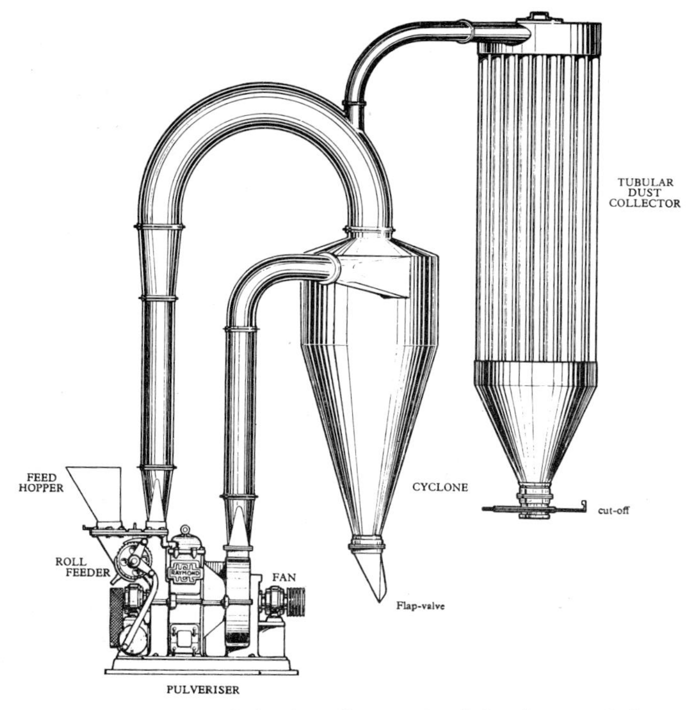

Fig. 51 *Impax Pulveriser* (*International Combustion Ltd.*)

The Raymond Impax Pulveriser (Fig. 51) introduces material through a roll feeder to the grinding chamber, which is swept through by an air stream. The air entrains finished material and dust and carries them to a collecting cyclone and dust-collecting chamber. A further refinement in fine grinding by hammer milling is reported by Robertson[2]. This is a two-stage mill, in which the runners in the second chamber have a higher peripheral speed at which they complete the work transferred from the first chamber.

The jet pulveriser or microniser carries a feed of $-\frac{1}{8}''$ material in air or steam at a pressure of about 100lb./sq. in. This streams out through suitable circular expanding chambers from its tangential delivery. Extremely fine grinding results, partly by mutual jostling between the solid particles, partly by contact with the chamber walls and by pressure release. Finished fine product is discharged from the centre at sizes varying down to one or two microns. The capacity is good, but wear is heavy and the mill is limited to specialised work.

Edge runners resemble Chilean mills with the mode of motion reversed, since the rollers remain stationary while the disc on which they bear revolves. The rollers are spring loaded. They, and the disc, may be made of ceramic material. Plain iron and perforated iron are also used. These mills are used to treat clays and ceramics. One form is the German Loesche Mill,

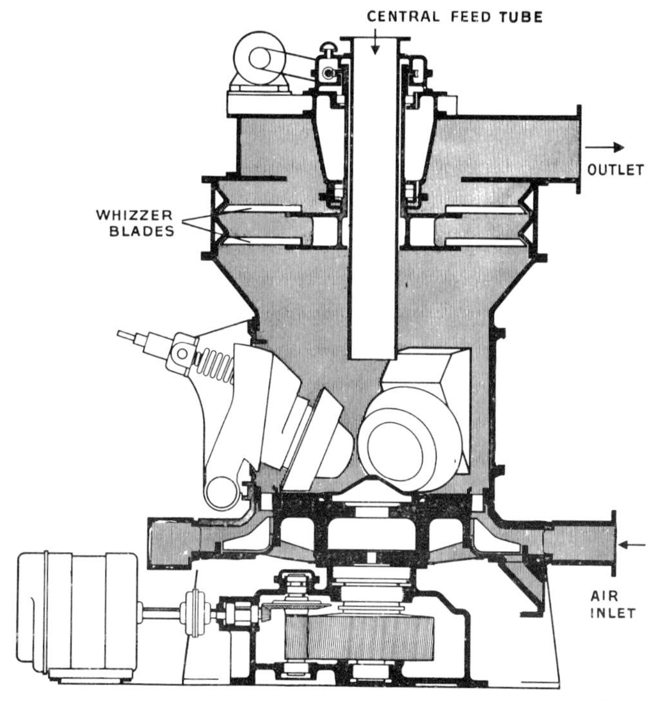

Fig. 52 *The Lupulco Mill (International Combustion Ltd.)*

from which the Hardinge disc roll mill has been adapted. In this two conic-section rolls ride above a revolving horizontal disc. This disc table, on which the ore arrives centrally, runs at a speed just below that at which peripheral discharge of crushed material begins. A dam forms round the circumference over which the discharging material is pushed by arriving feed, and falls into a classifying air stream which lifts finished mineral and returns anything coarser to the grinding disc. In the Lopulco mill (Fig.52) there are either two or three spring-loaded conic rollers. These mills range in output from 1 to 50 tons/hour and are designed for low, medium or high speed. There is external provision for adjustment of clearance between the rotating table and the rolls, which cannot make direct contact. The feed receives both loaded crushing and shearing attrition. An exhaust fan maintains air sweeping and removes finished product. In addition to its wide use in producing powdered coal this mill grinds a variety of softish minerals substantially through 100 mesh. The list includes gypsum, lime, phosphate rock and various industrial earths.

The older pendulum, or roller mills include the Huntington and the Griffin. In both, one or more pendulums revolve inside a wearing ring against which they bear owing to the centrifugal force set up by their rotation. Material trapped between roller and ring is ground till it escapes through guarding screens, set peripherally. In the Williams mill three to five rollers are pressed outward in similar manner, but the mill is swept through by a current of air which carries to a collecting cyclone or air-filtering arrangement all particles small enough to be borne along. The air or gas can be preheated in a furnace, and natural draught is aided by an exhaust fan above the grinding compartment. In the Raymond bowl mill, which is used for producing pulverised coal, the bowl rotates against spring-loaded mullers and finished material is removed by a current of air. The Babcock and Wilcox machine has balls rotating in a horizontal grinding ring, where they press on and pulverise material fed down into the ring, the finished product dropping by gravity to an external classifier. In a variation, the mill is swept by air or gas, hot if necessary, discharge being upward. Where air-sweeping is used, hot air can can be used for the purpose of drying the feed.

The Vibrating Mill

Intermediate between the fixed-path and the fully tumbling mill is the vibrating ball mill. This has not changed importantly since its prototypes were developed in the United States and Germany in the pre-war period. An industrial model, now in increasing use, as described by Paricio[3] is nearly 6′ high, 9½′ long and 7′ wide. It weighs six tons and is vibrated by the unbalanced rotation of eccentrics driven by two 50 H.P. motors at 1,200 r.p.m. The layout consists of a grinding cylinder rigidly attached to eccentric mechanisms in independent horizontal cylinders parallel with it. This assembly is mounted on four sturdy vertical springs and vibrates with an amplitude of 3/4″. In action the grinding media have a vibrating period of about 1,160 r.p.m. and occupy about 80% of the mill's volume. The load tends to rotate. For suitable ores this system has proved versatile and cheap in standing space, cost of installation, weight and maintenance. A 30″ diameter mill reduces *minus* $\frac{3}{4}$″ soft and friable feed to 99% *minus* 325 mesh at a rate of $2\frac{1}{2}$ tons/hour. Feed and make-up media enter through a dust-tight flexible spout and are discharged *via* a retaining grate. This mill can also be worked wet, and the retention time can vary from one minute up. When used for batch grinding two receiving tanks are used and the material is worked through from one to the other and then back until the required fineness is reached.

Vibration milling has met the need for grinding metallic powders to sub-micronic sizes in an inert atmosphere, and handles such inflammable elements in this way as aluminium and magnesium. Capacities quoted range, on limestone and using steel balls, from 1 ton hourly in the 15″ diameter mill to 15 tons in the 42″, an 80% *minus* 4-mesh feed being reduced to 80% *minus* 200 mesh. Where iron in the product is inadmissible, alumina balls and special linings are available. Materials handled commercially include tungsten carbide, aluminous nickel, silicon carbide and iron oxide.

Tumbling Mills

Dry ball mills have much in common with those described in Chapter 5, but there are marked differences in the method of operation. Continuous and batch grinding are practised, sometimes with rods as grinding media, but more usually with balls or autogenous crushing bodies.

Since batch-treatment mills are not limited in design to types which can

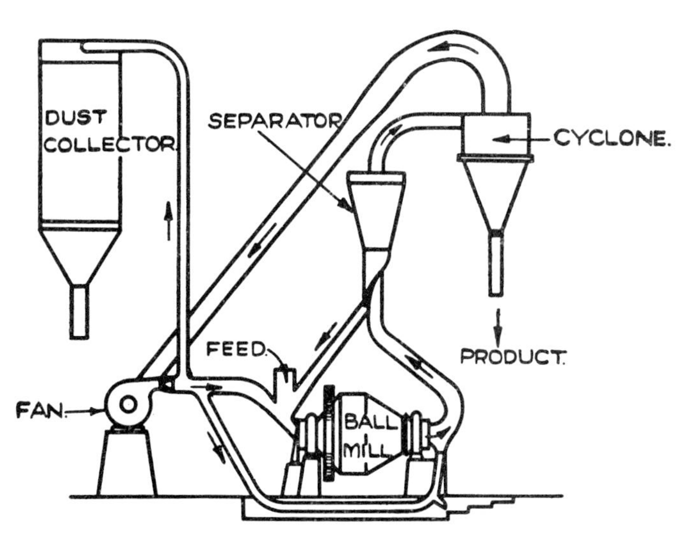

Fig. 53. Dry Grinding in Closed Circuit with Air Classifier (International Combustion Ltd.)

be fed at one end and discharged from the other, a variety of shapes are in use—cylindrical, conico-cylindrical, oval, polygonal, and even cubic. The liners and crushing media can be made of iron or ceramics, according to the requirement of the work. Feed and discharge are usually made *via* an aperture in the shell, a screen being placed over it to retain crushing media during discharge. The mill is charged, run for a suitable period, stopped, opened, and emptied.

In continuous work, the rod mill is sometimes used. Mineral tends to move sluggishly through, and can be aided in its progress by the highest possible gradient from feed to discharge end. This latter usually takes the form of a grate with peripheral discharge. If the feed end is shaped to a flattish cone, a reservoir of new material builds there and is pressed toward the rods. The rods are not displaced by this action, because they are steadied by the combined effect of rotation and of bearing against the discharge liners.

The ball mill is the most favoured machine for dry grinding of hard rock. A critical factor in good working is the moisture content of the feed. In dry grinding the discharge product may have to satisfy a maximum moisture specification. If the feed enters with a water content much above 1% the air circulated through the charge by the air-classifying and sweeping arrangements has increasing difficulty in drying the load during its retention period. By the time some 5% of moisture in the feed is reached, this air is unlikely to pick up enough heat from the frictional action of the crop load movement to do much drying unless it is given external heating. Some moisture-saturated air may have to be bled off from the closed air-sweeping circuit in addition to that which escapes under working conditions. A typical air-swept arrangement is shown in Fig. 53. Details of one type of air classifier are given in Fig. 54. Dry ore has a high angle of repose, making it desirable that progress through the mill should be assisted. Feed is often forced in by a screw conveyor, and a steep gradient through the mill is obtained by the use of a large diameter, short length, and peripheral discharge. Another critically important factor in operation is the feed rate. Owing to its low mobility, partly finished material tends to accumulate over the toe of the crop load, where it cushions the falling balls and uses their kinetic energy to redistribute the ore instead of to crush by impact and shock-wave propagation through the mass. It is therefore important to restrict loading so that the ore in the crop load remains in the interstices. The ball soon becomes coated with a layer of ore. This robs the blow of its crispness, and damps down transmission of shock. On the other hand, the coating layer protects the metal parts of the mill against wear, and considerably reduces this item of cost, thus enabling dry grinding to compare economically with wet in suitable fields. One of the best ways of avoiding loss of efficiency is "air-sweeping", in which a current of air is blown continuously through the mill. This removes the finest material and thus takes out of circuit the principal cushioning and coating fraction of the ore.

The ball load is kept lower in dry than in wet grinding—below 40%—to avoid "over-carry" at cataracting speeds, which would cause the flying balls to hit the down-running side.

One of the older methods of ensuring fineness in dry grinding is to divide

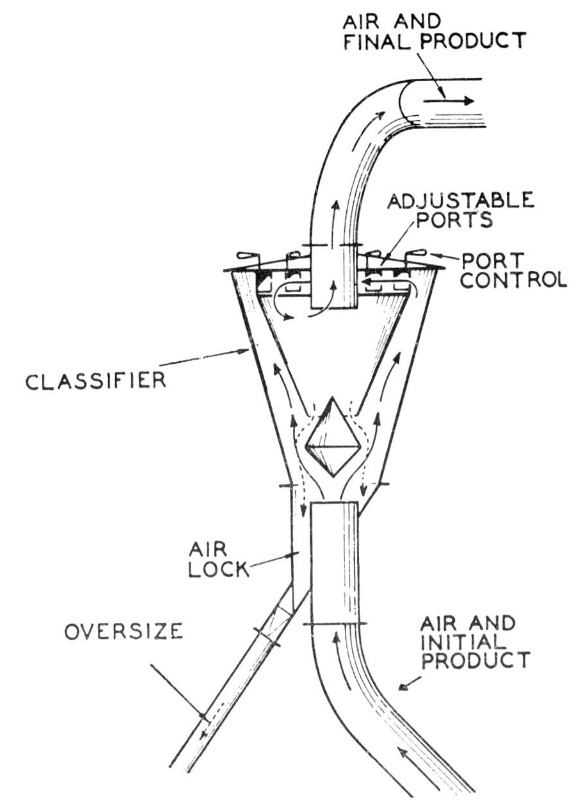

Fig. 54 *Basic Elements of the Hardinge Double-Cone Classifier*

the ball mill into two or more compartments. These are separated by grates which retain the material in the compartment for which crop-loading is most suitable until it can pass through the grate and be elevated by scoops to the next (Fig. 55). The Hardinge Tricone Compartment mill uses the back slope of it first (conic) section to promote media segregation and makes it possible to dispense with the third section shown in Fig 55.

In an air-swept single compartment mill, which has gained considerable favour in the preparation of pulverised coal, the difficulty of maintaining a gradient through the long cone of the Hardinge is partly overcome by using a retaining grate toward the discharge end. Behind this, lifters pick up the undersize and drop it into the classifying stream of air sweeping through the mill.

Another class of mill has peripheral discharge through the shell, the openings being guarded by screens. This mill is also used for wet grinding. The Krupp mill (Fig. 33) guards the fine screens, first by heavy grinding-plates which take the full liner wear, and next by coarse punched-plate screens

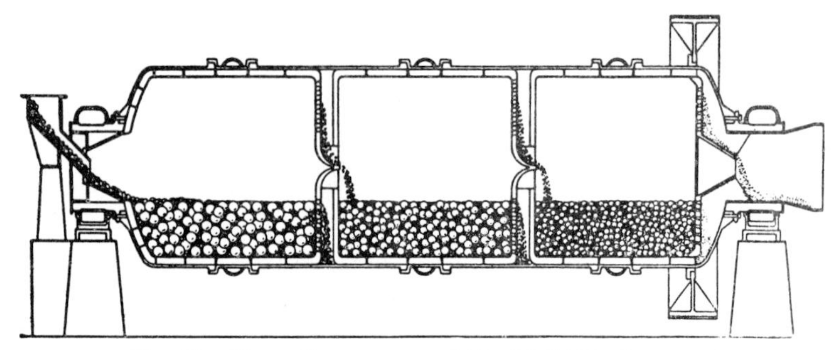

Fig. 55 Three Compartment Tube Mill (Hardinge Co.)

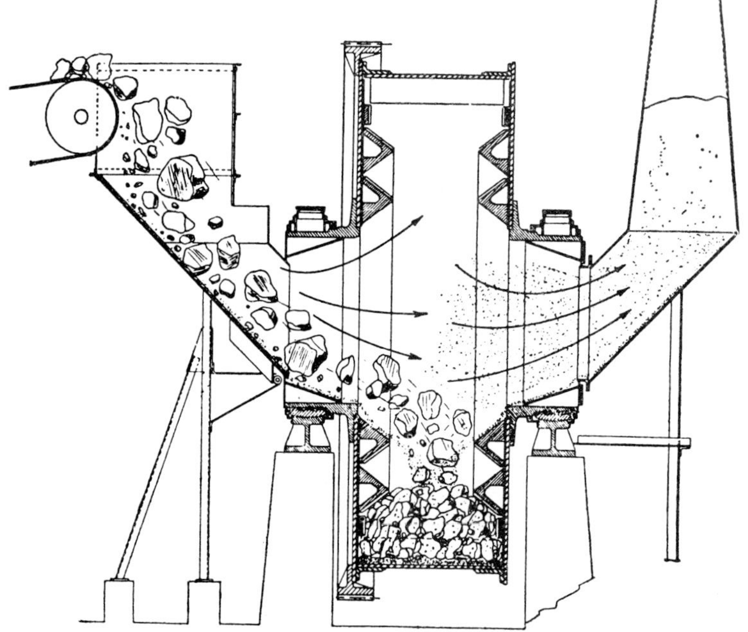

Fig. 56. Air Sweep through Aerofall Mill (after Waspe)

which hold back material that might damage the more delicate fine screens.

The Aerofall mill (Figs. 56 and 57) is either fully autogenous or run with a small percentage of balls as part of the grinding media.[4,5] It receives a wide-ranged dry feed, anywhere up to 18″ in size, and reduces it to a fine sand. One mill works on —42″ ore. A ball charge varying up to $2\frac{1}{2}$% of 5″ or larger balls, can be used where the natural feed does not contain an adequate proportion of heavy rock, but usually none are needed. The mill varies in diameter from 5′ to 28′ and in width from 2′ to 8′, and works at peripheral speeds between 8′ and 20′ per second. It can be used in open circuit or closed either by screens or pneumatic classification, finished material being removed by air-sweeping and sorted in cyclones. Successful use is reported with asbestos, iron ore and gold.[6] As with the pebble mill in wet grinding, ore is used to break ore, but the dynamics of the crop-load are very different. Shape and speed of the mill are designed to carry its —1″ contents high on the rising side, and the bulk of the free-falling action takes place near the 6 o'clock position. The full kinetic value of fall is developed, since no pool of water exists to cushion impact.

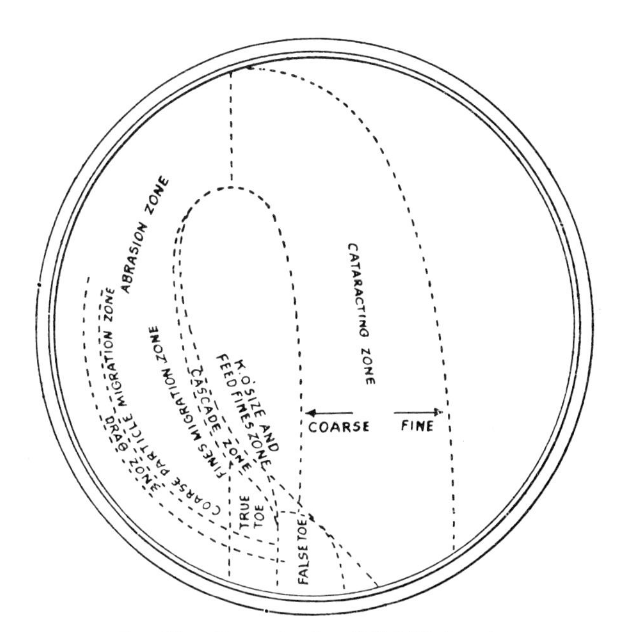

Fig. 57. Zones in Aerofall Mill Interior

Pilot tests[5] show one-step comminution to use less power than a conventional crushing and grinding sequence and to favour medium and fine-grained ores which require finer finished grinding, where air-sweeping is more effective. Ore tends to disintegrate selectively along grain boundaries and to produce mineral species at their natural grain size. Any such effect assists subsequent concentration. Liberation appears better than with rod-milling and wear is less. If selective grinding accompanies quick passage through the mill the later treatment benefits by having less over-ground particles to deal with. The Swedish tests suggest that when working for a coarser liberation mesh, air-sweeping is not fully effective. Tests at Doornfontein[4] showed higher recovery both of gold and uranium after 24-hour laboratory leaching of *minus* 200 mesh banket ore crushed direct from run-of-mine feed than of various samples drawn from the conventional flow-line. With a steady increase in the chemical extraction of values from their ores this improvement, which again suggests disintegration along grain boundaries, attracts the interest of process research.

Under normal weather conditions Weston considers a moisture content in the feed tolerable up to 3½%, but in sub-zero temperatures external heat to be needed from 1½%. Installed capital costs are lower than for comparable wet grinding and maintenance is much lower.

The Cascade mill (Fig. 34) was described in Chapter 5. It is used in dry grinding in the same general way.

Operation

The moisture content of the feed is an important factor in dry milling. If nothing is done to classify a circulating load, a very small percentage of water (½% to 1¼%) ruins fine grinding. If the circuit is closed through fine screens, a little more moisture can be tolerated without clogging the screen apertures, unless the material is soft and clinging. The temperature of the circulating air rises as the ore is milled, and its humidity increases when it picks up moisture from the passing feed. Enough air must be bled off to avoid saturation and thus ensure that this transfer of moisture can continue. The heat requirement for drying changes with the season, and must be supplied by warming the air when necessary. The water content affects grinding, cling in the crop load, cushioning, cataracting, coating of crushing media, and mobility and classification of partly finished material in the closed circuit.

As with wet-grinding work, the mill capacity required for a given rate of throughput depends on the grindability of the ore and on the sizing analyses of feed and finished product. The finer the grinding is taken, the lower will be the tonnage treated. Removal of finished material is very important to high capacity. If it is allowed to remain in the mill it deadens the grinding force by packing the interstices. It diverts what should be grinding energy to the task of re-distributing the crop load and of overcoming its frictional resistance and cling. This, though it adds to the heat of milling, contributes little or nothing to the comminution of the larger particles. Where a fairly coarse final grind is wanted (in, say, the 20-mesh zone), screening affords positive separation since at this size a lively open load can be worked over screens without "blinding". (Blinding connotes the wedging of an oversized particle into a screen aperture, thus putting it out of service, and it becomes serious with clinging material on small screens.) This coarseness calls for strong air-currents to sweep finished material out of the mill, and a mill design which includes retaining grates followed by lifters to the discharge overflow may be needed.

When the finished material is required at *minus* 60 mesh, reasonably gentle air-currents can be cheaply introduced into a circuit without giving rise to a serious dust or abrasion problem, so that between 10 and 60 mesh screening gives place to air-sweeping. In many cases tests have shown that the introduction of one or other of these methods into what had been open-circuit practice has resulted in substantial improvement at lower cost.

Where screens are used, the ore leaving the mill is usually elevated to them, so that the oversize can fall and join the incoming feed.

Small variations in size and moisture content of feed have a disturbing effect in dry grinding, and automatic control has been developed in order to govern feed rate and thus keep the amount of ore in the crop load at the most effective ratio. Sound control, thermostatic control actuated by the discharge-end temperature, and pneumatic control based on changes in the load carried in the air stream, are available,

Application

Industrial dry grinding is concerned mainly with homogeneous materials, whether naturally occurring or synthetic. In such work the problem of subsequent removal of a selected fraction does not arise. Increasing use is now being made of dry comminution as a liberating stage in ore processing, even where subsequent treatment is a vital factor. If chemical or magnetic treatment is to follow grinding there is little risk of trouble, but when froth-flotation is to be employed the delicate surface-active forces which this process exploits may rule in favour of wet grinding. This is discussed in later chapters. Flotation following dry grinding is already in use.

One manufacturer (Hardinge Co. Inc.) has over 5,000 commercial installations currently at work on nearly a hundred kinds of material. These include ores of chromium, gold, lead, manganese, molybdenum, phosphate and platinum. Other substances ground vary from such hard and abrasive materials as carborundum and cement clinker to soft or clayey feeds like coal, talc and ball clay. A common factor is the need to avoid the use of water.

Mill Capacity

This section applies to tumbling mills generally, wet or dry. The capacity is related to the mill volume, and to that portion of it occupied by the crop load. In a cylindrical mill the diameter can be reduced by interposing wood battens between the shell and the liners. The length can be shortened by inserting a timber packing behind the end-liners.

If only temporary reduction of capacity is desired, reduction of the volume of the crop load is a simple expedient. Big pieces of ore can be introduced deliberately to replace part of the feed. They grind down very slowly, thus producing the effect of a reduced crop load. It is also possible to arrange for a mill to run at more than one speed, so that when needed it is held to its slower rate of turn. Reduction of capacity by such methods is accompanied by lower operating efficiency, shown either as higher cost or reduced recovery of value.

Capacity can be increased by building up the crop load till unbalance reaches its maximum. It is sometimes feasible to increase the mill speed.

The circulating load can be increased if the classifier and mill trunnions are able to handle a higher tonnage, thus reducing the amount of over-grind and thereby increasing throughput. In the primary circuit, the liner contour can be accentuated so as to give higher lift and greater impact-grinding, and the diameter of the mill can be increased a little if thin alloy-steel liners are used. In most cases the simple practical course is to throw more of the work of comminution on to the crushing section, by closing the set of the machines until a finer product is delivered. It may be justifiable to introduce a further stage of intermediate crushing, by rolls or rod mill, which can be put in circuit to further reduce particle size when the grinding plant is intermittently overloaded.

General Conclusions

What follows applies particularly to wet grinding after dry crushing, and summarises the discussions of grinding up to this point.

The finer grinding is carried, the lower is the capacity of the plant and the greater the wear and tear. It is therefore of major importance to control all comminution in such a way that the products are released as close as possible to the required release mesh. Most dry-crushing machines can be worked as "arrested" crushers. A distant approach to arrested crushing can also be achieved in the low-discharge ball mill by working with a large re-circulating load. High-discharge ball mills cannot release finished material so easily, since it must be displaced by entering feed and water, and not sluiced out down a gradient.

Dry-crushing plants are subject to bearing wear if dusty conditions prevail. This can be avoided by using an air-swept sealed circuit. When crushing rock, safety precautions are sometimes needed to guard the health of the personnel. For humane reasons, it should be borne in mind that a mill management drawing its labour from a drifting native population is unlikely to be directly in touch with evidence of any bad after-effects of dust inhalation.

Metal sulphides, which comprise the bulk of values in ores subjected to treatment, are usually far more friable than their gangue rock and are therefore liable to receive selective comminution. This can be bad for concentration, and may make it advisable to remove some of these sulphides as a rough concentrate in the grinding circuit itself. If gravity methods of concentration are being practised, stage grinding can be interlinked with stage concentration to minimise such harmful over-grind of values. With flotation to follow, the problem may not arise.

When the mill is receiving an ore carrying two or more valuable minerals, it may happen that one of these is liberated at a coarser mesh than the other. This could be an occasion for stage treatment. The usual practice calls for a high-grade finished concentrate and an impoverished tailing, with regrinding of insufficiently liberated *middlings*. In gold treatment, metallic particles are frequently trapped out at the earliest possible stage in ore to avoid locking them up in the interstices and quiet crannies of the milling circuit—sometimes ironically called "gold-plating the mill". With extremely fine gold distributed through tiny particles of auriferous pyrite special arrangements are sometimes made to concentrate this pyrite and then subject it to intensive grinding or other specialised treatment.

The plant should be kept clean. Whenever power fails or a breakdown occurs, ore may have to be dumped on the floor. If this floor is covered with pools of oil, blobs of gear grease, etc., the ore will be contaminated. When the mill restarts, all this rubbish may be shovelled back into the circuit, upsetting the chemical condition of the particle surfaces, and introducing hydrocarbons which can seriously reduce the efficiency of flotation. Grinding must not be thought of only as a breaking treatment liberating to separation sizes. It is also a method of scouring tarnished old surfaces, and of preparing new ones for treatment by the methods of the surface chemist. It paves the way to good recovery, while bad grinding militates against it.

References

1. Taggart, A. F. (1945). *Handbook of Mineral Dressing*, Chapman & Hall.
2. Robertson, R. H. S. (1960). *Chemical Age,* June.
3. Paricio, R. (1960). *Mining World*, June.
4. Weston, D. (1960). *I.M.P.C.* (*Lond.*), I.M.M.
5. Fagenberg, B., and Ornstein, H. (1960). I.M.P.C. (Lond.), I.M.M.
6. Waspe, L. A. (1955). *Mining Magazine,* Nov.
7. Waspe, L. A. (1956). *Mining Magazine,* June.

CHAPTER 8

LABORATORY SIZING CONTROL

Preliminary

The response of a particle to mineral processing is influenced to an important degree by its shape — i.e. its ratio of surface to volume. If one gram of a solid of S.G. 1.0 could be crushed into equi-dimensional particles just passing through a 200-mesh sieve these, if cubic, would have a combined area of 674 cm^2; if oblong 562 and if composed of platy flakes some 2,700. A typical mineral sand would have an area of 922 cm^2. The surface of a particle is, so to speak, the door to its interior. The bigger the surface *vis-à-vis* volume, the faster is the rate of reaction and the greater its potential (surface) energy. A sphere has the minimum surface for its volume, and a thin flat plate the minimum volume for its surface. This surface is available for:

(a) Frictional retardation of the particle.
(*b*) Acting as a transition area between solid substrate and fluid surroundings of particle.
(*c*) Introducing heat or chemicals to the interior of the particle.
(*d*) Displaying the contents of the particle.
(*e*) Contributing to the *pseudo-viscosity* of a pulp (total friction per unit volume).
(*f*) Stoichiometric reaction.
(*g*) Protecting the contents of the particle it encloses.

These possibilities are exploited under controlled conditions during the separation of different kinds of particles into "concentrates" and "tailings" when treating an ore by physical and chemical methods. Other things being equal, a cube would react with 73% of the vigour of the "typical mineral sand", an oblong with 61%, and a "platy flake" 270% where surface-to-volume ratios were concerned. Characteristic breaking mode and particle shape are therefore statistically important factors for each specific mineral or ore, and their study is an important branch of the technology of processing.[1,2]

Of the many techniques employed in laboratory sizing and surface investigation, those chiefly used are:

(*a*) Screening.
(*b*) Elutriation.
(*c*) Sedimentation.
(*d*) Infra-sizing.
(*e*) Microscopic examination (including electron microscopy and electron diffraction).
(*f*) Turbidimetry.

(*g*) Permeability.
(*h*) Gas or liquid sorption.

The more important routine methods employed in mineral dressing are described in this chapter.

Laboratory screening can be carried down to 37 microns. The micron (μ) is 1/1000th of a millimetre. The millimicron ($m\mu$) is 1/1000th of a micron, or 10 angstroms (Å). Routine sizing tests on laboratory screens usually finish at about 70μ. "Sub-sieve" sizing is practised in the $-70\mu + 5\mu$ range. Microscopic inspection is applied at all sizes from sands down to 0·25 micron. Centrifuging of extremely fine suspensions is used down to 10 millimicrons, and the electron microscope can be employed in suitable cases down to a millimicron ($1m\mu$).

Purpose

The purposes of laboratory sizing control are to check the quality of the grinding, the extent to which the values are liberated from the gangue at various particle sizes, and to aid specific examination of ore constituents[3]

Size analysis of the feed transferred from the crushing to the grinding section shows whether the former is doing its work correctly, and how much comminution remains to be done by grinding. In the wet grinding section samples from the mill discharge show the size distribution of the delivered particles. Samples taken from the closed-circuit return show what sizes of material are being sent back for further grinding. Those from the closed-circuit discharge check whether properly liberated particles are being released to the concentrating section of the plant. In addition, sizing tests at selected points help in checking the progress of the material through the various stages of treatment. Study of the sizing analyses of all these samples gives the plant manager clear information as to the way in which grinding power is being used and how grinding affects recovery of the values from the ore.

The concentration processes applied to the pulp exploit selected differences between particles of valuable mineral and gangue. If two particles of the same size are produced, one being composed of heavy mineral and the other of light gangue, the heavy one has more potential gravitational energy than the light one. If both start to fall together, the heavier one will convert more of this potential to kinetic energy in a given distance. This difference in behaviour is used in "gravity" methods of separation, which exploit differences in mass and shape of mineral particles. By screening, the size-variable can be removed, and the density-variable accentuated. In other concentrating processes size control is used to find when the optimum size had been reached, when to stop grinding, and when the pulp is ready for treatment. In the mill the problem usually takes the practical form of adjusting working conditions in order to obtain the maximum tonnage at the required liberation size as cheaply as possible, while using the full capacity of the grinding plant. This problem is solved by empiric alterations, guided by laboratory sizing analyses of samples drawn from the mill circuit.

Influence of Particle Shape

For convenience in relating surface to volume, much research literature discusses particle behaviour in terms of the sphere, or sometimes cube. Particles are rarely spherical or cubic. If one gram of powdered mineral can just pass the apertures of a square-meshed sieve in which the linear distance between successive parallel wires is *A* mm., the surface (in square cm.) is $\frac{K}{A.\Delta}$. The constant *K* is 60 for cubic particles, 50 for oblongs, 240 for laminar ones in which thickness is about 10% of length and breadth, and 82 for the "average" mineral particle with dimensions 1·0, 1·2 to 1·5, and 1·6 to 1·8.

If screening motion is 100% efficient, the particle twists about on the sieve meshes until its smallest cross-section has been presented to the screen aperture in the manner most likely to facilitate its penetration, and if small enough, its passage through. A flaky particle may in such a case be presented diagonally, so that it has the chance of passing although actually wider than the distance between screen wires.[4] The longest dimension of the particle would not, in such a case, affect the result. A needle-like particle would, if presented end-on, be "threaded" through the mesh.

Taggart[5] has estimated that a particle having half the cross-section of the sieve mesh should pass once in four presentations, but if it is 95% of the mesh area only once in 500 presentations. With an acicular shape, the particle has increased difficulty in finding its way through, as it tends to lie along the screen, not to stand on end. Hence, the shape influences the accuracy of screen analysis.

When a particle is immersed in a fluid it is subject to the pull of gravity. If its density exceeds that of the fluid, gravitational force acts to convert potential energy into kinetic energy by setting the particle in downward motion. At the same time, retarding forces are applied to the particle surface. These are due to friction between the fluid and the surface and to viscous shear by the cross-section of the particle as it pushes asunder the molecules of the fluid. Thus, the reaction of a particle moving in a quiet fluid is determined by the relative strength of the gravitational force pulling it downward, the surface forces (total surface, and roughness of particle surface), and the fluid's viscosity. Surface area is a function of particle shape. The relation between the surface and the surrounding fluid influences the situation. A liquid does not "wet" a particle (that is, enter into direct contact with no intervening film of, say, air) until its surface tension is overcome. For example, steel is much heavier than water, yet an ordinary sewing-needle gently lowered will float on water if it is first drawn once or twice through the fingers, so that it acquires an impalpable coating of grease. Here the mutual "antagonism" between grease and water helps the latter to resist penetration. The laboratory methods used to differentiate between particles settling through fluids exploit differences of specific gravity, volume, shape, and surface area.

Particles on a microscope slide tend to lie on their most stable surface

(Fig. 58), as is shown by the photograph taken in plan and cross-section of such a field.[6] The thickness of such a settled particle can be estimated by using special techniques. The vertical distance represented by a given shift of the fine focusing screw of the microscope can be found and used for transparent particles. For opaque ones, the material can be displayed on a mirror instead of an ordinary glass slip. Alternatively it can be viewed through a prism, or specially mounted. Micro-observation of particle size is tedious and open to personal error, though appliances are available which take over some of the drudgery of counting and comparing.

The meshes of a sieve form a rigid reference framework, arresting or passing a particle in accordance with its cross-sectional area. In *elutriation* the fall of the particle under the influence of its mass is opposed by its rubbing contact with a rising column of fluid. Sieving or screening therefore measures size, and elutriation a combination of size, shape, and density. The former is a "*sizing*" operation and the latter "*sorting*" or "*classifying*" Microscopy gives data with regard to cross -sectional area.

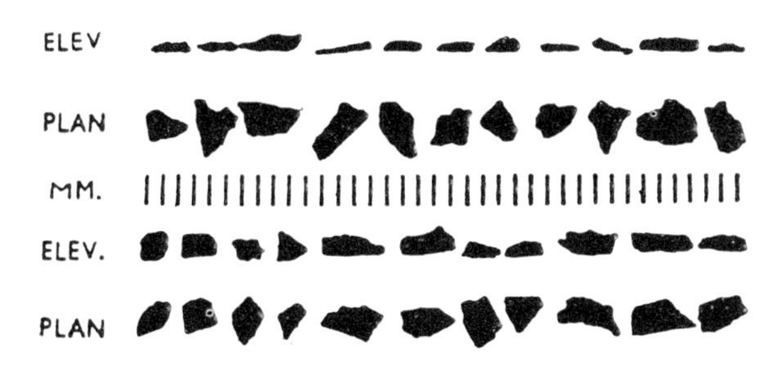

SHALE PARTICLES.

Fig. 58 *After Heywood and Pryor*

Laboratory Screens

Sieves used in ore testing are square meshed. The wires of the warp are woven successively over and under those of the woof (Fig. 59). The "*mesh*" is defined as the number of wires (or of openings) per inch, measured along either warp or woof (not diagonally). Since the main systems used in the English-speaking world differ in the thickness of the wires used, the apertures framed by these meshes also vary appreciably. The chief American systems are the Tyler and the U.S. Series (A.S.T.M.). The leading British systems are the I.M.M. and the B.S. In many laboratories an essential operation is to separate the sample being tested into a graduated series of sizes. This is done by assembling a "nest" of screens of diminishing mesh-size. A widely used reduction ratio is provided by the Tyler series (Table 10, col. 1) which is characterised by an arithmetical progression ($\sqrt{2}=1{\cdot}414$). The 200-mesh Tyler screen has a distance between wires of 0·074 mm. or

74μ (Table 10, col. 4). The next larger screen in the series is therefore of 104μ aperture (74 × 1·414) and the next smaller 53μ $\left[\frac{74}{1 \cdot 414}\right]$. The wires used to form the screen "cloth" as it is usually called are of a diameter complementary to this spacing. A $\sqrt[4]{2}$ series is also made for close sizing. The A.S.T.M. series uses the $\sqrt{2}$ ratio, but begins from 1 mm. or 18 mesh. The new U.S. sieve series (A.S.T.M. E–11–61) has the relationship of openings, wire diameters and mesh sizes shown in Fig. 60. Dimensional details are listed

Fig. 59 *Mesh of Woven Screen*

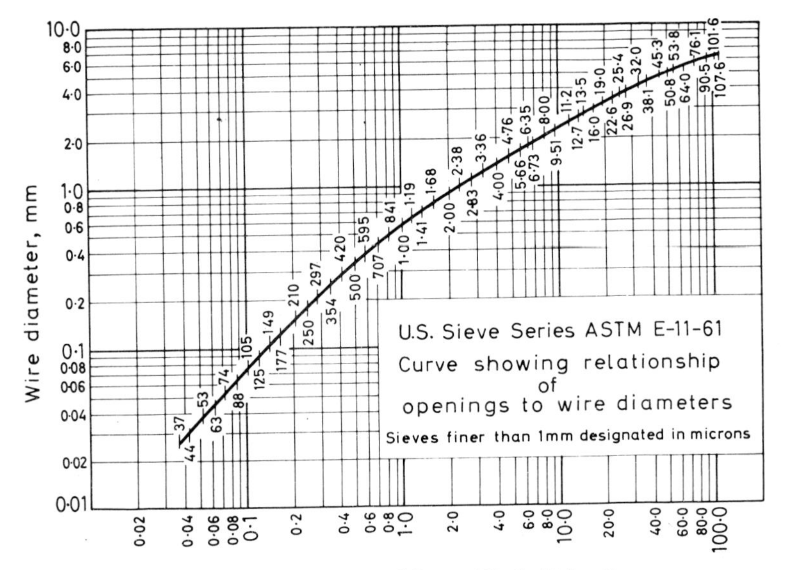

Fig. 60. *By courtesy Messrs W. S. Tyler Co.*

in Table 8. It corresponds fairly well with the Tyler series, with which it can be used interchangeably. The I.M.M. series is sometimes considered as obsolete and superseded by the B.S. series, but it is in still wide use in the British Commonwealth. The important difference from the Tyler system is that in the I.M.M. series the wires have the same diameters as the spaces. Thus, a 200 mesh I.M.M. sieve has 200 wires and 200 openings in one inch, measured along either the warp or the woof, so each wire must have a diameter of 1/400″. This gives a spacing of 63 microns. It follows that the description "200 mesh" can refer to 74μ or 63μ screen cloth, unless the system is specified. The I.M.M. system has the convenience of allowing the user to work out the mesh spacing by mental arithmetic. A series can be selected which gives a rough approximation to a Tyler series. The B.S. series which is taking the place of the older I.M.M. conforms to the Tyler series, with slight variations in wire diameter. The smallest woven-wire screen is 400 mesh or 37 microns, and a tolerance for weaving or wire-drawing error is permitted. The old German system (D.I.N. 1171 Standard) designated sieves by the number of meshes per cm^2. This has been superseded by D.I.N. 4188, the mesh dimensions for which are listed in Table 9, together with those for France.

A source of error with the wire-cloth laboratory screen is irregularity of weaving. Any displacement of wire or change of wire diameter results in a number of meshes of the wrong size. Unfortunately, a row of meshes too small (which would not matter much) is offset by a corresponding row of oversized meshes when a wire is out of position. Given time, every particle small enough could find its way through a very few such holes. Sieves, particularly the finer sizes, are delicate instruments of precision and should be carefully handled. When used for wet screening, they should be dried gently, immediately after use. If they become "blinded" by particles stuck in apertures, they should be freed by light brushing of the underside with a camel-hair brush. The use of metal discs and rubber balls to force material through a laboratory sieve is thoroughly bad, as is any other overloading or rough handling. A master set of sieves should be used for periodic checking of the work. Wear and tear will thus be revealed, and damaged sieves can be removed.

Although described above as square-meshed, the effect of the weave (Fig. 59) is to produce a curved set of wires. Until a near-mesh undersized particle falls centrally with its minimum cross-section normal to the aperture it will bounce off repeatedly, unless it becomes wedged. "Blinding" is minimised by taut mounting of the cloth, which should be laid on a wooden block and carefully soldered into its eight-inch frame, the soldering being finished with a smooth down-slope to the sieve so that no particles are liable to lodge in any crevice between cloth and frame. The fixing of a new cloth is best left to experts, as a slack screen gives bad results, while wrong handling in fitting may distort the meshes into parallelograms.

For most test purposes a series of screens such as is given in Table 10, col. 1, or the nearest B.S. equivalent, is adequate. If particles are more closely sized, prolonged shaking on the screen is needed, as many of them will be so near the aperture size as to become wedged, or fail to find a way

TABLE 8

U.S. SIEVE SERIES AND TYLER EQUIVALENTS

ASTM—E-11-61

Sieve Designation		Sieve Opening		Nominal Wire Diameter		Tyler Screen Scale Equivalent Designation
Standard	Alternate	mm.	in. (approx. equivalents)	mm.	in. (approx. equivalents)	
107·6 mm.	4·24 in.	107·6	4·24	6·40	·2520	....
101·6 mm.	4 in. (a)	101·6	4·00	6·30	·2480	....
90·5 mm.	3½ in.	90·5	3.50	6·08	·2394	....
76·1 mm.	3 in.	76·1	3·00	5·80	·2283	....
64·0 mm.	2½ in.	64·0	2·50	5·50	·2165	....
53·8 mm.	2·12 in.	53·8	2·12	5·15	·2028	....
50·8 mm.	2 in. (a)	50·8	2·00	5·05	·1988	....
45·3 mm.	1¾ in.	45·3	1·75	4·85	·1909	....
38·1 mm.	1½ in.	38·1	1·50	4·59	·1807	....
32·0 mm.	1¼ in.	32·0	1·25	4·23	·1665	....
26·9 mm.	1·06 in.	26·9	1·06	3·90	·1535	1,050 in.
25·4 mm.	1 in. (a)	25·4	1·00	3·80	·1496	....
* 22·6 mm.	⅞ in.	22·6	0·875	3·50	·1378	·883 in.
19·0 mm.	¾ in.	19·0	0·750	3·30	·1299	·742 in.
* 16·0 mm.	⅝ in.	16·0	0·625	3·00	·1181	·624 in.
13·5 mm.	·530 in.	13·5	0·530	2·75	·1083	.525 in.
12·7 mm.	½ in. (a)	12·7	0·500	2·67	·1051	....
* 11·2 mm.	7/16 in.	11·2	0·438	2·45	·0965	.441 in.
9·51 mm.	⅜ in.	9·51	0·375	2·27	·0894	.371 in.
* 8·00 mm.	5/16 in.	8·00	0·312	2·07	·0815	2½ mesh
6·73 mm.	·265 in.	6·73	0·265	1·87	·0736	3 mesh
6·35 mm.	¼ in. (a)	6·35	0·250	1·82	·0717	....
* 5·66 mm.	No. 3½	5·66	0·223	1·68	·0661	3½ mesh
4·76 mm.	No. 4	4·76	0.187	1·54	·0606	4 mesh
* 4·00 mm,	No. 5	4·00	0·157	1·37	·0539	5 mesh
3·36 mm.	No. 6	3·36	0.132	1·23	·0484	6 mesh
* 2·83 mm.	No. 7	2·83	0·111	1·10	·0430	7 mesh
2·38 mm.	No. 8	2·38	0·0937	1·00	·0394	8 mesh
* 2·00 mm.	No. 10	2·00	0·0787	·900	·0354	9 mesh
1·68 mm.	No. 12	1·68	0·0661	·810	·0319	10 mesh
* 1·41 mm.	No. 14	1·41	0·0555	·725	·0285	12 mesh
1·19 mm.	No. 16	1·19	0·0469	·650	·0256	14 mesh
* 1·00 mm.	No. 18	1·00	0·0394	·580	·0228	16 mesh
841 micron	No. 20	0·841	0·0331	·510	·0201	20 mesh

TABLE 8 (*continued*)

Sieve Designation		Sieve Opening		Nominal Wire Diameter		Tyler Screen Scale Equivalent Designation
Standard	Alternate	mm.	in. (approx. equiva-lents)	mm.	in. (approx. equiva-lents)	
*707 micron	No. 25	0·707	0·0278	·450	·0177	24 mesh
595 micron	No. 30	0·595	0·0234	·390	·0154	28 mesh
*500 micron	No. 35	0·500	0·0197	·340	·0134	32 mesh
420 micron	No. 40	0·420	0·0165	·290	·0114	35 mesh
*354 micron	No. 45	0·354	0·0139	·247	·0097	42 mesh
297 micron	No. 50	0·297	0·0117	·215	·0085	48 mesh
*250 micron	No. 60	0·250	0·0098	·180	·0071	60 mesh
210 micron	No. 70	0·210	0·0083	·152	·0060	65 mesh
*177 micron	No. 80	0·177	0·0070	·131	·0052	80 mesh
149 micron	No. 100	0·149	0·0059	·110	·0043	100 mesh
*125 micron	No. 120	0·125	0·0049	·091	·0036	115 mesh
105 micron	No. 140	0·105	0·0041	·076	·0030	150 mesh
* 88 micron	No. 170	0·088	0·0035	·064	·0025	170 mesh
74 micron	No. 200	0·074	0·0029	·053	·0021	200 mesh
* 63 micron	No. 230	0·063	0·0025	·044	·0017	250 mesh
53 micron	No. 270	0·053	0·0021	·037	·0015	270 mesh
* 44 micron	No. 325	0·044	0·0017	·030	·0012	325 mesh
37 micron	No. 400	0·037	0·0015	·025	·0010	400 mesh

* *These sieves correspond to those proposed as an International (ISO) Standard. It is recommended that wherever possible these sieves be included in all sieve analysis data or reports intended for international publication.*

(a) *These sieves are not in the fourth root of 2 Series, but they have been included because they are in common usage.*

TABLE 9

COMPARISON TABLE OF U.S., BRITISH, FRENCH, AND GERMAN SIEVE SERIES WITH TYLER EQUIVALENTS

TYLER (1)	U.S. (2)		BRITISH STANDARD (3)		FRENCH (4)		GERMAN DIN (5)		TYLER (1)
Equiv. Mesh	Size	Opg. mm.	No.	Opg. mm.	No.	Opg. mm.	Opg. Microns	Opg. mm.	Equiv. Mesh
	4·24"	107·6							
	4"	101·6							
	3½"	90·5							
	3"	76·1							
	2½"	64·0							
	2·12"	53·8							
	2"	50·8							
	1¾"	45·3							
	1½"	38·1							
	1¼"	32·0							
1·05"	1·06"	26·9							1·05"
	1"	25·4						25.0	
·883"	7/8"	22·6							·883"
·742"	3/4"	19·0						20·0	·742"
								18·0	
·624"	5/8"	16·0						16·0	·624"
·525"	·530"	13·5							·525"
	1/2"	12·7						12·5	
·441"	7/16"	11·2							·441"
								10·0	
·371"	3/8"	9·51							·371"
·312"	5/16"	8·00						8·0	·312"
·263"	·265"	6·73							·263"
	1/4" No. 3	6·35						6·3	
3½	No. 3½	5·66							3½
					38	5·000		5·0	
4	4	4·76							4
5	5	4·00			37	4·000		4·0	5
6	6	3·36	5	3·353					6
					36	3·150		3·15	
7	7	2·83	6	2·812					7
8	8	2·38	7	2·411	35	2·500		2·5	8
9	10	2·00	8	2·057	34	2·000		2·0	9
10	12	1·68	10	1·676	33	1·600		1·6	10
12	14	1·41	12	1·405					12
					32	1·250		1·25	
14	16	1·19	14	1·204					14
16	18	1·00	16	1·003	31	1·000		1·0	16
20	20	0·841	18	·853					20

Table 9 (*continued*)

TYLER (1)	U.S. (2)		BRITISH STANDARD (3)		FRENCH (4)		GERMAN DIN (5)		TYLER (1)
Equiv. Mesh	Size	Opg. mm.	No.	Opg. mm.	No.	Opg. mm.	Opg. Microns	Opg. mm.	Equiv. Mesh
					30	·800	800	·800	
24	25	0·707	22	·699					24
					29	·630	630	·630	
28	30	0·595	25	·599					28
32	35	0·500	30	·500	28	·500	500	·500	32
35	40	0·420	36	·422					35
					27	·400	400	·400	
42	45	0·354	44	·353					42
					26	·315	315	·315	
48	50	0·297	52	·295					48
60	60	0·250	60	·251	25	·250	250	·250	60
65	70	0·210	72	·211					65
					24	·200	200	·200	
80	80	0·177	85	·178					80
					23	·160	160	·160	
100	100	0·149	100	·152					100
115	120	0·125	120	·124	22	·125	125	·125	115
150	140	0·105	150	·104					150
					21	·100	100	·100	
170	170	0·088	170	·089			90	·090	170
					20	·080	80	·080	
200	200	0·074	200	·076					200
							71	·071	
250	230	0·063	240	·066	19	·063	63	·063	250
							56	·056	
270	270	0·053	300	·053					270
					18	·050	50	·050	
325	325	0·044					45	·045	325
					17	·040	40	·040	
400	400	0·037							400

(1) *Tyler Standard Screen Scale Sieve Series.*

(2) *U.S. Sieve Series - ASTM Specification E*-11-61.

(3) *British Standards Institution, London BS*-410.

(4) *French Standard Specifications, AFNOR X*-11-501.

(5) *German Standard Specification DIN* 4188.

TABLE 10 (*After Heywood and Pryor*)

SIEVE MESHES IN RELATION TO PARTICLE SURFACE AREAS

Sieve Mesh and Apertures in mm.			Aperture Dimensions in mm. based on Tyler Sieves				Surface Area in sq. cm. per gramme at Unity S.G.				Weight in grammes at Unity S.G. to expose one sq. metre of Surface				Surface relative to Unity at 200 Mesh	Weight in grammes at Unity S.G. to cover 8-in. Diam. Sieve one particle deep
Tyler $\sqrt{2}$ Series	Nearest B.S.	Nearest I.M.M.	Distance between Wires	Area sq. mm.	Diagonal	Mean Aperture	Cube	Oblong	Plate	Average Particle	Cube	Oblong	Plate	Average Particle		
(1)	(2)	(3)	(4)	(5)	(6)	(7)	(8)	(9)	(10)	(11)	(12)	(13)	(14)	(15)	(16)	(17)
1·05 in.	1·00 in. 25·40 mm.	—	**26·67**	710	37·7	32·18	1·86	1·55	7·5	2·55	5,360	6,440	1,340	3,920	0·003	—
0·742 in.	0·75 in. 19·05 mm.	—	**18·85**	355	26·7	22·76	2·64	2·20	10·5	3·60	3,790	4,550	950	2,780	0·004	—
0·525 in.	0·50 in. 12·70 mm.	—	**13·33**	177	18·9	16·09	3·73	3·11	14·9	5·10	2,680	3,220	670	1,960	0·006	—
0·371 in.	0·375 in. 9·53 mm.	—	**9·423**	89	13·3	11·38	5·27	4·39	21·1	7·20	1,900	2,275	474	1,390	0·008	—
3 mesh	0·25 in. 6·35 mm.	—	**6·680**	44·6	9·45	8·052	7·45	6·21	29·9	10·2	1,340	1,610	335	982	0·011	170
4 ,,	0·1875 in. 4·76 mm.	—	**4·699**	22·1	6·64	5·690	10·55	8·80	42·2	14·4	948	1,140	237	694	0·016	120
6 ,,	5 mesh 3·353 mm.	—	**3·327**	11·1	4·70	4·013	14·9	12·4	59·8	20·4	668	803	167	490	0·022	85
8 ,,	7 mesh 2·411 mm.	5 mesh 2·540 mm.	**2·362**	5·6	3·34	2·845	21·1	17·6	84·3	28·8	474	569	118	347	0·031	60
10 ,,	10 mesh 1·676 mm.	8 mesh 1·574 mm.	**1·651**	2·63	2·33	2·007	29·9	24·9	119·5	40·8	335	401	83·6	245	0·044	42
14 ,,	14 mesh 1·204 mm.	10 mesh 1·270 mm.	**1·168**	1·36	1·65	1·410	42·6	35·5	170	58·2	235	282	58·8	172	0·063	30
20 ,,	18 mesh 0·853 mm.	16 mesh 0·792 mm.	**0·833**	0·69	1·18	1·001	60·0	50·0	240	81·9	167	200	41·7	122	0·089	21
28 ,,	25 mesh 0·599 mm.	20 mesh 0·635 mm.	**0·589**	0·347	0·833	0·711	84·4	70·4	338	115	118	142	29·6	86·7	0·126	15
35 ,,	36 mesh 0·422 mm.	30 mesh 0·421 mm.	**0·417**	0·174	0·590	0·503	119	99·4	477	163	83·8	100	21·0	61·4	0·178	10·6
48 ,,	52 mesh 0·295 mm.	40 mesh 0·317 mm.	**0·295**	0·087	0·417	0·356	169	141	674	230	59·3	71·2	14·8	43·4	0·25	7·5
65 ,,	72 mesh 0·211 mm.	60 mesh 0·211 mm.	**0·208**	0·043	0·294	0·252	238	199	952	325	42·0	50·4	10·5	30·7	0·35	5·3
100 ,,	100 mesh 0·152 mm.	80 mesh 0·157 mm.	**0·147**	0·022	0·208	0·178	337	281	1,350	461	29·7	35·6	7·42	21·7	0·50	3·8
150 ,,	150 mesh 0·104 mm.	120 mesh 0·107 mm.	**0·104**	0·0108	0·147	0·126	476	397	1,910	650	21·0	25·2	5·25	15·4	0·71	2·7

200 ,,	200 mesh 0·076 mm.	150 mesh 0·084 mm.	**0.074**	0·0055	0·105	0·089	674	562	2,700	922	14·8	17·8	3·71	10·9	1·00	1·9
270 ,,	300 mesh 0·053 mm.	200 mesh 0·063 mm.	**0.053**	0·0028	0·075	0·064	938	782	3,750	1,280	10·7	12·8	2·67	7·80	1·40	1·4
400 ,,	—	—	**0.038**	0·0014	0·054	0·046	1,300	1,090	5,220	1,780	7·66	9·20	1·92	5·61	1·94	1·0
End of Sieve Series.			Microns **30**	—	—	0·034	1,760	1,470	7,060	2,410	5·66	6·80	1·42	4·15	2·63	—
			20	—	—	0·025	2,400	2,000	9,600	3,280	4·16	5·00	1·04	3·05	3·57	—
			10	—	—	0·015	4,000	3,330	16,000	5,460	2·50	3·00	0·625	1·83	5·96	—
			5	—	—	0·0075	8,000	6,670	32,000	10,900	1·25	1·50	0·312	0·915	11·9	—
			1	—	—	0·0030	20,000	16,700	80,000	27,300	0·500	0·600	0·125	0·366	29·8	—
			0.5	—	—	0·00075	80,000	66,700	320,000	109,000	0·125	0·150	0·031	0·092	119	—

Note.—These figures for surface area take no account of internal fissuring, or compacted aggregates. Surface area, as measured by gas-adsorption methods, may show much higher values.

(1) Tyler $\sqrt{2}$ Sieve Series, apertures in inches or mesh.

(2) Nearest corresponding B.S.I. (British Standards Institution) Perforated Plate or Fine Mesh Test Sieve, and aperture in mm.

(3) Nearest corresponding I.M.M. (Institution of Mining and Metallurgy) Test Sieve, and aperture in mm.

(4) Distance between wires or aperture dimension of Tyler $\sqrt{2}$ Sieve Series. (*Note.*—Figures in all subsequent colums are based on the Tyler Series.)

(5) Area of sieve aperture in sq. mm. Defines the smallest cross-sectional area of the particle.

(6) Diagonal of sieve aperture in mm. Defines the upper limit of the breadth of intermediate particle dimension for a tabular particle.

(7) Arithmetic mean of the aperture of the sieve in the same line and of the next sieve immediately above. This average is expressed by the symbol A and is in mm.

(8) Surface area in sq. cm. per gramme of cubical particles of which the side dimension is A mm., specific gravity assumed to be unity. These figures equal $60/A$. (*Note.*—**Divide** the figures given by the actual specific gravity of the mineral concerned; this applies to columns (8) to (11) inclusive.)

(9) Surface as above for particles having dimensions A by A by $2A$ mm. long. Figures equal $50/A$.

(10) Surface as above for flat particles A by $A/10$ mm. thick. Figures equal $240/A$.

(11) Surface as above for particles of average shape. Figures equal $82/A$.

(12) Weight of material in grammes to expose surface of one square metre, if particles are cubes of side dimension A mm. and specific gravity is unity. This column equals 10,000 divided by column (8). (*Note.*—**Multiply** the figures given by the actual specific gravity of the mineral concerned; this applies to columns (12) to (15) inclusive.)

(13) Weight as above for particles having dimensions A by A by $2A$ mm. long. Figures equal 10,000/column (9).

(14) Weight as above for flat particles A by A by $A/10$ mm. thick. Figures equal 10,000/column (10).

(15) Weight as above for particles of average shape. Figures equal 10,000/column (11).

(16) Surface areas expressed relative to the surface area of material of the same shaped particles retained between 150 and 200 mesh Tyler. These figures are equal to 0·089 divided by the respective values of A from column (7).

(17) The approximate weight of material that will cover an 8-in. diameter sieve to a uniform depth of one particle; the voidage between particles has been assumed to be 35% and the specific gravity unity. These figures are equal to 21·1 multiplied by A in mm., and must be multiplied by the specific gravity of the mineral concerned.

through. The mechanical friction during shaking, though gentle, may produce some attrition and thus falsify results.

A more serious danger arises when ores containing metallic flakes or wires of gold are screened. One such particle wrongly in, or missing from, the small sample finally handled by the assayer would seriously distort the results. Extra care is needed when there is any possibility of the temporary lodgement of a "metallic". This applies in lesser degree to native silver and copper.

In place of the finer cloths, which are easily damaged, it is possible to obtain sieves with square apertures manufactured by electroplating methods. Apertures flare outward and downward, so a particle that passes the "mesh" falls clear.

Despite their robustness, and the high precision with which they can be made,the author has not found them so useful as had been expected. The roughness imparted by the weave is missing, and the irregularly shaped particles do not appear to receive the vigorous tossing action which aids "near-mesh" particles to present their minimum cross-section to the sieve apertures.

Screening is in universal use for control of mill operations where precise knowledge of the state of the material down to –200 mesh is required. Below this size, routine control should preferably use methods based on the rate of fall through liquid. The change from a gauging method, such as screening, to a hydrodynamic one like sedimentation is not smooth, so it is good practice to let the two systems overlap in order to adjust them to continuity of record. Thus, screening can be carried down to 270 mesh on an unknown ore, and sedimentation commenced at 150 "mesh", in order that the overlapped portion of the test can be reconciled, using graphical methods.

In describing material, fine gravel sizes are thought of as ore broken to between 3 and 10 mesh; sand as 10 mesh down to 250 mesh; silt 200 mesh to 5 microns; molecules (inorganic) from 4 millimicrons down and unit crystals from about 5 angstroms. The following definitions of particle size are suggested in a British Standards publication[7], as giving a general descriptive term for finely crushed material:

Powder. Discrete particles of dry material with a maximum dimension of less than 1,000μ.
Grit. Hard particles, usually mineral . . . retained on 200 mesh B.S. test sieve.
Dust. Particulate material which is or has been airborne and which passes a 200 mesh B.S. test sieve (76 microns).
Agglomerate. Assemblage of particles rigidly joined together, as by partial fusion (sintering) or by growing together.
Aggregate. Assemblage of particles which are loosely coherent.
Particle mean size. Dimension of a hypothetical particle such that if a powder were wholly composed of such particles, such a powder would have the same value as the actual powder in respect of some stated property (sometimes called "average particle size").
Size range. Lower and upper limits of particle size of a powder.
Size fraction. Portion of a powder composed of particles between two given size limits. It may be expressed in terms of weight, volume, surface or numerical frequency.

Size distribution. Distribution of size fractions in a powder.
Fines undersize. Portion of a powder composed of particles which are smaller than some specific size.
Oversize. Portion of a powder composed of particles which are larger than some specific size.

In addition to the mesh-dimensions in Table 10 (cols. 1-7), factors are given which permits conversion of a "screen analysis" of ore into surface area/weight relations.

The total surface involved is a major factor in comminution, classification, flotation, and the handling of ore pulps. Table 11 records the computation of percentage surface in the undersize for each screen fraction of ground silica subjected to a sizing analysis.[6] In view of the importance of particle surface as the threshold across which all the reactions of that particle are transacted, it is illuminating to observe that such a rock ground "to 75½%–200 mesh" has 95·6% of its total surface in the undersize and less than 5% in one quarter of the volume or weight ground to "+200". If we assume that chemical reagents added to the pulp distribute themselves equably over the

TABLE 11

DISTRIBUTION OF WEIGHT AND SURFACE IN GROUND SILICA
(After Heywood)

Size	*Medium Grade*		*Fine Grade*	
	Weight in Undersize	*Surface in Undersize*	*Weight in Undersize*	*Surface in Undersize*
	%	%	%	%
100 mesh B.S.	97·8	99·8	99·9	100·0
150 ,, ,,	90·0	98·7	—	—
200 ,, ,,	75·5	95·6	99·7	100·0
240 ,, ,,	69·2	93·9	—	—
300 ,, ,,	60·3	91·0	99·4	100·0
40 microns	49·0	86·5	94·4	99·8
30 ,,	37·6	80·3	87·1	99·3
20 ,,	24·8	70·5	74·5	98·2
10 ,,	10·8	52·8	50·2	94·8
5 ,,	4·3	36·3	32·3	89·8
2 ,,	1·0	18·5	16·0	79·8
1 ,,	0·25	8·9	9·1	70·1
0·5 ,,	0·05	3·8	4·9	58·2
0·2 ,,	—	—	1·7	38·9
0·1 ,,	—	—	0·6	23·4
0·05 ,,	—	—	0·15	10·6

Note.—Percentage weight of undersize measured down to 2 microns size; subsequent percentages obtained by extrapolation.

solid surface, then practically the whole of such reagents are directed toward the ordinary grinding mill undersize and, in the example cited in this Table, more than half the reagent is consumed by the "—10 micron" fraction of the ore. The significance of this will appear more fully later, when flotation physics is studied.

Again, the —10 micron fraction, by providing more than half the surface, is responsible for more than half of the pseudo-viscosity and internal friction of the pulp. In some circumstances this is a valuable property, in that it aids the transportation of pulps by retarding settlement of the coarser particles which might otherwise choke the travelling pipes and launders. At other times a preponderance of extremely fine material interferes seriously with the dewatering or thickening of the pulp, and with its filtration through porous media which become "blinded" by flocculated slimes. Primary slimes are the worst offenders in these cases, and are most likely to be encountered when treating weathered ore from the upper zones of the ore deposit, or during dump reclamation. The very fine slime produced from virgin ore-rocks during grinding usually gives but little trouble in its physical handling.

Routine laboratory sizing is rarely carried below 200 mesh, and therefore leaves the greater part of the ground ore-sample unfractionated. From this point "sub-sieve sizing" must be used. The term is applied to methods of grading particles too small to be readily sieved on even the finest laboratory screens, which in normal practice finish at 200 mesh (♯). Save for measurement under the microscope, sub-sieve methods sort out equal-settling material under specified conditions (sedimentation, elutriation, infra-sizing, centrifuging) and the phrase "sub-sieve sizing" is somewhat inaccurate since the specific gravity and shape of the fractionated particle are the main factors in its segregation.

Particle Terminology

In studying the behaviour of a particle it is often useful to compare it with a sphere, or to define some dimension common to a group. The mean size, or "average particle size" may be defined as the dimension of a hypothetical particle such that, if the comminuted material being examined were composed entirely of such particles, this particle would have the same value in respect to some stated property as is exhibited by the actual powder.

The projected diameter of a particle is defined[7] as the diameter of a circle which has the same area as the projected profile, which will be greatest when the particle is examined from a direction perpendicular to its most stable plane. The equivalent surface diameter is that of a sphere having the same surface area as the particle. The equivalent volume diameter is that of a sphere of like volume. Particles are open-pored when they contain cavities open to the surface and *close-pored* when such cavities are internal only. The true density of a porous particle is that of its $\frac{\text{mass}}{\text{volume}}$ excluding pores. Its apparent density is that which excludes open pores but makes no allowance for closed ones. Specific surface is an important characteristic of a mass of particles. It is usually expressed as cm^2/g, of mass for a dry powder, or as cm^2/l, at a concentration of n g/l or n ml/l for mineral pulps or slurries. A powder consisting of particles of differing sizes is said to be polydispersed, and one of uniform sizes monodispersed.

Particles are said to be *acicular* when needle-shaped: angular when sharp-edged and roughly polyhedral; dendritic when of branched crystalline shape;

fibrous when thread-like; flaky when plate-like; granular when irregular but roughly equi-dimensional; *irregular* when non-symmetrical; nodular when rounded but irregular; and spherical when globular.

Laboratory Screening Methods

If no mechanical appliances are available for shaking the sieves, hand screening is used. A "nest" of screens is assembled with the coarsest above and the finest below. A tight-fitting pan receives the final undersize. Usually not more than three sieves are "nested" for hand screening, and they should fit closely so that no "dusting losses" are incurred. Before assembling a nest the sieves are examined and if necessary unblinded by gentle brushing from below with a soft camel-hair brush. The examination should disclose displaced wires, ruptures in the screen cloth and, when malleable metallic material may be present, whether any particles of gold, silver, copper, etc., have become wedged or entangled in the cloth. This can be done quickly and accurately by making a contact print of the screen cloth on photographic paper and scanning it with a magnifying lens. Direct observation of the meshes may be misleading.

For rough work, up to 200 grammes of dry ore are next placed on the uppermost screen. In the older textbooks the use of coins, washers, soft rubber balls, etc., as "beaters" is sometimes recommended. The idea is that such aids will slide above the particles and help to press reluctant ones through the meshes. This practice is thoroughly bad. If a particle cannot find its way through the mesh without adventitious aid it should report with the oversized fraction. If comparatively large objects are allowed to hammer small and sharp particles through the meshes the sieve is damaged and the sample undergoes attrition during screening. It thus ceases to be a true sample. For good work large samples are not recommended. Not only does the competition for a position next to the screen cloth vitiate the accuracy but particles are "damped down" by superincumbent layers and prevented from turning until their minimum cross-section is presented to the gauging action of the mesh. The ideal load is one particle deep, but tests show that up to four particles deep can be screened without serious loss of accuracy. *Minus* 150–mesh material of S.G. 1.0 covers a 200-mesh 8″ circular laboratory screen one particle deep when a weight of 1·9 grammes is worked (Table 10, col. 17). With sand of S.G. 2·7 four particles deep the equivalent loading would be about 20 grammes. If the Tyler $\sqrt{2}$ series 100 mesh, 150 and 200 mesh is nested, it is desirable to limit the charge to one which arrests all but the last 30 grammes or so ahead of the 200-mesh sieve. A good proportion of this 30 grammes will probably fall through readily, leaving ample screen surface available for the thorough searching of the fraction arrested. If the sieves are old and the cloth has sagged into pouches, this loading would be excessive. A sieve should be taut and springy. The decision as to the size of the sample includes other factors. The degree of accuracy used in testing a critically important sample on a special set of screens is not required in a routine sizing-analysis, where the tendency is for the operator to use a large sample, screened with moderate efficiency, rather than a small one with possibly a high sample-cutting error. Laboratory screening methods must be adapted to the general rule that the degree of accuracy at which they aim should be co-ordinated both with the accuracy required by the work in hand and with that possible in sample cutting.

The nest of screens is assembled, with the largest-mesh sieve on top, and a close-fitting pan at the bottom to receive final undersize. The sample is placed on the top screen and a lid is fitted closely to minimise dusting loss. The nest of screens is then shaken, and gently jarred. The point at which the pan is bumped is changed every half minute for one 60° further round the rim. Shaking, with jarring, is continued until very little further material falls through the bottom screen. The time required for this is then made standard for the particular class of material or work. When friable material is being tested, care must be used not to work with such violence, or for so long a time, that the sample is degraded on the sieves.

Most laboratory work is done on mechanically operated shaking appliances. One widely used appliance is the Rotap shaker (Fig. 61). Up to six standard

Fig. 61 *Rotap Laboratory Screen Shaker* (*International Combustion Ltd.*)

8″ round screens can be nested on this, and on most mechanised devices. Vertical reciprocation of the sieves is accompanied by a rotatory and a tapping motion. The length of screening time can be controlled by an automatic timer. Another six-nester is the Sherwen shaker, operated by A.C. current (Fig. 62). The vertical displacement of the table at each vibrating stroke is controlled by an electrical damping device, so that the amplitude of the

Fig. 62 Vibrating Screen Shaker

vibrating stroke can be varied to suit the material. Dusting loss is thus minimised and the best kind of shaking action is selected for the job in hand. Since no circular distributing motion is applied, the sieves must be taut, to assure good use of the whole surface. In the Russell shaker (Fig. 63) the nest is agitated by an out-of-balance coupling connected to a small electric motor. The result is a "panning" motion which floats and churns the material with good searching, mainly over the central part of the sieve.

With the Russell shaker, attachments are available which allow the nest of screens to be sprayed by water, the undersize being flushed to a receiving vessel.

Inaccuracy in screening arises from several causes. If the sample is properly dried, friction on the sieve leads to some electrostatic "bunching" of the particles, so that several which are truly undersized to a given aperture report as part of the "on" fraction. Such seizure increases with temperature, and screening in a cool room therefore minimises it. Dampness is the worst enemy of accuracy. It leads to the sticking together of particles, or their clinging to the screen cloth when they should fall through. When dry-crushed rock is being tested, and also when the test sample has come from a pulp containing much soluble material care should be taken that no dried salts are adhering to the particles. These would increase their effective cross-section. In very exact work, microscopic examination may be used to check whether particles have been aggregated by impact during grinding, so that several are clustered together in such a way as to report as one. Dispersal and free movement are aided by limiting the size range of the sample. This can readily be done by removing the sub-sieve particles (say those finer than 200 mesh) before making the screen analysis.

The connexion between surface friction and mass mobility can easily be observed. If a level teaspoonful of dry and wide-ranged sand (say – 65 + 400 ♯) is dropped from a height of about five feet it falls with but little scattering. If a similar quantity of closely sized sand is similarly dropped its particles will scatter evenly and over an appreciable area. In the second case the specific surface has been reduced and with it the binding effect of inter-facial friction. This simple experiment shows one of the many ways in which solids moving through a fluid (gaseous or liquid) are influenced by the

Fig. 63 *Russell Screen Shaker*

specific surface involved. In panning, tabling, sluicing, thickening, pulp transport, *dense media separation,* jigging, and the building of flotation froths the effects of *pseudo-viscosity* are at work, as in screening. The degree of individual mobility of particles is a major operating factor in mineral processing, and this simple experiment with a sand sample after screen tests are complete and before it is thrown away helps to bring home the fact.

The surface/volume ratio can be seen at work in its effect on the behaviour of agricultural soils or engineering foundations. Following the presentation in Fig. 48 and considering each grain as a cube, the following figures show the tremendous differences between various forms of soil. The mesh sizes, stated in millimetres, for sand, silt, etc. are arbitrary but commonly used.

	Average cube side		*Area per cubic cm.*
Sand	1·0 mm	(1,000 μ)	60 sq. cm.
Silt	0·05	(50 μ)	1,200 sq. cm.
Clay	0·005	(5 μ)	12,000 sq. cm.
Colloidal soil	0·0005	(0·5 μ)	120,000 sq. cm.

Anyone familiar with the land will appreciate the change in the amount of work involved in cultivating soils of varying textures. In mineral processing a lot of work goes into the essential business of fluidising large tonnages of ore in connexion with its treatment. The specific surface involved in transport, gravity treatment, chemical extraction and froth flotation is an important element in any problem.

For precise work, the wet-and-dry screening method is to be recommended. It removes a substantial amount of interfering undersize and also re-dissolves dried-on salts and disperses any caked clay. The dried and weighed sample is dispersed in water containing a little sodium silicate, (about 0.02 g/l) and washed gently on the finest screen of the series. The undersize is caught in a bowl below, settled (perhaps with the aid of a flocculating agent), dried, and retained. The oversize is also dried and then sieved over the desired series. More ore will work through the finest sieve. These two final undersizes are mixed for final weighing.

A screen analysis can be reported as in Table 12. It is usual to add to

TABLE 12

SCREEN ANALYSIS

Messina Copper Ore

B.S. Mesh	*Mesh Aperture* (μ)	*Direct % Wt. Retained*	*Cum. % Wt. Finer*
+18	853	7·0	93·0
—18+25	599	10·4	82·6
—25+36	422	14·2	68·4
—36+52	295	13·6	54·8
—52+72	211	9·2	45·6
—72+100	152	8·1	37·5
—100+150	104	8·2	29·3
—150+200	76	5·1	24·2
—200		24·2	
		100·0	

this a graphic presentation of results, which illustrates more clearly the particle size distribution.

The graphical methods most commonly employed in mineral dressing are the weight frequency or direct plot, and the cumulative weight or integral plot. The direct weight frequency plot (Fig. 64) is essentially the derivative or slope of the corresponding cumulative weight curve, and the cumulative plot is a summation of the area under the direct plot curve.

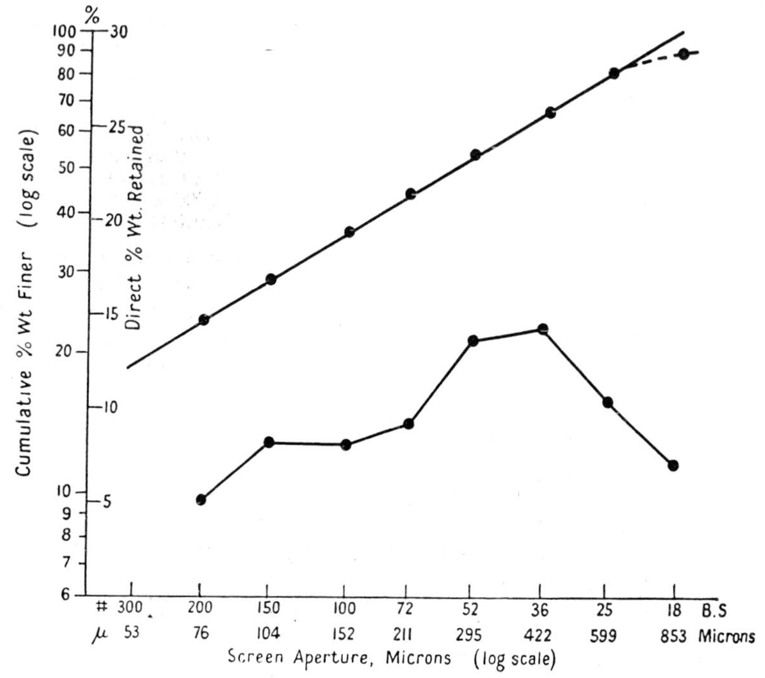

Fig. 64. *Graphs of Screen Analysis in Table* 12
Lower, *direct plot*. Upper, *cumulative plot*

To avoid distortion and possible misinterpretation, of the direct plot curve, values of the weight increments should be plotted at equal geometrical intervals on the size scale. To give equal prominence to all parts of the size range, a logarithmic size scale is used, and the intervals taken normally correspond to a $\sqrt{2}$ sieve series. The same considerations apply to results which are obtained by sub-sieve sizing methods, and the common practice of plotting some arbitrary relationship between the increments on the size scale should be avoided. If the graph is developed as a histogram, the areas show the relative proportions of the material in each size interval.

The cumulative weight, or integral plot has the advantage that a constant geometrical relationship is shown, even in the absence of the $\sqrt{2}$ succession of screens, since the logarithmic scale is plotted as apertures in microns, and not according to any labelled sieving system.

The particular integral plot, in which log scales are used on both axes of the graph, and in which percentage weight finer than a given size, as ordinate, is plotted against particle size as abscissa, is known commonly as the Schuhmann[8] plot (see also Fig. 64). As shown by Schuhmann and others,[9,10] the size distribution of fine powders produced by comminution tends to follow certain empirical relationships. In the Schuhmann plot the distribution approximates to a straight line over the greater part of its length, and may be represented by the following equation.

$$y = 100.\left[\frac{x}{K}\right]^{m} \tag{8.1}$$

where, y = the percentage weight finer than size x.
m = a constant. The distribution modulus.
K = a constant. The size modulus.

The arbitrary constant m, the distribution modulus, gives a measure of the size distribution of the material. The larger the value of m, the narrower is the distribution of the material over the whole size range. For unclassified ground ore, the value of m is usually close to $\frac{1}{\sqrt{2}}$. The size modulus K, which is the value of x when $y = 100$ (as determined by extending the approximately straight portion of the distribution until it intercepts the $y = 100$ ordinate), represents the theoretical upper limiting size of the material. It must be emphasised that the above equation is only an approximate empirical relationship which is extremely useful in interpreting size distributions.

Study by means of graphs of the patterns or modes of disintegration taken by an ore in response to comminution at various stages of reduction and in various ways will frequently yield information of value concerning the crushing characteristics, and mineralogical composition of the ore, provided the work has been done carefully on accurate screens.

Laboratory sizing is often followed by analytical tests made on a sample of mineral from each screened fraction. This shows the distribution of values through the various sizes in the head feed. It is of special value in tailings control, when it shows the amount of loss in each sized fraction—information of vital importance in process control.

Elutriation, Infra-sizing, and Sedimentation

The principles underlying these methods of size control are considered in later sections of this chapter. When correctly applied, they result not in sizing but in the *sorting* of particles into groups which have equal rates of settlement through a fluid medium. Two forces act upon a particle when it is free to move in a quiet fluid. One is the force of gravity, exerting a downward pull proportional to the mass. The other is frictional, proportional to the rubbing of the total surface of the particle against the fluid through which it is moving. This frictional retardation is of a complex nature. One factor is the nature and degree of wetting of the surface by the ambient fluid/s. Wetting is only complete when the air which was in contact before immersion is displaced, and this is not necessarily complete. The relative surface tensions of surface/air and surface/water (or other liquid used) affects the balancing forces of attraction, and indirectly the flow pattern of the liquid as it streams past the particle. Another retarding factor is the spinning and swerving effect produced by unbalanced braking as the solid moves through the liquid. At any moment the particle is drawing on its gravitational energy in order to momentarily force its maximum horizontal cross-section through the molecularly bonded liquid. On each side of its axial plane of movement it has two equal masses in balance about a vertical axis, but is simultaneously subject to frictional unbalance to the extent that the corresponding surface areas are not similarly equal. Save for a perfect sphere there can be no such balance. Differential braking therefore either deflects the particle from a vertical path, or sets it spinning. This leads to collision with other particles and loss or gain of momentum and kinetic energy. Further, the path thus elongated calls for more conversion of gravitational energy if the particle is to continue to fall then if it could have moved vertically. In other words, gravitational effect is weakened, and the decelerating effect of friction is greater than can be accounted for by surface braking alone. Thus, the determinants of the response to a given fluid medium of a single particle are the resultant of (*a*) its specific gravity and volume, (*b*) its total surface (result of shape *plus* texture) and (*c*) deviation from linear drift. In screening, the selection into size grade depends on the gauging effect of a system of rigid wires applied to the particle's cross-section. The relative densities of heavy and light minerals do not enter into consideration. When "equal-sorting" methods are applied, the criteria are mass and shape. In dealing with the settling behaviour of homogeneous material, volume and shape alone need to be considered. With ores, the fact that more than one mineral is present, and that each may have been liberated completely or may still only be partially unlocked, must be remembered in assessing results.

In practice, excellent reproducible results are obtainable. Simple methods can be improvised to allow size-control investigations to be applied to particle sizes down to 15 microns or less. More complicated methods are needed when dealing with very fine sizes. It is significant that several mill laboratories find it profitable to include routine tests down to the 5-micron zone of investigation. Since a "jog in the data" is inevitable when two so diverse techniques as sizing and sedimentation are applied to a sample, tests are usually overlapped so that a curve can be constructed which smooths out the differences.

In distilled water at 20° C., a quartz sphere 200 microns in diameter accelerates till it reaches a terminal velocity of 29·4 cm./min. At 15° C. the rate is 25·8 cm./min. If instead of being allowed to fall unhindered through static water it were introduced into a smoothly rising stream, this sphere would, in theory, "hang" or remain stationary when the rising rate of the water equalled the failing rate of the sphere. In practice, difficulties arise through the setting up of small eddy-currents between the fluid and the sides of the tube and through convection effects, so that the sphere would "dance" rather than remain stationary. Certain anomalies therefore arise when a rising current of water is used to lift small particles while larger ones drift down, although elutriation, which exploits this difference of motion, is used in sub-sieve sorting.

A sphere is convenient for use in research work, since it has a calculable area which is the minimum possible for a given volume, and which always presents the same cross-section to the fluid through which it falls. When an irregularly shaped particle falls through a fluid under similar conditions, the total surface rubbing against that fluid is greater than for an equal spherical mass and the rate of fall is therefore slower. Fall is also erratic owing to changes in the cross-section as the particle spins during its descent.

When a large number of spheres of equal specific gravity move in a rising column of water the smaller ones can be made to rise, the medium ones to *teeter* or dance about in a dilated layer (called a teetering zone) and the larger ones to fall through. The behaviour of any single sphere under these circumstances is random, depending on the chances of collision, eddying of the water, and of transfer of kinetic energy acquired during falling or rising when combating such eddies or when taking part in such collisions. A generalised behaviour can, however, be predicted. If instead of spheres a collection of particles of homogeneous material but of random shape is used, the individual variations become more complex, but a generalised behaviour sufficient for some sorting separations can be obtained. When the further complication is added of differential density between the particles, there is further loss of accuracy. Nevertheless, this behaviour of particles in rising currents of water permits of a very simple and convenient industrial application called *classification* which is widely used in the large-scale control of fine sizing or sorting. In the laboratory classification or elutriation of solids in a rising column of liquid has limited use. With patience and care, even better work can be done by sedimentation, a method based on the falling rate of small particles through static liquids. The elements of the elutriator (Fig. 65) include:

(*a*) a steady-head source of supply of the liquid;
(*b*) a metering arrangement which controls the flow rate of liquid rising through the elutriator;
(*c*) a sorting tube with smooth and parallel sides;
(*d*) receptacles for catching the overflowing and sinking particles separately.

Elutriation can be practised in water or any appropriate fluid and recirculation of the elutriating fluid can be obtained by the use of a small laboratory pump. A very accurate instrument is the Blyth elutriator (Fig. 66[12]).

The sorting tubes d to d_6 have a progressive increase in internal area of cross-section of the order of $\sqrt{2}$. The elutriating liquid is syphoned from the beaker below each tube, the valves a to a_6 being used to start the flow.

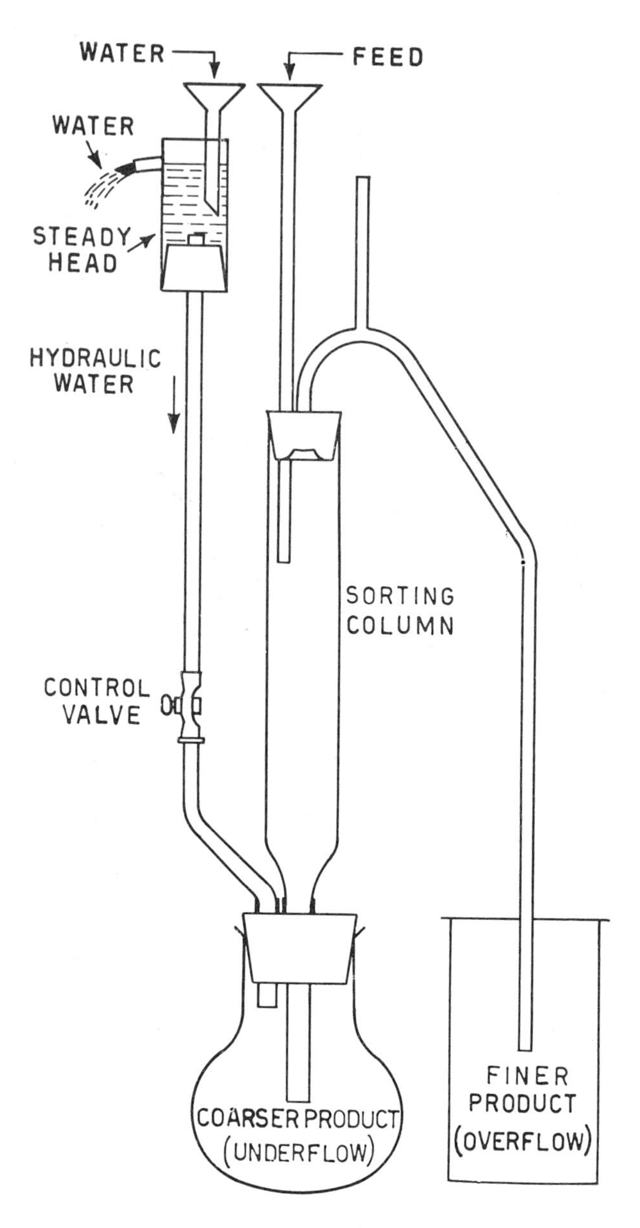

Fig. 65 A Simple Elutriator

Overflow from the sorting tubes is delivered by the syphon legs b_1 to b_6. Glass stirrers driven by a fractional motor agitate the sample, which is introduced into the first beaker after the rate of flow of liquid has been adjusted.

An elutriator which incorporates an element of *cyclone* action is the Kelsall "cyclosizer"[13]. It accelerates the movement of fine particles and can produce up to 100 g. of sorted material in the *minus* 50 μ *plus* 8 μ particle range in half an hour's running time.

When sedimentation is practised, a simple method, using only some beakers and a laboratory stirrer, can be used. For treating small samples the Andreasen pipette gives excellent precision. Another simple apparatus, the Palo-Travis tube[14] has been developed for rapid estimation of coal sizes by sedimentation in the sub-sieve region. The sample is gently introduced into a tall glass cylinder and part is collected below in a graduated glass tube. Rise of sediment shows the percentage of each size in a timed period, the coarsest fraction settling first. To minimise sampling error the author prefers the beaker method (Appendix B), which treats larger quantities. In all sedimentation methods care must be taken to disperse flocculated particles. The differing specific gravities of the products which settle equally

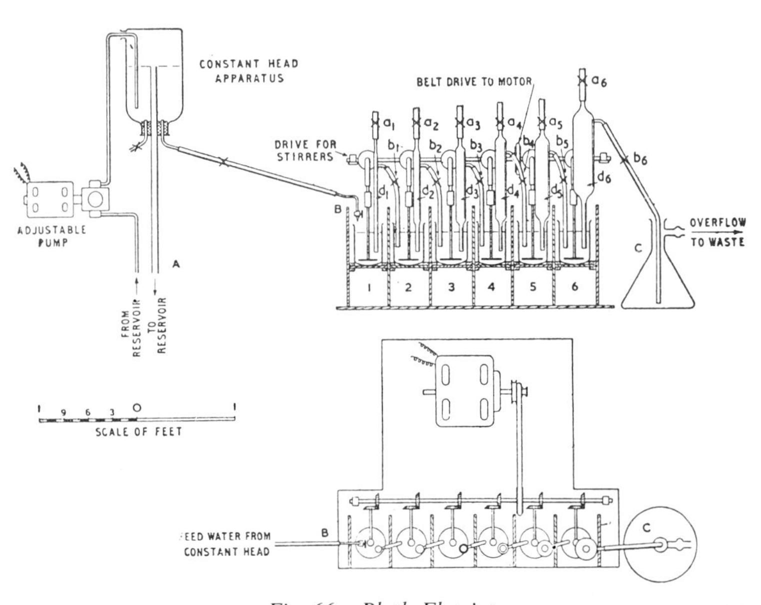

Fig. 66 Blyth Elutriator

may necessitate tedious sorting of them under a microscope if no convenient method of separation according to physical difference is possible. Sedimentation can be practised down to 5 microns or in some instances to as low as 2 microns.

When the fluid used is a liquid, surface friction is relatively high and appreciable viscous resistance is offered to the movement of particles in any direction. When a gas is used as the elutriating fluid, its molecules are not bonded and viscous resistance is negligible. Effective opposition to fall comes from bombardment of the particle by rising gas molecules. Horizontal cross-section of the particle varies from moment to moment as it spins downward through the ascending current of gas. If, then, enough gas molecules bombard a particle, they can transfer to it sufficient kinetic energy

Fig. 67 The Haultain Infra-sizer

to lift it "in teeter", or to slow it as it drifts down. Infra-sizing employs this principle by causing a current of air to be sent with minimum turbulence through a series of tubes of increasing diameter, the sample of ore being introduced into the air stream. The heaviest particles drop out earliest, and the lightest retained in the system stay in the largest tube where the bombardment rate of rising air molecules is lowest.

In the Haultain Infra-sizer (Fig. 67) a sample of 50 grammes can be treated with but little need for skilled supervision, and satisfactorily large samples of products are made with excellent reproducibility. The sample and air must be dry. Electrostatic seizure is avoided partly by the use of conducting rubber in the flexible connexions, and partly by jarring the tubes regularly. Separation can be carried down to sizes of a few microns. The use of air elutriation in specialised fields of size determination is given in some detail in B.S. 3406,[15].

In the permeability method of sizing the specific surface is derived by measurement of resistance to flow of a fluid (usually air) through a settled bed of particles. The original method of Carman[16] used a filtering system and measurement of liquid flow. Lea and Nurse[17] developed an apparatus using air which has been widely adopted. For coarse material (say between

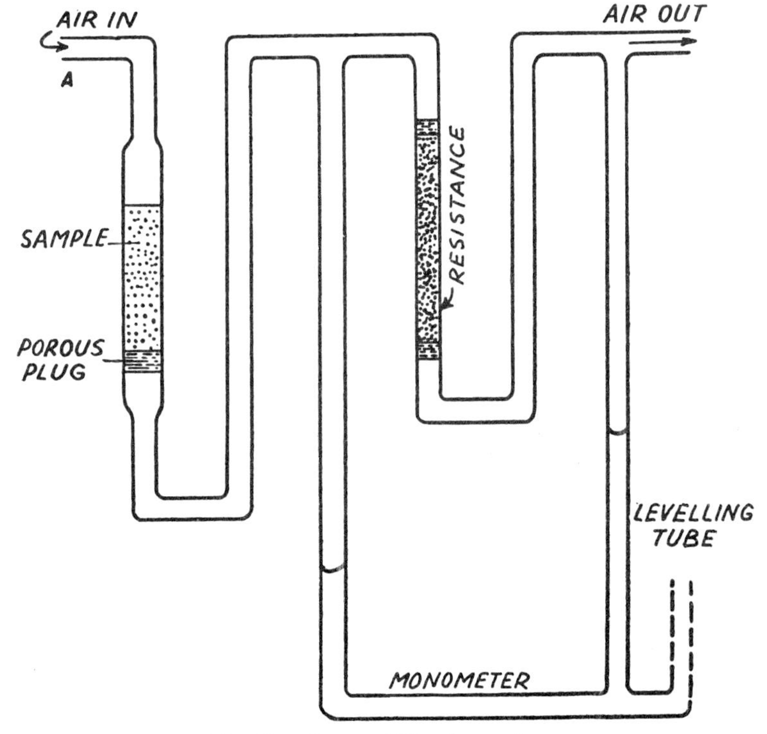

Fig. 68. Permeability Apparatus

2μ and 50μ the Gooden and Smith apparatus[18] has proved suitable for routine work. The formula used in air permeability measurement at the Royal Institute of Technology, Stockholm,[19] based on the rate of flow of a gas through a compacted bed of powder, is:

$$Q_m = \frac{1}{k_0.k_1} \cdot \frac{1}{\mu} \cdot \frac{1}{S^2} \cdot \frac{\epsilon^3}{(1-\epsilon)^2} \cdot \frac{A}{L} \Delta pt \frac{8\sqrt{2}}{3\sqrt{\pi}} \cdot \frac{1}{k_1} \cdot \sqrt{\frac{RT}{M} \cdot \frac{1}{S} \cdot \frac{\epsilon^2}{1-\epsilon} \cdot \frac{A}{L} \cdot \frac{\Delta p}{P_m}}.$$

It gives reproducible values within 2% + or − on powders with specific surface above 500 $\frac{\text{cm}^2}{\text{cm}^3}$.

Here

p_m = middle pressure of the gas in the bed (dynes/cm.^{2}),
Q_m = volume rate of the gas flow through the bed measured at the pressure p_m (cm.^{3}/sec.).
Δp = pressure drop across the bed (dynes/cm.^{2}).
A = cross-sectional area of the bed (cm.^{2}).
L length or depth of the bed (cm.).
ϵ = porosity or the fractional void space in the bed.
S = specific surface of the powder (cm.^{2}/cm.^{3} firm measure).
μ = viscosity of the gas (poise).
R = gas constant per mole = 0·8315 . 10^8 erg/deg. mole.
T = absolute temperature of the gas (degree Kelvin).
M = molecular weight of the gas.

k_0 and k_1 are numerical constants of which k_0 is supposed to be a real constant and equal to 2* while k_1 is supposed to depend upon the shape of the particles. To be able to calculate the specific gravity one usually takes $k_1 = 2{\cdot}5$†.

Movement of Particles in Liquid

It was observed in the last section that a particle moving in a liquid is influenced by gravitational pull and by the braking force of friction at its surface. If the particle is sufficiently massive, gravitational force dominates its behaviour. If sufficiently small, the major influence is exerted by its total surface. In the intermediate zone between these two types of reaction a reasonably consistent changeover is observable.

First, consider a perfect sphere of, say 50μ diameter, light enough to fall slowly through water. If this sphere falls through a large mass of water, so that no waves generated during its passage are reflected back, then the nature of the flow of water past the surface of the sphere is viscous or laminar. At the surface, the sphere is seized to a wetting layer of molecules of water.

* Value for a bed of glass beads, the surface of which has been directly measured.

† Modified constant after determining bed permeability at two pressures, found to be $k_1 = 2{\cdot}25$—$3{\cdot}0$

These molecules are associated with the water molecules beyond them against which they drag in a viscous manner. Thus, layer by layer, the zone surrounding the falling sphere can be pictured as being disturbed by its passage, the effect of this disturbance being calculable as loss of kinetic energy through surface friction, or laminar flow. The effect has been embodied in a classical formula by G. G. Stokes.

$$v = \frac{d^2 g\,(\sigma - \rho)}{18\eta} \tag{8.2}$$

where v = velocity (free falling).
σ = density of particle
ρ = density of fluid
g = gravitational acceleration
η = absolute viscosity
d = Stokes's diameter of particle
R = Reynolds number

in c.g.s. units.

The free-falling velocity for a particle moving through a still fluid is that at which its effective weight is in balance with the drag exerted by its surroundings, so that its rate of movement is steady. The Stokes's diameter is the equivalent free-falling diameter of a sphere moving at such a rate that its Reynolds number is less than 0.2. The Reynolds number is a dimensionless group of parameters which defines the flow pattern of a fluid surrounding a particle

$$R = \frac{v d \rho}{\eta} \tag{8.3}$$

These definitions, together with the following example of a calculation of Stokes's diameter, are taken from B.S.2955:[9]

EXAMPLE OF CALCULATION OF STOKES'S DIAMETER

To calculate the Stokes's diameter of quartz particles (density 26·5 g/ml) corresponding to an observed time of sedimentation of 2 minutes over a distance of 20 cm. in water at 15°C. (absolute viscosity 0·0113 poise).
Free falling velocity = 10 cm. per minute = 0·1667 cm. per second

Stokes's Law (8.2) above, is in this case

$$\therefore\ 0{\cdot}1667 = \frac{d^2}{18} \times \frac{(2{\cdot}65 - 1)\ 981}{0{\cdot}0113} \tag{8.4}$$

$$d^2 = \frac{18 \times 0{\cdot}1667 \times 0{\cdot}0113}{1{\cdot}65 \times 981} = 2094 \times 10^{-8} \tag{8.5}$$

$$d = 45{\cdot}76 \times 10^{-4}\,\text{cm} = 45{\cdot}76 \text{ microns.} \tag{8.6}$$

NOTE. For explanation of symbols see Stokes's Law. (Definition 328)

To check whether the fluid flow round the particle is of the streamline type and the motion within the range of Stokes's Law, the Reynolds Number is calculated as follows:

Reynolds Number (8.3) is in this case

$$R = \frac{0{\cdot}1667 \times 45{\cdot}8\ 10^{-4} \times 1}{0{\cdot}0113} = 0{\cdot}0675 \qquad (8.7)$$

As this is less than 0·2, the motion is stream-line and the above calculations are valid.

A sphere falling through water moves in accordance with Stokes's Law if its mass is insufficient to accelerate it to a steady rate of fall at which the surrounding water is agitated to incipient turbulence or eddying. A change begins when true viscous or laminar flow is exceeded (in the Allen zone), at a diameter for quartz of about 100μ, or correspondingly less for heavier minerals.

Sir Isaac Newton enunciated a law which can be applied to particles which have sufficient mass to make them fall so fast that they leave eddies and turbulence in their wake.

It is not borne out experimentally unless modified by the use of a coefficient Q, which is approximately 0·4 for spheres 0·2 cm. in diameter falling through water.

$$= v \frac{8}{3Q} . g \frac{\sigma - \rho}{\rho} r \qquad (8.8)$$

In the intermediate zone between Stokesian and Newtonian progress through a liquid there is a steady change from laminar flow to turbulent flow. Photographic study has shown this to commence with the setting-up of "anchored" tail eddies which collapse behind the particle or sphere. These increase as the rate of fall is accelerated until drifting eddies separate themselves in the wake of the sphere and dissipate turbulently the kinetic energy which it has transferred to them. Thus it is seen that in each case the fall of the sphere is slowed down by the transfer of part of its kinetic energy to its surroundings. This transfer is orderly and quiet at slow falling rates and can be calculated with high precision, but as the particle gathers speed, shocks and buffetings are set up through the liquid. The speed of the sphere determines the manner of the transfer of kinetic energy to the fluid and the quantity which is transferred. As the speed rises, the rate of dissipation of kinetic energy also increases.

This intermediate state has been given empiric form in the Allen equation, in terms of the resistance of the fluid (Rf) to the movement of the particle.

$$Rf = Kr^{n} \rho \mu^{2-n} v^{n} \qquad (8.9)$$

where K is a constant modified by the shape and velocity of the particle, n a coefficient of velocity and μ the kinetic viscosity b/η, η being the absolute viscosity. When conversion of potential energy of the falling sphere is balanced by its transfer to the surrounding liquid in the form of kinetic energy, the maximum falling speed has been attained. Under Stokesian conditions of fall single layers of molecules of liquid are affected. With Newtonian fall masses of molecules are set in motion as waves and eddies. There is complex rebounding between such masses and collapse of the vortexes. The Q factor is consequently arbitrary.

In mineral processing the conditions are far removed from those postulated for the perfect expression of these laws. Particles are not spherical, but of random shapes. A turbulent mass is in motion. Its component particles vary in their densities. They collide with each other and with the walls, bends and mechanisms immersed in the vessels and pipes through which they flow. The transporting liquid is not straight water of S.G. 1·0 but a more or less opaque pulp carrying finely suspended solids which raise its density and *pseudo-viscosity*. This mixture is accelerated or decelerated as the cross-section of its flow path changes, and it may receive kinetic energy from the external devices at work on it as it flows. Pseudo-viscous flow is typical of the relatively quiet movement used in the froth-flotation process and in dense-media separation. Turbulent flow is the rule in gravitational methods of concentration which depend on fast movement of heavy particles through a relatively light fluid. The essential difference between the formulas of Stokes and Newton in their practical significance is that with the former, falling velocity varies as the square of the particle's diameter, (due to braking proportional to area), while in the latter it varies as the square root (in proportion to mass). Most gravity concentration is performed in the interzone between the full application of these two laws.

Quicksands, Teeters, and Slurries

When a single particle falls through still water, it displaces its own volume, thus creating a rising current. The effect is indefinitely small. If it falls from a relatively warm upper layer it transports heat and increases any convection at work. When a continuous stream of particles falls through still water a definite but minute rising current is set up. (For the process to continue it would be necessary to withdraw the settled solids and water would be carried out at the same time, thus more than offsetting this rise). Each falling particle presses upon the water below it, and generates a current at the cost of part of its own kinetic energy, which is thus transferred to the surrounding fluid. To this loss of energy must be added that used in disrupting the bonding of the liquid molecules through which it cleaves its passage.

Consider next a vertical cylinder of uniform horizontal cross-section through which a current of water is rising. Ignoring the effects of convection and wall friction, the water rises everywhere at an even rate. (In mineral processing vertical currents are called "hydraulic water", and are not necessarily steady.) If a particle is introduced horizontally into the body of the liquid it is carried up and out provided its falling rate through still water is less than the rising rate of the hydraulic current. If the rates are equal it will hover. If its mass causes it to accelerate still more it will drift downward.

The energising forces which determine the direction in which it moves are contributed in part by the particle's mass and in part by the hydraulic water. Each makes its contribution to the frictional heat generated as the water rubs against the surface. The amount of energy thus transferred from the hydraulic current to the particle is minute, though the statistical general effect where a multitude of particles is concerned is shown as a back pressure on the movement of the energising source. The effect on the movement of a single drifting particle is observable. Since both water and particle contribute to the resulting friction, both lose kinetic energy.

Consider next a stream of particles falling steadily under the above conditions. The broad general picture at any cross-section through a vertical tube through which coarse grains are dropping turbulently, either under still conditions or against a rising current of water, is this. The total area at a given cross-section is composed partly of liquid and partly of solid. The greater the proportion of solid, the more restricted are the channels between the particles composing that solid. Hence, the greater the percentage of solid, the faster is the rising current because it has less area through which to rise. At the same time, the greater the proportion of solid the heavier is the displacement and the greater the volume of water seeking to escape, a factor increasing the back-pressure below the falling particle. The other important component of this back-pressure is the loss of hydraulic energy due to friction, or "loss of head". From thermodynamic considerations the loss of head should ideally be equated with gain in heat, but the practical engineer is more likely to observe it in terms of wear on metal work or abrasive degradation as the particles thrash their way along the flow line.

The above conditions are true for a moment at a cross-section. Unfortunately, there are further complications in what is already an involved state of affairs. All the particles are attracted downward by gravitational forces, but not all the particles have sufficient mass to enable them to travel against the rising current of water. Thus at any moment the cross-section contains falling particles, rising particles, and particles which are more or less in balance. The overall effect is the sorting of all these particles into well-defined zones. At the bottom are the ones with the largest supply of retained kinetic energy. These are the particles of high specific gravity and the relatively large particles of lower specific gravity. Above these there is a steady transition in accordance with the downward pull of gravity and the "arresting" or braking force of a particle's surface friction against the hydraulic stream.

Assume that the whole tube containing these particles is carrying a steady rising current of water, introduced above a still zone. The particles which can spend the necessary kinetic energy in overcoming the resistance to fall offered by the contents of the tube drop to the still zone and leave the system. (See Fig. 65.) At the overflow end the smallest ones are flushed away. Those neither quite heavy or light enough to persist downward or upward are held above the entry area of the hydraulic water. Those that drift downward fall into ever denser crowding conditions. They consequently are subjected to an increasingly severe rising current, augmented by loss of kinetic energy through the rubbing together of the surfaces of the particles. This arrangement is called a "teeter bed" from the behaviour of the particles in it, which dance or "teeter" a little up or down as they alternately acquire and lose kinetic energy *vis-à-vis* the surrounding fluid and particles, and in spinning vary the cross-section which they present to the rising stream. At the top of this teeter bed the lighter particles dance and occasionally break away upward, rise a short distance into the unobstructed full bore of the tube, and then sink back because the rising velocity is no longer sufficient to bear them upward.

The density of the teeter column depends on the proportion of liquid and solid of which it is composed. "Density", when considering a teeter, is a word needing some qualification. Though the words are often considered interchangeable "teeter" and "quicksand" are not quite similar states. Electrical dispersion works in the latter, so that the quicksand density which would seem to be available for support of any object is not a true guide. Only the liquid component is fully available to resist penetration of the quicksand by a heavier body. A man can be engulfed though his S.G. is far below that of the solid/liquid mixture in which he is trapped. The appellation "quick" is also misleading as the particles of sand, though quick to move apart when pressed on by an intruding body, are not otherwise particularly mobile. A teetering column of dancing particles is different. It corporately bombards a larger body pressing down upon it and acts with the full value of its solid/liquid density *plus* that of the kinetic energy of the particles in collision with the body. A man, though lower in density than a mixture of sand and water, can be engulfed in a quicksand. If he throws himself flat and so spreads the area of contact he can usually scuffle back to safety, though the experiment is not recommended. Similarly, where the teeter is partly composed of coarse grains, it can be penetrated by an object cleaving its way in, because the teeter opens and reacts at the density of its fluid component, water. When however, the same object is presented over a large area, the teeter bed reacts homogeneously and supports it. The qualifying consideration in this is that there must not be much silt or very fine sand, since teetering columns of the type under consideration are being dilated by rising water, which bears fine particles up and out of the system. A quicksand may contain silt and an electrolytic fluid such as brackish water. It is not at its *iso-electric point.*

The density of a teeter, under conditions where the solid fraction is small enough in particle size to act homogeneously but not so small as to act viscously, is given by the equation:

$$\text{Effective Density} = \frac{(\%\ \text{solid} \times \text{S.G. solid}) + (\%\ \text{liquid} \times \text{s.g. liquid})}{100} \tag{8.10}$$

If homogeneous particles close-ranged in size form the teeter bed, then the behaviour of any mineral presented to it can be calculated.

This discussion is limited to the consideration of a batch of particles charged into an elutriating column. In plant practice new material arrives continuously and part is trapped in the teeter column. The added complication thus caused is considered in Chapter 9.

Thus far conditions have been considered for coarse particles which require rising currents sufficiently strong to maintain a state of turbulence in the suspension, although this turbulence is damped down by the crowding and colliding of the solids present. When the particles are much smaller, turbulence is no longer needed for the maintenance of a reasonably open fluid condition, or "pulp". A thin solid-liquid mixture acting as a fairly stable fluid (for a short enough time to allow settling to be ignored) is a pulp. A thicker mixture, with more solids than liquid, and flowing viscously, is termed a "slurry". The vast increase in the total surface per unit volume of finely ground material, as compared with that for a similar ratio of coarse particles and water, so increases friction that the solids behave as an integral part of the fluid for minutes at a time, and even for hours if these solid particles are very small, even when no hydraulic or other stirring forces are at work on them.

It will be seen that there is a great difference between the behaviour of teeter beds, which require agitation to prevent their settlement into compact layers, and slurries, which have high pseudo-viscosity and a slow-settling character. A teeter, or a coarse sand, tries to settle out at any quiet point in the flow-line while a slurry flows steadily and maintains its fluidity even in a moderately quiet zone. A teeter discriminates as to whether it will accept a particle of lower density than those of which it itself is composed on the basis of the mass of that particle. A slurry acts more homogeneously because of its higher pseudo-viscosity. It accepts a particle denser than its own effective density, provided that particle can bring into the system enough kinetic energy to overcome the frictional resistance of the slurry, but it will not permit entry of *any* falling particle which is below it in effective density.

If a teeter collapses (as when rising water ceases to flow), it settles down and forms a tightly packed sand. Even a large and sharp object of comparatively high specific gravity now fails to penetrate it. Thus the slurry acts mildly like a mudbank in which a boat can stick tenaciously, and the collapsed teeter as a sandbank on which the same boat rests without appreciable penetration.

The physical laws which govern the manipulation of pulps and slurries are considered in connexion with industrial sedimentation, thickening and dewatering in the next chapter.

Measurement with the Microscope

Direct observation and measurement can be made down to less than one micron (1μ) by use of a metallurgical microscope. For measurement of particles, a linear scale engraved on glass is placed in the eyepiece (ocular) and the instrument is then focussed on a stage micrometer. This is a transparent engraved scale which shows either squares or other figures of known dimensions, or better, one millimetre divided linearly into tenths and hundredths (1000μ, 100μ, and 10μ). It is placed on the microscope stage and the two scales can then be correlated for the ocular-objective system in use. The stage micrometer is then removed and the particles are examined and measured. If they have been placed loose on the glass slide, they will settle in their most stable position and the two major dimensions are presented. For routine size evaluation special graticules such as the Patterson-Cawood[4] are useful. The use of a drop or two of a dilute solution of Calgon helps to disperse the material on the counting slide and to display the smaller particles which might otherwise be masked by the big ones. Where a tedious amount of particle-counting is involved various aids are available. A projection microscope is less tiring to work at than a stand machine. Particles dispersed in water can be made to flow through a small aperture and work a counting mechanism as they interrupt a beam of light. The Zeiss particle size analyser is semi-automatic.

"Liberation-Mesh"

Ore consists of at least two minerals intimately inter-attached. Before one mineral species in the ore can be segregated by mineral processing, its particles must be adequately detached from the other species present by suitably controlled comminution.

Probably the most important work in connexion with laboratory sizing

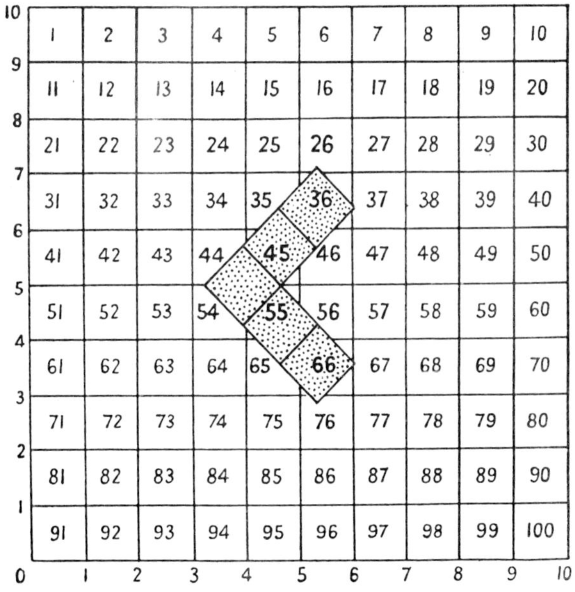

1	2	3	4	5	6	7	8	9	10
11	12	13	14	15	16	17	18	19	20
21	22	23	24	25	26	27	28	29	30
31	32	33	34	35	36	37	38	39	40
41	42	43	44	45	46	47	48	49	50
51	52	53	54	55	56	57	58	59	60
61	62	63	64	65	66	67	68	69	70
71	72	73	74	75	76	77	78	79	80
81	82	83	84	85	86	87	88	89	90
91	92	93	94	95	96	97	98	99	100

Fig. 69 Phase Liberation

control has to do with liberation-mesh and the mesh of grind at which optimum unlocking of the desired mineral from those accompanying it in the ore is reached. Thus far this chapter has discussed methods of sorting the particles either by size or by settling characteristic. For some specific purpose such a grading may be an end in itself, but comminution control is more usually concerned with economic liberation of values from the crude ore.

Liberation is not achieved by grinding the ore down to the grain size of the desired mineral particles. If Fig. 69 is taken as representing a stylised cross-section through a cube of ore in which the values are 5% and the gangue 95%, then it will be observed that if the rectangle shown is cleft into 100 square subdivisions, 87 represent clean gangue (i.e. completely valueless mineral) and 13 partly liberated material in which gangue and value are still tightly bound. Though in Fig. 69 the value is 5% and the L-shaped flake should theoretically be completely liberated from gangue with a 20 to 1 reduction by comminution, the diagram represents a 100 to 1 crushing. The incompletely separated squares or "middlings" (Nos. 26, 35, 36, 37, 44, 45, 46, 54, 55, 56, 65, 66, and 76) contain varying amounts of value from the tiny fleck in 37 to the high loading of 55. Despite the theoretical heavy overgrind (100/1) instead of 20/1 no value has been completely liberated.

If the grain sizes of an actual ore specimen had been measured, they would not correspond with the grinding treatment necessary for adequate liberation of any specific mineral, as Fig. 69 shows. "Middling" is the name given to a particle intermediate in composition between one of clean concentrate and one of clean tailing. Usually it contains fractions of both minerals and can only be treated after further grinding has liberated them. A particle which contains firmly joined value and gangue is called a "locked middling". It can only be separated into its constituent minerals by further comminution. A variant is the middling composed largely of a third mineral species not sought as a concentrate, but intermediate in its response to a given method of treatment. The position in Fig. 69 is artificial, but the principle illustrated by it is very important. Grinding commences by liberating, or freeing, part of the major constituent at a relatively coarse *mesh-of-grind* (usually abbreviated in our literature to m.o.g.). This may be above the maximum particle size of the valuable constituent if the latter is the minor mineral present. Even when grinding has brought all the ore down to the grain size of the minor constituent, it is unlikely that any of this mineral will have been cleanly liberated, though a good quantity of the major one is now ready for separation. Grinding must be continued well below the measurable liberation-size of the minor constituent before any substantial quantity of it is freed.

Consider, for example, particle 44 in Fig. 69. If a further 4 to 1 size reduction is made squarely one quarter-sized new particle of gangue will be freed, two "locked middlings" carrying but little value, and one high-grade middling. Yet, no completely gangue-free value has been liberated as the result of this considerable (and costly) further grinding.

Occasionally the ore may have a weak boundary between gangue and value, along which breakage occurs preferentially. In such a case, liberation may be satisfactory at mineral-grain size. Usually the adherence of the minerals in the crystalline matrix of the ore is strong, and during crushing and grinding the various constituents are cleft across. This leaves the resulting particles in the form of locked middlings, composed partly of gangue and partly of value, with the more abundant phase (usually the gangue) far freer than the scarcer one.

Loosely bonded gravels are easily disintegrated. A pebbly phosphate, or a clay carrying nodules of manganese or iron, can be broken up to release the value with but little comminution. Fracture planes (cleats) in coal and schistose weaknesses in hard rock can be preferentially opened up by gentle comminution appropriate to the material. If there is marked difference in the hardness or toughness of the constituents of the ore, some differential grinding may be possible using forces sufficient to break up the more friable mineral while leaving the tougher one unaffected. In the case of coarse sands or gravels which have become encrusted with some mineral of special value, this can sometimes be washed off by tumbling treatment in log washers or removed by gentle grinding action. These are special cases. Normally, the ore must be ground down to an economic liberation point determined by test work. Sizing control provides a valuable means of isolating the various mesh-grinds for such testing.

It was noted above that the more abundant gangue is freed at a far coarser mesh than the values. In order to avoid the cost of grinding all the ore to the mesh where the "values" are fully liberated, it may be possible to arrange for primary coarse grinding to be followed by a stage of treatment in which some of the free gangue is removed. This reduces the quantity of ore needing finer grinding and thus cuts down the grinding cost. If tests show that the cost of such a treatment is lower than the cost of grinding all the ore to a fine mesh, then this fact is considered during the construction of the flow-sheet and the planning of the treatment plant. Again, primary crushing may develop enough difference between the clean gangue and the partly liberated value for the latter to be concentrated into a small bulk. This is a form of stage concentration used in dense-media separation. The best method of solving the problem of optimum economic and technical liberation depends less on the facilities at hand than on the nature of the ore and the way in which its values are distributed through the gangue. For convenience, consideration has been confined in this discussion to an idealised squared area through a simple ore with only two components or phases. In mineral dressing, a series of valuable products must frequently be separated and the problems of liberation are then correspondingly complicated.

It is now clear that the crystal size of the desired mineral in the ore is not the complete guide to the liberation-mesh or m.o.g. (mesh-of-grind) needed for its concentration. This is arrived at by empiric laboratory testing, described later. The optimum degree of liberation of the value is called the "break".

References

1. Dallavalle, J. M. (1948). *Micromeritics*, Pitman.
2. Herdan, G. (1953). *Small Particle Activities*, Elsevier.
3. Cohen, E. (1957). *I.M.P.C.* (*Stockholm*), Almqvist & Wiksell.
4. Heywood, H. (1946). *Trans. I.M.M.* (*Lond.*), 55.

References—continued

5. Taggart, A. F. (1945). *Handbook of Mineral Dressing,* Chapman & Hall.
6. Heywood, H., and Pryor, E. J. (1946). *Trans. I.M.M.* (*Lond.*), 55.
7. British Standards Inst. (1958). B.S.2955.
8. Schuhmann, R. (Jnr.). *A.I.M.M.E.* (T.P. 1189).
9. Austin, J. B. (1939). *Indust. Eng. Chem.* (*Anal.*), 11.
10. Gaudin, A. M. (1926). *Trans. A.I.M.M.E.,* 73.
11. Rosin, P., and Rammler, E. (1953). *J. Inst. Fuel,* 7.
12. Pryor, E. J., Blyth, H. N., and Eldridge, A. (1953). *Recent Developments in Mineral Dressing,* I.M.M.
13. Kelsall, D. F. (1962). "R & D", Nat. Trade Press (June).
14. U. S. Bureau of Mines Report (1961). No. 5838.
15. British Standards Inst. (1963). Part 3, B.S.3406.
16. Carman, P. C. *Trans. Inst. Chem. Eng.,* 15.
17. Lea, F. M., and Nurse, B. W. *J. Soc. Chem. Ind.,* 58.
18. Gooden, E. L., and Smith, S. M. *Indust. Eng. Chem.* (*Anal.*), 12.
19. Svensson, J. Royal Inst. Tech., Stockholm.

CHAPTER 9

INDUSTRIAL SIZING AND SORTING

Preliminary

Sizing, as performed on screens, grades material according to the minimum cross-section presented during the time of passage across the meshes of the screen cloth. The regularity of the industrial product is dimensional, and takes no account of differences between the weights of the particles in a given grade. "Sorting", or as it is more usually called classification, discriminates between the behaviour of particles in a fluid and grades them according to their surface, volume, and density. Since ores contain particles of varying densities, this is not a sizing operation. The fluid mostly used is water, though high-density salt solutions are employed for special purposes. Material required in a dry state may be sorted by floating it in air currents of controlled strength. Screening is only used for comparatively coarse material, as the rate of treating large quantities of ore becomes slower when fine-meshed screens are used. Wet screening is practised commercially down to 65 mesh, but industrial dry screening is rarely carried below 20 mesh. Classification can be used from coarse-sand sizes down to well below 200 mesh. Fine particles (say —20 mesh) must have fluid mobility if they are to be sorted, and these conditions cannot be contrived with a long-ranged dry feed subjected to screening.

Ore may be screened for any of the following purposes:

(*a*) To retain oversize in a given section or circuit, and thus prevent it from being fed to a machine not suitable for dealing with it.
(*b*) To remove undersize from the feed to a crushing machine set to treat bigger lumps.
(*c*) To grade rock into specified sizes.
(*d*) To present a correctly sized feed to a concentrating process.

Classification is used to:

(*e*) Separate ore into relatively coarse and fine fractions by exploiting differences in settling rates.
(*f*) Split a long-ranged feed into fractions settling equally.
(*g*) Close grinding mill circuits so that no particles escape from them into the concentrating section of the plant until they have been reduced to the desired sizes.
(*h*) Remove or segregate slimes.
(*i*) Regulate size-range fed to a process.

Action on the Screen

The purpose in screening is to hold as oversize all particles too large in their minimum cross-section to pass through the apertures, and to let all smaller ones drop through as undersize. The separated products are then sent to their next processing point by independent routes. The behaviour of a particle upon the screen depends chiefly on the relationships listed in Table 13. Consider first the ratio of its cross-section to that of the meshes (*a*). A relatively small particle has no difficulty in falling through, but the nearer it approaches "retaining mesh", the more difficulty it has in hitting an unoccupied mesh centrally with a suitable presentation. Shape plays a part, the fairly equidimensional particle having a better chance than the acicular or tabular one, unless the screens have been chosen to assist such shapes.

TABLE 13

(*a*) Ratio between cross-section of particle and of mesh.
(*b*) Percentage of screen area open.
(*c*) Angle of incidence of feed.
(*d*) Efficiency of spread of feed over screen area.
(*e*) Kinetic energy of particle approaching screen opening.
(*f*) Moisture of ore feed.
(*g*) Stickiness of particle and of aggregated particles.
(*h*) Pressure of particles riding above those next the screen cloth.
(*i*) "Blinding" of screen apertures.
(*j*) Corrosion of screen material.
(*k*) Electrostatic "bunching".
(*l*) Shape of particle.
(*m*) Percentage of "near-mesh" particles in the feed.
(*n*) Rate of feed, thickness of layer, tautness of screen.
(*o*) Shape of screen apertures.
(*p*) Motion imparted to particle by screen vibration.

Square-meshed cloth does its best work when set horizontal, but the shaking movement imparted to it must then have a forward transporting component. Oblong meshes are often used for feed which is moist or clayey, and for needle-like particles. If the feed tends to "blind" the cloth, the long axis of the rectangle is set in the direction of flow. Square mesh is usual with tabular material. The percentage of screen area open (*b*) depends on weave, diameter of screen wires, and shape of aperture. For a given mesh various ratios of opening are available. (*c*) is concerned with the mode of arrival on the screen. Ideally, a particle would fall with its minimum cross-section normal to the aperture and at negligible velocity. In practice, it competes with a crowd of other particles of random shape and size, falling along various trajectories. Hence item (*d*) is most important, since the wider the entering feed is spread, the easier will it be for a particle to find unobstructed passage to the screening surface. Since this is so (*e*) should be low. A particle flying nearly horizontally toward the screen is most likely to hit the layer of feed with its broadest dimension and to slide down on top, with no early chance of burrowing its way down. (*f*) and (*g*) vary with climate and working conditions and to some extent can be compensated by using slightly larger screens in the monsoon season. (*h*) can aid the work where the screen cloth has little or no vertical motion, but when the action is designed to "dance" the feed along, it exerts what is on the whole an undesirable damping pressure. (*i*) arises partly from near-mesh material and is worst in closed-circuit work (*m*) where the return particles tend to be close to "release-mesh" and so to be retained. Such material prevents the passage of true undersize and cuts down the capacity of the screen, so that it is usually found desirable to provide more generously for screening in closed circuit than in grading open-circuit ore, or to use a screen aperture greater than the set of the crusher. (*j*) introduces roughness along the wires, and greatly increases the proneness to "blind" and to resist clearance of the blinded apertures. Corrosive pitting is lessened when the ore can be neutralised with lime, or if a resistant alloy is used for the screen wires. (*n*) affects the resistance the upper particles must overcome in penetrating the feed bed, while the tautness of the cloth decides how vigorously and widely the shaking and tossing action of the agitating mechanism is transmitted to this bed. Finally (*p*) can be modified on some mechanical screens to allow any variation in the vibratory orbit of a free particle from counter-current to concurrent. With electrically vibrated screens the upward movement of the cloth, as it vibrates normal to the direction of flow, can be terminated by abrupt arrest, thus "unblinding" the cloth at each stroke. This action is not possible on mechanically shaken screens, which can, however, be assisted in keeping open by the slapping action of rubber cords stretched below.

Types of Separating Surface

The grizzly (Fig. 17) was described in Chapter 3. Robustly built screens (Fig. 70) are also available. The one shown can handle up to 1000 tons of large rock hourly with far less loss of head than would be possible on the static surface of a fixed grizzly. The machine must stand up to rough treatment and the impact of heavy rock.

Punched screen is used for many purposes (Fig. 71). A variety of shapes is manufactured, circular openings being recommended for coarse work and slotted ones for fine.

Woven-wire cloth is widely used in the range between $\frac{3}{4}''$ and 200 mesh. Various shapes of aperture, crimps and weaves are available (Fig. 72). Steel, stainless steel, monel metal, copper, and bronze are the chief metals from which cloths are made. For delicate work at fine meshes, dry material is sometimes sieved through silk or nylon. For special purposes very fine square-meshed screens are made by electro-forming instead of weaving. They are sold in 100′ lengths three feet wide, with hole diameters in the range 25μ down to $2\frac{1}{2}\mu$. The screen surface is nickel, on a copper base. Conventionally woven materials of construction include mild steel, brass, phosphor-bronze, copper, copper-nickel, nickel-chrome, austenitic stainless steel, galvanised and tinned mild steel. In milling, steel screen wire should not be used to treat acidic and corrosive ores, unless it is first protected by being

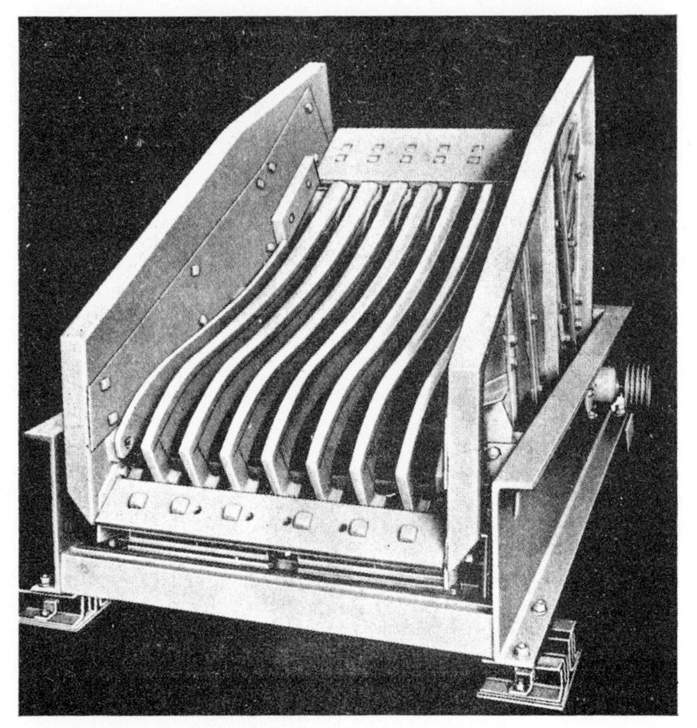

Fig. 70. Vibrating Bar Grizzly (Nordberg Manufacturing Co.)

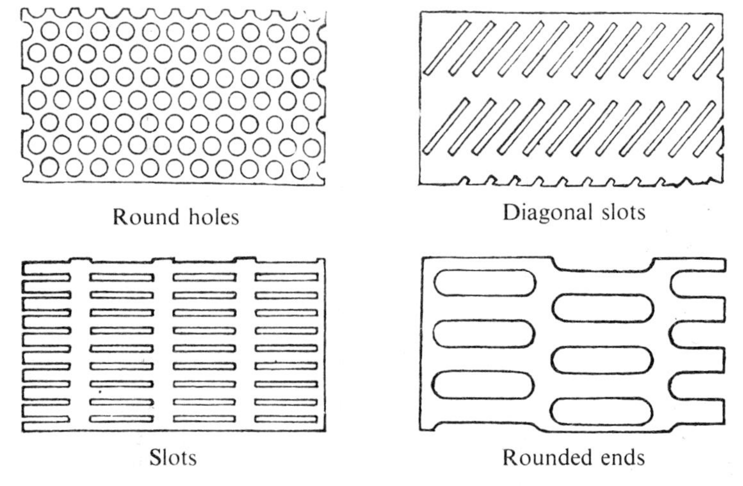

Fig. 71. Types of openings, Punched Screens

given a suitable plastic coating. Where corrosion is not serious, high carbon steel is suitable, being strong and hard wearing. Maximum capacity of screens with oblong apertures is obtained by using them with the long side of the mesh set across the flow.

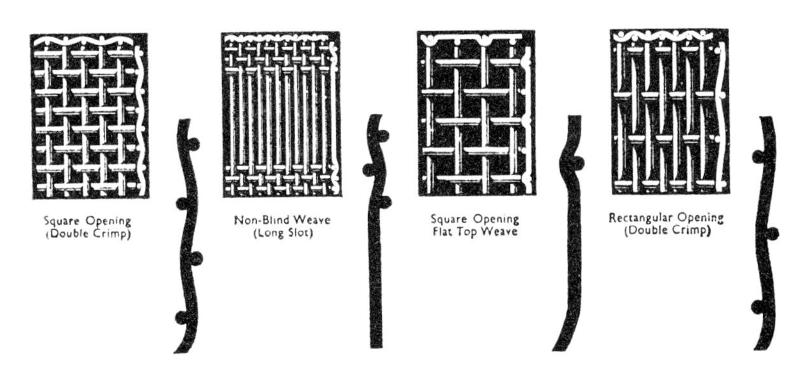

Fig. 72. Types of Woven Screens

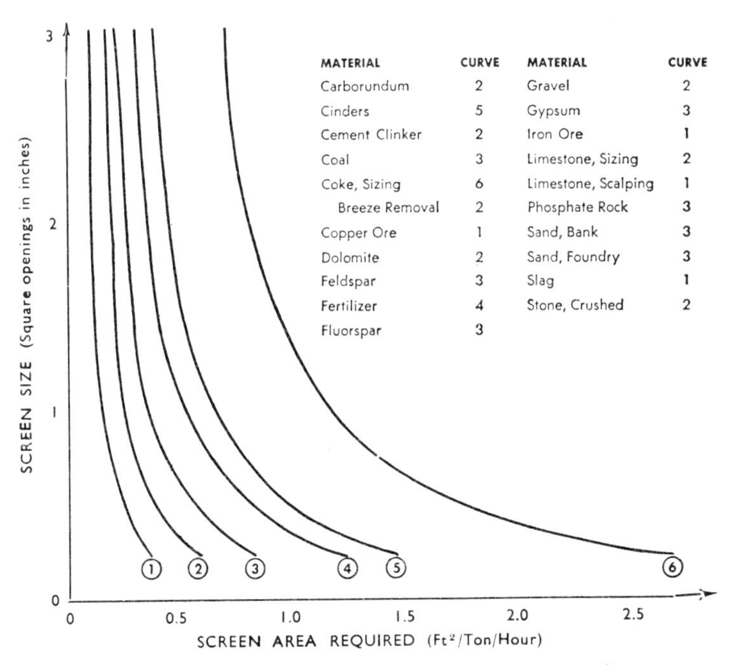

Fig. 73. Screen Capacity. (Denver Equipment Co.)

In Table 14 the recommended wire diameters for screens with openings from ⅛″ to 4″ are given, together with percentage open area. The screen area required for a given feed tonnage can be estimated by use of the graph in Fig. 73. The appropriate curve is selected and the screen area (sq. ft. per ton per hour) is read for the required linear opening (ordinate) or square-meshed screen-cloth. The area figure (abscissae) is then multiplied by the tons per hour to be screened. The curves are representative where the feed contains less than 65% of oversize and where about 50% of the undersize is about half the screen-size opening.

TABLE 14

STANDARD COARSE SCREEN SPECIFICATIONS

Recommended by the Division of Simplified Practices, U.S. Department of Commerce, for screening of mineral aggregates

Opening	*Wire Diameters*							
	Medium Light		*Medium*		*Medium Heavy*		*Heavy*	
	Wire Diameter Inch	*% Open Area*	*Wire Diameter Inch*	*% Open Area*	*Wire Diameter Inch*	*% Open Area*	*Wire Diameter Inch*	*% Open Area*
4″	·500	79·0	·625	74·8	·750	70·9	1·000	64·0
3½″	·4375	79·0	·500	76·6	·625	72·0	·750	67·8
3″	·4375	76·2	·500	73·5	·625	68·5	·750	64·0
2¾″	·375	77·4	·4375	74·4	·500	71·6	·625	66·4
2½″	·375	75·6	·4375	72·4	·500	69·4	·625	64·0
2¼″	·375	73·4	·4375	70·1	·500	66·9	·625	61·2
2″	·3125	74·8	·375	70·9	·4375	67·3	·500	64·0
1¾″	·3125	71·9	·375	67·8	·4375	64·0	·500	60·5
1½″	·250	73·4	·3125	68·5	·375	64·0	·4375	59·9
1⅜″	·250	71·5	·3125	66·5	·375	61·6	·4375	57·5
1¼″	·250	69·4	·3125	64·0	·375	59·2	·4375	54·8
1⅛″	·225	69·6	·250	67·0	·3125	61·0	·375	55·7
1″	·225	66·6	·250	64·0	·3125	58·0	·375	52·9
⅞″	·207	65·3	·225	63·3	·250	60·5	·3125	54·3
¾″	·192	63·4	·207	61·4	·250	56·3	·3125	49·8
⅝″	·177	60·7	·192	58·5	·225	54·0	·250	51·0
½″	·162	57·1	·177	54·5	·192	52·2	·207	49·8
7/16″	·148	55·8	·162	53·2	·177	50·7	·192	48·3
⅜″	·135	54·1	·148	51·4	·162	48·7	·177	46·1
5/16″	·120	52·2	·135	48·8	·148	46·0	·162	43·4
¼″	·105	49·6	·120	45·6	·135	42·2	·148	39·4
3/16″	·080	49·1	·092	45·1	·120	37·2	·135	33·8
⅛″	·054	48·7	·072	40·2	·092	33·4	·105	29·5

Heavy wire recommended for trommels.
Medium heavy for high speed vibrating and shaking screens.
Medium light and medium for other vibrating screens.

In addition to the foregoing, rod-deck screens are used for coarse work. The rods are sprung into place and can be changed individually. Wedge-wire screening is employed for some purposes. It is strong and can be made with small apertures. The blunt sides of the wedge strips which form the separating surface are upward, so that material passing falls clear without "blinding". Many other wire shapes are obtainable.

When a long range of sizes is being fed to a comparatively weak or light screen, a robust coarse screen should be mounted above it to form a double-decked system in which the heavier pieces do not reach the second deck. A delicate screen can be supported by a coarser backing screen beneath. The apertures in punched-plate screening may be round, square, rectangular, or oblong with rounded ends, the last-named being less prone to blind than holes completely circular. These screens are made of steel, steel alloy, brass, monel, copper, or bronze. If openings are disposed in an equilateral triangular pattern, the maximum ratio of opening to total surface is obtained. The stouter the plate, the closer can be the openings and the longer the service life. Against this must be set the fact that the thicker the plate the greater the proneness to blind, and the higher the initial cost. This blinding with increased thickness is still more noticeable with woven wire.

The selected aperture depends on the working requirement, while constructional strength depends on the nature, size, and loading of the feed. The ratio of aperture to screen area is not, therefore, a function of the mesh in commercial screens. For instance, an 8-mesh screen woven from thin wire can have 24% more screening area than a 6-mesh weave in thick wire. Hence, the opening required dominates specification as regards size of product, but the total amount of screen capacity available at any point in the flow-sheet depends partly upon the thickness of weave of the selected mesh. Screens undergo such rough treatment that their working life may be measable in hours in extreme cases. Flow of ore must be interrupted while a broken cloth is being replaced. This consideration affects the choice as regards robustness of wires.

Screening Machinery

Since separation must be made of all sorts of feeds, varying in condition from the completely dry to the sticky, clayey, or "porridgy", each ore presents its own screening problems. Capacity, efficiency, mesh size, and wear rate are relative to the specific ore and should dictate the choice of appliance. The flat grizzly is used to retain large oversize which might cause trouble if allowed to continue to the next point in the flow line. It is sturdy enough to act as an anvil if such oversize is to be sledged down by hand, or can operate in conjunction with any required breaking arrangement.

The inclined grizzly is sloped between 25° and 39° if it delivers its oversize to a crushing machine, the gradient being chosen to allow sliding control by means of hand or chain feeders. If the oversize is to run to the crusher free of control, a slope well in excess of 35° from the horizontal is usual. For dry quartz 45° should suffice, while sticky or moist rock might need 50° or more. Grizzlys are simple and strong, but are wasteful of headroom.

The general types of separating device are presented schematically thus:

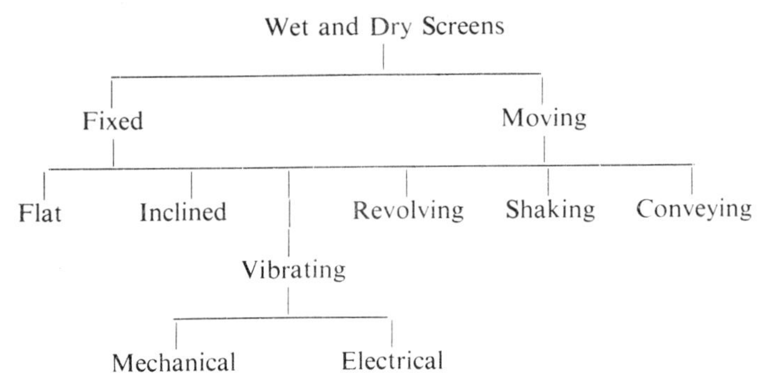

The roll grizzly consists of a series of grooved rollers driven by sprocket and chain unidirectionally in a supporting frame, the speed increasing from entry end to discharge roll. The undersize drops between the grooves while oversize moves flatly along. With this arrangement loss of head is minimised, but power is needed for the roll drive.

Grizzlys can be vibrated mechanically, electrically, or by the impact of falling rock. They may also be shaken, or alternate bars may be moved by eccentrics. They can be used as sorting tables or to control the rate of feed. The mechanically vibrated bar grizzly illustrated in Fig. 70 is designed to separate ore at 1½″ bar spacing or more, and handles up to 1000 tons/hour.

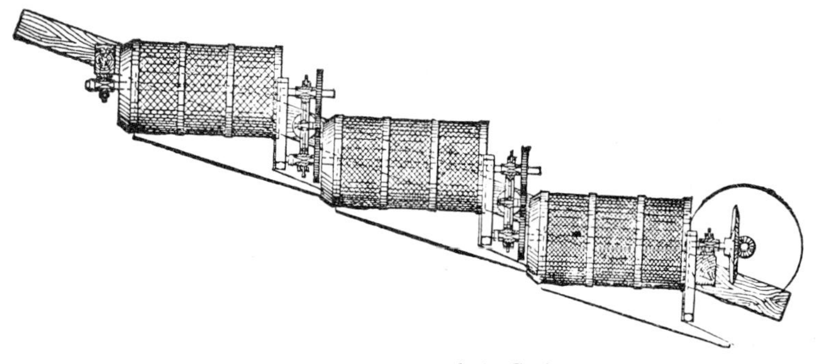

Fig. 74. *Trommels in Series*

The curved bars tend to tumble the feed, and the vibrating action, powered by a 7½ h.p. motor, is effective on wet and sticky ore. The jar-bar feeder grizzly is illustrated in Chapter 3.

One of the oldest screening devices is the trommel, which can be used wet or dry. Its usual form is cylindrical, with the screening plates forming the side walls, and a downward slope from feed to discharge end. Trommels may be arranged in series (Fig. 74), with the coarsest discharge at the start, in order to remove heavy oversize at the earliest point and by means of the most robust screenplates. When "washing trommels" are used, the feed is picked up by internal lifters made of angle-iron, and is sprayed. The water runs out with the undersize and slime. A variation of the series of diminishing-mesh trommels is the compound trommel with concentric screens, the coarsest inside, and separate discharge launders for the products. The disadvantage of this arrangement is that failure of a screen is hard to observe and difficult to deal with quickly. Other variations include the polygonally-sided trommel which gives positive lifting action to its contents and permits a complete flat screen to be replaced as a unit. To overcome the problem of providing a skewed bevel drive to sloping trommels, the conical trommel, set horizontally and flaring down from feed to discharge, has had some use. Trommels are chiefly used as sizers in gravel plants and stone-breaking work, and on tin and gold dredges where they remove boulders and clay from the gravels brought up by the buckets. In tin dredging a serious source of loss of cassiterite is that due to embedding of the mineral particles in nodules of clay, either because the richest alluvial ore lies directly on clay bottom in the deposit, or through the jumbling of gravel and clay together during digging and discharge from the dredge buckets. "Disintegrating trommels" have been used, with cutting blades rotating inside the revolving cylinder. These meet and slice up clay lump, so that the trapped cassiterite is released and can run through as undersize. Trommels are simple, vibration-free, cheap, strong, and economical of head loss through a series. Against this must be set the fact that they "blind" easily, have poor capacity and cannot be repaired speedily or changed rapidly to a different mesh. Apart from the uses mentioned above, the vibrating screen has replaced the trommel for most ore-dressing purposes.

Shaking screens are usually worked dry, and chiefly in the sorting of coal. A typical arrangement consists of an oblong box of which the screen forms the bottom. This may be hung by chains or links, or supported from beneath. In the latter case the Ferraris motion may be used. The Ferraris truss carries a loaded deck by means of flexible battens set at a calculated angle. When the deck is pressed forward the battens move through a rising arc, lifting the load and throwing it forward. On release, the deck falls backward to its stop-point, the effect being to toss the ore in the air, or at least to reduce its clinging contact with the screen after the forward stroke has imparted to it kinetic energy in the direction of travel. This loosening aids in the stratification of the ore, and leaves the largest particles on top, where they press upon the smaller ones which are trying to work through the meshes of the screen. The upward tossing motion can aid in unblinding (a little), while the jarring arrest as the deck falls back tends to loosen particles wedged in the meshes. The vibration of shaking screens is a disadvantage, and they are today but little used for hard-rock work.

Travelling-belt screens are typified by the Callow screen. In this, usually made duplex, the screen cloth is bound along the selvedge to rubber strip which forms a retaining lip on each side. Ore pulp is fed on, washed through by sprays, and removed as undersize, while oversize is discharged at the far end. They are little used today.

Vibrating Screens

These screens dominate modern sizing practice. They handle dry to moist or sticky material as coarse as 10″ in ring size, and as fine as 65 mesh. In special cases they can work on dry feed down to far finer sizes. For most mineral-dressing operations, screening stops at the point where the crushing section delivers ore to the fine-ore bins for wet milling (say between ¾″ and ¼″ size), though an important tonnage is handled by gravity concentrators after screening down to 20 mesh. Once wet milling has begun, sizing usually gives place to sorting in classifiers though wet screening is being increasingly used on sands.

Vibrating screens can work at low slopes and need but little headroom. Though loss of height as material drops through a machine is not an expensive item, it influences plant layout and the choice of appliances and should be minimised. Other important advantages of vibrating screens are accessibility, easy visual control, crisp transmission of the input power, and (given good design) avoidance of transmission of vibration to the mill structure.

The vibrating screen has one or more decks, usually plane and kept in sprung tension. The screen forms the floor of a box which is vibrated mechanically or electrically. The electrically vibrated screen (Fig. 75) uses an electro-magnetic device (Fig. 76), usually a solenoid arranged to set up a reciprocating motion. This solenoid is activated by alternating current, and a striking block or anvil may be incorporated in the design. The rising motion is communicated to the screen-cloth through a rod. Lift can be made to terminate with a jarring blow adjusted so as to counteract "blinding" of the meshes. In other variations, the electrically induced vibration may be resisted by adjustable springs, thus modifying the severity with which the screen is vibrated. A slight variation in the frequency of the A.C. supply has a magnified effect on the motion.

If the pushing and pulling of the solenoid acts direct *via* transmitting rods upon the tensioned screen-cloth (and this is the usual arrangement), the ore particles dance normally to the surface unless (as in conveying screens shaken in this manner) the movement is applied at an angle. There are no rotating parts, and but little that calls for maintenance. Metallic dust occasionally gathers around the striking anvil and leads to sparking. It is removed by blowing with compressed air. The screens are mostly used on —½″ feed.

The most widely used screens for coarser sizing are mechanically vibrated. The motion impressed upon a particle is not necessarily a simple straight-line one. Anything between this and a circular orbit spinning either counter-current to the feed (giving the particle a tendency to climb back toward the feed end) or concurrent (accelerating its progress toward the discharge end) is in theory feasible. One advertised motion shows a counter-current ellipse at the feed end, a reciprocation normal to the screen at the centre, and a concurrent ellipse toward the discharge. The oncoming feed is thus stated to be checked and searched during stratification, then sent down to an area of screen unusually free from blinding, and finally accelerated off the screen. These variations are produced by balancing the forces producing vibration, the movements of the tensioning springs, the yield of the screen cloth, the

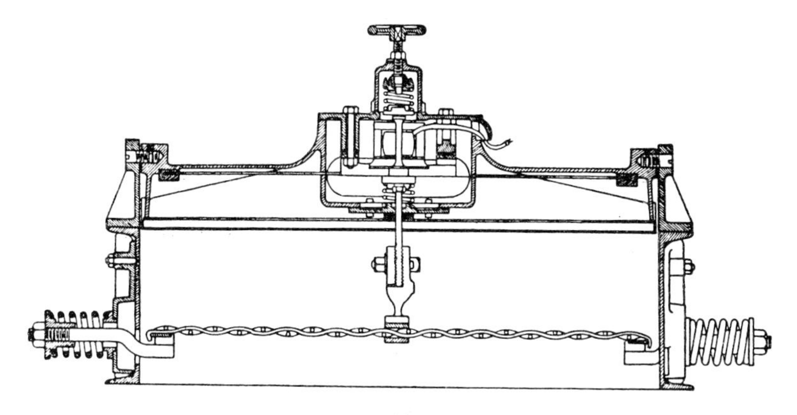

Fig. 75. The Hummer Screen

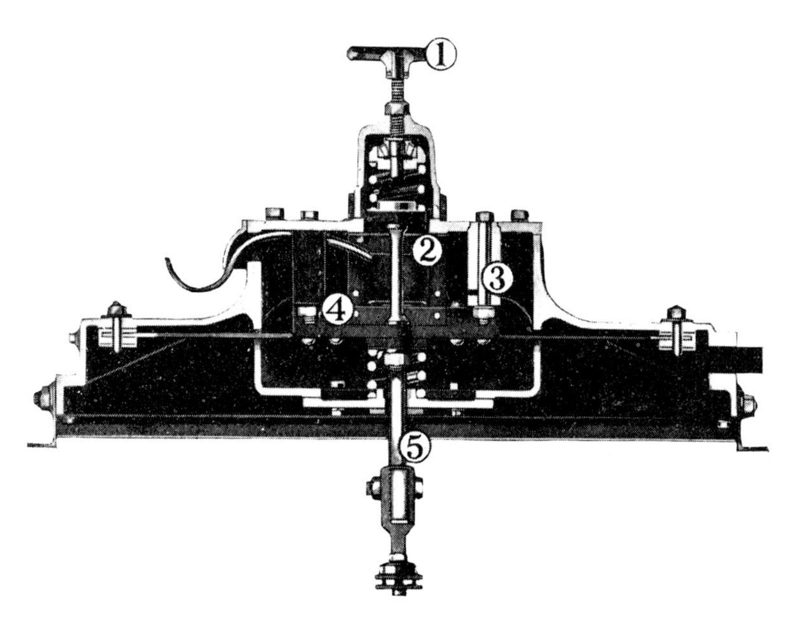

Fig. 76. Cross Section of Hummer V16 Vibrator (International Combustion)

1. Hand wheel for regulating intensity of vibration
2. Coil and magnet
3. Striking block, wearing plate and shims
4. Armature
5. Armature post and bracket

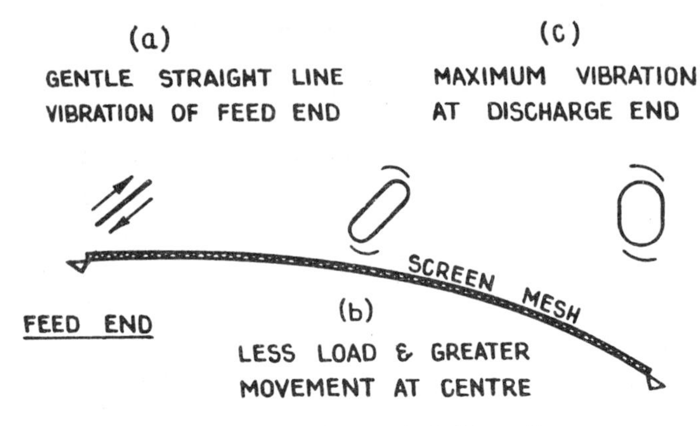

Fig. 77. Compound Particle Orbits on Screen
(Mining Bulletin, King's College, Durham, No. 11)

inertia of the framing, and the weight of the passing load. A motion developed in connexion with the screening of coal is shown in Fig. 77.

There are two main methods of producing vibration. For feed coarser than a limiting size of the order of $1\frac{1}{2}''$, eccentric motion is preferred. Below this, and increasingly down toward a retaining mesh of $\frac{1}{8}''$, the unbalanced pulley is favoured.

Eccentric motion imparts a circular orbit. The typical arrangement con-

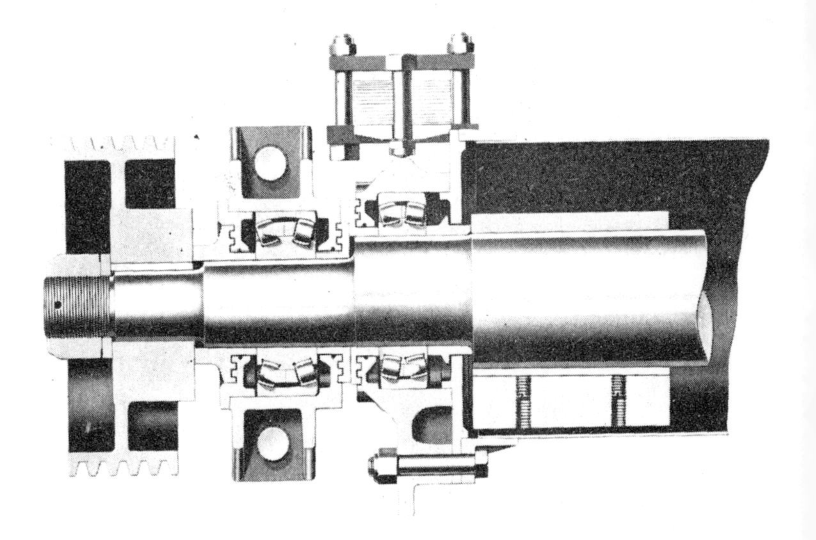

Fig. 78. Floating Eccentric Drive Unit (Nordberg Manufacturing Co.)

sists of a floating drive unit of which one end is shown in Fig. 78. From left to right are (*a*) the drive unit, (*b*) a concentric bearing which carries a side bar attached to the balance deck (*c*) a follower bearing connected to the screen deck, and (*d*) the main shaft. When the revolving shaft attains its operating speed the total throw of the eccentric (*c*) is divided between the screening and balancing deck in inverse proportion to their weights. Many variations in detail have been developed by the manufacturers, all being designed to lesson the mutual strain between shaft and follower and to avoid transmission of vibration to the mill structure by feeding back such impulses in the form of useful work. Boxes can be tilted, sometimes while running, between horizontal and 30°.

Where an unbalanced weight is revolved, vibration results. This can be

Fig. 79. Russell Screen

made to shake a screen box. In the simplest form an unbalanced rotating shaft is mounted across the screen box. In more developed systems the driving shaft is balanced and spins two sets of unbalanced flywheels. One set is keyed to the shaft while the other set can be locked in any desired relation to the keyed set, thus giving neutralisation or reinforcement of the out-of-balance force generated at each revolution. In a third form of development there are two driving shafts, rotating in opposite directions at the same rate, each carrying unbalanced weights. These weights then pull in the same direction twice per revolution, and oppose each other twice. The screen box is mounted on springs, or in flexible rubber blocks. The Russell screen (Fig. 79) incites a gyratory motion generated by an unbalanced weight.

In dry screening, dust protection of the moving parts is desirable. It should be a simple and speedy matter to change a cloth, to adjust its tension, and to ensure that feed is being evenly spread over the whole area.

The vibrating action can conduce to local flexing and premature fracture of the screen wires. The mode of attachment of the cloth to the holding frame or bars must be designed to minimise this effect, and also to permit rapid change of screen-cloth.

Mine ore is often moist or sticky. One method of reducing the blinding and clogging of the meshes by such feed is to heat the screen wires by gas or electricity. Below about 20 mesh, the rate of efficient dry screening becomes increasingly uneconomic. Classification, which takes over when this happens, has some drawbacks when used on sands much coarser than those in the 65–100 mesh range. Further, sorting action in a classifier does not give size discrimination. There is constant pressure on the industry to extend the practicable range of wet screening.

"Wet" Screening

If the ore can be simultaneously held in suspension and given screening action the adverse effects of specific surface friction are reduced. Several ways of doing this have been worked out, but the difficulty in keeping a pool of hydraulic water where it is wanted while at the same time allowing the undersize to be screened away is considerable. Sprays can be directed on the passing layer of ore with turbulent strength, but the water thus used rapidly drains out, leaving the fairly fine ore in the form of a wet sand or silt in which screening motion is almost *nil*, since the wet particles are clinging tightly to each other. Reciprocation of the screen surface under water in a pool has been tried, but is subject to the difficulty of "dashpot" damping of the vibration, and to the fact that a particle free to move in air has a far greater effective thrust downward than one which is waterborne and colliding with many others.

An approach, which has been claimed as economic in some plants, uses a series of sprays (Fig. 80) to repulp the ore as it passes along the screen. If new water is used, this entails considerable dilution and an ample supply of fresh water, but this repulping water can be clarified and recirculated. A refinement of this method for which good economy in use of water is claimed,

Fig. 80. *Screen with Sprays* (Deister Concentrator Co.)

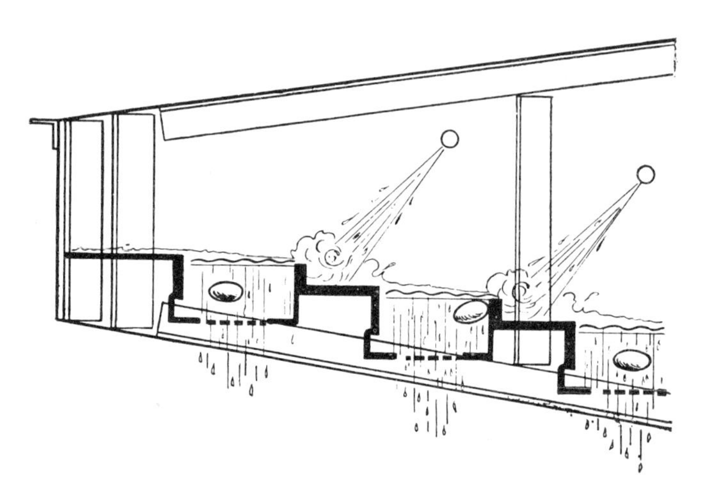

Fig. 81. *Horizontal Wet Screens* (Allis-Chalmers)

is the "pool washing screen" (Fig. 81). The deck is interrupted at intervals by a transverse "pool" into which water is sprayed. The feed arriving from the previous section of screen-cloth is pulped in this pool, and the undersize readily drains through the screen on the next section. This wet screen is made with stepped series of decks, inclined deck, or a pool-interrupted flat deck.

A stationary screen, the Dutch State Mines "Sieve Bend"[1] has gone into considerable industrial use in the past few years. It handles satisfactory volumes in a small mill space and makes good separation down to 100 mesh or finer, with throughput as high as 50 tons per hour. The separating surface

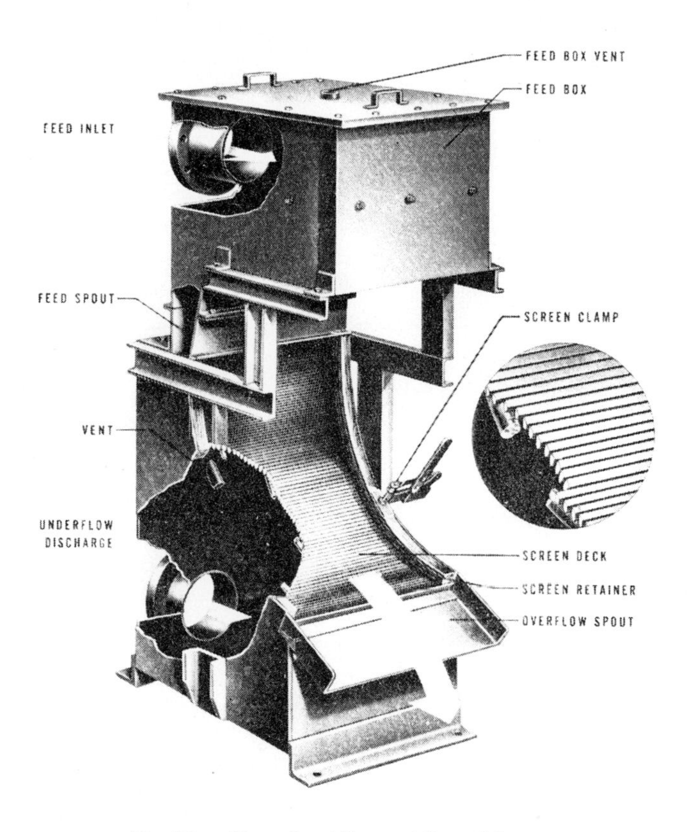

Fig. 82. Sieve Bend Screen (Dorr-Oliver)

is a stationary concave (Fig. 82) formed of wedge-wire bar screen set across the pulp flow.. Modified shapes are marketed in which the concaves extend over 60°, 120°, or 300° of arc. In the last of these feed enters vertically from beneath and is either delivered *via* an internal baffle or a flat nozzle, so as to sweep up and round the interior of the partial cylinder formed by the horizontal wedge bars. These are spaced from 50μ to 150μ apart and have a standard length of 63″. Construction material is stainless steel and maintenance is light since there are no moving parts. Power cost is limited to the pumping needed to present the pulp at the required height and velocity.

A novel wet-screening method has been developed by Hukki.[2] The experimental arrangement (Fig. 83) consists of a cubic box which receives the feed at one side, a stirring arrangement deep in the box, and inclined vibrating screens through which the undersize overflows above. Baffles in the box direct the flow, and sand oversize is removed below *via* an adjustable discharge valve.

Efficient Operation

The simplest expression for efficiency of screening is the weight of undersize actually obtained as a percentage of the weight of undersize actually in the feed. This expression is not of great practical value, however, since it ignores the effect on efficiency of particles of near-mesh size. Particles just too large to pass the limiting mesh, or so large as only to pass with difficulty, are far more prone to blind the available separating meshes than others in the feed. Obviously, the efficiency of the screen is related to the dwelling time of transient material and to the openings available during its passage, hence this near-mesh material defeats a simple formulation. When the screen is closing a crushing circuit, the tendency is for a near-mesh circulating load to build up, so that a progressive falling-off in screen efficiency and capacity is to be expected as this increases. The criterion of efficiency used by one manufacturer is stated as 100 minus the percentage of true undersize in the rejected oversize, and this is a better practical figure. Efficiency varies between 60% and 80%, and increases with:

(*a*) the percentage of the screen open to passage of undersize;
(*b*) the smoothness and freedom from pitting of the mesh wires;
(*c*) the suitability of the shape of aperture to the average particle shape under treatment;
(*d*) the time taken in transit.

Efficiency is adversely affected by:

(*e*) increasing the rate of feed;
(*f*) increase in percentage of near-mesh grains;
(*g*) thickness of bed which hinders presentation of particles;
(*h*) lack of "liveliness" of the screen cloth in responding to the vibrating impulses;
(*i*) moisture in the feed (this can be serious).

Efficiency can be calculated by means of the two-product formula

$$R = \frac{100c(f-t)}{f(c-t)}. \tag{9.1}$$

where R is recovery, and f, c, and t are percentage sizes of feed, concentrate and tailing (Taggart, *Manual*, 19-191, Eq. 133).

This formula, together with that for ratio of concentration

$$K = \frac{(c-t)}{(f-t)} \tag{9.2}$$

can also be used to show the relative amounts of sand discharge and overflow. K is the ratio of weight in feed to the weight of concentrate. In this case sand takes the place of the valuable product usually shown.

Square mesh exercises a restraining effect in two dimensions, whereas an oblong mesh gauges the passing particle in one principal dimension. It thus increases screen capacity at the expense of accuracy. The choice of mesh shape must take into account three interrelated criteria—precision of sizing, permissible tolerance of wrong sizes, and effect on overall operating profit. Usually in mineral dressing, optimum liberation is finally regulated at the classifier overflow, and it suffices in the screen-controlled sections if material too large for efficient comminution in the next grinding section is held back at any designated points. In this case, square-mesh accuracy is rarely of vital importance. Rectangular mesh is favoured for acicular particles, and for moist or clayey feed smaller than $\frac{1}{2}''$. Slabby particles are best handled on square mesh. Material prone to blind the meshes should be treated on oblong screens set with the long axis in the direction of flow. If the screen product is to be delivered to mineral jigs the tighter size control possible with square meshed cloth may be found important.

Suitable tensioning of the cloth in its securing frame is needed. The vibrating strokes should be distributed fairly evenly over the whole area, (*a*) to avoid overstress at a point, line or node and (*b*) to ensure adequate tossing of the passing stream of ore.

Good tension of the screen cloth is desirable to give efficient transfer of the vibrating strokes from mechanism, *via* cloth, to load. Backlash and slackness of the assembly are bad for efficiency. The combined effect of vibration speed and amplitude, together with slope of screen, must be such as will keep the material well stirred and running freely. If the amplitude is too great, stratification will be upset and near-mesh particles will not be adequately "ridden" into the meshes. If it is too feeble, the apertures will blind. The moisture of the incoming feed may vary seasonally, in which case several cloths of varied mesh can be kept ready, and changes made in aperture to suit the altered condition of the feed.

Capacity of a section monitored by screening is higher when oversize is not returned for retreatment but is sent to a different crushing system. This is probably due to the lessening of re-circulated material not crisply dealt with by the crusher from which it has already escaped to the screen. In a large mill the cost of an extra crushing stage may be justified, but a small plant would return the oversize in closed circuit for another pass. This arrangement could be aided by the use of a screen rather larger than the set of the crusher so as to keep down the volume of near-mesh circulating load.

The possibility would arise with spring-loaded crushers rather than those having a rigid set.

The area round a dry screen is usually dusty, and may require hooding and an extracting system. The main running cost in screening is for replacement of cloth, subsidiary items being for power, labour, and loss of gravity head through the appliance. Where the screens are set to gauge the size of an important product it may be desirable to take special precaution against delivery of oversize owing to the unnoticed rupture of a screen. This may be done by duplicating the same mesh on a double-decked machine. If the upper cloth ruptures, oversize will commence to come over from the lower one, and this change can be caused to actuate an alarm.

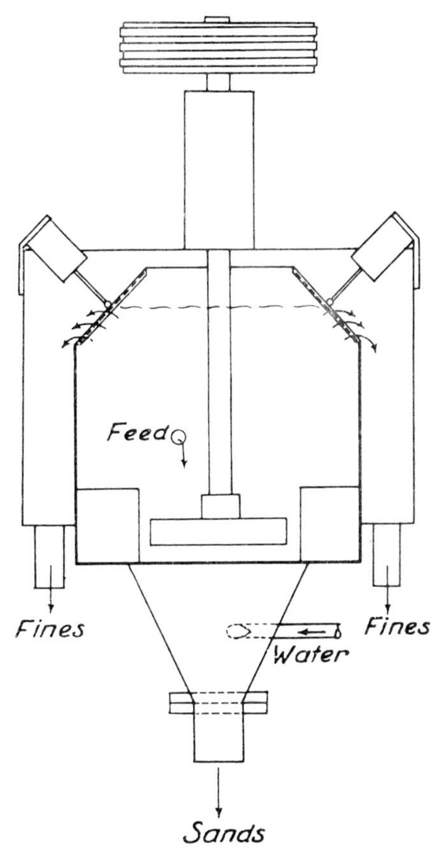

Fig. 83. *The Hukki Screening Cell*

Screens may be classified in terms of the path of a point on the vibrating surface. The five types listed in a Paper by Kuenhold[3] include full or modified circular throw (vertical or horizontal), and lineal reciprocation (vertical, tilted, or horizontal). In coarse sizing vertical circle motion is most widely used, either from an eccentric drive or an unbalanced shaft. The size of screen used is governed by duty, available space, headroom, and position in relation to other appliances.

There is no single formula for capacity, but a basic point is that capacity decreases as oversize fraction increases. This is not always appreciated. Capacity is more related to width than to length, but efficiency of separation improves with the repeated opportunities for passage as the loading lessens toward the discharge end, and travelling particles become more free to move. Screens are usually suspended (preferably on flexible cable) or mounted on a base bolted to the supporting structure or on vibration dampers. The natural frequency of the supports should be at least $1\frac{1}{2}$ times that of the screen at running speed. Optimum slope is that at which the maximum amount of oversize is handled while removing the required percentage of true undersize. To aid this, the bed thickness must allow stratification of fines down to the screen-cloth with adequate mobility. Feed should be delivered across the full width of the screen with sufficient gentleness to avoid wear. If the feed is dry, gravity chutes work efficiently, but should incorporate a stone-shelf to check the entering velocity. Sticky feed is best handled by mechanized feeders.

Where wet sizing is practised, the feed should be slurried at a liquid-solid ratio of 2 : 1 and flushed on gently but uniformly. Skirtings should not be attached to a screen unless the manufacturer can confirm that this dead weight will not upset the balance and bearings. When installing, thought must be given to convenience of access for maintenance and replacements, and this should not be obstructed by badly planned fixtures such as chutes, hoodings, and skirts. Hoods should be mounted separately from the vibrating body, and in fine screening the suction used to remove dust should maintain a downward flow of air.

Principles Governing Classification

In commercial sizing on screens, the particles are presented to a rigid system of gauging meshes which causes separation in terms of one or two dimensions of their cross-section. In classification no such physical restraint is at work. Instead the rate of fall of each particle through a fluid medium is exploited under controllable conditions so as to direct it into either the "oversize" or "undersize" class. The terms "oversize" and "undersize" thus used are not truly descriptive, since classification is a sorting operation, each particle reacting to the resultant effect of the gravitational force pulling it downward and the frictional and kinetic arresting forces generated during this fall.

If it is heavy enough a particle can fall to the discharge gate at the bottom of the classifying vessel. If of intermediate mass it is held in the teeter column. If light it is swept out with the overflowing fluid. The vertically acting force is hydraulic and is provided by the velocity of rising water. The strength of the force is determined by the speed at which this water passes upward through the horizontal cross-section of the classifier at a given point. This in turn is a function of the volume of rising water and the free area at the cross-section, part of which is occupied by grains of mineral in teeter. If a particle is to fall under these loading conditions it must overcome frictional drag and collision in the teeter zone. Under these circumstances it is said to be separated by *hindered settling*. If the classifying vessel also imposes horizontal flow on its contents, a falling particle is displaced from the vertical to a distance proportional to the time it has taken in passing through the current. A pulp of water and fine particles is the classifying fluid normally used in mineral processing. Fine, dry powders can be classified in vertical or horizontal air currents.

Classification deals with a mass of small particles in movement varying from slight drift in parts of the mass to turbulence in other parts. An individual particle is constrained by the packing density in its immediate neighbourhood. If this is thought of in terms of the number of particles in a unit volume of pulp (termed by the author the *specific population* in order to tie up with the frictional factor of *specific surface*) it can be seen that a pulp has a critical concentration below which unimpeded motion of individual grains occurs and above which increasing intergranular interference is encountered.

The factors which influence movement of particles relative to the surrounding fluid may be summarised:

(*a*) The relative velocity of particles of the same S.G. and shape varies as their sizes, a larger falling faster than a smaller one.

(*b*) With two particles of the same size and shape, but of different densities, the heavier falls faster.

(*c*) With two particles of the same S.G. and size (displacement), but of different shapes, the fall is retarded by skin-friction relatively to their surface areas. (Maximum falling rate is developed by a sphere, minimum by a thin plate.)

(*d*) Resistance to fall depends on the velocity of the falling particle (Newtonian, intermediate, or Stokesian) and varies directly as the velocity when slow-falling through an intermediate zone of change till it varies as the square of the velocity when falling rate is higher.

(*e*) Other things being equal, the velocity of fall varies as the squares of the particle diameters when these are small, as the square roots of the diameters when larger.

(*f*) Resistance to fall increases with the S.G. and viscosity of the fluid medium through which fall occurs.

(*g*) Anomalies in behaviour may arise through flocculation, or the presence on the particle of minute air-bubbles.

(*h*) The degree to which individual particles can develop their shape and mass properties is conditioned by the *specific population* and *specific surface* of their environmental pulp.

Though this matter is discussed more completely in the chapter dealing with concentration by gravity methods, it is helpful at this point to note certain ways in which gravitational forces act selectively in classification. When grinding has liberated a heavy particle and a similarly sized light fragment, item (*b*) of the above list provides a simple means of separating them. If the two particles are introduced into a vertical current of water, which flows upward faster than the light particle would fall through still water but slower than the falling rate of the heavy particle, then the one will drift upward with the current while the other will fall slowly downward. Here, if a feed were sized on screens and then presented to some such system, efficient concentration would result. Removal of the variable of *size* would develop a maximum difference in behaviour due to *density*. If the order were reversed, it might be possible to cause small heavy particles to fall at the same rate as big light ones, and then to separate them on a screen of intermediate mesh, or by other methods if they were too small to screen. In this case a *classifying* difference would be removed in order to develop maximum response to a *sizing* difference.

Since classification depends partly on frictional retardation, it cannot be applied effectively to coarse material. It takes over from screening somewhere below 20 mesh, and is used on free-falling or "hindered-settling" particles down to sizes of a few microns, the range being extended further when required by the application of centrifugal force.

The mixture of fine particles with water acts as a heavy liquid. Its density depends on (*a*) the specific gravity of the ore from which the fine particles have come and (*b*) the solid-liquid ratio or percentage by weight of solids in this fluid mixture. When a particle of relatively coarse size falls through this fluid, it converts its potential energy to kinetic, its motive power being the difference between its weight and that of an equal volume of classifying fluid. Hence, the higher the ratio of solid to liquid the smaller becomes the gravitational effect. In one form of classifier (the free-settling mechanical type) fluid density is an important controlling factor in maintaining the desired separation of undersize from oversize.

The kinetic energy of the particle, as it is generated, is used not only to overcome viscous resistance in the fluid but also (if the particle attains sufficient speed) to start vortexes and to displace other particles during collision or frictional contact. It has been convenient to consider a single particle dropping through the fluid, but myriads are moving in all directions in the classifier, and their individual collisions and reactions are far too complex for mathematical resolution. Fortunately, the resultant of all these collisions and rubbings is a generalised effect sufficiently controllable to be of the greatest possible value in the fine grinding of ore. There is no precise cut-off point in classification such as can be obtained in careful screening, but at a given separation point the bulk of the particles will respond in the desired manner. At the point where the separating cut is being made, some particles will be diverted by drifting vortexes which throw them into the wrong stream, but in milling the scheme of concentration is kept sufficiently elastic to allow for this.

It was noted that large particles fall with some strength while small ones fall gently. Two types of classifying treatment are available for exploiting this difference, and when sorting an ore the one chosen is that which will select the most appropriate product for further treatment. For film sizing or "stream action" (discussed under gravity treatment in sluices and on shaking tables) the greater the difference between the sizes of two equal-settling particles, the better will be their separation. Hence, if a sufficiently coarse *break* or liberation grind to give Newtonian settling is called for, classifiers which give maximum size differentiation in their sorting work are used. These are hindered-settling classifiers. When the break point of the ore must be taken at a mesh so fine as to bring the particles toward or into the realm of Stokesian settlement the crowding of the particles through a dense teeter bed, characteristic of the operation of hindered settlers, is too violent to be practicable. A change is then made to a system which permits free settling through a fairly quiet fluid pool, in which the specific population is much lower. This form of separation is used in mechanical classifiers and thickeners. Hindered settlement is used with separating "fluids" carrying from 40% to 70% solids, while free-settling conditions operate with between 3% and 35% of solids in the separating medium. Hydraulic classification is applied to the sizing of homogeneous fine gravels and sands, to settle a relatively fast-settling coarse fraction from a slower one. Classifiers (including *thickeners*, which exploit a more slow-moving variation of the same principles) are used for a wide variety of purposes, including sizing, sorting, desliming of foul effluents, dewatering muddy pulps, adjusting the solid-liquid ratio of a pulp, and development of greater response of ground ore to concentrating processes.

Spitzkasten and Settling Cones

Many special developments of the general principles of classification exist, but it is possible to separate these into groups, despite some obvious overlaps.

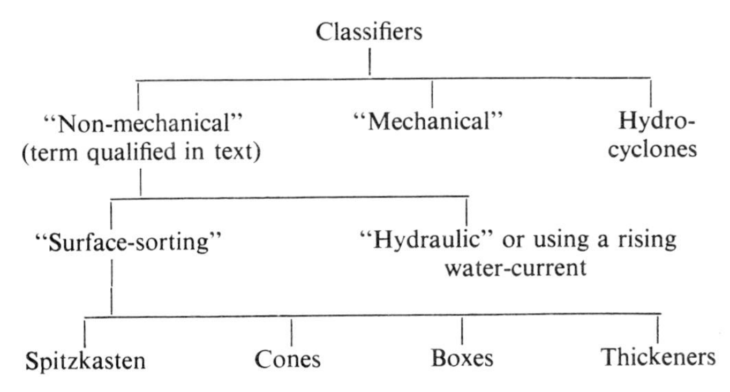

Since powered mechanisms are applied to specially developed sub-types of the so-called "non-mechanical" classifiers, this term is not strictly accurate. In common technical usage, "mechanical classifiers" are inclined troughs in which part of the pulp settles and is continuously withdrawn by raking, spiral, or other suitable gear, while all other types are classed as non-mechanical, whether motorised or not.

Truscott[4] groups non-mechanical classifiers into

> ". . . two main types, namely, Surface Classifiers, wherein the sizing is effected at the water surface, by the water which brings the material to the classifier; and Hydraulic Classifiers, wherein it is accomplished in a restricted passage by fresh or added water introduced below. Surface classifiers are employed for finer material, the discriminating velocity being that which wells upward across the relatively extended surface at the level of overflow; hydraulic classifiers are used for coarse material, say above 80 mesh, the restricted passage permitting the requisite rising-velocity to be obtained with no great amount of added water."

The elementary form is the spitzkasten (Fig. 84), a pyramidal box. A stream of pulp flows in at one side, drops part of its solid charge with some water to an aperture at the pointed end, and discharges the remainder *via* an overflow lip. The entering pulp transforms part of its horizontally directed kinetic energy into downward-acting eddies and vortexes. Particles caught up during these shifting and whirling movements acquire centrifugal

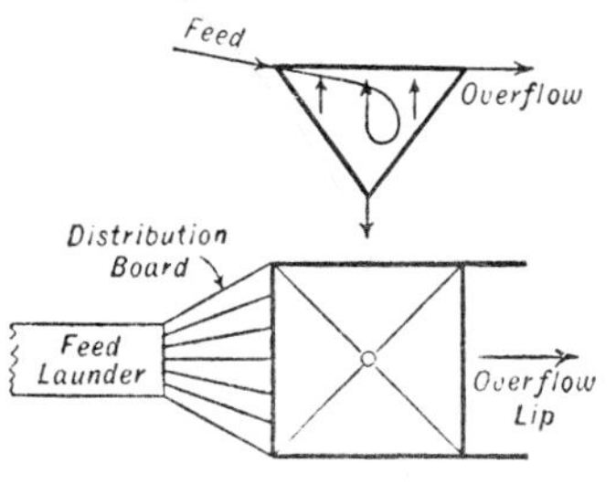

Fig. 84. *The Spitzkasten*

acceleration, which may throw them out of the vortex in any direction from vertical to horizontal, and this force acts without discrimination on large and small, heavy and light particles. On the whole, the heaviest and coarsest sands gravitate downward to the bottom discharge while the lighter material is crowded back to the top overflow. In its crude form, the spitzkasten does not make a particularly satisfactory separation, but it illustrates the fundamental principles at work, and is therefore worth examination.

First. As with all classifiers, only two products are delivered. Everything entering must either depart above or below. The greater the proportion of solid withdrawn below, the greater will be the amount of fine material included in it. With feed and discharge rates steady and equal the total number of particles, and their size distribution, is definite. All that can be done in classification is to vary the ratio between the quantity of coarser sand leaving below (*via* the spigot discharge) and the relatively fine overflow. If the spigot were closed, all the pulp would eventually overflow without classification, and this overflow would contain the maximum possible percentage of coarse particles. The greater the fraction removed as relatively coarse oversize the finer must be the average size of both overflow and underflow, and *vice versa.* Separation does not mean that only coarse particles will underflow and only fine ones overflow. This pitfall sometimes confuses the learner, who imagines that if the coarsest particles are being removed from one point the result must be to give a much finer product at the other withdrawing place. What happens is that classifier adjustment simply varies the splitting conditions. Overflow and underflow are respectively fine and coarse relative to this splitting.

Second. Since the volumetric capacity of the spitzkasten is constant, the settling rate available to a particle varies as the volume of feed. Other things remaining unchanged (including the ratio of solid to liquid in the feed), the greater the fed volume, the less will be the dwelling time of each particle, hence the coarser the overflow fraction since it has less time in which to settle.

Third. If the rate of feed is constant, but the solid-liquid ratio varies, the specific gravity of the pulp varies with the increase or decrease in content of solids. Since the rate of settlement varies inversely as the solid fraction, the more watery the feed, the finer will be the average size of overflowing particles.

Fourth. It is customary to consider that classifiers separate particles by virtue of a combined vertical and horizontal movement. To this end, the maximum cross-section is usually provided in the horizontal plane of overflow, and in the surface area immediately adjacent to the overflow lip all particles are deemed to move horizontally. Hence, the argument continues, if in this area the particle is swimming or drifting in the overflowing current, it is carried out of the classifying system. If its physical relation to the system at this point causes it to sink, it is retained and may sink to the underflow zone for discharge in the coarse fraction. Teetering particles are discussed later in this chapter. Hence, the capacity of a classifier is related to its horizontal area in the plane of overflow, and the concept of "surface classification" or "surface-sorting" is closely bound up with pulp behaviour in this plane. In theory, it should be possible to control the density of the pulp at this surface, and also the horizontal speed toward the overflow which influences the drifting rate of a particle. Operation based on these considerations, together with those of mass and surface friction, should lead to an accurate sorting system. In practice, only rough sorting takes place in the spitzkasten. The entering particle undergoes random acceleration, in a confused mixture of interfering vortexes. It passes from one solid-liquid ratio to another in various parts of the box, with varying freedom of packing. Finally it is delivered to the sorting surface with an unpredictable kinetic energy directed at any angle by the vortex from which it is separating at the moment.

Fifth. The shape of the spitzkasten provides quiet zones down the corners, and beds of sand pack into them. Some particles work their way over these beds to the overflow without undergoing sorting action. Toward the continuous underflow, periodical collapse of the packing chokes the discharge orifice, causing surging and abrupt change in the downward flow of pulp.

Appreciation of the foregoing considerations will aid in understanding the physical limitations of classification in the more elaborate appliances discussed later. The first improvement in simple spitzkasten work is to use a series of boxes, increasing in volumetric capacity, so that the coarsest sands are removed first and the finest last (Fig. 85). To avoid settlement of material on the sides, these are made steep. With the larger boxes this would lead to high pressure on the underflow and the production of too watery a discharge there. This is avoided by using the "gooseneck" discharge pipe shown, with suitable provision for clearing it should "tramp oversize" from a preceding spitzkasten settle in the pipe and choke it. At one time spitzkasten series, either in box or trough form, were widely used ahead of gravity separation, but they are not much seen today. Baffles must not be used with spitzkasten as they would interfere with surface selectivity.

To overcome the surging effect produced by the periodical sliding down of settled sands from the corners of the box, cones are used. Feed is central and overflow peripheral into a launder. If provided with a diaphragm to prevent very fine material from dropping through with the coarser sands, the appliance is called a diaphragm cone, and still further regulation of the solid-liquid ratio at the bottom discharge may be provided by a float system

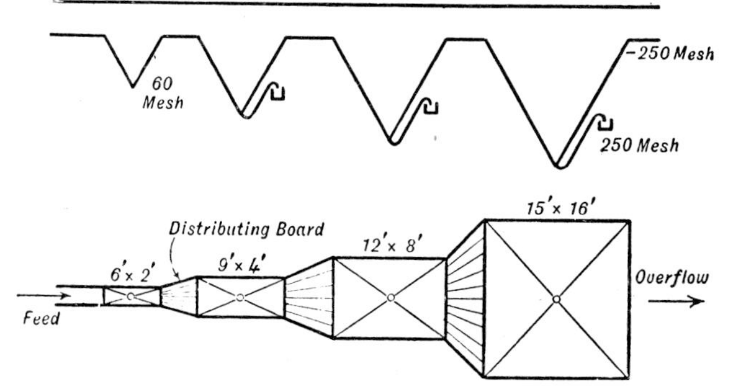

Fig. 85. Series of Spitzkasten (Section and Plan) (after Truscott)

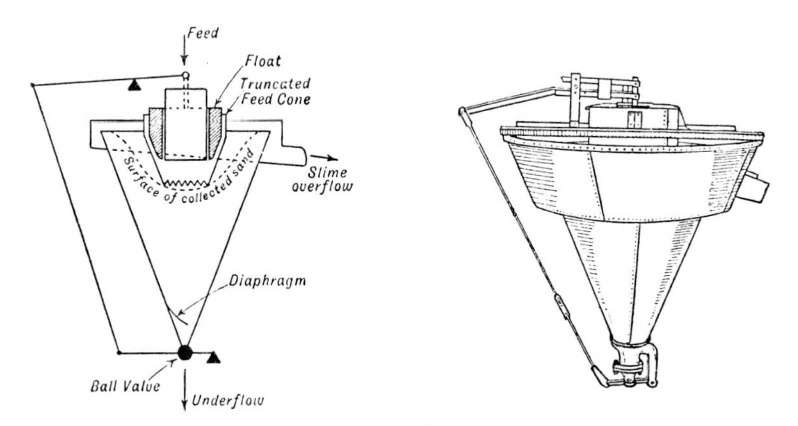

Fig. 86. The Settling Cone

which varies the aperture in accordance with changes in the feed (Fig. 86). Very little classification is done in these cones once sand has settled to a fairly compact mass. The selective action occurs in the fluid layers of swirling pulp above this bedding. Since feed is introduced centrally and the pulp is then swept through a rising path toward the peripheral overflow, the coarsest and heaviest particles tend to drop straight through or to settle out from this eddying stream, while those smaller than a certain mesh are lifted to the overflow lip. The solid-liquid ratio of the underflow depends on the closeness of packing of the settled sands.

Under favourable conditions, the cone compares inefficiently in performance with the far more costly mechanical classifier, save that it is unable to restore the loss of head of the coarse sands dropping through it, and cannot therefore close a milling circuit without the addition of a pump of some kind. It can be used for desliming or dewatering. Simple devices based on the spitzkasten find little application in modern plants, but have a limited field of use in small operations.

Nests of boxes, with either pyramidal or V-shaped settlement zones, were developed for thickening dilute suspension of fine sand, and for reclaiming mill water. They offer a large quiet zone with a gentle motion over the surface-classifying area. They afford a gentler application of the sorting principles at work in the spitzkasten, in the same way that the settling cone does in comparison with the cone classifier. Their place today has been taken by thickeners of the type described in the next section.

The Thickener

Thus far the conditions which affect particles settling quickly from random vortexes and under crowded conditions have been considered. In the thickener, settlement is free and the particles are given hours to gravitate downward. The classifiers hitherto considered were either trying to split a long-ranged feed into coarse and fine fractions in a short time, to trap out coarse sands, or to remove slimes from a fast-settling pulp. When the particles have been very finely ground a prolonged dwelling time under quiet conditions is needed for their settlement. The thickener is constructed with sufficient volumetric capacity to give this time. It can be used:

(*a*) to reclaim water from a muddy effluent by allowing the silt to settle;
(*b*) to decant fouled water or chemical solutions;
(*c*) to change from one chemical wash-solution to another;
(*d*) to remove a dissolved mineral product from a pulp;
(*e*) to thicken (i.e. increase the solid-liquid ratio of) a pulp;
(*f*) to reclaim some mill water before discarding the solids from a tailings pulp.

The essential features of a thickener are shown in Fig. 87. In a typical operation, mill pulp carrying finely ground solids in suspension is fed in centrally, through a "trash screen" which holds back any debris that has accidentally entered. The entering pulp displaces part of its volume as a

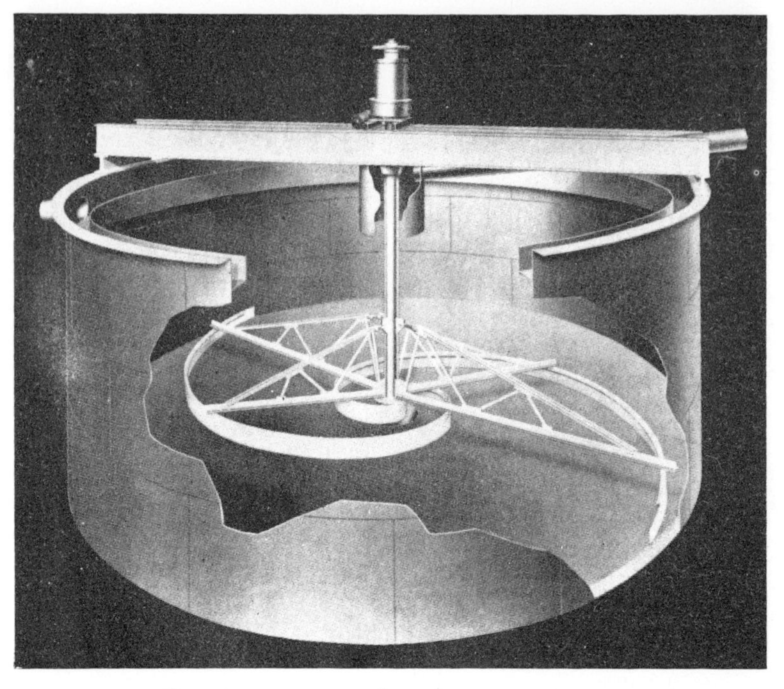

Fig. 87. Cutaway View of Hardinge Spiral Rake Thickener

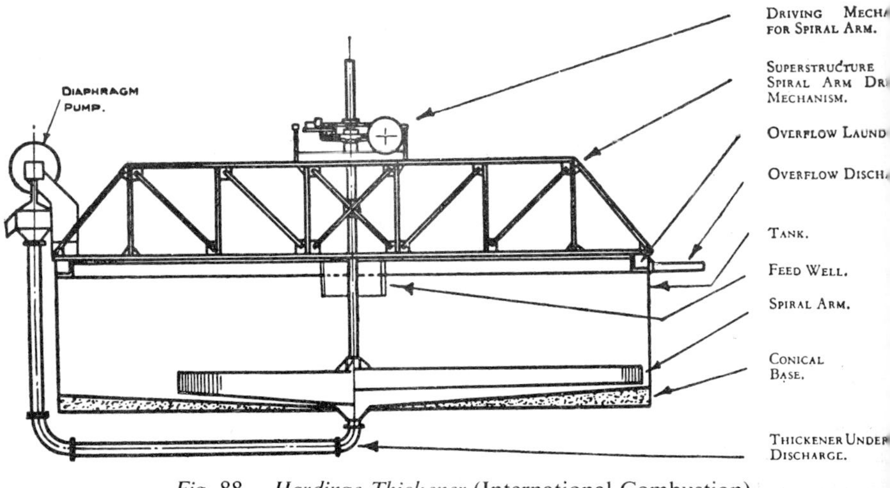

Fig. 88. Hardinge Thickener (International Combustion)

peripheral overflow of moderately clean water. During the very gentle radial drift of this overflowing water from centre to sides the solids fall slowly downward, perhaps individually, or more usually with some degree of flocculation into aggregates of particles. Material sufficiently coarse falls rapidly to the raking zone below, while the rest of the solid fraction settles and leaves a zone of clear water toward the upper periphery, followed by one similar in consistency to the feed (less its coarser particles). This is succeeded by a transition stage through which the pulp steadily increases in solid-liquid ratio as it settles downward until it reaches the compression zone where the particles, or more probably the floccules, are being squeezed together by the weight of fluid above. Through this compression zone the rakes (in Figs. 87 and 88 the spiral gathering arms) of the thickener are very slowly revolving, gathering and sweeping the settled slurry or slime toward the central discharge well. The rake arms may be revolving once in from two to eight minutes, and as they move, their superstructure cuts through the billowing floccules, opening channels through which clear water can be squeezed upward.

In thickening, flocculation of the pulp is usually an important factor. The subject is considered later. At this point it suffices to remember that the more dilute the entering pulp, the slower it is to form flocs and hence the longer its settlement time. Thus, when a thickener is showing signs of being overloaded (by discharging insufficiently clear water at its launder) the trouble may be due, paradoxically, to insufficient solids in the feed. This is because flocculation depends partly on the opportunity given to particles to collide, which is proportional to their concentration.

The thickener may be a very large round tank, or a cylindrical excavation lined with concrete. It must be capable of containing the continuously entering feed-pulp for the number of hours required for efficient settlement and compression down to the required solid-liquid ratio in the well at the centre of the discharge zone. The bottom of the tank usually has a gentle slope inward to this well. The rakes which gently press the slurry and gather it to the centre may be driven from a shaft, or be towed by an electric motor running round the periphery on a monorail.

In operation the feed rate and feed-pulp condition must be such that there is ample time for a protective clear zone to form and to be maintained toward the launder, while the settling fraction has adequate time to consolidate. The rate of discharge from the well is regulated by means of a diaphragm pump which is run at a rate allowing some two feet of fully thickened slurry to be maintained in the compression zone. This layer holds back the insufficiently compressed pulp and ensures that only a completely settled slurry is withdrawn. If the zone is allowed to become too thick, there is danger of burying the rakes or of overloading them and injuring or distorting the mechanism. Alarm and trip mechanisms are fitted to indicate the advent of such overloads. The overflow can be monitored by an *electric eye* so that warning is given if its turbidity rises unduly. Since the slurry being pumped from the underflow is dense, it is carried through pipes of small diameter to the pump, thus maintaining it in motion sufficiently vigorous to reduce any tendency to settle out and choke the withdrawal system. Flushing points are also provided through which water or compressed air can be injected in the event of choking. Instead of pumping, some large installations use bottom valves to run slurry off. The piping discharge is then of ordinary diameters. The thickened discharge is commonly led to a continuous filter which must periodically be shut down for servicing. During such a period, which may last several hours, the thickener continues to receive feed and must store its slurry. Provision is made for raising the rake mechanism to prevent overstrain under such conditions. Since this is a safety precaution, the rakes must be gently lowered as soon as normal running has been restored and the loading has been reduced.

Choking around the well of the thickener is a serious matter that may lead to the shutting down of the whole mill. Solid objects fall in, either through sabotage, a kink in human nature, or through failure of workmen on the thickener to tie loose tools to a safety belt. It is therefore a wise precaution to have a run-off tank available into which the contents of the thickener can be sluiced in emergency. If the thickener is "stalled", speedy repair is essential, as there is rarely standby capacity to which feed can be diverted.

Where space is cramped, or where the risk of freezing entails protection, tray thickeners having from two to six compartments are often used. The pulp is divided into equal streams, and each is fed centrally to a compartment. In one type each compartment rakes the settling slurry to a common well-discharge. In another type the slurries can be kept separate. The water overflow rises naturally from the lower trays to join that from the top peripheral launder, or, if desired, these overflows can be kept separate.

Failure of the settled slurry to come away from the well may be due to a solid object obstructing the outlet, a choke in the piping system or a defect in the diaphragm pump, such as a stuck valve or a ruptured membrane. Thickeners receiving a flotation concentrate sometimes build a thick scum of floating froth on their surface. To minimise this, the feed should run in gently, since splashing is likely to entrain air-bubbles.

Little power and attendance are normally required, but when a thickener breaks down it can very seriously upset the running of the plant, since it usually constitutes a "bottle-neck" in the flow-line. To simplify maintenance and avoid a lengthy shut-down thickeners are sometimes built with an approach tunnel below, ending in a pump room. Alternatively, underflow may be pumped up through a central column large enough to permit a man to enter.

Among recent developments in thickener design[5] are a two-stage raking zone. The peripheral area has a relatively shallow gradient, while coarser material which tends to settle more centrally falls to a steeper central portion. The inner rakes which sweep this zone are attached to posts below the trusses. These posts cut through the deposit and open channels through which water can squeeze upward. The trusses themselves are not subjected to the strain of shearing through this material. Another design is a flat-bottomed thickener with peripheral discharge.

Where continuous thickening is coupled with periodic filtration so that storage capacity inside the thickener is required, automation has been successfully used to raise or lower the rakes in accordance with changes in torque signalled to a motor which lifts or lowers the raking mechanism. For sub-zero working of exposed plant, electric heating of driving gear and pump room can be used. For a thickener which is used to clarify a thin, but finely divided and slow-settling feed, hydraulic disturbance where the feed enters can be avoided by delivery through vertical screening which divides the flow into a number of thin streams and checks these by baffling.

Thickening Theory

In 1916 Coe and Clevenger[6] produced a formula for calculating the thickener area required in handling a known rate of loading. Their observations were based on the zone sequence seen during the settlement of pulp—clarification of the uppermost layer and increasing pulp density down to the final stage of compression (the critical point) below which no further settlement occurred. Thus at any horizon in the transition zone there was a change in pulp density which affected the subsequent settling rate and the specific population at that horizon. Their calculation of the required thickener area was formulated in respect of a transitional rather than a final or saturated concentration, and within that limitation was dependable. It is

$$A = \frac{1{\cdot}33\,(F-D)}{R\delta} \qquad (9.3)$$

where A is the area (ft.^{3}/ton) of dry solids per 24 hours
F is wt. of liquid/wt. of solid in feed
D is wt. of liquid/wt. of solid in discharge
R is the settling rate and
δ the specific gravity of the pulp.

In 1952 a new approach was made by Kynch[7] in which the original basic assumption (that the settling velocity of a particle is a function of the local concentration of surrounding solids) is retained. Overloading, in the sense that solids arrive faster in the feed than in the fully compressed zone, changes the density values through the compression zone and thus reduce the rate of delivery to the discharge zone. The thickener area required by Kynch's formula is

$$A = \frac{t_u}{C_0 H_0} \qquad (9.4)$$

where t_u is the time in days, C_0 the concentration of feed as tons of solid per ft^3 of pulp, and H_0 the height in feet.

Both formulae bring out the fact that (ignoring any effects of flocculation or need for clarification to completely clean water) sedimentation is governed by pool area rather than pool depth. One element in thickening is, however, overlooked by both formulae, as is pointed out by Fitch.[8] As pulp density increases the particles tend to be locked into a plastic structure the yield value of which rises as the percentage of solids increases. Above this fairly narrow zone particles still have some freedom of movement but the zone itself, called by Fitch the "steady state thickening" phase, controls the volume of arriving solids which can get through it. An elusive compression factor is at work, connected with the shearing of the plastic structure of this zone by the revolving rakes of the thickener. These, in addition to moving settled material to the discharge point, cut channels through which clear water can be seen to well up as it is squeezed out by the heavier settled pulp. This zone is called "the zone of rake action" and the final solid/liquid ratio possible in thickening takes this into account in testing for design of a thickener for a given pulp by using slow stirring of the bottom layer of a sedimentation column in order to provide a channel for this final squeeze-out of water.

As the result of laboratory tests Cross[9] concludes "that the use of a deep, slotted feed well will increase the capacity of a thickener and that the principles originally put forward by Messrs. Coe and Clevenger are no longer tenable. . . The re-designing of raking mechanisms is called for in order to both cope with the increased tonnage and impart greater movement to the compacting sediment."

A mathematical model of the thickening process has been developed by Gaudin and Fuerstenau.[10]

Hydraulic Classifiers

These appliances are also called "spitzlutten" and upward-current classifiers. In its simplest form (Fig. 89) the hydraulic classifier resembles the spitzkasten, with the essential difference that, in addition to horizontal sorting at its maximum cross-section where the lighter particles overflow the discharge lip,

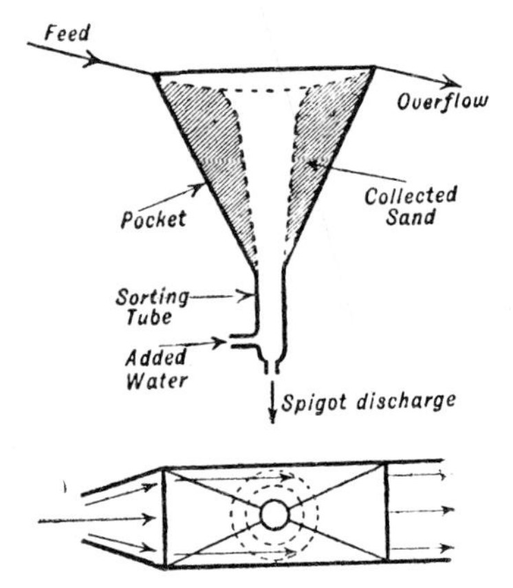

Fig. 89. The Hydraulic Classifier (after Truscott)

there is now added a column of water rising at a controlled rate. A single particle falls through a still column of water at a speed resulting from the gravitational pull downward as moderated by consumption of energy in overcoming shear or turbulent resistance set up by its motion.

Consider three particles which have accelerated to their maximum steady rate of fall through such a column, and which differ in terminal velocity. If the water is now caused to flow upward, a rising velocity can be chosen which neutralises the downward rate of the medium of these particles so that it hovers. The slowest one now rises slowly while the fastest one drifts slowly downward. This is the principle applied in hydraulic classification. The sorting column is usually of even vertical cross-section so that there is no variation in the rising velocity. When the cross-section of the sorting column varies from a maximum at the upper discharge end to a minimum at the lower, it is possible to arrange the flow rate so that similarly energised particles hover on reaching a point where their falling velocities are balanced. Actually, the water does not flow smoothly, so such particles will dance or "teeter" in the column. Thus, two types of construction are possible. In the "free-settling" hydraulic classifier a particle heavy enough to fall goes right through to the bottom discharge. In a "hindered-settling" classifier it may fall to a certain distance and there teeter at the entrance to a more restricted column through which water is rising faster.

In the classifier many millions of particles are simultaneously being sorted. All are jostling against one another to some extent, and at any cross-section of the sorting column the liquid is rushing through the interstices between the tossing and tumbling grains. Thus, the kinetic energy of the falling particle is being dissipated by friction and collision, and the liquid cross-section is constantly varying, giving rise to spinning motions which may equally well impel a particle upward or downward. Only particles decidedly heavy can force their downward way through such a scouring crowd. A heavy particle can be "rafted" clear across and out of the system, by some accident of momentary turbulence. It is easier for a particle to become arrested and entangled in the teeter bed than for it to get out again. In consequence such a bed grows as the hindered-settling classifier works, till the point is reached where the sides are comparatively packed by slowly circling grains moving in vertical ellipses or occasionally sliding down to the discharge. Meanwhile, the centre is occupied by a more mobile teeter column dilated by the rising water. This is indicated in Fig. 89 by the clear central portion of the classifier, which in operation would be occupied by a teetering column of sand giving way (shaded portion) to packed sand at the sides. Here the bulk of the sorting of new feed is done. The central column is a pseudo-fluid of density corresponding to its solid-liquid ratio, which works against the attempt of a heavy particle to fall through. If the feed can travel downward at this pulp density and against the hydraulic rise through the teeter, the jostling it encounters may retard, but will not stop, its descent.

Hydraulic classification may be used as a sizing method, if applied to particles of the same specific gravity. In this case the bigger and heavier ones fall through, given suitable adjustment of the restraining forces, and the lighter ones are carried up and out. Such work can be done with less precision, but far more cheaply, than on the delicate fine screens which would otherwise be needed. With sufficiently fine particles, classification can be performed in columns of air instead of in liquid.

In mineral processing the main purpose of classification is to sort a feed into two classes, settling and rising. If this feed contains heavy and light mineral, small heavy particles will have the same falling characteristics as larger light ones. They can therefore be dropped together and then separated on a screen of an appropriate intermediate mesh.

Many shapes and varieties of hydraulic classifier have been developed. Truscott[4] observes in his text-book:

> "Hindered-settling classifiers are those wherein . . . water rises through the sorting tube or orifice into a chamber so dimensioned that the material collected there is brought into a quicksand suspension or 'teetering' movement. The (horizontal) area of the teetering chamber should not be greater than about four times that of the sorting passage or some of the material will lie quiescent; nor much less, because then the conditions of the ordinary free-settling classifier would be approached. With these conditions fulfilled and where particles of varying density and size are present, another specific advantage of hindered-settling is realized, namely the small dense particles fall with much larger less-dense particles than under condition of free fall; with quartz and galena, for instance, the diameter ratio of equal-falling particles under hindered-settling condition is about 6:1—as though the specific gravity of the medium in which fall took place were about 1·5—instead of 4:1 which obtains in free-falling conditions. Of such an increased difference in particle-diameter, advantage may be taken in the processes of separation. . . ."

The teeter column thus heightens the separating effect based on exploitation of differences in mass and S.G. It also has a scrubbing action which aids in cleaning the discharged product that runs the gauntlet. This scouring arrests fine particles which otherwise might be carried, attached to larger particles, to the lower discharge. The hydraulic column must be steady, and is therefore fed from a constant-pressure source of water. Some water is withdrawn with the solids leaving at the bottom discharge, and the rest is discharged as the overflow, together with the water in the original feed. Classifiers can be arranged in series, to give progressively finer underflow products. They may receive feed from a desliming cone, or send out a final overflow product for quieter sorting in such a cone. The coarser the desired underflow, the more strong must be the rising current.

The teeter bed is intermediate in settling character. It tends to build up so as to alter the character of the products discharged through it. To minimise this defect, a class of hydraulic appliances having the general distinguishing name of "hydrosizers" has been developed in which this teeter bed is controlled as to composition. As the bed builds up, the weight of its component quicksand also increases. This can be measured by providing a water connection in the form of a hydrostatic head which shows changes in the back pressure exerted upon the entering hydraulic water (the "added water" of Fig. 89). Appropriate control mechanism can be actuated by pressure changes as in the Stokes hydrosizer (Fig. 90) in which the bottom discharge aperture is opened or restricted in response to changes in the load of teetering sands. This classifier is one of several made in multi-spigot form, to deliver a series of graded products.

Another form of hydrosizer takes the shape of a miniature thickener. Its rakes rotate fast enough to loosen up the settling sands and upward currents may be imposed upon the contents of the circular tank. By suitable use of these combined forces very fine material is caused to overflow, while denser particles drop and are raked to a central well-discharge.

Mechanical Classifiers

In the so-called "mechanical" classifier the pulp is fed into a rectangular tank under conditions which allow the heavier and coarser solids to gravitate fairly freely downward, while the lightest particles flow to a weir discharge.

The "mechanical" part of the classifier is usually a drag belt, a set of rakes, or a spiral screw. Its functions are to stir the pool of pulp and remove settling solids. In Fig. 91 a cross-section through the type of tank usual with a Dorr (rake), Akins (spiral) or drag-belt classifier is shown, without its mechanism. The pulp overflowing from the grinding mill carries 70%–80% solids, and flows to the classifier through a short launder, or possibly after elevation through a centrifugal pump. *En route* or in the tank more water is added, so as to thin the pulp to an operating density at which overflow of finer sands and settlement of coarser ones are most effectively produced. In the launder the pulp stream is split into two or more channels which distribute the feed fairly equably across the pool, approximately two-thirds of the distance from the weir in direction of its "V" end.

Ignore for the moment the stirring effect produced in this pool by the raking gear. The pulp, diluted to say 30% solids, falls from a slight height into the shallow end of a wedged-shaped body of fluid. The height of the feed launder above the pool determines the kinetic energy with which this plunge is made. Rate of settlement through the pulp results from the interplay of complex forces. These may include *thixotropy*, adhesion and cohesion, which receive further discussion in Chapter 15. These forces influence shear in the body of the pulp under stated operating conditions (mainly concerned with temperature, *specific surface* and turbulence). For convenience of operating control this complex is treated as a simple S.G. effect for which the classifier pool is held at a specified operating density. This determines the resistance to settlement which is offered at successive horizons to the particle as it seeks to gravitate downward. There are two channels of escape from the pool. A particle having sufficient mass to force a passage to the bottom, or raked zone, falls and is quietly withdrawn by the mechanical withdrawing device (not shown in Fig. 91). A particle which has insufficient mass to descend can drift with the current from the feed launder to the overflow weir through the transporting zone. Equal-settling particles will report together. A small heavy particle will settle with a larger one of lower density.

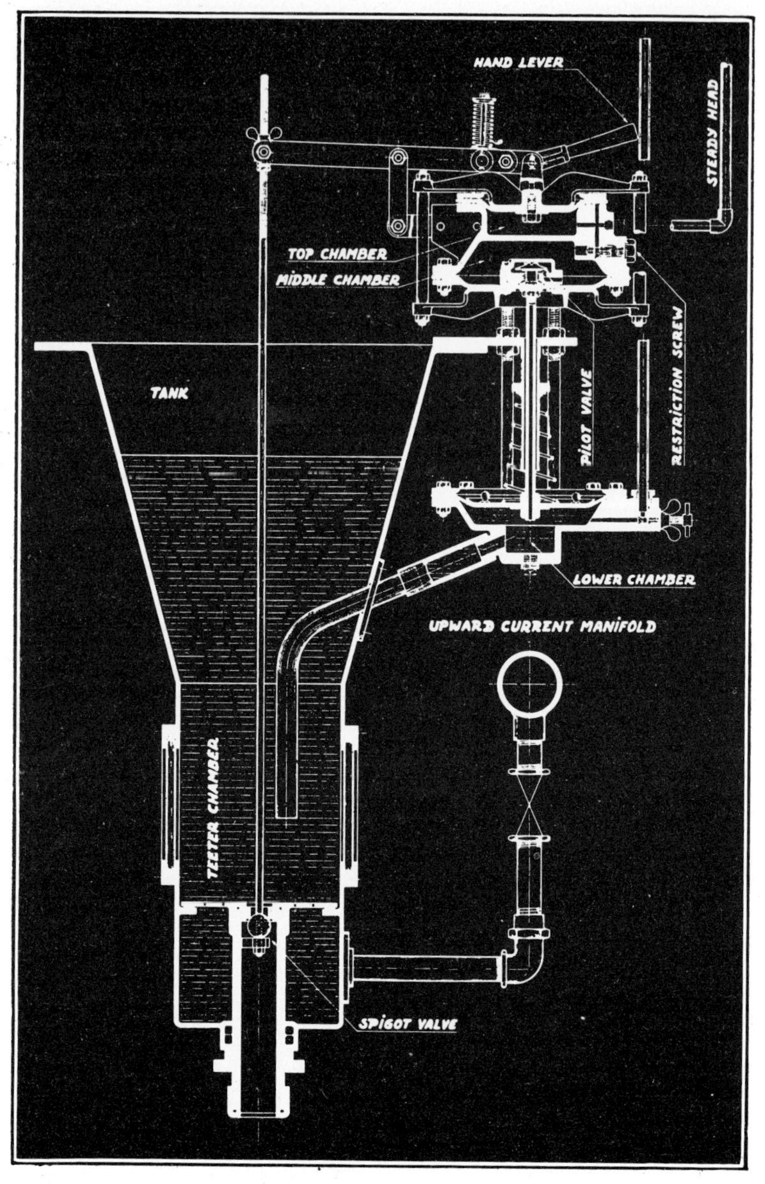

Fig. 90. Section through Stokes Hydrosizer showing Relay Valve

If all particles in the classifier were equal in size, a clean separation into heavy (sinking) mineral and light (overflowing) mineral would be possible. In fact, the mechanical classifier tends to retain the heavier mineral in the ore and return it, *via* the raking mechanism to the feed end of the mill with which it is close-circuited even when the operator desires it to be sent on. The particles comprising the solid fraction of the feed cover a wide range of sizes,

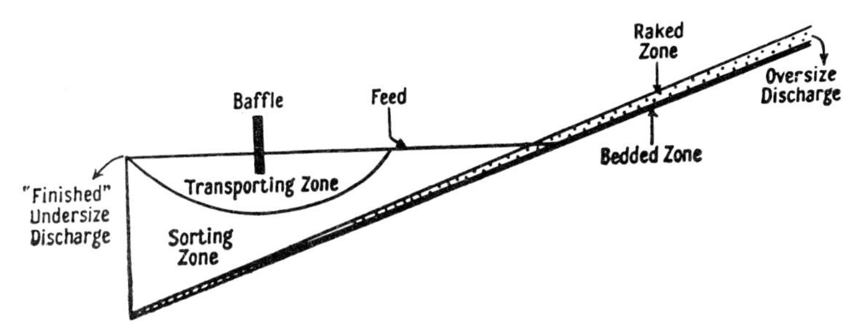

Fig. 91. *Cross-section through Classifier Tank*

and are in various stages of liberation. Consequently some are swift to drop, some slow, while others build up in the sorting zone. This zone is not homogeneous but thickens from the most watery layer at the top of the pool, by increasingly dense strata, to the densest part of the pool immediately above the raking zone. These strata are stirred by the rakes or spirals as they move in the pool of the classifier. Below the rakes lies an undisturbed stratum which packs the clearance space down to the steel bottom of the tank.

It follows from this sorting of the pulp into zones that each layer continuously receives new particles from above, and either (*a*) lets them fall through, (*b*) rejects them back to the layer above it, or (*c*) retains them. So long as a layer only sorts the entering particles according to (*a*) or (*b*) it retains its integral composition and performs consistent and predictable work. When, as is inevitable, it captures more particles, it increases its density and any further entering particles must be correspondingly more dense in order to fall through. To some extent this continuous rise of pool density is offset by the stirring action of the rakes, by the downward drift of the sinking particles, and by the slow horizontal displacement of the rest toward the weir discharge. These movements alone are not able, in the ordinary classifier, to maintain a constant distribution of strata densities in the sorting and transporting zones, though they can be helped considerably by skilled operation.

The bed below the rakes often traps the heaviest particles fed to the classifier. If these constitute a valuable product such as free gold, an undesirable concentration of material will result. In some grinding circuits small concentrating devices are set between mill discharge and classifier to "*scalp out*" such heavy values.

The rate of movement of rakes or spirals can be varied to produce the required turbulence in the pool. With a mechanical classifier set to yield a fine overflow this might be as low as 9 rake-strokes per minute when making a separation at —200 mesh, varying up to 32 or so for 28 mesh, rapid-settling sands. The rakes have the further major function to perform of clearing the settled sands up-slope to the mill feed launder, and they must therefore be run at a speed which deals adequately with the load falling to them. A section through the Dorr rake classifier is shown in Fig. 92. The rakes (1) are set

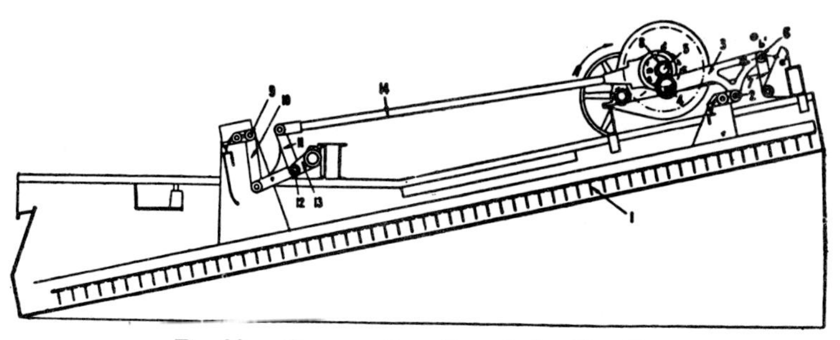

Fig. 92. *Cross-section, Dorr Rake Classifier*

in an inclined tank and are actuated *via* an eccentric motion (8) and two sets of link motion, 9 to 13, and 3 to 6. These links cause the rakes to moves through an elliptical orbit, almost flat on its long axis which corresponds with the slope of the classifier tank. The rakes start their climb at the lowest setting, and gather settled sand. At the rising end of the stroke they lift sharply. They then return, drop, and repeat the cycle. These classifiers are described as simplex when the trough has one compartment, duplex and quadruplex when there are two and four divisions separately raked. Provision (not shown) exists for raising the rakes. Spiral classifiers (Fig. 93)

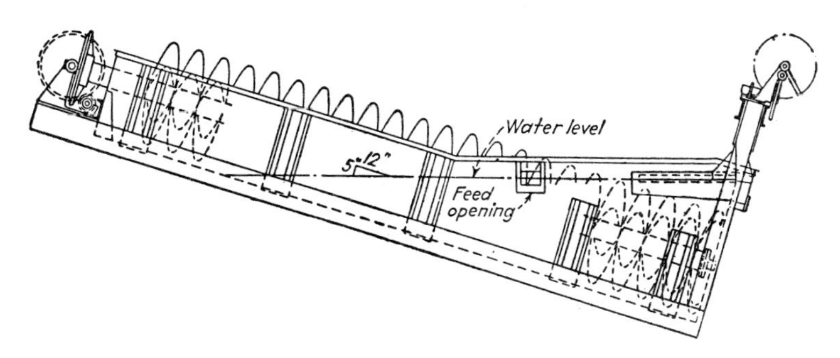

Fig. 93. *The Akins Classifier*

do not normally set up a turbulent stirring action in the classifying pool, but cross-pieces can be bolted to the spiral shoes to produce this. The helix is run at a speed varying with its maximum diameter, the shaft usually turning at between 3 and 6 r.p.m. for a large spiral, and up to 20 r.p.m. for a small one.

As settled sands emerge from the "V" end of the pool, they bring with them some trapped or "drag-out" slime which should properly have overflowed at the weir. Washing sprays are often used to return this material to the pool. They offer a simple means of adjusting the pool density by providing a region where extra water can be mixed in, but are not very efficient as deslimers. With the rake classifier the emerging sands usually take an undulating contour, and if the sprays are allowed to play over a length of rising sand the slime is trapped in the depressions, ploughed under at the next dragging up-stroke, and thus smuggled through. To be effective the sprays must be directed forcibly on to sands close to the point of emergence, and other methods must be used for making final additions of water to the pool as the sprays must run at full pressure. In this respect the spiral classifier is at an advantage. It carries its sands upward on the rising side of the helix, so that sprays can be used at any point and there is a good channel for running their washed product straight down to the pool. The churning of newly emerged sand by sprays causes it to slip back and interferes with the elevating function of the rakes. Sufficient height must be gained for the sands to run by gravity back to the feed scoop of the mill, after they fall from the classifier into the return launder. The continuous and more gentle action of a spiral makes possible a steeper slope in the classifier tank, settings of from 3″ to 4″ per linear foot being common as against 2″ to 3″, with an occasional 3½″, for reciprocating rakes. Removal of slimes reduces the tendency to slip, but a certain amount of deliberate inefficiency is liked by some millmen, who consider that the slime has a valuable lubricating effect in the feed trunnion of the ball mill, and that without it the mill could not handle so large a circulating load. In the "Overdrain" classifier a drag belt moves between stationary longitudinal walls (called shrouds) so as to carry material trapped between successive moving sections in bottomless compartments. From these compartments supernatant slimes overflow through side openings and back to the pool between shrouds and sides of classifier.

In the event of power failure, or shut-down, care must be taken that the raking gear does not become buried as the sand settles compactly down. Reciprocating rakes can be raised in some models, and spirals can be swung clear. If a classifier is restarted with buried rakes, the mechanism may be injured. Rakes and helixes wear, and are renewable.

If tramp oversize or steel from the mill overflows to the classifier, it may lodge so as to project from the bedded zone. This causes rough action and undue wear of the sand-moving mechanism. Spiral edges can be notched, or given welded projections to dig such material up and get it back to the mill.

The separating point can be up to 28 mesh, or anywhere between this and *minus* 200 mesh. Several factors are used in setting the machine to deliver at and around the required mesh. Since a particle must rise from the body of the pool in order to overflow, the upward current in the vicinity of the weir is one determinant. This weir, which is adjustable for height, is at the deep end of the pool. The upward (hydraulic) current can be varied by the introduction of a baffle (see Fig. 91). The deeper this is set, the deeper the transporting current is forced before an upturn in the flow stream of pulp can take place. The distance between baffle and weir determines the cross-section of the rising column, and therefore, the rising speed of the transporting current. For a coarse-grained overflow the baffle may be set as close as 1½″ from the weir, and for finer separations as far back as 24″. It may be removed altogether for very fine work.

The finer the *release mesh,* the longer must be the settling time given to the entering feed and the gentler the stirring. The volume of pulp available for sorting purposes is therefore an important factor. It is affected by the slope of the tank bottom, the height of the weir, and the width of the classifier. Usually the slope is made as steep as is possible without risking slip-back of raked sands, in order to obtain maximum elevation. This aids their drainage and allows the use of a steeper fall to the feed box of the ball mill from rake discharge into the return feed launder.

Consider a classifier set with a slope of 3″ in the foot, and with vertical walls. The enclosed pulp has a wedge-shaped volume and a triangular longitudinal section. When the weir overflow is raised, the progressive increase in the volume of the pulp is proportional to change in longitudinal area (Table 15).

TABLE 15

Weir Height	*Longitudinal Section Area*	*% Increase*	
8 Units	128	Unity	On previous figure unity
9	162	26	26
10	200	56½	23
11	242	89	21
12	288	125	19
13	338	164	17½

The effect is to give each particle entering the pool a corresponding overall increase in settling time.

The weir height affects the capacity in the pool. It also influences the total area of horizontal cross-section in the plane of overflow. When the weir is high, it damps the surging motion set up by the plunging movement of the reciprocating rakes of the Dorr-type classifier. The greater the tonnage the classifier is handling, the bigger should be the pool, and hence the higher the weir setting.

Lowering the weir helps when a denser overflow is needed. Care must be taken, particularly when fine pulp is being sorted, not to reduce the capacity of the pool too severely or cyclic surging may become excessive. This would allow sands to escape from the closed circuit before they had been reduced to the proper overflow-mesh, and lead to trouble in the concentrating section of the plant. Consider the classifying zones shown in Fig. 91. When no baffle is used, the transporting zone merges layer by layer into the sorting zone.

Think of the pool as it would be under conditions of smooth and steady flow, with no stirring action from rakes to break up the stratification. It would soon arrange itself in a series of thin layers, sorted from the vertical cross-section of the feed and becoming more individualised as they flowed toward the weir. The surface layer would have a density approaching that of water and the bottom layer that of a high solid-liquid ratio. Further, the grain size of each mineral would be highest at the feed end and lowest at the weir end, because of the combined effect of gravity and rate of drift upon the particles moving in or through the layer. (This discussion temporarily ignores two factors—the upthrust of the layers at the weir end and the accumulation of newly arriving particles at the bottom of the pool.)

An individual particle moving in such a layer can drop through, if its gravitational force is sufficient to overcome its frictional retardation, during the period of its travel from feed to weir. It can rise into a less dense layer through eddying or if it is squeezed up and out by heavier particles. Each of the lower layers is continuously receiving such particles as they drop downward, and receives a greater load than it can part with by rejection upward, or downward, of a corresponding weight of solids. Thus the near-release particles accumulate, and the sorting zone becomes too dense to work properly. The generalized effect is that a rise in the overall density of the pool occurs, accompanied by an increase in the *pseudo-viscosity* so that new feed finds it increasingly difficult to fall through. This can be seen in operation particularly in a circuit set for fine grinding such as is frequently closed by a bowl classifier. The level of the pulp in the main tank rises and falls in a cyclic rhythm, corresponding to a gradual rise in the overall specific gravity of the pool. This is periodically relieved by a surge of heavy pulp over the weir and a drop in the overall S.G. The word "overall" includes normal running density and the heightened one of the high-density surge period, since there need be no change in that of the overflowing pulp during the period of build-up. This would be particularly the case if little or no use of a baffle was made, as the lightly loaded surface layers of the pulp would be disproportionately represented in the samples used to check the specific gravity. For this surging to occur, more solids must arrive in the pool than are departing. During the rise, the rakes become more lightly loaded although no change has been made in the rate of new feed to the mill. If at this stage vigorous action is taken to dilute the body of the pool (say, by directing a jet of water into it), the rakes load up to normal, or beyond, and proper recirculation is restored.

This phenomenon only becomes serious when the mill has insufficient classifier pool capacity for its input of ore. These layers can to some extent be mixed by raking, or by setting up turbulence in other ways, e.g. by allowing the feed to plunge into the pool from a slight height. When water is added to reduce the density of the pool it should be mixed in, not simply used as a spray wash on the raked-up sands. Unless it is mixed in, it may run as a lightly loaded layer above the sorting zone, simply washing out some fine material by its streaming action. The use of the baffle forces all surface water down into the body of the pool and aids such mixing. Provided the diluting water is properly mixed in—partly in the feed launder and partly in the body of the pool toward the weir end, the density of the overflowing pulp becomes the vital operating control. If this mixing is not assured, the value of the control is weakened.

There is a certain ratio of solid to liquid, called the critical density or the critical dilution, above which the size of particle overflowed depends upon the pulp density in the upper part of the pool. If the pulp is diluted below this critical density, the particles no longer respond significantly to this form of control. The critical density varies with the ore, temperature, and the flocculating effect of chemicals such as lime which are frequently added in the grinding circuit for reasons explained in the chapters dealing with flotation. Broadly, the percentage of solids at which change is to be expected is somewhere between 3% and 5%. The most important factor in this critical density is pseudo-viscosity or specific particle surface. A given ore has a critical density proportional to the square of the particle mesh for a given solid-liquid ratio since this governs the total area of surface in a given volume of pulp. If the feed contains clay or other primary slime, the critical density is lower, because the pseudo-viscosity is higher.

Provided the solids which should be returned to the mill for further grinding are able to fall through to the raking zone, the speed at which the rakes are run depend on:

(*a*) the degree of turbulence desired in the pool to break up the layers as they try to form;
(*b*) the amount of raking needed to withdraw settled material.

Some confusion exists in the minds of learners with regard to this. It is sometimes argued that classifier capacity can be increased by increasing the rake speed and hence the rate of withdrawal. Any increase thus obtained would be temporary. The material raked up returns to the mill, where it displaces an equal volume of crop load. All of this comes straight back to the classifier, so the only effect of faster raking could be to increase the classifier feed by the same amount. It is unlikely that this would in fact happen. The rakes can only remove the sand falling to the raking zone, and this is partly regulated by the sorting rate in the pool. In practice, the new feed to the mill is so regulated as to keep the rakes well, but not excessively, loaded. Since the circulating load usually exceeds the tonnage of new feed, any change in the raking speed affects the dwelling time of material in the grinding zone of the mill. This is controlled to an optimum time, and changes in rake speed are therefore only made as the result of tests. This also applies to changes in the weir height and position of the baffle, if one is used. Such changes should only be made after studying screen analyses made in the laboratory.

All Dorr classifiers can be fitted with a shallow bowl (Fig. 94) in which gathering rakes revolve slowly—2 r.p.m. being usual—so as to move settled material gently toward the centre. The feed from the mill is introduced centrally, and spreads radially, thus giving a greater surface area from which the particles can drop from the outward streaming pulp. The bottom of the bowl is a shallow cone dropping $2\frac{1}{2}$ in./ft. The return sands are drawn to a slot, through which they drop to the raking zone of the standard Dorr classifier below. Owing to the loss of head in this arrangement, it is not normally possible to close a circuit containing a classifier bowl unless a low-head centrifugal pump is used to elevate the mill discharge ahead of the classifier. A modification, the "bowl desiltor" has been described by H. W. Hitzrot.[1]

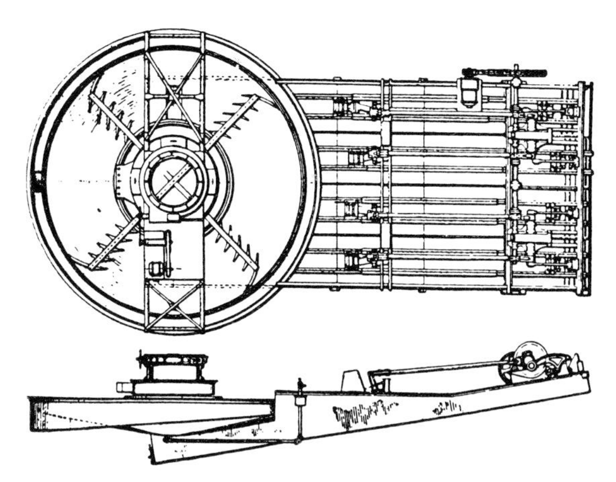

Fig. 94. *Dorr Bowl Classifier*

In another proprietary make (Fig. 95) when fineness of grinding necessitates the use of more surface settling area than can be provided in the ordinary mechanical classifier, a bowl somewhat like a miniature thickener is used. This, the hydroclassifier, works on lines similar to those of the Dorr bowl.

Fig. 95. *Denver Bowl Classifier*

Such a machine can also be used independently of mechanical classification. Hydraulic water may be introduced where the hydroclassifier is acting as a deslimer.

Except when the mechanical classifier is overloaded and is improperly surging part of its undigested load over to the concentrator in an unfinished condition, this class of machine makes no middling. The feed is split with moderate efficiency into two fractions. These are either fine plus moderately fine, or coarse plus moderately coarse. They cannot be fine and coarse, since the use of controls to separate a fine product at the weir means that the bulk of the entering feed must have been ground fairly fine. The higher the specific gravity of the ore treated, the higher must be the pool density for a given discharge mesh.

Other classifiers include the Dorr multizone, the hydro-oscillator, and the Hardinge (Chapter 12).

Centrifugal Classifiers

The rate of fall of a particle varies as its effective mass. If centrifugal force is applied, the effective mass is increased and, provided nothing happens to offset this effect, settling rate is higher. As particles are ground smaller they reach a size where the surface drag against the surrounding fluid almost neutralises the gravitational pull, with the result that the particle may need hours, or even days, to fall a few inches through still water. This slowing down of settling rate reduces the tonnage that can be handled and increases the quantity of machinery and plant required. A 10μ particle of silica settles through water at speed varying round 6 mm./min. which, for many purposes, would be too slow. By superimposing centrifugal force the gravitational pull can be tremendously increased.

The Hydro-Cyclone

During the past few years cyclones have replaced mechanical classifiers in many grinding plants. The liquid-solid cyclone (this name, and the prefix "hydro" usually being omitted) was introduced by Driessen in 1939, and the references[11-16] indicate only a fraction of the literature already published concerning its principles, applications, and performance. When a pulp is fed tangentially into a cyclone (Fig. 96) a vortex is generated about the longitudinal axis. The accompanying centrifugal acceleration increases the settling rates of the particles, the coarser of which reach the cone's wall. Here they enter a zone of reduced pressure and flow downward to the apex, through which they are discharged. The percentage of feed leaving as coarse product depends on the aperture of the inlet and vortex finder provided the underflow does not exceed some 30% of the feed. At the centre of the cyclone is a zone of low pressure and low centrifugal force which surrounds an air-filled vortex. Part of the pulp, carrying the finer particles, moves inward toward this vortex and reaches the gathering zone surrounding the air pocket. Here it is picked up by the tube called the vortex finder, and removed through a central overflow orifice. The vortex finder is so adjusted as to project into the cylindrical section of the cyclone, and short-circuiting of newly arriving pulp is thus minimised. The main controlling factors in cyclone operation are:

1. Feed Inlet Diameter.
2. Feed Pressure.
3. Feed Rate.
4. Solid-Liquid Ratio.
5. Position of Vortex Finder.
6. Diameter of Vortex Finder.
7. Diameter of Apex.
8. S.G. of Solids in Feed.

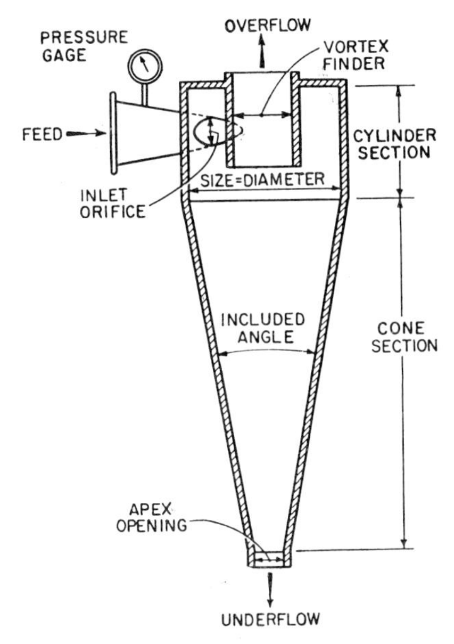

Fig. 96. *Cyclone Nomenclature*

Dahlstrom has produced an empirical equation.

$$D_{50} = \frac{97(be)^{0.68}}{Q^{0.53}\sqrt{P_s - P}} \qquad (9.5)$$

D_{50} = 50% particle diameter (μ)
b = cyclone inlet diameter (inches)
e = cyclone overflow diameter (inches)
Q = Imperial gals./min. of feed.
P_s & P are S.G. of solid and liquid in gm./cc.

The 50% particle diameter is further defined as the equilibrium particle size at which centrifugal and centripetal forces in the cyclone are so balanced that half the solids are discharged as the coarse fraction (apex discharge or underflow) and the rest *via* the central overflow.

Q in this equation is a function of inlet diameter, overflow diameter, pressure drop in transit and a constant *K* which varies with the included angle of the apex.

Tarjan's equation[17] for the size of particle which revolves in equilibrium at the circumference of the cyclone cylinder is

$$d = \frac{42e^2}{\sqrt{(P_s - P)\,h\,Q}} \qquad (9.6)$$

where h is the height of the cylinder in cm., e is given in cm. and Q in l/s. son Fahlstrom observes that the hydro-cyclone has two sets of characteristics. First are the ones fixed by construction—diameter, area of feed entry, length of drum and vortex finder, and cone angle. Second are the operating variables which include pulp concentration, feed pressure, diameter of overflow pipe and of vortex finder. The effect of the classifying action can be defined in terms of weight of yield and of percentage of solids in the underflow. Separating size is given by his equation

$$\delta = k_o\,(1 - g_u)^{1/n} \qquad (9.7)$$

and maximum sharpness of separation by

$$\eta\ \text{max.} = 1 - v_i\,(10 - 16{\cdot}7\ v_u) \qquad (9.8)$$

where δ is the size of separation, g_u the weight yield at underflow, k_o and n constants characteristics of feed size, η max. the sharpness of separation and v_i, v_u pulp density and parts by volume of solids in the feed i and apex discharge u. Precision of separation (η max.) is highest when v_i and v_u are related to the diameter of the cyclone in respect of δ. Overflow size distribution is a simple function of the diameter of the apex orifice. Distribution of solids between underflow and overflow is a function of the ratio between the areas of vortex finder and apex orifice, and either of these may be modulated for the purpose of regulation, other conditions being held steady.

The advantages which have led to the widespread adoption of cyclones in Rand practice are, according to Krebs:

1. Sharper classification.
2. Saving of floor space.
3. Less power consumption.
4. Less maintenance.
5. Ability to shut down the mill immediately under full load.
6. Ability to bring the circuit rapidly into balance.
7. Elimination of cyclic surging.

In addition, operators of flotation plants claim operating benefits due to the higher percentage of solids in the overflow, and reduced dwelling time in the closed circuit. This latter consideration will be better understood when the effect of oxygen on newly sheared surfaces has been considered.

The cyclone is increasingly used for classification in the finer grinding ranges, between 150μ and 5μ, although coarser separations are possible. Separating efficiency is measurable as the percentage of misplaced product in either the overflow or underflow. Confusion can arise in assessing operating efficiency if the apex discharge is not under proper control. With a sprayed discharge the issuing solids should not exceed some 70% by weight of the total amount leaving at the apex. It is sometimes set to give a thick underflow carrying a higher percentage of solids. This means that the cyclone is overloaded and that some oversized material is unable to report with the apex discharge,

Fig. 97. *Multi Cyclone* (Liquid-Solid Separations)

and consequently is overflowed with the finer fraction. The correct adjustment can be made by varying the diameter of the apex discharge. This can be automated by pneumatic or hydraulic adjustment of the apex, signals being initiated in the vortex chamber or feed zone.

The cyclone does not act as an effective substitute for the thickener in dealing with material below about 5μ in size. It can, however, concentrate the feed to a thickener or, alternatively, remove the bulk of the solids in the underflow. The overflow then carries a relatively small percentage of the finest solids, and can be led to any convenient settling area for further thickening. The tonnage of solids deposited in the re-thickening of the cyclone overflow would be too small to present serious handling difficulties.

An assembly of cyclones fed from a central distributing point is illustrated in Fig. 97. A testing unit for laboratory use is shown in Fig. 98, and a large cyclone in Fig. 99.

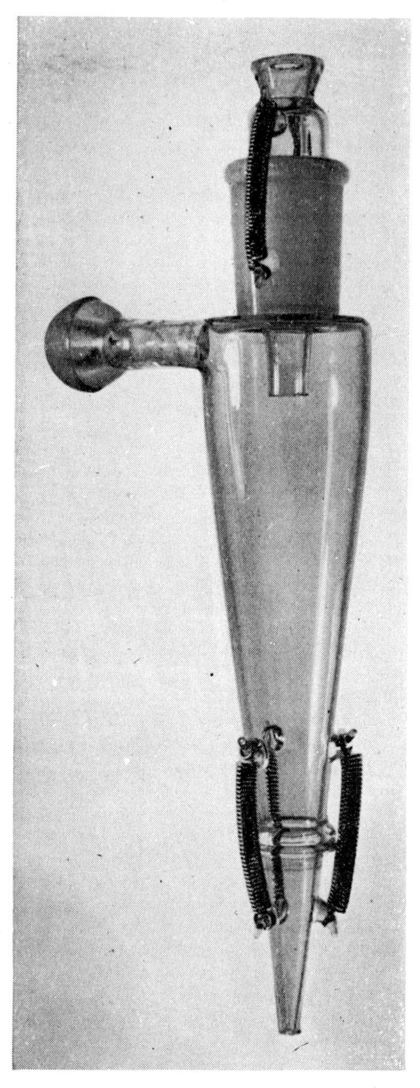

Fig. 98. *Pyrex Hydrocyclone* (Liquid-Solid Separations)

When classification of much finer material is needed, mechanical spinning of the separating vessel can be used as in the Dynocone (Fig. 100). This consists of the conical revolving shell in which a screw conveyor rotates at a slightly higher speed. The solids settling to the inner wall of the cyclone are

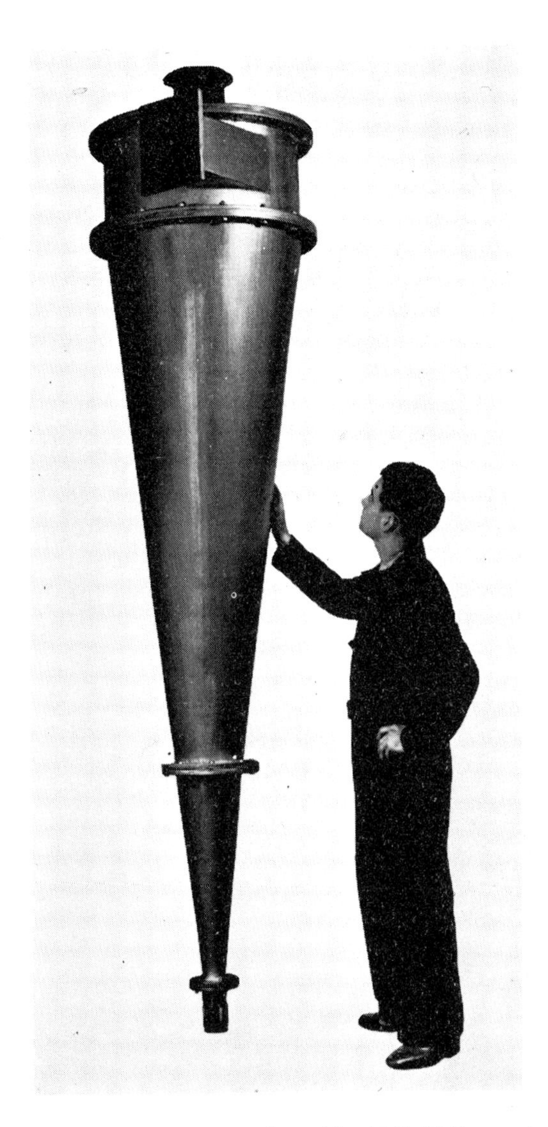

Fig. 99. *Rubber-Lined Hydrocyclone* (Liquid-Solid Separations)

discharged by means of the screw, while the finer fraction overflows at the other end.

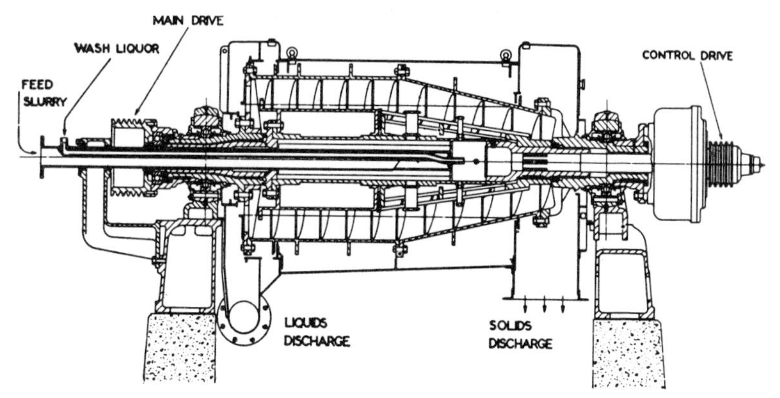

Fig. 100. *Section through Dynocone Centrifugal Classifier* (International Combustion)

Air Sizing and Dust Control

When a particle is in movement in a liquid, part of its kinetic energy is expended in overcoming the viscosity, or molecular adhesion, of that liquid. If it moves through the liquid quickly enough to generate vortices, a substantial transfer of kinetic energy must occur in order to give these vortices sufficient power to thrust their way outward from the particle and maintain a temporary existence. When a particle moves through a gentle current of air viscous resistance is negligible. The only effective resistance encountered by the moving particle is the pressure-effect due to collision with gas molecules, which is a function (*a*) of the cross-section of the particle and (*b*) of its speed of motion relative to the air. Where the number of particles in a given volume is low, collision between these particles is negligible and they are "free-settling". As the crowding increases, collision multiplies and a change in settlement behaviour occurs. The interested reader will find mathematical treatment of the subject in Taggart[18] and Dallavalle.[19] Air is used for classification in dry grinding, and air currents are also so manipulated as to collect dust and aid in its deposition in suitable containers. In the first of these two cases, a split is made between larger and smaller particles as part of ore preparation. In the second, all possible dust is removed from the air in order to mitigate an industrial nuisance.

When air is the selective fluid in a sorting operation, reliance is placed on the use of an equable current of air, moving with as little unplanned turbulence as possible. If the air stream is vertical, the particle is either carried up or down as an elutriation. If it is horizontal, particles entering at one point fall out of the stream in accordance with their mass (or the time they take to fall through the stream) (Fig. 101).

The largest of the three spheres starting from rest at the point marked by an arrow falls fastest. It is therefore exposed for the shortest period to the displacing effect of the horizontal air stream. The smallest drifts the furthest under the same conditions.

The sheltering, and the frictional drag upon air and particles touching the containing walls of a vertical vessel, is sufficiently pronounced to require special precautions. Build-up of particles on these containing walls may occur, and machines used in air sizing and dust control sometimes contain devices for preventing this from becoming serious. Friction in a dry atmosphere causes electrostatic forces to be generated upon the particles, and aids their adhesion to one another and to the walls of the vessel. Apart from effects of this kind, the physical characteristics which determine how a given particle will behave in an air stream are its size, shape, density, and liability to collide with other particles (dispersion or concentration). The effects produced by the air stream depend on its velocity, humidity, viscosity (a pressure effect), and the way in which it is constrained to move by the containing walls of the appliance in which it operates. The settling velocity of a particle in still air is quoted by Taggart for irregular grains of S.G. 2·5 as 3140 ft./min. for 5000μ particles; 470 ft./min. for 500μ; 50 ft./min. for 24μ and 0·5 ft./min. for 0·0032μ or 3·2mμ.

If the velocity of the transporting stream of air is suitably controlled, relatively coarse particles are dropped while finer ones are carried onward. Baffles and deflecters may be placed in the stream to sort out the coarse particles in accordance with their inertia. In the cyclone a centrifugal sorting action is set up, the dust finding its way to the sides and then falling down while a vortex of comparatively dust-free air rises at the centre. The air

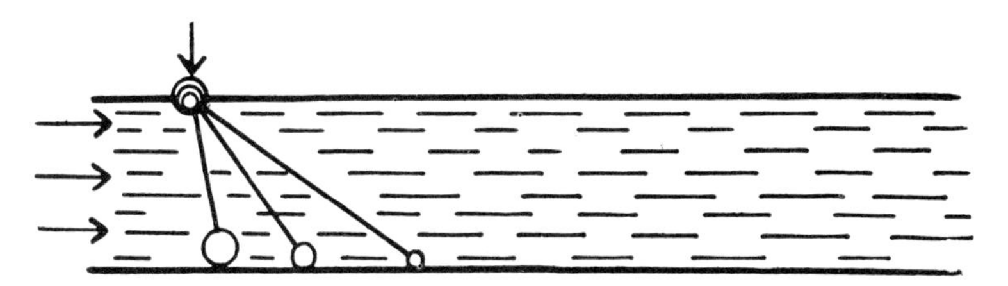

Fig. 101. *Relative Fall through a Horizontal Air Stream*

is usually moved by a fan, working either to push a clean column of air forward or to draw dusty air through the system. To avoid wear of the blades, fan-power should be applied to clean air, or, alternatively, after the coarse particles have been trapped. A modified form of air classifier—the "gravitational-inertial"—for removing *minus* 100 mesh material from crushed limestone has been developed.[20] (Fig. 102.)

Despite the application of centrifugal force, the finest particles of dust cannot be trapped in a cyclone. Such dust might be a menace to the health of the workers. Various methods are used to minimise this risk. Cyclones can be used to remove dust coarser than 5μ. Parts of the mill operations which set up contamination should be enclosed and kept under a slight vacuum, so that dusty air does not leak out.

Electronic dust precipitation is used industrially to deal with particles too fine to settle by gravity. When a wire of small diameter receives a high charge of electricity a corona of gas molecules forms around it and these

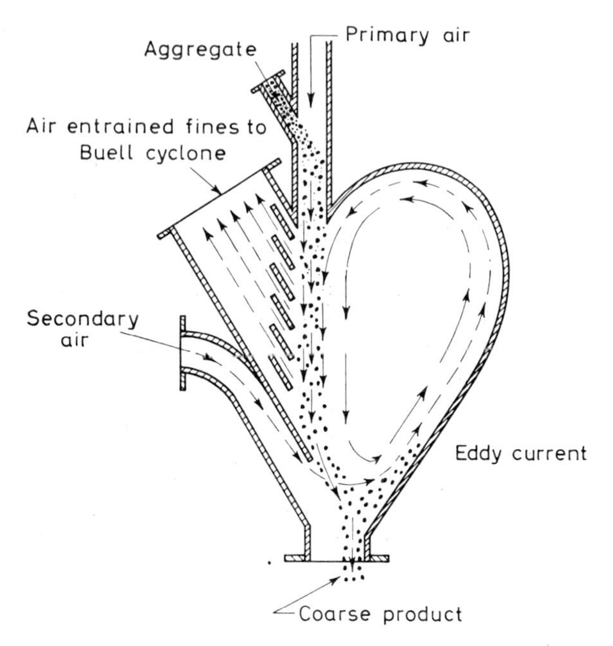

Fig. 102. The Gravitational Inertial Classifier

molecules collide and ionise. The ionised molecules can then be attracted toward a near-by grounded conductor. As they migrate they charge any passing dust particles they may collide with, and influence them to travel toward the grounded electrode. Dust can also be precipitated by passing it through a duct in which supersonic waves are vibrating at a sufficient intensity to cause the particles to flocculate into rapidly settling clots.

Closed Circuit Concentration

If two particles of the same shape and size, but of different specific gravities, are fed into a mechanical classifier, the heavier one may sink and be returned for further grinding while the lighter one overflows at the weir. Under such circumstances the mechanical classifier becomes a concentrating machine. This possibility was noted above when classifiers were referred to as sorting devices, not sizers. It is often undesirable that particles should remain in the closed circuit in this way. A metallic sulphide, once properly liberated, can be caught in the concentrating section of the plant more efficiently when it is still somewhat coarse than after it has been ground to an impalpable slime. A particle of free gold is usually malleable and merely changes its shape as the result of further passages through the mill. When necessary, effects of this kind are partially countered by the introduction into the closed circuit of a moderately efficient concentrating device appropriate to the working conditions. This is used to remove as much as possible of the desired value in the form of a rough concentrate, in order to minimise its over-grinding or accumulation in the closed circuit. Such appliances are colloquially termed "scalpers". Jigs, flotation cells, hydraulic classifiers, and corduroy strakes are among the devices used for this purpose.

Classifiers v. Cyclones

The closed circuit in wet grinding is rarely controlled by screening. The choice lies principally between mechanical classifiers and cyclones, though flow-sheets using gravity processes of concentration may have hydraulic classifiers for reasons which are discussed in Chapter 14. The growth in use of the cyclone has been rapid during the past few years, and data for adequate comparison with the older appliances are still inadequate. The mechanical classifier has the advantages over its rivals of tolerating and smoothing out surges, of returning its oversize to the mill launder without need for an extra machine, of being robust, long wearing, and easy to control. The screen fails because it cannot handle fine sand efficiently, and is therefore unsuitable for liberation of most of the low-grade and fine-grained ores which are treated today. Use of the cyclone owes its phenomenal growth to several facts. First, by its use of centrifugal force it can speed up settlement rate and therefore either handle larger tonnages with light equipment in a small space, or make a separation at finer meshes than can the classifier. Next, running costs are comparable but capital cost and installation are far cheaper. Third, by returning its oversize direct to the feed trunnion, it does away with the need for a scoop and feed box. Next, a point which specially affects flotation, it only keeps a small tonnage in circulation and hence reduces oxidising effects in the grinding circuit. Last, the limitation in fine grinding when the circuit is closed by a bowl classifier is that only a moderate circulating load is possible, the limit being dictated by the free-settling speed of the near-release particles. Using the accelerated settlement due to centrifugal force the cyclone makes possible a large circulating load. Repeated passage through the secondary mill, as with the primary one, lessens over-grinding.

Classifier Efficiency

This is checked by making screen analyses of properly collected samples of (*a*) material entering the classifier, (*b*) coarse discharge, (*c*) fine discharge.

These sizing analyses, plotted as "direct" graphs, should show a marked peak at the desired mesh of maximum separation, while the relation of the undersize and oversize to this peak mesh is a guide to behaviour in the circuit under the conditions ruling when the sample is taken. If operating changes are called for, they must be made one at a time, empirically. It must be remembered when judging efficiency that each change made anywhere in the closed circuit affects the whole circuit. Classifier efficiency cannot safely be considered in isolation from the conditions in the grinding mill and/or concentrator.

The control used by the shiftsman during the hour-by-hour running of the wet-grinding closed circuit corrects the density of the overflowing pulp. Provided this pulp is truly representative of the conditions in the top few inches of the classifier pool and is not upset by the presence of an overrunning stream of diluting water, its two sorting components—pulp density and hydraulic transporting energy—exercise effective control over the mesh sizes in the overflow. The accurate control of pulp density is therefore a major factor in operating efficiency. Haultain[21] observes that most bowl classifiers are worked with a cyclic surge of between 15 min. and 30 min. duration, during which the pulp in the rake compartment rises anywhere up to 9″ above the normal working level, while corresponding variations in the sorting action takes place.

References

1. Hitzrot, H. W. (1957). *I.M.P.C.* (*Stockholm*), Almqvist & Wiksell.
2. Hukki, R. T. (1960). *Aufbereitungs-Technik*, 12.
3. Kuenhold, N. (1957). *Min. Eng.*, June.
4. Truscott, S. J. (1923). *Textbook of Ore Dressing*, MacMillan.
5. King, D. L., and Schepman, B. A. (1962). *Trans. S.M.E.*, 223.
6. Coe, H. S., and Clevenger, G. H. (1916). *Trans. A.I.M.E.*, 55.
7. Kynch, G. J. (1952). *Trans. Faraday Soc.*, 48.
8. Fitch, B. (1962). *Trans. S.M.E.*, 223.
9. Cross, H. E. (1963). *J. S. Af. I.M.M.*, Feb.
10. Gaudin, A. M., and Fuerstenau, M. C. (1960). *I.M.P.C.* (*Lond.*), I.M.M.
11. Bradley, D. (1960). *Ibid.*
12. Peachey, C. G. (1960). *Ibid.*
13. Kelsall, D. F., and Holmes, J. A. (1960). *Ibid.*
14. Cohen, E., and Isherwood, R. J. (1960). *Ibid.*
15. Stas, M. (1957). *I.M.P.C.* (*Stockholm*), Almqvist & Wiksell.
16. Fahlstrom, P. H. son. (1963). *I.M.P.C.* (*Cannes*), Pergamon.
17. Tarjan, G. (1950). *Acta Tech. Hung.*, No. 1.
18. Taggart, A. F. (1945). *Handbook of Mineral Dressing*, Chapman & Hall.
19. Dallavalle, J. M. (1948). *Micromeritics*, Pitman.
20. Anon. (1960). *Min. Eng. A.I.M.M.E.*, Nov.
21. Haultain, H. E. T. *Can. Min. J.*, 67.

CHAPTER 10

GRINDING CIRCUIT CONTROL

Preliminary

This chapter is concerned with the routine control of the closed circuit in the wet-grinding section of the plant. Sampling control, as designed to check the state of the pulp at critical stages in its treatment and end-product disposal, is discussed more fully in Chapter 23.

Control of the closed circuit has three aspects:

(*a*) Immediate running control.
(*b*) Long-term quality control, as it affects assay grades.
(*c*) Capacity control, which considers wear rates and economic through-put.

The first of these is usually limited in its scope, and is designed to give the shiftsman discretion to vary a few simple settings in order to maintain the correct mesh-of-grind and solid-liquid ratio within permissible limits. With the increasing use of automatic control, possibly linked by computer with the overall operation, the scope and character of human shift-working are changing. In some new plants an analogue computer integrates aspects (*a*), (*b*), and (*c*) and restricts local intervention which might interfere with the overall control plan. Subject to such considerations, which may include operating elements outside the concentrator, items (*b*) and (*c*) include controls only applied by the mill superintendent. They are rarely of a type which can be entrusted to the workmen to vary.

Every alteration in the character of the ore pulp escaping over the classifier weir into the concentrating section of the plant results in a definite change in the working conditions there. It is therefore not possible to give an operator in the grinding section discretion to make changes in his circuit which can lead to such effects later on. The only person who can properly initiate such major changes is the one who also can watch and foresee (or be prepared for) their effects all along the flow-line. Since most of these effects are slow to show themselves and are only visible as assay-return changes, the mill super-intendent cannot delegate initiative in introducing changes in the grinding section. After the bulk of the concentrate has been removed, and the pulp approaches the final stages of treatment, greater discretion can be delegated to the shiftsman. The grinding requirement in all good plant practice is a constant tonnage rate of pulp held closely to its optimum grind and solid-liquid ratio. This is in control of the shiftsman until the pulp overflows the final classifier weir on its way to the concentrating section, and it is his job to make sure nothing happens ahead of that weir which results in the overflow of unsuitable pulp.

It may be helpful at this point to recall the idea of the sectional or block flow-sheet discussed in Chapter 1. Transfer from grinding to concentration is the most important sectional step in the whole series of operations. Mistakes made in the crushing section can be retrieved at the cost of efficiency by grinding adjustment. Mistakes in grinding affect quality and quantity of value recovered, and must therefore be appropriately guarded against. Once the main work of treatment has been completed, risks may be taken so toward the tailings end the shiftsman can be given more freedom to vary his controls. Hence, the slightly paradoxical position is that the best workmen may be needed in the grinding section, yet they may be given the least discretion as to their activities. Vigilance in this case is of a higher order, concerned to check trouble at its source and not to restore order or take chances. For such reasons as these automatic control is in increasing use in the grinding circuit. The bulk of this chapter deals with control as a human activity used to apply basic principles and maintain running norms.

Running Control

By "running controls" are meant the adjustments of the closed circuit which can be made by the shiftsman in order to hold the delivery of finished pulp at the required mesh-of-grind. Consider the closed circuit (Fig. 103).

The new feed varies in size, grindability and moistness. The returned sand varies in grain size and tonnage rate. If the operator's instructions are that the crop load is to be held to a solid-liquid ratio of, say 70:30, then he must adjust the feed water (2) so that 30 units by weight are scooped in for every 70 units by weight of ore.

Hence:

$$\frac{\text{New ore (dry)} + \text{Return ore (dry)}}{\text{Moisture } (1+2+5)} = \frac{70}{30}.$$

If the specific gravity of the 70 units of mineral is 2·8, then the volume of 100 weight units of pulp is $\frac{70}{2 \cdot 8} + 30 = 55$ and the specific gravity $\frac{\text{(weight)}}{\text{(volume)}}$ is $\frac{100}{55}$ or 1·82. The density of the mill discharge must be held at this figure. To check this, the shiftsman periodically weighs a known volume of pulp, and if the weight is wrong he varies the feed water (2) slightly in order to restore the correct solid-liquid ratio. The next control he may vary is the rate of new feed. Since the total load of balls in the mill is constant and the total weight of ore in the crop load is a definite fraction of this, the total weight of ore ((1) + (5)) fed to the mill should be constant. The amount returning at (5) is the amount in the mill discharge (4) less the amount overflowed (6). It therefore varies with the grindability of the ore, and adjustments are made to (1) to compensate these variations. The shiftsman may be guided by the power intake of the mill (ammeter reading); the height at which the return sands stand on the rakes: or the grinding noise of the mill. Any variation in grindability, or in fineness of (4), alters the rate of settlement through the

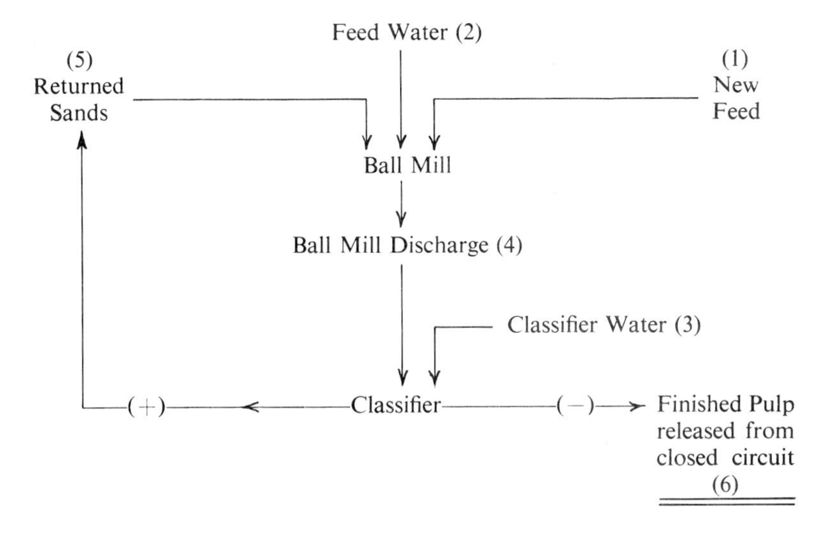

Fig. 103. *Control of the Grinding Circuit*

pool of pulp in the classifier and in due course shows at the overflow as a change in pulp density. The third control is therefore (3) the volume of diluting water mixed intimately with the mill discharge pulp to form the pool pulp. This must be varied so as to maintain the correct pulp density at the weir overflow. The figure should never be allowed to vary when overflow pulp is being sent to the concentrating section, no matter what emergency may arise. Summarising, the shiftsman may vary any or all of the following:

(*a*) (1) Rate of new feed.
(*b*) (2) Flow rate of mill water.
(*c*) (3) Flow rate of classifier water.

As these rates react on one another, he must make any variation called for with due regard to its effect on the others, and must hold the pulp density of mill discharge and weir overflow at the specified values while maintaining the optimum load of ore in the mill crop. When these controls are integrated with reliable monitoring devices which check continuously on such running signals as noise level in the mill, power draft, density and flowrate changes in mill discharge and classifier overflow, tonnage rates of new and returned feed, it is inevitable that the swifter and more comprehensive coverage thus afforded must exceed human ability in detecting variance, and in aiding corrective measures.

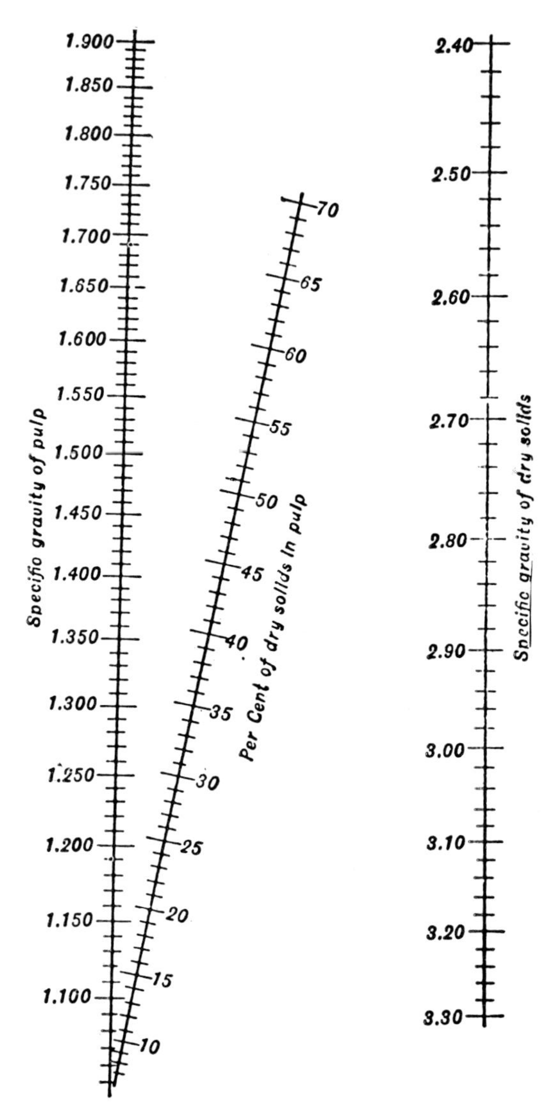

Fig. 104. *Pulp Density Nomogram* (after Heinz)

The effect of varying each of these controls is set out in Table 16.

TABLE 16

Controlled Variable	*Effect*	
	Increase	*Decrease*
(*a*) Rate of new feed (1).	Average particle size larger, coarser discharge and faster throughput. Higher circulating load.	Longer dwelling time, finer grind, less return sand, more over-ground particles in weir overflow.
(*b*) Mill water (2).	Flushes fines through mill. Increases wear of steel.	Increases coating on balls, slows progress through mill. May choke mill.
(*c*) Water to classifier (3).	Finer release mesh. Greater circulating load.	Coarser overflow. Lower circulating load.

Handling the Controls

When balance must be restored to a closed circuit which has become upset, the shiftsman begins by adjusting the pulp density at the weir overflow. The response is rapid, and it safeguards the release mesh while he finds what has happened and decides what to do next. When a working condition has changed slightly the circuit as a whole is slow to respond, and may even take hours to show that it is upset, so the cause of the trouble is not necessarily recent. Provided the classifier is not discharging tramp oversize in cyclic surges, there is time to think before taking further action. With experience, the shiftsman is unlikely to lose control. He watches the new feed, and if it is coarser than usual he is prepared for the extra work being thrown on the mill. Perhaps he has noticed a slight rise of the height of ore on the classifier rakes, showing that the circulating load is building up, and he may have to cut down the new feed a little to avoid an overload. He may see that the ore—a friable sulphide, say—is richer than usual, and therefore knows that it will grind more speedily. He will be alert to take advantage of this by raising the feed rate a little.

How does the circuit become unbalanced when all controls are in working order? The most likely cause is a change in the physical state of the new feed. Coarse lump, and usually also low-grade ore, takes longer to reduce to mesh-of-grind and must be fed in more slowly. Otherwise one of two things will happen—a circuit overload, or a discharge of tramp oversize to the concentrator. Finely crushed ore, and high-grade sulphide, is more quickly finished and can be fed in faster. Since the sulphide is a better lubricant in the crop load than sand, the experienced operator listens to the sound of the mill and may decrease the solid-liquid ratio a shade. At the same time he watches his weir overflow in case this dense sulphide is failing to rise, and needs a slightly higher pulp density to help it. He must, however, have special authority if he is to be allowed to vary this control.

The unbalancing of the circuit is therefore due to a change in the amount of work required to produce a given end-condition. Since the work input is constant, the quantity of ore must be varied in accordance with its varying grindability. Where the scale of mining and the nature of the ore body justifies the cost, blending of the ore ahead of grinding may be practised.

One of the most serious emergencies that can arise is the plugging of the grates of a low-discharge mill. If they have become partially clogged with rock or small balls, the new feed is stopped, the mill kept running until the return sands are negligible, and the mill is then stopped while the grates are freed. It may be desirable to raise the classifier rakes at the same time, to prevent an accumulation of return sands in the feed launder and scoop box. If the feed rate has been too high or the mill water has failed, the remedy is obvious. The worst position arises when the grates have become completely choked or "plugged". The mill runs silently, as the balls are no longer free to move, and as the scoop goes round in the overfilled launder, return sands are being thrown out on the floor. New feed must be stopped immediately. If the mill is discharging at the central orifice, classifier sands are still being returned, and for the moment nothing can be done about it. The problem is to get the grates unplugged without starting a torrent of pent-up material which may rush over the classifier weir and through to the concentrator. If the layout of the circuit permits, the safest thing to do is to open the discharge port at the bottom of the classifier tank so that its contents run out to the floor. This "emergency exit" is not shown in Fig. 92. It is a simple orifice, closed by a valve or plug. When the classifier has been emptied, the water supply to the head of the mill can be gradually restored. If there is a violent outrush of mud, the mill bearings should be inspected before returning to normal working. When things have been put straight, the classifier port is closed, rakes are lowered, and all spilt sand and rock is returned to the circuit before restoring new feed. The cause of this rare trouble may be failure of feed water, but is most likely to arise from the shiftsman's inexperience or inattention.

Symptoms and their Correction

When the grinding circuit is becoming unbalanced, the alert shiftsman should be able to see what is happening before things go seriously wrong. In Table 17 some of the main discernible phenomena are set out in Column 2, the causes or effects which are probably directly associated with them in Column 3, and the appropriate counter-control in Column 4, while the point in the circuit at which the phenomenon is first seen is given in Column 1.

All these adjustments are dictated by the need to balance the varying condition of the newly arriving ore till it is in adjustment with the fixed amount of energy supplied to the crop load. The returned sands (Fig. 103) have a steadying effect, but cannot directly modify the load on which the work acts.

TABLE 17

Item (1)	*Phenomenon* (2)	*Possible Associated Changes* (3)	*General Nature of Control* (4)
NewFeed from Ore Bins.	Ore finer or richer than usual.	Less grinding needed to produce finished *m.o.g.** Pool density may be upset by over-grinding.	Increase feed rate. Watch pool overflow density.
	Ore coarser or tougher (lower grade) than normal.	Longer grinding needed. Large pieces of ore may accumulate in crop. Circulating load may rise.	Reduce feed rate, also feed water. Pool may need less water.
Mill Discharge.	Solid-liquid ratio has dropped, and pulp is too dilute.	Feed rate too low, or feed blocked. Return sands too low. Feed water too high.	Check feed and feed water. See if classifier is surging, and pool water on.
	Solid-liquid ratio too high.	Feed too fast, feed water low, or failed. Ammeter may drop if mill overloads, classifier may send out oversize.	Check feed rate. Perhaps stop new feed. Increase feed water cautiously. Check pool density.
	Stopped.	Grates plugged. Return sands spilling from feed launder.	Stop new feed. Run out classifier. Then check feed water and restore it if failed. Be prepared for surge of mud.
Mill Noises.	Noisy, rattling.	Not enough feed.	Check feeding.
	Quiet, choking.	Too much feed. Too high solid-liquid ratio.	Correct feed. Cut pool water if safe, to decrease return.
Ball Mill	Amps dropping.	Mill either overloaded and too quiet or underloaded and too noisy.	Check feed and pool water.
	Amps rising.	If abnormal, mechanical defect developing.	Check bearings and lubrication.
Classifier Weir Overflow.	Density rising.	Mill discharge finer. Too little feed water. Pool dilution insufficient.	Gently increase water to mill and/or classifier.

* m.o.g.—*mesh-of-grind*

TABLE 17 (*continued*)

	Density falling.	Mill discharge coarser. Load rushing through too fast. Mill water too abundant. Classifier water too high.	Lessen mill and/or classifier water gently.
Classifier Circulating Load.	Rising.	Pool too dilute. Mill overfed. New feed too coarse.	Watch rakes. If they overload, reduce new feed.
	Falling.	Pool density too high. Classifier surging. New feed fine, rich, too slow.	Find which item is abnormal, and act accordingly. Perhaps dilute pool.

These are the chief control measures, and it will be noticed that they can be reasoned out once the shiftsman has noticed at which of the three points the quantity or quality of the new addition—new ore, feed water, pool water—has changed or become unbalanced. A steady deterioration in, say, throughput of correctly finished ore would be due to maladjustments of a less temporary nature, such as incorrect ball loading or a misplaced baffle.

Starting-up and Stopping

In a well-controlled plant, the grinding mill rarely stops for any unexpected reason except power failure. Routine maintenance overhaul is foreseen, planned, and arranged to fit in with the mill superintendent's schedules for such work.

Stoppages of a grinding plant are of two kinds, planned and involuntary. The involuntary stoppage arises either from breakdown somewhere along the flow-line which involves the mill, from power failure, or from a mechanical defect in the grinding circuit itself. Planned stoppages can be divided into two types—long stoppages for major overhaul, and brief stoppages for minor repair and check-up. Take first the sudden and unpremeditated shutdown. If the whole power supply to the system has failed, the first step is to see that all switches are "off" so that a sudden restoration will not strain the motionless machinery by throwing on full power. The place where trouble is most likely to develop in the circuit is the pool of the rake classifier. The moment the rakes stop, sand begins to settle, burying them. If this happens, the rakes must be freed before restarting the classifier. If power failure has been general there can be a standing instruction that everyone available must start to turn the classifier over by hand until its load has been raked up and out. The alternative is that the pool's contents be run out to the floor and the rakes raised. With spirals, and some modern rake classifiers, the lifting gear alone need be used and the rakes can be gently lowered after restarting. If the mill discharge is being pumped to the next point in the flow-line, the power failure may lead to settlement of sands in the pumping line, tanks and pump casings, which must therefore receive appropriate attention. If part only of the grinding circuit has broken down it is probably necessary to shut down the rest. New feed should be stopped at once. The water can wait, but solids must not be allowed to pile up. The sections receiving pulp from the grinding circuit should be warned of the interruption and kept advised of developments.

Emergencies apart, it is sometimes necessary to shut down for a brief period in order to make some running adjustment. A liner bolt may have worked loose, a broken classifier rake may need removal, or some other small matter may enforce a short stoppage. Before stopping, the necessary tools and spare parts are made ready and anybody concerned is assigned his job. This is important because the repair must be finished before the classifier has had time to rake up a big load of return sands to the return launder. A good shiftsman should be able to shut down, complete his repair, make a rapid inspection of pinion, crown wheel, and shell, and have the mill restarted in less than one minute. If he needs much longer than this a proper circuit shut-down is necessary. This, unless the classifier is of a type which can be restarted under load, is a more detailed business, and starts with the "grinding out" of the circulating load until mill and classifier are only lightly charged with ore. If the circuit is closed by a cyclone, the load undergoing classification is small and shut-down is consequently simpler. The sequence of operations is:

(*a*) Warn sections affected.
(*b*) Cut off new feed.
(*c*) When return load is small enough, stop the mill (this may be 20 minutes or longer after (*b*)).
(*d*) Raise the classifier rakes.
(*e*) Stop the classifier.
(*f*) Stop mill and classifier water.
(*g*) Stop classifier overflow pump.

If reagents are being fed into the circuit, they should be shut down with (*c*). At this stage stores required should be assembled and fitters ready. A stop of this kind is usually made on the instruction of the mill superintendent and arranged to take place during the day shift when the Company stores are open. The supervising engineer should make a list of points he proposes to have checked, including the condition of rakes, bearings, drives, submerged classifier bearings, liners, and the wearing part of the scoop feeder. When the work is completed, he should check over his list, see that loose tools have been removed and that all bolts are secured. Before starting up, it may be wise to make sure nobody has taken advantage of the occasion to take a siesta in a dangerous dark corner. The usual order for restarting is:

(*a*) Warn the concentrator sections.
(*b*) Put on the lubricators, if shut off during repair.
(*c*) Start the classifier.
(*d*) Start the mill ("inching" it round to begin).
(*e*) Start feed and mill headwater.

(*f*) Lower the rakes.
(*g*) Start flow of reagents.
(*h*) Start the pool water, roughly judged.
(*i*) Start the classifier overflow pump as overflow begins.
(*j*) Build up to a normal circulating load.
(*k*) Adjust weir density to normal.
(*l*) Check mill discharge density.
(*m*) If the mill has a capacitor to control the power factor of the plant, check its setting.

Efficiency Control

The controls dealt with thus far are operated by the shiftsman as he steadies the grinding circuit. There are several other running adjustments beside that of feed and water, but these should only be made under the instruction of the mill superintendent. They have one or more of the following aims:

(*a*) Reduction of tailings loss.
(*b*) Improvement in concentrate grade.
(*c*) Reduction of ball and liner wear.
(*d*) Saving of power.
(*e*) Increase in throughput of properly liberated ore.
(*f*) Lowering of grinding costs.

All these items are interdependent. The increase in throughput of properly liberated ore can best be achieved by reducing over-grinding to its minimum and increasing the inevitable slight rise in oversized particles to its safe limit. Since ball and liner wear are proportional to power used and tonnage ground, each improvement in item (*e*) at the same time improves (*c*), (*d*), and (*f*). The approach to higher efficiency in an operating circuit is usually made by trial-and-error methods, and is therefore made cautiously. Any variation which causes a serious loss of values is expensive, and should be avoided. The golden rules in bringing up the grinding circuit to peak efficiency should be remembered. They are:

(*a*) Go slowly.
(*b*) Only change one thing at a time.
(*c*) Give each change plenty of time to make its effect felt right through the plant.
(*d*) Observe the effect of one change for a substantial period before making another.

It is cheaper in the end to wait a little than to rush enthusiastically ahead. An alteration which seems completely logical and safe may have ignored some detail in the layout of the plant which makes it impracticable. Multiple variations can be made in test-work in a pilot plant where a limited tonnage is exposed to experimental hazard, but are out of place in a full-scale commercial operation. The methods used in bringing a circuit up to higher efficiency are dealt with more fully in Chapter 20. They enable the mill superintendent to select the best conditions for each important factor in the combined operation. In closed-circuit grinding these factors include:

Place	*Detail*
Ball Mill	Liner Contours
	Mill speed
	Solid-liquid ratio
	Ball load
	Ball size ratio
Classifier	Rake speed
	Weir height
	Baffle depth
	Baffle distance from weir
	Circulating load
	Water addition technique

to name some of the more important. It is obviously not possible to arrive at accurate conclusions simply by varying any of these conditions in a commercial circuit, since their mutual reaction would make complete observation impossible and false conclusions would be drawn. Efficiency control in the working circuit is obtained by holding all the selected factors (or elements) interacting in that circuit to the optimum values laid down as the results of tests made under laboratory control. The methods of maintaining this control depend on careful sampling of the pulp passing through the circuit at key points, in conscientious accordance with a routine procedure. This is followed by a routine laboratory testing of these samples, devised to disclose any shift-by-shift variations.

Such sampling may include:

Place	*Material*	*Information Given*
Mill Head	New Ore	Assay value. Size analysis, moisture.
Mill Discharge	Pulp	Solid-liquid ratio, *p*H. Size analysis.
Classifier Overflow	Pulp	Solid-liquid ratio, *p*H. Size analysis, assay value.
Classifier Returns	Return Sands	Ratio of circulating load to new feed. Sizing analysis.

After studying the returns for a given period, and correlating them with the results reported from the concentrating section for the same lot of pulp, information is gathered which suggests useful lines of process research. These are carried out by laboratory methods where possible, and can then be introduced into the plant in an orderly manner.

Automation in the Grinding Circuit

A number of running and radical adjustments have now been considered in terms of human control. These interact, and change in one is liable to throw the circuit out of balance in a way which misleads the operator into applying the wrong corrective. This unbalancing proceeds slowly, and correction can be tedious. The final objective is usually a steady flotation feed correctly ground and conditioned, and this can only be achieved through steady grinding and release. Here automation can help. Fuller discussion of methods and economic considerations is deferred to later chapters, but some general observations are relevant at this point.

The mineral processing industry must meet four main challenges.

(*a*) Rising cost of treatment.
(*b*) Lowered grade of ore.
(*c*) Increasing complexity of both ore bodies and treatment.
(*d*) More stringent specifications on quality of mill products.

The metallurgist must provide industry with highgrade ("advanced") metals and to do this he passes part of the technical problem back to the concentrating plant. This in its turn demands new standards of liberation in the grinding circuit, only to be met by higher operating skills. Some of these are best applied by automation and closed-loop computer control.

A short list of conditions in the closed circuit which can be monitored and/or automatically controlled is given in Table 18. Save for the entries marked *, all these items are thus handled in some plants. Each success-

TABLE 18

Operation	*Method/s*
	(A) THE FEED.
1. Pre-blending	Computer-controlled mine valuation and stoping
2. Blending from bins	Constant-weight delivery to a common blending belt
3. Blending into bins	Mines' controlled delivery
4. Preliminary *p*H control	*Lime to ore bin
5. Sorting	(*a*) Radio-active check and rejection
	(*b*) Colour check
	(*c*) Conductivity check
	(*d*) Detection of tramp iron
6. Sizing	(*a*) Oversize warning by electric eye
	(*b*) "Pebble" selection for autogenous grinding, by screening
7. Feed rate	See under "Mill"
	(B) MILL
8. Crushing bodies	(*a*) Power draft change
	(*b*) Sound-level check + pebble feeder
	*(*c*) Check on crop load level
9. Solid-liquid ratio	(*a*) Flowmeter for mill water controlled by discharge density
	(*b*) Flow rate and density check at mill discharge, controlling new feed and mill water
	(*c*) Pulp temperature change, feed to discharge
	(*d*) Sound-level check linked to new feed
	*(*e*) Size analysis at mill discharge
	(*f*) Power draft variance signalled to new feed: with D.C. mill varying mill speed
	(C) CLASSIFICATION
10. S.G. in pool or overflow	Back pressure on bubble pipe
11. Rake load	Ammeter, electric eye and feeder link
12. Pulp *p*H	*p*H electrode system controlling reagent addition
13. Assay grade check	X-ray fluorescence scan of classifier overflow linked to reagent feeders, etc.
14. Cyclone load	(*a*) Vortex probe controlling apex aperture
	(*b*) Automatic weighing of underflow
	(*c*) Gamma-ray, radio-activity or ultrasonic check on rate and density of overflow

ful incorporation of a warning or correcting device has a twofold effect. *First,* it improves steady operation and may pave the way for further progress on similar lines as hitherto unsuspected variances from the operating norm are exposed. *Second,* the shiftsman is freed from some routine work, and is aided by information displayed by the detecting device.

The economic aspects which influence decisions as to installing mill automation take into account the quality and cost of available labour, expected improvements in throughput, recovery, etc., in relation to capital and running cost, dependability of the proposed system, and the problem of adaptation to an existing plant. Automation is proving its value right through the mining field, but like all new technologies is liable to teething troubles which call for expert nursing. Present developments tend to interest plants which handle large tonnages, highly complex treatments, or operations where special labour difficulties are met. In the rapidly evolving use of automation many ideas now undergoing research will duly win industrial application. Notes on instrumentation are given in Chapter 21.

Keeping Down the Costs

The economic side of mill operation is constantly scrutinised by the administrative branch of the enterprise. Grinding is the most costly single item in most plants. The costs in grinding are fairly evenly divided between those for power and those for steel replacement. Labour is usually a minor item provided the mill is running continuously at full capacity. If the mill is shut down, labour, depreciation, and a percentage of the total overhead charges of the undertaking continue unchanged. The only way to keep these three from becoming a serious item of cost is to avoid loss of operating time. Two forms of criticism are applied to the work being done—financial and technical. The mill personnel may be pressed to use methods which continuously increase the mill capacity. By now, the reader will appreciate that there is no magic formula for doing this. Grinding produces new surface, and the most technically efficient grinding produces it at the desired particle size. If this grinding is done perfectly, then 100% separation of value from gangue is theoretically possible. The cost would be prohibitive, so it becomes necessary to compromise between the theoretically possible and the financially permissible. Tonnage milled per shift is easily estimated, and if the mill runs without stopping, the obvious "financial" effect of putting through more ore than could be properly ground would be to send inadequately liberated material to the concentrating section. This would result in higher tailings losses. A compromise must be arrived at. This is reached, so far as the mill alone is concerned, when the cost of further treatment (grinding or otherwise) is not repaid by a corresponding increase in recovery. Since the mill is only one part of the whole enterprise, external considerations might override this criterion.

In addition to the technical efficiency controls discussed in the previous section, several practical things can be done which do not need laboratory guidance. Round balls are the most efficient crushers, and any suitable opportunity for rejecting those which are broken or badly shaped should be seized. The power factor in the main motor-circuit should be held at 1·0. Tramp oversize in the new feed of a carefully ball-rationed mill will reduce the capacity and a little care taken to avoid its delivery into the fine-ore bin is justified. The mechanical state of the circuit affects the power bill, so good maintenance and efficient use of the proper lubricants are called for.

Summary

At this point preparation of the ore has been sufficiently discussed. With the completion of preparation, the run-of-mine ore should have been subjected to some, or all, of these operations:

Sorting of true ore from detritus.
Washing to remove barren slimes.
Crushing.
Weighing.
Head sampling.
Grinding to desired m.o.g. at required solid-liquid ratio of discharge.
Smooth delivery to the concentrating section.

These operations should have been performed in accordance with a system which has developed the solids in the pulp to the state in which they respond most profitably to the treatment selected for their separation into the various products made in concentration. Within practical limits such sampling controls should have been imposed upon the work as will give the management sufficient information to enable the efficiency and financial value of the operation to be assessed.

Reference

1. Ramsey, R. H. *Eng. Min. J.*, 146.

CHAPTER 11

METHODS OF SEPARATION

Introductory

In previous chapters consideration was given to the operations necessary for the unlocking of the constituent minerals in ore as mined. Comminution was studied:

(*a*) as a liberating treatment preliminary to separation into concentrate and tailings;
(*b*) as an economic operation;
(*c*) as a stage in developing individual qualities needed in further treatment.

Economically, there is a limit to the amount which can prudently be spent in grinding. Technically, there are degrees of liberation which directly affect the efficiency of separation and the purity of the resultant products. The word *concentration* denotes the selective separation of the head feed into characteristic products. In the simple case of an ore containing galena (PbS) and calcite ($CaCO_3$), comminution, followed by concentration, might produce three fractions (Fig. 105).

Here a concentrating process has been applied which has segregated the broken ore into three products, two of which are "finished"—the valuable galena and the valueless calcite—and one in which the two constituents remain interlocked as particles of *middlings*. Processing methods are normally concerned with physical separation only, so nothing more need be done to products (1) and (3). If further treatment is decided on, product (2) can be crushed finer and given another separating treatment.

This intermediate middling, being composed partly of the sought concentrate, galena, and partly of the waste product, calcite, should neither be accepted as a concentrate nor rejected as a tailing. If accepted, it would lower the grade and thus prejudice sales. If rejected, its galena would be lost. In all concentrating processes the decision as to further treatment of middlings includes technical and economic considerations specific to the ore.

When the ore body contains more than one economically recoverable mineral treatment is more complicated. Not only does each species contribute a middling product, but these middlings may contain more than one valuable mineral and thus lead to contamination of concentrate A by A–B type middlings. Treatment must sometimes be elaborated in order to reach a specified degree of purity in the end-product. In general terms, a multi-product flow-sheet provides for series concentration, as at Rammelsberg, where *differential flotation* is used to produce a series of high-grade concentrates serially (Fig. 106).

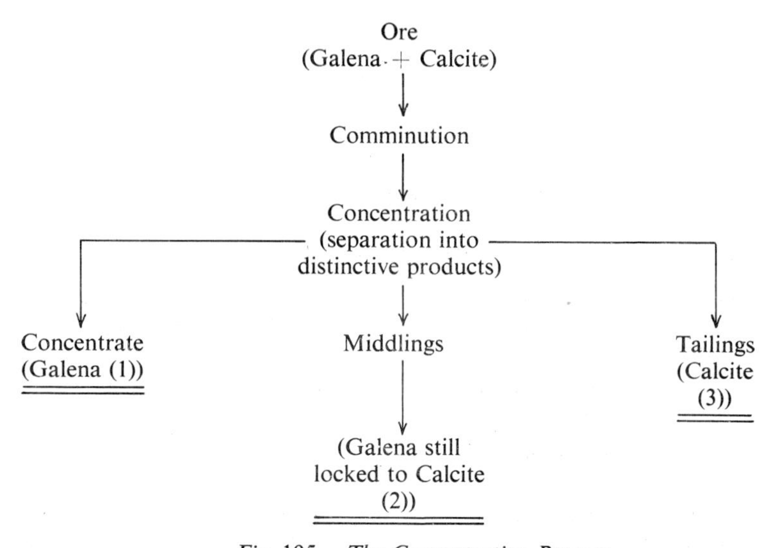

Fig. 105. The Concentrating Process
(Note. *Double underline shows end-product with respect to flow-sheet*)

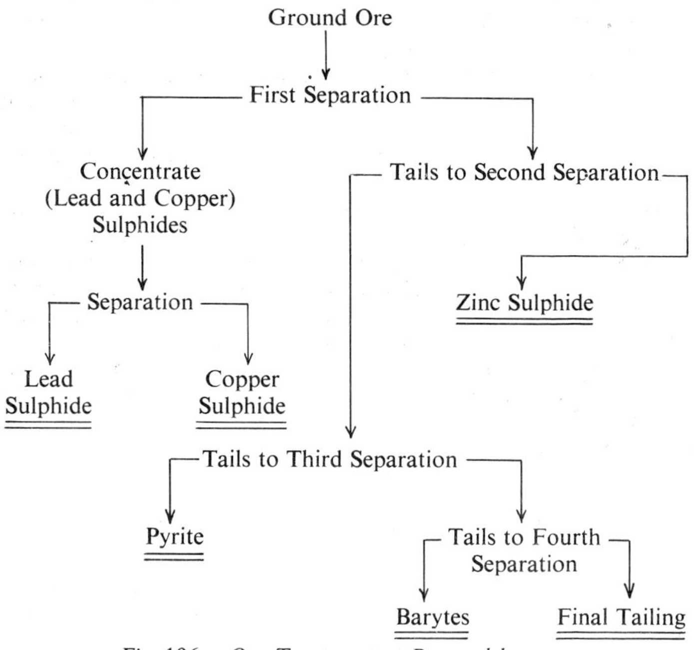

Fig. 106. Ore Treatment at Rammelsberg

In a third type of problem it may be sufficient for the concentrating process to segregate all the minerals containing some desired element without regard to the associated elements. Thus, a copper ore containing a variety of copper compounds (Fig. 107) might be treated thus:

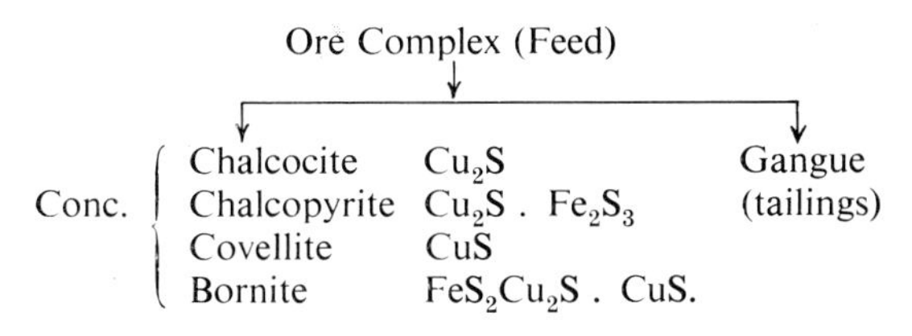

Fig. 107. *Bulk Concentration*

Particle Characteristics

If comminution of the ore is efficiently performed, each of the multitude of resulting particles acquires distinguishing characteristics which can be exploited by a suitable separating (concentrating) process. The treatment chosen in the case of the mixture of galena and calcite (Fig. 105) is applied to the pulp to make as complete a separation between the two minerals as their degree of liberation allows. It is not necessarily completed in a single step, because of the presence of incompletely liberated middlings. It may produce

(*a*) finished galena (1) and calcite (3),
(*b*) finished galena (1) and middlings (2),
(*c*) middlings (2) and calcite (3),
(*d*) rich middlings (2) and poor middlings (2).

These (*d*) products could be disposed of as a low-grade concentrate and a rich tailing, re-treated, or stockpiled, if better treatment facilities or higher realisation prices were expected in due course.

The characteristics of the particle undergoing treatment may combine to make the work of separation easy, or they may interfere with each other, making concentration difficult. To be usable, they must be such that at the same time and in the same pulp the two kinds of particles (concentrate and tailing) do not, to any important extent, respond in the same manner to the chosen treatment.

The physical, electrical and chemical properties of minerals which are most commonly exploited in their concentration are listed in Table 19.

Middlings

Provision is usually made in concentrating appliances for the separation of the feed into three products: (*a*) *concentrate,* (*b*) *middlings* and (*c*) gangue or *tailing*. In gravity separation a particle sufficiently coarse to move principally in accordance with its gravitational pull can be moved in one direction if clean and heavy, and another if clean and light. If it consists partly of heavy mineral, and partly of light gangue, it either remains neutral or moves feebly in the direction either of the light or the heavy particles. This depends on which of its two component minerals predominates in the *locked* (incompletely liberated) middling. If the separating appliance has three exit channels, this indeterminate middling forms a dividing band between the two finished products (concentrate and tailing) as shown in Fig. 105.

If the middling, instead of being run through the appliance in open circuit, is returned and mixed with new feed the total quantity of middling in the circuit increases until the position shown in Fig. 108 is reached.

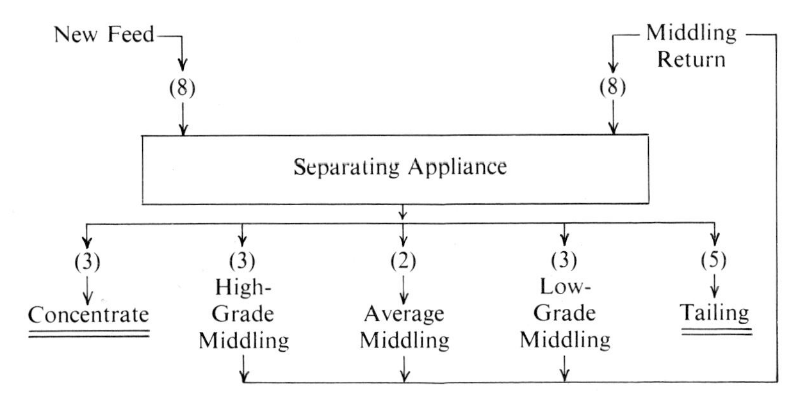

Fig. 108. *Separation with Returned Middlings*

It is thus possible to give these hesitant particles more chances to sort themselves out. If they are made to issue from the appliance in a band lying between concentrate and tailing, the separation is at the same time made more effective.

A 100% circulating load has been built up at the moment illustrated. For each eight units of new feed eight are being recirculated *via* the middlings band, while eight leave the system, three as finished concentrate and five as finished tailings.

While this happens, new feed comes to the separating appliance, and the middlings leaving is therefore increased beyond the 100% circulating load. This cannot continue indefinitely. A balance must be struck between the amount returned and the amount of new feed. When the operation has steadied to this state of balance, the position is again as depicted in Fig. 105 save that the return circuit of middlings now provides an excellent dividing zone between concentrate and tail. As the separating appliance does not alter the degree of liberation of the middlings, something more must be done outside the circuit if further separation is to be obtained. Failing that, the

TABLE 19

Exploitable Characteristic	*Exploiting Channels, Processes and Machine Types*
Particle density Liberation mesh Particle size Particle shape	Separation based on differential specific gravity of sized or sorted particles, using *dense media separation, sluicing, jigging, shaking tables, screening, classification.* All movement of solids in fluids is affected, hence in some degree all mineral processing.
Chemical reactivity Ion exchange Electrolysis	*Chemical extraction* (hydro-metallurgy, cyanidation, *leaching*). Redox roasting, electro-extraction, corrosion prevention. Chemical and physical superficial change (ad- and deeply sorptive, modifying natural reactivity.) Flocculation, dispersion, flotation, detergency. Amalgamation.
Ferro-magnetism Electrical change Electronic induction Radio-activity Optical quality Thermal effect	Magnetic separation (natural or induced ferro-magnetic quality). High intensity separation (*electrostatic sep.*). Radio-signalling and control. Dust precipitation electrostatically. Fluorescent sorting. Electric conductivity. Colour sorting, light reflection. Heat retention devices.

lower-grade middlings will be crowded out with the tail and lost, and the higher grade particles will join the true concentrate, lowering its purity. Since the cause of this is incomplete liberation, the remedy is to give the middling a regrinding treatment. Fig. 109 shows how this can be done.

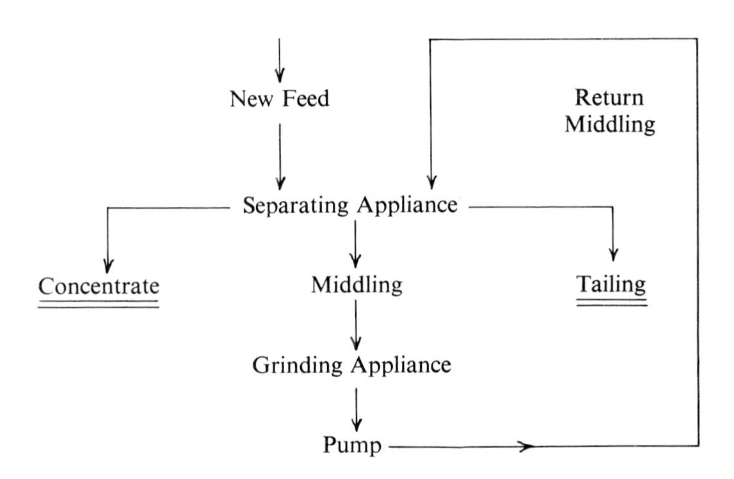

Fig. 109. *Separation, with Regrinding of Middlings*

If the separating appliance is unsuitable for the treatment of the ground middling, the latter is sent to a more appropriate machine.

Three important points in mill practice should now be apparent. *First,* it is not necessary to begin by grinding *all* the minerals to their fully liberated state in order to procure clean separation. Results can frequently be achieved in stages. First comes grinding, next separation into clean concentrate, clean tailing and "locked" middling (as incompletely liberated particles are called). Finally, the middling is unlocked by grinding, and retreated.

Second, the middling, or any fraction of it, can be held in a closed circuit if by doing so the work of separation is made more efficient.

Third, a true middling always needs special treatment not provided for in the appliance which has sent it out as a middling. After this special treatment it may be unsuitable for return to the sorting appliance. In the case just considered, if treatment depended on the mass of the particle, the ground fragments would probably be sent to a machine specially adapted to deal with their smaller size.

Types of Middling

Table 19 lists a number of mineral characteristics of which advantage may be taken to achieve separation. The behaviour of a middling can sometimes be strongly influenced by the characteristics of one of its minerals while the other part of the binary system offers little or no opposition. The coarser the grinding, the cheaper it is and the greater the amount of middling made. Where binary particles are sufficiently responsive, advantage can be taken of their breaking behaviour to cheapen the working costs without losing valuable mineral. Middling particles can be associated in various ways (Fig. 110).

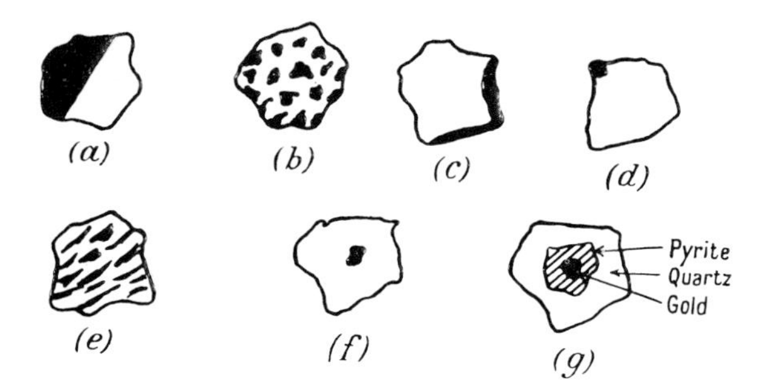

Fig. 110 (*magnified*). *Types of Middling Particle* (after Taggart)

A particle of type (*a*) would need breaking before it could be correctly graded. Since half of its surface is gangue and half value, it will respond well to surface attack such as is employed in chemical or flotation treatment. Particle (*c*), in which the value is deposited as a shell on a core of gangue, is likely to behave as a tailing in gravity work and as a clean concentrate in flotation; (*d*) is usually lost in flotation, but satisfactorily relieved of its value in a solvating process; (*a*), (*c*), and (*d*), if the value is ferromagnetic, should respond to electro-magnetic pull. Particles (*b*), (*e*), (*f*), and (*g*) might act as gravity middlings. When the values are almost or quite masked by gangue, they will be completely lost in flotation or chemical attack. Particle (*g*) can be a special case. Suppose the gangue to be quartz, the outer enclosed mineral pyrite and the centrally enclosed mineral gold (the last being the value sought), then the grinding treatment must expose the gold. Regrinding must therefore be far more elaborate than in usual cases, particularly if the gold is segregated in the pyrite in specks only a few microns in size. Here roasting treatment is usually preferred to further comminution.

Staged Concentration

Older textbooks show methods of treatment (Fig. 111) in which concentrates are removed at definite stages of comminution.

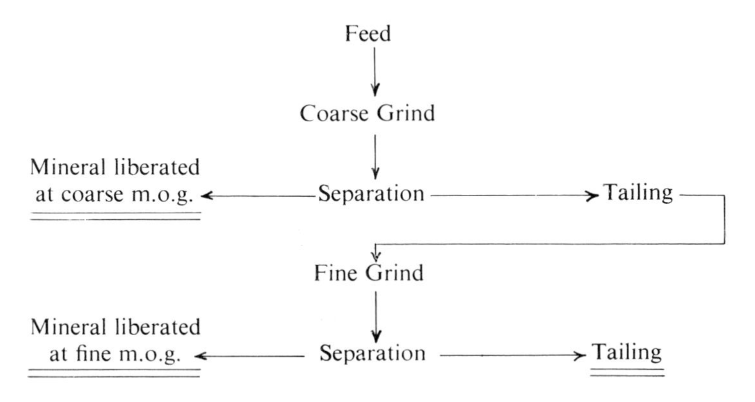

Fig. 111. *Staged Treatment*

Coarsely mineralised ores suitable for such treatment are now rare. Further, most minerals can be cheaply concentrated at a fine m.o.g. by flotation, so the need to avoid fine grinding which dictated gravity-concentration methods before 1920 no longer exists. If flotation is used, as it is today in most mineral-dressing plants, instead of removing finished concentrates by stages it may be feasible to discard liberated gangue at each grinding stage, as in Fig. 112.

This principle may be used to give selective treatment to the middlings product at each grinding stage, as in Fig. 113.

This arrangement is a variation of the "cascade" principle, which is used in chemical engineering to enrich or deplete an original feed by gentle stages (Fig. 114).

A number of identical concentrating devices are arranged in series, new feed being introduced midway. Each unit divides its feed into a slightly

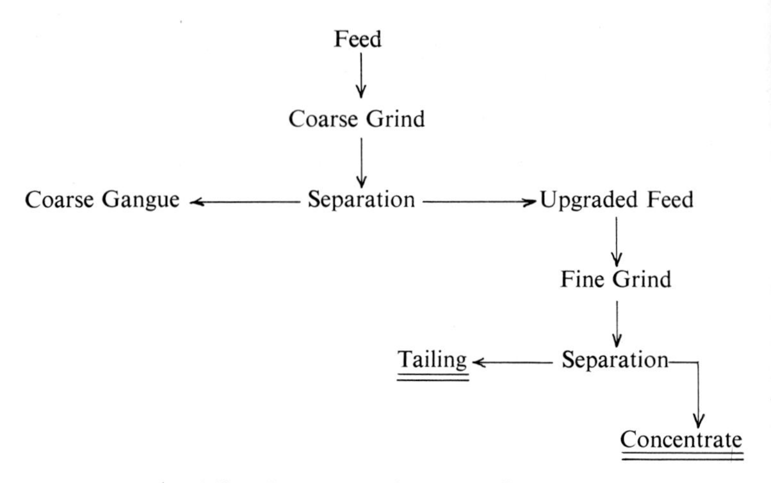

Fig. 112. *Rejection at Successive Stages*

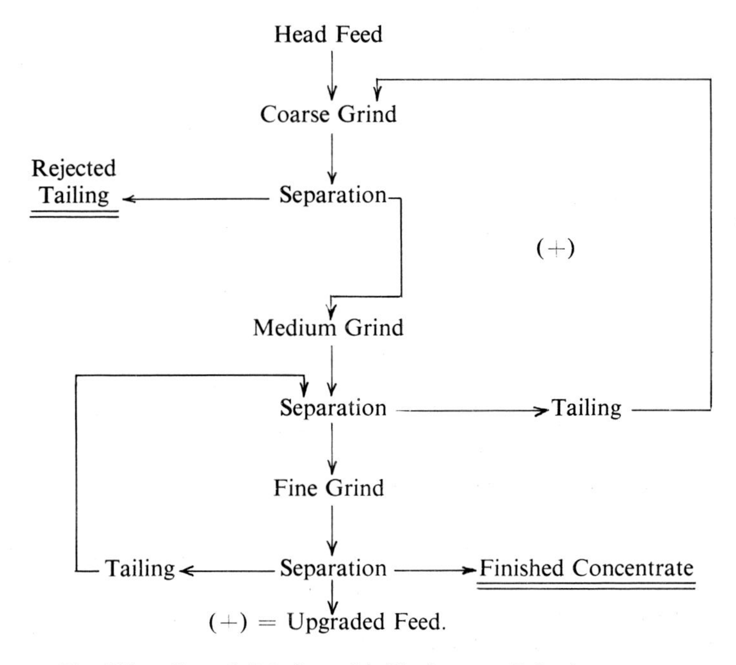

Fig. 113. *Staged Grinding with Single-stage Rejection*

richer product C, which is sent upstream or counter-current (to the left in

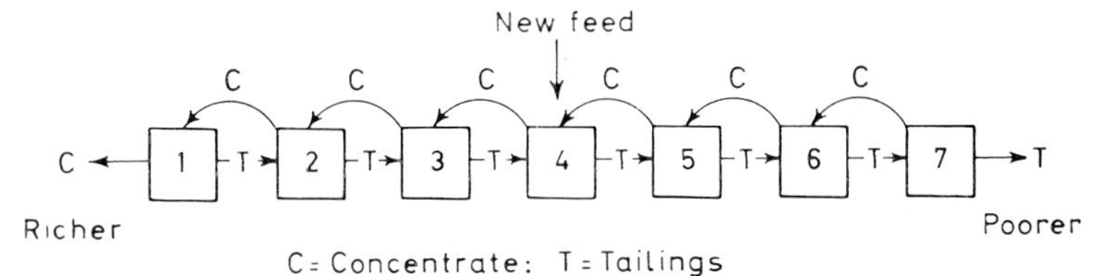

Fig. 114. *"Cascade" or Counter-current Principle (C = Concentrate; T = Tailings)*

Fig. 114), while an impoverished tailing flows concurrent to the next unit below (to the right in the figure). Thus concentrate C from unit 3 joins the tailing from unit 1 in unit 2. The result is progressive enrichment in one direction and corresponding impoverishment in the other. In mineral dressing the usual arrangement is more compact (Fig. 115), the separating machines being marshalled into blocks, called roughers, cleaners, and scavengers.

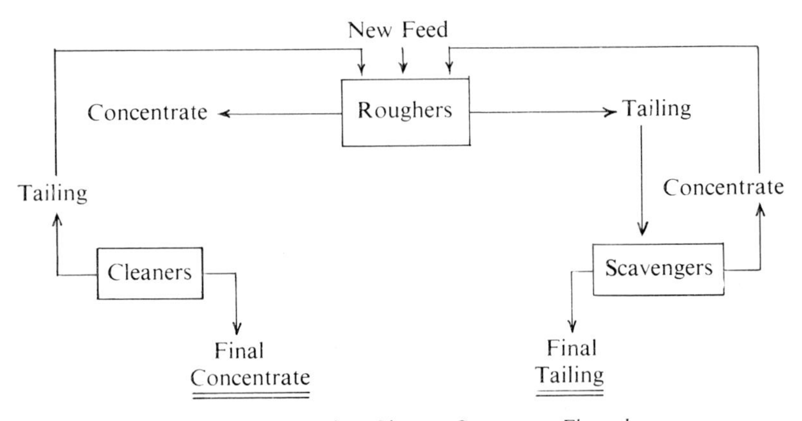

Fig. 115. *A Rougher-Cleaner-Scavenger Flow-sheet*

The extent to which this principle of gentle upgrading is used in a given case depends on the operating difficulties which must be overcome and on the cost warranted by extra treatment.

Panning

At this stage the student is recommended to practise separation of a heavy mineral from a light one by panning. This will familiarise him with the "packing" of sands, the transporting power of a slurry, the difference in behaviour of value, middling and tailing particles, and the difficulties encountered in attempting to treat a long-ranged feed in a single operation. A clean prospecting pan should be used. Scouring with sand usually suffices to remove old rust. The pan should not contain oil or grease. If a gold ore is to be tested, the pan should be dark in colour so as to show up the golden specks clearly. Since a great deal of panning is done in pools and streams, the beginner should learn to squat on his heels and pan from one gold pan into another full of water, so that he can save and retreat the discarded material until he has become expert.

An excellent practice material is a —20–mesh mixture of sand and galena. A few hundred grams of this material are wetted down into water and worked into a running pulp. This is next deslimed by gentle decantation, the pan being held with its double riffle away from the operator. At no time should pulp be allowed to stream over the riffle. It should be floated out into the pool of water with a gentle swirling or rocking motion.

From time to time the pan should be tapped with the heel of the hand to aid the heavy particles to burrow down to the bottom of the fluid body of pulp. The top strata should then be panned off, using a jigging or swirling motion. As soon as heavy particles show, the material still in the pan should be repulped and rethumped. Successive barren strata can thus be removed without loss of values. When the point is reached where most of the sand has been rejected, the decision must be made as to whether a low-grade concentrate and a clean tailing is to be produced, or a high-grade concentrate and a middling. In the latter case, rejected sands which now carry heavy mineral should be panned out into another holding vessel. It is not possible to make a clean concentrate and a clean tailing in one operation. The rejected tailings should be repanned to see how efficiently the work was carried out. With practice, it is possible to use panning as a rough guide in assessing efficiency of gravity treatment of sands.

The plaque, a white-enamelled concave disc 11″ in diameter, is used in a similar manner in the examination of fine sands.

Gravity Separation

In gravity separation the combined effect of mass and shape of the particle determines its movement relative to flowing water. In one development of this effect the water flows vertically either continuously (classification) or in oscillating motion (jigging). In a second method the ore is fed into a fairly quiet pool of dense media (water mixed with slow-settling heavy minerals to form a fluid which can be maintained at a high specific gravity). Ore entering this dense medium either floats or sinks. A third type of gravity treatment uses flowing streams to effect separation. Here the pulp is carried horizontally or down a slope, and separation depends on the rate of fall, and resistance to displacement after the particle reaches the floor of the appliance.

Methods which exploit differences of gravity require that there shall be a marked difference between the specific gravities of the value and the gangue. The material must be sufficiently coarse to move in accordance with Newton's law. Particles so small as to settle in accordance with Stokes' law are unsuitable for concentration by simple gravity methods. If centrifugal force is applied, such fine particles can sometimes be treated. Gravity methods of separation become cumbersome and inefficient when the average particle is so minute that its surface friction dominates its movement through the surrounding fluid. Exploitation of differences between the specific gravities of particles below 150 to 200 mesh is difficult and usually avoided when alternative methods of treatment are available.

Chemical Methods

These depend on adequate exposure of the valuable ore mineral to the action of a solvent. Solvation attacks the surface displayed. The new surface developed as the result of comminution is doubled for each halving of the mesh size of the mineral. Hence the time taken for dissolution of a particle bears an important relation to its comminution. The steps taken in separation by chemical methods are:

(*a*) Grinding for optimum exposure.
(*b*) Solvation of value.
(*c*) Separation of solute from residual solids.
(*d*) Re-precipitation of the value.

Step (*d*) may produce the value in a new compounded form, or as metal.

Flotation

Flotation is the term used generally to denote "froth-flotation" rather than processes in which gravity provides the main separating force. In this book the prefix "froth" is not employed. This extremely important process is used in most of the mills at work today, and frequently provides the main separating method. It must not be confused with "sink-and-float" processes which are termed by the author *dense-media separation* (DMS).

Flotation is usually applied to minerals (sometimes values, sometimes gangue) that cannot be liberated at a size suitable for treatment by all-gravity methods. It is chiefly practised on feeds ranging between 60 mesh and a few microns in particle size. Under the influence of appropriate chemicals, the surfaces of some mineral particles can be physically and chemically modified. Thus treated, some particles in the pulp remain attached to water (hydrophilic) while others become air-avid or aerophilic. The latter cling to air bubbled through the pulp. Loaded bubbles rise to the surface where they form an unstable froth which overflows, carrying off the mineral thus concentrated.

Magnetic and Electric Methods

All minerals are to some extent able either to concentrate or to repel magnetic flux. When the former characteristic is strongly exhibited, it is called ferromagnetism. If this ferromagnetism is specific to one or more minerals liberated by comminution of the ore, it can be used. This is done by conveying the ore through a regulated magnetic field in such a way that the responding particles are able to move to collecting devices associated with the magnets.

Fine particles may be made to deviate from their normal line of fall through quiet air, if they carry an electrostatic charge so that they are attracted or repelled as they drop past a charged electrode. This principle is used in electrostatic or high intensity separation.

Amalgamation

At the turn of the century *amalgamation* was the main method used for the separation of gold and silver (in their "free" or "clean metallic form") from their ores. It is now subsidiary to cyanidation, with rare exceptions, and is sometimes omitted altogether. Amalgamation exploits the fact that mercury readily "wets" the surface of clean gold, silver, and electrum-metal. Once particles of these metals have been absorbed into the mercury, the resulting amalgam can easily be trapped and removed from the ore pulp. The mercury is then removed, and the residual gold is refined.

Mixed and Minor Methods

The flow-sheet developed for treatment of a specific ore may combine some of the foregoing methods, or may reinforce their efficient use by special local techniques. Mixed and specialised methods of treatment are best understood after the basic principles have been studied. A selection has been included in Chapter 23, where specific treatments are considered under the heading of the main mineral being recovered. Such a "combined operation" as, for instance, the leach-precipitation-float process (L-P-F) is not, therefore, dealt with in Chapters 16–19.

Among the mixed and minor applications which receive brief attention in Chapter 23 are thermal segregation, radio-active picking, reduction roasting as used in iron ore beneficiation, mixed gravity and flotation such as is at work in some coal cleaning plants, and the segregation process.

Exploitable Factors

In order to influence or control changes in a balanced system the basic laws which govern its composition must be understood if they are to be utilised efficiently. A mineral pulp, whether in a more of less "held" state or in steady flow, can be regarded as such a balanced system. Each concentrating treatment can be seen as an unbalancing activity designed to segregate and remove selected fractions by separation into two new balanced systems. Instability is created in the original system by the application of an external force sufficiently powerful to overcome the opposing forces of stabilisation it encounters, or one able to enlist their aid. Whether this (or these) applied forces are called physical, chemical or electrical one or more of them is the disturbing factor. The general descriptive term "physical chemistry" embraces the study of these forces and their scientific use in such fields as thermo-dynamics, electro-chemistry and electro-magnetism. In the following chapters the role of physical chemistry is considered in many aspects. The forces at work are sometimes mutually antagonistic and able to reduce efficiency or even prevent the desired reaction unless they are recognised and correctly dealt with. Basic and applied research, the latter specific to the ore which is to be treated, should therefore isolate and study such interfering factors. It should ascertain the strength and nature of both the forces relied on for reaction, and of those inimical thereto. Only thus can successful concentration (separation of the previously balanced pulp into two differentially balanced streams) become a science rather than an art or empiric skill.

When these matters have been thoroughly looked into, the flow-sheet is worked out. Larger-scale tests in the *pilot plant* usually follow. The full-scale plant is then designed and built and worked up to its operating norms by the permanent staff. Since their work, like that of all who must deal with an empirical element in the daily variations encountered, is "the art of the possible" it is invaluable to the mineral engineers concerned to understand something of the forces at work, so that process controls can be made to strengthen their effective use.

Concentration Formulae

The basic two-product formula used in mill control was given in Chapter 2 under "Acceptance Operations". To this weight balance ($F = C + T$) and ingredient balance ($Ff = Cc + Tt$) for weights FCT and assays fct of feed, concentrate, and tailing, may be added the following:

$$F = C \frac{(c-t)}{(f-t)} \tag{11.1}$$

$$F = T \frac{(c-t)}{(c-f)} \tag{11.2}$$

$$C = F \frac{(f-t)}{(c-t)} \tag{11.3}$$

$$C = T \frac{(f-t)}{(c-f)} \tag{11.4}$$

$$T = F \frac{(c-f)}{(c-t)} \tag{11.5}$$

$$T = C \frac{(c-f)}{(f-t)} \tag{11.6}$$

Another important figure is that for ratio of concentration K. This is the ratio of weight of feed to that of concentrate.

$$K = \frac{F}{C} = \frac{(c-t)}{(f-t)}. \quad (11.7)$$

For two-product calculation the percentage recovery R is formulated

$$R = \frac{100Cc}{Ff} = \frac{100c\,(f-t)}{f(c-t)}. \quad (11.8)$$

These formulae, derived from the weight and ingredient balance, permit a solution where four of the six quantities, including two assays, are known.

Where the mill operation is set to produce a concentrate rich in mineral a reporting in C_1, followed by one rich in mineral b reporting in C_2, three-product formulae can be used. That for weight balance is

$$F = C_1 + C_2 + T \quad (11.9)$$

and for ingredient balance with respect to minerals a and b:

$$Ff_a = C_1C_{1a} + C_2C_{2a} + Tt_a \quad (11.10)$$

$$Ff_b = C_1C_{1b} + C_2C_{2b} + Tt_b \quad (11.11)$$

These simultaneous equations can be solved for C_1, C_2, or T (Taggart, *op. cit.*, 19–192, eq. 149–158). Alternatively the assay values and known weights can be written into 11.10 or 11.11 for direct solution. Ratio of concentration for mineral a is

$$Ka = \frac{F}{C_1} \quad (11.12)$$

$$= \frac{(c_{1a}-c_{2a})(c_{2b}-t_b)-(c_{1b}-c_{2a})(c_{2a}-t_a)}{(f_a-c_{2a})(c_{2b}-t_b)-(f_b-c_{2b})(c_{2a}-t_a)} \quad (11.13)$$

and recovery

$$R_a = 100\left[\frac{C_1c_{1a}}{Ffa}\right]$$

$$= \frac{100c_{1a}(f_a-c_{2a})(c_{2b}-tb)-(f_b-c_{2b})(c_{2a}-t_a)}{fa[(c_{1a}-c_{2a})(c_{2b}-t_b)-(c_{1b}-c_{2b})(c_{2a}-t_a)]}. \quad (11.15)$$

For many practical purposes it is more convenient to treat each of a series of concentrating operations independently and to consider all values not then removed as tailings with respect to the immediate work. For most concentrating purposes recovery and ratio of concentration are the important criteria, together with concentrate grade as shown by assay. Efficiency of operation must be considered at the same time in order to maintain a process in which product yield and grade are satisfactory. Diamond[2] uses the arithmetical average of each main constituent recovered in the formula

$$E = \sum \frac{RnN}{n} \tag{11.16}$$

where R_{nN} gives the recovery of constituent n in the N product. Gaudin[3] discusses this. He then puts forward a Selectivity index (S.I.) as a convenient measure of two-way separation. As formulated it gives a criterion of trend in a continuing operation, by which current results can be compared with previous ones.

$$\text{S.I.} = \sqrt{\frac{R_a J_b}{R_b J_a}} \tag{11.17}$$

where R_a is recovery of mineral a in A and J_a its rejection in B; R_b is the recovery of b in A and J_b its rejection in B. Since $J_a = (100 - R_a)$ and $R_b = (100-)J_b$

$$\text{S.I.} = \sqrt{\frac{RaJb}{(100-(Ra)(100-Jb)}}. \tag{11.18}$$

The subject receives detailed consideration by Douglas[4] who proposes this formula for calculating efficiency.

$$\mathrm{E} = \frac{(\mathrm{R}-\mathrm{C})}{(100-\mathrm{C})} \cdot \frac{(\mathrm{c}-\mathrm{f})\ 100}{(100-\mathrm{f})} \cdot \tag{11.19}$$

in preference to an empirical one[5] which is more applicable to a low-grade ore, and which gives efficiency of concentration (Ec) as:

$$\mathrm{Ec} = \frac{(\mathrm{c}-\mathrm{f})}{(\mathrm{c_{max}}-\mathrm{f})} . \mathrm{R} \tag{11.20}$$

General

In addition to the foregoing main methods of separating ores into value (or concentrate) and gangue (or tailing), there are several specialised techniques which lie beyond the general purpose of this book. The processes used in ore treatment may combine more than one method in a given flow-sheet. In gold milling, amalgamation can be followed by flotation, cyanidation, or both. The lead of a lead-zinc ore can be concentrated on jigs and tables, and the zinc by flotation. The method, or combination, chosen depends on the break point of the values, the ore characteristics which can most satisfactorily be exploited, and the cost of treatment. Such considerations as early elimination of part of the *gangue,* simplicity of operation, specific reaction of the sought mineral and a brisk rate of throughput are factors which influence the choice of method.

References

1. Taggart, A. F. (1945). Handbook of Mineral Dressing, Chapman & Hall.
2. Diamond, R. W. (1928). Trans. A.I.M.M.E., 79.
3. Gaudin, A. M. (1939). Principles of Mineral Dressing, McGraw-Hill.
4. Douglas, E. (1961/2). Trans. I.M.M. (Lond.), 71.
5. Stevens, J. R., and Collins, D. N. (1963). Trans. A.I.M.E., 226.

CHAPTER 12

DENSE MEDIA SEPARATION

Introductory

The processes considered in this chapter are variously called "sink-float", "sink-and-float", "dense-medium separation", "heavy-medium separation" and "heavy-liquid separation" in response to their variations on the basic principle of dense media separation (usually abbreviated to DMS).* They may include:

(*a*) Separation in a quiet single bath
(*b*) Separation with controlled agitation of the bath
(*c*) Separation in a series of baths
(*d*) Separation in a single bath changing in density between top and bottom
(*e*) Use of hydraulically expanded separating media (e.g.Chance and Stripa processes)
(*f*) Use of autogenous media
(*g*) Use of blended heavy liquids
(*h*) Centrifugal aid, e.g. by cyclone bath.

If a fluid of suitably high density can be maintained in a sufficiently quiet and stable condition, the relatively heavy particles fed into it sink while the light ones float. The principle is illustrated in Fig. 116.

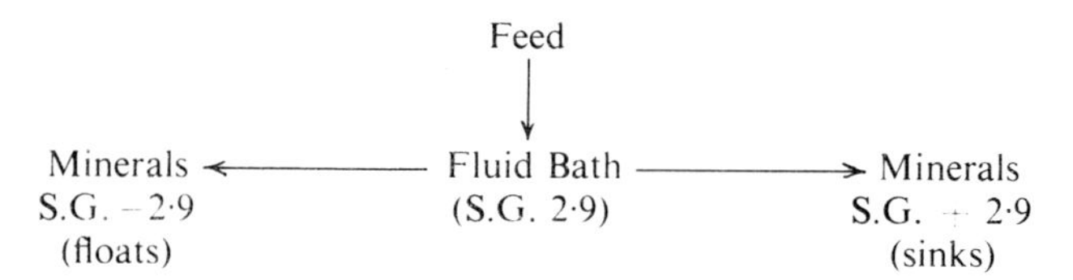

Fig. 116. *Dense Media Separation*

Heavy liquids have long been employed for laboratory separations of this kind, but have so far only limited industrial use. Several decades back the possibilities of using a mixture of finely divided magnetite and water were explored. The fluid made from this mixture is cheap and settles slowly enough to act like a true heavy liquid. Commercial development failed at that time for lack of an effective method of keeping it in clean working condition. Bessemer, in 1858, patented the use of solutions of metal chlorides, and in 1911 Du Pont developed the use of chlorinated hydrocarbons to obtain high densities. The first important success was achieved by the

* Following usage, the term "dense media" is used in this book, though in some contexts the plural is less appropriate than the singular word.

Chance process, patented in 1917. This is not strictly a simple DMS process, since it employs a hydraulically dilated bed of sand. In 1928 Lessing re-developed the use of calcium-chloride solution. In 1931 clay, gypsum, and pyrite slurries were used in water, and in 1932 G.J. de Vooys established a coal-cleaning process based on use of a barytes-clay medium. Mixtures of such earths with water produce fairly stable fluids with specific gravities ranging between 1·4 and 1·6. When run-of-mine material is fed into a separating bath maintained at some such density, coal floats and the accompanying shale sinks. The density range is, however, insufficient for treating the great majority of ores which require a "*parting density*" of 2·7 or more. In 1935 the possibilities of a suspension of finely ground galena in water were investigated and matured by Huntington, Heberlein and Co., in a pilot plant which in due course led to a successful application at the Halkyn lead mine in Wales. Meantime, DMS was being pioneered by the American Zinc, Lead and Smelting Co. at Mascot, Tennessee, which started work with a galena-medium in 1939. Later, magnetite and ferro-silicon took the place of galena in the Mascot plant. The change was made after methods had been developed for using the ferro-magnetic qualities of these substances to keep the dense media fluid clean.

The DMS principle is extremely simple. The high specific gravity of the bath may be produced by the use of heavy liquids, solutions of salts, or suspensions of slow-settling solids in water. The bath may be almost static, gently agitated, or in moderately brisk motion.

Two main applications are possible. DMS can remove lightweight rock at an early stage in crushing. It is then a sorting treatment, used to remove barren waste or ore of a grade too poor to warrant treatment in the concentrator. The second application of DMS produces a commercially graded end-product. It is chiefly used to produce clean coal as a "float", and a low-grade "sink" consisting of shale or of coal too high in ash for sale.

The various methods used in cleaning coal embody specialised applications of mineral-dressing techniques, having the common factor of low specific gravity in the mineral recovered as a "float".

Amenability

In mineral dressing the term "*amenability*" refers to the manner in which a given ore responds to a given method of treatment. Broadly, DMS is applicable to any ore in which, after a suitable degree of liberation by crushing, there is enough difference in specific gravity between pieces of rock to separate those which will repay the cost of further treatment from those which will not. If this liberation is achieved at coarse gravel size or larger, a difference in specific gravity of 0·1 or even less can usually be exploited commercially. In the laboratory, where longer time can be given to small quantities of material to move, much smaller differences in specific gravity are sufficient to ensure separation. Parting densities above 3·4 are difficult to maintain, and limit the applicability of the process.

If the values are irregularly distributed in the ore body, either as coarse crystals or aggregates, DMS may be possible. If the values are finely disseminated through the ore, an adequate difference of density between the crushed particles cannot be developed by coarse crushing. For the process to be applicable the heavier fraction must be sufficiently liberated from the lighter fraction at a mesh coarser than 48 mesh (preferably coarser than 5 mesh). Ore can be separated from waste if there is a sufficient difference in specific gravity between the two minerals. This can be 0·1 or even less for separation at 5 mesh or over, and at still finer grain sizes if the viscosity of the media is low. If the specific gravity of richer lumps of ore is higher than that of poorer ones it is frequently possible to upgrade run-of-mine ore which otherwise would not repay the cost of treatment, by using DMS to reject the low-grade material.

The Economic Criteria

Consider the hypothetical case of an ore in which the average assay value is 4s. 6d. per ton. Ignoring the cost of DMS treatment and taking cost of subsequent treatment (fine grinding, concentration and disposal) as 5s. per ton milled, the results of discarding a low-grade fraction might be as shown in Table 20.

TABLE 20

	Tons	*Value/ ton in Contents*	*Total Value*	*Treatment Cost*	*Profit*	*Loss*	
Float	10	1s.	10s.	—	—	—	Case 1
Sink	90	4s. 11d.	440s.	450s.	—	10s.	
Feed	100	4s. 6d.	450s.	—	—	—	
Float	20	1s. 3d.	25s.	—	—	—	Case 2
Sink	80	5s. 4d.	425s.	400s.	25s.	—	
Feed	100	4s. 6d.	450s.	—	—	—	
Float	30	2s. 8d.	80s.	—	—	—	Case 3
Sink	70	5s. 3½d.	370s.	350s.	20s.	—	
Feed	100	4s. 6d.	450s.	—	—	—	

If this ore is treated as mined there is an inevitable loss of 50s. This can be mitigated or even turned into a profit. Case 1 shows that by removing ten tons of almost barren rock, 10s. worth of valuable concentrate is lost which could only have been recovered by milling those ten tons at a cost of 5s. per ton, or 50s. Thus the effect of this amount of pre-treatment is to reduce the total loss. In Case 2 a more severe sorting has been applied, with the result that more of the valuable mineral has been lost. The overall effect is that a profit of 25s. is made on treating the tonnage. Case 3 shows the result if too severe a DMS sorting is applied—the net profit is reduced to 20s. As the result of tests it might be found that for maximum profit a specific gravity should be chosen at which more than 20% and less than 30% of the feed would be discarded as the floating fraction.

This illustrates an important principle in mineral processing. Normally, the plant must be worked to produce maximum profit, *not* maximum recovery. Note the qualifying word "normally". Under the stress of war the profit criterion may be ruled out. Again, it may be the case that the mill is short of ore, but costs nearly as much to run at 90% capacity as at full load. The foregoing type of calculation might not then apply.

In suitable cases, dense-media separation can sort a wide size-range of ore cheaply. The tonnage of suitable feed must be sufficient to justify the cost of installing a plant. Easily movable plants can be rented. They are used to treat old dumps, upgrading their contents to a value which justifies the cost of more expensive processing. They can also be used to make a comprehensive field test on a few thousand tons of development ore while the question of a permanent installation is being considered. The exploitation of low-grade properties which otherwise were too poor to be mined has repeatedly been made possible by the use of this cheap up-grading method. Where the ore mineral occurs in a lode too narrow to mine selectively, the DMS plant can be placed underground, so as to sort out barren wall or "*country*" *rock* for return as filling material, thus saving transport cost.

Forces at Work

In DMS in a quiet bath the essential separating force is gravitational, and the essential counter-restraint is viscous shear.

Equations 8.2 and 8.3 show how these forces are at work. Consider five spheres equal in volume but differing in specific gravity, being respectively S.G. 2·0, 2·1, 2·2, 2·3 and 2·4. Let them be placed in a still bath of homogeneous heavy liquid of S.G. 2·2. Provided that the bath is truly static, with no convective currents, streaming or vertical flow or other turbulence, the behaviour of these spheres can be accurately predicted. The potential energy due to its S.G. is neutralised for sphere 2·2 so that it remains poised at the point of release into the bath. In the absence of frictional restraint the lighter spheres float upward under pressure from the heavier liquid, the lighter one being the first to reach the surface. Similarly, the two heavier ones sink, sphere 2·4 being the first to reach the bottom. Figuratively, the liquid can be thought of as weighing each free-moving particle introduced without kinetic energy or shock.

If these conditions remain unchanged the experiment can be repeated with smaller spheres of the same S.Gs, until their mass has been reduced to the point where gravity drag is cancelled by the resistance to viscous shear of the liquid media. Spheres 2·1 and 2·3 will be the first to hover, and with further size reduction all five will remain suspended. (It would, of course, be equally possible to leave the largest 2·2, the medium 2·1 and 2·3 and the small 2·0 and 2·4 in this hovering state together, thus producing a sizing effect). The important point to remember is that any of these spheres which lacks sufficient gravitational push or pull with respect to the heavy liquid to start moving upward or downward, has become a middling. In a media of given viscous resistance there is a minimum size limit for a sphere of given S.G., below which it lacks sufficient potential energy to move up or down. This may be true viscosity (molecular shear) or *pseudo-viscosity*. DMS on the commercial scale rarely uses true heavy liquids, but suspensions of appropriate heavy minerals in water, sufficiently finely ground to have a slow rate of sedimentation. Hence the retarding force in such a bath is largely a function of the *specific surface* involved. The main contributor to this specific surface is the heavy mineral used in the media, but slimes, clays, and detritus from the ore also exert an influence. (Physico-chemical influences associated with the *p*H of the media will be more apparent when Chapter 17 has been studied. They affect the dispersive and coagulative tendencies of the solids in the bath as a whole). A further important factor arising from the effect of specific surface is the size and shape of the particles used to constitute the dense media. Spheres were considered above, in order to simplify the essential picture, but a working bath is filled with angular mineral particles under roughly controlled turbulent conditions, with a slow rise in media density from top to bottom due to the tendency of the particles to settle. Despite the complex of working forces it is helpful to start this study with its essentials, the "weighing" of the ore particles in accordance with the amount of media each of them displaces, and the generation of force in each ore particle according to the amount it is out of balance with the media.

It now becomes clear that the particle size and shape of the mineral used as dense media, together with such chemical effects as may originate with *p*H of the water in which it is slurried, control its viscous resistance. This is modified by the presence of contaminants and by temperature effects. These therefore are control factors, since they help to determine the minimum size of particle which can move sufficiently fast through a bath at a given *parting density*. Leaving the hypothetical case of spheres, it is now clear that for ordinary fragmented particle of ore this minimum size is also influenced by shape. The main effect of shape is to develop a rubbing surface over which the braking laminal flow of the media acts. A secondary effect is that, since a particle cleaves its way relative to the media, the disturbance it generates is proportional to the cross-section it presents.

Thus far consideration has been confined to a few pieces of ore or waste rock. Continuing with the theoretical concept of a quiet bath (and ignoring the disturbance due to removal of separated products from a plunging feed), consider next a continuously arriving stream of dry ore which has been sized by screening. Assume that of a hundred such particles forty-five will sink and be withdrawn, forty-five float and overflow, and ten will form an equipoised teeter. At first separation is clean, but deteriorates as the teetering population grows with the further arrival of middlings. The sinking fraction is obstructed and dissipates part of its gravitational force in collision and jostling through the crowded middlings and then tends to be retained. The problem has some analogy to that considered in Chapter 9 under Mechanical Classifiers. Left uncontrolled, the bath would soon choke with near-middlings and separation would cease. The problem is handled by inbuilt features of design in the various DMS systems, and by suitable operating controls described later.

The Process

The stages in DMS are:

(*a*) Presentation of a suitable prepared feed.
(*b*) Separation into "floats" and "sinks" in a bath containing dense media.
(*c*) Withdrawal of products, and removal from them of adherent dense media.
(*d*) Cleaning, reconstitution, and return to bath of clean dense media.

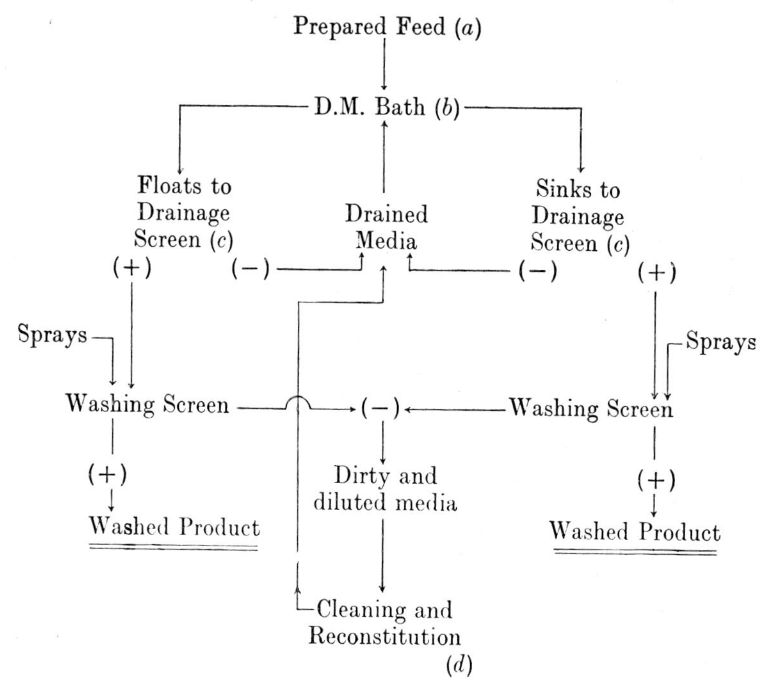

Fig. 117. *Stages in Dense Media Separation*

These steps are merged into a continuous process on the lines shown in Fig. 117.

In operation, new feed and reconstituted media are fed continuously into the bath. Floats and sinks are withdrawn, and drained of adherent or "drag-out" media which may either be pumped back into the bath direct, or given a cleaning treatment before return. The sinks and floats are next washed by sprays which remove almost all the residual media, after which they are sent to the next stage of treatment. The medium removed from the finished products by washing is collected, cleaned, adjusted to the working density, and returned to the bath.

Feeding the Ore

The feed must not contain colloidal material, primary slimes, or fine ore. These are removed, as far as possible, before bath treatment. Colloidal slimes and "fines" reduce the efficiency of treatment because they increase the specific surface beyond the amount provided by the dense media. This random variation in the pseudo-viscosity of the bath is undesirable if it increases unduly. A controlled amount of such fine material may be needed in order to aid stability of the dense media by decreasing its rate of settlement. When the ore has been sufficiently washed it must be drained before entry to the bath, to avoid undue dilution of the medium. Provided the feed rate is fairly steady and its moisture content does not vary, the effect of such added water can be compensated by adjusting the density of the re-circulating media.

Normally dense media separation is applied to a feed ranging from about 3″ down to + 10 mesh. For larger or smaller sizes special equipment is needed.

A dense media composed of magnetic particles suspended in water is easy to clean, and can tolerate a limited amount of slime from the ore without serious deterioration of the separating action. New feed should, however, be free from slime and undersize in order to aid the selectivity of the separating fluid. This new feed should not include material liable to break down and thus produce contaminating slime and fine sand during treatment.

The Dense Medium or Media

In this section three types of dense media are discussed:

(*a*) Solutions of salts in water.
(*b*) Organic liquids.
(*c*) Supensions of solids in water (dense media).

Autogenous media, provided by the ore pulp as it passes through the system, are mentioned later.

Solutions of salts. If a solution of a salt such as calcium chloride is used, it must not react with the material treated in such a way as to depreciate the selling value of the product. At one time three English plants based coal separation on the Lessing process in which a solution of $CaCl_2$ in water, at S.G. 1·35, was used.[1]

The de-dusted coal was washed in to the separating bath, over which it floated to the "deliquoring screen" while the heavy fraction sank to the bottom, whence it was withdrawn. Weak liquor was reconcentrated by evaporation. The process was abandoned owing to high cost. Other methods using $CaCl_2$ include the Bertrand and the Belknap. This last depends partly on the density of a $CaCl_2$ solution, and partly on upward currents induced by an impeller submerged in the bath. It treats sized coals.

Organic liquids. Little commercial use of heavy liquids is at present made in mineral processing, although they are in routine use in laboratory research and process control. Since chemical engineering methods have overcome most of the operating difficulties, and a useful range of fairly stable liquids is today available, an expansion in the use of heavy liquid separation seems probable. [1-4]

Liquids in the 1·4–3·0 range of specific gravity, with viscosities well below those of aqueous heavy media make possible sharper separation and treatment of smaller feed sizes. This permits more complete liberation crushing. Most gangue minerals are siliceous and lie in the 2·6 to 3·5 range of S.G., the rock-forming silicates (bar pyroxenes, amphiboles and chrysolite) being below 3·0 in density. The economics and technical advantages of elimination of these gangue minerals at an early stage are therefore of lively interest in mineral processing.

Taggart's *concentration criterion* gives a rough guide to the size ranges in which various methods of gravity separation are practicable on a large scale. If two minerals, respectively 3·0 (Sh) and 2·5 (Sl) in S.G., were to be separated in water the ratio

$$C = \frac{(\text{Sh}-1)}{(\text{Sl}-1)} = \frac{3-1}{2{\cdot}5-1} = 1{\cdot}33 \qquad (12.1)$$

would be too low for efficient treatment on *shaking tables.* If (though no method has yet been achieved) a heavy liquid could be used at S.G.2 the ratio would be

$$C = \frac{3-2}{2{\cdot}5-2} = 2{\cdot}0 \qquad (12.2)$$

and tabling would be in theory efficient. Health hazards from inhalation of fume, danger of contact with the operator's skin, loss by splash, drag-out and evaporation, corrosive action and chemical breakdown have all been reduced but the economics of cheap and abundant water *versus* chemicals costing some 3/– per lb. remain at present prohibitive to general use of methods involving the circulation of a large volume of heavy liquid. A break-through is probable in the use of such DMS appliances as drum washers for coarser ore and cyclones for sands.

Some commercially available heavy liquids are listed, together with their key properties, in Table 21. If a liquid is adjusted by dilution, the diluent must be miscible and chemically compatible. It should have a similar vapour pressure and b.p., or form an azeotropic mixture. Heavy organic liquids are lower in surface tension than water, and therefore "wet" dry ore more readily. This, by reducing the tendency of smaller particles to float regardless of their density and so to overflow without being treated is a minor point in favour of heavy liquid separation, when the feed is dry. Working losses on the pilot scale and in the small operations now at work is of the order of 1 lb./ton of

TABLE 21

(1) *Compound*	(2) *Formula*	(3) *S.G.*	(4) *Visc.*	(6) *S. Ten.*	(6) *b.p.* / *f.p.*	(7) *Sol. of water %*	(8) *Stability at b.p.*	(9) *Miscible with*	*Notes*
85% Thallous formate solution	$HC_0O.Tl$	3·39	2·7	74·5				Water	(4) Viscosity in centigrade at 25°C.
Methylene iodide	$CH_2 I_2$	3·31	2·6	62·5	182 / 6·1		Not	Carbon tetrachloride Benzene	
Tetra-bromo-ethane (TBE)	$C_2H_2Br_4$	2·96	3·4	48	239 / 0·1			Alcohol CCl_4 Chloroform	(5) Surface Tension at 20°C.
Bromoform	$CH Br_3$	2·89	1·8	44	149 / 8		Not	Alcohol	(6) b.p. and f.p. boiling and freezing points °C.
Methylene bromide	CH_2Br_2	2·48	1·0	39·9	97 / 0·2	·07	Yes		
Ethylene dibromide	CH_2Br, CH_2Br	2·17	1·6	38·5	131 / 9·8	·07	Yes		(7) Solubility of water in heavy liquid %
Pentachloro-ethane	$CCl_2 CHCl_2$	1·67	2·3	35·6	161 / –22	·24			
Tetrachloro-methane (carbon tetrachloride)	CCl_4	1·5			76·7 / –22·9		Yes	Most organic liquids	
Trichloro-ethylene	CCl_2CHCl	1·46	·6	29·3	87 / –87	·02	Yes		

material treated. Ore structure affects this, granular and compact particles being best for treatment. Those with voids raise the drag-out loss. Micaceous, fibrous or acicular small particles are unsuited to treatment.

Dense media. One of the earliest modern processes to use a quiet bath for upgrading coal was the Conklin (1922). The solid employed was magnetite ground to —200 m.

Although magnetite is today widely used in dense media baths, it failed to establish itself at that time, since a simple method of keeping it clean had not been found.

In the 1930's the Barvoys process was developed in Holland. It continues to enjoy a substantial success. The separating fluid is a mixture of clay (S.G. 2·3) and finely ground barytes (S.G. 4·2), mixed in a ratio of 2:1 and diluted with water to any desired density up to 1:8. The barytes, at —200 mesh, together with the clay, forms an almost stable pulp in which great accuracy of "cut" between sink and float is practicable. The viscosity of the medium is an important operating factor. If too high, it impedes the movement of particles in the separating bath. Froth-flotation is used to regenerate fouled dense-media, by removing fine coal.

In another Dutch process, Tromp uses a less stable fluid, prepared from finely ground magnetite or specially treated pyrite. The settling characteristics of the medium cause the bath density to be somewhat lower at the top than at the bottom. This aids in separation.

Another substance used in the Netherlands is loess, which differs from clay in having a lower viscosity at the same pulp density. The processes using this medium are discussed under "Coal" (Chapter 23).

In the treatment of ores the lightest mineral is usually quartz, at a density exceeding 2·63. The substances used to form the dense media must be far denser than for coal treatment, as the working density of the bath is twice as high. Any substance used is chosen for the following main qualities:

Hardness. It must not easily break or wear down into a slime under working conditions.

Chemical Stability. It must not be chemically corrosive, or liable to react with the ore minerals undergoing treatment.

Slow Settlement at Tolerable Viscosity. It must form a fairly stable pulp without having to be ground very fine, or the medium will be too viscous.

Specific Gravity. This must be high enough to give the required bath density under reasonably non-viscous conditions.

Regeneration. In working, the dense media becomes foul, and must be easy to clean before further use.

Non-fouling. A certain amount of media lodges in cracks in the lumps of cleaned and washed ore. It must not be of such a composition as to upset the subsequent treatment of the ore.

Among the dense media best able to meet these requirements are those listed in Table 22.

Galena was used in the dense-media process at Halkyn[4] following its development in a pilot plant. It was also extensively used in North America in lead-zinc plants. The density of PbS is 7·4–7·6. When pure it can be used to maintain a bath density of 4·3, but the practical working limit is of the order of 3·3. Above this, movement of the ore is slowed down by the viscous resistance to the medium. This leads to difficulties in dealing with small particles of feed, and of middlings having a specific gravity close to that of the bath. Galena is a soft mineral, and is prone to slime readily. Removal and recovery of this slime by froth-flotation is somewhat difficult. The material itself is expensive. The tendency of recent years is toward an increased use of magnetic materials rather than galena. These can be more easily regenerated and are lower in primary cost.

TABLE 22

Material	*Method of Regeneration*
Galena	Froth-Flotation
Magnetite, Mill Scale, Ferro-Silicon	Magnetic Separation
Pyrites	Used autogenously
Copper Pyrites, Hematite	Obsolescent

Magnetite (S.G. 5·0 $<$ 5·2) is used where bath densities below 2·5 are suitable. This is rather low for ores, but magnetite is increasingly used in cleaning coal. It is regenerated by magnetic treatment.

Ferro-silicon is now the most widely used substance. With a silicon content of 10% its S.G. is 7·0, and at 25% Si it is 6·3. If the silicon content exceeds 22% it is only feebly magnetic, and if below 15% the compound is prone to rust. A 15% ferro-silicon can be used to produce dense media with specific gravities up to 3·5, but the more usual working range is 2·5–3·2. Where the separation is between 2·65 and 2·9, the ferro-silicon is usually ground to —100 mesh and blended with between 10% and 20% of Fe_3O_4. At higher working densities (2·8 $<$ 3·0), —100-mesh ferro-silicon is used alone, or a mixture of —65-mesh ferro-silicon with the Fe_3O_4. For bath densities above 3·0, —65-mesh FeSi is employed. FeSi containing 15% Si is non-rusting and has good magnetic qualities. A little lime is sometimes added to the slurry, particularly when carbonate ores are treated. A dispersing agent, such as tri-sodium phosphate is occasionally employed to reduce any tendency for slimes in the bath to flocculate. Ferro-silicon is either prepared by grinding lump material, usually in a wet ball-mill, or by air-blasting the molten compound as it is tapped out from the furnace. In the latter case, the resulting particles are roughly spherical, and it is claimed that higher bath densities can be used because the specific surface (and hence the pseudo-viscosity) of these spheroids is less. A further advantage claimed for atomised FeSi is that there is reduced loss through drag-out, the spheroids being less liable to cling to the ore products.[5] One contributing factor to loss through such adhesion is roughness of the ore surface.[6]

It is further claimed that rusting tends to commence at sharp corners and the spherical ferro-silicon is chemically more stable than that produced by crushing and grinding. Provided due care has been taken in manufacture to avoid trapping of air into the molten globules and thus reducing their specific gravity, this claim is interesting.

Mill scale is used as an alternative to magnetite. Flue dust from steel smelters is used in Australia. Research has been active in seeking suitable heavy powders to increase the commercial working limit of bath density. Such metals as tungsten have had limited laboratory use in bath make-up. Monitoring and automatic regulation of media density and viscosity are practised in some installations.

In the Stripa process[7] an autogenous DM is segregated from the mill feed. It is a fine magnetite-hematite sand, and its S.G. is regulated by rising hydraulic water which keeps it in teeter. In the Chance process, used in coal preparation, a similarly dilated bed of sized sand is used.

Dense Media Separation

The first Huntington-Heberlein sink and float plant came into use for ore treatment in 1937, at the Halkyn Mine in North Wales. In its early stage the method used was similar to that of the Barvoys coal cleaning process, save with respect to type of medium and method of regeneration. As now worked, the media is controlled for rate of flow through the open topped pyramidal bath partly by use of several "steady head" flow tanks and partly by use of an internal "steady head" outflow from the elevator casing, through which the settled fraction of the ore is removed. At the top of the tank (Fig. 118)

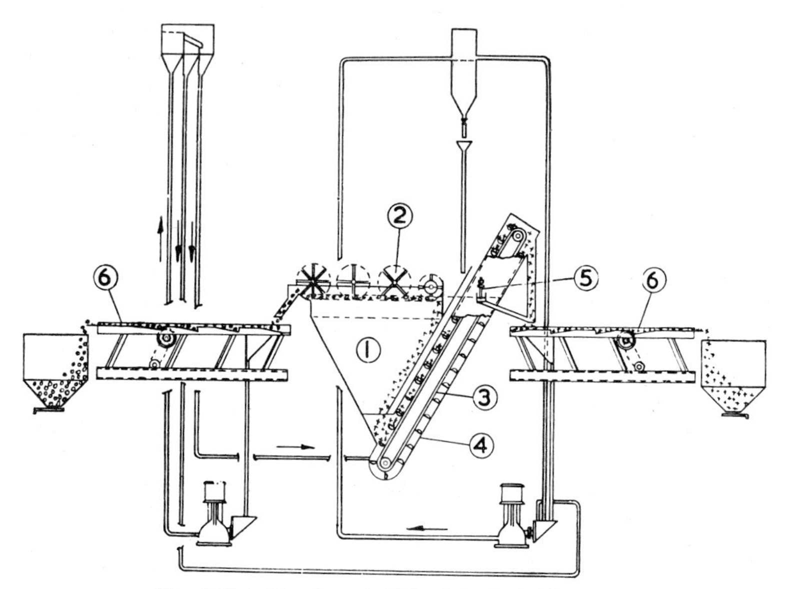

Fig. 118. *Huntington-Heberlein Sink-Float Unit*

rotating paddles (2) submerge the entering feed and impel the floating fraction to the discharge. Rate of paddle rotation is controlled, a dwelling time for float ore of between 15 and 25 seconds being normal. Sinking ore is accelerated by flowing media through the pyramidal part of the tank to the slowly moving elevator (3) where it is raised by perforated buckets which allow non-adhering media to drain back down the casing (4) to the outflow point (5). Further drainage is performed on the shaking screens (6) as they convey floated and sunk products separately out of the system. This drained media returns to the system. Wash-sprays remove the last of the adherent media, which is dewatered, reconstituted and returned. An essential feature of the process is its quiet bath, no mechanical agitation being used to stir its contents. Media can be cleaned between overflow and return (for example by froth-floating galena away from the entrained waste).

The Cyanamid Process

This is also called the "differential-density" process, because under ordinary working conditions there exists, with the media used, an appreciable rise in specific gravity from the top to bottom of the separating vessel. The standard flow-sheet (Fig. 119) starts by removing the fines (usually —10 mesh) at (1). These should by-pass the process. Oversize is sent to the cone (2) where it joins the dense-media fluid and either floats to a discharge or sinks and is then pumped up by an airlift. Float and sink move separately to two parallel drainage screens (4)[8] along which the sinks and floats proceed in their separate channels, under the transporting influence of a vibrating mechanism. These products then traverse two parallel washing screens (5) and are discharged to their respective following sections. All drained medium is pumped by (6) back to the cone (2). A diverting chute (15) can be used to divert part of this drainage into the wash-water sump. This receives all the washings from (5), and sends it along for reclamation. First the drainings pass through a magnetic field (7). Here the magnetite and ferro-silicon used as media receive a magnetic charge which causes the particles to flocculate into aggregates. These behave as though they were large single particles, so when they reach the thickener (8) they settle rapidly, having temporarily lost the physical qualities which enabled them to remain dispersed in the separating bath. The thickened sediment is now pumped up to two wet magnetic separators (11 and 12) in series. Here the magnetic flocs are washed, separated from non-magnetic particles in the feed, and dropped into the densifier (13). This is an adapted Akins classifier, made more sturdily than is needed for ordinary pulps. The thin medium is fed in over what is normally the Akins weir, and water overflows from what in classification would be the feed launder entry. The magnetic flocs are discharged through a demagnetising coil (14) back to the medium drainage sump, whence they are pumped up via (6) to the cone (2). The densifier (13) can be used to regulate the density of separation. If the spiral is raised a little, some solid DM is withdrawn from circulation. Since all the water (spillage, evaporation, and drag-out water excepted) is in continuous closed circuit, this results in a decrease of bath density. Similarly,

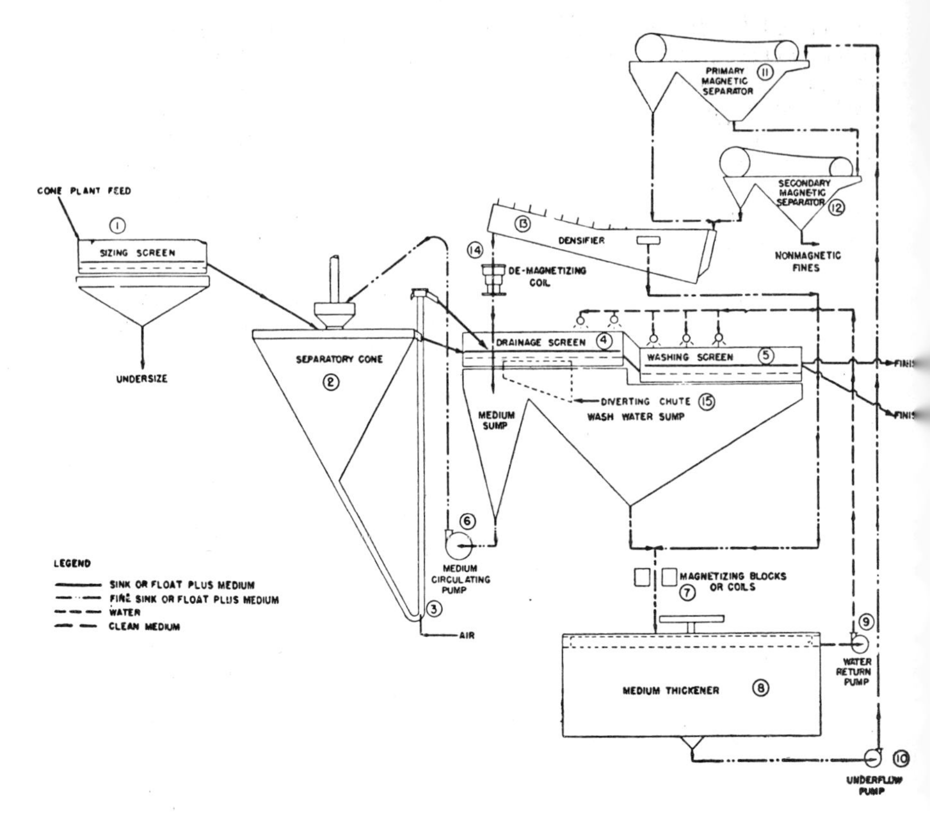

Fig. 119. *DMS with Magnetic Reclamation*

a slight dropping of the spiral introduces more D.M. solid into circulation. The operator thus has excellent control of separating density. The thickener can be used as a storage tank when shutting down, as dense medium must not be allowed to settle solidly in the cone, where it would be difficult to restore it to a fluid condition. It is usual to provide a run-off sump to which all circulating dense medium can be drained in an emergency, and from which it can be returned to circuit either by pump or bucket elevator. In recently constructed plants no run-off sump is provided. Instead, the bottom floor is of concrete, and is surrounded by a curtain wall. The floor slopes gently to a small gathering sump. This arrangement serves to catch spilt medium and provides temporary storage if the contents of the cone must be dumped. In another arrangement the contents of the cone can be piped to the thickener.

Medium of any required operating density can be introduced in any desired volume at any depth or set of depths down the vertical axis of the cone. Part escapes freely at the overflow launder and part with the underflow. The distribution of the outflow depends on the force with which the air-lift (Fig. 119 (3)) is worked, this air-lift being, in effect, a controlling pump which varies the intensity of downdraw of the medium. A stirring mechanism works inside the cone, but its action is indiscriminate so far as this hydraulic effect is concerned. Since in operation a light ore fraction must overflow, while a heavy one drops to the underflow, the inflow must always exceed the underflow.

Recirculating media enters the cone gently at the required depth, and then flows either toward the upper discharge, which is situated on the periphery, or the bottom (apical) discharge. If the air-lift is run gently, the upward current will be relatively strong, becoming gentler as the horizontal cross-section of the cone widens. In its standard form the cone finishes above with a cylindrical section. Some of the older designs had "closed" cones, with a frustrum inverted upon the lower cone to give acceleration to the rising fluid before discharge, but this is no longer favoured. If the entering media is released near the top of the cone, much of it streams across to the overflow after being distributed by the revolving paddle. This gives the upper part of the cone a strong streaming force, while toward the apex there is an increasingly heavy downward pull, controlled by the strength with which the air-lift is being caused to withdraw fluid.

If the media is fed low down, streaming force gives place to lifting force. Two simple controls are therefore available for adjusting the vertical and horizontal components of the fluid motion to suit a given type of ore. Their effect is intensified by the slight instability of the media, which by giving a gentle density rise from top to bottom of the cone makes it more difficult for a particle to fall the further it descends. Here two forces are acting against one another. If the cone is imagined to be quiet, so that the only operative force is fluid density, then a number of particles lying very near to the average density would arrange themselves horizon by horizon with the lighter above and the heavier below. If, however, the added effect of the continuous withdrawal of medium at overflow and underflow be now considered, each particle, at its particular horizon, is being moved by the frictional drag of fluid running past its surface. The lighter material will be at the top and tends to receive streaming friction, which moves it horizontally to the discharge of the floated fraction; the heaviest will be well down in the drag toward the underflow, where it is accelerated throught he continuously reducing section to the apex. Other particles will fail to generate a sufficiently strong characteristic alignment and will build into a teeter bed, according to their mass and surface.

In cone operation this teeter bed can be used as a controlling factor. It has some points of resemblance to the teeter zone in a hydraulic classifier. It "scrubs" the material passing through it, and its added resistance affects the sorting action in the cone. The operator can see the level of fluid in the sump from which medium is being pumped back. Since this sump is the only "surge" or holding tank, and the total amount of fluid in the system is constant, any increase of teetering ore in the cone displaces an equivalent volume of media and causes a rise in the sump level. This can be used as a teeter control by a skilled operator.

Normally the weakness of all two-product concentrating devices is that they cannot simultaneously produce a clean concentrate, a middling, and a clean tailing. If set to make a good concentrate, they must put middling with the gangue, and *vice versa*. Using the sump level to indicate the volume of teetering material (middling) in the cone, the operator can let this accumulate while he makes satisfactory light and heavy fractions, until the time when he decides to dispose of this middling. If he wishes to treat it as a low-grade concentrate he now increases the air-lift, thereby strengthening the downward flow. The teetering particles are drawn down and can either be discharged with the heavy fraction or separately removed from the circuit. If the teeter bed is being worked as a "scrubbing zone" and is to be rejected, it can be switched to waste by the same method. When the sump-level has dropped to normal, the air-lift is reduced and the building-up process is repeated. Control of the bottom discharge has been proposed by use of a "magnetic valve" through which, by varying the magnetic flux, non-magnetic sinking ore can fall while part of the magnetic media is held back.[9]

The working balance in the cone can also be varied by other methods. Where high operating densities are needed, −65-mesh ferro-silicon is favoured. Such a coarse media leads to a higher density differential from top to bottom of the cone, and aids in the quick formation of a teeter bed. For lower densities—100-mesh ferro-silicon is used, or a blend of this with magnetite. Frictional resistance to fall of ore could in theory be exploited by controlling the grain sizes blended together, but this is not favoured in practice, anything which introduces an element of viscous resistance tending to reduce efficiency of separation. Plants can handle suitable ore down to fine gravel size, but if the viscosity and friction were high, the retardation produced upon the smaller particles would be so serious as to reduce capacity. An essential feature of DMS is the cheapness and high capacity of this method of upgrading low-grade ores. A low-viscosity medium is therefore desired.

It has already been noted that the level of media in the hopper below the drainage screens varies only with the teeter volume. This needs some qualification. If the ore fed in is wet, it introduces water which may thin the media and upset the bath density. If this input of water is constant, it can be corrected by adjusting the input density of regenerated media returning from the magnetic cleaning circuit. The balance between water entering the system on ore, and that leaving the system, must be maintained. Any desired fraction of the recirculating media can be diverted from the drainage hopper to that below the washing screens, for cleaning purposes.

A 14-ft. diameter cone working at a capacity of 2000 tons per 24-hour day holds a ferro-silicon charge of about 60 tons. If the "drag-out" loss of FeSi were of the order of $\frac{1}{2}$ lb./ton of ore treated, this loss of about $\frac{1}{2}$ ton of media daily would continue for some time before it became apparent. New media can be fed direct to the plant. Occasionally it is preconditioned for several days. The reason given is that a week or two of wet tumbling serves to wear off grain angles and produce uniformity. It is also possible that when ferro-silicon is produced from pigs of alloy by dry crushing and dry grinding its surface is grease-contaminated and aerophilic as shipped, so that the particle surfaces need some cleansing. When a new charge is put into the plant, some clay or slimed ore should be added, so as to produce a media having a consistency similar to that which characterises normal operation. Clean ferro-silicon in a new bath usually settles rather too quickly without this addition of slimes.

An interesting development of dense-media service is the Mobil-Mill (Fig. 120). This is a compact cone plant, with single-stage magnetic cleaning, fed direct from the washing hopper. Flocculation of the media takes place in the magnetic separator itself. Sufficient permanent magnetism remains with the material pulled out on the belt to ensure its settling during progress through the densifier. The Mobil-Mill takes only 36 man-days to erect or dismantle and stands on an area of 13′ 10″ wide by 29′ 3″ long and is 16′ 2″ high. It is sectionalised, on skids, and has a duty of between 5 and 15 tons per hour; it may be leased or bought in some countries. This ease of installation opens interesting possibilities to the young concern, or the one which is considering the possibilities of sink-float treatment but wants a run on a pilot-plant scale before finally committing itself.

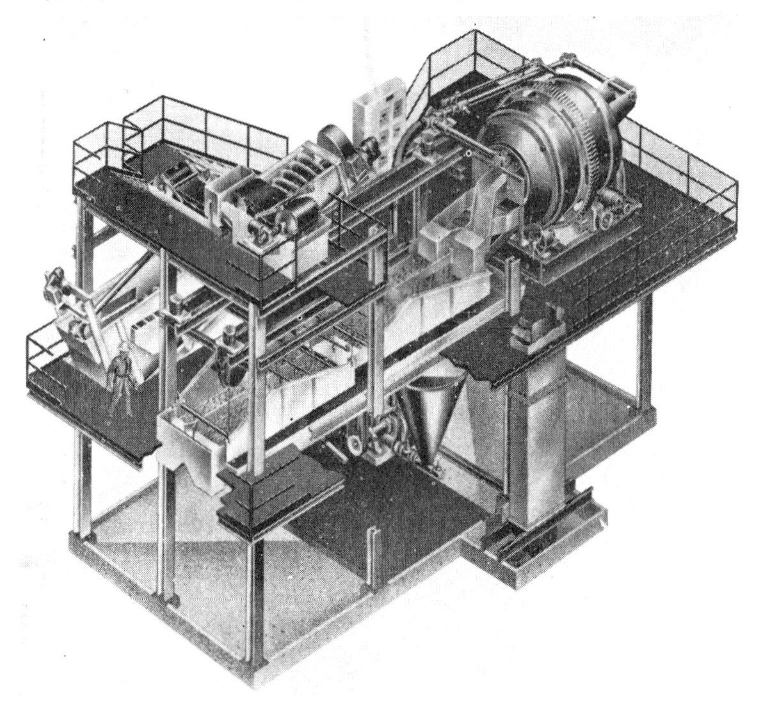

Fig. 120. *Mobil-Mill* (Western Machinery Co.)

Some Other Systems

One of the oldest appliances was the puddling machine developed at

Kimberley for breaking down the diamond-bearing "blue ground" into a slurry through which diamonds settled while lighter semi-colloidal mud overflowed. The machine is a round shallow trough, the contents of which are stirred by knives rotating at a linear speed of 300′ to 400′/minute. The feed enters (in the modern form) through a submerged aperture in the outer wall, and tailings overflow an adjustable weir in the inner wall. Tangential entry of the feed, *plus* the retarding and swirling motion imparted by the stirring knives, produces a quicksand of a density around 2·0, which allows a substantial volume of valueless solids to overflow at the inner weir, the settled heavy fraction being discharged from below.

In another process the Akins classifier is used as the separating vessel. It is built into some such arrangement as that shown in Fig. 121. The system

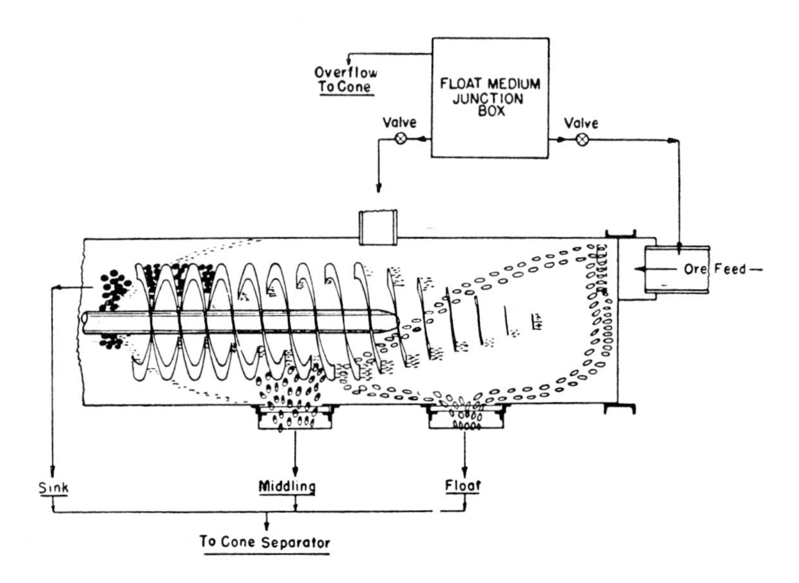

Fig. 121. *The Akins DMS System*

has the advantage in use that is can be stopped under full load and readily re-started, since the Akins spiral is available for working settled medium up to a suspended state. Where the DMS plant is required to work only during hours of delivery of ore this can be an important aid to the operator.

A type of machine well suited to the gentle handling of large material such as coal is the Link-Belt Co.'s Float-Sink Concentrator (Fig. 122). The separating drum principle is also incorporated in such appliances as the Wemco (Fig. 123) and the Hardinge (Fig. 124). Wemco machines are also made with two separating drums (Fig. 125), operating at slightly different densities so that a true middling can be continuously removed.

In coal preparation, which pioneered modern DMS, the feed is much lighter and the economic product is floated away from a minor sinking fraction of shale and "dirt". Operating requirements therefore differ sharply from those in which the same principles are applied to the treatment of heavy minerals. Among the substances which can be used in the bath are clay, barytes, loess, magnetite, blast furnace soot and ferro-silicon. The

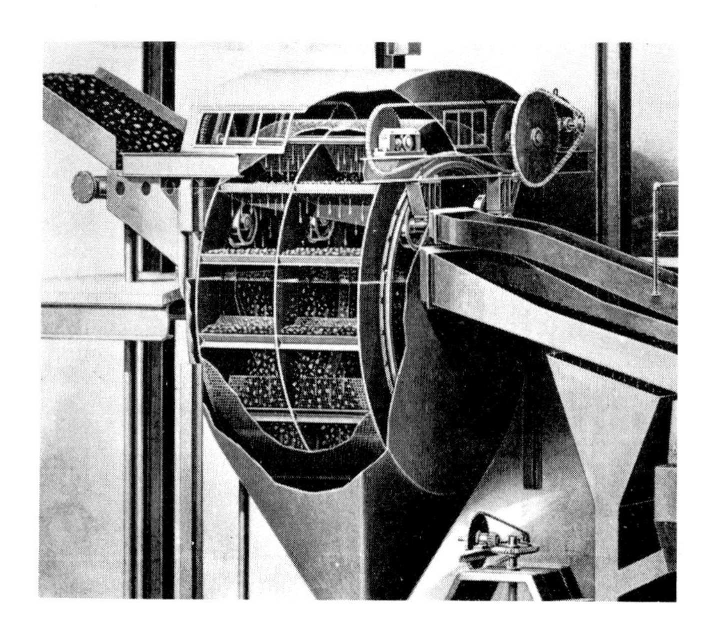

Fig. 122. *Link-Belt Separator*

tendency is toward finely ground ferro-magnetic substances, which in their normal state settle slowly, but can be magnetically flocculated for quick settlement, recovery, and cleansing. One such medium has a settling rate of 4 minutes/inch in the presence of 10% of clay slimes in a bath working at density of 1·5 and over 21 minutes/inch at 30% slimes. The slight tendency to settle is compensated by the continuous flow media through the separating bath. It carries one or more of the separated products out of the system. Separating baths may be designed to make a two-product division into "floats" and "sinks", or may also deliver an intermediate middling in one pass. Alternatively, three products can be made by operating two baths in series at different densities.

In the Drewboy separator (Fig. 126) the feed crosses the bath, floats being helped out by travelling flights while sinks are lifted out by the vanes of the slowly revolving dirt wheel. It treats material from 24″ down to $\frac{1}{2}$″. In the shallow-bath separator (Fig. 127) raw coal is fed centrally, and floats are moved to one end by the top part of the separator conveyor while sinks are carried up to be discharged at the opposite end.

Both the Tromp and the Barvoys separators have provision for three-product working. One modern form of this system is the Simcar (Fig. 128). Raw coal enters at A and first-stage separation (clean coal from the dirt *plus* middlings) is made in the low-density zone B at the top of the bath. The float is moved by the scraper conveyor C to drainage screen D while the rest of the feed sinks to the boot of the elevator F and out *via* G. Middlings are checked between low- and high-density zones and carried up the compart-

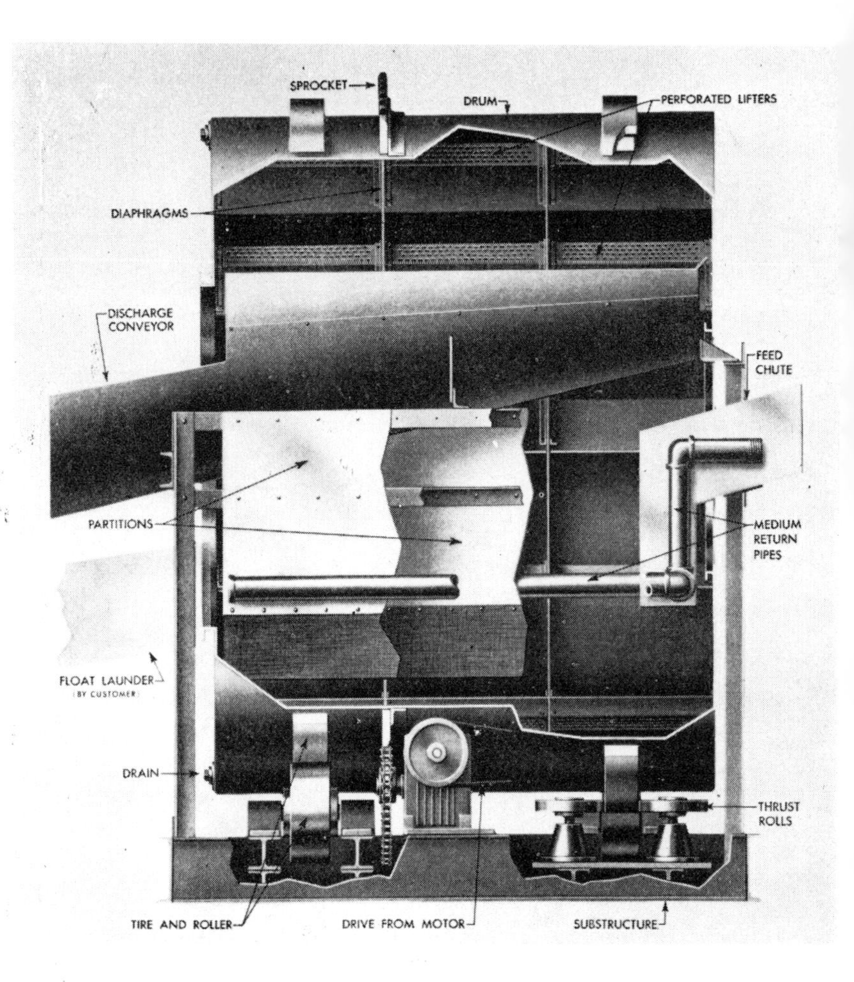

Fig. 123. The Wemco Drum Separator (Western Machinery Co.)

ment H by gently rising medium, the flow rate of which is controlled from the overflow box L. They are then moved by the middlings scraper conveyor K to the middlings chute J, and delivered to their drainage screen. This machine treats up to 200 tons per hour of $-8'' + \frac{1}{4}''$ feed.

In the Barvoys process the density of the clay-barytes suspension only varies slightly from top to bottom of the bath. Middlings are lifted out by a propeller-induced upward current. The Tromp process uses a horizontal flow to yield a top float, a middling, and a sink discharge of dirt. In the Ridley-Scholes process float and middling are withdrawn at points along a V-shaped bath and sinks are removed on an endless belt.

A comparatively new type of coal washer, the Norwalt[10] has been designed

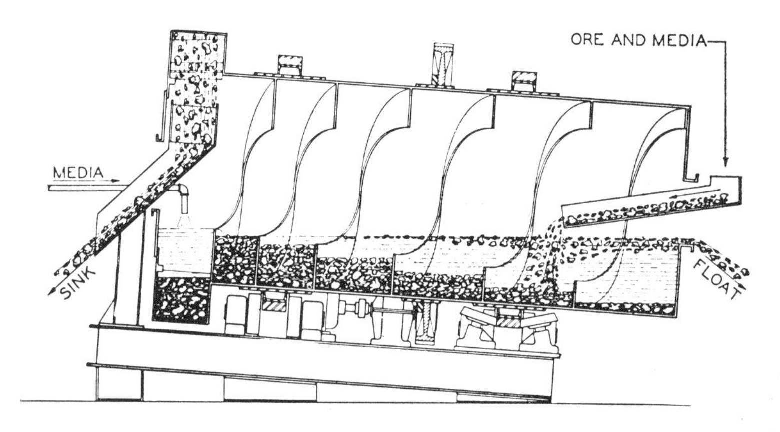

Fig. 124. *Counter-Current Heavy-Media Separator* (Hardinge Co.)

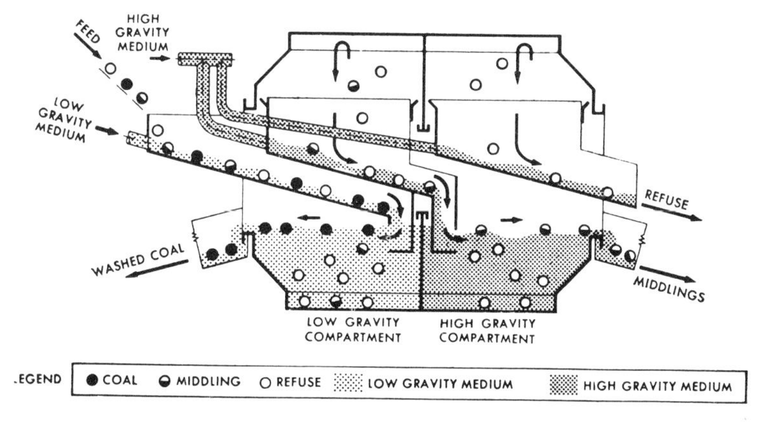

Fig. 125. *Wemco Two-Drum Separator*

to unite the best features of both "deep" and "shallow" baths. The bath is a cylindrical drum in the centre of which is the frustrum of a cone. Feed enters at the apex and is forced into the separating fluid via an annular curtaining hood. The heavy fraction sinks and is swept to the discharge section by slowly rotating plates.

The Chance process differs from the DMS methods in which a relatively stable slurry flows through the system. It more nearly approaches hydraulic classification, but the shape of the cone and the method of introducing the hydraulic water is designed to maintain a dilated bed of sand which for the most part remains at work in the separating zone. The large volume of water overflowing with the clean coal (Fig. 129) is fed to the upper part of the cone and has not been used to dilate the main separating zone as would be the

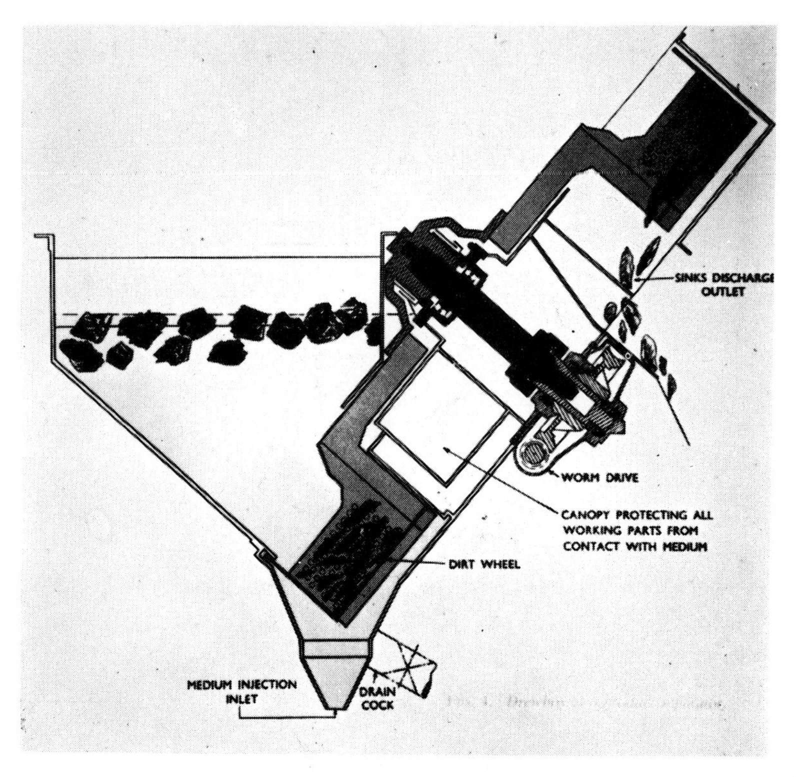

Fig. 126. *The Drewboy Separator* (Automatic Coal Cleaning Co., Ltd.)

case in straight hydraulic classification. The sand used is silica sized between 30 # and 100 #. When this is dilated by water rising between $\frac{1}{4}''$ and $\frac{1}{2}''$ / second (according to packing voids of the sand), the quicksand density is

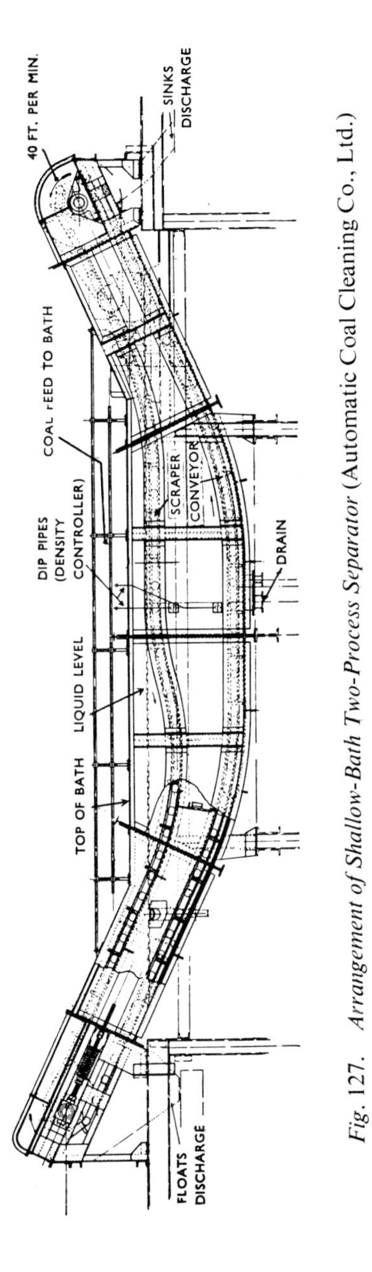

Fig. 127. *Arrangement of Shallow-Bath Two-Process Separator* (Automatic Coal Cleaning Co., Ltd.)

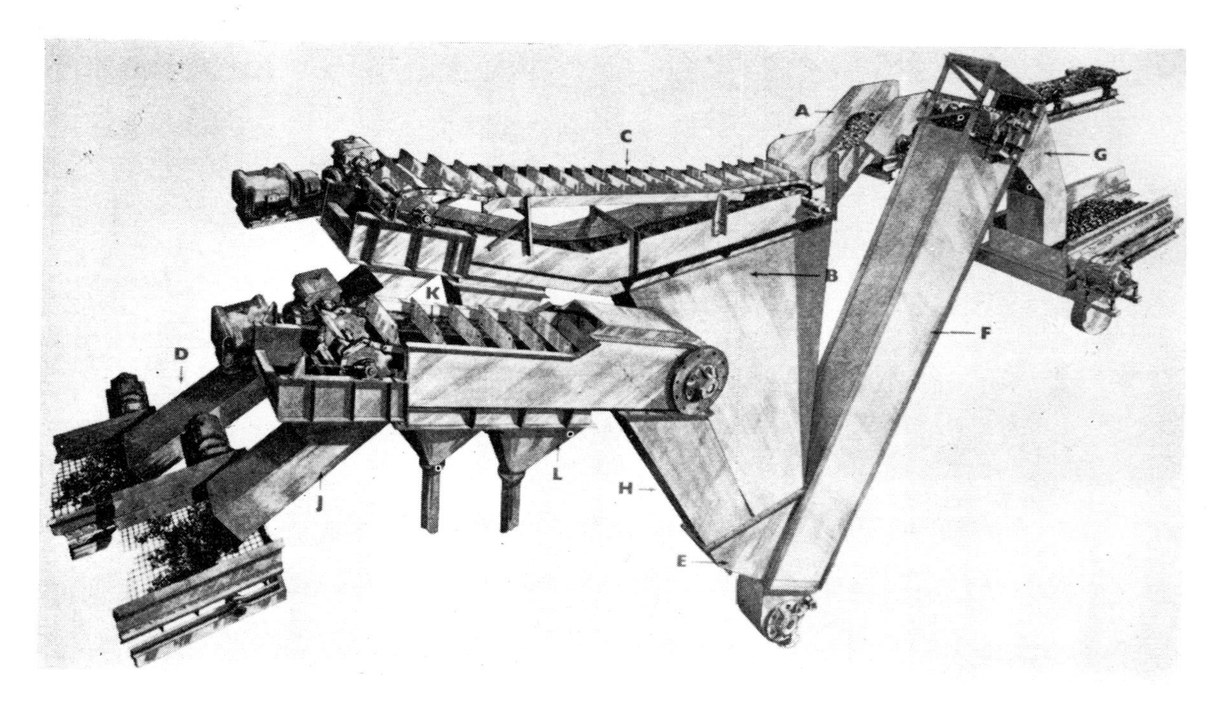

Fig. 128. *Simcar Three-Product Separator* (Simon Carves)

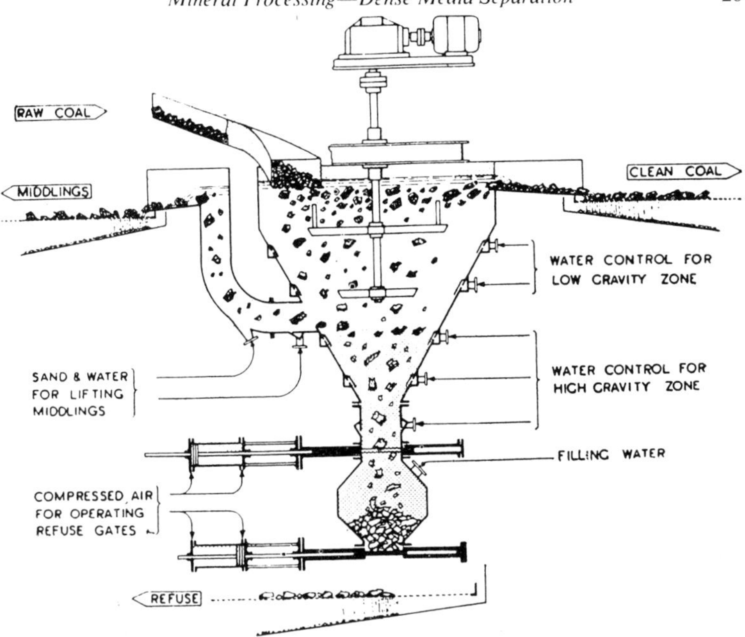

Fig. 129. The Chance Cone (G.E.C.)

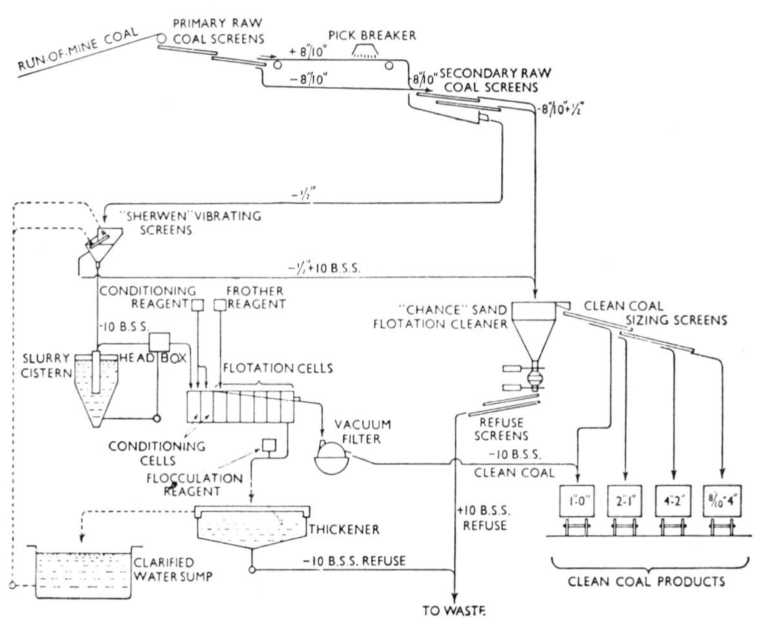

Fig. 130. Diagram of Coal Preparation Plant

easily held at a point between 1·4 and 1·7. Gravity control is effected by varying flow-rate in the high-gravity and low-gravity zones. A particle of coal 1·676 mm. in diameter (10 ♯ B.S.S.) and S.G. 1·55 has a terminal falling velocity of about 6″/second in a liquid of S.G. 1·5. It thus has no difficulty in sinking through a current rising ½″/second. The stirring mechanism in the cone rotates slowly and assists the coal outward to the overflow weir. Overflow coal and shale discharged through the automatically controlled refuse gates are desanded, the sand being returned to the top of the cone. The normal cone does not have the additional middlings discharge shown. Operation is checked by testing the products with heavy liquid of the correct separating density. Too dirty a float is corrected by increasing the proportion of water to sand, or by reducing the rate of feed.

In Fig. 130 the flow-sheet for a two-process treatment is shown. Dry or wet screening removes the —10 ♯ material from the raw feed for froth flotation, and sends the −10″ + 10 ♯ coal to the Chance cone.

A still more pronounced step away from "weighing" the feed into a quiet bath is applied to the separation of heavy minerals in the Stripa process, shown in Fig. 131. As in the Chance process a hydraulically dilated quicksand is used, water rising into the shaking trough (1) through the valves (F). In a typical Swedish operation a fairly coarse magnetite sand is concentrated on shaking tables from ore undersize screened through, say, 20 mesh, and passes along the trough at a teeter density of 3·4. Coarse feed entering at (9) either floats on this teeter or sinks into it, the two fractions being separated by (D) into floats (7) or sinks (8). The media drains through (5) and is either recirculated via (6) or sent to further treatment, since in this particular case it is an economic mineral. With a copious supply of autogenous media no regenerative treatment is needed, though this can be achieved simply if desired, since the grain size is coarse enough for slimed material to be screened away. There is no problem of dilution of media by wet feed, and the process tolerates unscreened run-of-mine feed.

DMS in cyclones. In the Chance and Stripa processes an ascensionary hydraulic pressure is imposed in place of a quiet bath. In the cyclone used for DMS, though there is a minor movement of ore roughly parallel to the long axis, the main driving force is centrifugal. Consideration of the formula for centrifugal force (mv_2/ρ) shows a complicated work-pattern at any cross-section in the sorting zone. The intensity of g-acceleration on an individual particle in an evenly swirling system is a function of its radial distance from the axis. This is modified in practice by the fact that the particle is entrained in an environment which is itself being subjected to sorting action. The swirling system increases in viscosity and S.G. outward, but the swirl decelerates toward the periphery because of drag against the containing walls of the cyclone, thus introducing a stirring element due to shear.

Since the feed in a typical cyclone treatment is smaller than 10 mesh (say *minus* 2 mm.) its area-volume ratio is much higher than that of coarse material fed to a quiet bath. At first sight the viscous restraint exerted on each particle might appear a major factor in dictating its response. Such braking force is, however, being applied in a spinning system in which both

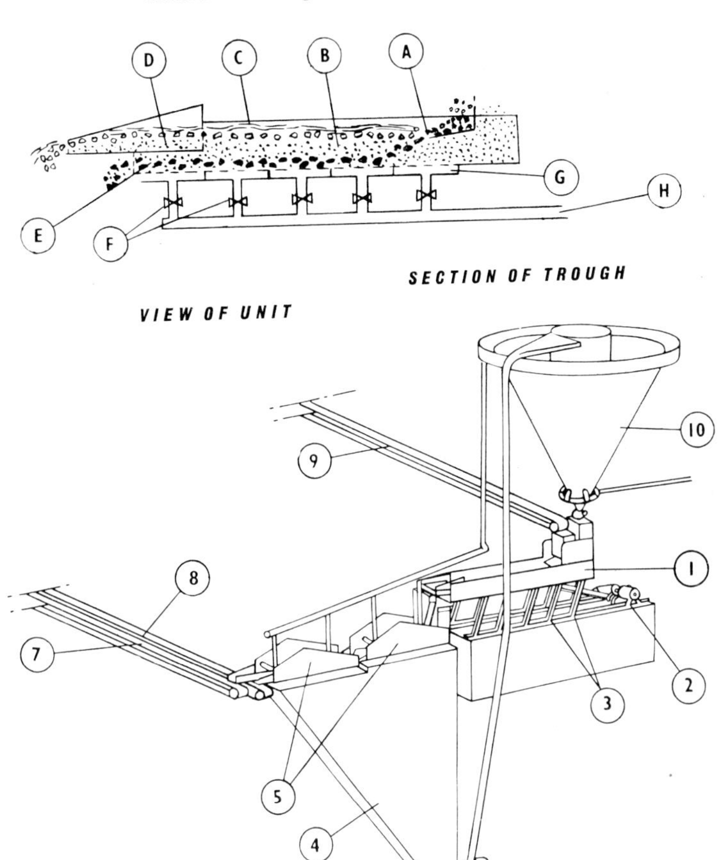

Fig. 131. *The Stripa Process*

A. Feed plate; B. medium bed; C. floats; D. adjustable splitter plate; E. perforated bottom plates; F. valves; G. water boxes; H. bottom water pipe; 1. Stripa shaking trough; 2. driving mechanism; 3. flexible rods; 4. pump sump; 5. washing screens; 6. medium pump; 7. floats; 8. sinks; 9. feed; 10. medium thickener - cone or cyclone.

brake and particle are moving at roughly equal radial speed. Hence, the viscosity of the medium only affects the situation while the particle moves inward or outward along its radius, and is therefore a minor influencing factor.

To simplify the approach to our study of the complicated dynamics of cyclone separation we can start by assuming the media to be a heavy liquid. This defers the extra problem of centrifugal distortion of the S.G. of the bath from its lightest central core to its heaviest outer zone. Consider next the distribution diagram in Fig. 132.

Feed F (heavy liquid plus ore) enters peripherally at a constant rate R, constant solid-liquid ratio cF and pressure *p*F. Part leaves the cyclone as the heavy product H at consistency and pressure *c*H and *p*H, and part as the light product L, at cL and pL, *via* apex and vortex finder respectively. With steady operating conditions the position is as shown in Fig. 132 (b) but with no fluctuating pressure between *p*H and *p*L. Assume that the volume and rate of feed F have been adjusted until a satisfactory split into ore products H and L has been reached.

Such an ideal condition is never achieved in practice. The main disturbing factors which must be compensated or controlled in each specific ore treatment are

(*a*) Change in S.G. of heavy liquid (or dense media).
(*b*) Change in solid-liquid ratio of F
(*c*) Change in (*p*F, *p*H and/or *p*L).
(*d*) Change in feed rate R
(*e*) Change in constitution of ore (particle shape, size analysis of feed, mass distribution of heavy, medium (middling) and light particles.
(*f*) Inhomogeneous feed velocity.

Reduction of F lessens overflow L and lowers the concentration of correctly placed heavy ore in the H product. If carried far enough sorting ceases, and all F leaves at H in its original condition. In the other direction, an increase of F raises *p*F, because the system which at the moment of change was operating at full load with a correct pressure on both H and L must now adjust itself to an increase. The power input to the pumping system rises to overcome the cyclone's resistance to the higher load. Ignoring pump slip and frictional loss, this reports in the system as increased *p*F. In turn the centrifugal force is increased. On theoretical grounds this might increase the volume of ore departing correctly separated *via* H and L. Practically, sorting in the cyclone drum is unbalanced and a disproportionate volume of solids leaves at H, lowering the efficiency of performance by misplacement. Since the H-exit is, at this stage of our study, assumed to be of fixed area, the passage can also be seen as congested so that an undue proportion of the proper sinking ore is forced up to the L-exit. Exploratory tests show that either or both these things occur, leading to decreased efficiency of separation.

Many variables affect the separating action, and they are not adequately comprehended in such formulae as those given in Chapter 9 (Eq. 9.5, 9.6, and 9.7). Even with heavy liquid as the parting fluid there is, with the finely crushed feed characteristic in operation, some autogenous addition to the bath density, derived from the size-weight distribution of ore, and particularly of its middling fraction in the D_{50} zone. From this point, discussion refers to a cyclone using a standard dense media such as a water-ferro-silicon pulp.

Three types of particle, including ore and FeSi, can be classified:

(*a*) Those with sufficient mass to be centrifuged to the wall and discharged via the apex.
(*b*) Those light enough to remain more or less stably in the parting fluid passing through both apex and overflow discharge.

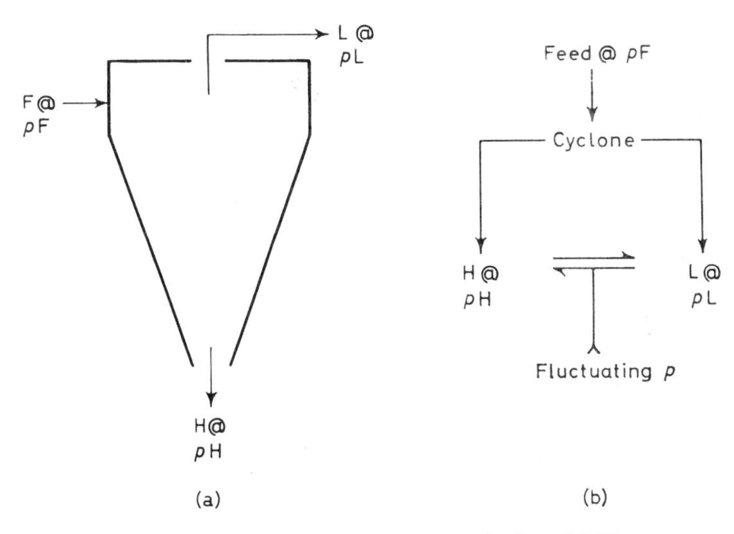

Fig. 132. *Distribution Diagram, Cyclone DMS*

(*c*) Indeterminate particles which change their route in response to minor operating variables.

In research on these variables it has been concluded[11, 12, 13] that with a ferro-silicon media the cyclone deals simultaneously with the sorting of light gangue, heavy values and the heavier particles of FeSi, each material dividing at its own D_{50} (Eq. 9.5, Chapter 9) under prevailing conditions. A FeSi medium sufficiently coarse to react to the centrifugal force in the cyclone would be distributed irregularly, outside the area round the apex discharge and interfere with the desired sorting effect. The settling characteristics of the ferro-silicon under normal operating conditions (size analysis and shape) are therefore important.

In DMS only enough stable (i.e. finely ground) media should underflow to maintain a fluid apex discharge. The ratio between the desired underflow and the volume (not weight) of dry solids in the ore should be maintained at the figure specified for the particular treatment, as arrived at by testing the ore. This is possible if a constant-volume pump delivers feed at a controlled solid-liquid ratio. Alternatively, the apex area can be so controlled (by hand or instrument) as to maintain a steady solid-liquid ratio in its discharge. An essential relationship exists between the proportion of true underflow ore in the feed, the feed rate and the apex discharge area. Feed pressure is of secondary importance, and if low can be neglected. A system worked at a high pressure is in danger of using power uneconomically and of suffering undue wear and maintenance cost. Provided the centrifugal acceleration brought to bear on the size range of particles undergoing treatment is adequate, a relatively large apex and low operating pressure is desirable.

Centrifugal pumps are less satisfactory for feeding in the ore pulp than are those which deliver a constant volume at a steady rate, though they are in wider use. The centrifugal pump is easily upset in delivery rate by wear in any part of the system, since a change in the pressure head against which it works varies the slippage of pulp passing through its impeller system. When the feed rate is steady the vortex finder accepts delivery of the float fraction without altering the pressure in the cyclone. If this pressure rises, the apex discharge flow is increased and float product is being mishandled and wrongly delivered. Where feasible, a constant-head tank between pump and cyclone could be used. The volumes of solids rather than their weights in the pulp are the key consideration in reaching a constant rate of properly blended feed.

Smooth running is often aided by instrumental control based either on pressure-sensitive devices which control the apex area or on radio-active gauges which measure the pulp consistency and tonnage rate at a suitable point. One such device measures changes in the vacuum in the vortex zone and then controls the annular area at the apex by varying the hydraulic pressure on a rubber ring. Another arrangement, best suited to large-scale treatment where close sorting is not vital, is the "umbrella discharge",[12] in which only a selected core of apex underflow is removed, the peripheral portion being re-circulated.

Tests in a 6″ cyclone working in the 2·4 to 3·5 range of density at a 15′ pressure head[13] confirm the importance of using ferro-silicon of a size appropriate to the required parting density of the bath. The finer grade used in these tests contained 50% *minus* 20μ and only 2% *plus* 43μ, the minerals tested being in the *minus* $\frac{3}{16}$″ *plus* 28 mesh range. Somewhat larger sizes were successfully concentrated in a 12″ cyclone, the test ores including such minerals as limestone, silica, magnetite, phosphate rock, iron ore and chromite. Commercial plants accepting feed up to 1″ in size are made, but in our present state of knowledge ores exceeding some 10 mesh (2 mm.) are better handled in quiet baths, provided the required degree of liberation can be reached in these larger sizes. An exception might be a case where the bath's parting density was just insufficient in a quiet bath. Here, the centrifugal effect on a DM circuit charged with ferro-silicon could simulate a moderate extra bath density.

Cleaning the Medium

This, together with the regulated return of cleaned media adjusted to the correct bath density, is the critical control factor in DMS processes. The usual sequence of operations is:

(*a*) Drainage of media from ore leaving the bath.
(*b*) Washing of ore products to remove balance of adherent medium.
(*c*) Collection of foul media, and cleaning it by a suitable process. Since this is a mineral-dressing operation the methods are applications of those described later under:—
  (1) Flotation. (For galena, and for barytes fouled with coal.)
  (2) Magnetic Separation. (For ferromagnetic media.)

(3) Hydraulic Separation. Tabling, mechanical classification. (For sands.)

(*d*) Reconstitution and return to circuit of reclaimed media.

After cleaning, the media usually leaves this operation as a thick slurry or sludge and goes to surge storage (e.g. as in item 13 of Fig. 119). From this it is steadily returned to the separating bath, passing on its way through magnetic de-flocculation in the case of a magnetic or FeSi medium. The amount thus returned is slightly in excess of that bled off for cleaning, as there is a small drag-out loss due to incomplete removal of medium from the departing ore. En route from storage the medium is adjusted to the entering density required to maintain steady parting conditions. As ore usually enters somewhat wet from the washing screens ahead of the bath, this density is somewhat above that of the bath media.

Bath Density

The effect of varying the specific gravity of the dense media during a test for *amenability* of ore to DMS is shown in Table 23. The purpose of the test was to find whether a low-grade ore could be made worth treating. At a separating density of 2·7 just under 54% of the feed was retained as an enriched sink, while by stepping up the cone density to 2·8 the enriched sink was reduced to 7%, a float of 93% of material too poor for treatment being discarded.

If the cost of DMS treatment was of the order of 1s. per ton and of further treatment to recover saleable cassiterite and wolframite 5s. per ton, the cost would be, for 100 tons:

(*a*) Treating all the feed £30
(*b*) DMS at 2·7 S.G. £18½
(*c*) DMS at 2·8 S.G. . . . . . £ 7.

The feed for this test was −1″ + 10 mesh, and the ore minerals included cassiterite, wolframite, quartz, felspar, muscovite, and biotite.

TABLE 23

	Bath S.G. = 2·70					*Bath Density* = 2·80				
	Wt %	*Assay*		*Distribution*		*Wt* %	*Assay*		*Distribution*	
		Sn	WO_3	*Sn*	WO_3		*Sn*	WO_3	*Sn*	WO_3
Float	46·25	·09	·01	17·52	4·05	92·93	·09	·03	37·6	16·89
Sink	53·75	·36	·26	82·48	95·95	7·07	2·06	1·74	62·4	83·11
Feed	100	·23	·15	100	100	100	0·23	0·15	100	100

In older installations the bath density is manually checked at hourly or half-hourly intervals. Instrumentation and automatic control are in increasing use in newer plants.[14, 15] With continuous monitoring, and consequent immediate detection of drift from the operating norm, variance can thus be reduced considerably. When this is coupled with corrective adjustment—for example, by automatic lowering or raising of the densifier rakes and suitable variation of the amount of make-up water—sorting efficiency is correspondingly improved. Although viscosity control is still largely experimental, some progress has been made in automatic check. The methods used are discussed further in a later chapter.

References

1. Bird, B. M., Mitchell, D. R., and Smith, F. E. (1943). "Coal Preparation", A.I.M.M.E.
2. O'Connell, W. L. (1963). *Trans. S.M.E.* 226.
3a. Roe, L. A., and Tveter, E. C. (1963). *Ibid.*
3b. Baniel, A. M., et al. (1963). *Ibid.*
3c. Baniel, A. M., and Mitzmager. A. (1960). *I.M.P.C.* (*London*), I.M.M.
4. Pearson, A. *Trans. I.M.M.* (*Lond.*), 48.
5. Rodes, F., and Cremer, J. (1960). *World Mining*, March.
6. Geith, G. (1958). *Trans. I.M.P.C.* (*Stockholm*), Almqvist & Wiksell.
7. (1955). "*Das Stripa-Schwimm-Sink-Verfahren" Erzmetall*, 4.
8. American Cyanamid Co. *Ore Dressing Notes*, No. 14.
9. Moiset, P., and Dartois, R. (1960). *I.M.P.C.* (*Lond.*), I.M.M.
10. Anon. (1961). *The Mining Magazine*, May.
11. Cohen, E., and Isherwood, R. J. (1960). *I.M.P.C.* (*Lond.*), I.M.M.
12. Dreissen, H. H., and Fontein, F. J. (1963). *Trans. S.M.E.*, March
13. Davies, D. S., Dreissen, H. H., and Oliver, R. H. (1963). *I.M.P.C.* (*Cannes*), Pergomon.
14. Nesbitt, A. C., and Weavind, E. G. (1960). *I.M.P.C.* (*London*), I.M.M.
15. Oss, D. G., and Erickson, S. E. (1962). *S.M.E.*, May.

CHAPTER 13

SEPARATION IN VERTICAL CURRENTS

Introductory

The laws governing the motion of a particle immersed in a fluid show its behaviour to be determined by its mass and its surface friction. The gravitational force necessary to initiate and to sustain its movement is modified by the specific gravity of the circumambient medium, as was noted in Chapter 12.

The net force acting on the particle is

$$f = \frac{v\ (\Delta' - \Delta'')}{g} \qquad (13.1)$$

where v is its volume and Δ' its density, Δ'' is the density of the fluid and g the acceleration due to gravity. If $\Delta' > \Delta''$ the particle gravitates downward, if $\Delta' < \Delta''$ it tends to float. If the densities are almost equal it may lack sufficient energy to move relative to its fluid surroundings. The greater the excess of Δ' over Δ'', and the larger the value of v (the bigger the particle) the more readily and speedily it will fall. The potential energy of the immersed particle is converted to kinetic energy, and used to overcome the retardation (viscous, eddying and colliding) which opposes the attempts of the particle to move. In commercial ore treatment by gravity methods it is essential that the particles move briskly through the separating appliance. When they settle so slowly that the flow of the pulp past them is laminar, or viscous, the terminal velocity (in accordance with Stokes' law) varies as the square of the diameter d. With fast and turbulent flow (Newtonian settlement) it varies as $\sqrt{d}$.

This rules out the application of most gravity-based processes to solids settling according to Stokes' law, because surface friction dominates their reaction to the fluid medium. They therefore move too sluggishly to be treated at the speed required if a reasonable tonnage is to be efficiently handled. For gravity to act as the selecting force, there must be a marked difference in settling rate between the heavy and the light particles in the feed. The closer the specific gravity of the minerals being separated, the coarser must be the minimum size treated. For example, a 150-mesh particle of quartz (S.G. 2·7) has a mean diameter of about 0·104 mm., and a falling rate in still water of about 0·7 mm./sec. A quartz particle of 65 mesh (0·208 mm. diameter) falls about 26 mm./sec. The tremendous increase in the falling rate due to a doubling of the diameter might make it possible to use treatments based on gravity for the coarser-sized material while ruling it out for the finer mesh.

Finally there is the hindered settling effect. The particle under consideration is one of millions, each moving individually according to its physical nature, but subject to the constraining influence of the mass movements in the separating machine. The total number of particles for a given solid-liquid ratio varies inversely as the average particle size. Consider a single cube of quartz, 1 cm. long and weighing 2·7 gm., dropped into a vessel containing 1 ml. of water. If this cube is reduced to smaller cubes by successive halving of length of side, the changes in settling conditions take place in accordance with Table 24.

TABLE 24

	Solid-liquid Ratio	*No. of Particles*	*Length of Side mm.*	*Total Surface sq. cm.*	*Available Colliding Bodies*	*Effective Weight in Water (grammes) of each Particle*
(1)	1:1	1	10	6	Nil.	1·7
(2)	1:1	8	5	12	7	·2125
(3)	1:1	64	2·5	24	63	·0266
(4)	1:1	512	1·25	48	511	·0033

This suggests that with each halving of mesh size there is an eightfold increase in opportunity for collision and a similar decrease in the kinetic energy of the individual particle, together with a doubling of its surface area/size ratio, although there has been no change in the solid-liquid ratio. Only approximate prediction of the settling characteristics of particles moving under constraint is possible, but empirically the retardation of fall is of the order of 50%–65% of the free-settling rate.

Applied Forces

At first glance the use of gravity in mineral separation may appear somewhat elementary. In practice it can be extremely complicated. Process control must always be kept as simple as possible, but a number of mutually contending forces are at work. Though this chapter is concerned with the use of vertical currents in mineral processing, streaming action is also occurring. Study can usefully begin with the vertical component of separation, exploited in classification and jigging.

Consider a mixture of grains of galena (S.G. 7·5) and quartz (S.G. 2·7) which have been limited in size-range by screening through 14 mesh and on 20 mesh. Particles of similar shape and having a diameter of 1·1 mm. have terminal velocities of 260 mm./sec. and 96 mm./sec. when falling freely through water. If, instead, they are placed in a column of water rising at a speed of 150 mm./sec., the particle of galena appears to fall 110 mm./sec. while the quartz appears to rise 54 mm./sec. Because the size variable has been eliminated by the use of screens, the settling variable can thus be used to make these particles pursue opposed paths. (Fig. 133).

The more closely the feed to a rising current has been ranged by sizing upon screens, the greater will be its accuracy of response to the combined influence of the rising water and its own mass. The main conditions which determine whether an ore can be treated by gravity methods can be represented as a *concentration criterion* (Taggart). This is a ratio

$$\frac{(\Delta h - \Delta f)}{(\Delta \text{L} - \Delta f)}$$

where Δ_h is the S.G. of the heavy mineral, Δ_f that of the fluid medium, and ΔL that of the light mineral. When the ratio exceeds 2·5, effective concentra-

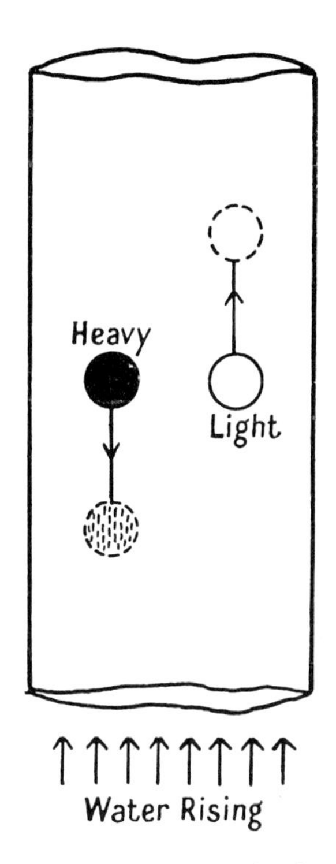

Fig. 133. *Separation in Hydraulic Current*

tion is possible down to the size of fine sands. With a ratio of 1·75, commercial separation is possible down to 100 mesh. At 1·5 the limit of fineness is around 10 mesh, and at 1·25 only gravel sizes can respond satisfactorily. There is a big difference in sensitivity between the forms of gravity separation now to be considered and the quiet ones of DMS in which accurate splitting is possible at very small differences of density, provided the feed is moderately coarse.

Despite some viscous retardation in DMS, the particles have space in which to move independently. They are, broadly, free to settle or to rise. In jigging and tabling this is no longer the case. The restraints of crowded conditions are at work. These include collision, friction, localised turbulence, and drag against machine surfaces.

In the study of classifying action (Chapters 8, 9 and 10) it was shown that a dominant factor was the mass of the individual particles involved. Those above a certain weight would require an extravagant use of water if they were to be lifted. This limits the practical use of steady elutriating currents to the treatment of moderately small ore particles. These then become an alternative to the methods of treatment considered in Chapter 14, where streaming currents are the main dynamic agents.

The limitation is reduced when the current of water is "jigged" up and down through a loosely compacted bed of coarser ore, with sufficient force to dilate this bed during the rising portion of an up-and-down cycle. The same jigging motion aids re-settlement during the return portion of this cycle, (drain-back of the water probably with positive suction added) If the ore is held on a screen in an open-topped box a column of water can be pulsed through it at a desired rate, with a suitable force and volume in movement. The useful power entering the system is that used to energise the water actually dilating and contracting the bed of ore, though this is only part of the total power consumed.

Consider next the action in a jig-box loaded with ore sized between appropriate limiting and retaining meshes, the latter being larger than that of the retaining screen on which the ore is held. These meshes have been chosen so as to allow all the particles to teeter during the pulsion stroke and re-settle on the return. Smaller and lighter material has been screened off. In a working plant this tends to limit the use of jigs to + 20 mesh feed, since screening of smaller sizes is slow and expensive. There are important and increasing exceptions to this generalisation, and if the sizing problem can be handled jigs can be considered as an alternative to *shaking tables* for treating ore sands.

The particles in our jig box become stratified after repeated displacements (Fig. 134). The hydraulic water accelerates during pulsion from zero to its full rate of flow through the opposing teeter produced while the bed opens and its constituent particles stir and jostle. Upflow then diminishes to zero and direction is reversed on the return half of the stroke. The teeter collapses by the end of the jigging cycle.

The individual particle finds its way as the result of these repeated teeterings to its most stable horizon in the mass and becomes stratified there with others which have reacted similarly. It does this in response to the effect on it of the following forces.

(*a*) The size-range of the total load in the jig-box.
(*b*) Its own size, shape and specific gravity.
(*c*) The average density of the total load, including voids.
(*d*) The average particle shape in the load.

(*e*) The "neighbourhood" density in its stratum.
(*f*) The pulsing and accelerating rates of the dilating fluid (usually water) and distance risen.
(*g*) Rate of sorting and stratification of the load.
(*h*) Back pressure bearing on its horizon and restraining its freedom of movement.
(*i*) Voids and their distribution, horizon by horizon.
(*j*) Specific surface of load.

To these may be added, simply for convenience of reference, certain streaming effects not present in the batch-loaded jigbox here considered, but which arise as soon as feed is introduced and products withdrawn. These include:

(*k*) Horizontal stratifying effect of cross-streams.
(*l*) Progressive changes of load of density between the feed and discharge sides of the jigbox.
(*m*) Rate of lateral displacement of the jigged strata.
(*n*) Percentage of heavy mineral removed during passage across jigbox.
(*o*) Density ratio between sinking (withdrawn) and rising (overflowed) mineral particles.

The fifteen factors thus summarised do not complete the possible list, but they indicate the complexity of gravity separation and explain the need for a largely empiric approach to the treatment of each ore.

Elutriation of a single particle stages a conflict between its mass and the upward urge exerted upon its area and cross-section by rising water. In jigging this force is pulsed and new conditions enter. The hydraulic current's rhythm takes the form of a sine curve (Fig. 136). The inertia of the particle is overcome at some point on the upward part of this curve, when it starts to rise and to acquire kinetic energy by transfer from the water. If light and small, movement begins while the hydraulic urge is building up; if large and heavy, as the water is decelerating toward zero from the top of the pulsion stroke. Thus, as long as the water rises faster than the particle the latter continues to receive energy. When the situation is reversed the particle uses any excess energy it has stored in continuing its rise, and then starts its fall, tracing a sine curve of smaller vertical amplitude which lags behind that of the pulsing water. The point can be reached for a sufficiently weighty particle where it is still descending while the water is rising on its next pulsion. The point at which its reversal occurs, and the direction and strength of water flow at that instant varies according to the particle's mass. One might start to fall during the dying away of pulsion while another might still be carried by its stored energy over to the suction stroke.

When consideration moves from the single particle to the loaded jig-box a complex situation is revealed. There is graduated teetering, with the top of the box well expanded and the bottom rather close-packed. As stratification proceeds the lightest particles in the upper strata are the first to loosen and rise, and therefore travel furthest. Energy transfer from hydraulic water to load is probably fairly even for all particles during rise, but as the teeter collapses on the return half of the cycle loss by collision and friction is heaviest in the lower strata. The hydraulic force consumed by the system must equate the back pressure of the teetering load and overcome frictional resistance. During pulsion the course of the water is turbulent, because it bursts through the channels which offer least resistance as the packing of the load changes. This causes momentary increased flow at such points. The general picture must therefore include local disturbance in the teetering load. At a late stage in the pulsion stroke the overall condition in the jig-box is one of a system being pressed upon from below by water and from above by the over-riding weight of the suspended solids. Since this storage-pressure is released during the return half of the cycle, pulsion and suction are not strictly equal, the suction stroke being reinforced by the downward movement of the collapsing load.

In continuous jigging, with the load working across the jig-box so as to introduce a streaming effect, extra water is usually added under the screen (Fig. 137, *water service*) in order to increase the loosening action during pulsion and/or reduce it during the suction half of the cycle. New feed is continuously added to the top of the load on one side, and two products are continuously withdrawn. These are the heavy fraction, either removed through the screen or via bottom gates as shown in Fig. 137, and the light fraction which works its way across the box. As this travels, it is repeatedly dilated and resettled by the pulsing stream of hydraulic water.

Two jigging cycles are illustrated in Fig. 134, one at the first stage of treatment and one when stratification is complete. In the pulsion stroke (*a*) water has been forced up through the screen. The mixture of closely sized

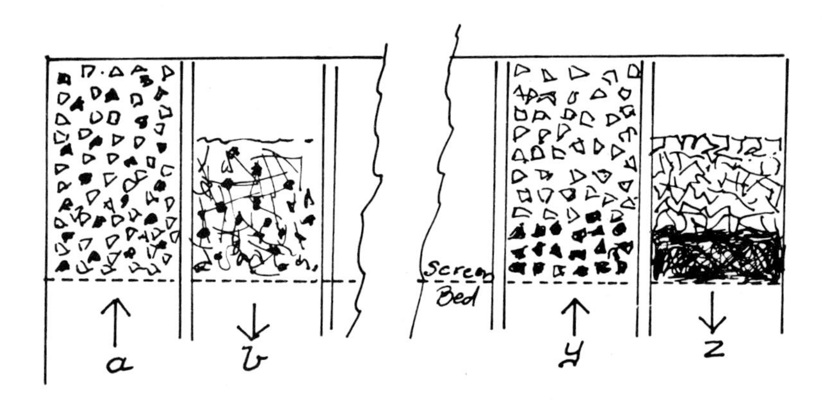

Fig. 134. *Stratification during Jigging*

particles (galena and quartz, for example) is dilated and teetering. Next the pulsion decreases, dies out and the suction stroke commences. The heaviest particles are the first to be affected by the reduced hydraulic lift, and they therefore fall sooner and faster. As the teeter collapses it forms the bed (*b*) with the galena beginning to concentrate at the bottom and the quartz at the top. This stratification is helped by the fact that the galena particles have a greater kinetic energy than the quartz and so are able to thrust aside the lighter grains while the bed is still closing. (y) and (z) show the position after a number of cycles, when the particles have arranged themselves in strata determined by their mass and shape.

Instead of equal-sized and equal-shaped particles of galena and quartz, consider next a mixture of various shapes and sizes, some being completely liberated and some locked middlings, such as would be found in a typical feed of crushed ore which had been screened to a size-range such as, say $-6 + 14$ mesh.

Ignore for the moment the new feed and consider only a cycle of pulsion and resettlement of the coarse grains. As water is forced up through the bottom of the screen, part of its kinetic energy is transformed into an equivalent static pressure. The back pressure at any depth in the dilated bed is that of the overlying particles. If now the current is stopped (or reversed), the bed packs tight, the action beginning at the bottom where the back pressure is highest, and where any change in flow rate or direction is first felt. Thus, the bottom particles interlock and present a barrier to those above, regardless of density. During the change from dilation to interlock, the first heavy particles to be thus barred are the largest, and the last the smallest. This explains an apparent anomaly in jigging—that small heavy particles move downward more readily than large ones when the feed is long-ranged. These small particles next pack the voids between the larger ones and thus increase the bed's resistance to penetration. This packing effect must be considered when determining the size range which is to be used when screening the feed. The greater the dilation of the bed the easier is the movement of particles in jigging. This dilation is adjusted by controlling both the quantity of hydraulic water used and the force applied to it.

Solid particles press down upon supporting media proportionally to their immersed weights. Since the immersed weight of the particle is proportional to its mass (volume $\times$ S.G.) this leads to the following effects:

(*a*) Particles of the same S.G. but of different sizes fall at rates varying as size (or mesh).

(*b*) Particles of the same size but of different S.G. fall at rates varying as specific gravity.

(*c*) Modifying (*a*) and (*b*) above, rate of fall is inversely as surface presented—equi-axial particles falling fast and plate-like particles slowly for a given volume.

Jigs are run at speeds varying from 57 to 330 cycles/minute. With streaming action added to the expanding and contracting teeter the movement becomes one of plastic flow. With impulses arriving from two to five times each second while the bed is being pushed toward the discharge side by incoming feed, individual particles tend to dance sideways as they stratify. The effects of k and l in the above list of forces now appear. The stirring effect of pulsion helps the particles to drift with the streaming water and to be pushed along by the new feed working its way through and across. At any vertical cross-section the load density has fallen steadily from feed toward discharge. Since the hydraulic lift is applied equably over the whole of the containing screen, there is somewhat greater expansion of load during pulsion toward the discharge end. It is usual to arrange a series of jig-boxes so that the tailings climb slightly as they move to the next box in series. This ensures gentle entry and minimum disturbance of stratifying action, and the pulsion effect aids the transfer.

Research regarding the movement of particles in the jig load[1, 2, 3, 4] and on the physical requirements for efficient operation deals *inter alia* with feed regulation, pulse control, de-sliming, packing of the bedding and load dilation.

Uni-Directional Separators

In Chapter 12 two methods which combined the use of hydraulic water with a heavy medium were discussed—the Chance and Stripa processes. Comparative newcomers to separation by more or less direct classifying methods are the Lavodune and Lavoflux systems.[5] The former resembles, broadly, a straight-sided hydraulic classifier working at a tilt of 45°. Feed ranging between 0·5 and 20 mm. in size is treated, in a pulp diluted below 50 g./l. in tubes 2 to 3 m. long, each of which handles a ton hourly per square decimetre of its cross-section. The closer the size-range the more efficient is the action. Over 4,400 gallons of water per ton of ore are kept in motion at a water-gauge pressure of from 6 to 13 feet. This consumes from 0·2 to 0·8 kWh./t. of ore. As the feed moves upslope the lightest particles are swept up and out of the tube while the heavier material forms a rotating dune from which entrained light particles are freed as it performs an oval roll. During this rotation part of the load breaks away periodically and falls to a separate discharge. The Lavoflux, described in the same Paper, treats 0·1 to 0·5 mm. feed and forms a fluidized dense medium instead of a dune.

The Hand Jig

A description of one of the oldest operations in mineral dressing—hand jigging—is now given, to aid in visualising the essential features of the work. In primitive situations, or where it is desired to test a limited quantity of ore with "home-made" equipment, the hand jig has its place. It consists of a watertight box or hutch in which an ore-box can be jigged up and down. The ore-box has screen cloth for its bottom, strong enough to retain a bed of ore, open enough to allow water to run freely through, and small enough in mesh to hold as oversize substantially all the ore fed into the box.

The feed should be sized, and filled into the ore box, which is then jigged up and down through the hutch-water until the particles have stratified with the lightest above and the heaviest below (Fig. 135). The barren top layer is then removed and new feed is added. From time to time the accumulation of heavy mineral is removed after a series of such stratifications. The descent of the ore box causes an uprush of water (the pulsion stroke in jigging) which dilates the bed in accordance with the speed at which water rushes

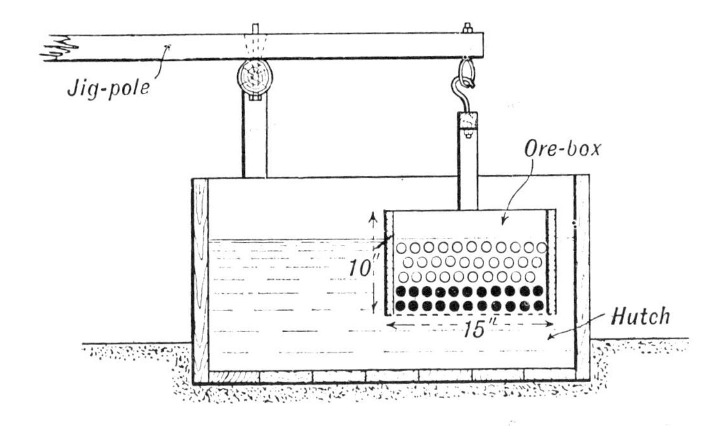

Fig. 135. *The Hand Jig* (after S. J. Truscott)

upward through the loaded box. The stroke direction is now reversed, usually more gently in order to avoid making too compact a bed.

Hand-operated jigs include the Willoughby, the Cornish Kieve, and a Chinese method of jigging on hand-held screens. Since the jig motions can be duplicated with mechanical precision, hand jigging is not much used today. The stratifying effect of the up-and-down movement of the ore box in a static pool of water is like that produced when a vertical column of water is pulsed upwards and downwards through a stationary ore box.

Application of Jigging Forces

It is helpful when studying a method of treatment to consider how the working force is applied, and the way in which the energy thus introduced to the ore is consumed. In dense-media separation the comparatively quiet media was used to discriminate between the specific gravities of the particles. In separations based on the use of moving water the hydraulic current is so manipulated as to exploit (*a*) the particle mass, which depends on its size and specific gravity, and (*b*) its shape, which affects its frictional associations and cross-sections when moving in a crowded teeter bed. In a vertical current the hydraulic force exerts a general pressure equal to the back-pressure of the load pressing upon it *plus* the energy needed to replace the frictional loss during flow. The pressure at any given point in the hydraulic flow varies slightly from the general pressure, since the water seeks the easiest channel. The picture is, therefore, one of momentary turbulence at ever-changing points in the teetering column, with immediate local rise in pressure as a result. This is followed by a rapid change of course as the water finds new and easier channels, which in turn immediately develop increased resistance. The power transferred from the hydraulic current is converted partly to friction and is doubtless finally dissipated as heat. Part of it goes, however, to overcome the inertia of individual particles and the general mass of the solid load. In "hindered settling" the teetering action produced by a hydraulic current was considered in terms of the gravitational force of the particles and the surface friction in the system. In jigging, a third factor is in the acceleration from rest several times a second of the whole mass of particles in the bed.

Consider next the contents of a jig box. Here the mass of quick settling particles rests on a screen through which water is pulsed. This bed of material could be thought of as though it were a leaky plug. If a sudden thrust of water is given it can be pushed up almost as a solid mass and will then re-settle on the down stroke of the same pulsating water column. This for the purpose of jigging would be useless. What is wanted is to bring the pressure of water to bear in such a way as to disperse the bed of material into individual teetering particles and then to reverse this action so that they start to settle back into a solid mass.

With each pulsion strike sufficient energy must be imparted to the separating fluid to lift all but the heaviest particles, and thus expand the contents of the jig box to teetering point. It follows that the heavier the particle, whether because of its size or its specific gravity, the more powerful must be the thrusting action during pulsion.

The effective components of the jigging column of water are:

(*a*) The volume of water in motion above the screen.
(*b*) The area of horizontal cross-section through which this volume rises or falls in each jigging cycle.
(*c*) The rate of build-up and reversal of the rise.
(*d*) The maximum speed attained in each phase of the cycle.
(*e*) The point of the half cycle at which this maximum rate is reached.

The energised volume of water is produced by the interaction of several factors. These include the length of stroke of the piston or other mechanised driving force mechanically applied at the input end of the column. Next there is the leakage passing the plunger at this end. Third there is the effective area of the energising unit (piston, diaphragm, etc.). Finally there is the effective area of the jig screen. By "effective area" is meant the open space after allowing for the solid components of the screen and its stationary supports. These factors determine the volume of water which rises and falls through the jig screen in each complete cycle. The speed of rise or fall varies from *nil* at the commencement to *nil* at the end, through a maximum which is a function of the volume and the cross-sectional area. This area is modified by the porosity of the jig bed at any horizon and at any moment. This is changing continuously throughout the fraction of a second during which the bed expands or contracts. Thus a jig running at 180 r.p.m. would have three cycles per second, each half-cycle taking one-sixth of a second from zero to zero. Such a jig bed probably consists of a mass of particles varying from half-teeter to full teeter, and never having quite enough time to settle into a locked mass. In this discussion the term "plastic" is applied to this bed to indicate the fact that this rapid alternation of strong thrust and weaker return action is keeping the whole of the material more or less mobile. Many jigs are driven by eccentrics or linking motions which accelerate the stroke in accordance with the harmonics of an angular motion. Thus the jigging cycle is seen to be far more complex than a uni-directional teetering column.

The fluid normally used in jigging is water. Air has had limited application in the pneumatic jig, used in desert countries. Attempts have been made to jig in dense-media fluids, but no industrial process of importance has been reported apart from the jigging element in the Stripa DMS machine. The water used can be set in motion in several ways. Gentle oscillation between the limbs of a U-shaped box is used in the Baum coal jig, the pulsing force being applied by low-pressure air introduced through valves at intervals, so as to sustain the oscillating action as the water rises and then gravitates back through the material resting on the screen. For heavy minerals a loosely fitting piston or a flexible diaphragm is used to impart more rapid pulsion strokes, followed by positively powered return (suction) strokes. Abrupt short strokes are possible because water used in this way exerts a plastic thrust rather than a streaming action, and the particles with the least mass and inertia are most readily set in movement. Before a particle thus urged into motion has time to settle in response to the ensuing suction stroke, the next thrust is delivered. The jig bed is thus kept in partial teeter, only the upper layers being fairly free to move. By this relatively violent method of applying force the work is done with a moderate quantity of water in motion. The Denver, Bendelari and Pan-American jigs use compression in the water chamber instead. An alternative is to move the jig box up and down in a tank of water (Hancock and Halkyn Jigs).

Kirchberg and Hentszchel[1] have derived this formula for resistance of a bed of spherical grains to loosening (Δp)

$$\Delta p = \psi \frac{V^2 m}{2g} \gamma_M \frac{h}{d_m} \frac{m}{V^4{}_R} \text{ g./cm}^2., \qquad (13.2)$$

where ψ = coefficient of resistance.

V_m = velocity of flow of medium in non-channelled parts of the bed.

g = acceleration.

γ_M = specific weight of medium.

h = height of bed.

d_m = average diameter of the spherical grains.

V_R = voids in bed.

γ_F = specific weight of solid material.

Such a bed resists loosening while the general flow rate of the jigging fluid (V_m) lies between Reynolds Nos. 20 and 3000. The maximum rate of flow such a bed can resist without opening is:

$$V_m = \sqrt{\frac{(\gamma_F - \gamma_M)\,(1 - V_R)\,2gd_m V^4{}_R}{\psi\gamma_M}} \text{ cm./sec.} \qquad (13.3)$$

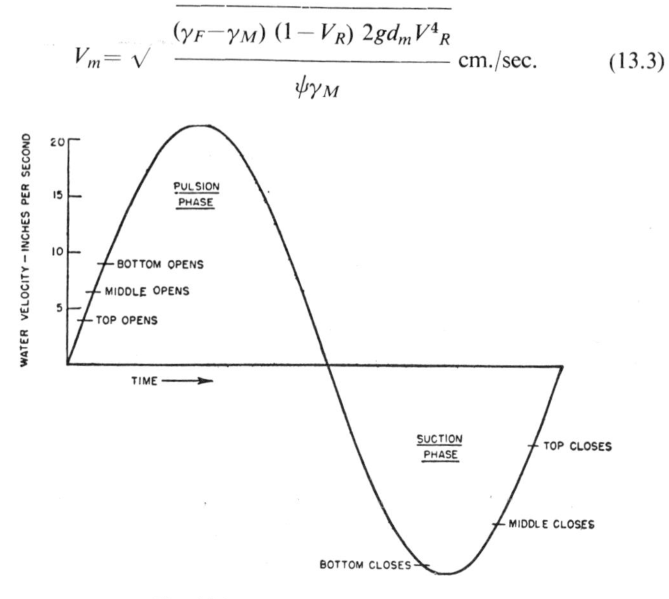

Fig. 136. Jigging Cycle for Coal

Consider first the case of coal jigging, where there is a comparatively small density difference between mineral and jigging water (say 1·5:1·0). The gentle pulsion stroke needed to dilate such a bed permits greater refinement in the work than is practicable in jigging minerals with a density ratio of 3:1 or higher. The jigging cycle for coal may be represented as a sine wave (Fig. 136) in which water velocity is plotted against time.[6] Starting from rest, the water commences to rise through the closely packed bed of coal with increasing velocity. Dilation of this bed begins with the loosening of the uppermost stratum. (In jigging coal it is possible to apply water so as to toss the whole bed solidly upward and then to allow it to open out upward and downward, but this is not feasible with heavy minerals.) As the water column attains a rising velocity of 100 mm./sec., the upper part of the coal bed opens and begins to teeter. Velocity continues to increase, and layer after layer opens until the whole bed is dilated and teetering. Then, as the pulsion stroke passes its peak, the velocity of the water lessens and the suction stroke begins. At this point each particle in the loosened bed has some degree of mobility, and has acquired both kinetic and potential energy in accordance with its movement and vertical displacement in the teeter bed. These it dissipates as the suction stroke develops. The particles fall downward in the falling column of water and repack into a tight bed. First there is momentary free settling, then hindered settling with the teetering mass rapidly consolidating to the packing density of the solid phase, less voids.

The bottom of the bed is the first to pack tight, then the middle shuts down upon it, and last the top closes. Jigging is thus seen to be hindered settling carried out under oscillating conditions during which the material passes from fully fluid conditions (S.G. 1) to those of a bedded deposit at each cycle. In coal jigging the pulsion-suction cycle is usually about 57/minute. In treating heavy minerals, where more violent action is needed, speeds up to 280 cycles/minute are employed.

If a continuous upward current were used, stratification of the jig bed would result, with the smallest and lightest particles uppermost, the largest and heaviest ones at the bottom and a mixture of largish light particles and smallish heavy ones in the middle of the teeter bed.

If the material had been close-sized, this would result in clean separate strata (heavy below and light above). Such close sizing would not be economic in commercial practice.

When, instead of as a continuous rising current, the water is applied in a succession of up-and-down pulses, a new factor appears. As the lifting current becomes gentler, all the particles cease to rise, hover, and then begin to fall. The heaviest are the first to change direction, and the lightest the last. Soon the falling movement is accelerated as the upward pulsion stroke of the hydraulic water dies out and downward suction commences. Hindered settlement becomes rapidly more severe and the bed closes tightly. As the spaces between the particles shrink, the bigger particles are arrested in their fall. The smaller ones can still burrow down while this is happening, and are the last to be stopped as the bed packs tight.

Since the heavy particles have the greater falling rate, stratification is effected with the largest and lightest particles above and the smallest and heaviest below, the remainder being disposed intermediately.

This accomplished, the rest of the work of the jigging appliance consists of the orderly reception of new feed and its dispatch as separate stratified products. A list of requirements for efficient jigging of coal has been suggested by Bird.[7]

Two methods of withdrawing the heavy product are practised. In the original "German" (on-the-screen) jigging the supporting screen is smaller than the mesh of feed. Concentrate collects above the screen and is withdrawn *via* side ports adjusted so as to maintain a bed of desired thickness (Fig. 137, *concentrate discharge*). A later development, more widely used, is "English" (through-the-screen) jigging, in which the screen mesh is larger than that of the feed, so that the ore reaching the screen falls through to a receiving chamber, the hutch. An advantage of the English method is that withdrawal can be more closely controlled by use of an oversized layer of selected material (broken ore, steel shot, etc.) called the *bedding* or ragging. This bed is maintained in a suitably thick layer, and its component particles are chosen for shape and specific gravity.

The controllable factors in through-the-screen jigging include:

(*a*) Amplitude of jigging cycle.
(*b*) Strokes per minute.
(*c*) Mechanical unbalancing of stroke-producing device.
(*d*) Hydraulic unbalancing over selected part of stroke.

(*e*) Size, shape and specific gravity of component particles in bedding layer.
(*f*) Bedded thickness.

The bed serves several purposes. It evens out the upthrust during pulsion, thus preventing a break-through of water which would ruin stratification. It provides a controlling barrier through which too light an ore particle cannot pass and join the concentrates, or hutchwork. The overall density of the bed, as modified by back-pressure from the load above it, determines this controlling action. This density derives from the mass of its component particles *plus* that of the water in its interstices. Bedding selected from oversized ore is not necessarily the best choice. In use, shape tends to wear to a roughly spheroidal form and this gives the most smooth distribution of rising water, beside reducing entanglement to a minimum. The further a bed material departs from a well-rounded shape the less precisely does it control the quality of concentrate worked down through it. Metal discs, rivets, steel punchings and broken ore are commonly used in jigging. It is possible to do fairly good work with them, but testing which leads to the choice of bedding of the best shape and size yield better results. In all processing work it is possible to correct a wrong working condition to some extent by over-compensation. "To some extent" implies acceptance of higher tailings or lower grades of concentrate than could be produced, or inflated processing costs and, perhaps, some working instability. This point may escape notice where an autogenous bed is used, composed of oversized concentrates or middlings of the mineral which is being concentrated. With wear the original angularity of the material is reduced, and the resulting improvement may be mistakenly attributed to some other cause. The particles in this bed act rather like inefficient valves, opening on the violent pulsion stroke and shutting (packing tightly) under the combined force of a gentler suction stroke and the overriding gravity of the mass above them.

The Jigging Impulse

The two main types used in industry are "moving-screen" and "fixed-screen" jigs. In the former the screen box is moved upward and downward in a tank of water. In the latter, the screen box is stationary and the jigging fluid is pulsed upward and downward through it.

Water is normally the pulsing fluid. When it is desired to aid selectivity by applying the dilating impulse more gently, the density of the water is sometimes increased slightly by permitting it to carry solids in suspension. Little commercial development of this possibility is known, as the method creates problems which offset its advantages. In the treatment of dredged tin gravels de-sliming before jigging has proved its worth.[8] Extra power would be needed to handle and circulate dense media. The need for removing drag-out material would add to working costs.

Pneumatic jigging has a definite field of application. In arid country, and in cases where it is desirable to avoid wetting the feed, air is a good pulsing fluid within somewhat severe limitations of particle size. It has the further virtue of being available everywhere and of needing no effluent equipment beyond that which may be used to remove dust.

The control of the application of the jigging impulse is discussed in the following paragraphs. To some extent it is specific to the type of jig used.

Fixed-screen Jigs

The oldest type of jig in use is the Harz, which takes its name from the Harz Mountains in Germany, where it was developed for treating the lead-zinc ores which have been mined in that region for several hundred years. In the Harz jig (Fig. 137) an ore box, 18″–24″ wide, 24″–48″ long, and 6″–8″ deep, occupies about half the plan area of a compartment in a water-filled tank, called the "hutch". In the other half of this compartment a loosely fitting plunger moves vertically, so as to pulse water upward and downward with a force proportional to:

(*a*) The displacement caused by the variable eccentric actuating the plunger.
(*b*) The speed at which the plunger is reciprocated.
(*c*) The leakage past the plunger.

The body of water thus set in motion reverses its direction of flow after passing the centre board, and alternately dilates and closes the bed of material resting on the screen which forms the bottom of the jig box.

One reason for using a loosely fitting plunger is to reduce "water hammer" in the reciprocating system. A second is to enable the operator to differentiate between the strength of the pulsion and the suction stroke by applying "hydraulic water" at a controlled rate. In Fig. 137 it will be noted that a water service is provided above the plunger. All water introduced at this point either finds its way out with the tailings or to the concentrate discharge after passing through the screen. If no hydraulic water is used, only the water entering the hutch with incoming feed is available. Since this arrives above the screen, the pulsion and suction strokes are equal and opposite, save for water lost through leakage from the hutch and periodic withdrawal of concentrates at the spigot of the hutch-work discharge. When hydraulic water is added, the pulsion stroke is fortified and the suction stroke weakened, proportionately to the amount so added. Thus, hydraulic water is employed to aid in controlling the tightness of the closed bed and the differentiated speed and violence with which it is opened and closed.

Several successive compartments—usually four—are placed in series in the hutch. The sized feed is introduced gently at the head of the first compartment, care being taken not to upset the stratifying work by plunging it in violently. The feed should be distributed evenly over the full width of the box. In the first compartment the controls are manipulated so as to produce as the bottom layer a high-grade concentrate. The lighter fraction of the feed stratifies upward and passes over the far end of the compartment into the second section, and so on down to the final discharge. Concentrating conditions are made progressively less discriminating from head to discharge so that a lower-grade concentrate or "middling" is removed from the later

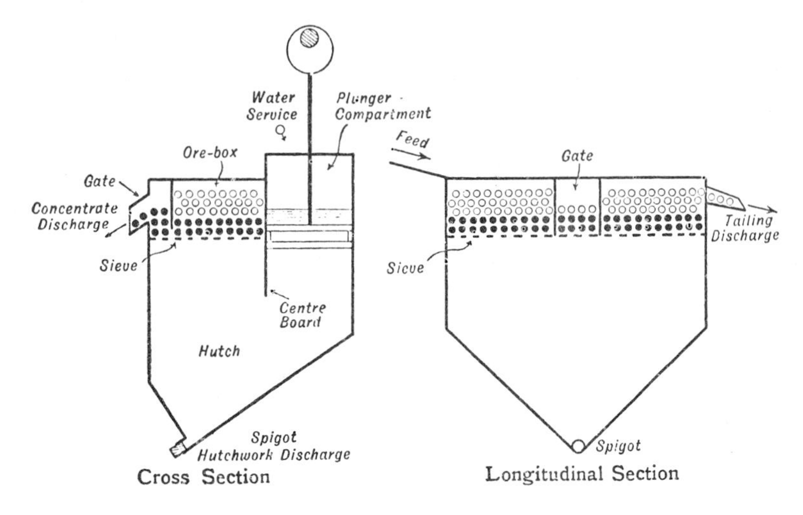

Fig. 137. *Harz Jig* (after Truscott)

boxes, the top discharge from the last box being the jig tailing.

The concentrate can be withdrawn from the Harz jig in one of two ways. If the feed particles are larger than the aperture of the supporting screen, they cannot find their way down into the hutch. Jigging "over the screen" originated in Germany, and is sometimes called German jigging. It is characteristically used with fairly coarse feed and moderate suction, and the grade of concentrate produced is partly regulated by the thickness of the bottom stratum, which is determined by the rate of withdrawal through the undergate of the concentrate discharge.

A later modification, the English method, was introduced when it was found possible to treat finer material. The feed is finer than the apertures in the supporting screen, so the concentrate is mineral which has succeeded in working its way down through the screen during the jigging cycle, while the tailings consist of particles unable to descend under the prevailing operating conditions. The hutchwork discharge is delivered to a receiving vessel in such a way that concentrates can be removed when the hutch valve is closed periodically, without upsetting the water-level in the jig-box.

A number of plunger mechanisms and product take-offs have been used. The Baum jig is described below.

The amount of water used in jigging varies considerably from one mill to another. If a round figure of 5:1 per compartment be taken as a reasonable average, twenty tons of circulating water are used per ton of ore treated in a four-compartment plunger jig. This water is mostly reclaimable, but must be kept in hydraulic action by applied power and transported through a conserving system. Instead of plungers, many modern jigs use flexible diaphragms to produce the stroke. These tend to be more economical of water and allow higher running speeds than are practicable with loosely fitting plungers.

In the Bendelari jig (Fig. 138) the diaphragm (A) is reciprocated from below and hydraulic water is admitted at H. A modified form of this jig is widely used for concentrating the values from alluvial gravels in dredging.

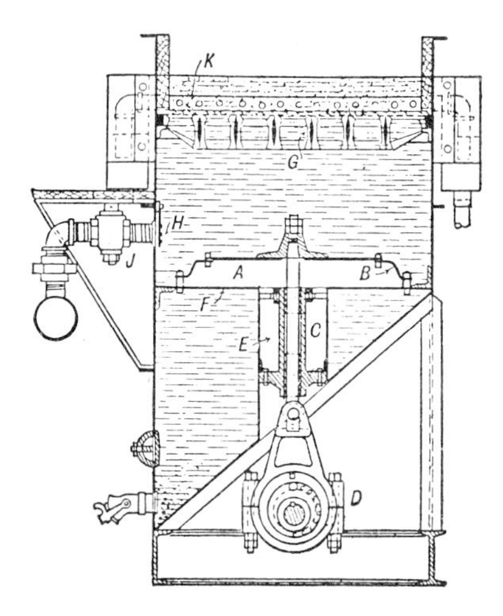

Fig. 138. *Bendelari Jig*

The Pan-American jig, frequently used to remove coarse metallic gold at an early stage in milling, is arranged so that the conical bottom of the cell is reciprocated by an eccentric. The moving bottom is attached to the square jigging compartment above by rubber strip.

The Denver jig (Fig. 139) was designed in the first place as a unit cell working in the closed circuit of a ball mill. It can take a long-ranged feed, and involves but little loss of height between mill discharge and delivery of its tailing discharge to the classifier. The rotary valve (*c*) can be adjusted so as to open during the required portion of the jigging cycle, thus admitting hydraulic water in phase either with the suction or the pulsion stroke, which

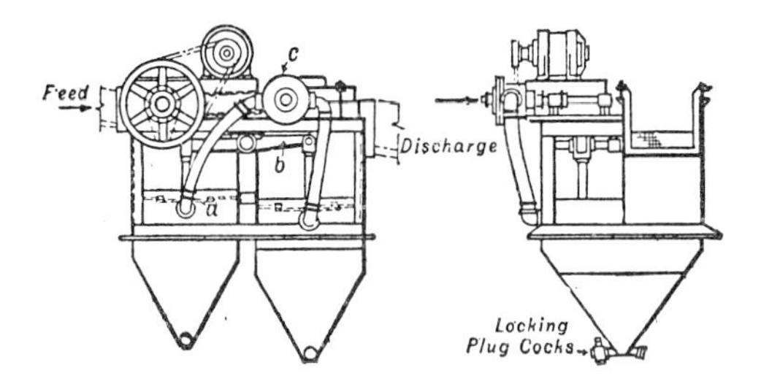

Fig. 139. *Denver Jig*

is applied by means of a flexible diaphragm *a*. This modification in jig design has led to increasing applications of the machine, which can be adjusted so as to produce fairly sharp separation of material down to coarse sand sizes. Duplex jigs embodying this phased control of the hydraulic water are now used outside the closed circuit, one important application being in the "roughing" of diamond gravels. By suitable adjustment of the rotary water-valve any desired variation can be arranged, from complete neutralisation of the suction stroke with hydraulic water to a full balance between suction and pulsion. Jigging is through a wedge-wire screen.

Jigging

The jig is widely used in coal cleaning. Reversing the conditions normal with most minerals, coal is the abundant and light fraction. There is also a marked shape factor, since the impurities are tabular. The range of size fed to the jig determines the efficiency of separation. The closer the sizing, the higher this becomes. If a jig is set to operate at a density of, say 1·5, this means that it should produce washed coal with the same ash content as that from a heavy-liquid float at a specific gravity of 1·5. Part of the sample, and of the product, would have floated at a lower density. Of the several types of coal jig, that in greatest use is the Baum. Various developments from the original jig have been made. Where there is a marked difference in the breaking characteristic of high-ash and low-ash coal it is possible, by splitting the feed at two sizes to separate jigs, to obtain better results by operating each size at its most appropriate density-equivalent. Normal modern practice avoids such a complication and treats all the −6″ feed in one circuit.

The Baum jig (Figs. 140, 141) has a stroke rate of 55–57 strokes/minute, the water being pulsed by means of compressed air. The bed recloses by unassisted gravitation, no suction stroke being used. This closing can be modified by the use of hydraulic water, or by slowing down the release of the pulsing air during the return stroke of the jigging cycle. The jig box has a U-shaped cross-section and each of its two compartments forms a complete jig. The screen of the first compartment slopes backward toward the feed end. The feed is introduced gently, perhaps below a very shallow baffle to ensure wetting, but not with sufficient disturbing influence to upset the stratification taking place in the box. The weirs between the boxes in the older types have been abolished for the same reason, so the feed proceeds with, if anything, a slight climb through the jig. Air pressure is $1\frac{1}{2}$ to $2\frac{1}{2}$ lb. at least. Most of the dirt is dropped in the first compartment. Part of it falls through the screen to a spiral conveyor which moves it out. With one type of control the oversize dirt builds up on the screen until it has lifted a controlling float, which responds to change in bed density at the interzone between dirt and coal. This actuates a switch that starts the elevator motor and dirt is withdrawn through the refuse gate until the falling density causes the float to switch off the motor. In a variation, float motion controls secondary pulsion under the discharge gate. Thus the thickness of bed is automatically controlled. The float setting correlates the jig discharge products with the required average density

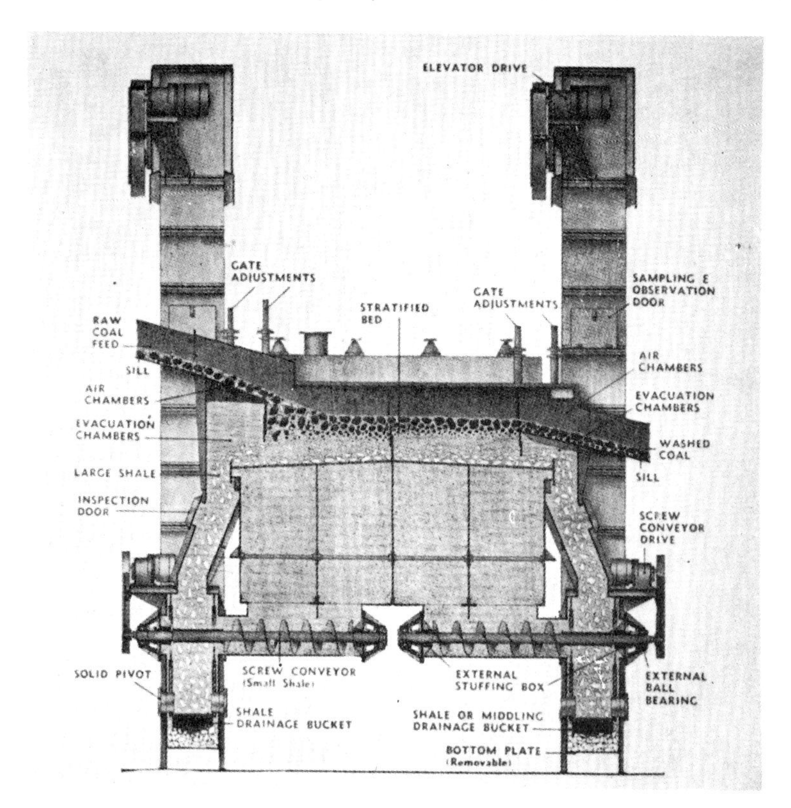

Fig. 140. *Longitudinal Section of ACCO Baum Jig* (Automatic Coal Cleaning Co. Ltd.)

of the washed coal.

With feed coming intermittently, as can happen when the washery is fed by trucks through tipplers, the bed may be upset until restratified by new feed. One type of Baum jig has an automatic idling control which minimises this upsetting influence.

Specialised research led to the post-war redesign of the automatic control. Rotary air valves replace the older slides and plungers, and detailed mechanical improvements raised the feed rate of the fully automatic machine to hundreds of tons hourly. Idling and overload performance have been improved. The size range of feed is 6″ – 30 #. Automatic control of the thickness of the retained bed of shale is essential to success under these conditions. To obtain this the shale discharge control (Fig. 142) works by means of a float which responds to changes in the height of water in the tube in which it is suspended. The thicker the layer of shale on its supporting

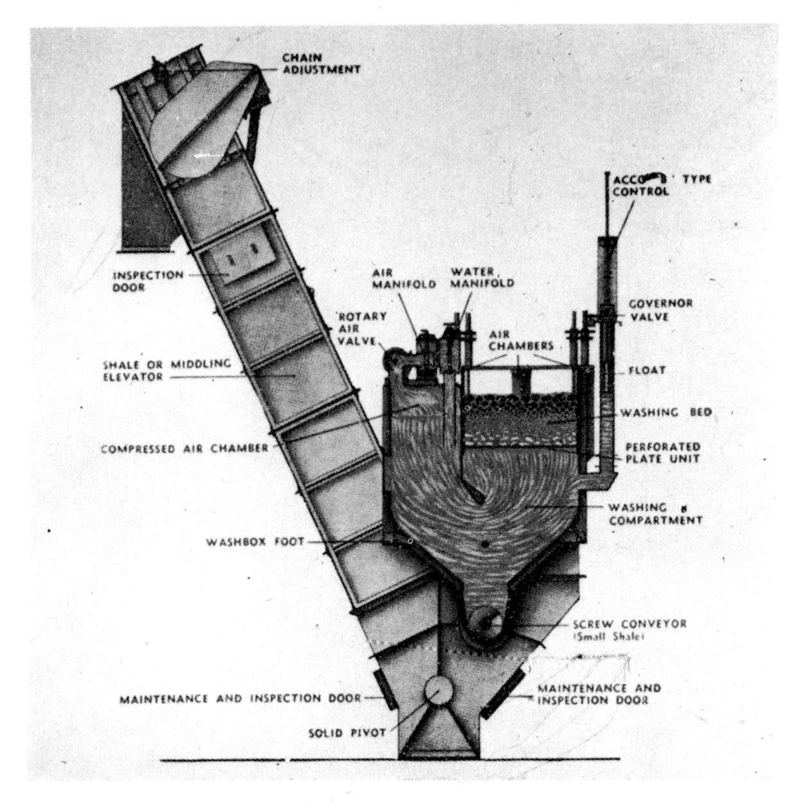

Fig. 141. *Lateral Section of ACCO Baum Jig* (Automatic Coal Cleaning Co. Ltd.)

perforated plate, the greater the overall resistance to the water being pulsed upward. Taking the line of least resistance, water surges more strongly through the tube and raises the float at each pulsation. Above the shale gates (A and B, Fig. 142) is an air chamber C which traps air when the release-valve D is closed by the float. This air-cushion prevents water pulsation near the end of the plate and shale discharge stops. When D opens the air-pressure is released, pulsation occurs and discharge takes place.

To check performance the shiftsman is given a supply of heavy liquid. He judges by the amount of floating coal shown in a sample whether the setting is right. The second jig box delivers a coal middling which is either suitable for use as a low-grade fuel or can be given a liberation crushing and returned to circuit.

The Remer jig (Fig. 143) has a single long box. Impulses are produced by dual eccentrics which give differential acceleration to the three hutches. This

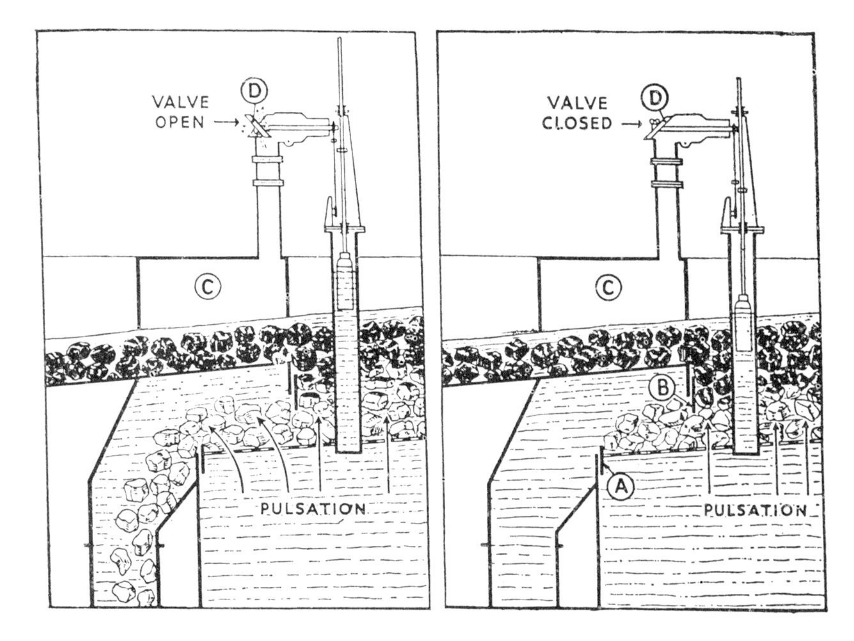

Fig. 142A. *Automatic Shale Discharge Control Operating with Thin Shale Bed* (Automatic Coal Cleaning Co. Ltd.)

Fig. 142B. *The Same Operating with Thick Shale Bed*

jig is used in roughing work to remove an upper layer of deleterious material and deliver a middling of gravel and a hutchwork of finer sands.

The Hancock jig (Fig. 144) consists of a long tray, closed at the bottom by screens which increase in aperture from feed to discharge. The tray hangs in a hutch divided into five compartments, each with intermittent bottom discharge for the concentrates it receives. The tray is lifted by cam gear so as to throw its load of ore forward and upward. The jigging cycle thus produces stratification and positive assistance of the bed from feed to discharge, and the return stroke can be softened by the admission of hydraulic water into the hutch below the tray.

The Halkyn jig was developed to treat a coarsely crystalline lead-zinc ore occurring in a limestone gangue. It has a tank divided into several compartments and a screen box closed with a bottom of wedge-wire screening, which, like the Hancock, receives vertical and horizontal motion that stratifies the feed and helps it to move toward the discharge end. Hydraulic water can be applied below the screen. *Fines* fall through to the hutch, but the concentrate is taken *via* an adjustable gate in the discharge end of each section of the screen box.

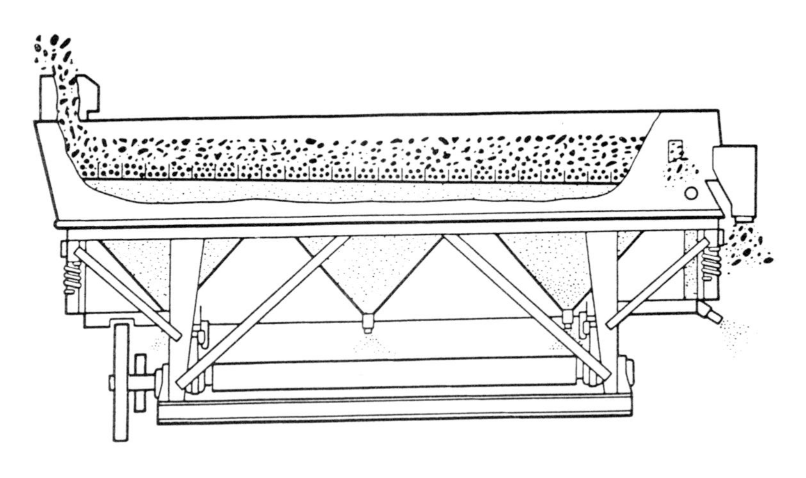

Fig. 143. *The Remer Jig*

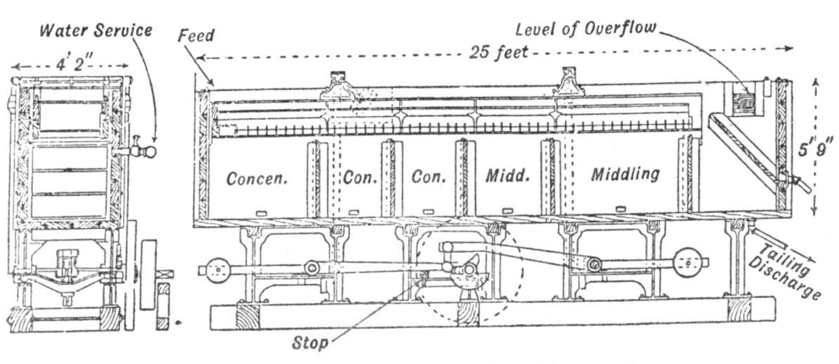

Fig. 144. *Hancock Jig* (after Truscott)

Practical Application

Since jigging exploits the difference in density between particles of similar size (and to a minor extent differences in shape), the process is mostly suitable for ores which are adequately liberated at a fairly coarse m.o.g. It is theoretically possible to jig fine sands provided there is a substantial density difference, but the fact that screening is desirable to control size range limits treatment range to sizes which can be commercially screened—say a minimum of 65 mesh. Jigs such as the Denver can be used down to 120 mesh, but, broadly speaking, separation of particles in size ranges finer than about 20 mesh can be done better by other methods.

Jigs are widely used in beneficiating coal, iron, barytes, galena, sphalerite, and fluorite, but many of the coarsely crystalline deposits which relied on jigging for their concentration have been virtually exhausted. Ores mined today usually require fine grinding and are therefore unsuitable for jigging. Another factor in the decline of this method of treatment is the increasing use of dense-media separation for the treatment of long-ranged feeds. This process is competitive in cost and is more sensitive to small differences of specific gravity between the minerals separated. DMS cannot readily be used, however, in the treatment of minerals so heavy that the maintenance of a suitable bath would be difficult, a consideration which does not affect jigging.

In certain specialised fields jigging is of great value. In the closed circuit between the mill discharge and the mechanical classifier, a jig of the Denver type removes a heavy fraction from the ore pulp which would otherwise tend to recirculate and be over-ground. In the case of malleable particles of gold, to take an extreme example, repeated passage round the closed grinding circuit would not lead to the overflowing of the liberated gold from some ores. Removal in the closed grinding circuit by some such device as a jig prevents an undesirable building up of the precious metal.

Another application is to the rough concentration of diamondiferous gravels by jigging. One plant in East Africa uses two Duplex Denver jigs in series, with first $\frac{3}{4}''$ screens and next $\frac{1}{2}''$ ones limiting the entry-size of the feed. The jigs have 2-mm. bedding screens which accumulate a bedding stratum of ilmenite. The diamonds smaller than 2 mm. work through to the hutch and are then recovered by feeding the hutch-work down a greased plane, to which the diamond adheres while the non-values slide. The diamonds larger than 2 mm. are held in the ilmenite bed, and periodically recovered. These jigs are run at 280 r.p.m., water being drawn from a 20′ head, and used at full suction. At this unusually high speed the contents of the ore box hardly get a chance to pack down, and the settling conditions are intermediate between normal jigging and those in a hindered settling classifier. Research which led to modifications in the setting of Denver jigs used at Premier Mine has been described by Weavind and McLachan[9]. These jigs treat diamondiferous kimberlite at 20 tons/hour per jig, the feed being *minus* 2 mm. The rotating water valve restricted the original flow rate to 7,600 gallons/hour. Separation at this high rate of feed improved when this valve was removed. The kimberlite has a density of about 3·5 and the valueless associated minerals 2·7. Gravel with the intermediate S.G. of 3·2 was found to be the most suitable bedding medium, at a bed thickness of from 2″ to 3″ and a size range *minus* 6 *plus* 10 mm. Stroke frequency of 270 cycles per minute at an amplitude of $\frac{1}{2}''$ gave best recovery of radio-active test diamonds used to check recovery.

In the earlier days of tin dredging all the preliminary concentration was done in sluices carried longitudinally on the dredge pontoons, and jigs were sometimes used to clean the rough concentrate thus produced. Today most dredges send the tin-bearing gravels straight to jigs after de-sliming through hydrocyclones.

These are but a few examples of the operations in which various types of jigs are used. They are also employed in prospecting, field testing, and the early development of mines.

Operation

Usually the size range of the feed is limited by screening ahead of the jig, the spread of this range depending chiefly on the differences between the specific gravities of the particles which are to be separated. When this difference is small, the feed should be close sized and the suction stroke should be gentle, as separation of equal-sized particles can best be achieved by stratification according to density.

Sometimes—though this is not good practice—the feed to the jig is hydraulically classified, so that it lies in a range of particles which settle more or less equally. Jigging action to be effective must now assist the small, heavy particles to burrow downward through the interstices in the bed as it closes after the dilating stroke. In this case the jig should be run with strong suction, thus aiding penetration of the small sizes.

When jigging "through the screen", the apertures must not become blinded. If the discharge of concentrate from the hutch is continuous, water in excess of that required to modify the stroke cycle is needed to compensate that lost with the discharge. It is added together with that used to modify the suction-pulsion cycle.

Feed is usually sized by screening in an appropriate series of fractions from −2″ downward. Heavy-mineral jigs are run at speeds varying from 100 r.p.m. on coarse feed, with a stroke amplitude as high as 2″, down to 300 r.p.m. with a $\frac{1}{8}$″ stroke on fine feed. The relationship between speed and "throw" must take account of the working stresses at the point where the jigging impulse is given to the water column. As was noted earlier in this chapter, this relationship does not by itself ensure correct hydraulic conditions in the jig box. Coal jigs, such as the Baum, work at 57 cycles/minute. Here gravity alone supplies the return half of the cycle and the speed is chosen to suit the harmonics of the oscillating column of water.

The capacity of the jig is proportional to the transporting effect of the horizontal current from feed to discharge end, as modified by the restricted mobility of the bed. Two factors determine the maximum efficient rate of feed. The first is the time needed for the finest desired particle of concentrate to settle down and be withdrawn from below during its progress from feed to discharge end of the box. The horizontal speed is determined by:

(*a*) The volume of feed (ore + water/minute).
(*b*) The cross-section normal to flow.
(*c*) The nett volume of jig-water (excess of pulsion over suction)/minute.

The second factor is the smoothness of working conditions in the ore box.

If the feed is run in so rapidly as to churn up the bed, it takes longer to secure stratification, and separation suffers. If the feed is such that heavy use of hydraulic water, or violence of stroke is needed to procure adequate dilation, then there is risk of local breakthrough of excessive quantities of water. This, by churning up the bed, upsets stratification and leads to loss. The balance between total mass of material in the jig box, masses of individual particles and blend of particle sizes must be such as will lead to even diffusion of the pulsed water through the interstices of the jig bed. There must neither be danger of local weakness leading to a breakthrough nor insufficient dilation to achieve stratification. To aid transporting rate, a slight drop from one compartment to the next is better than the use of streaming water. Some operators, particularly those working on light material such as coal, hold that the turbulence set up as ore falls over to the next box in a series is undesirable, and prefer to run the compartments either level or at a slight upward climb. The capacity of an ore box is proportional to the screen area, and particularly to efficient delivery of feed over its whole width.

In addition to variation of the length and rate of the strokes, and the addition of hydraulic water, there is another important variable. This is the bedding or "ragging" layer, which rests immediately on the jig screen. Under working conditions the contents of the ore box sort themselves into three fairly distinct layers (Fig. 145). At the bottom is the separating layer, which should be maintained at a grain size and density suited to the severity with which it is to oppose penetration by particles from the feed. In jigging through the screen the concentrate must fall through this bed, which may be autogenous, but is usually composed of particles of a material chosen for its specific gravity, of a size larger than the screen openings. Various materials are used, among which are steel punchings, shot, galena, and other heavy

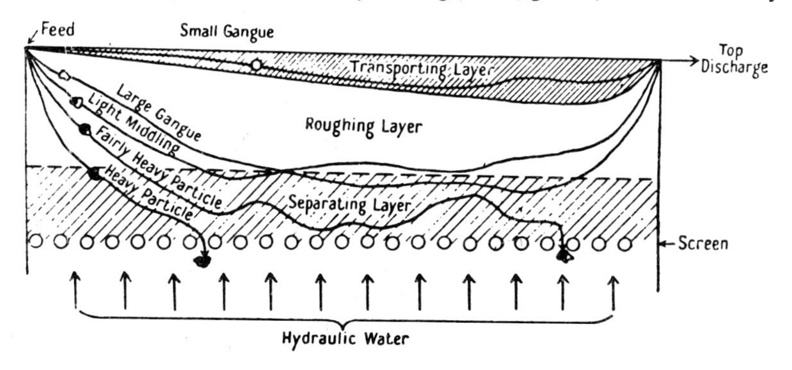

Fig. 145. *Particle Paths in Jig*

sulphide minerals tough enough to resist attrition. At each pulsion stroke the bed dilates, but provided it has been made sufficiently thick it distributes the uprushing water equally through its interstices and damps out any tendency to "boil". The specific gravity of this bed is not that of its component particles, but of these *plus* the interstitial water in a given volume—a solid-liquid ratio. This ratio changes from a maximum when the bed is close-packed at the end of the suction stroke, and a minimum when it is fully dilated shortly after the peak of the pulsion stroke. The route taken by an ore particle (also in up-and-down motion) is relative to the changing bed density and to the residual kinetic energy of the particle after it ceases to receive hydraulic thrust. In jigging "through the screen" heavy mineral from the feed must pass through this bed. It is drawn down during the suction stroke and in due course falls to the hutch, since it is undersized to the screen. At the same time, lighter mineral seeking to penetrate the separating layer lacks the necessary dynamic thrust, and is rejected upward. Thus, the thicker and heavier the separating layer the cleaner will be the concentrate. The main factors determining whether the feed particle is rejected, held in the bed, or passed down through it, are:

(*a*) Bed density (solid *plus* liquid).
(*b*) Size, shape, and size-range of bed particles.
(c) Depth and uniformity of bed.
(*d*) Size, shape, and density of feed particles.
(*e*) Balance between pulsion and suction strokes.
(*f*) Bed dilation.

Too slow a stroke, by failing to dilate the bed, makes it almost impossible for particles to work through to the screen. Too sharp a stroke may upset the stratification and evenness of the bed.

Above this bed is the roughing layer (Fig. 145). This also has sorting work to do, though of a less severe nature. The layer is composed of partly liberated particles of ore, which have stratified downward from the passing feed, in accordance with the forces already discussed. The roughing layer varies its composition somewhat in accordance with changes in the feed, the setting of the jig, and the rate of withdrawal of concentrate which it allows to fall through. It rejects any coarse gangue presented to its upper surface, and if the jig is making a high-grade concentrate it either retains or rejects (upward) any middlings. When the jig is set to make a "concentrate" of middlings, as would be normal practice in the final compartments, the adjustments are such as to allow these particles to drop through. The roughing layer must be thick enough to press down on the separating layer with moderate force, thus helping to resist any "boiling" through of water on the pulsion stroke. The feed to the compartment must not arrive with sufficient kinetic energy to scour the roughing layer.

Uppermost is the transporting layer (Fig. 145). Here the feed is received and spread across the ore box, and from this layer of light and fairly mobile gangue particles the heavier material in the feed drops down to the roughing layer. Much of the gangue streams across to the discharge end without leaving the roughing layer. Thus, part of the original feed is withdrawn below from the system as concentrate, and part, after rejection from the roughing layer, emerges and is lifted out at the discharge end.

In a series of four ore compartments there is progressive diminution of the weight of entering feed and, consequently, a slowing down of the rate of horizontal movement of the solids, proportional to the weight leaving as concentrate. Against this, some increase in the total water probably occurs in each compartment, because of the excess of pulsion over suction, so that the streaming effect produced by the added water increases from one compartment to the next.

Working Control

This accords with the conditions analysed in the preceding section. Changes can be made in:

(*a*) The density of the oversize material in the separating (ragging) bed.
(*b*) The thickness of the separating bed.
(*c*) The addition of hydraulic water.
(*d*) The rate of withdrawing concentrates (German jigging).
(*e*) The feed rate.
(*f*) Speed and length of jig stroke.

When concentrate is drawn from above the screen, the slower the rate the more severe is the selective action and the higher is the grade of the product. Provided the final tailing is not too high in mineral value, the draw-off from successive gates can be adjusted on some such lines as this (Fig. 146)

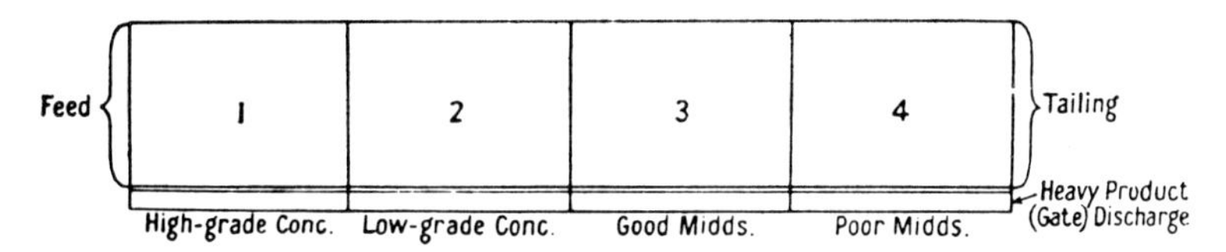

Fig. 146. *Successive Hutch Products in Jigging*

When jigging through the screen, the corresponding control is made by adding or removing some bedding material. The thicker the bed, the higher is its effective density and resistance to passage, and the looser are the layers above it (since more force is needed for the hydraulic water to reach them). In order to raise the grade of concentrate, or to loosen the roughing zone, the bed can be thickened.

If other details are in good adjustment but the tailings are too high in value, the feed rate should be decreased and the hydraulic water reduced, so as to give the material a longer dwelling time in the system.

Some Specialised Jigs

One of the most elementary jigging motions is that sometimes used on primitive mines in Malaya by Chinese tin-dressers. The final concentrate from the sluices or "palongs" may be contaminated with heavy minerals such as ilmenite and monazite. These minerals often occur with cassiterite and are so near to it in specific gravity that it would be unwise to attempt their separation during the crude operation of sluicing. As the price paid by the smelter for cassiterite penalises impurities and encourages production of a concentrate assaying well over 70% Sn, upgrading calls for skill. In the most elementary process the worker half fills a round fine-meshed sieve with sluice concentrate and gives it a circular panning motion, at the same time jigging it up and down in a tank of water. The heaviest mineral stratifies downward and outward, while the impurity collects in a central "eye" and as a top layer. This layer is scraped off, replaced by new material, and the process is repeated. From time to time the "eye" is removed and clean concentrate taken from the sieve for drying and dispatch. The rejected material is collected and given further cleaning treatment. The process is called "dilluing".

A somewhat similar principle is embodied in the Cornish kieve. In its simplest form this is an open-topped barrel into which fine tin concentrate is shovelled by one worker, while another keeps the semi-liquid mass swirling. When the barrel is full, and its contents have been paddled round until segregated, they are allowed to settle down, the outside of the barrel being

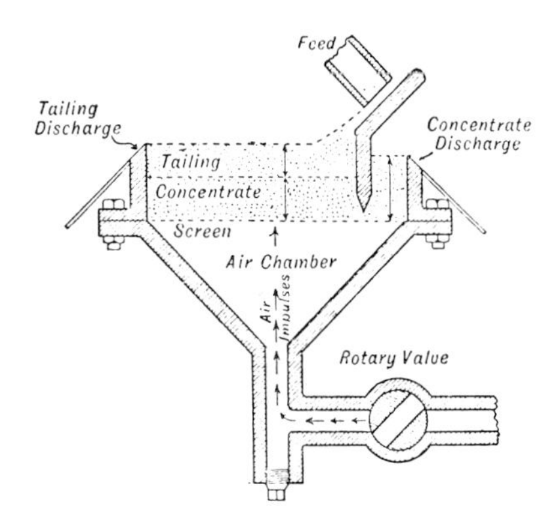

Fig. 147. *Pneumatic Jig* (after Truscott)

vigorously rapped. Both swirling and rapping can be mechanised. The upper part of the settled solids (particularly toward the centre) carries a low-grade fraction, and the outer lower part is relatively high grade. The rapping of the barrel transmits a jigging vibration through the material, and aids the heavier particles to burrow down while the mass is still mobile.

The Willoughby jig is a simple elutriating device. The bottom of the ore box is a finely punched screen. The box is filled with rough concentrate through which water from a tank is allowed to flow upward. At first the flow rate is high and the particles stratify under teeter conditions. with the lightest and smallest on top, the largest and heaviest below. As the water head falls, the teeter bed collapses, the uppermost layers being the last to lose their mobility. The impurities are scraped from the top, more concentrate is added, the tank is refilled. and the process is repeated until a satisfactory shipping grade of cassiterite is obtained. The apparatus, though primitive, is easy to make and there is little difficulty in learning to use it.

The pneumatic jig (Fig. 147) in various forms is used in arid country. Its operation is self-explanatory.

References

1. Kirchberg, H. and Hentzschel, W. (1958) *I.M.P.C.* (*Stockholm*), Almqvist & Wiksel.
2. Robinson, H. Y. *Coal Cleaning Research*, 9, Bull.1.
3. Michell, F. B. and Swarnapradip, P. (1963). *I.M.P.C.* (*Cannes*), Pergamon.
4. Batzer, D. J. (1962). *Trans. I.M.M.* (*Lond.*), 72.
5. Condolion, E., Hoffnung, G., and Moreau, C. (1963). *I.M.P.C.* (*Cannes*). Pergamon.
6. Bird, B. M., Mitchell, D. B., and Smith, F. E. (1943). "Coal Preparation". Seeley-Mudd Series, A.I.M.M.E.
7. Bird, B. M. (1943). *Ibid.*
8. Chaston, L. R. M. (1960). *I.M.P.C.* (*Lond.*), I.M.M.
9. Weavind, R. G., and McLachan, D. F. C. (1961). *J.S. Af. I.M.M.*, Jan.

CHAPTER 14

SEPARATION IN STREAMING CURRENTS

Streaming Flow

The term "film sizing" has been defined as "Reverse classification; sorting of mineral particles on such flattish surfaces as sluices and shaking tables in accordance with the sizes of the particles moved by a flowing film of water, which exercises transporting force proportional to the cross-section exposed to flow".[1] The thrust exerted by the streaming water can be used to size homogeneous sand, to wash particles of a light mineral away from those of a heavier species, or to separate particles of similar mass but different shape. In ore treatment all three effects are produced together unless pre-treatment has modified one or more of the operating factors—particle size, density and shape. This streaming force is manipulated in various ways and in several types of machine. The appliances used to separate heavy minerals from relatively light ores depend mainly upon the work done by water as it encounters ore particles while flowing down an inclined plane. This plane may be smooth, rough, or riffled; gently or steeply inclined; stationary or mobile. Instead of a plane, concave or convex surfaces may be used, or compound planes. Flow may be quiet or turbulent, steady or intermittent, and elements of vertical or centrifugal motion can be introduced. This section discusses the forces that act upon particles moved by a layer of water as it flows down an inclined plane.

The kinetic energy carried by this moving water provides the dynamic force. Constraints, modified cross-sections, obstructions, changes of direction, eddies and turbulent zones can be introduced in order to check, accelerate or produce a centrifugal effect. Essentially, however, the separating effect on particles of ore introduced into a streaming current results from their reaction to two successive sets of circumstances. The first is their lateral displacement or drift, which is determined by the time taken by each particle to fall from the point of entry to the floor of the sluice (the channel through which the water into which it is fed is streaming). The second is the resistance offered by each particle to further lateral displacement after it has reached this floor. The mode of presentation of kinetic energy to the resisting particle is different from that at work in a vertical current. Efficient use of flowing streams as a means of applying separating force to ore particles therefore requires a different preparation of the feed.

Consider first a simple sheet of water flowing gently down a smooth plane. The rate of flow and depth of the layer depend on:

- (*a*) Volume of water available.
- (*b*) Width of plane.
- (*c*) Speed of water in direction of flow at moment of arriving on plane.

(*d*) Inclination of plane.
(*e*) Obstruction to smooth flow.

Such obstructions (*e*) to smooth flow are introduced by:

(1) Multi-directional currents in flowing layer.
(2) Wind blowing along surface.
(3) Viscous resistance to flow.
(4) Surface friction between water and bed.
(5) Bad distribution at entry to flowing layer.

When a plane surface is wetted by a liquid, a film of that liquid a few molecules thick becomes seized to the surface with appreciable tenacity. If the surface is set vertically, excess fluid drains down with increasing reluctance, but final rejection of its wetting effect is only obtained when sufficient force has been applied to disrupt the attraction between solid and liquid (e.g. by heat of evaporation). On an inclined plane an indefinitely thin layer of water is attracted to the surface of the supporting plane, where it becomes almost static. (Surface tension effects are discussed in a later chapter.) Above this another tenuous layer is formed, the molecules of which continuously interpenetrate the seized layer so that they must be pulled away again by the general gravitational force causing the water to flow down the plane. Thus a zone of shear exists between the truly solid plane and the fully mobile water. Across this, from the solid outward, there is a change from complete rigidity to completely unobstructed flow. If the plane is nearly horizontal, there are very few gravitational forces to upset the mating and keying of the water molecules. Since the mobility of the water molecules increases with temperature, the viscous keying at the interface varies inversely as temperature.

These forces operate over extremely thin layers, but they affect every particle moving through a wetting liquid. With coarse material such as constitutes the feed in jigging it was possible to ignore the indefinitely small retardations introduced by these interfacial zones, but they begin to affect the behaviour of extremely fine particles when film-sizing methods of separation are used on material finer than about 200 mesh. The transition from the seized layer of wetting liquid at the surface of the solid, over to the fully mobile mass of liquid molecules, applies to particles of ore as well as to separating planes.

The flowing film can be pictured as in Fig. 148.

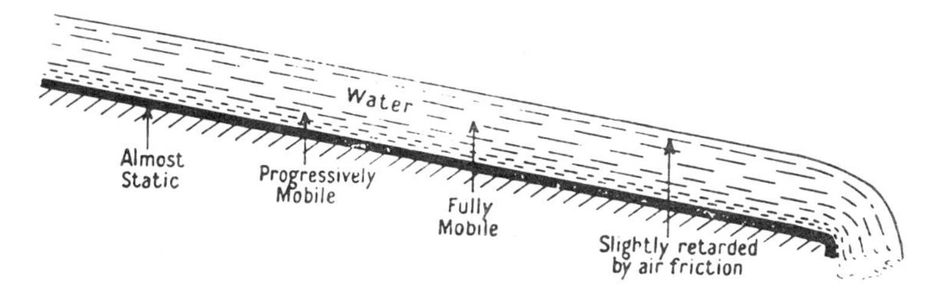

Fig. 148. *Retardation of Flowing Water*

A short distance up from the plane the current of water runs with a swiftness depending chiefly on the slope (i.e. the gravitational pull). Near the top of the layer the water is slightly retarded by friction with the layer of air above it. This may set up wavelets or interfering currents.

The Liquid-Particle System

Consider next the behaviour of a number of spheres introduced into a thick layer of water running gently down a smooth plane. If these spheres are composed of two kinds of mineral, one heavy and one light, and are of different sizes, they will be separated during their fall through this layer. Each sphere will have a downward gravitational pull corresponding to its mass, which determines its rate of fall. It is, however, carried along by the stream through which it is falling. The distance displaced depends on the time it takes to fall. This results in a separating effect based on mass (Fig. 149).

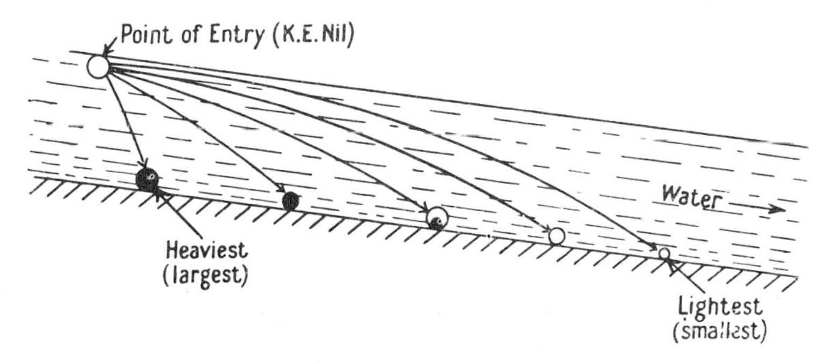

Fig. 149. *Particle Drift in Flowing Water*

At the moment of reaching the plane the biggest high-density sphere, having fallen fastest and consequently been least affected by the current, lies nearest downstream to its point of entry. The smallest low-density sphere has drifted furthest downstream. The others are to some extent overlapped.

It would obviously be possible to base a separating process on this displacing effect, with a clean coarse concentrate upstream, a clean light gangue downstream, and an overlapped mixture between which could be separated by screening. The practical difficulties, however, outweigh the advantages.

On arrival at the surface of the supporting plane, the spheres become subject to a new system of forces, which may be represented as in Fig. 150.

The pull of gravity anchors the sphere to the plane if this is horizontal, but aids rolling if it is inclined. The flowing water presses against the cross-section of the sphere and encourages it to move downstream. The differential rate at which the laminae of water are flowing causes pressure to be lowest next the plane and highest at the top of the sphere, thus exerting a slight overturning moment which strengthens the tendency for rolling downward to commence.

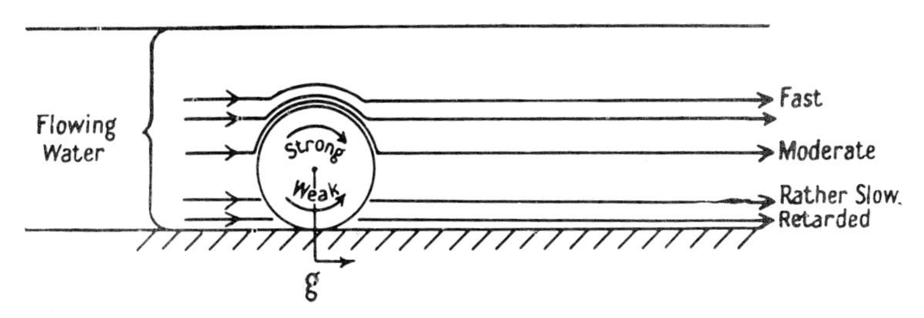

Fig. 150. *Thrust of Flowing Water on Particle*

If the combined influence of plane slope and streaming velocity is sufficient to keep all the spheres in rolling movement, they rearrange themselves in response to the combined effect of thrust and mass. If a system is used which now allows all the spheres to come to rest (a multiplane surface, flattening out, or a broadening of the stream), they will have changed their arrangement as in Fig. 151. This is one form of reverse classification.

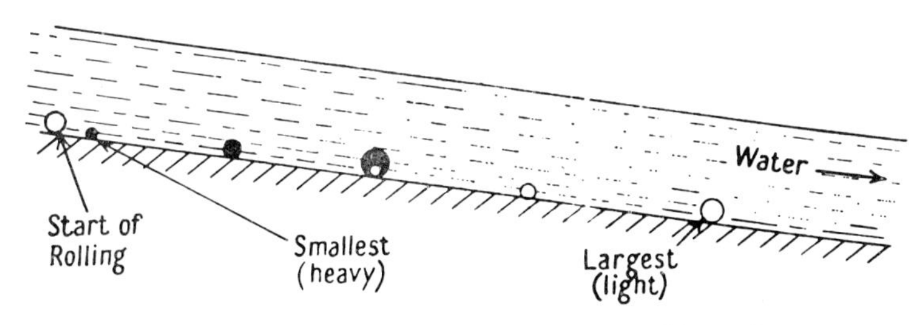

Fig. 151. *Classifying Effect on Streamed Particles*

Application

It is now clear that three physical characteristics of the mineral particle play a dominant part in determining its behaviour in film sizing or film separation. These are

(*a*) Shape.
(*b*) Volume.
(*c*) Specific Gravity.

In the previous section it was convenient to simplify the picture by considering a sphere. In practice, the shape determines the cross-section presented to the push of the stream and the ease with which rolling can be induced. If not packed together, particles settle in their most stable positions, each on its flattest available base. A single particle exposes a minimum cross-section to the stream and offers maximum resistance to overturning movements. Since it has achieved a maximum of frictional hold it is to that extent reluctant to skid. If it lies close to the plane, the passing water rises over it and presses it down yet more firmly. In fact, wherever it is in contact with an unmoving surface it is to some extent shielded from streaming action. Shape is therefore an important discriminating factor in film separation, and judicious manipulation of hydraulic pressure can be caused to set cubic shapes in motion while flakes of similar mass remain quiescent. If in Fig. 151 irregular shapes were substituted for spheres, and all the particles had the same mass, then the flattest flakes would be the first to stop, the cubes would come later and rounded particles would roll furthest. Because of its shape, a flake of mica has the greatest stability in film sizing and a marble the least.

Under practical working conditions it is not feasible to provide the large area which would be needed for a full development of these conditions. The separating appliances must carry a bed of material many particles deep. The angle of repose of an individual particle depends from moment to moment upon its adjustment to the voids between particles. A quicksand is produced which permits the heaviest particles to gravitate downward, and which at the same time tends to accept edged light particles and reject the larger rounded ones. There is no simple generalisation from which the angle of repose of a specific particle can be predicted.

A smooth separating plane is rarely used in the early stages of sluicing. Obstructions vary from rough-textured floors to the use of cross-*riffles* ((bars set across the stream). These obstructions introduce turbulent flow, vortexes, and obstacles against which the bottom strata lodge while the upper ones climb over by creeping, rolling, or jumping. A rough plane shields the lowest particles and tends to immobilise them. The stationary particles then become in turn the bed along which the upper strata move. Thus, if an increased dynamic pressure is desired along a sluice it can be obtained by:

(*a*) Increasing the flow of the water (rate, volume, or both).
(*b*) Increasing the slope.
(*c*) Increasing roughness.
(*d*) Increasing and combination of (*a*), (*b*), and (*c*).

Method (*c*) is usually found to give a maximum throughput for a minimum use of hydraulic water (i.e. expenditure of power). This roughening of the bed is most effective when produced by cross-riffling.

In Fig. 152 the effect of placing a riffle, such as a short length of 2″ × 2″ wood, across a sluice is illustrated. The smallest and heaviest grains have burrowed their way to the bottom, and are covered by heavy particles of increasing size mixed with small particles of the lighter minerals. At the top of the bed are the largest light particles, which are unable to force their way through the layers below. This applies only where the feed to the sluice has been restricted to a suitable size range. Preliminary classification of the feed within suitable equal-settling limits is a pre-requisite to efficient gravity concentration by film sizing.

The main control factors in sluicing are:

(*a*) Slope of bed.
(*b*) Thickness of water film.
(*c*) Roughness of riffling bed.
(*d*) Relation between maximum and minimum settling rate (in water) of particles fed.

(*e*) Ratio of solids to water in pulp.

Other factors, not subject to direct control, include relative density of particles of concentrate and tailing, particle shape, and tightness of particle interlock along the bed. Control of the partly locked middling is considered later.

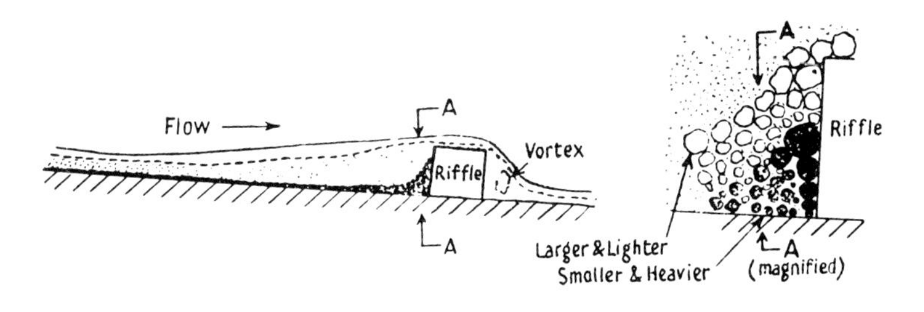

Fig. 152. *Riffle Effect on Streaming Sands*

Simple Sluicing Systems

One of the most elementary systems of concentration by sluicing may be seen in most mineralised countries after heavy rain. The surface waters have washed down whatever sand they can shift, leaving grains of heavy minerals concentrated in natural depressions lying along their course. Advantage used to be taken of tidal action by Cornish "tinners", who would select rich patches of sand for removal and sluicing, and would rove the beach between tides in order to locate spots where nature had done part of the work for them.

In addition they supplied riffles in the form of strong boards anchored into the sand parallel to the waves. As the tide ebbed the lighter sand was washed down while the cassiterite and other heavy minerals remained trapped above these timbers.

Where water is plentiful, and pioneering conditions favour the use of manual labour, ground sluicing (Fig. 153) may provide a cheap method of "muck-shifting". Usually the deposit consists of a barren overburden underlain by an alluvial gravel rich in such heavy and inert minerals as cassiterite or gold. By diverting part of a river into an upper leat and by sluicing down from this, considerable tonnages have been cheaply removed, and the same process has been repeated with greater care to produce a rough concentrate when the pay-gravel is exposed. Such methods denude the country of its agricultural soil and choke the natural drainage system, and would not be permitted in developing areas.

Where an adequate fall exists, and there are no social objections, concentration can be combined with extensive earth removal by the use of hydraulic monitors. A powerful jet of water washes down the overburden and then is used to move the pay-dirt into sluices excavated along the ground, or constructed of timber. Variants of this technique use a jet of water to elevate the gravels to a sluicing height from which they can run gently down to a tailings dump, the values being trapped along the sluice behind riffles. The Chinese "palong", used in Malaya, may be fed thus or from a centrifugal sand pump which elevates the tin sands (karang) from an opencut or lombong.

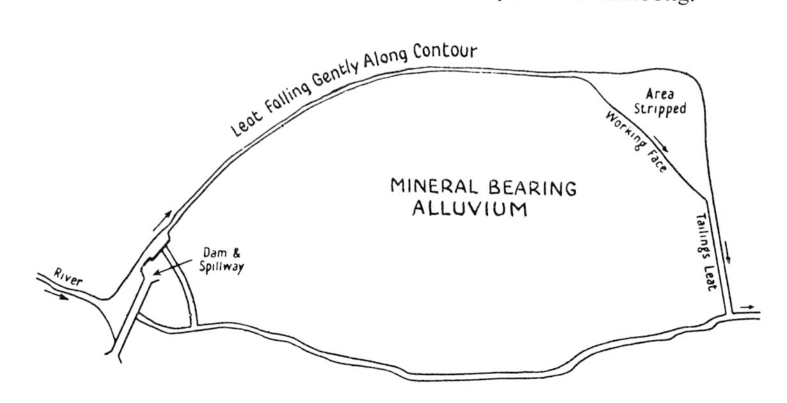

Fig. 153. *Ground Sluicing*

In Fig. 154 one method of concentrating cassiterite from rich tin gravels is shown. First the barren overburden is removed—perhaps by hydraulic monitors and jet pumps if abundant pressure water can be cheaply led to the workings. The mineralised gravel is next washed down to a sump cut in the barren bed rock, and from here is pumped to one of the two or more long parallel sluices. As the sands stream down this sluice they are kept from packing too tightly by hand raking against the current. From time to time riffles are set across the sluice, and existing ones are heightened. Thus the light and valueless sands are washed down and run to waste while the heavier ones accumulate and are trapped behind the riffles. When a full load of low-grade concentrate has been thus produced, feed from the opencast (lombong) is switched to a parallel sluice, and clean-up commences. Wash water is streamed down, the sands being raked vigorously against the flow and the riffles removed cautiously working upward from the discharge end of the sluice. When the concentrates have been sufficiently upgraded in this way they are removed for final cleaning by more delicate methods, and the sluice is again ready to receive run-of-mine feed.

The simple but laborious operations of the lone prospector or "fossicker" still produce a substantial yield of tin, gold, and alluvial diamond. His tools include the gold pan, with such variants as the batea and calabash. The cradle rocker (Fig. 158) allows a quantity of gravel to be washed, if necessary with some recirculation of water. Feed is placed on a punched screen at the bottom of a small feed box. The mineral-bearing sand is washed through this, large pebbles being held up and discarded. Clay which might trap valuable mineral is either rejected, or puddled and disintegrated by hand. Water is either piped in or caught below the rocker and recirculated by the use of a

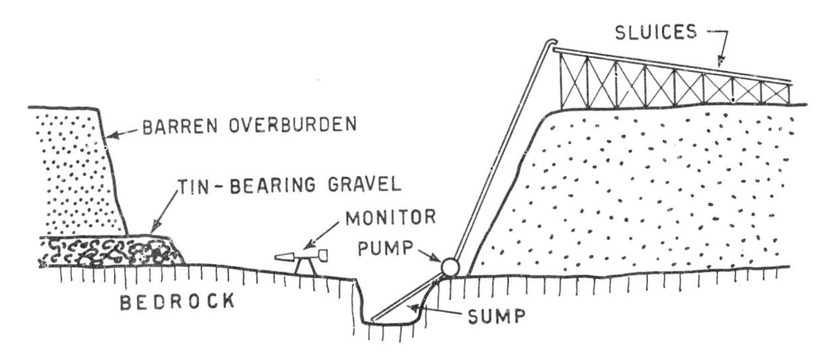

Fig. 154. *Malayan Lombong*

Fig. 155. *Hydraulic Monitor at Work* (Tin Industry Board, Malaya)

dipper. Particles of gold may be trapped upon canvas screens below the hopper, or they may be washed through into the sluice box. This is carried upon rockers, so that it can be oscillated from side to side, thus giving some fluidity to the sand as it gravitates downward. The gold stratifies and is arrested by the retaining riffle at the bottom of the sluice, while the impoverished sand climbs over and is discarded.

Another method uses a stationary sluice, recirculating the water when necessary by means of a hand-operated diaphragm pump. The bed of material is stirred by means of rakes, rabbles, or hoes to prevent it from consoli-

Fig. 156. *General View of Malayan Open-Cast showing palong sluice middle distance* (Tin Industry Board, Malaya)

Fig. 157. *Malayan Sluice* (*palong*). *Left compartment is being cleaned up; middle and right are being loosened by raking to aid settlement of cassiterite between riffles running across sluices* (Tin Industry Board, Malaya)

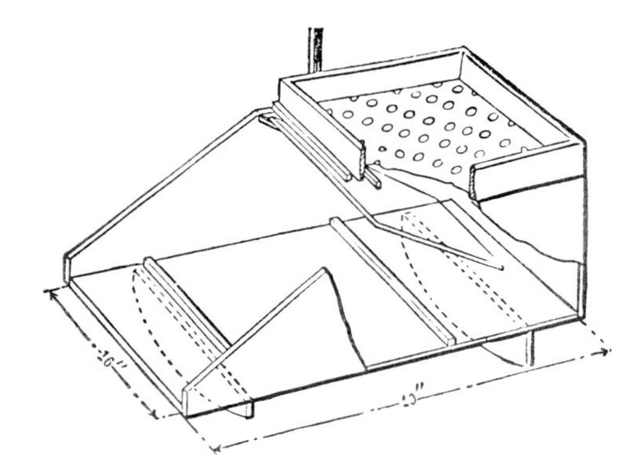

Fig. 158. *A Cradle Rocker*

dating too tightly to permit stratification. The sand bed must be kept sufficiently open to trap values if concentration is to result.

The gold strake is used in milling to entrap coarse particles of gold at an early stage in the treatment of ore.

The simplest form consists of a stationary inclined plane, sloping downward with just sufficient inclination for a film of pulp to flow gently along. The light mineral particles roll down, while the heavier ones stop on a removable cloth lying on the surface. The pulp is distributed evenly over the strake at entry, and slope, feed size, solid-liquid ratio and feed rate are adjusted to give smooth flow with no channel formation. At one time rough blanketing was used as the catching fabric. The modern straking system uses strips of corduroy cloth laid with its ribs across the direction of flow. The nap of the pile points upward, so as to entangle the gold falling to it. Corduroy is sold in bolts 75 yards long and 28″ or 36″ wide. Three pieces of cloth are laid on a typical strake, the upper sheets overlapping the lower ones. The head cloth may be held down by means of a strip of flat-sectioned iron. Two strakes are set in parallel so that pulp can run over one while concentrate is being removed from the other. If several strakes are set in series, the head strake catches most concentrate, and the uppermost cloth needs removal and washing more frequently than the others. Pulp is fed on at a size below 10 mesh, and the table slope is adjusted till the largest gangue particles just manage to progress over the corduroy ribbing. The solid-liquid ratio of the pulp may be from 20% up to 50%. The tonnage fed varies from 0·2 ton/sq. ft. of cloth to 2·0 ton/sq. ft. per day, 0·3 being a normal figure. Concentration ratio is 1:2000 or less.

The cloths are washed regularly, at intervals of two hours or more. Where this work is done by hand, the pulp is diverted to the parallel strake, and the cloth is folded on itself, refolded, and carried to the concentrates tub. This is a vessel filled with water, with a metal grid locked in place below water-level. The cloth is opened out face downward on this grid and shaken so that the concentrate falls through. When the values have been scrubbed and sluiced away, the cloth is refolded and returned to the strake. Bottom cloths may be left in place for four or eight hours. Old cloths contain an appreciable amount of gold. They are burned, and the ash is sent to the cyanide plant.

Under the rough-and-ready conditions of elementary sluicing, little attention is given to limitation of size as a variable. A coarse screen is commonly used at the feed end to remove stones, the undersize from the screen passing on to the sluice.

Sluicing Fine Sands

Thus far hand-operated sluicing in quiet beds has been considered, with hand rabbling or gentle agitation by rocking. In addition to the need for maintaining a sufficiently loose texture in the bed, it is also important to maintain even distribution of the feed. Left to itself, the stream of pulp settles irregularly and cuts channels along which the material scours too violently for good separation to ensue. At the point where the riffle is set across the stream, the sand tends to pack so tightly that newly arriving particles cannot embed themselves and may therefore flush over and be lost. Such tendencies are watched and corrected by the sluice operator. Most rabbling movements can be performed mechanically if the cost is justified. Stirring arms have been invented which apply steady strokes against the current, with short traversing movements that break up channels. Instead of a quiet bed, moving belts have been developed, which quietly carry their load upstream. Since the values commence to concentrate on the bed near the head of the sluice, this arrangement allows continuous withdrawal and the bringing up of moderately enriched sands from further down the belt for final upgrading. Entering feed is liable to disturb this material. A gold particle (density 19·2) partly entangled in the pile of a corduroy belt can usually withstand such a displacing inrush. Lighter minerals have less clinging power and need more gentle withdrawal. Stationary sluices are cleaned by sluicing with water, feed being cut off. Sometimes only a few riffled sections at the head of a long sluice are cleaned up regularly, during a short stoppage of feed. In this case a complete clean-up is carried out at, say, weekly intervals. Such a complete clean-up is usually worked counter-current, clean water being sluiced downward while the deposited sands are raked up against it. The streaming action is adjusted so as to sweep away light sands while leaving heavy particles to be worked upward. When a rough concentrate has been accumulated at the upper end of the sluice it is removed and brought up to shipping grade in appropriate concentrating appliances.

In the section "Working Principles" in Chapter 12 the concentration criterion was briefly noted. The finer the mesh of the minerals being separated by gravity treatment, the greater must be the differences between their specific gravity. Where an alluvial deposit has been laid down as the result of natural forces working through geological ages, the valuable mineral usually exists as moderately coarse liberated particles, since wind, flood, and erosion have sized and classified the sands. Very different considerations arise when an ore is broken to liberation-mesh by crushing and grinding, since the particle sizes produced include much material which, despite all the skill used in its comminution, is far too small to respond efficiently and quickly to treatment based on differences of specific gravity. Later it will be seen that this defect in grinding technique is compensated by the development of the flotation process which makes a completely different approach to the problem of concentration. Certain important minerals, however, are not readily amenable to flotation. The outstanding one is cassiterite, SnO_2, the main source of tin. It has a S.G. varying between 6·4 and 7·1 according to its purity, and typically occurs interlocked predominately with quartz (S.G. 2·7). Other minerals are associated in the ore, but these are the two which must be separated by gravity treatment. As a result, much attention has been given to the problem of making a separation, however inefficient, of the fine "slime tin" produced in grinding.

The first essential is good classification of a pulp which has been ground under conditions which avoid over-grinding. In the flow-sheet shown in Fig. 184 some concentrate and tailing has been removed ahead of the first ball mill, and the ball mill discharge is sent to a series of hydraulic classifiers and thence to tables (discussed later in this chapter). This lessens over-grinding. At a certain stage in the flow-sheet a classifier overflow (slimes) is produced containing mineral so finely ground as to be slow to settle. This still contains valuable tin, much of which is recovered on buddles and similar devices. The dominant requirements in treating this fine material are:

(1) Very gentle action is essential.
(2) Upgrading must proceed by easy stages.
(3) The area of the concentrating devices must be generous.
(4) Feed must be evenly distributed in thin and slow-moving layers.
(5) Water must be copious and reasonably clean.

Mechanised Sluices

With the decline of the Cornish mining industry the use of devices which incorporate mechanised movements has lessened, but a brief description of the buddle is given since it illustrates certain important principles in pulp handling where gravitational force is used. In the convex round buddle (Fig. 159) the pulp is distributed radially from the centre head which is usually of concrete and about 5′ in diameter, and drops some 12″ to 15″ on to an annular cement bed, the overall diameter being about 15′. As the pulp spreads the carrying film becomes thinner and moves more quietly, so that the particles in it, though of subsieve size, are arrested and begin to build up into a layer of mineral, the heaviest nearest the centre and the lightest out towards the periphery, or even overflowing. Channelling is avoided by the light drag of brushes which rotate slowly over the settling material. The word "buddle" is believed to be derived from the building up of the ore bed as flow proceeds. Periodically, feed is diverted to a duplicate buddle and the bed is sampled, the products either being retained for further upgrading or rejected.

In the concave buddle (Fig. 160), feed is peripheral and material not settling immediately is flushed away as the pulp-stream accelerates towards the centre. A modern improvement on the concave buddle uses the same peripheral feed and brushing gear. Instead of allowing build-up and periodic removal it is,

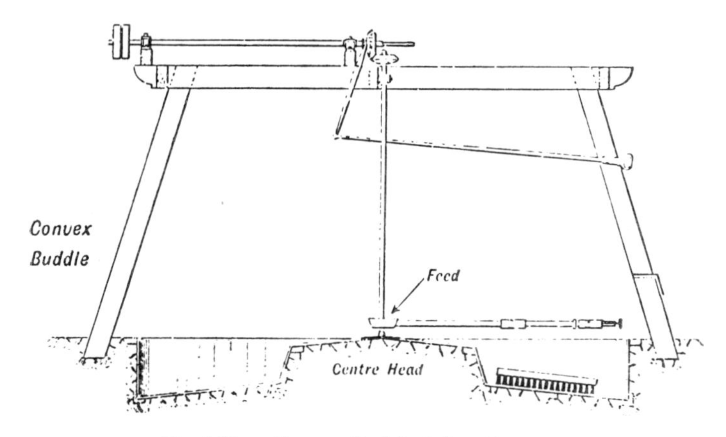

Fig. 159. *Convex Buddle* (after Truscott)

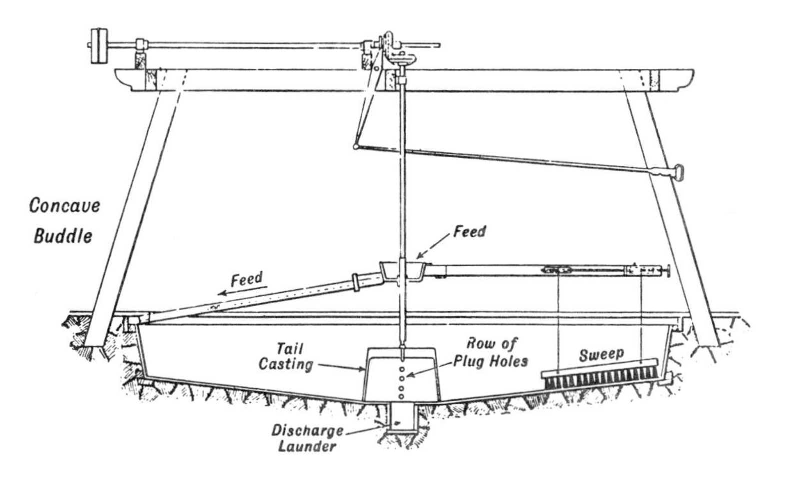

Fig. 160. *Concave Buddle* (after Truscott)

however, continuous in action. A flush of water is carried from the outside of one of the brushing arms, and a short laundering bridge carries the displaced material flushed down by this water over the tailings ring in which the buddle terminates and into an inner concentrates ring, from which the enriched pulp is delivered to the next machine in the series. The principle is adapted from the Cornish round frame, a circular convex table made of wood, which is rotated slowly and from which tailings and concentrates are removed by separate channels.

One drawback to the use of sluices for treating fine sands is the speed with which tightly packed beds are formed. These beds should not be smoothed by raking, as the force used would interfere with the gentle separating action

Fig. 161. *Denver-Buckman Tilting Concentrator in Feeding Position*

needed. As with the sluicing of coarser sands, efficient trapping of the heavy particles calls for even flow of pulp over loosely settled sands, with no channelling to introduce scour or reduce the working width of the separating plane.

For the first minute or two after flow of pulp has commenced a correctly sloped plane forms a level bed of good trapping consistency. Mechanised sluicing systems have been developed which exploit this, by feeding for a short period and then flushing the settled material into a special launder for further upgrading. In the old Cornish tin-dressing plants a large number of "ragging frames" were used to catch "slime tin" in this way. These were automatic in action. The feed came gently on to a sloping wood table for a few minutes, and at about the time when the bed was packing too tightly, water-actuated balance weights tipped the table sharply and flushed its load into a concentrate launder. Considerable space was needed in order to feed these tables at the gentle velocities required for separation of particles, most of which settled in accordance with Stokes' law. At a Canadian mill the rag-frame principle has been rediscovered and developed in a space-saving form, for the recovery of cassiterite from a lead-zinc tailing. This Buckman tilting table has since been used successfully for other heavy minerals. Feed is distributed over five hinged decks (Fig. 161) and after a short time is switched to a parallel unit. The loaded decks are tilted sharply backward and washed with spray water which removes the concentrate. The cycle is then automatically repeated. Embossed rubber matting is favoured as the mineral-catching surface on these tables. The vital point for success is a timing sequence in which the period of sluicing is ended before the layer of settled material has packed down hard or become channelled. If under this operating condition only a low ratio of enrichment is advisable, the concentrate can be given further upgrading by running two or more groups of tables in series.

Details of the operation of a test installation have been described by Chaston.[2]

Strakes and open tables are open to pilferage when used to catch gold. In some African plants their place has been taken by the Johnson concentrator (Fig. 162). This is a 12′ steel cylinder of 3′ diameter, lined circum-

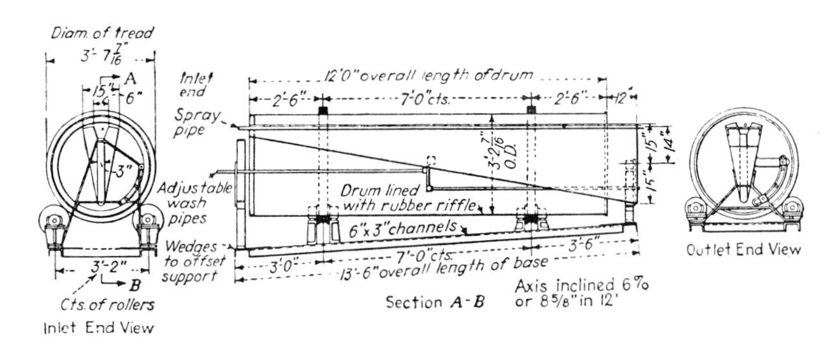

Fig. 162. *Johnson Concentrator*

ferentially with corrugated rubber. It rotates at 7 r.p.m. with a 5° slope. Feed flowing downward deposits its heavy mineral in the corrugation. As the shell rises, this mineral is washed off by a spray into an axial launder.

In coal cleaning by means of "trough washers" a stream of water carries the raw coal down an incline. The lightest fraction stratifies at the top, the dirt below and the middling forms an intermediate layer. Mobility of the sluiced stream increases from the bottom and sides inward and the coal, being least impeded, travels fastest. Shape factor aids the differentiation since shale is tabular and slides quietly without projecting into the stream. Four types of separator exploit these possibilities:

1. Troughs, in which coal is sluiced away from the dirt.
2. Endless belts, with dirt dragged up and out while coal runs down and over.
3. Troughs with undercurrent discharge (Rheolaveurs).
4. Riffled shaking tables.

In the Rheolaveur the feed runs down a steep incline and acquires kinetic energy as it commences to stratify. The trough then flattens abruptly to a gentle slope, and in the stirring and dissipation of kinetic energy stratification is completed. The dirt stratum is removed through apertures in the bottom of the trough, guarded by a rising current of water (Fig. 163). Middlings are recycled, and washed coal runs on and out.

The Pinched Sluice

In its elementary form the pinched sluice as described by Stewart[3] is an inclined launder 2′ to 3′ long, narrowing from about 9″ in width at the feed end to 1″ at discharge. Pulp with a consistency between 50% and 65% solids enters gently and stratifies as it descends. At the discharge end these strata are separated by splitters (Fig. 165). In one marketed arrangement, the Cannon, forty-eight pinched sluices are set as a circle which discharges inward on to two concentric rings which divide the product into concentrate, middling and tailing. In another the Carpco Fanning concentrator (Fig. 164) delivery is on to either one sluice or two set back to back, and delivery is made to a plate (the fan) set at a slight angle across the discharge. This causes the pulp to spread before it reaches the splitters set at the edge of the fan.

The systems are used, together with variants developed from local practice in Florida and Australia, in the *minus* 10 mesh *plus* 200 mesh size range where the concentration criterion exceeds 1·7. Capacity varies from half a ton hourly for fine sands up to 2 t/h for coarser ones. A typical flow-sheet is shown in Fig. 166. Operating controls are concerned with

(*a*) Pulp density (held between 50 and 65% solids).
(*b*) Splitter positioning.
(*c*) Sluice slope (16–20°, depending on that needed to keep the feed gently moving.)
(*d*) Fan orientation (Carpco).

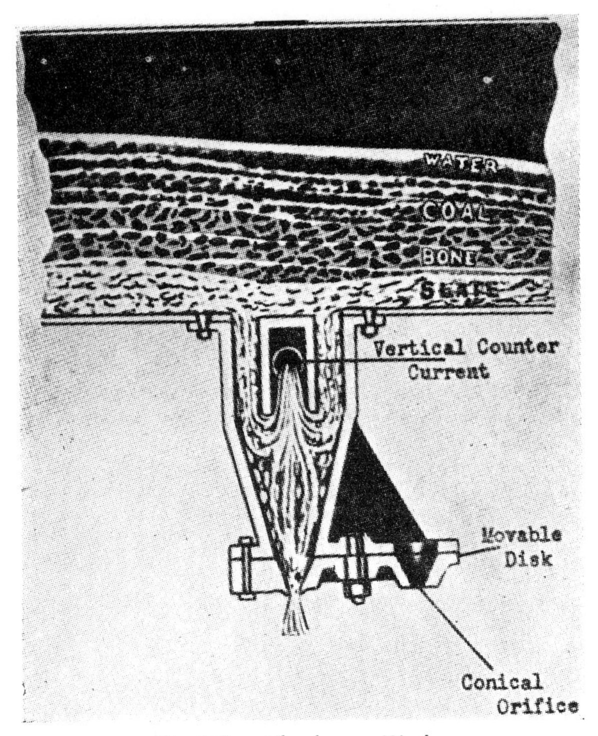

Fig. 163. *Rheolaveur Washer*

These sluices are very simple. Light sheet metal, rubber coated against wear, is adequate for construction. They are cheap to buy and to run, and take but little space. Pullar[4] in reporting on their growing use in Australia, mentions operating difficulties which stem from fluctuations in pulp density or grade of feed. He also describes a type of sluice in which a transverse slot of adjustable width with a movable "tongue" is used for splitting. An element of flotation has been introduced in the treatment of flake graphite and of rock phosphate. Suitable flotation reagents are added to the feed and air is blown in at the head of the sluice. The splitter then works in a water layer between the sunk fraction and the floating mineral/s.

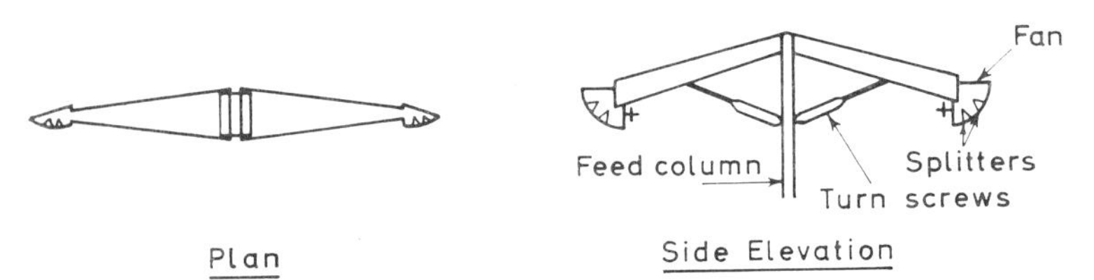

Fig. 164. *Carpco Fanning Concentrator*

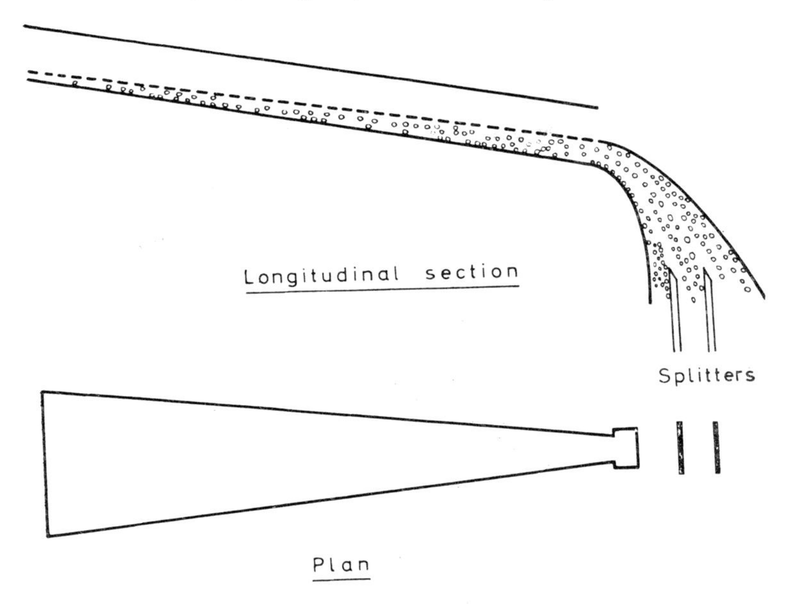

Fig. 165. *The Pinched Sluice*

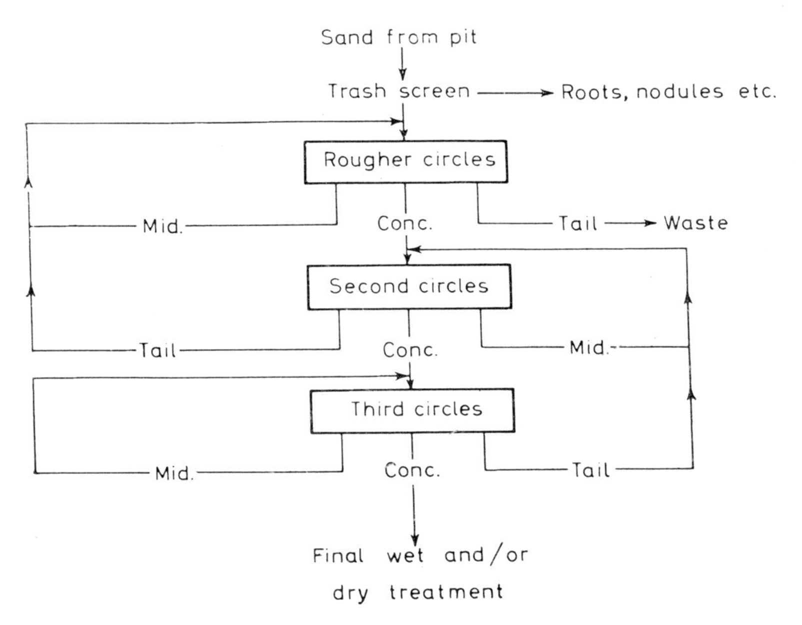

Fig. 166. *Cannon Flow Sheet*

The Humphreys Spiral

This appliance consists essentially of five turns of troughing, with a cross-section roughly following the quadrant of a circle. The mean radius is 5″ with a 13″ fall per turn. Spirals used for coal treatment have six turns and a more gentle gradient. Wash water runs down a small trough alongside the inner (horizontal) part of the cross-section, and can be deflected into the portion carrying the ore-pulp wherever required. A single-spiral testing assembly is shown in Fig. 167. At intervals down the inner part of the spiral withdrawal ports are provided. These are either sealed off by flat discs or filled with half-discs having deflecting wings, which can be adjusted so as to remove part of the inner pulp stream to the central concentrate collecting pipe.

The separating action in the spiral is complex. It combines centrifugal action with multiplane sluicing in a partially controlled sink-float medium autogenously provided by the pulp. Viewed from above, the descending ribbon of pulp can be seen to spread (Fig. 168), the heaviest and coarsest particles remaining nearest the centre and on the flattest part of the cross-section, while the lightest and finest material climbs well up the sides. Wash-water is fed in at an angle and assists this rising action. The innermost part of the ribbon is diverted by the deflecting discs or "splitters" at appropriate intervals. As the outside edge of the heavy fraction concentrated in this section is partially concealed beneath middlings and slimes, more than one such cut is needed, but the material thus removed can be held in separate channels.

This ribbon forms shortly after delivery from the feed box to the first bend after which rearrangement, in accordance with the film-sizing effect shown in Fig. 168 commences. It is sometimes worth while to take advantage of this when treating a long-ranged feed, by removing coarse sands high on the spiral. Heavy minerals should be removed as soon as a steady band has formed. Pulps are treated at between 20% and 30% of solids by weight, unless they are handling fine sands when the dilution is greater, or coarse material, when up to 50% solids may give better results. In practice a battery of spirals is used, feed gravitating from the highest block of "roughers" and gradually being upgraded to the economic end-point. Each spiral must be vertical in order to ensure regular development of its concentrating ribbon. The feed is distributed evenly over each block of parallel spirals from a feed box which must be kept in good adjustment to avoid unequal loading and the consequent loss of efficiency. Ores vary in their reaction, but a throughput between one and two tons/hour of feed at 25% to 50% solids and *minus* 20 mesh is normal. Tests are necessary to determine whether this material should be treated at all −20 mesh or broken down by classification into two or more fractions. On the one hand, the very fine mineral may aid separation by introducing a slight dense-media effect, but this may be outweighed by an increase in the accompanying retardation.

For a feed of 100 tons/hour discarding 50% at the first roughing stage, recirculating 25% twice as middlings and ending with a 25% final concentrate a total of between 150 and 180 spirals would probably be needed. Material of

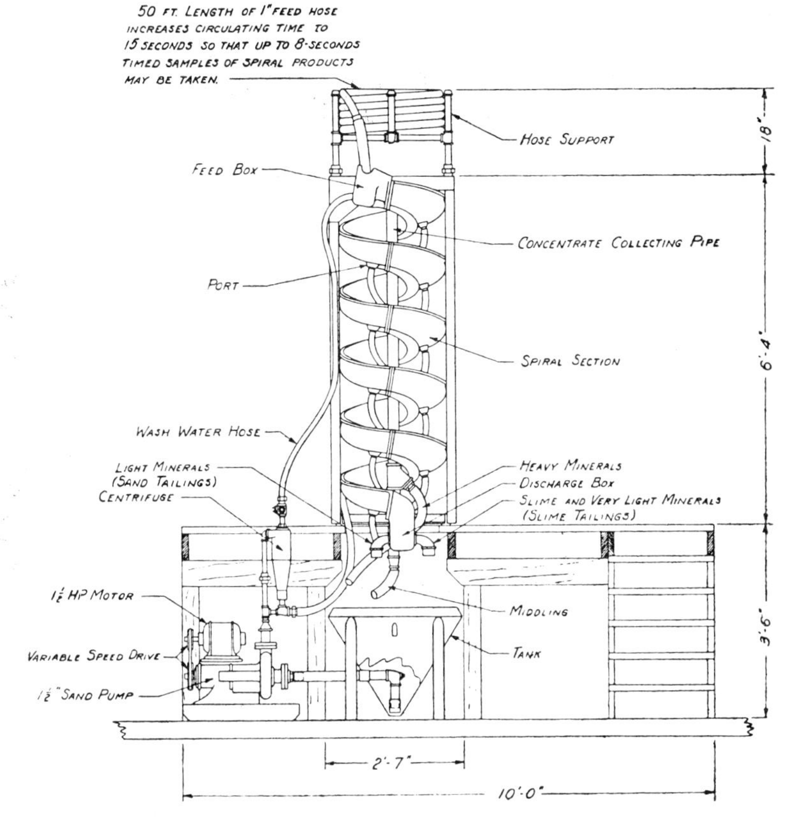

Fig. 167. A Laboratory Test Spiral

construction can be iron, a Ni-hard or similar corrosion-resisting alloy or iron covered with rubber.

Despite the complexity of the principles utilised, the Humphreys spiral is very simple to operate, one or two men taking care of a large installation in which the only mechanical parts are the centrifugal pump, feed and distributing arrangements. The gentle centrifugal force allows material to be separated at sizes too small to be handled on a simple sluice. Spirals are in increasing use for the recovery of heavy minerals such as zirconium, chromite, tin, tungsten, garnet, rutile, monazite, etc., from beach sands and tailings.

Developments from the standard Humphreys spiral have been described by Pullar.[4] Concrete and old lorry tyres have been used instead of cast iron. Distribution through tubes instead of by deflection may improve precision of use of wash water, provided these tubes are not allowed to become blocked by dirty water. Moulded plastic reinforced by fibre-glass has proved excellent, the sections being moulded with a 24″ outer diameter and a pitch which can be adjusted between 13½″ and 15½″ with this flexible material.

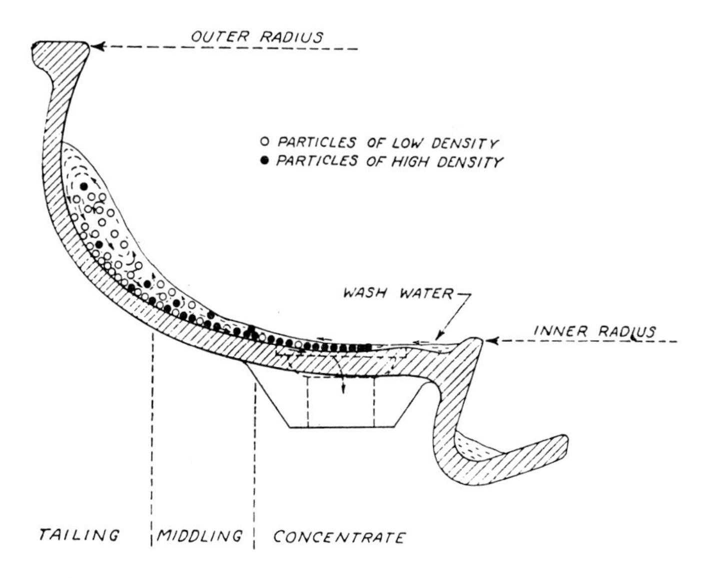

Fig. 168. *Cross-Section of Spiral Stream* (Humphreys Investment Co.)

Vanners

Thus far separation above quiet planes has been considered. In the vanner an endless belt with flanged edges moves slowly up a slightly inclined plane, and over pulleys at the ends (Fig. 169). Feed is distributed gently in a thin layer across the belt, and wash water runs down from above. As it travels, the whole belt system is given a rapid shaking from side to side. This keeps the feed spread and prevents channelling, and it helps the heavy mineral to stratify down through the light gangue. The latter is washed down-slope by wash water adjusted to a suitable velocity for the purpose, while the heavier concentrate continues to cling to the rubber surface of the belt, until it is carried over and washed into the concentrate hopper of the vanner.

Various oscillatory movements have been incorporated in the vanner, of which those in the Frue and the Isbell are the most widely used. A few vanners survive to this day in tin concentrators. They treat slimes too fine to respond to ordinary tabling methods, and produce an impure concentrate with a low recovery (50% or less). In the treatment of "tin slimes", as finely ground cassiterite is usually called, the vanner lingers where low through-put, moderate working costs, and ample space are to be had.

The feed to the vanner is first passed through settling cones or hydraulic classifiers. It is therefore not an equal-sized but an equal-settled material, comprising relatively coarse light gangue and relatively fine heavy concentrate. This discriminating sorting effect, produced by proper use of classification of the feed, is important in the efficient running of shaking-bed sluicing devices. Equal settling is valuable in simple sluicing, but cannot be used with full effect unless the bed is sufficiently disturbed to permit the fine

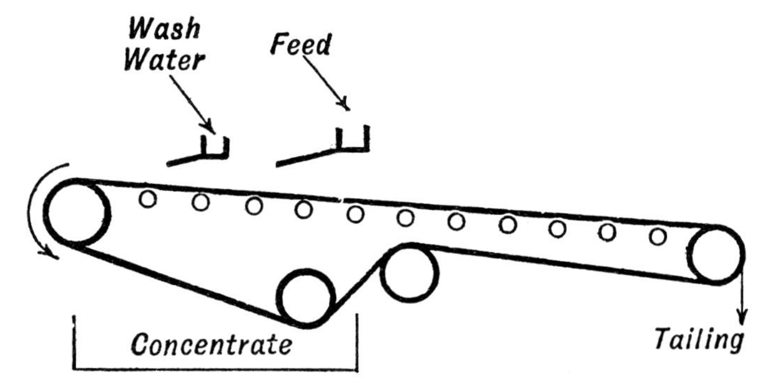

Fig. 169. *Frue Vanner* (after Truscott)

particles to burrow down, thus forcing the coarse ones up to where they meet the full weight of the sluice water.

The difference between feed preparation for jigging and tabling is now clear. Before jigging the feed is screened, the size variable is removed and the density variable developed. Before tabling, which to an important degree includes simple sluicing, the feed is hydraulically sorted into equal settling fractions, the mass variable is removed and the size variable developed. Stratification on the bed completes this phase of preparation for separatory action.

The Frue vanner has an adjustable amplitude of side shake. This is varied from 1″ when treating very fine cassiterite (say *minus* 270 mesh) up to 2″ for sand (at −30 mesh), at 200 shakes/minute.

Shaking Tables

The shaking table is widely used for the gravity separation of sands too fine to treat by jigging. The physical principles utilised in tabling must be understood if preparation of feed and application of control are to be efficient.

Consider a number of spheres rolling down a slightly tilted plane under the urging influence of a flowing film of water. Some of the spheres (shaded) in Fig. 170 represent heavy mineral and others (white) light gangue. The largest sphere travels fastest and the smallest one slowest, under the combined influence of streaming action and gravitational pull. Of two spheres having the same density, the larger moves faster. Of two having the same diameter, if the slope is relatively gentle and the hydraulic urge relatively strong, the lighter sphere travels faster. If during the otherwise free downward travel of these spheres the whole plane is moved sideways, then the horizontal displacement of the spheres varies in accordance with the length

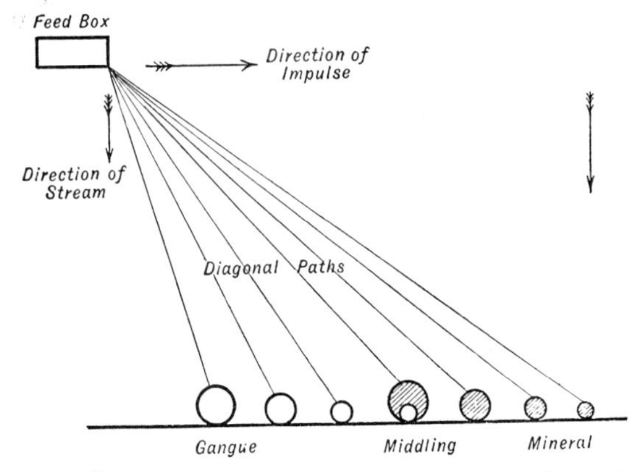

Fig. 170. *Effect on Streaming Particles of Lateral Displacement*

of time they take to roll down. This is represented in Fig. 170, which shows that the largest light sphere has undergone the least horizontal displacement because it travelled fastest, whilst the smallest heavy one has been carried furthest to one side. From this it is seen that if a suitable displacing movement can be applied to a plane, the feed can be spread into bands according to the size and density of its constituent particles. If these bands are collected into separate vessels as they leave this deck, the feed will have been segregated into three main products:

(*a*) Fastest-moving. Coarse light mineral (gangue).
(*b*) Medium-moving. Fine light mineral plus coarse heavy mineral plus partly unlocked particles (middling).
(*c*) Slowest-moving. Fine heavy particles (concentrate).

A particle light enough to respond mainly to the hydraulic influence of the flowing film of water moves down-plane with little horizontal displacement. A typical particle, unlike a sphere, will either slide or "skip" downward, rather than roll, provided it is reasonably free to move. Apart from the limited use of the automatic strake in concentrating metallic gold, continuous lateral displacement across the sorting plane cannot handle an adequate tonnage and is not used in the mill.

With the shaking table a reciprocating side motion is applied to the sloping surface or "deck" down which the pulp is streaming. If this shaking action was applied symmetrically in both directions across the stream, each particle would move an equal distance in each direction, and separation into bands would not occur. The displacing stroke must be applied gently, so as not to break the grip between particle and deck. The deck accelerates, and in doing so imparts kinetic energy to the material on it. Then the deck motion is abruptly reversed so that it is snatched away from under the particles resting immediately above it. These continue to skid sideways (across the flow) until their kinetic energy has been exhausted. It is therefore essential to provide a differential side-shake which builds up gently and then breaks contact between deck and load. This is provided by the shaking mechanism or head motion of the table. The slower the particle travels downstream, the further it slides sideways under the influence of the shaking motion.

Thus far discussion has been limited to a series of individual particles fed to the deck from one starting-point. If, instead, a layer several particles deep is fed from a starting-line, it becomes possible to handle a greatly increased load on the deck. The operating conditions have now changed. In the cross-section through such a layer (Fig. 171), as seen normal to the direction of shake, the mixed feed first stratifies itself under the disturbing influence of the shaking action. The smallest and heaviest particles reach the deck, the largest and lightest stay uppermost, with a mixture of large heavy and small light grains between. This arrangement exposes the large, light particles to the maximum sluicing force of the film of water as it streams down the table, a force that can be controlled in intensity by varying the volume of water used and the slope of the deck. It is thus possible to exert some degree of skimming action to accelerate the downward movement of the uppermost layer without disturbing those below.

The particles next to the deck are pressed to it by the material above, and therefore can grip it with greater firmness than would be given by their own unaided weight. They thus are able to cling during fast sideways acceleration, and are only freed and set skidding by the sudden reverse action.

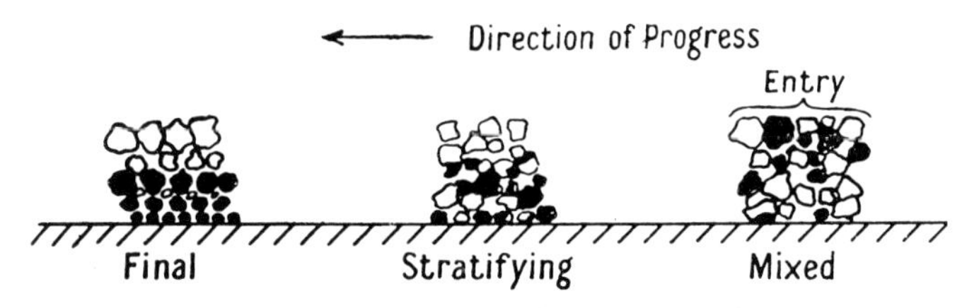

Fig. 171. *Progressive Stratification along Riffle*

The overlying particles have only a precarious hold. This aids the discriminating action of each stroke. The bottom particle travels furthest, breaks free at stroke reversal and is the first to skid. Those above it sway backward and forward and consequently receive less lateral movement. This accentuates the separating action by giving the bottom (heavy mineral) particles the maximum horizontal displacement per stroke and the upper (light gangue) grains the least. This aids the sorting discrimination. If the feed has been properly prepared by hydraulic classification, ensuring that all the grains have similar settling characteristics through vertical currents, film sizing can now take advantage of the variation in cross-section between the heavy and light particles in each stratum, sweeping down the lighter and leaving the heavier untouched. The particles thus segregated are then removed in separately discharged fractions, called bands, at the far end of the table's deck.

It would not be possible to form and maintain an evenly distributed thick bed of the kind called for by the foregoing considerations if a smooth plane deck were used. Riffles are therefore employed to provide protected pockets in which stratification can take place. They are usually straight and parallel with the direction of shake, but may be curved or slanted. The deck, instead of being plane, may be formed to provide pools in which the feed can stratify (see Fig. 152). The riffles must

(*a*) Arrest and spread the entering feed.
(*b*) Aid transmission of shaking action to their enclosed load.
(*c*) Expose the top layer of sand, after it has stratified, to the cross flow of wash water down the table deck.
(*d*) By suitably gentle tapering, promote delivery to succeeding riffles, wash planes or discharge points.

Thus (*a*) rules out as bad practice the use of "stopping" riffles set high above the rest, sometimes used to arrest and spread entering feed. If all riffles are not of similar initial height the stratifying action and transfer between them is upset. Smooth delivery is best achieved with a feed box integral with the moving deck, and aligned with the vibrator. It should let the feed down gently to the head riffles. Items (*b*), (*c*), and (*d*) are arguments against the use of curved riffles, which increase wall friction and upset stratifying action. A badly maintained mechanical action and deck coupling may mislead the engineer into re-designing his riffle plan, just as an incorrect stance may cause the unwary golfer to modify his swing instead of standing correctly.

In the standard Wilfley table (Fig. 172) the riffles run parallel with the long axis, and are tapered from a maximum height on the feed side (nearest the shaking mechanism) till they die out near the opposite side, part of which

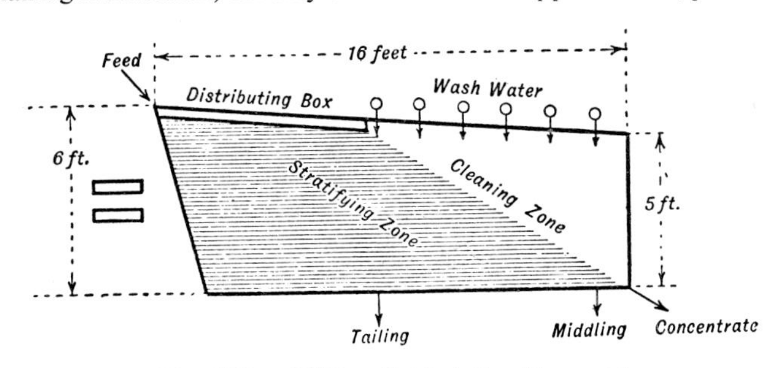

Fig. 172. *Wilfley Table* (after Truscott)

is left smooth. Where the riffles stand high, a certain amount of eddying movement occurs, aiding the stratification and jigging action in the riffle troughs. As the load of material is jerked across the table, the uppermost layer ceases to be protected from the down-coursing film of water, owing to the taper of the riffle. It is therefore swept or rolled over into the next riffle below. In this way the uppermost layer of sand is repeatedly sluiced with the full force of the current of wash water, riffle after riffle, until it leaves the deck. This water-film is thinnest and swiftest while climbing over the solid riffle, and the slight check and downpull it receives while passing over the trough between two riffles helps to drop any suspended solids into that trough.

At the bottom of the riffle-trough, then, the particles in contact with the deck are moving crosswise as the result of the mechanical shaking movement. At the top they are exposed to the hydraulic pressure of a controllable film of water sweeping downwards. In the trough of the riffle the combined forces—stratification, eddy action, and jigging—are arranging them according to density and volume. Provided the entering particles have been suitably sorted and liberated, good separation can be achieved on sands in any appropriate size range from an upper limit of about $\frac{3}{8}''$ to a lower one of some 300 mesh. The difference in density and mass between particles of concentrate and gangue determines the efficient size range which must be maintained by hydraulic classification or free-fall sorting of the feed. A further separating influence is applied hydraulically along each riffle as the water in it gathers energy from the deck's movement. As it gathers speed in the forward half of its cycle, the water flowing along the trough parallel to the axis of vibration is accelerated. When the deck's direction is abruptly reversed this flow is only gently checked relatively to the more positive braking force exerted on the skidding particles in the riffle. There is thus a mildly pulsed sluicing action across the table, in addition to the steady stream at right angles to it, down-slope. This cross-stream helps the particles to travel along the riffles.

Since separation depends to a large degree on the hydraulic displacement of the particle, its shape influences its reaction. Flakes of mica, though light, work down and cling to the deck, and may be seen moving nearly straight across, even at the unriffled end where they meet the full force of the stream.

Where there is no marked influence in density between the constituent minerals of a pulp, the shape factor aids a flat particle to move along the deck to the concentrates zone, and under like conditions helps an equi-dimensional one to move down-slope toward the tailings discharge. Shape factor can therefore help tabling in some cases, and be disadvantageous in others, depending on whether it reinforces or opposes differences in size between the classified particles of value and tailing.

Applied Principles

The principles which underlie separating action, together with the methods by which the combined energy of the sluicing water and the mechanical shaking can best be exploited, may now be summarised.

(*a*) Feed must be in a size range chosen with regard to the particle size, density and shape of each mineral being separated.

(*b*) Riffle spacing must promote stratifying action and reasonably uncrowded sluicing by the transverse currents set up by the deck's vibration.
(*c*) Shaking must accelerate smoothly from a gentle start in the direction of axial discharge.
(*d*) The forward part of the stroke must end with enough abruptness to promote some skidding of the particles in the direction of discharge.
(*e*) The return part of the stroke must commence abruptly, to snatch the deck from under the loosened or skidding particles.
(*f*) The return stroke must die away gently.
(*g*) Stroke amplitude and frequency must be chosen to ensure adequate grip of the load at the start, sufficient transfer of energy to the riding particles, and enough shock at the forward end and gentle reversal at the return end of the cycle.
(*h*) Riffles must be designed, spaced and tapered so as to promote stratification in the chosen size range of feed, and prevention of indiscriminate down-stream coursing of the pulp (streaming across the table).
(*i*) Riffles must not set up side-wall restraint sufficient to prevent free sluicing action across the table.
(*j*) The solid-liquid ratio of the pulp must be kept steady, and auxiliary water service must be suitably and smoothly distributed so as to maintain a correct balance between cross-streaming and down-streaming.
(*k*) The requirements of (*j*) may be aided by correct table tilt about its long axis.

In amplification of (*a*) the effect of specific surface must be reckoned with. If the voids between particles in the feed are allowed to become congested by undersize material and slimes, the solids tend to form a cohering mat on the deck. Stratification is then impeded and sluice action only ruffles the surface of this mat. When the feed is properly prepared and the table correctly controlled sluicing action should be delicately selective, not forced. Increased use of hydraulic water, tilt and violence of shaking action are not efficient substitutes for careful preparation of the table feed.

It will be seen when the chapters concerned with the physical chemistry of pulps has been studied that the rheology of a fluent suspension of finely ground minerals in water can to some extent be considered as that of a plastic system undergoing stresses and strains. As such, its behaviour is influenced by such factors as mine water acidity, temperature, and the interacting surface tensions of the components. Development of the most efficient scheme of tabling (deck design, vibrating mode and cyclic acceleration) should, but rarely does, take these factors into consideration. Shaking tables were developed empirically before much was known of surface physics, and practice in their use varies widely. A draughtsman's design cannot be expected to envisage the flocculating effect of mine water in another country. Hence, empirical claims for the special virtues of this or that variation in tabling need critical examination when new selection is being made. The basic principles outlined in this section should be checked in these matters by test work subjected to the local conditions which will be encountered.

Mechanical Details

Shaking action on the table is controlled through an adjustable head motion, which converts the rotary drive of the motor into a differentiated horizontal stroke. In the Wilfley table (Fig. 173), one toggle (*B*) is seated against a fixed

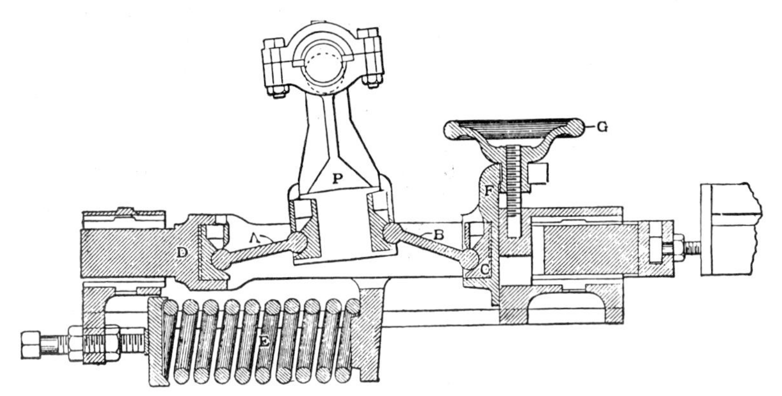

Fig. 173. *Wilfley Table. Head Motion*

Fig. 174. *Head Motion of Holman Table*

mounting (*C*) which is bolted to the foundations. The other toggle (*A*) bears on a yoke (*D*) which is connected to the table deck. The toggle system is held together by the spring (*E*), and driven by the eccentric and pitman (*P*). At the beginning of the forward stroke the toggles are at their flattest and the spring in maximum compression. As the pitman rises, the toggles steepen out. At the top of its stroke they are at their most acute angle and the table has reached its maximum speed. The pitman now descends, flattening the toggle angle and abruptly reversing the direction of the table. The abrupt reversal occurs at the end of the forward thrust of the deck, and at this moment the particles resting upon this deck skid forward. The decompression of the spring aids the forward stroke to some extent. This spring is adjusted until its increased compression just quietens the knocking sound of too loose a set-up. For coarse feeds the Wilfley table is run at 240 full strokes/minute and a stroke length $< 1''$. For finer sands speed can be raised (by changing pulleys) to 300 r.p.m., but the stroke must be shortened. This adjustment is made by means of the handwheel (*G*) which moves the sliding piece (*F*) (Fig. 173). The toggle (*B*) in its seat (*C*) can thus be raised or lowered, shortening the stroke (deck movement) from the full $1''$ down to a minimum of $\frac{1}{2}''$. In operation the table uses $\frac{3}{4}$ h.p. but may have a starting torque of twice this amount.

Various types of head motion are found on James, Deister, and other tables. The Fraser & Chalmers table is moved by an electro-magnetic motion (Sherwen).

Various materials of construction are available for the deck and its riffles. A linoleum deck with hardwood riffles secured to it by copper tacks, is most commonly used. Rubber decks with rubber riffles have a good working life and freedom from leakage, but must be protected from full sunlight. The rubber is bonded to the softwood underdeck with adhesive. On slimes tables, brass riffles are used for the shallow section, above linoleum. Some use of aluminium decking with metal riffles has been made. Where a large tabling area is required, double-decked tables have been successfully employed, suitable strengthening of the structure and increase of driving power being provided.

Triple-deck tables are also manufactured. Lightness of the deck, which is shaken 15,000 or more times hourly, has a direct bearing on working life. In laboratory tests, the honeycomb aluminium sheet used in aeroplane construction is excellent, though costly.

Some of the many riffling plans are shown in Fig. 175. The variations in feed preparation from one mill to another are such that sound judgment regarding the relative merits of the systems is difficult. Broadly, a riffle plan in which the feed box moves with the deck and lets down the pulp smoothly on to the top row of riffles is best. These top riffles should be high enough to check any tendency of the newly arriving feed to scour its way down the table without having time to settle and to begin its stratification in the sluicing valleys between the riffles. Some operators check this tendency of the pulp to rush down-slope by making every other riffle (or fourth one) at the feed end twice the height of those between as in Fig. 175 (*d*). The desired effect can usually be obtained more easily by careful control of the classifying and feeding arrangements, thus ensuring that all the riffles are equally loaded instead of overloading part of the deck at the expense of the low riffles.

Again, some operators consider that in place of the standard riffle (*a*) which runs parallel to the shaking motion of the deck, efficiency is gained by forcing the pulp to climb slightly on its way across. If the true function of the sluicing space between the riffles is carefully helped by good feed preparation, and by proper adjustment of deck tilt and wash-water the author considers that deviations from straight-through sluicing have little to justify the complication, wear and loss of sorting capacity they introduce. If, for any over-riding reason it is not possible to provide a steady rate of feed at a consistent solid-liquid ratio, of correctly classified sands, then deviations from the standard riffle plan (Fig. 175 (*a*)) acquire some rather doubtful justification. On the whole, however, an expert application of basic principles is better than such compensating action.

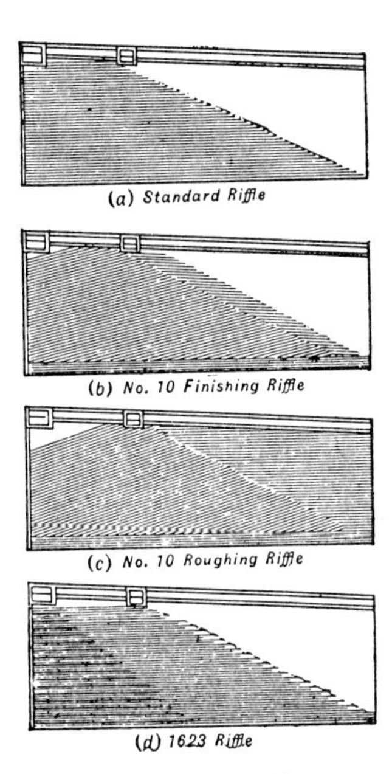

Fig. 175. *Riffle Plans*

The table deck is supported on wood carried on steel framing and a running adjustment is provided for altering the tilt. In some tables a truss mounting is used instead of rocker action. With this trussing, the whole deck rises very slightly during the forward stroke and this may have a slight tossing effect upon the load of sand. It is important for good efficiency that a table be set so that its trusses are accurately aligned with the vibrator, that they move in a vertical arc, and above all that they do not reach a perpendicular position at the forward end of the longest stroke employed. In installing a table to comply with these requirements due allowance should be made for wear and back-lash.

Full table efficiency is only possible when the mechanisms, linkages and slopes are accurately set and maintained. It is usual to have the long axis horizontal, as this best helps sluicing action along each riffle. The cross-tilt mechanism should work freely so as to maintain an even slope under working conditions and avoid warping strains. There must be no resistance to vibrator action from avoidable friction. Splash, rust, seepage and neglect can produce cumulative wear in the under-deck supports and thus damp down the desirable vibration with its compound cycle of flow and return. If there is loss of alignment at any pivot, rocking bearing or truss the deck is distorted in the vicinity. There is usually enough abrasive material under the table to regrind such a point smooth but this is a poor substitute for regular check to ensure that moving parts are truly parallel to the vibrating action. Backlash dulls the thrust either by slack coupling which substitutes a jerk for the correct steady build-up and final snatch of the forward stroke, or by letting the return stroke over-run so as to produce a jerk as the slack returning deck is hit by the forward push of the next forward stroke. This sloppy motion cannot be correctly compensated merely by tightening the compressing spring in the head motion or by increasing the inclination toward the vibrator of the springs in a truss-supported deck, useful as these aids may be.

An indication of bad alignment is a high starting torque. With the table in good trim this is moderate. With misalignment there may be fracture of the connecting bolts or blown fuses. The probability that once the table has started up it runs smoothly may mislead the operator. Another useful indication of local trouble is obtainable by laying the unloaded table horizontal and observing the progress of a small coin along selected riffles. If the coin zigzags from wall to wall or progresses erratically, a misaligned, binding or worn support is indicated. Though the deck is thought of as a rigid body, the effect of millions of rapid impulses during its working life produce local "soft spots", if maintenance is neglected.

A standard 6′ × 15′ single-deck table can handle from $\frac{1}{2}$ ton/hour to 2 tons/hour, the safe limit rising with coarseness of feed and with decrease in the percentage of concentrate to be removed.

Table Controls

These are of three kinds. The first are:

(*a*) The type of riffling used;
(*b*) the material of which it and the deck surface are made;
(*c*) the acceleration and deceleration during one revolution of the drive;
(*d*) the mode of presentation of the feed.

These are major points of design, and can only be varied by reconstruction. They should therefore be decided by tests made on the material which is to be treated, before the table is installed.

Next come running speed and stroke amplitude. Speed depends on motor and pulley system. Stroke is adjusted by varying the toggle spread, or in the Sherwen, by altering the amplitude of the solenoid vibration. With most tables, stroke can be varied while running. It is desirable that these adjustments should be settled and fixed as early as possible in a working plant, as constant "fiddling" with the controls leads to inefficiency and working trouble. If a small alteration is needed, it is easily made at any time during the working life of the table.

Third there are the operating controls which must be made by the shiftsman as the need arises. These are:

(*a*) The tilt across the table.
(*b*) The solid-liquid ratio in the feed distributing box.
(*c*) The wash water.
(*d*) The position of the product cutters.

The interplay of these controls will be understood if the operation of a typical shaking table in a plant is considered.

Assume that the table is operating at half-tilt (for a Wilfley, which can be adjusted from flatness to 1″/foot, this is ½″ per foot run from feed to discharge side) and that the bands are spread as shown in Fig. 176.

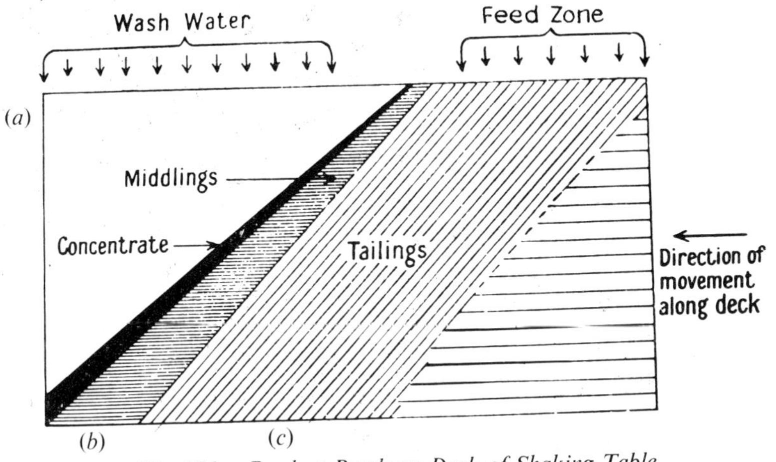

Fig. 176. *Product Bands on Deck of Shaking Table*

As the shiftsman may have to supervise a large number of tables, the controls must be of such a general nature that they will deal with any changes which may occur during his absence. The only thing which can change during that time (accidents excepted) is the nature of the feed. This feed is composed of sand and water having the following variables:

	Variable
Sand . . .	Grain size, degree of liberation, proportion of concentrate to gangue.
Feed Water . .	Volume.
Water-Solid Ratio.	According to steadiness of discharge from classifier and/or feed to classifier ahead of table.

The shiftsman must operate his table so as to produce a concentrate of specified grade, a gangue from which as much value as possible has been removed, and a middling. The concentrate and the gangue are being withdrawn from circuit and must be kept to the specified standard. The shiftsman can hold for retreatment the material in the middlings band, and he therefore tries to set his table so that any irregular running during his absence will only affect that band.

His chief preoccupation is with "band wander". If the volume of feed changes, the work done in separating it also varies and this causes the concentrates and middlings bands to rove slightly, either uphill or down. If the solid fraction of the feed increases, or if the sands become richer in value, these two bands broaden correspondingly. The middlings band tends to push the concentrates band up and a little away from the ends of the riffles and further over along the cleaning plane from (*b*) toward (*a*) (Fig. 176). The richer part of the middlings now begins to report with the concentrates product. This lowers shipping grade but does not lead to loss. If the bands wander in the other direction, from (*b*) toward (*c*) the poorer middlings fall with the tailing, and this results in a definite loss of value.

Similar considerations acting in reverse would affect operation if the feed values dropped, or the rate of solid feed was reduced. The concentrates and middlings bands would shrink and wander toward the tailing cut. Concentrate grade would improve, but a certain amount of true concentrate would report as middlings (whence it would later be recovered) and some good middlings would be lost as tailings.

To guard against this, the shiftsman keeps his bands spread during normal operation, to a comfortable working width. The flatter the tilt of the deck, the longer the stroke and the gentler the downward wash of the wash water, the wider will be the bands. If, as may be the case, more than one heavy mineral is being removed separately, a fairly wide spread is needed to assist cutting of two or more types of concentrate as well as middlings and tails. The shiftsman must provide enough wash water for adequate cleaning at all times, and this wash water may have to aid an inadequate supply of feed water if any classifying or distributing irregularities take effect during his absence. When he has adjusted the wash water, he varies the tilt until the bands are adequately displayed. This upsets the film action a little, and may call for a secondary adjustment of wash water and tilt. Nothing now remains to be done, but to set the product cutters in position in the *a* to *b* and *b* to *c* launders surrounding the table . Whatever happens, the concentrates must not wander completely through the middlings band to the tail, and *vice versa*. The middlings is the steadying, protecting factor in the scheme. The cutters are therefore set so that a certain amount of gangue is included with the middlings cut on the lower side, and a low-grade fringe of concentrate on the upper side, the middlings band being therefore set somewhat too widely. The amount of safety setting thus made depends on the overall operating efficiency of the system and the period between inspections.

Stress has been laid on the importance of classification. This produces coarse, medium, and fine sands which are treated on decks with standard riffles, the main variation being that the finer the feed, the faster and shorter is the stroke used. Somewhere below the −150 to 200 mesh sizes decks with a different catchment plan are used. A typical *slimes table* as these are called (Figs. 177/178) has a series of planes rather than riffles, on its linoleum-covered deck. Plane *A* (Fig. 177) is the major settling section, some ten feet along and four feet down the table. Here the finest particles settle and stratify. There is no check of flow till the area *B–B* is reached, the settled rough concentrate being pushed toward the discharge from here. Gangue overflows to plane *C*, where middlings are held and sent progressively to *D*, *E*, and *F* for a repetition of the original action. The concentrate and/or middlings produced on the slimes table may require further treatment on buddles or in kieves before shipment.

One possible variation which should be mentioned must be guarded against. If water used in treatment is caught and recirculated, it may become contaminated with colloidal iron and aluminium hydrates. These could, by coating the deck and the particles with a tenuous layer of slippery slime, upset the frictional grip of deck on particles, and also the stratification betweeen riffles.

Slimes Treatment

Here the important particle size is that lying between true Newtonian and Stokesian movement, in the zone covered by Allen's equation

$$\rho = \mathrm{K r^n} \, p \mu^{2\text{-}n} \, \nu^{n} \qquad (14.1)$$

where ρ is the resistance of the fluid to the motion of the solid, K a constant modified by shape and velocity of the particle relative to the fluid, r the radius of the equivalent sphere, n a coefficient of velocity v, p the fluid density and μ the kinematic velocity. Practice varies considerably as regards rate of feed and range of size fed to a given type of concentrating appliance. Table 25 summarises a cross-section representative of reported practice.

TABLE 25

	Average lb./ft.² *per hour of feed*	*Particle size-range*	*Types of Concentrate*
(*a*) Newtonian	60	20–200 mesh	Shaking tables
(*b*) Allen	30	150–400 mesh	Slime tables, vanners, buddles, strakes
(*c*) Stokesian	down to 7	270– 30μ	As (*b*)

The basic principles which should be observed in slimes treatment may be summarised:

(1) Horizontal area required varies inversely as grain size of feed.
(2) This area is mainly determined by the economic mesh-of-grind.
(3) The finer the feed, the gentler and slower must be the action.
(4) Feed distribution must be even and channeling avoided.
(5) The coagulative-dispersive characteristics of the feed affect its response to gentle flowing action.

The concentration criterion for a mixture of quartz and cassiterite is 3·5 and at less than some 50μ of particle size is not sufficient to give crisp separation. That for a quartz-gold mixture is nearly 9, and provided the gold has not become flaky during grinding good action is still possible at this size.

In assessing delivery from any shaking table mistakes are possible if control is based entirely on visual appearance. This is particularly noticeable with the common practice of running slimes tables with their vibrators in reverse. The reason usually given is that much more efficient separation is thus obtained, as judged by the width of the concentrate band leaving the table. If, however, the actual amount of concentrate leaving the table under this condition is compared with that when normal vibrating action is used, there is little or no difference in weight or grade. All that reversal has done is to retard delivery rate and hence broaden the retained band.

Sampling Check

Although this is concerned with tabling control, it was not discussed in that section, since it requires laboratory tests not possible in shift running. Where a major tabling operation justifies the cost of sampling check, it becomes important to have detailed information of the rate of change in concentration along the discharge edges. Samples may be collected for a timed period in a trough divided every few inches into compartments, a small head sample being taken at the same time. Scrutiny, size analysis and assay then present a picture of the value-change along the discharge area, and of the quantity represented at the prevailing discharge-rate. It is not uncommon for the most serious loss of readily recoverable value to be disclosed as coming from feed running straight across the table from entry to the tailing discharge nearly opposite, because it has not been properly arrested and bedded into the riffles. Poor classification of the feed is also disclosed, when it is seen that material which cannot be handled at the setting appropriate to the main-size range of the feed is present. Various further improvements become possible once the technical supervisor has reliable facts to work from. Cutters can be positioned better, and the amount of middling best retained for re-circulation brought under control.

Tabling—General

The maximum particle in the feed must not be thicker than the riffling or more than one-third of a riffle in width. If tables are used in series, the first group handling a given grade is set to produce a low-grade concentrate containing all recoverable values, and a stripped gangue for discard. This is called *roughing* and the low-grade concentrate is then passed to a smaller group of tables for finishing, or cleaning, the tailings from these tables being given further treatment, possibly by returning them to the rougher feed, as in Fig. 179.

This is called a "rougher-cleaner" operation. Roughing tables have for their primary task the rejection of barren tailing and operation can, at need, be simplified to two-product work (a rejected gangue and retained low-grade values). This last will then go on to cleaning tables to be separated into two bands—a final concentrate and a middling requiring further treatment. Tables worked thus require less skilled attendance than where band-wander in a three-band separation must be checked, since most of the gangue is removed at the start, and operation such as that shown in Fig. 179 is steadied by a wide-spread return middlings band, particularly if re-grinding is performed between cleaners and classified re-distribution (not shown in diagram.) With single-pass treatment any change in ore texture or richness calls for

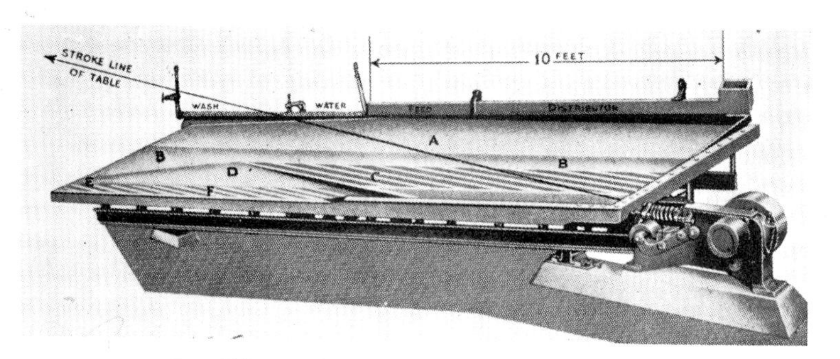

Fig. 177. *Holman Slimes Table—Right Hand*

Fig. 178. *Holman Slimes Table at Work*

trimming adjustment by the shiftsmen. Where there is a substantial re-circulation of middlings fluctuations in the feed tend to be taken up without calling for such close supervision.

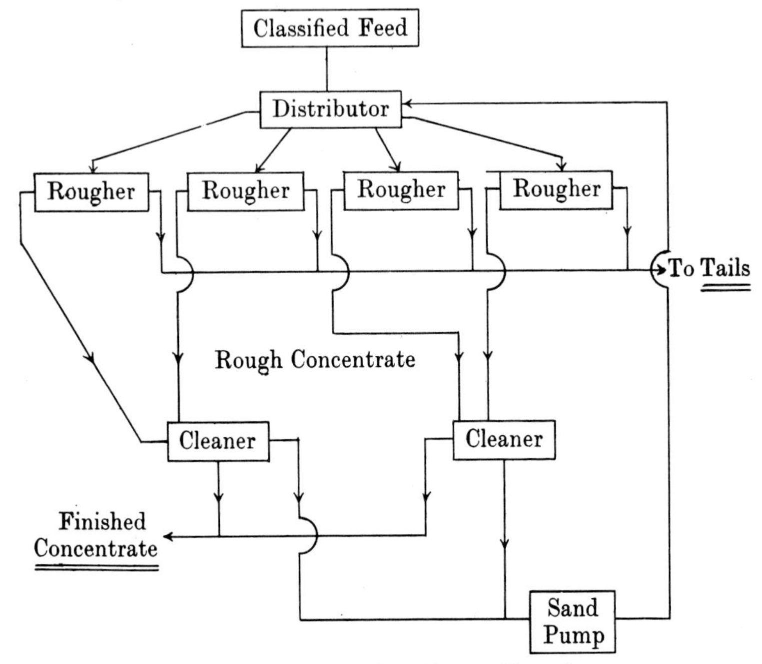

Fig. 179. *A Rougher-Cleaner Flow-sheet*

Tables set for roughing work are fed more heavily than when a complete operation is carried out in one pass since the operator has only to make sure that one product (usually the tailing) is finished efficiently. More water is used, and a greater tilt and stroke length, thus moving the sand vigorously along and across the deck.

A further development usually found in tabling such ores as cassiterite is the grouping of the film-sizing appliances, each group being fed with ore which has been graded (probably through a series of hydraulic classifiers), in some such manner as in Fig. 180.

Each group of tables in such an arrangement can either be operated in parallel (one-pass treatment) or series (rougher-cleaner).

A further refinement may be practised with the coarse tabling group. In order to avoid excessive production of "slime tin" (a cassiterite so finely comminuted that much of it is lost with the tailings), the feed may be ground and tabled in stages. Sliming loss occurs when fine particles settle very slowly and are washed over the table without being trapped. It is therefore sometimes advantageous to remove liberated concentrate early in the treatment, using some such form of stage concentration as shown in Fig. 181.

It is also possible to use this scheme to make a stage rejection of barren tailings.

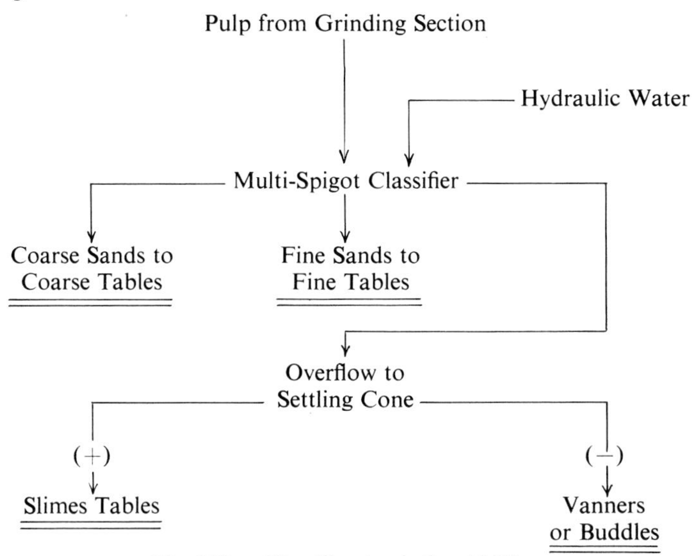

Fig. 180. *Classification before Tabling*

The coarser the feed to a table, the longer and slower should be the stroke. One operator can take care of between ten and one hundred tables. Tables used for roughing should have more riffling than those used for finishing the concentration. They are required to hold the feed in teetering columns between their riffles while the motion causes the particles to be jigged and stratified. The deck of the table must be watertight.

Riffles are made of hardwood such as oak, of soft pine, or of plastics, e.g. rubber. The maximum particle size usually treated by tabling is $\frac{3}{8}''$ and sands as fine as 300 mesh are handled. Frue vanners are given even finer feeds of tin-bearing pulp, there being no simple alternative to gravity concentration for separating this mineral.

The general tendency in modern mills where tables play an important part is to "rough" a run-of-mill ore pulp which has been released from the grinding circuit at a controlled upper size limit. The tailings from this roughing are discarded, and the concentrates are classified and then given a cleaning treatment.

Beside their use on cassiterite, tables are employed to treat free-milling gold ore, auriferous pyrite, barytes, fluorspar, foundry dross, coal "smalls", etc. A single miniature table is sometimes used to receive part of the tailings from some other concentrating process (e.g. froth flotation) to aid the shiftsman by showing him what minerals are leaving the circuit. Because tabling depends on a good gravity difference between the minerals being separated, care must be taken to avoid any unnecessary grinding of values which are sufficiently unlocked to respond to treatment. Particles which are so fine that they do not fall rapidly through still water cannot be treated quickly or in quantities on tables, and even under the most favourable circumstances poor recovery of a low-grade concentrate is probable. In all treatment based on gravity the rule is to remove the desired mineral at the coarsest mesh of liberation which can economically be justified.

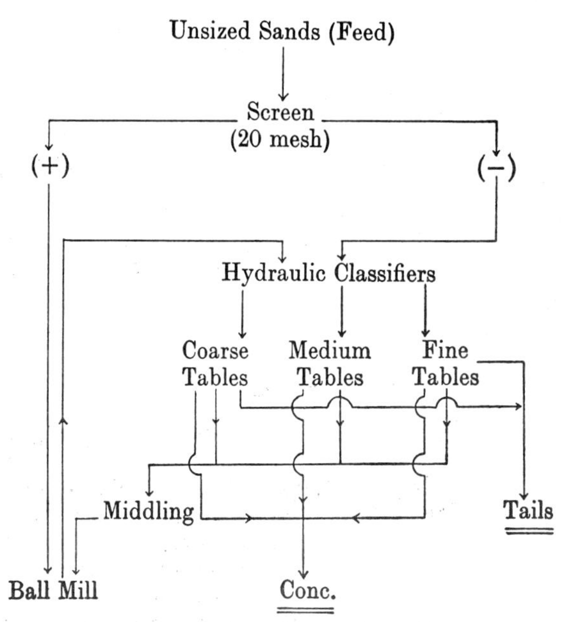

Fig. 181. *General Flow-sheet. Stage Concentration*

Tables make a middling product which, as has been noted, may contain true tails and true concentrates. If this middling is recirculated, it grows in volume, and the time arrives when it crowds over to join one of the other products. Simple recirculation is not, therefore, the best way to handle a true middling. If it consists of incompletely unlocked particles, part gangue and part concentrate, these can only be freed by further grinding.

If the middling is partly composed of some mineral or minerals of intermediate density, it should either be cleaned up and marketed, stockpiled, or sent to waste. This involves separate retreatment of the middling, not simple recirculation to the feeding head of the table producing the middling.

It is rarely feasible to make a high ratio of concentration in a single pass without paying the penalty either in the form of dirty concentrate or of tailing losses.

The ratio of concentration R is expressed in the simplest forms as weight of feed F divided by weight of concentrate C. In cases where it is not convenient to weigh feed and concentrate, a formula based on the assay grades of feed

(f) concentrate (c) and tail (t) can be used.

$F = C + T$ (T being weight of tailing) (14.2)

and $Ff = Cc + Tt$, (14.3)

from which is derived $R = \dfrac{c-t}{f-t}$. (14.4)

The importance of this concept arises in the assessment of recovery Q, which can be expressed as a percentage thus:

$$Q = 100\,\frac{Cc}{Ff}, \qquad (14.5)$$

where the actual weights of C and F are known or

$$Q = 100\,\frac{c(f-t)}{f(c-t)}, \qquad (14.6)$$

where assays only are known.

So far as general considerations permit, a series of operations in which grade of concentrate (or leanness of tailing) is reached in steps is preferable to high-ratio treatment in which throughput must be kept low if grade of product is to be maintained.

Friable minerals, powdery oxides, and light flaky material are not amenable to table treatment. Shape difference between desired products is a help when the particles of the heavy mineral tend to be flattish and those of the gangue equi-axial. Even then the difference in density should not be less than 1·0 between concentrate and tailing and more in the case of fine sand, for gravity separation to be both efficient and cheap.

Dry Tabling

However carefully water used in gravity concentration is conserved, any quantity up to fifteen times the weight of material undergoing treatment may be in circulation. This water becomes foul with use, and if slimy, upsets frictional grip of the sand to the deck.

Methods of separation based on gravity use least water in dense-medium separation and most in the treatment of very fine slimes. They can only be used where a plentiful supply of unfrozen water is available during the operating season.

Sometimes the product would be injured by wetting, or may contain soluble salts which preclude water treatment. Again, the cost of drying a wet concentrate may be prohibitive. In such circumstances the possibility of dry jigging, "dry-blowing", and dry tabling can be considered. In the pneumatic jig air takes the place of water as the pulsating medium used to dilate the bed of ore and promote stratification. Various forms of jig have been developed for desert use.

Pneumatic tables use a throwing motion to move the feed along a flattish riffled deck, and blow air continuously up through a porous bed. The general principles of separation are similar to those applied in wet tabling. Use is limited, and the recorded data of performance are somewhat confusing.[5] A few coal plants use pneumatic methods, which have the advantage of producing a dry product. The upgrading of asbestos by pneumatic tabling is another important commercial application.

Hudson[6] has observed that whereas in wet tabling the particle size increases and the mineral density decreases from the top of the concentrates band towarc the tailing (in the absence of marked shape differences), on an air table both particle size and mineral density decrease from the top down, the coarsest particles in the middlings band having the lowest density. Thus, air tabling is similar in effect to hydraulic classification (the air blown through the porous deck exerting an effect similar to that of the rising water in the classifier), while wet tabling does not develop such an effect. This has, of course, beer developed by pre-classification of the feed in the case of the wet table. Logically, in air tabling a similar pre-treatment in reverse is indicated (sluicing before tabling). In experimental work Hudson did this by sizing the feec through a launder. The dry sands stratified during descent to the table anc fanned out as they fell from the launder, the coarser particles being thrown further horizontally during fall than the finer ones.

In dry blowing, a horizontal air current is used to displace particles falling vertically, each of which is displaced laterally in accordance with its rate o drop.

A recent development of the Lavodune principle[8] uses air instead of wate as the transporting and dune-forming agency.

Some Gravity Flow-sheets

The methods upon which the consulting engineer designs an all-gravit flow-sheet are dictated by the behaviour of the ore under test. Despite a apparent wide variety of treatments, he must always consider certain requirements for efficient work. These may be summarised:

1. Minimum further grinding after liberation.
2. Sizing before jigging (to limit size variable and emphasise densit variable).
3. Classification (usually) before tabling (to develop rolling response t film-action).
4. Grading of sands for separate handling on coarse, medium, and fin tables, possibly followed by vanners, buddles, or spirals.
5. DMS rejection of barren rock at the earliest possible stage.
6. Table rejection of roughed-out gangue at earliest possible stage.
7. Appropriate special handling of true middlings.
8. Use of middlings bands to keep true concentrates well separated fro true tailings on the table.
9. Stage comminution with stage concentration wherever the extr handling justifies its cost.

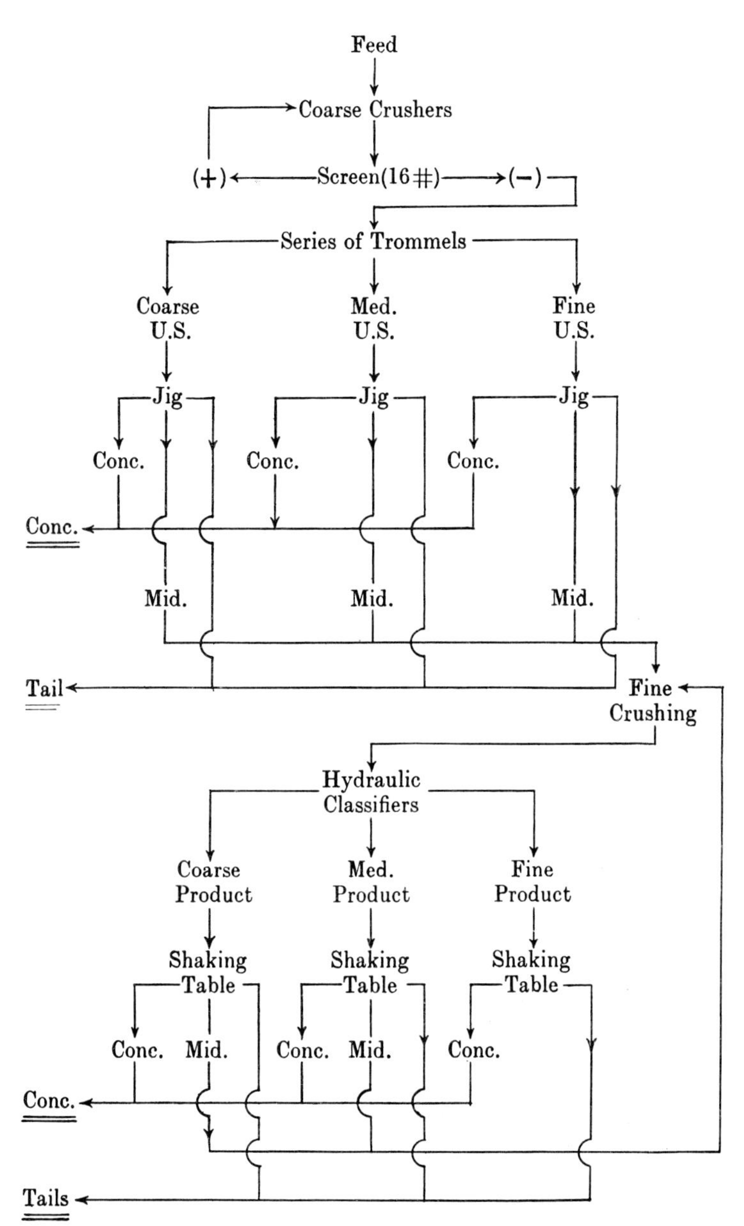

Fig. 182. *Jigs followed by Tables*

Fig. 182 gives the flow-sheet for the combined treatment of a sulphide copper ore using jigs and tables. The main minerals are chalcocite, bornite, and a siliceous gangue. For a number of years students at the Royal School of Mines have worked parcels of this ore through the old pilot plant, and have obtained excellent recovery of high-grade products, mostly in the jigging section. This flow-sheet exemplifies the value of stage crushing with stage concentration for material breaking free at a coarse enough mesh to permit jigging.

Fig. 183 shows the series up-grading arrangement of groups of Humphreys spirals.

Fig. 184 illustrates a Cornish sand treatment plant, in which rehandling must be unusually thorough because an important part of the value resides in the slimes, and there is no accepted alternative to gentle gravity separation for its treatment.

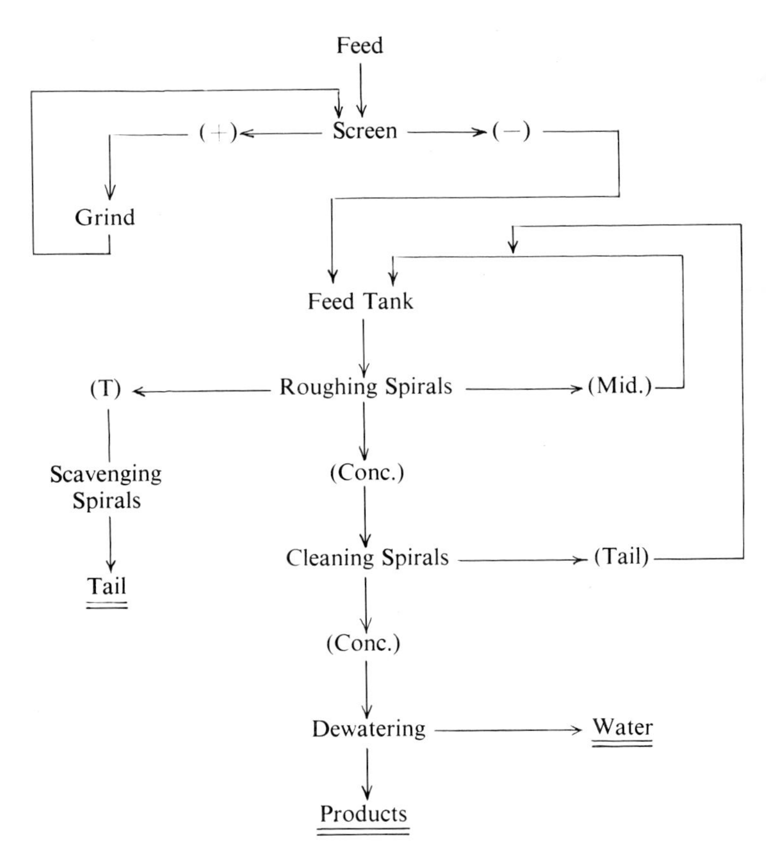

Fig. 183. *Stage Concentration on Spirals*

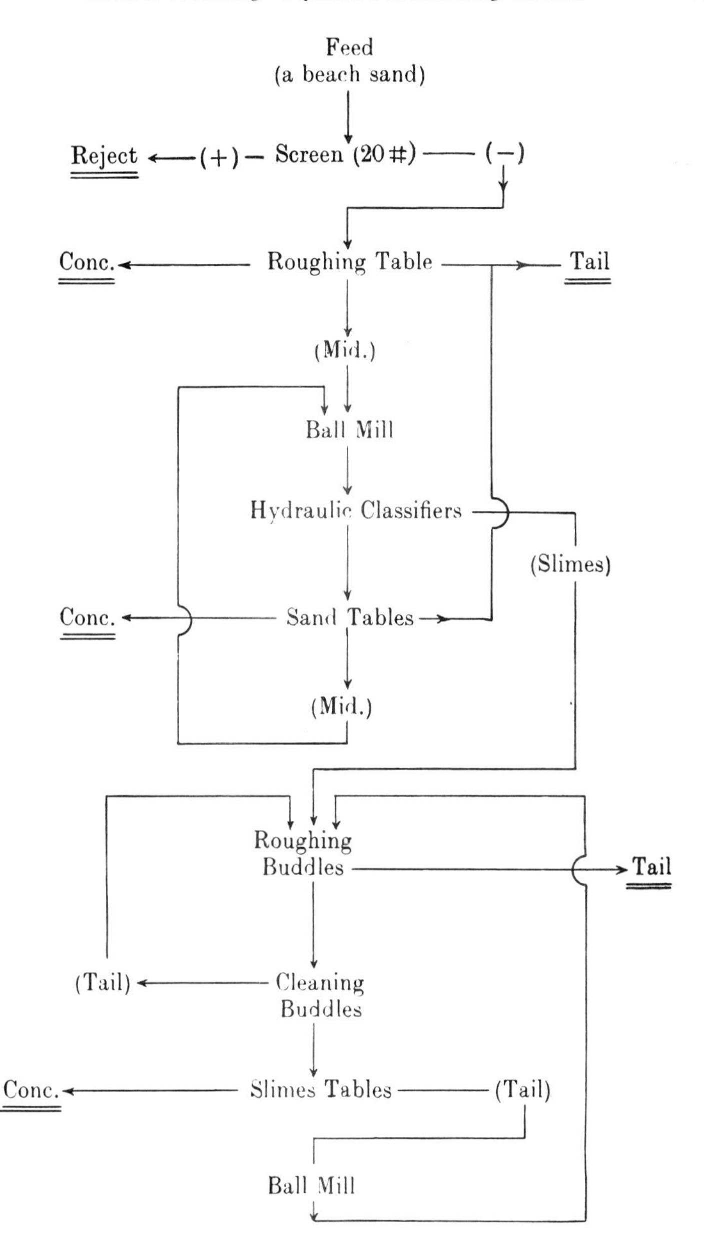

Fig. 184 *A Tin-bearing Beach Sand*
(*Dewatering circuits and surge arrangements not shown*)

References

1. Pryor, E. J. (1963). *Dictionary of Mineral Technology*, Mining Publications.
2. Chaston, I. R. M. (1961/62). *Trans. I.M.M.* (*Lond.*), 71.
3. Stewart, A. L. (1961). *The Mining Magazine*, Sept.
4. Pullar, S. S. (1963). *Proc. Aust. I.M.M.*, March.
5. Knapp, E. A. (1953). *Recent Developments in Mineral Dressing*, I.M.M.
6. Hudson, S. B. (1962). *Aust. I.M.M.*, Dec.
7. Knapp, E. A., and Sweet, C. T. (1953). *Trans. I.M.M.* (*Lond.*), I.M.M.
8. Conta del Fa, M., and Ferrara, G. F. (1963). *I.M.P.C.* (*Cannes*), Pergamon.

CHAPTER 15

PHYSICS AND CHEMISTRY IN ORE TREATMENT

Introductory

During the past few decades a radical change in the approach to inorganic chemistry has occurred. This is due to an increasing knowledge of the valences, structure, and bond strength of crystalline compounds. The forces which integrate and largely determine shape, behaviour, and the energy used in producing a change of compounded structure have been worked out for many substances by precise mathematical methods. Physical and inorganic chemistry overlap.

The work of such pioneers as Sidgwick[1] and Pauling[2] brought new concepts of structural order to the chemical speculations of earlier workers who had not the aid of radiography and spectrography which are today in common use in research. The nature of a chemical bond can now be classified on considerations of valence, ionic, covalent, and other forms of coupling, strength, molecular shape, and crystal packing. The thermodynamics of reaction, synthesis, and dissociation, the rate of crystal growth or degeneration, and the chemical reactions possible in a given system are governed by forces assessable in mathematical terms.

The segregation of valuable minerals into ore bodies during geological aeons does not finish with the production of neat and formal end-products. A metal will balance its hydrate, silicate, carbonate, and sulphide radicals in various ways, and intruding ions of other elements have time, opportunity, and sufficient directing energy to distort the lattice structures of the almost insoluble compounds that go to make up an ore complex. A general name for a given mineral species has not, therefore, the same identifying significance as would apply in the case of a fairly pure chemical compound. Classification by name is not a guarantee of identical behaviour of, for example, galena drawn from different deposits. Chemical analysis is helpful in the study of a mineral, but is only a starting-point in the investigation of its nature. The distribution of the elements of which it is mainly composed gives some guide to its bond strength, but this is modified by any non-conforming ions in the lattice and by the regularity or random dispersion of their distribution. The particles produced by comminution have been severed from a rock formation which for millions of years has been reducing its potential and free energy to a minimum. Grinding has given these particles a new stimulus and disturbed their structural bonds. Concentrating processes can be used to exploit the energy differences between the various mineral species thus liberated. At the fundamental level, valency theory shows where bonds are weak or strong, and this suggests promising lines of attack where changes of state must be induced as a prelude to concentration or rejection. Among the techniques by which mineral structures can today be studied are:

Thermodynamics. Osmosis, vapour pressure, boiling and freezing points, adsorption isotherms.
Electrical. Conductivity, e.m.f., capacity, magnetism, electrostatics, electro-kinetics.
Spectroscopic. Emission or sorption of electro-magnetic radiation.
X-ray study. Diffraction, etc.
Radio-activity and radio-tracing.

Nearly all treatments leading to the concentration of desired ore constituents are applied to a finely ground suspension of ore in water. Consideration must now be given to the manner in which individual particles are dispersed in this pulp, and to the way in which their chemical and physical nature may be affected by wet grinding, exposure to air, and other factors. In the course of treatment it is essential to use, and if necessary, to develop exploitable differences between particles of valuable mineral and gangue. Chemical extraction calls for a working knowledge of chemistry, reaction rates and elementary thermodynamics. An understanding of physical chemistry is important to the scientific study of particle dynamics and the control of delicate physical and chemical forces acting at the surfaces of finely ground particles. It also explains the dispersion or agglomeration of particles, their treatment and settlement, smooth transport, separation of clean water from settled solids, and effective display of the true particle surface involved in a chemical or physical reaction.

The available energy (free and potential) at the surface of a mineral particle governs its methods of attaining equilibrium with its surroundings. The expression "surface tension" is used to aid the mathematical expression of this surface energy.

A characteristic of the ordinary mineral particle is its almost complete insolubility in water. (Exceptions such as alkali salts mined in desert regions do not enter this discussion.) The discriminating controls used in leaching and flotation are applied at the interphase between the solid particle and the surrounding fluid, whether the latter be liquid or a froth. Electro-magnetism is discussed in Chapter 18. It is therefore important to know something of two-phase and three-phase systems made up of air, water, and rock particles.

This chapter is designed to describe and define the main forces used in mineral technology to separate (or concentrate) mineral species by methods in which the use of gravitational force is subordinate to chemical, electrical and surface-physical discrimination.

In gravity concentration the particles are but little affected by the chemical "*climate*" in the transporting fluid. Their direction of movement is manipulated by methods which exploit mass and area. In comminution prior to chemical extraction adequate exposure of the species which is to be dissolved dictates the optimum particle size. For flotation (the prefix "froth-" is dropped from this point on) the particle is rarely as coarse as 60 mesh (say 250μ) and may be only a few microns in diameter. Direct gravitational forces play a minor part. Differentiation of their surface energies provides the discriminating force used to separate particles of concentrate from gangue.

Reactions affecting the surfaces of "insoluble" particles—nothing is, of course, completely insoluble—take place across the interphase, the transition zone between the characteristically solid and liquid phases. In the case of a gas-liquid-solid system, this is a complex drawn from all three components or phases. It is therefore necessary to know something of the physical chemistry of each of the bodies meeting at an interphase, because these bodies are the reservoirs from which the reacting forces are drawn, sustained, and replenished. If a given state is to be maintained for a given period at an interphase, there must be a correct concentration of forces drawn from each reacting phase lying behind that interphase. Knowledge of this interplay and of the steps necessary to hold it in controlled balance is essential for efficient control of the forces used in flotation.

The governing principles cover a broad field in which both fundamental and applied research continue to modify our concepts and hypotheses. Some terms are defined in the Glossary and others are briefly set out below, perhaps with further expansion in later chapters. A few of the excellent textbooks which deal at length with special aspects of physical chemistry are referred to in the text. The author's teaching experience has also led him to define many words and phrases of special significance in mineral processing (see References).

As an example of blurred precision, the terms "interface" and "interphase" are often used as though they carried the same meaning. This can be confusing, since in our present state of knowledge no definition of the word "surface" has been produced which can satisfy the rigorous requirements of physical chemistry. When a particle is surrounded by a gas, liquid, or gas-liquid mixture there exists a zone of transition between the phases concerned. The word "interface" suggests a sharp change of character. This abrupt and absolute change does not in fact exist. The physical and chemically induced changes in the behaviour of a particle dictate its reaction in the flotation process. These may or may not be influenced dominantly by its substrate structure and composition. They are, however, critically dictated by reactions in the interphase, as is later shown.

Gases, Liquids, and Solids

A gas is a molecular dispersion of matter as an elastic fluid, in which individual molecules are in free and independent motion, distributed fairly evenly over the space in which they are contained. The gas with which flotation is normally concerned is air, which maintains its full gas-molecular mobility under operating conditions, even when lightly compressed, since high pressure and a temperature of $-140°$ C. are needed for its liquefaction. Oxygen and nitrogen are both slightly soluble in water, this solubility increasing with pressure. On relief of such a pressure these gases separate from a supersaturated solution, and emerging bubbles are often negatively charged. This change of state may be of importance in flotation reaction, but at present the effect (if any) of gas ionisation is not well understood. The function of gas (apart from any ionising effect) in mineral processing is two-fold. It conveys oxygen or any other gas molecules in the air to a reacting interphase. Since it conveys all its molecules, the possibility of undesired ones could exist in special cases (e.g. an atmosphere heavily charged with sulphur dioxide from an adjacent smelter). Second, when air is blown through a pulp under suitable conditions it acts as a general stirring and transporting agent and perhaps by picking up particles into the walls of individual bubbles, as a selective transporting agent for mineral specifically attracted into the air-liquid interface.

The liquid of dominant interest in mineral dressing is water. In the research and testing laboratory this may be distilled or deionised water, to which controlled additions of reagents are made. In the plant it is the natural water of the local catchment area, mine water, river water, or piped water from a distance. Normally it is used as received, but for some processes it must be partially "softened" or otherwise prepared. Physically, a liquid can be defined as a fluid which flows readily in such a way as to seek the lowest available level and there moulds its underside to the shape of the containing vessel and its surface to the mean curvature of the terrestrial globe. A liquid differs from a gas because its molecules are bound together more firmly by forces of adhesion, so that its particles settle in a closed vessel at normal temperatures and do not distribute themselves evenly through the container. If heat were applied to water in such a closed vessel, the liquid molecules could be stimulated to speeds at which their group-bonds were destroyed, after which the entire contents of the vessel would become gaseous. Hence, water is the liquid form of a substance which can exist in more than one state or phase, according to the temperature and pressure to which it is subjected.

Since water is not the only liquid used in flotation, a fuller definition of the liquid state is needed. Liquids occur in three forms. Associated liquids of polar type have polar molecules, which seek to form oriented groups in accordance with their polarity.

A polar molecule is one which has an electric moment, its positive and negative charges being disposed at opposite ends. Its polar bond is the electrostatic union of atoms due to the transfer of one or more electrons, as with sodium chloride

(1) Na. ⟶ · C̈l: (outer electron shells represented by dots).

(2) Na:C̈l: (Cl octet completed by transfer of one electron).

(3) Na^+Cl^- (resulting electrostatic union and disposition of charge).

A polar molecule is chemically active and either soluble in water or hydrophilic. A polar group is a portion of a molecule which has polar characteristics.

In a non-polar molecule the electro-magnetic union is produced by the sharing of electrons, as with carbon dioxide:

(4) :O: ⟵ ·Ċ· ⟶ ··O··

(5) ··Ö:C:Ö:

(6) O=C=O (a covalent bond).

The non-polar molecule has an equal number of + and − charges with coincident centres of gravity. It does not ionize, is symmetrical and chemically inactive when compared with a polar molecule, and tends to be hydrophobic.

A heteropolar molecule combines non-polar and polar groups, the latter including such hydrophilic ions as hydroxyls, carboxyls, thios or amidons. Hence, part of such a molecule is hydrophobic and part ionizing. If the molecule adsorbs to a mineral the non-polar portion orients outward from the surface layer. This property is utilised in the *conditioning* of minerals prior to their separation in the flotation process.

The polar number or valence number of an atom expresses its valence mathematically. That for sodium, with one excess electron, is one and is positive. Sodium is thus monovalent and on losing its electron when becoming part of a molecule is positively ionized. Oxygen, which lacks two electrons to complete its outer octet, is divalent. Since it receives two negative charges in bonding it is electro-negative. When an atom forms an ionic compound with oxygen it therefore parts with electrons or becomes de-electronated (oxidised). In reverse process, it is reduced or electronated when it acquires electrons during chemical reaction in which oxygen is removed. Electronation reactions are those in which the valencies of the participating elements change through the exchange of electrons. They include oxidation-reduction (redox); nitridation; chlorination; sulphidation. In a polar molecule the oppositely charged ions are held together by electrostatic attraction. Such a molecule is one form of a *dipole*. It has electrical symmetry (neutrality) with its + and − charges sufficiently separated to orient it relative to the electrical system in which it moves. It has a dipole moment or molecular constant μ showing the distribution of its electrical charges which vary from zero if these are symmetrical up to 5×10^{18} e.s.u. (electrostatic units). This moment is the product of the magnitude of the dipole charges and their distance apart.

In a liquid system such as water the polar molecules H.O.H. tend to be constrained by their dipole forces into a regularly spaced lattice structure. (This trend is opposed by electrical, thermal and mechanical activity.) The general equation of state

$$pv = \mathrm{RT} \tag{15.1}$$

where p is pressure, v is volume and T the absolute temperature, is modified in the Van der Waals equation to

$$(p + a/v^2)(v - b) = \mathrm{RT}. \tag{15.2}$$

b is a volume factor corresponding with $\sqrt{4}$ of the molecular space, and a/v^2 expresses the mutual attraction between the molecules. This equation is valid for the solid, liquid, and gaseous state if pressure, volume, and temperature are expressed by their critical constants Pc, Vc, and Tc. under defined conditions.

If the molecular agitation due to temperature of an associated liquid system is sufficiently low, its polar groups can combine into a regular lattice structure, as when water cools below 4° C. and the molecules orient and finally lock into the expanded lattice pattern of solid ice (Fig. 185). (This orienting expansion, which is almost unique, is the cause of pipe-bursts in freezing.)

There are two sub-types of association—Newtonian liquids and thixotropic liquids. The former are true liquids, and undergo no change of viscosity when agitated. The latter are able to strengthen their molecular orientation in zones in the liquid, so as to produce a jelly-like or semi-liquid structure which does not flow. Agitation breaks up this orientation and temporarily restores the fluent behaviour characteristic of a true liquid.

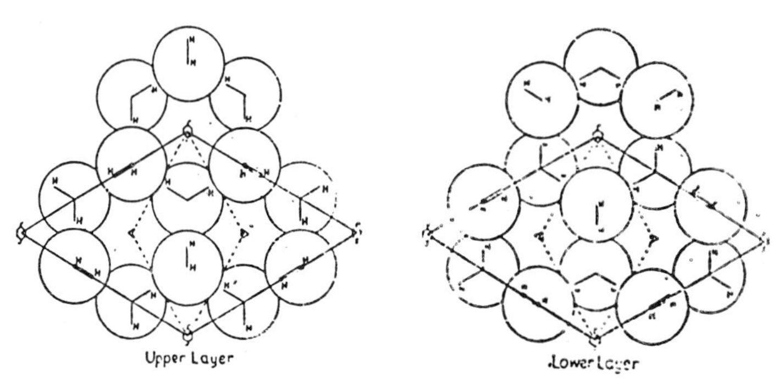

Fig. 185. *Lattice Patterns of Ice* (after Evans)

The second form is the non-associated or "normal" liquid. Here the molecules are non-polar and independent, and they are not held together by such forces as co-ordinate bonding or sharing of electrons. Benzene and carbon tetrachloride are liquids of this type. Intermediate between polar and normal comes the third type, the semi-polar liquid such as alcohol, which contains both a non-polar and a polar group in its molecules. The three types are illustrated in Fig. 186.

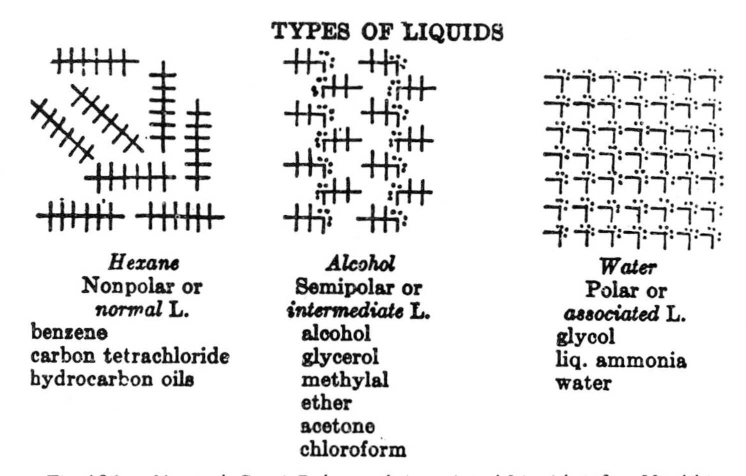

Fig. 186. *Normal, Semi-Polar and Associated Liquids* (after Hackh)

The dividing line between the liquid and solid state is not always clearly drawn. As the energy of the molecules in a substance becomes reduced through loss of heat, they move more and more slowly. The "temperature" of a molecule indicates the kinetic energy due to its thermal agitation. As this falls the molecule becomes less able to force its way through the surrounding molecules, and tends increasingly to become bound into a rigid structure or lattice. At some stage of cooling, for all short term purposes, the substance holds a definite shape and is no longer able to spread itself quickly at the lowest available level. The molecules may become strongly locked in the regular pattern of a crystal lattice, or there may be a *titer*-range of several degrees between the truly liquid and solid states, as with most commercial fatty acids. Pitch, though solid over a period of hours, is flowing plastically, as may be seen by leaving a lump of it for a year and then noting how it has spread.

The atoms of a substance may be bound ionically along one or more planes, producing considerable strength in the planes so bound, with weakness planes when the mode of crystallisation is such that no ionic forces unite them in other directions (cf. graphite, mica, and asbestos). Such symmetrical binding can be found both in pure crystal lattices and in solid solutions, where more than one element or compound has crystallised in an ionic pattern.

In addition to heat and pressure changes, which can make the same compound occur as gas, liquid, or solid—e.g. steam, water, and ice—solids can be built up from discrete molecular dimensions to colloidal aggregates and thence, *via* micro-crystals, to the macroscopic state. Sometimes single molecules associate into a colloidal concentration; sometimes they grow into large molecules (molecular colloids). Colloids can be considered as a stage in the growth or, if produced by detrition, the breakdown of a substance.

The structure of a substance may be determined partly by its rate of growth or solidification from the discrete molecular state of sub-colloidal disperse phase. If barium sulphate is precipitated from a solution of barium chloride by the addition of sulphuric acid, a flocculent precipitate is produced when the reacting substances are fairly concentrated. The newly formed molecules rush together so swiftly that they entangle water in the growing aggregates and the true micro-crystalline form of the compound is not built up. If extremely weak solutions are used, the molecules precipitate down upon suitable solid nuclei slowly enough to become arranged in a crystalline pattern ionically bound, and no water is trapped into the precipitate.

A particle may be homogeneous, in which case the atoms of which it is composed are bound together in a regular order. The pattern thus formed in a crystal is termed the *space lattice*, and a straight line drawn through two adjacent *lattice points* can be prolonged to pass through successive points equally spaced. Parallel lines do the same and the crystal structure is referenced to its x, y, and z distribution of such points with regard to a rectangular tri-axial set of co-ordinates. The order will be repeated throughout such a particle if it is a complete crystal, or a fragmentary portion of such a crystal. It can be composed of an intergrowth of true crystals, in which case there will be abrupt changes in lattice co-ordination at the boundaries between the individual crystals forming the particle. A particle composed of minute crystals, or crystallites, is usually tougher and harder than one derived from a large single crystal. At the boundary of the particle its crystal lattice is interrupted, so that the atoms at the surface are only partially linked to those in the body or *substrate.* The surface atoms are therefore in a less stable condition than are the atoms deep inside the substrate, and are more sensitive to the "climate" surrounding the particle. The electrical disturbances created by the disruptive force which creates new surface produce polarisation of the surface ions of an ionic compound.[11] At the surface newly created during comminution a reticulated and presumably uneven pattern of cations and anions is developed, and their electrical disturbance affects both the substance and the interphase. Considerations rising from this affect the consideration of comminution as a process which, though in itself mechanical, has important electro-chemical consequences. Throughout Nature all systems tend to arrange themselves in a position of maximum stability, and the atoms lying uneasily at the surface of a newly sheared piece of rock are no exceptions to this rule. Whether such a particle is surrounded by air or by a liquid, its incompletely linked surface atoms are attracted toward any atoms in the external system which can help to compensate them for the loss of the bonds sheared away when the new surface was formed during comminution. If the forces holding such an atom in the crystal lattice are weaker than those attracting it through the interphase, it may detach itself from the lattice and enter the fluid phase. If charged atoms or groups in the fluid phase are sufficiently mobile to respond to the attracting force, a charged lattice point on the surface of the crystal may reduce its excitation by the capture of an external atom carrying opposite charge. Several forms of sorptive attraction are possible, and these vary in intensity. The essential requirement for reaction is that there shall be an unstable area at the surface of the particle due to incomplete balance of the uppermost layer of partly bound atoms, and that action across the interphase shall be capable of reducing this instability.

Most of the large particles of ore dealt with in mineral dressing are not homogeneous, but consist of a mixture of crystalline compounds distributed through the particle. If liberation of such an aggregation is completed by comminution, each resulting fragment is homogeneous. Its newly sheared surface lattice is in a state of unbalance characteristic for the given compound. At the surface the unsatisfied valencies produced by disruption (*lattice discontinuities*) immediately commence to satisfy their instability by seeking new bonds, and this process does not necessarily go forward at a uniform rate all over the particle. It is therefore essential to the study of flotation physics that even when a pure compound is being considered, the difference between its surface condition and that of its substrate be borne in mind.

Dominantly Two-phase Systems

The extreme complexity of surface physics can readily be appreciated if a list is made of the possible two-phase systems resulting from the three basic forms—gas, liquid, and solid, considered above. These are shown in Table 26.

TABLE 26

SOME TWO-PHASED GAS-LIQUID-SOLID SYSTEMS

Group	System	
I. GAS-LIQUID SYSTEMS .	Mists and fogs	*a*
	Foams or froths	*b*
II. GAS-SOLID SYSTEMS .	Aerosols	*c*
	Occlusions or deep sorptions	*d*
	Pumices	*e*
III. LIQUID-LIQUID SYSTEMS .	Phases, Newtonian-normal	*f*
	Emulsions, Newtonian-normal	*g*
	Phases, normal-normal	*h*
	Emulsions, Newtonian-Newtonian	*i*
IV. LIQUID-SOLID SYSTEMS .	Unstable dispersions	*j*
	Peptised dispersions	*k*
	Flocculations	*l*
	Solvating systems	*m*
	Slurries	*n*
	Gels	*o*
	Thixotropic gels	*p*
	Moist solids	*q*
V. SOLID-SOLID SYSTEMS .	Two-component particles	*r*
	Multi-component particles	*s*
	Solid solutions	*t*
	Low dielectric systems	*u*
	High dielectric systems	*v*
	Zoned systems	*w*

in addition to others of no interest in mineral dressing.

Systems such as (*a*) usually commence as three-phased, since mists and fogs have their water droplets condensed upon an electrolyte suspended in the atmosphere. When it is remembered that this condensation is an essential first step toward the formation of a droplet, and that fog sufficiently dense to obscure vision contains only a few pints of water, something of the tremendous potency of these seemingly minute forces of the interphase can be glimpsed. Foams (*b*) can be of several types, depending on the tenacity with which the gas phase is stabilised in the liquid phase, and the proportion contributed to the foam by each phase. Later it will be seen that the interphase is complex and almost always adulterated. Systems of Type III can exhibit varying types of segregation. Sometimes oil floats on water, and sometimes it is dispersed as a stable emulsion or cream. The more thorough the dispersion the greater is the total surface area developed in the system, a fact which becomes important if any changes of state are influenced by forces residing at such interfacial surfaces. The eight systems listed in Group IV are all of significance in the physical chemistry of ore pulps. They affect settlement rate (*j*), (*k*), and (*l*), electrical state (*k*) and (*l*), deep sorption and chemical reaction (*m*), handling of fluids in the plant (*n*), (*o*), and (*p*), and sorption and drying (*q*). The solid-solid systems possible in Group V lead to many variations in the surface physics of a particle when it is brought into phase-contact with liquids or gases.

Ore treatment is especially concerned with the following two-phase systems:

(*a*) Colloids
(*b*) Emulsions
(*c*) Floccules
(*d*) Froths
(*e*) Micelles

These are rarely, if ever, unadulterated two-phase systems but are considered as such in the following discussion.

The Colloidal State

The special significance of the colloidal state may be seen as a transition stage rather than a semi-permanent condition of the liquid/solid system. In this state solids are *either* becoming part of the liquid by molecular dissociation *or* precipitating from solution and grouping their molecules densely as threshold solids. These steps in chemical solution or its reversal can produce an intermediate phase neither truly solid nor liquid. The forces which vectorise its trend (toward either increased solidity or liquidity) are at work in many mineral processing operations even when a distinct colloidal condition does not appear to exist.

Many colloidal phenomena and all electro-kinetic ones are primarily electro-chemical[10] and can be understood if colloids are regarded as electrolytes having surfacial electric tensions due to the presence of substances electrolytically dissociated. Graham's original definitions of substances entering into solution, made in 1861, divided them into two types. Those which diffuse rapidly he called crystalloids, and those diffusing slowly colloids (from the Greek word for glue). Today this difference is seen as one of state rather than nature, since most substances can be rendered colloidal. In 1907 Ostwald re-defined colloids in terms of particle size, between 10^2 and 10^7 cm. in diameter. Anything smaller is called a molecular dispersion.

A colloidal solution has two phases, the dispersed (particles) and the continuous (the dispersion medium). A solid dispersed in a liquid is a sol, and if the liquid is water, a hydrosol. A lyophil is an attracting liquid, or a colloidal system in which the dispersed phase is a liquid with an attraction for the dispersing medium. It includes hydrophils. Lyophilic sols attract, and lyophobic ones repel, a solvent. A sol may remain in suspension for months, if its solid particles are only a few μ in diameter, and a cubic centimetre of material reduced to this size has a surface area of an acre or more. A gel is a plastic lyophilic sol in which the continuous phase has been adsorbed by the particles.

The colloidal state can also be defined[10] as that in which the number of surface atoms approximates the number of internal atoms in each particle and the general structural disorder lies between monophasic solution and bi-phased suspension. The main characteristics of the colloidal state are:

(*a*) It cannot be dialysed.
(*b*) It has a very low diffusion velocity.
(*c*) There is a low tendency to crystallise.
(*d*) It is a transition state between solution and suspension.

Arbitrarily (due mainly to the difficulty of micro-observation) systems containing $10^3 - 10^9$ atoms (10^{-7} to 10^{-5} cm in diameter) are considered colloidal.

In the transitional state of a colloidal particle physical and chemical properties are blended. The colloid is stably knit electrically and may be helped in this stability by affinity between its molecules and those of the ambient liquid, exceeding those of the liquid molecules for one another.

This leads to a coating of the colloid by a layer of liquid. Such an incipient solvent state is characteristic for lyophiles, which form spontaneously when a disperse substance and dispersing medium are in contact and free enthalpy is decreased. Lyophobic colloids are maintained by an energy barrier of electrostatic repulsive forces. Hence there are two types of colloid with distinguishing features. Lyophobes such as metallic colloids are thermodynamically unstable, readily precipitated by electrolytes, migrate in an electric field and have a viscosity and surface tension comparable to that of the disperse phase. Lyophiles, such as proteins, are thermodynamically stable, insensitive to electrolytes, non-migratory in an electric field and of high viscosity and low surface tension. Due to the complexity of colloidal systems these distinctions may be blurred but in mineral processing, where the main interest in this state is centred on lyophobes, electric charge is essential to stability.

This electrical charge is explained by two theories. That of Helmholtz and Smoluchowski postulates an almost rigid electrical double layer formed by selective adsorption on each granule of one ionic species from the electrolyte, while the counter-species remains in the disperse medium and gives it opposite charge. Gouy considers this double layer to be diffuse, the charge density decreasing outward from the colloid into the ambient liquid.

The Malfitano-Duclaux theory, elaborated by Pauli, considers the colloid to be a true electrolyte dissociating into normal ions and ions of colloidal size. Here again an electrical double layer results. Broadly, the Pauli dissociation theory fits the observed facts better than the Helmholtz adsorption theory, though neither is completely satisfactory. Where a dipole moment is induced between a polarizing ion in solution and a dipole in the colloid stable adsorption follows and the colloid acquires charge. Though it is then become an electrolyte it is fundamentally different from a solution of true electrolytes. It is much larger, has the same electrical sign as its neighbours, is of heterogeneous composition and for the most part neutral, being largely composed of molecules insoluble in the solvent phase. The iogenic part of the colloid lies at its surface and contains soluble compounds capable of dissociating into ions. One species of these ions remains adsorptively fixed while the counter-species goes into solution. There are three types of iogenic complex—the iso-molecular colloid, the heteromolecular and the valence-linked. Silicic acid, colloidal ferric hydroxide and starch are of these respective types.

Electrophoresis indicates the electrical sign of a colloid. If it migrates toward an anode it is negative and *vice versa*, the phenomena being anaphoresis and cataphoresis. If, due to electrophoresis or other influence the electrical charge is sufficiently diminished, the colloidal state breaks down and particles may coagulate in response to surface attraction.

Coagulation is induced by the addition of electrolytes. Colloids then adsorb counter-ions and if the concentration of electrolyte is sufficient the number of free charges at the colloid surfaces diminishes together with the charge density on the surrounding zone of shear (Fig. 187). Electrokinetic

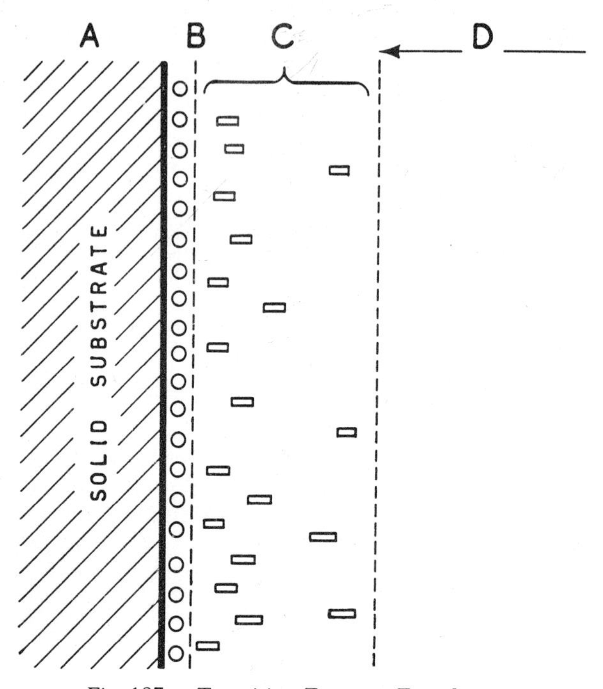

Fig. 187. *Transition Zone, or Zeta Layer*

potential then falls as the counter-ions become de-activated. Valence-bonding, ionic radii, hydration and other influences are also present. The reader not yet familiar with the aspects of physical chemistry summarised in this section is recommended to re-read it after completing his study of this chapter, when the further examination of electro-kinetics should have shed more light on this very complex subject, which plays an important part in controlling reaction rates in chemical extraction, flotation and all operations which are made possible by reaction through an interphase.

Emulsions

A true emulsion is a dispersion of one liquid in another (usually oil and water) under conditions in which they are practically immiscible, forming a fairly stable suspension with little or no solubility of the one in the other. If the dispersed droplets are of oil the emulsion is oil-in-water or O/W. If oil forms the continuous phase it is W/O. This two-component system is thermodynamically unstable and reverts to fully separated phases either by sedimentation, flocculation or coalescence. For coalescence to occur any stabilising energy barrier surrounding the droplets must be overcome. This can be achieved by water molecules in a W/O emulsion as they pass between neighbouring hydrocarbon chains or by localised displacement. The result is irreversible for the coalescing system unless new shearing force is applied from outside. In the flocculation of a dispersed phase the effect is reversible.

Stabilizing agents can be used to interpose a film in the interface between the emulsified phase and the continuum, when the system ceases to be two-phase. Among these reagents are sulphonates, quaternary ammonium compounds, soaps and gelatine. A non-reactive barrier can also be set up by finely divided solids which adsorb lyophilically. An emulsoid is a colloidal sol with the colloid as one phase and the liquid as the other. In liquid-liquid extraction, a technique used to transfer a solute from one phase to the other (e.g. in uranium technology), the liquids are emulsified and then allowed to separate. The speed of disengagement of the phases is termed the emulsion breaking rate.

Floccules

In an emulsion the two-phase system is dominantly liquid-liquid, though it was observed that a solid interphase could part these liquids. In typical flocculation the two-phase system is liquid-solid. The liquid of main interest in mineral processing is aqueous and the solid particles are too small to settle down with appreciable speed from an unagitated system.

The surface potential of a particle tends to repel it from others similarly charged. Opposing this, its surface tension promotes cohesion, whereby the overall solid area of the cohering particles is reduced, and with this the total surface energy. If repulsive force is dominant these minute particles remain dispersed (peptised). If on close approach the mutual attraction offered as reduced tension can overcome this electrical energy barrier they may flocculate. Floccules may consist largely of finely divided solids, or of particles loosely clustered in an aggregate predominantly aqueous—perhaps micellar in form. Forces disposing the particles to flocculate include magnetism (used in the dense-media process to manipulate ferro-silicon through its cleaning process); ionic forces of attraction; and secondary entropic forces. Each particle has an unmeasurable but considerable surface tension. This can be partially satisfied, as has already been observed, by reactions between it and the aqueous phase of the pulp. Further reduction of surface tension can also be achieved by adherence to other particles. If a number of particles can bind together so as to reduce their total free surface, the conditions for flocculation are propitious. If, however, each particle has an ionised surface and the bulk of the ions are of the same electric sign in each case, then a repulsive force will be exerted between approaching particles. Should this exceed the attraction offered by a reduction of total surface energy, the particles remain dispersed in the pulp. The most favourable conditions for flocculation therefore require electric neutrality between particles of similar composition. This neutral state is called the *iso-electric point,* or the *zero-potential.* It denotes the *p*H below which a given substance is acid, and above which it is alkaline. At or near its iso-electric point a colloid coagulates while a pulp tends to flocculate. The iso-electric point is also the *p*H of the liquid phase at which there is minimum movement into it of ions from the solid phase, or minimum solubility. Movement of a species from the solid into solution as an ion increases the residual electrical charge at or near the solid surface by the same amount, but of opposite sign. Electrostatic tension is thus set up between ionizing particles and the system remains dispersed.

The adherence of particles of a given species as floccules must not be confused with the phenomenon of slime-coating. In this an extremely fine layer (e.g. of lime) adheres to the surface of a relatively large particle of sulphide and impedes true surface-reaction.[15] A pulp being prepared for flotation may need dispersive conditioning to clean the particle surfaces, whereas one containing the end-products of treatment may require flocculation to settle and remove solids. In the first case the presence of two mineral species which have been conditioned to develop strong differences in surface excitation may produce ionic coupling; in the second the particle population in the pulp is in the same general state of excitation and a tendency to flocculate may be achieved near the iso-electric point. Modification of *p*H, to the extent that it increases surface charge, is a dispersant, but so far as it neutralises charge is a flocculating agent. The principal dispersing agents used are weak-acid salts of the alkali metals, such as sodium silicate, carbonate, sulphide, cyanide, and tri-phosphate. The most used flocculating agent is lime, others including sulphuric acid, sodium aluminate, and ammonium chloride. Trivalent ions appear particularly effective in inducing flocculation.

If the floc is considered as a particle the effect of flocculation is to increase particle size and facilitate sedimentation and filtration. The main stabilising factor in a sol is its electrical charge, and flocculation is aided by the addition of an electrolyte capable of discharging the component particles. Subsidiary forces which influence flocculation include the physical state of the dispersion medium, the temperature, population density (affecting opportunities for contact between particles), particle shape and the viscosity of the system. The Schulze-Hardy rule is that the ion of an electrolyte which causes a sol to coagulate is of opposite electrical charge to that on the colloid particles, and that its flocculating power increases greatly with an increase in ionic valency.

The concentration of particles in the pulp has an important effect on rate of settlement due to flocculation since collision must precede adhesion, and the denser the population the higher will be the rate of collision. Once floccules have formed, they act as a membrane when they drift downward, trapping particles and filtering the pulp.

Before particles can flocculate from a pulp, they must come into contact. When particles are already flocculated, some force is needed for their dispersion. This is provided by thermal motion. It can be observed with small particles in a pulp as a vibratory effect, due to their ceaseless bombardment by moving molecules of water. The energy transferred in each such collision by a moving molecule or ion is about 6×10^{14} erg at room temperature. The higher the pulp temperature and the lower its viscosity, the stronger is this stirring effect. If the particle surfaces are nearly electrically neutral, the movement will favour flocculation, as will also be the case if a flocculating agent having bonding counter-ions has been dissolved in the water. If the surfaces have similar electric charge but are still in contact, this Brownian vibration may suffice to shake them apart. Since a very small gap suffices to reduce the attractive forces across the interphase between surfaces, they will then tend to remain dispersed. The thermal agitation is impartial, but will aid any change which decreases the total surface energy in the system. Brownian movement does not directly affect rate of settlement. A "stable" suspension of colloidal particles in water falls in accordance with Stokes Law unless it is peptised by electrical charge. Molecular bombardment *per se* cannot oppose gravitation, since it is directed at random on all sides of each particle. The rate of fall may, however, be extremely slow.

Froths

The froths of technical concern in mineral processing are those produced in the flotation process. A specialised texture is aimed at, in which an unstable liquid-air foam increases its stability by becoming loaded partly by particles of select mineral species and partly by the addition of a chemical reagent (frother). The mineralised froth is a three-phase system and as such is more closely considered in connexion with the flotation process in Chapter 17. A foam may be defined as a dominantly gaseous mixture of gas and liquid, in the form of more-or-less stable bubbles. The liquid phase may be colloidal. Here again a foam is possibly three-phased in character. Such a system may be stabilised by soaps which emulsify to form a binding network which encloses water in each bubble, or by capillary-active substances.

The work required to produce a foam[9] is the foam surface multiplied by the surface tension. This foam is thermodynamically unstable since its collapse is accompanied by a decrease in the total free energy. When a heteropolar surfactant is dissolved in water its molecules preferentially concentrate at an air-water interface with their hydrocarbon groups oriented toward the air phase and their hydrophilic polar groups immersed in the aqueous phase[3]. The surface tension of the liquid decreases toward a minimum value as the concentration of these molecules rises from zero to the saturation limit in accordance with the Gibbs adsorption equation

$$\Gamma = \frac{1}{RT}\,\frac{\delta\gamma}{\delta \log_e a} \tag{15.3}$$

where γ is the surface tension (dynes/cm.)

R is the gas constant
T is the absolute temperature
a the activity, related to concentration of solution.

The lowering of surface tension on addition of pine oil (a widely used flotation frother) to pure water is from 72·3 dynes/cm. at N.T.P. to 63 dynes/cm. for an addition of 0·2 g./l. and 58 for 0·4 g./l. Commercial pine oil contains an erratic quantity of borneol and fenchyl alcohol (15–20%) and some 50% of α-terpineol, and has a maximum solubility of 1·28 g./l. at 25°C.

For all pure liquids, including water, the life of an emerging bubble is about 0·01 second. Addition of any soluble substance increases this duration. A highly stable bubble (e.g. one produced from water containing saponin) is not usually able to pick up mineral particles.

Sutherland and Wark[18] observe "a correlation between the size of bubble in the froth and the concentration of the frother. In very dilute solutions the most stable bubbles are large, but as the concentration of the frother is increased the average size of the stable bubble diminishes." They also discuss the bearing on froth stability of such factors as surface viscosity, liquid-phase viscosity, rate of adsorption of frother and surface tension gradient. Another factor which affects stability is the size range of the bubbles in a blanket of froth emerging continuously from an aerated body of water. Other things being equal, when these bubbles are sheared into the water by a mechanical propeller a wide range of sizes is formed, but when it enters through a porous septum a more closely sized and unstable foam is produced. (See also Chapter 17.

Micelles

A micelle is an aggregation of organic surface-active ions or molecules so clustered that their hydrophilic ends are oriented outward and their hydrocarbon chains tightly packed inside the cluster. The size and shape of the micelle[8] is determined by the equilibrium between the attractive forces among the hydrocarbon chains and the repulsive ones among the electrically charged heads of the ions. When the concentration of surfactant ions in a solution exceeds a certain point (the critical micelle concentration or CMC) the independent ions aggregate in these complex groups. Three structural shapes have been postulated.

Electro-Kinetics in the Pulp

From the specialised conditions in the transition zone which surrounds the individual particle consideration now turns to the overall forces operating in the pulp. These may be in part thermal, mechanical, magnetic and chemical but include an important amount of electrical activity. Such localised factors as sorption, surface activity and ion exchange are dealt with later. This section is concerned with generally accepted definitions of the forces which react selectively on the mineral species in the pulp.

Electrochemistry has been defined[10] as the branch of physical chemistry which studies the relationship between chemical transformations and energy in reactions which involve electrical energy originating outside the affected system. Although it is convenient at this point to consider pulp in a vessel insulated from random external influences, the normal working conditions under which ore is processed are fluent and unprotected from such forces. Although process control in the mill must be applied in the context of these conditions, causal forces are considered in isolation in this chapter.

Electrochemistry uses electrolysis and other electrical stimuli to promote chemical changes. They act when a current flows through an electrolyte, when potential difference is set up between electrodes, and when ionizing solvents are added to the pulp. An electrolyte may be defined as a compound which when melted or dissolved conducts electricity by the movement of the ions into which it has dissociated. If strong, it is considerably dissociated in solution and if weak, requires much dilution to effect partial dissolution. The ionisable conducting liquid used in electrolysis is an electrolyte. Electrolysis is the transfer or transport of matter through a medium by means of conducting ions. The medium may be fused salts or conducting solutions. Electrical mobility has been defined[19] as the constant velocity of an ion (in cm. sec.$^{-1}$) under the influence of IV/cm. The electromotive force (e.v. or e.m.f.) is the quantitatively measured free energy change during electrical reaction as ions migrate under the influence of potential difference between the two poles of a cell.

$$\text{e.m.f.} = \frac{\text{Amperes}}{\text{Ohms}} = \text{Volts} \qquad (15.4)$$

e.m.f. is not synonymous with potential difference (the drop of voltage across an open or resistant electrical system) since it measures the flow conversion of electrical energy.

Electrokinetic phenomena include electro- or cataphoresis. This is the movement of colloidal particles through a liquid toward an oppositely charged electrode. The mobility (μ) of a charged spherical particle is proportional to the potential gradient E/T (volts/cm.), the zeta potential ζ and the di-electric constant ϵ of the medium. It is inversely proportional to the viscosity η of the medium.

$$\mu = \frac{\left[\frac{E}{1}\right]^{\zeta\epsilon}}{6\pi\eta} \qquad (15.5)$$

Electro-osmosis is an effect arising from the partial separation of mobile from fixed charges in an electrical double layer. Akin to it is the electro-osmotic effect produced by the flow of liquid through a relatively fixed porous solid mass or aggregate under the influence of an externally applied electric field. Counter-ions surrounding the charged solid surfaces are attracted to the area of applied charge of opposite sign while the relatively immobile solid remains in place.

Three-phase Systems

The number of components capable of contributing to reaction in a so-called three-phase system can be considerable. Air, with its content of water vapour, CO_2 and noble gases is itself at least a nine-component mixture. A lump of ore may contain several discretely crystallised species capable of reacting to the stimulus of the forces at work in a mineral pulp. Detailed study of all possible interactions and reactions would be impossible. In plant practice, which is the commercial battleground of these forces, the operator can only control a few dominant factors in order to produce the most favourable "climate" for the required development of physical effects. It is, however, desirable to recognise the general nature of the uncontrolled factors which are present as well as those which are under control.

Gas-liquid-solid systems can be developed in mineral pulps in various forms, according to the percentage of each phase in the mixture. These percentages affect such characteristics as viscosity and rate of separation of an evenly distributed mixture into separate components. The systems include mixtures such as those shown in Table 27.

TABLE 27

THREE-PHASE SYSTEMS

	Type	*Example*
(*a*)	Mostly gas	Froth
(*b*)	Mostly liquid	Flocs
(*c*)	Mostly solid	Sediments
(*d*)	Much gas and liquid, little solid	Bottom of a froth column
(*e*)	Much liquid and solid, little gas	Thickener underflow
(*f*)	Much gas and solid, little liquid	Moist filtercake

When the physical chemistry of a three-phase system is receiving scientific consideration, an exact experimental approach is possible, provided each phase is rigorously pure. The engineer in charge of industrial operation cannot control his process with such meticulous precision, nor is he allowed to experiment freely. His raw material, once it has become part of a pulp, can only be brought into process control in terms of its generalised reaction. In research work all the known elements of a problem are recognised and as far as possible held constant. One only of them is then selectively varied for experimental purposes. In this case, the phases are combined in the simplest effective manner possible for control, test, and observation. Care is taken to start with a consistent and reproducible system. Variations are then introduced methodically and singly, the effect of each such variation being studied and, if possible, comprehended before passing to the next. In studying interphase reaction, which critically determines the efficiency of treatment, this painstaking empirical approach is vitally important because the number of possible variables makes theoretical reasoning too speculative to be reliable. Much misconception has crept into flotation literature because step-by-step variation of single conditions, together with the check of reproducibility of phenomena, is so hard to achieve even under precise laboratory conditions. Chemical extraction and flotation acts through selective change at relatively inert mineral surfaces. Reagents used must penetrate the interphase between the mineral and the surrounding fluids, liquids and/or gaseous. The words "interface" and "interphase" are sometimes considered to be interchangeable, but the latter is used in this book when there is an appreciable zone of change between the two faces of phases concerned. The chemical, physical, and electronic state in the interphase is determined by forces contributed by, and originating in, the true phases surrounding it. Solvation of a species from an exposed mineral surface requires access of the reacting electro-chemical forces. These must be drawn from the surrounding liquid phase and, when spent, replaced by fresh supplies. In facilitating these movements viscosity, pulp temperature, turbulence in the system, sorption, electrophoresis, osmosis and capillarity affect both rate and magnitude of reaction. Any of these may find expression in the rate determining step, expressed in the equation

$$\frac{dn}{dt} = \frac{DA(C-C_0)}{\delta} \tag{15.6}$$

where D is the diffusion constant, A the area from which diffusion proceeds, C the concentration of reactant in the liquid, C_0 that of reactant at the surface of the solid and δ the thickness of the diffusion layer or transition zone surrounding the product. In chemical extraction the leaching steps include agitation (which keeps the solids moving through the liquid and promotes blending in of air or other gases essential to reaction), transport of reactants to the interphase, adsorption reaction at the solid surface, desorption of the solubilised product, transport of this from the interphase into the liquid, and replacement in the interphase of exhausted reactants. An apparatus designed to facilitate study of leach kinetics is the modified Levitch rotary disc.[26]

The surface chemistry of flotation, though including some of the above elements of reaction, has for its objective the physical transfer of a virtually unchanged and undissolved mineral particle of selected species from an aqueous pulp into passing air bubbles which are rising to form a mineralised froth. Its chemistry is therefore less concerned with solvation through the interphase than with those reactions at the interface which cause water to be displaced in favour of air to which the particle will cling when it makes contact with a "coursing bubble" (one rising through the pulp). Here special study of surfactants and of changes in surface tension is needed, and is deferred to Chapter 17.

Zeta Potential

No known solid is completely insoluble in an electrolytic liquid. Consider a monovalent metal immersed in water. It tends to dissociate into the aqueous phase as its atoms or molecules dissolve to become cations in the surrounding solution. With each such departure of a cation an unsatisfied negative charge (electron) is added to the residual solid mass. These charges tend to attract dissociated cations in the liquid back toward the surface of the solid so that a concentration of them is produced in the interphase. This is strongest close to the metal and increasingly tenuous through the diffusion layer and outward. When a solid particle of any species swims in an aqueous solution (or any other electrolyte) the latter gains ions until saturation-point is reached. While this is happening the interphase between the true solid surface and the incompletely saturated ambient liquid develops as a zone through which only ionically charged species common to both solid and solution pass. These species are called potential-determining ions.

The Helmholtz concept of the electrical double layer thus created was of a partially seized layer of ions separated from the solid surface by a thin film of water molecules, held to it condenser-wise by electrostatic force. Gouy, in 1910, added the concept of thermal motion which showed this layer as more mobile, diffused or non-uniform. A moderately rigid part of the electrical double layer is thus pictured as held in the liquid close to the solid surface while the balance of the ions, and those of counter-charge (counter-ions or gegen-ions) are diffused outward until their attenuation is that of the ions in the characteristic liquid phase.

The gradient between the potential-determining ions, shown as anchored circles at the interface in Fig. 187, and the normal concentration in the liquid of fully hydrated ions and counter-ions from D outward (these hydrated ions not being shown in the diagram) is the transition zone and the electric potential difference across it is called the zeta-potential, ζ. At a very low temperature the anchored charge B is embedded in the liquid phase, its ions being rigidly adsorbed to the solid. As the temperature rises the system becomes increasingly mobile and electrostatic coupling is more strongly opposed by molecular movement until the characteristic situation is reached—the diffuse double layer. The Gouy concept, which pictures a rigid inner layer of adsorbed or structural ions diffusing for a few milli-microns outward from this interface, is generally accepted in our present understanding. The presentation here given is over-simplified, since electric charge on the surface of the solid can arise from other causes than straight ionization—for example, from preferential ad- and de-sorption at selected sites on the surface, or the co-existence of oppositely polarised surface areas. The double layer can be either ionic or dipolar. The first of these can be several Angstrom units deep and the second only a few molecular diameters. In the kinetics of the anchored "wall", plane of shear or Stern layer (Fig. 187B) dispersion due to Brownian motion is at work. Electro-kinetic effects also active include the streaming potential (that of the plane of shear between the wall and the fully mobile phase D).

The defects of many hypotheses concerned with the diffuse double-layer theory arise from the concept of dimensionless counter-ions and hence from preoccupation with electrical charges rather than mass-containing species. It is not possible to follow this special branch of electro-kinetics further in a textbook concerned with the practice of mineral concentration, but it must be remembered that the surface of the particle is the doorway to its interior in chemical extraction, and the selecting factor in flotation. As such, the surface must be reckoned with if process control is to be scientific. In connexion with traffic to and from the surface the significance of the iso-electric point must be remembered—the point of minimum solubility, minimum electrical attack and transfer, minimum ionization. It is represented by a critical *p*H value for each species in an ore pulp.

Change in the System

The three laws of thermodynamics define changes in the forms of energy shown, among other manifestations, in chemical and physical reactions.

The first law of thermodynamics (the law of conservation of energy) states that energy may be transformed within a closed system but that its total quantity remains unchanged. The net amount of work yielded by a given system while changing from a higher to a lower energised state is its free energy, if the change occurs at constant temperature and pressure. This applies to mechanical, thermal, electrical, and chemical aspects of energy. Energy is heat, or whatever can be changed into heat, and is measurable as potential energy, kinetic energy, surface energy, latent heat, calorific value, and electrical energy. In most forms it has intensity and capacity.

The second law deals with the degradation of energy. It can be stated in several ways. It is impossible, for example, to transfer heat from one body to another body at a higher temperature without external aid—hence in all changes the entropy (amount of irreversible energy) in the closed system increases. Since the initial excitation of the system is reduced by internal rebalancing without any emission of energy the energy thus rendered unavailable is degraded. It has changed from the free to the entropic state. It is sometimes easier to see a chemical reaction in terms of free energy loss, this loss being the measure of the force used to drive the reaction, and hence being the maximum work obtainable from the system undergoing change.

The third law states that every substance has a finite positive entropy. At absolute zero this is *nil*. This entropy is defined[16] as the unavailable energy of a substance due to the internal (irregular and compensating) motion of its molecules. It is unlike free energy, because it cannot be used for mechanical work. Entropy tends to a maximum, while free energy tends to a minimum. Hence, when a reaction in a closed system reaches equilibrium, the free energy is at its minimum and the entropy its maximum.

Every closed system seeks to reach the maximum available stability. It reaches this condition by rearranging its components until they form the most stable system possible under the circumstances. The potential energy of water pent up behind a dam becomes kinetic energy when the gates are opened. The water falls to the lowest available level, thereby reducing its potential energy to a new minimum. A hot piece of iron gives out heat until its temperature is in balance with that of the surrounding air. Its molecules now dance less violently in the mass. In other words, at this lower temperature they are in a less excited state. When this lower level has been reached, no further change can take place unless something is done to the whole system to upset its equilibrium. More heat could be conducted into the iron, or the air could be cooled. A further readjustment of the relations between iron and surrounding air would then begin. These are entropic processes, in which the system is at rest when its entropy has reached a maximum and the free energy a minimum. Because the system is achieving its equilibrium in isolation from all external influences this entropic energy, which is exclusively due to mutually compensating reactions of its component particles (electrons, atoms, ions, molecules, etc.), is not available for doing external work. In accordance with the requirements of the second law of thermodynamics "all naturally occurring processes are accompanied by an increase in the entropy of the system" (Taylor). "System" in this sense is the portion of the universe being studied in isolation from its surroundings. The energy of such an (ideally) isolated system is constant, while its entropy tends toward a maximum. Entropy can also be thought of as "the measure of the capacity of a system to undergo spontaneous change".[20] Clausius formulated the fundamental principles of thermodynamics thus:

(*a*) The total amount of energy in the universe is constant.
(*b*) The total entropy (S) is increasing, therefore entropy shows the direction taken by a spontaneous natural process, which is irreversible.

For a stable or ideal crystal system[21]

$$S = \int_0^n \frac{dQ\ rev}{T} \tag{15.7}$$

where T is the absolute temperature, rev indicates the reversibility of reaction, Q is quantity of heat and $\int_0^n dQ$ the quantity needed to bring a macroscopic system from state 1 to state 2. The equation does not hold for micro-states in which there is interaction between atomic groups.

Entropy defines unavailable energy since it is due to internal molecular and other motion which is compensated inside the *closed* system. Free energy (F) only is available for doing work, in accordance with the equation

$$F = (H - TS), \tag{15.8}$$

where H is the heat content (internal energy *plus* pressure × volume), T the absolute temperature and S the entropy. Entropy is directly proportional to the heat of a body and inversely as its temperature. At absolute zero temperature S is nil. In general terms entropy tends to a maximum and free energy to a minimum. Reaction tends to equilibrium and in equilibrium entropy is at its maximum. It is important to understand something of the meaning of this intimidating word "entropy" because in mineral processing we are dealing with rocks which have spent millions of years in attaining equilibrium. They have achieved such external signs of stability as great physical strength, chemical inertness, and sluggishness in reacting to environmental change. Each mineral particle presents a substantially closed system, and the only area which can be made adequately receptive to environmental influences is its newly disturbed surface. Thus, the effect of comminution in producing new surface is again seen as going far beyond mechanistic concepts. Molecular agitation (measured as heat) is only one form of energy tending to equilibrium in a closed system. The act of compression as part of crushing disruption of a rock mass leaves its mark on the products of comminution. Lattices are distorted and recover slowly, while incipient fissures inside the particle set up local areas of special strain and availability to penetration. This has been proved by radio-tracer methods which show that infiltration takes place along selected avenues rather than by generalised absorption. The large piece of ore is normally in such a state of equilibrium as to be but slowly affected by environmental change during its severance and transport to the grinding mill. The −60# particle detached from that lump is still relatively inert. It can, however, be unbalanced at its surface by electrochemical forces in the aqueous phase of the pulp which forms its new environment. This unbalancing must be guided and controlled in order to render selected particles floatable by surface-modifying treatment.

In the closed systems here being studied (air-water-mineral particle), maximum stability is attained when the movement of energy outward from one phase, through the interphase, and into an adjacent phase occurs at about the same rate in both directions. There is still plenty of activity, with packets of energy going individually to and fro, but the total inward traffic is balanced by the total outward traffic. Potential energy has settled at its lowest available level, free energy is at its minimum and nothing happens to disturb the tranquility of the system. This could be the state of affairs in a vessel filled with settled ore pulp before concentrating forces were applied.

In order to segregate one constituent (such as the metal-sulphide particles in a mixture of galena and quartz), steps are taken to disturb this equilibrium. Forces are introduced from outside the system which affect one of the solid constituents of the mixture in some special way. This constituent next commences to regain its equilibrium in the disturbed system. To do so it may send out more packets of energy to the other phases than it takes in, or it may do the opposite. There is excessive one-way traffic, of a volume and nature determined by the external force which has been introduced. At a suitable point in the readjustment it is possible to attract this one-way traffic along a new route, again by external interference. Examples of this have already been discussed in the previous chapters on concentration, when comparatively crude manipulations were used to bring gravitational forces to bear on mixtures of light and heavy particles in jigs. They are used when the unsettling force of sodium cyanide is exerted in a gold-bearing pulp to prise the atoms of metal loose from the solid system and attract them into the liquid phase. Flotation physics is concerned with delicate electrical forces which can be generated in pulps in such a way that only one mineral or group of minerals is thrown out of equilibrium. When this unbalancing has been performed an alternative system must be created in which this upset mineral can again settle stably—a system which has no attraction for the other minerals in the pulp. The new system is then separated from the old one, and with it goes the concentrate. Froth-flotation works in just this way. There are points of similarity between a manipulation of this sort and such chemical processes as cyanidation. In the latter a system is set up in which gold dissolves in the liquid phase, after which the solution is filtered off from the solid residual phase. Flotation deals with larger particles of valuable mineral, and with far larger quantities. It would not normally be economically possible to apply solvation methods to such resistant compounds as the metal sulphides. It is, however, possible to unsettle the physical state of most minerals at their interphase with water, and to do so cheaply. This is followed by the use of appropriate chemical treatment, which completes the surface change needed to ensure the required reaction in the flotation process or leads to selective solvation.

Since the rate of change toward greater stability is conditioned by the resistance met with when energised packets (ions, colloids, etc.) move through the interphase between particle and liquid, mechanical and thermal agitation of the system may be used as rate-controlling or rate-determining steps. Viscosity of the liquid phase is its internal friction, or resistance to flow. With the exception of certain silicones, it varies inversely as the temperature and directly as the pressure. It is measurable as the tangential force per unit area of either of two horizontal planes unit distance apart, of which one is fixed while the other moves at unit velocity, the intervening space being filled by the viscous fluid under test. It is measured in poises (c.g.s. units of viscosity—dynes $\times$ seconds $\times$ cm^{-2}). Anomalous viscosity is that which varies with either a shearing stress of a rate of shear. This latter is typically a thixotropic effect produced during mechanical agitation. In its most pronounced form a thixotropic system changes from a gel when at rest to a sol when shaken. The change is isothermal but can be produced by warming, its mechanism not being fully understood. In it particles associate into loose networks and the viscosity increases until gelling point is reached.

Chemical equilibrium is reached when a reversible reaction proceeds at the same rate in each direction, and may be indicated by the use of split arrows instead of the = sign, thus

$$H_2 + Cl_2 \rightleftharpoons 2HCl \tag{15.9}$$

Where ions are concerned an ionic equation can be written

$$Ba^{++} + SO_4^{=} \rightleftharpoons BaSO_4 \tag{15.10}$$

the arrows in each case indicating two-way traffic tending toward equilibrium. Metastability is a steady but incompletely satisfied state of equilibrium, which will undergo further change on the introduction of something which aids reaction tending to improve or complete stability.

Changes in the system may be self-initiated or may start with additions from outside. The rate of change is expressed in terms of first, second and third order reactions:

$$\frac{-d(A)}{dt} = k_1(A) \tag{15.11}$$

$$\frac{-d(A)}{dt} = \frac{-d(B)}{dt} = k_2\,(A).\,(B). \tag{15.12}$$

$$\frac{-d(A)}{dt} = \frac{-d(B)}{dt} = \frac{-d(C)}{dt} = k_3\,(A.\,B.\,C.) \tag{15.13}$$

according to whether reactant (*A*) only changes at constant volume (A + B ...) react, or (A + B + C ...) react, k being a velocity constant.

Bonds, Lattices, and Surfaces

An aerophilic (or "hydrophobic") particle, in flotation, is one which is attracted strongly to an air-water interphase. A hydrophilic particle under the same conditions tends to remain completely in the water. This differentiation of behaviour is to some extent inherent, but is mainly induced by the use of small quantities of surface-active chemicals. These are added during a process called pulp *conditioning*. After conditioning, air is bubbled through the pulp. The aerophilic particles are carried up by the rising bubbles and stabilised at the surface in a mineralised froth from which the mineral can be recovered as a concentrate. The preference shown during the actual separation of aerophilic from hydrophilic particles is a surface phenomenon. Each particle has two aspects—its surface and its mass. For flotation purposes the surface is to the mass as an envelope is to its contents—a means to an end. If the surface (to continue an imperfect analogy) has been correctly "addressed", the contents will be delivered according to plan. "Addressing" in this case consists of the addition of suitable reagents to the pulp, so as to develop the desired aerophilic quality on the surface of the desired particles. Pursuing the analogy, the address on the envelope and not its contents determines the place of delivery. Similarly, an aerophilic surface ensures flotation, regardless of the nature of the enclosed mass. We float surfaces, and only incidentally do we also float their underlying substrate.

The old definition of surface—a magnitude having length and breadth but no thickness—is inadequate. A naturally sheared mineral particle consists as regards its substrate, of repetitions of its unit cell. This cell has atoms arranged in one of the following patterns. It may be held together by electrovalent (ionic, polar, or hetero-polar) bonds as in the case of sodium chloride, which is an electrovalent compound, Na^+Cl^-. Here, the sodium atom, possessing eleven positive nuclear charges (protons) has the original shells shown in Fig. 188.

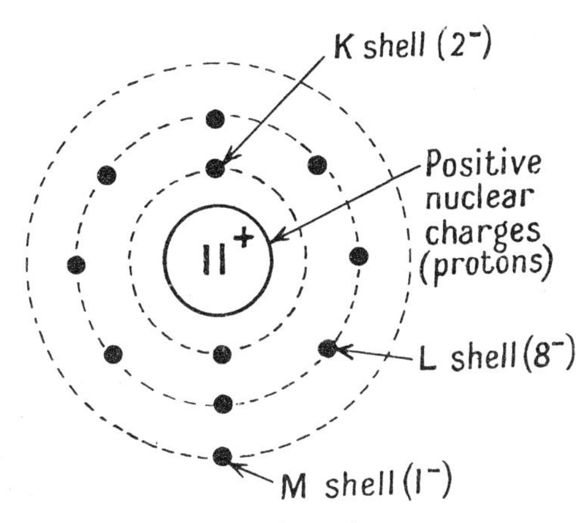

Fig. 188. *The Sodium Atom*

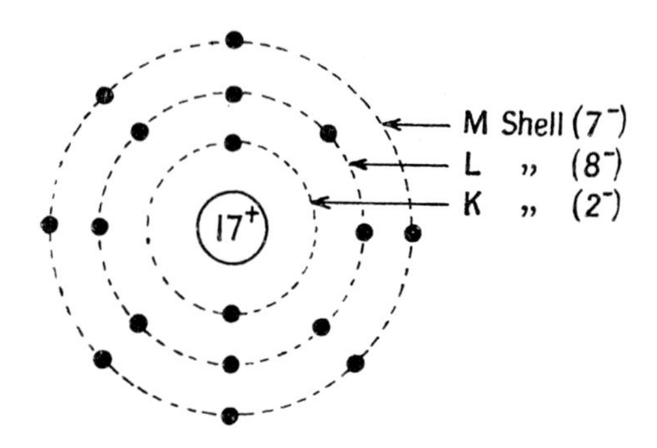

Fig. 189. *The Chlorine Atom*

Its eleven electrons spin round the nucleus to which they are held by the attraction exerted by the eleven positively charged protons at the centre, round which they whirl in balanced orbits. The one M electron is unbalanced. The chlorine atom has seventeen protons, balanced by seventeen electrons (Fig. 189). In this case, one unbalanced electron on the outermost shell leaves the atom dynamically unstable. The lone M electron (valence electron) of the sodium atom is transferred to the "hole" in the chlorine's M shell to complete the latter's octet. There is now one unneutralised proton on the sodium, which gives this atom positive charge, making it positively ionised. Similarly, there is now one excess negative charge on the chlorine atom, due to its capture of an electron from the sodium, making it negatively ionised. The two ions couple electrovalently to form

$$\left[\mathrm{Na}\right]^{+} + \left[:\ddot{\underset{..}{\mathrm{Cl}}}:\right]^{-} \quad \text{or}$$

sodium chloride. Each ionised element has a perfect octet on its outermost shell of electrons. The work done in removing this electron from the sodium atom is 5·12 electron volts—the ionising potential. To form an ion by the removal of two electrons in Mg^{++}) requires 22·58 e.v.; three (in Al^{+++}) need 53·0 and four (in Sn^{++++}) 101·9 e.v. Because the monovalent atoms can be ionised at the cost of less work than those with more valence electrons, the "alkali metals" are more active than the "alkali earths".

The valencies available at the active mineral surfaces offer four levels of reaction to the attacking chemicals and, consequently, four possibilities for differential surface modification. It is therefore of fundamental interest to study the valencies offered by a newly cleaved surface in relation to the ionising potentials required to attract ions from that surface into the liquid phase. Even with the clues thus afforded the research worker, many further considerations enter, and the overall picture is obscure. Electrical reaction proceeds at electrical speed, and there is no established evidence that new surfaces are immune from such rates of change. The author's hypothesis as to events during comminution and cleavage is very tentative, but it may be found helpful. When mechanical impact and abrasion have created a new separate particle it proceeds at electrical speed to neutralise the majority of its charged lattice points from immediately available sources. Counter-charges are available in the pulp and in its own substrate, and there may be some lattice structure rearrangement. After an extremely short period of intense activity on these lines, equilibrisation slows down until it reaches rates which are measurable by laboratory methods. Much of the residual unbalanced energy at this later stage is the result of lattice distortion, diffusion between liquid and solid phases and chemical readjustment. The picture ends with quiet relaxation, restoration of distorted lattices, and penetration of ions from the aqueous phase where zones of high strain have been developed inside the particle by comminution and chemical action. With moderately good conductors such as the metal sulphides which are readily floatable, electro-valence is an important factor in the general surface reactions during which the sulphur lattice points appear to undergo a measure of hydration as a necessary conditioning step. The nature and extent of this is neither fully clear nor fully agreed. Electro-valence is defined by Hackh[16] as "a fundamental type of atomic linkage, corresponding with ionised or polar linkages . . . (and) due to the transfer as distinct from sharing of electrons". Enough free energy must be made available in the pulp to initiate and sustain each desired change. The energising forces are supplied through hydrogen-ion concentration (by *p*H control); anion concentration (hydroxyl, carbonate, silicate, cyanide, etc.); available oxygen (either in the mill water or entrained); and surface-active chemicals. A simple example of ionic structure is provided by crystals of sodium chloride.

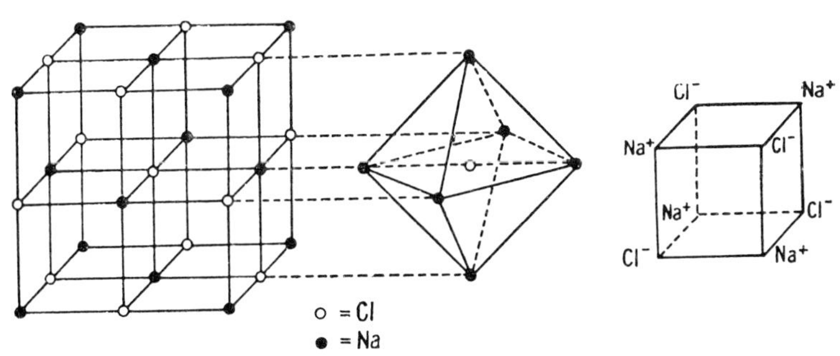

Fig. 190. *Sodium Chloride. The Unit Cell*

The unit cell of NaCl has the form shown in Fig. 190. It can grow (repeat itself) in all directions, provided sodium and chlorine ions arrive slowly enough from a pure solution of these ions in water. If they reach the nuclear lattice too rapidly, they have not sufficient time to find their ordered place in the lattice, and a new oriented growth commences. For this reason, large crystals are rare, whereas crystallites, which are intergrown systems of minute crystals of similar composition, are common.

The second type of bond is the covalent, in which atoms share two electrons in each valence bond. An example of this is the gas methane, CH_4. The hydrogen atom consists of a single proton and electron (Fig. 191*a*) and the carbon atom of six protons and six electrons (Fig. 191*b*). These bond covalently as shown in Fig. 192. The closed K shell of the carbon is omitted, since only the valence electrons react to form an octet. Since no electrons are transferred in covalent bonding, no ions are formed. Some of the most-used organic reagents in flotation derive their aerophilic quality from covalent bonding.

In the third type (the co-ordinate bond), valence electrons are again shared, but all are contributed by one partner to the linkage. Examples are the sulphate and phosphate ions $(SO_4)^{--}$ and $(PO_4)^{---}$. Here, the group is ionised. The four oxygen atoms require eight electrons to complete their octets, and in addition to the six or five captured respectively from the sulphur and phosphorus atoms have acquired two and three respectively from external atoms, thus acquiring a corresponding ionic charge. The fourth type of bond, "metallic linkage", rarely affects flotation physics.

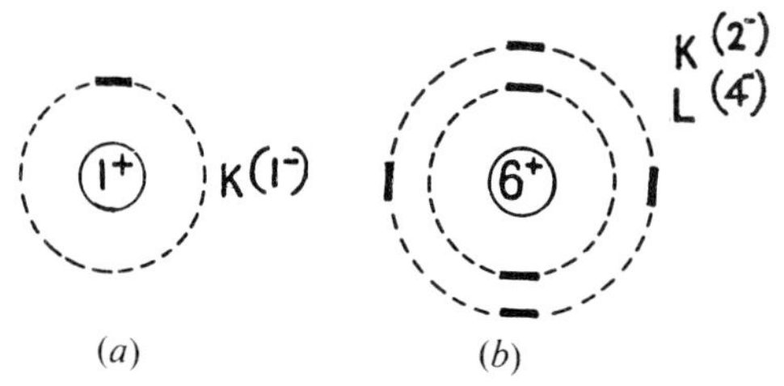

Fig. 191. *Atomic Hydrogen* (a) *and Carbon* (b)

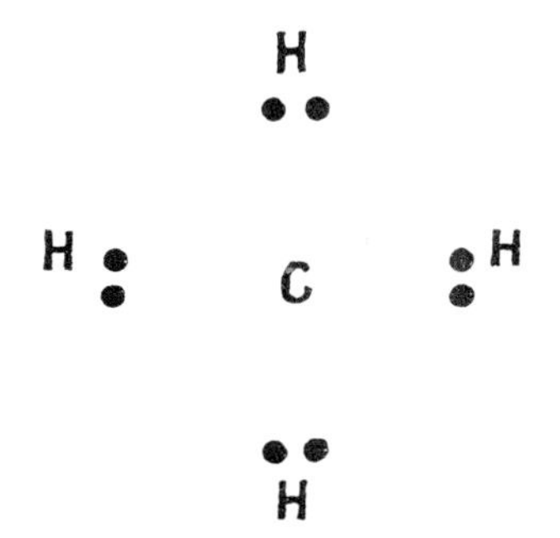

Fig. 192. *The Covalent Bond* (*Methane*)

Electrovalent compounds are ionised, even in the solid state. When melted, the ions are free to move, and are good conductors of electricity as compared with liquid covalent compounds. On solution in water the electrovalent compounds dissociate into ionised groups, and thus render the solution consuming them conductive or electrolytic.

Some substances occur in both the covalent and electrovalent form. The most important of these, in flotation, is water. In its covalent (non-ionised) form it has the structure

$$H : \ddot{\underset{..}{O}} : H$$

sometimes co-ordinated to dihydrol H_4O_2 or trihydrol H_6O_3. By ionisation, or through reaction with dissolved substances and solid surfaces of suspended particles, water molecules can change to the electrovalent forms:

$$3H_2O \rightleftharpoons (H\,.\,HOH)^+ + (:\ddot{\underset{..}{O}} : H\,.\,HOH)^-. \qquad (15.14)$$

The latter forms are polar and are attracted to oppositely charged lattice points. When an electrovalently bonded mineral particle is freshly severed, its surface monolayer is charged wherever the ionic lattice has been ruptured. It immediately begins to attract from the surrounding fluid the most suitable available substances for reducing its electric potential. This it does by sorption, hydration, oxidation (de-electronation), or exchange adsorption. Because of the great variety of reactions possible in a flotation pulp, the general word "climate", already used in this chapter, is often used. It refers loosely to the physico-chemical state of the flowing stream which is undergoing commercial treatment. This climate is partly influenced and controlled by means of reagents which vectorise the forces at work in desired directions. These are concerned to accentuate differences between aerophilic and hydrophilic mineral surfaces as a prelude to their separation by flotation.

A particle may be homogeneous, in which case the atoms of which it is composed are bound together in a regular order. The pattern thus formed in a crystal is termed the *space lattice*, and a straight line drawn through two adjacent *lattice points* can be prolonged to pass through successive points equally spaced. Parallel lines do the same and the crystal structure is referenced to its x, y, and z distribution of such points with regard to a rectangular tri-axial set of co-ordinates. The order will be repeated throughout such a particle if it is a complete crystal, or a fragmentary portion of such a crystal. It can be composed of an intergrowth of true crystals, in which case there will be abrupt changes in lattice co-ordination at the boundaries between the individual crystals forming the particle. A particle composed of minute crystals, or *crystallites*, is usually tougher and harder than one derived from a large single crystal. At the boundary of the particle its crystal lattice is interrupted so that the atoms at the surface are only partially linked to those in the body or *substrate*. The surface atoms are therefore in a less stable condition than are the atoms deep inside the substrate, and are more sensitive to the "climate" surrounding the particle. The electrical disturbances created by the disruptive force which creates new surface produce polarisation of the surface ions of an ionic compound. At the surface newly created during comminution a reticulated and presumably uneven pattern of cations and anions is developed, and their electrical disturbance affects both the substance and the interphase. Considerations rising from this affect the consideration of comminution as a process which, though in itself mechanical, has important electro-chemical consequences. Throughout Nature all systems tend to arrange themselves in a position of maximum stability, and the atoms lying uneasily at the surface of a newly sheared piece of rock are no exceptions to this rule. Whether such a particle is surrounded by air or by a liquid, its incompletely linked surface atoms are attracted toward any atoms in the external system which can help to compensate them for the loss of the bonds sheared away when the new surface was formed during comminution. If the forces holding such an atom in the crystal lattice are weaker than those attracting it through the interphase, it may detach itself from the lattice and enter the fluid phase. If charged atoms or groups in the fluid phase are sufficiently mobile to respond to the attracting force, a charged lattice point on the surface of the crystal may reduce its excitation by the capture of an external atom carrying opposite charge. Several forms of sorptive attraction are possible, and these vary in intensity. The essential requirement for reaction is that there shall be an unstable area at the surface of the particle, due to incomplete balance of the uppermost layer of partly bound atoms, and that action across the interphase shall be capable of reducing this instability.

A lattice structure (the Bravais, space, three-dimensional or transition lattice) repeats the unit cell. A net (lattice plane, two-dimensional lattice or two-dimensional translation group) repeats the unit mesh along a plane. A row (line lattice, lattice row, one-dimensional lattice or translation group) repeats the ordered occurrence of one line of lattice points. A lattice point has no periodicity. Crystallites are minute crystals bound together in a larger but imperfectly crystalline particle.

The physical properties of ionically bonded crystals are linked with the size of the metallic ion and the resulting change in the distance between it and the counter-ion. This, the A–X distance, has a striking effect on crystal hardness as measured on Mohs Scale (Table 28).

TABLE 28

ION-SIZE AND HARDNESS
(after Evans)[12]

	FeO	MgO	CaO	SrO	BaO
A–X	1·65	2·10	2·40	2·57	2·77
Hardness	9·0	6·5	4·5	3·5	3·3

The same effect is produced by increase in anion size.

TABLE 29

	CaO	CaS	CaSe	CaTe
A–X	2·4	2·84	2·96	3·17
Hardness	4·5	4·0	3·2	2·9

Change in ionic charge without substantial alteration of ionic distance also affects bond strength or hardness (Table 29).

TABLE 30

	NaF	MgO	ScN	TiC
A–X	2·31·	2·10	2·23	2·23
Hardness	3·2	6·5	7–8	8–9

Melting-point falls with increase in A–X (Table 30).

TABLE 31

	NaF	NaCl	NaBr	NaI
A–X	2·31	2·70	2·94	3·18
M.P. (C°)	988	801	740	660

In the solid state ionic compounds are non-conductors, their ions being bound in the atomic orbitals concerned. When melted, relatively high conductivity takes place by means of ion transport.

The effect of bond-type is summarised in Table 32.

A paper[22] deals with imperfections in the regular lattice structure of crystals and the methods of observing them.

Ions and pH

The ionic theory, modified from that propounded in 1887 by Arrhenius, has four points of interest in flotation:

1. When salts, acids, and bases dissolve in water they form positive and negative ions, and the number of unit charges on each ion is determined by the electrons it gains or loses when departing from atomic neutrality.

2. Ions are either produced by dissociation of the solid electrolyte,

$$NaCl \rightarrow Na^+ + Cl^-, \qquad (15.15)$$

or, by reaction with water,

$$SO_3 + H_2O \rightarrow SO_4^{=} + 2H^+ \qquad (15.16)$$

3. When an electric current passes through ion-containing water the positively charged *cations* move toward the cathode (negative electrode) and the *anions* toward the anode (positive electrode).

4. When an electrolytic compound dissolves in water, any acid properties are due to hydrogen ions (protons attached to water molecules $[H_2O.H]^+$), basic properties to hydroxyl ions (OH^-), and neutral properties to the ions of the salt concerned, save when these ions have reacted with ions of water (hydrolysed).

Acids in their pure state are unlike salts and bases, in that they are not dissociated. They react with water to ionise:

$$HCl + H_2O \rightarrow H_3O^+ + Cl^- \qquad (15.17)$$

$$H_2SO_4 + H_2O \rightarrow H_3O^+ + HSO_4^- \qquad (15.18)$$

$$H_3PO_4 + H_2O \rightarrow H_3O^+ + H_2PO_4^-. \qquad (15.19)$$

Ammonia acts as a base through reaction with hydrogen ion:

$$NH_3 + H_3O^+ \rightarrow NH_4^+ + H_2O. \qquad (15.20)$$

The neutral atom can acquire electrical charge. If it loses electrons (usually from its outermost shell) it manifests one unit of positive charge for each electron thus lost. If it captures electrons, it acquires one unit of negative charge for each such capture. Although changes of this kind are most

TABLE 32

PHYSICAL AND STRUCTURAL PROPERTIES ASSOCIATED WITH THE FOUR INTERATOMIC BONDS

(after R. C. Evans)

Property	*Ionic*	*Homopolar*	*Metallic*	*Van der Waals*
Mechanical	Strong, giving hard crystals	Strong, giving hard crystals	Variable strength Gliding common	Weak, giving soft crystals
Thermal	Fairly high m.p. Low coefficient of expansion. Ions in melt	High m.p. Low coefficient of expansion. Molecules in melt	Variable m.p. Long liquid interval	Low m.p. Large coefficient of expansion
Electrical	Moderate insulators. Conduction by ion transport in melt. Sometimes soluble in liquids of high dielectric constant	Insulators in solid and melt	Conduction by electron transport	Insulators
Optical and magnetic	Absorption and other properties primarily those of the individual ions, and therefore similar in solution	High refractive index. Absorption profoundly different in solution of gas	Opaque. Properties similar in liquid	Properties those of individual molecules and therefore similar in solution or gas
Structural	Non-directed, giving structures of high co-ordination	Spatially directed and numerically limited, giving structures of low co-ordination and low density	Non-directed, giving structures of very high co-ordination and high density	Formally analogous to metallic bond

readily made in the outermost shell where the electrostatic bond between electron and nuclear proton is weakest, the second shell may also play a part in ionisation. Many atoms can acquire more than one charge. Where an atom or group can have two valencies, the lower is by old custom called "-ous" and the higher "-ic" e.g. ferrous chloride ($FeCl_2$) and ferric chloride ($FeCl_3$). If the ionisation of the atom is negative, the attraction should be directly between a lattice point and a shell-electron. The hydroxonium ion of dissociated water is usually called the hydrogen ion, H^+, but is actually hydrated as $(H.HOH)^+$. If the H^+ breaks away this reverts to H_2O and H^+, a stripped single proton. This might be expected to have, in addition to its known high mobility and ease of transfer in the aqueous phase, considerable reactivity and ability to penetrate an attracting crystal lattice. It would thus act as a de-electronating (oxidising) agent.

The ionised lattice points at the surface of a particle suspended in water attract oppositely charged mobile ions (counter-ions) (see Fig. 187). The picture there given assumes a smooth surface plane, but it is improbable that this exists in fact. In the process of comminution the severance of structural bonds is believed to leave a pitted irregular area (in terms of molecular dimensions), and one in which the energised lattice sites are not necessarily regularly disposed. These matters are conjectural, but it is convenient to picture a Helmholtz mono-layer of ions strongly attracted to the counter-charges on the majority of the lattice sites.

The potential effective charge in electrokinetic action (the ζ-potential) is the work needed to bring unit charge from infinity (in this case phase D) to the surface of shear B. Hence ζ-potential can be modified either by changes in the constitution of the electrolyte, or in that of the strongly adsorbed compounds.

When a liquid moves relative to a solid, as when an electrolyte percolates through a porous aggregate or a capillary tube, loosely adherent ions in the diffused layer C are constantly removed, and a measurable potential (the *streaming* potential) occurs between the ends of the system, in accordance with the equation

$$\zeta = \frac{4\pi\eta K_s E_s}{PD} \qquad (15.21)$$

where E_s is the streaming potential;
P is the pressure forcing the liquid through the capillaries;
K_s is the specific conductivity of the liquid in the capillaries;
D is the dielectric constant of the liquid;
η is the viscosity.

ζ-potential may also be derived from the electrophoretic behaviour. The velocity with which a fine suspended particle moves is governed by its surface charge and the applied electro-motive force. An approximate equation is

$$\zeta = \frac{4\pi\eta}{D} u, \qquad (15.22)$$

u (velocity) being given in microns/sec./volt/cm. In these and related phenomena the general equation is useful.

$$\zeta = \frac{4\pi\gamma}{D} \qquad (15.23)$$

y being the dipole moment ($y = Se$), where S is the thickness of the double layer and e the charge per cm.^{2} of the double layer.

This "difference in potential" or potential gradient also provides the moving force which causes electro-osmosis (diffusion through a membrane, and capillary rise).

If changes are to be induced at the surface of a particle by ion-anchorage or reaction with agents which have been added to the aqueous phase of a flotation pulp there must be sufficient attractive force at that surface to overcome the inertia of the system. It must act across the interphase, and as its attraction varies as the square root of distance, the molar concentration of ions is important. If a lattice ion captures a counter-ion from this zone, the surface reaction at this point ceases to be polar, since the electrical tensions have been neutralised. If the captured ion carries a non-polar group, this forms a coating above the lattice point and the particle will react to flotation in accordance with the dominant coating thus formed at the surface. In an electrovalent compound such as a typical metal sulphide, there are positive and negative ions in the newly sheared surface lattice, and there can be traffic in both directions. It is possible to influence this traffic by adding desired ions to the water. Balance is reached when the traffic from the mineral surface to the water phase equals that in the opposite direction.

A major instrument of control of the ionisation in the pulp, and particularly at the particle surfaces, is the hydrogen-ion concentration. pH is the descriptive symbol. It denotes the logarithm of the reciprocal of the H-ion concentration.

$$\log \frac{1}{C_{H^+}}$$

Even pure water can conduct some electricity, because it is slightly ionised:

$$2H_2O \rightleftharpoons (H_2O\ .\ H)^+ + (OH)^-. \qquad (15.24)$$

For any given temperature of water this is a constant:

$$Kw = [H_3O^+]\,[OH^-]. \qquad (15.25)$$

With pH defining the negative log H-ion concentration and pOH that of the hydroxyl-ion concentration:

$$pH + pOH = pKw. \qquad (15.26)$$

At 25° C. the ion product for water is 10^{-14}. $(H^+) \times (OH^-) = 10^{-14}$. pK_w is then 14. The pH at this temperature is therefore $pK_w - p$OH. A pH of 7 indicates neutrality. Lower values denote logarithmic increase in acidity, and higher ones in alkalinity. Changes from the near-neutrality of pure water occur when compounds are added which dissolve to produce H-ions. As these increase, the hydroxyls ($OH-$) are correspondingly decreased.

An electrolytic compound becomes ionised on solution in water. If the compound consists of a strong acid and a weak base or *vice versa* it ionises fairly completely as regards the strong partner, but the relation between the ionised weak partner and the water molecules is only partial. It resists change in the *p*H of the solution such as would normally occur on the addition of a strong base or acid, and the solution is said to be "buffered". A salt such as aluminium chloride produces acidic buffering because the hydroxyls are neutralised by hydrolisation.

$$AlCl_3 + 3H_2O \rightleftharpoons Al(OH)_3 + 3H^+ + 3Cl. \quad (15.27)$$

In the case of sodium carbonate the hydrogen ion is withdrawn on dissociation, leaving the hydroxyl uncompensated and the solution alkaline

$$Na_2CO_3 + HO_2 \rightleftharpoons HCO_3^- + OH^- + 2Na^+. \quad (15.28)$$

This difference can be seen if water stained by the mixture of aniline dyes called a "universal indicator" receives first a drop of dilute sodium hydroxide (causing it to show the phenol-phthalein violet of *p*H 10+) and then small additions of dilute sulphuric acid, which will produce a sudden transition to the red of methyl orange at *p*H 4—. If a salt such as sodium acetate (strong base and weak acid) be now added and the experiment repeated, much greater addition of acid becomes necessary to effect the colour change, which proceeds gently through the range of colours of the mixture in use.

The surfaces of the particles in a pulp exercise buffering effect, and the opacity of such a pulp renders the use of dyes inconvenient for the measurement of *p*H. The usual method of measurement is electrometric. Glass and calomel electrodes are immersed in the pulp, and the potential difference between them is read direct on a galvanometer calibrated in *p*H units.

The "common ion effect" is the change in concentration of an ion in a saturated solution, through the addition of a different electrolyte which yields an ion in common with solid substance present in excess. The ion product remains constant, but with increase of one ion that of the other diminishes correspondingly. Since the solution was already saturated precipitation now occurs. The effect is to reverse the ionizing process.

Surface Activity and Sorption

In normal chemical reaction where the whole volume of each reactant is available precipitation, combustion, and other changes (particularly those involving ions) take place with some rapidity and visible alteration of state. The speed is affected by temperature, concentration, and catalysis. When reaction is reversible, it reaches equilibrium in a closed system after the free energy has been redistributed. When precipitation has occurred, a small amount of ionised material remains in solution. When an electrovalent substance is immersed in water, some ionises into solution, however low the solubility. This, for a given temperature, may be expressed as a constant

$$K_s = (B^+)(A^-), \quad (15.29)$$

but quantitatively it may be reduced by the presence of common ions or increased by the "salt effect" of other electrolytes. Salting out, in general terms, is the phenomenon in which a substance is precipitated from its solution by the addition of soluble salts.

The mineral particles involved in the chemistry of flotation have very low solubilities and their reactions take place either at the discontinuity lattice in a monomolecular surface layer or with decreasing force for a very short molecular distance inward. In electron-diffraction study of the boundary lubrication of metals zones have been shown to exist, starting from the substrate and proceeding outward *via* increasing metallic oxidation anywhere up to a thousand molecules deep to the lubricated interphase. No comparable picture of the depth into the ore particle through which change is taking place has yet emerged. When the wide variety of mineral lattice structures is considered, together with the lattice defects and distortions possible in any specific structure, it is surprising that research has led to so much agreement in the whole field of mineral surface-physics, rather than that there is still much that is controversial.

Surface chemistry is concerned with the specialised interface or interphase in which the substrate of each reacting phase is unbalanced ionically. This condition is induced by surface activity in which a surface-active agent (or surfactant) modifies chemical, electrical or physical characteristics at the surface of a solid, or changes the surface tension of either solids or liquids. Usually the heteropolar molecules of the chosen surfactant are attracted to a specific type of surface in the ore pulp so that only one species of mineral is affected. Change of state at the surface of the attracting particle follows. This may include either sorption or chemical change resulting in the formation of a surface film differing in composition from the substrate. Sorption is a general term for reactions at or near the surface. It includes the following:

Adsorption. The ability of the adsorbent (mineral surface in this case) to hold adsorbates (gases, liquids, or solutes) adhesively at its surface, without penetration. *Desorption* is the reverse action.

Absorption. Deep physical sorption by capillary action or superficial migration of the sorbate.

Chemisorption. Deep absorption with some degree of chemical change or lattice modification at the surface of the sorbing substance by means of valence bonds.

Persorption. Permeation into a porous solid.

In connexion with surface change, the term "orientation" is concerned with the position assumed by an adhering molecule, and the nature of the force concerned. Hydrolysis and electrification are among the possible forces at work on a zone (area of special interest on the solid surface) where molecules are oriented.

One group of the surface-active molecule may form a polar monolayer electrostatically attached to a counter-charged lattice point at the solid surface while the rest of the molecule faces outward and thus presents a new type of surface to the ambient phase at that point. If enough of these attachments occur the lyophilic-lyophobic balance of attraction between the surface of the particle and the ambient water and air decides whether it will remain wetted in an aqueous phase or cling to the air-water interface of a coursing bubble passing through the pulp. Other relationships modified by surfactants include surface tension, emulsification and micellar grouping.

When the effect is to reduce the interfacial tension (surface tension) a solute is said to be positively adsorbed. Such a solute's molecules contain both hydrophilic groups (which adsorb to the water phase) and hydrophobic groups (which are water-repellent). The three electro-chemical types of molecule are the un-ionised, the anionic and the cationic. An amphoteric surfactant is one which ionises in aqueous solution to become either anionic or cationic in response to the *p*H of the solution. The molecules of an ampholytic surfactant such as sodium lauryl sarcosinate contain both cationic and anionic groups.

Surfactants[7] adsorb to surfaces to produce (*a*) lowering of one or more interfacial boundary tensions in the system; (*b*) stabilisation of one or more interfaces by forming adsorbed layers which oppose further change. Except for solubilisation by some surfactants, action is either (*a*), (*b*) or a mixture of the two.

Specific adsorption is sometimes suggested as a better term than chemisorption. Change in the aerophilic quality of the mineral surface does not become apparent in a two-phase solid-liquid system strictly isolated, so methods such as contact-angle measurement described in Chapter 17 do not fully give satisfactory research information regarding this type of sorption.

Where adsorption is due to positive attraction or linkage between polar groups of opposite charge the surfactant is not necessarily amphipathic. Moilliet[7] suggests that is should be called a specific surface-active agent, and that it may be macro-molecular. Amphipathic adsorption is usually predictable, since the medium tends to expel one part of the molecule or ion and to retain the other. Specific adsorption is not thus predictable since it varies with the quality of the external attraction.

Surface energy can be stated in terms of free energy per unit area or surface tension. It is provided by those molecules in the outermost liquid layer of a liquid/air interface which are not compensated toward the substrate by the mutual attraction of surrounding molecules. Any change in the total free energy residing in the interfacial area depends on expansion or contraction of this area. The effect of a surfactant in producing this change can often be analysed thermodynamically.

In a liquid-liquid or liquid-solid interphase the hydrophilic groups of the surfactant may be bound to the solid discontinuity lattice and the hydrophobic groups oriented outward. Here there may be some chemisorption, as with the xanthates widely used in the flotation process to render a preferred mineral's surface aerophilic.

Taggart's[23] solubility theory for flotation postulated metathesis between the surfactant collector agent and the mineral surface, and has proved to be an over-simplification. It failed to take into account crystal lattice structure, surface electric properties and physical adsorption, but was a good practical guide.

An arbitrary definition of chemisorption regards it as an adsorption process in which the adsorbate attaches to the adsorbing surface with a molar free energy of 10 k.cal. or more.[8] Research has shown that with a few exceptions uncontaminated and unoxidised sulphide mineral surfaces cannot be attracted into a flotation froth. No fully agreed reason has yet been advanced but the evidence points to the need for oxygen-attack to produce threshold sulphatisation of the particle, which must be arrested far short of saturation. Plaksin[24] summarising part of a research programme, observes:

(1) Gases play an important chemical role in flotation. The behaviour of minerals in particular depends on the concentration of oxygen in solution.
(2) The effect of oxygen is not limited to sulphide minerals but, surprisingly, enters also into the behaviour of non-sulphides.
(3) The crystal structure of minerals, sulphides and non-sulphides alike, is a major factor in their response to variations in oxygen level.

It is not necessary that the whole surface of the mineral should be affected. Zones may develop on part of the surface. They may have special characteristics such as an attracted layer of oriented molecules from the surrounding fluid. A *zone* thus differs from a *phase*, the latter being homogeneous, while the former has random disposition of its molecules.

Adsorption takes place at the solid boundary of the solid-liquid interphase, some constituents of the fluid phase being attracted and held at the solid surface. The attracting force may be ionic or entropic. In the latter case the arriving molecules give a lower surface tension than that of the surface to which they adsorb, resulting in an overall lowering of the potential energy of the particle surface. If such a substance raises the surface tension, then it will be repelled from the surface, since entropy tends to a maximum and is irreversible. Among positive (adsorbing) additives used in flotation are fatty acids, short chain alcohols, amines, phenols, and terpenes. Adsorption can be caused by thermal or radiation energy, and can accompany hydrolysis or other solvating actions changing electrical surface-potential. It can result from residual valences when atoms at the surface of the adsorbent are only partly saturated at their inner ends and have unsaturated valency points at the outer ends capable of holding impinging counter-ions. An important characteristic of chemical adsorption is the formation of a reaction product which has the least solubility of all reaction products available for interfacial reaction. In a saturated solution of an electrovalent compound having limited solubility, the solubility product concentration is that of the ionic concentrations and is constant at a given temperature. For a substance BA which yields ions B^+ and A^- the equilibrium constant K is

$$k = \frac{(B^+)(A^-)}{BA} \qquad (15.30)$$

and the solubility product K.BA. A molecule, atom, or ion giving the most sparingly soluble and dissociable compounds with the adsorbent should be the most strongly adsorbed. The ability of ionised molecules to adsorb to a crystalline lattice is affected by:

(*a*) The distance apart of the centres of charge across which attraction must act.
(*b*) The spacing of the changed lattice points at the mineral surface.
(*c*) The size of attracted ions and their ability to pack themselves above the lattice points.

Broadly, all adsorption is accompanied by reduction in free energy. Physical (often called Van der Waal[15]) adsorption requires only weak adhesional forces between adsorbate and adsorbent while with chemisorption the activation energy can be high. The change in interfacial tension on charging an interface is sometimes called the electro-capillary effect. Adhesion usually refers to an interface between dissimilar substances and should not be confused with cohesion, which applies to the internal pressure due to the force of attraction between molecules or particles in a mass and varies inversely as temperature.

Changes in surface activity are accompanied by variations in surface tension. These can be measured for liquid/gas systems but for solids can at present only be estimated indirectly. For a liquid it is expressed in dynes/cm. as the force with which the surface on one side of a line 1 cm. long pulls that on the other side of the line. If stated in ergs instead of dynes it is the work required at constant temperature and liquid composition to increase the surface area by one cm^2 of free energy per unit area measured in ergs/cm^{-2}. Further consideration is given to surface tension in relation to flotation in Chapter 17.

A special type of sorption, indeterminate between ad- and chemisorption, is that of the amalgams. These are so-called alloys of mercury with various elemental metals. Though there is true chemical reaction in some cases (e.g. sodium amalgam), the chemistry of gold, silver and copper amalgams is more obscure. Some stoichiometric action takes place. Gold is slightly soluble in mercury (0·06% at 20° C.). The subject receives special reference under Gold in Chapter 23.

Ion Exchange

In IX, as this widely used reaction is often referred to in industrial practice, ions are exchanged at the boundary between two phases. Those of special interest in mineral processing are the liquid/liquid and the liquid/solid, the former being virtually immiscible save during agitation.

An ion exchanger is an insoluble solid carrying accessible exchange cations, anions or amphoteric ions. It differs from a sorbing solid in that stoichiometric action takes place, ions captured by the exchanger from the ambient solution being replaced in that solution by an equivalent amount of similarly charged ions. Thus, IX is preferential, selective electrostatic interaction leading to the capture of an ion by the IX solid in accordance with the relative ion sizes, valences and concentrations of the ions displaced by this capture. A liquid ion exchanger is a compound dissolved in a liquid immiscible with the one carrying the ions which are to be exchanged. If this liquid is aqueous, the immiscible phase may be kerosene, trochloroethylene, chloroform, etc., a high flashpoint kerosene being preferred industrially. The exchanging ions dissolved in the organic liquid contain hydrophobic groups which hold them in this phase, the rest of the molecule being available for capture. The compounds used include aliphatic amines, dialkyl phosphates and fatty acids.

Solid exchangers are solid electrolytes with inert matrices (R) to which are attached special polar groups (G) balanced by oppositely charged ions (X) which are available for exchange. The formula is R(GX)n. If G is negative X is a cation and the material is a cation exchanger.[25] Inorganic exchangers include zeolite, glauconite, natural montmorillonite and synthetic sodium aluminosilicates. Some coals can develop IX qualities on sulphonation and oxidation. The most important industrial IX materials are based on synthetic resins such as phenol formaldehyde and styrene-divinylbenzene.

The polar groups in cation exchangers are sulphonic acid ($-SO_3H$, strong), carboxylic acid ($-COOH$, weak), sulphonium ($-S.R_2OH$) and phosphonium (PR_3OH). The reaction is

$$R(GX)_n + nM\pm \longrightarrow R(GM)_n + nX\pm \tag{15.31}$$

The rate controlling step in IX is diffusion in the resin particle and across the film of solution at its surface, and this is aided by increase in active area, temperature rise, concentration of ions and their relative smallness. The diffusion of counter-ions appears to be the critical factor in rate determination. Any such ions which leave the IX media are replaced by an equivalent amount of other counter-ions in order to maintain electro-neutrality. This step can be subdivided into two rate determining steps—diffusion of counter ions in the media, and inter-diffusion of these ions in the adherent films or diffusion layers, and both may be at work simultaneously.

If a cationic resin is placed in an electrolytic solution there is cross-migration. Cations go into the ambient phase and anions into the resin. This sets up a positive charge in the solution and a negative one in the resin, and builds up a potential difference called the Donnan potential, which pulls cations back into the now negatively charged resin and *vice versa.* The counter-ion concentration remains higher and the co-ion lower in the resin than in the solution. With an anionic resin the effect is similar but of reversed sign. The co-ion has but little effect on exchange kinetics.

For IX to succeed the exchanging ion species must be soluble in the selected solvent. They must be at least partly dissociated and mobile, and the solid must not be soluble or destructible in the solvent. The solvent should have a high dielectric constant. That of water is ϵ 81.

Resins are built from monomeric units forming linear chainlike polymers, cross-linked so as to form flexible networks. The iogenic groups are carried on the chains. These coiled networks unfold to permit limited entry of solvent molecules and then swell till a balance is struck between the intruders and the elasticity of the resin. Thus the resins are typically gels with an irregular macromolecular matrix of hydrocarbon chains in a three-dimensional network. In cation exchangers this matrix carries such groups as $-SO_3^-$, $-COO^-$, $-PO_3^=$, $-AsO_3^=$, and in anion exchangers $-NH_3^+$, $>NH_2^+$, $>N^+<$ $\underline{\ }>S^{+3}$. The chemical nature of the groups affects the IX capacity. A weak acid such as $-COO^-$ is only ionised at a high *p*H, while a strong acid like $-SO_2^-$ remains ionised at a low *p*H.

For most purposes in mineral processing the resin is used in bead form. It can be formed into discs, plugs, mats and frames for special work, and belts and ribbons are also available. The use of IX in the mill is considered in Chapter 16.

References

1. Sidgwick, N.V. (1946). The Electronic Theory of Valency, Oxford University Press.
2. Pauling, L. (1945). The Nature of the Chemical Bond, Cornell University Press.
3. Adam, N. K. (1941). The Physics and Chemistry of Surfaces, Oxford University Press.
4. Gucker, F. T., and Meldrum, W. B. (1950). Physical Chemistry, American Book Co.
5. Harned, H. S., and Owen, B. B. (1950). The Physical Chemistry of Electrolytic Solutions, Reinhold.
6. Wells, A. F. (1950). Structural Inorganic Chemistry, Oxford University Press.
7. Moilliet, J. L., Collie, B., and Black, W. (1961). Surface Activity, Spon.
8. Fuerstenau, et al. (1962) 50th Anniversary Vol. Froth Flotation, A.I.M.M.E.
9. Osipov, L. I. Surface Chemistry, Theoretical and Industrial Applications.
10. Milazzo, G. (1963). Electro-Chemistry, Elsevier.
11. Weye, W. A. (1956). Trans. N.Y. Acad. Science, 12.
12. Evans, R. C. (1946). Crystal Chemistry, Cambridge University Press.
13. Gortner, R. A. (1937). Selected Topics in Colloid Chemistry, Oxford University Press.
14. McCarty, M. F., and Olsen, R. S. (1955). Eng. Min. J., Nov.
15. Del Guidice, G. R. M. (1934). Trans. A.I.M.E., 112.
16. Hackh (1949). Chemical Dictionary, Churchill.
17. Pryor, E. J. (1963). Dictionary of Mineral Technology, Mining Publications.
18. Sutherland, K. L., and Wark, I. W. (1955). Principles of Flotation, Aust. I.M.M.
19. Miall, L. M. (1961). A New Dictionary of Chemistry.
20. Thewles et al. Dictionary of Physics, Pergamon.
21. Fast, J. D. Entropy, Philips Technical Library.
22. Dash, W. C., and Tweet, A. C. (1961). Observing Dislocations in Crystals. Scientific American, Oct.
23. Taggart, A. F., Taylor, I. C., and Ince, C. R. (1930). Trans. A.I.M.E., 87.
24. Plaksin, I. N. (1959). Trans. A.I.M.E., 214.
25. Heftmann, E. (1961). Chromatography, Pergamon.
26. Kakovsky, I. A. (1963). Study of Kinetics and Mechanism of Certain Hydrometallurgical Processes. I.M.P.C. (Cannes)

CHAPTER 16

CHEMICAL EXTRACTION

Introductory

In the first edition of this book this chapter, then entitled "Gold Cyanidation", confined itself to the hydro-metallurgical process of cyanidation. With the steady extension of chemical treatment of ores and their concentrations the term "hydro-metallurgy" is being replaced by "chemical extraction", "chemical attack", and "chemical treatment". A change has therefore been made and elementary discussion of additional processes is now included.

It could be argued that the extraction of specific elements or compounds from their ores by dissolution is the province of the chemical rather than the mineral technologist, and is therefore out of place in a textbook on mineral processing. The counter-arguments are sufficiently cogent to outweigh this objection. They are economic as regards the handling and treatment of a moving tonnage of material, and technical in their use of the combined skills of chemist and mineral engineer. Broadly, the same principles are applied in

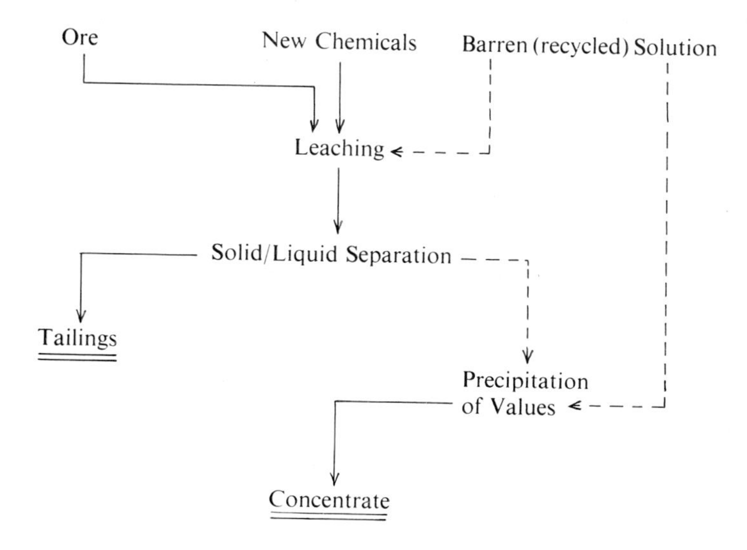

Fig. 193. The Basic Leaching Flow-sheet. Solids and pulps shown in line, liquids broken

the simple leaching by chemicals of all minerals, and even pressure-leaching can be thought of as a specialised form of this attack.

In concentration by means of gravity the treatment leaves the physical structure of the minerals unaltered. In the flotation process discussed later, only minute changes are made at the surfaces of specific minerals and the effect on chemical composition is indefinitely small. In chemical extraction, however, the desired metal or compound is separated from the finely ground ore by dissolution. This metal-rich or "pregnant" solution is then separated from the solids and receives further treatment to recover its value. It is then termed "barren" solution.

Because of the world-wide use of cyanidation in gold recovery this subject is treated here at some length. The "cyanide process", as applied to various types of gold ore on the industrial scale, receives further attention under Gold in Chapter 23.

The Cyanide Process

An understanding of the principles and practice of cyanidation establishes the basic knowledge of leaching. Probably this was the first important use of continuous chemical treatment in modern industry. It was forced upon the mining industry in the last decade of the nineteenth century by changes in the nature of the Rand banket ores. While these were being worked near the outcrop they were rich and oxidised. Even with moderate extraction by amalgamation (often below 50% being recovered) the operation paid. As the mines deepened, a poorer sulphide zone was reached and recovery by amalgamation suffered. By 1890 the financial future looked dark. The discovery that gold could be economically extracted by the use of cyanide salts transformed the outlook, and today a major industry works on low-grade material mined at great depths, with consistently high recovery. The basic process was patented by McArthur and Forest in 1887 and first applied successfully in 1894. A lecture[1] "The industrial chemistry surrounding gold" shows how a great chemical industry has grown in response to enlarged demand by the gold-mining industry.

The cyanide process proceeds in four stages:

(*a*) Preparation of the ore to expose its gold.
(*b*) Dissolution of gold from solids.
(*c*) Separation of gold-rich liquid from residual solids.
(*d*) Recovery of gold from pregnant solution.

Preparation has two main purposes in view. First, the ore must be ground to such a fineness that all the economically recoverable gold will be sufficiently exposed at the surface of a particle to ensure its dissolution during the optimum time of chemical reaction. Second, the other compounds in the ore which are liable to interfere with efficient recovery should, if possible, be removed or rendered harmless.

The solvation of gold by cyanide takes several hours, speed being a function of the total surface exposed to attack. Large particles of gold would need a very lengthy period of exposure. This would in turn require an inordinate volume of pulp-holding space if a big tonnage was treated daily. Fortunately heavy and large particles of gold can readily be trapped so the holding capacity need only be adequate to deal with particles small enough to escape from gravity concentrators. Contact is rarely less than for eight hours and may exceed twenty-four. During this period of solvation a large tonnage of gold ore, together with a much greater weight of solution, must be kept moving and thoroughly agitated. This is done in order to extract a few pennyweights of gold from each ton of crushed ore. Pulp handling is therefore important in the design and operation of the plant, in which the minimum amount of power-consuming movement must be aimed at, consistent with efficient extraction.

The weak cyanide solution circulates through the mill, becoming richer in gold. Periodically the dissolved value is removed by precipitation. During this circulation a little may be lost through spillage (though it is avoided as much as possible), and there is some evaporation. After a while the cyanide becomes "foul" with base-metal salts liable to lower the efficiency of recovery. It must then either be regenerated or discarded. Since new water is constantly coming from the grinding section, a balance must be struck between entering water, circuit loss, and discarded solution. The foul solution is still sufficiently poisonous to constitute a health hazard if allowed to run indiscriminately into the district drainage area, and the working arrangements of the plant must allow for safe disposal of waste liquor which cannot economically be regenerated and returned to use.

In the third stage of the extraction process the solids and liquids must be separated. It is customary to speak in cyanide plants of "sands" (relatively coarse beds of particles through which water can percolate) and "slimes" (true mineral particles which, when settled, would form a bed too closely packed to allow such easy passage of water). The reason for this distinction will be discussed later. The "slimes" are not a primary slime, such as a clayey or talcose material produced by the weathering down of alkali-earth minerals. Where such clays are mined with the ore, and are more or less non-auriferous (or "barren"), trouble in the separating section can be avoided if these primary slimes are removed before fine grinding is started. Some gold flow-sheets therefore incorporate a washing stage for this purpose. This has the further advantage of removing any soluble salts liable to interfere with cyanidation. "Cyanicides" and other substances which jeopardise or upset the gold-cyanide reaction include completely and partly oxidised sulphides of antimony, zinc, iron, and copper; mineral acids produced by this oxidation; some organic matter; graphite and charcoal (which can precipitate the dissolved gold prematurely), and chromium (which interferes with reprecipitation of the gold at the appropriate stage of the process).

Electro-chemical Concepts

Gold occurs in two forms, monovalent and trivalent. Monovalent gold takes a positive charge to form the aurous ion Au^+, according to the formula for a metal M of valence n

$$M + n(^+) = M^{n+} \qquad (16.1)$$

$(^+)$ being one farad (=96,500 coulombs). The atomic structure of gold (atomic number 79) shows the inner five electron shells to be filled (2, 8, 18, 32, and 18). This leaves one valence electron, the removal of which creates the aurous ion. In the electro-chemical series, with hydrogen = 0, the electrode potential of the aurous ions on a gold anode is $1{\cdot}36^+$. Gold is therefore highly resistant to attack by acids. Aurous ions can only exist in small concentrations, and their salts readily form complex ions or covalent (auric) compounds, or revert to metallic gold.

The trivalent form of gold (auric) has not at present been proved to take the Au^{+++} ionised form, despite the weakness of the outermost shell of 18 electrons. Two electrons from this shell readily accept covalent bonding. Thus aurous chloride hydrolyses back to the non-ionised forms:

$$3AuCl = 2Au + AuCl_3 \qquad (16.2)$$

in water or moist air. Auric chloride is not appreciably ionised.

When a metal is dissolved from its position at the anode of an electro-chemical system it must first become ionised. In accordance with equation (16.1) above, the mechanism for gold is:

$$Au + (^+) = Au^+. \qquad (16.3)$$

This unstable ion becomes open to corrosive chemical attack when the surface of the anode has a potential in excess of the equilibrium value for the metal ion system. If the solution in which the metal is immersed contains cathodic reagents, or if electrons are being otherwise conducted out of the system *via* the anode, the unbalanced condition required for production and solution of the aurous ions is maintained.

One such cathodic reagent is oxygen, which is slightly soluble in water and for this reason reacts so gently in the cyanide process as not to oxidise cyanide ions $(CN)^-$ to cyanogen (NC – CN) or cyanates. The cathodising power of oxygen is developed thus:

$$\tfrac{1}{2}O_2 = O^{--} + 2(^+) \qquad (16.4)$$

With water this becomes:

$$2O^{--} + 2H^+ + 2OH^- = 4OH^-. \qquad (16.5)$$

Pure gold is a bright yellow metal. It alloys with most metals. Silver whitens and copper reddens the colour. The element is most unlikely to be found in a pure state in ores. There is therefore a wide variety of possible alloys, lattice distortions and galvanic couplings. Thus a particle of gold, if of sufficient size, can develop anodic and cathodic areas at its surface in accordance with minute variations in local structure. With an extremely small particle of gold, the surface approaches a uniformity in which anodic and cathodic areas are not differentiated sufficiently to provide the poles needed for electro-chemical dissolution, and cyanidation becomes less efficient.

When gold anodes are ionised, the oxide-product is somewhat soluble in sulphuric acid, phosphoric acid, and alkaline hydroxides. If sodium or potassium chloride is electrolysed to the hypochlorite, chlorine attack on gold proceeds faster with such a salt in solution. Solution of gold in *aqua regia* proceeds according to the equation:

$$Au + HNO_3 + 4HCl = 2H_2O + NO + HAuCl_4. \quad (16.6)$$

Industrially, sodium or calcium cyanide is used to dissolve gold from its ores. The reaction (Elsner's Equation) is:

$$4Au + 8KCN + O_2 + 2H_2O = 4KAu(CN)_2 + 4KOH. \quad (16.7)$$

Oxygen is essential to the cyanide process, and the role played by it is indicated in Equation 16.5. The electro-chemical nature of the reaction was proved by Thompson.[2] He immersed a strip of gold in de-aerated cyanide jelly. When air diffused into the upper layer of the jelly corrosive dissolution of the gold commenced at the deepest point of immersion, and not at the air-cyanide interface.

Chemistry of Cyanidation

Elsner's Equation (16.7) is equally valid for silver. 130 gramme-molecules of KCN, with 8 of oxygen, dissolve 197 of gold. According to Bodlaender this reaction proceeds in two stages:

$$(1)\ 2Au + 4KCN + 2H_2O + O_2 = 2KAu(CN)_2 + H_2O_2 + 2KOH. \quad (16.8)$$

$$(2)\ 2Au + 4KCN + H_2O_2 = 2KAu(CN)_2 + 2KOH. \quad (16.9)$$

The rate of solution of pure gold was found by Barsky, Swainson and Hedley[3] to reach its peak in a solution containing 0·05% NaCN, and of silver at 0·10% NaCN. These workers also showed that when the alkalinity of a cyanide solution exceeded a *p*H of 10·3 (using sodium hydroxide) the rate of dissolution of the gold fell slowly. With lime as the alkali, the fall was more rapid.

Oxygen is essential to dissolution of the gold and the rate of reaction is proportional to the available dissolved oxygen. Unfortunately, other ore constituents in the treated pulp are consumers of oxygen, the supply of which must therefore be maintained by good aeration if extraction is to proceed with reasonable speed. The oxidation of metal sulphides is accompanied by the formation of acid. This, if ignored, would destroy the sodium cyanide used commercially. Hence a moderate excess of alkali, usually lime, is maintained in the pulp during aeration. This is called "protective alkali". In addition to sulphide minerals, organic substances such as wood, oil and grease and abraded iron from the milling machinery, are considered to slow down reaction by consuming oxygen. Various oxygen-yielding additives have been used experimentally to speed up the dissolution of the gold by cyanide, but none have been commercially adopted.

It is possible to remove abraded iron either by the use of magnetic separators or during the concentration by gravity devices of coarse gold ahead of cyanidation (for example, on corduroy strakes). Some operators consider that the exposure of thin films of pulp on strakes accelerates the oxidation of unstable sulphides such as pyrrhotite, and they defer addition of cyanide till this has been completed. Others consider that pre-aeration, as this is called, introduces soluble iron salts which "kill" part of the cyanide by converting it to thiocyanate and ferrocyanide, which are not solvents for gold. If the ore has been ground and given gravity treatment in water, the pulp must in any case be thickened before cyanide is added, in order to minimise dilution of re-circulating barren cyanide solution. Fouled water can be removed at this stage. When the gold-bearing fraction of the ore is concentrated by froth-flotation before cyanidation, the problem of oxygen consumption is especially severe. It may be dealt with by roasting such concentrates, sometimes using their sulphur content as the only fuel. Interfering soluble salts can then be removed by washing the resulting calcine before it is cyanided.

At one time bromine (in the form of cyanogen bromide BrCN) was used to aid dissolution, the reaction being supposedly:

$$BrCN + 3KCN + 2Au = 2KAu(CN)_2 + KBr. \tag{16.10}$$

This bromine salt is easily destroyed by excess alkali, iron, or pyrites, and although it was favoured for treatment of telluride ores, its use has virtually ceased in current practice. Improved grinding methods have led to better exposure of the gold, and where necessary a selected fraction carrying the troublesome mineral can be removed from the main stream of pulp by flotation, and given special treatment.

The rate of solution of gold from a pure specimen under ideal laboratory conditions was found by McLaurin[4] to be at its maximum with 0·25% KCN, a figure more than twice that obtained by Barsky, Swainson and Hedley[3] with NaCN. In plant practice the supply of oxygen is limited and a further factor must be considered—the rate at which dissolved gold moves away from the surface of the particle and is replaced by fresh cyanide solution. The "common ion effect" mentioned in Chapter 15 tends to hold solution ions in super-normal concentration close to similar ions still partly bound in their solid lattice. While it is true that this common ion attraction ceases with the completion of the chemical reaction that locks the gold into a non-ionised cyanide complex, there remains an envelope of partly spent enriched cyanide around the auriferous particle. This enveloping action retards further solvation unless overcome by such rate-determining factors as diffusion and brisk agitation of the ore pulp. Since the effect is influenced by viscosity, the rate of solution varies with temperature. In practice artificial heating is not used. Wet grinding introduces an appreciable amount of heat into the pulp. In sub-arctic conditions pulps must be kept from freezing.

The precipitation of gold from solution is effected by mixing the pregnant solution, after removing as much oxygen as possible, with zinc in excess of its gold content. The exact nature of the chemical reaction has not been agreed. McFarren favours the reaction:

$$\begin{aligned} &KAu(CN)_2 + 2KCN + Zn + H_2O \\ &= K_2Zn(CN)_4 + Au + KOH + H \end{aligned} \tag{16.11}$$

but Christy prefers:

$$2KAu(CN)_2 + 3Zn + 4KCN + 2H_2O$$
$$= 2Au + 2K_2Zn(CN)_4 + K_2ZnO_2 + 2H_2. \quad (16.12)$$

Bearing in mind that zinc and gold lie on opposite sides of the neutral point in the electro-chemical series, and that gold dissolves in the presence of oxygen, and is most effectively reprecipitated in its absence, there appears to be a case for considering the hypothesis that the action follows reversal from anodic to cathodic flow. Rose and Newman[5] observed that, in the presence of free oxygen, gold dissolves in free cyanide about six times as fast as zinc, whereas in the absence of oxygen zinc easily displaces gold from aurocyanide solutions.

Gold is also extracted from solution by carbon in the form of charcoal and by some types of graphite. When such carbonaceous material occurs with the ore, it can sometimes be removed by flotation before cyanidation, roasted out, or rendered innocuous by causing it to acquire an insulating film of oil. Some operators now use carbon to remove the gold simultaneously with its dissolution in cyanide by the use of activated charcoal. Activation is procured by heating the charcoal sufficiently in an oxidising atmosphere to remove the inert hydrocarbons at its surface. Activated charcoal can sorb up to 2000 oz. of gold per ton of carbon.

Introduction of Cyanide to Ore

It has been noted that the sequence of operations in cyanidation is:

(*a*) Comminution (perhaps with removal of coarse gold by gravity concentration).
(*b*) Dissolution of gold in cyanide solution.
(*c*) Separation of spent solids from pregnant solution.
(*d*) Precipitation of gold from pregnant solution.

Many plants mill their ore in cyanide solution. This method was used on the Rand in 1892, but discontinued, the technique involving difficulties which at that time could not be mastered. If amalgamation is to be used in the grinding circuit, there are objections to the presence of cyanide. The protective lime in the solution hardens the amalgam, and the supporting copper plates are liable to be chemically attacked. Again, if flotation is used to concentrate the gold and auriferous sulphides or to remove cyanicides, cyanide is not admissible at the same time. Gold-sorbing carbon in the ore, such as a graphitic shale, would also rule out introduction of cyanide before this carbon had either been removed or rendered innocuous. If grinding released cyanicides, it would probably be better to deal with them in a water pulp than in one containing cyanide. There is the further point that personnel working with corduroy strakes are exposed to some risk from the presence of the poison. Where the nature of the ore permits, the case for grinding in cyanide solution rather than in water is strong. It gives a maximum contact time, commencing the moment grinding has exposed a fresh, clean gold

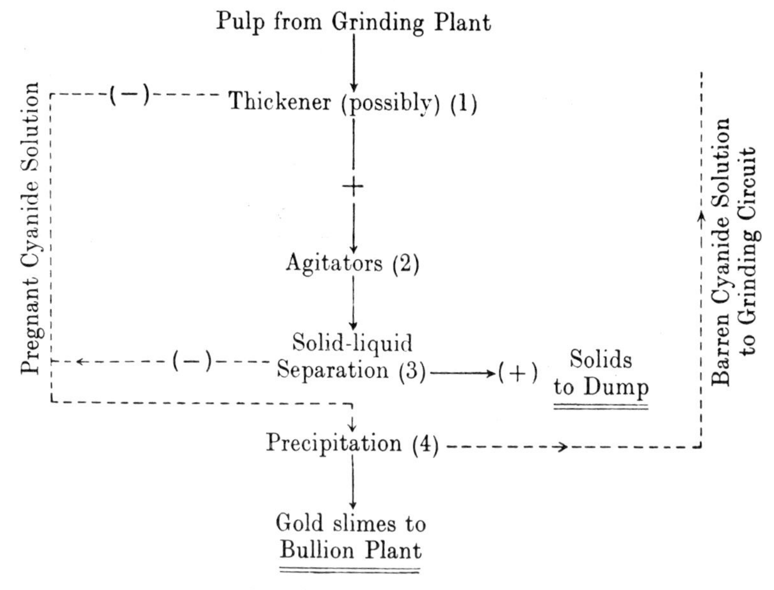

Fig. 194. *Grinding in Cyanide*
(*Liquids are shown by broken lines in this and subsequent flow-sheets*)

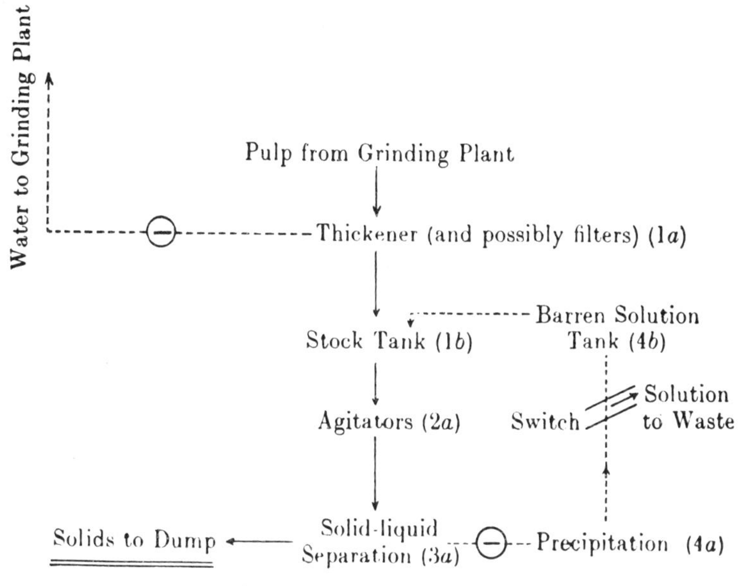

Fig. 195. *Grinding in Water*

surface and the problem of balancing the removal of foul cyanide solution against newly entering water is simplified. In ores in which the gold dissolves quickly there is a good case for grinding in cyanide, but where dissolution proceeds slowly, the extra time thus gained is a minor consideration. Gold in characteristically sulphidic Canadian ores dissolves much faster than that in typical Rand bankets. This may be due to plutonic as against alluvial genesis. Some Rand plants now mill their ore in cyanide solution.

Consider two forms of a simple flow-sheet in which all the ore has been finely ground and sent from the grinding section for cyanidation by agitation (the method is discussed later). Let the grinding sections operate respectively in cyanide and in water. The flow-sheets then are as shown in Figs. 194, 195.

In the cyanide grind, the only uncontrolled water entering the plant is the moisture in the head feed. Some water is lost by evaporation, and a certain amount of cyanide solution must be withdrawn when it has become too "foul" with base-metal salts to be efficient as a gold solvent. This make-up water can all be introduced at a point where it does useful work, and the quantity thus introduced is under the control of the operators. This point is reached after the pregnant cyanide has been displaced in stage (3) of the flow-sheet, but a wetting film of auriferous cyanide still envelopes each particle of the solids. If the water is used to wash or scrub these particles, the loss of dissolved gold thus carried over to the dump is reduced. Item (1) of the flow-sheet is shown as doubtful, since it might be better to agitate the effluent from the grinding circuit without preliminary thickening.

Consider next the alternative flow-sheet (Fig. 195). Item 1*a* (thickener and possibly filters) is no longer optional, as it is essential to remove most of the water before introducing cyanide. If a thickener is used for this purpose, and an underflow carrying 50% of solids is produced, each ton of ground ore entering carries with it one ton of new water. This water must be accommodated in the solution-holding system, and must therefore displace an equivalent quantity of cyanide solution, whether or no the latter is foul. The operator has thus lost a useful manipulative control. If to the inexpensive operation of thickening it is decided to add filtration, the new water is much less—say 15% entering on the filtered solids—but an extra element of complication and expense has been introduced. This thickened pulp (or repulped filter cake) must now be held in suspension in a collecting tank (1*b*, the stock tank) and thinned down with barren solution. Since the barren solution leaving 4*b* will now be diluted by fresh water, it must be adjusted in strength before use, or alternatively fresh solid cyanide must be added at 1*b*. Part of this addition would be needed in the first of these flow-sheets, since cyanide oxidises in use and must be replenished, but part of the addition now under discussion is needed to replace loss incurred at 3*a*. When the solids are separated at this point, any fresh water used to displace clinging auriferous cyanide adds to the amount of barren cyanide which must be withdrawn and run to waste, so the use of such water must be restricted.

In mineral dressing, the process should always be developed to suit the specific ore, and to achieve desired ends with the greatest simplicity and economy of effort. A strong case exists for milling in cyanide, but there are exceptions. One would be a plant relying on a transient population of unskilled labour, such as might be recruited in Africa from tribal reserves. If mistakes occurred ahead of 1*b* (Fig. 195), they could readily be put right. If the same thing happened ahead of (1) (Fig. 194), a heavy loss would result if pregnant cyanide were spilt. Derangement of the pulp transport system through inexpert handling of pumps and piping circuits could lead to such a loss.

Sand Leaching

Crushed ore from milling (together with auriferous tailing from gravity or flotation treatment) should be brought to its optimum technical condition before cyanidation. Cyanicides are removed, neutralised, or reduced to tolerable concentrations. Comminution liberates the values to the point where the greatest profitable recovery can be made, and the cost of further grinding would not be compensated by increased recovery of bullion. Since grinding is usually the most expensive item in treatment, the head value of the ore and the mode of occurrence of its gold will, in each case, determine the amount of grinding which should be done. At one extreme, ores such as the Mountain Copper Co.'s are treated after crushing to $-\frac{3}{8}''$, while at the other, the auriferous sulphides of a high-grade ore may be ground down to extremely fine slime before maximum profitable recovery is achieved.

When the crushed ore is sufficiently coarse, cyanide-bearing solution can be made to percolate through moderately thick beds, with sufficient speed and searching power to dissolve and remove the bulk of the exposed gold. The rate of percolation determines the number of days a bedded tankful of such crushed material must be treated, and hence the holding capacity (as capital investment) of a leaching plant for a given throughput tonnage. The porosity of the bedded material determines the rate of percolation. If an unclassified mass of sands were bedded, the associated slimes and fine sands would obstruct the interstices between the large grains and thus interfere with the percolation rate. This could in part be compensated by working with a thinner bed, by using vacuum to pull the leach liquors down through the bed, and by stirring these liquors into the upper layers of the bed. In the earlier days of cyanidation the feed was classified into two types of product—sands coarse enough to be leached and slimes too fine for percolation. With the improvement and cheapening of grinding methods it has usually been found better to do away with leaching of such ores and to reduce all the feed to a fine state of grind—called on the Rand "all-sliming". Static leach treatment continues to have value in a variety of special applications, notably with low-grade ores, dump retreatment and low-capital projects where simple home-made devices aid development finance.

The general scheme of leaching treatment can be pictured as in Fig. 196.

A certain amount of fine material can be tolerated in the sands classified to the leaching section without retarding percolation too seriously, but it is usually found undesirable to allow bad separation for another reason. This is that gold-bearing solution tends to cling film-wise to the surface of ore particles, and as the particle size becomes smaller not only does the total surface per

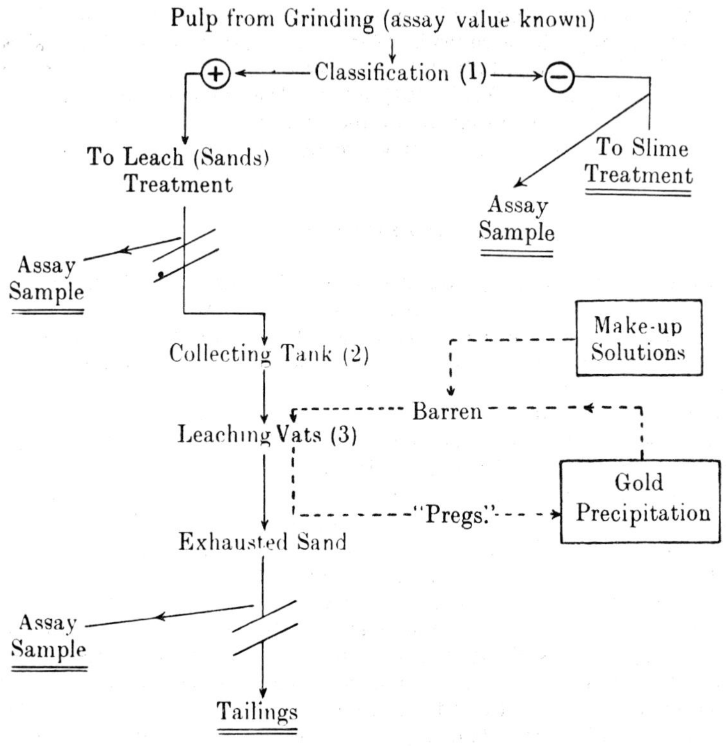

Fig. 196. *A Gold Leaching Flow-sheet (Solutions shown by broken lines)*

unit volume increase exponentially, but more violent displacing action must be used to remove these films than can easily be applied to leached sands. At Golden Cycle[6] better classification of −20 mesh material at a 200-mesh split raised percolation rate from about 1″ to 10″/hour.

The coarse fraction can be produced as an underflow from hydraulic classifiers or sand cones, or as the drag-out from rake, spiral, or bowl classifiers. It can be fed into steel collecting vats (25′ to 50′ in diameter and 6′ to 9′ deep), either through a hose directed so as to equalise the bedding, by turbo-distributors such as the "Butters and Mein" (Fig. 197) or by mechanical distribution. Sand settles, while slime overflows with the excess water through vents in the sides of the tank, which are plugged as the load rises. This gives some further classification, the overflowing slimes being led to the slimes section of the plant. (When calculating yield, correction must be made for gold thus transferred). Sand can be leached direct in the collecting tank, but in a large operation it is transferred *via* bottom discharge openings to leaching vats. These are large round tanks made of wood, concrete, or steel, up to 55′ in diameter and 15′ in depth, the largest holding 850 tons. One ton of Rand barren sand occupies 23 cu. ft. as collected, and 21½ cu. ft. when settled. If crushed ore is leached without preliminary classification, the practicable bedded depth rarely exceeds 3′.

The leaching tank has a false bottom through which solutions are discharged (and occasionally introduced) and through which vacuum can be applied. This bottom consists of an open wood framework of slats or perforated board. It is covered by twill, matting, or duck fabric, on which wooden boards are laid to protect the textile from damage when workmen are shovelling exhausted sands to the central bottom discharge. Fittings on appliances which handle cyanide solution are of iron, not of copper alloys. Leach rate varies from ½″ an hour up, 7″ being a fair speed with vacuum aid applied to the false bottom. The drained moist sands, still carrying up to 15% moisture, are levelled by hoes. Care must be taken that distribution of sand is reasonably uniform or during the application of successive leaching washes channels may form in the bed, through which the solutions will run without making the searching contact with all the grains which is necessary to dissolution. The other essentials are the presence of oxygen and of sufficient lime to prevent destruction of the cyanide. Some air is brought in with new solution, and some is drawn in during drainage. Air can also be introduced by top-raking, use of compressed air blowers, or by applying suction from below. A more thorough method is to transfer the whole charge from one tank to a second half-way through the leach, the movement breaking up channels and ensuring good exposure to air. Leaching tanks are sometimes built in pairs, one above the other, to facilitate this transfer and economise space. "Strong" cyanide (0·05–0·1% NaCN) is used for leaching, and is run on till the sand charge is covered, the solid-liquid ratio then being from 70:30 to 75:25. If the liquor is permitted to percolate continuously, a cover of up to 12″ being maintained above the sand, there is less risk of displacing entrapped air from the tank. This strong solution is sent for precipitation. From time to time the charge is drained, and when dissolution is nearing its end, weaker cyanide is employed (0·02 or 0·03% NaCN) to complete the reaction and displace pregnant liquor. This solution is not usually sent for precipitation but becomes stock for making up strong solution. Final washing is with water or waste liquor from the slimes plant. The treatment cycle takes from four to seven days. Tailing is then removed by shovel, sluice, or mechanical plough and transported to the dump.

The double handling of sands from one leaching tank to another, or when wetting with cyanide, begins in the collector tank. It aids aeration and counteracts loss due to the tendency of the sands near the bottom of the bed to become waterlogged and unaerated. The canvas bottom, which is held to the tank sides by a ring of caulking rope, must be kept free from clogging due to precipitated lime salts. A typical leaching sand is 50% *plus* 100 mesh in grain.

Leaching has been successfully applied to clayey ores and the small-scale retreatment of old dumps. The clayey material is worked by ploughing into balls, sometimes consolidated with a little lime, and varying from ¼″ to 3″ in size. Leaching solution is run gently on to, and off from, this material either in vats or in shallow trays constructed from vertical boardings on an impervious

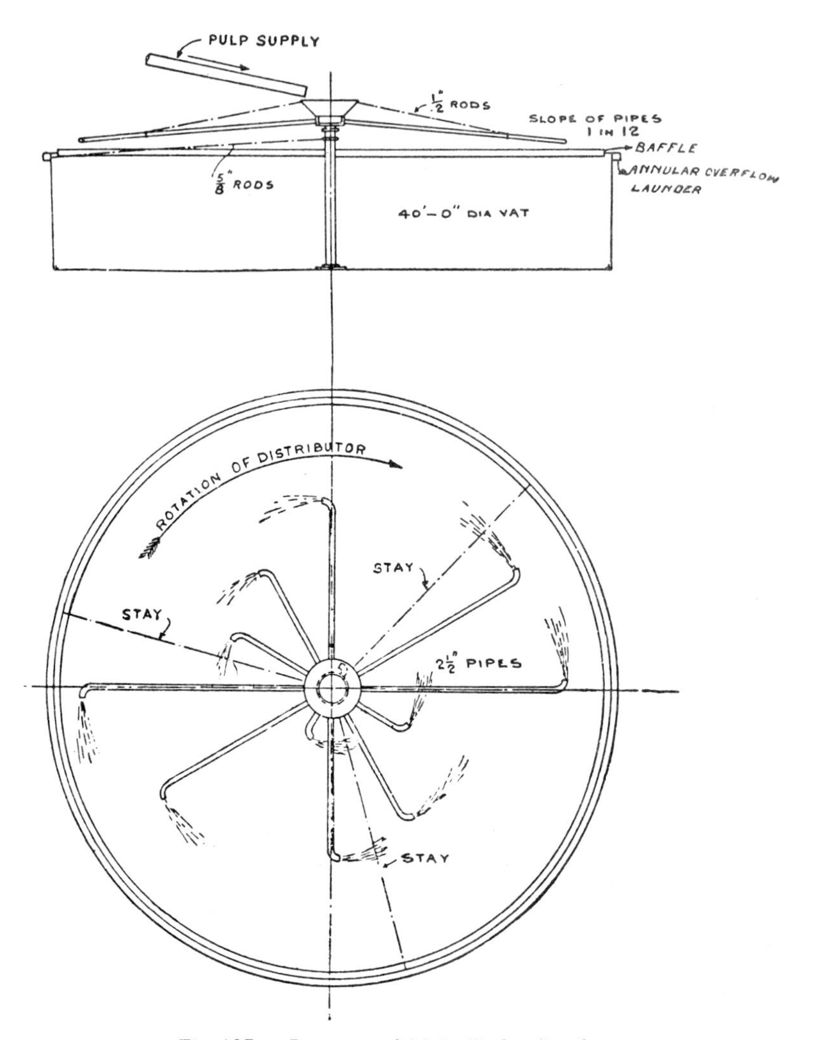

Fig. 197. Butters and Mein Turbo-distributor

clay flooring. In the case of reworking old dumps unusual care in needed to avoid adding more than a bare minimum of protective alkali, if trouble due to fouling of the solution by oxidised salts from the ore is to be minimised.

Slimes Agitation

"Slimes" is a description which has persisted in gold ore treatment from a less technically-minded age. It would be helpful if the use of the old term could be confined to colloidal clayey material associated with the true ore as the result of natural weathering, infiltration, or non-selective severance, and not applied loosely to a variety of minerals, including the finer mesh sizes of true ore produced during grinding. Primary slimes are usually loosely adherent to the ore, and can be removed by washing, without substantial loss of head value. When this operation can be justified, a nuisance is removed and subsequent treatment facilitated. Primary slimes are apt to coat surfaces which should remain open to chemical attack, to sorb expensive reagents wastefully, and to cause trouble in the thickeners and filters by their slow rate of settlement and ability to choke the interstices of separating media. The word "slimes", as used on the Rand, may apply to pulps containing a certain amount of such primary colloidal material, but is intended to describe ore too finely ground to be effectively infiltrated by gently applied leach liquors. The dominant characteristic of slimes treatment is its application to particles (say *minus* 100 mesh) too small to be permeable by a percolating leach-liquor when settled into a solid mass, but small enough to remain suspended in cyanide solution under conditions of mild stirring and agitation.

Treatment may be generalised as in Fig. 198.

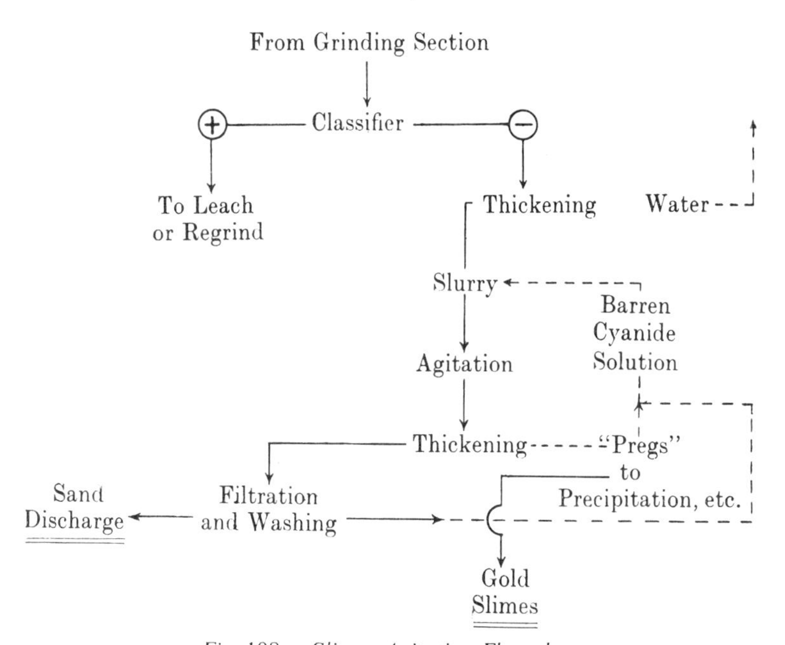

Fig. 198. *Slimes Agitation Flow-sheet*

As already seen, when milling is not done in cyanide solution, thickening is needed to remove as much water as possible before slimes treatment. This removal of water may be made the occasion for decanting off undesirable

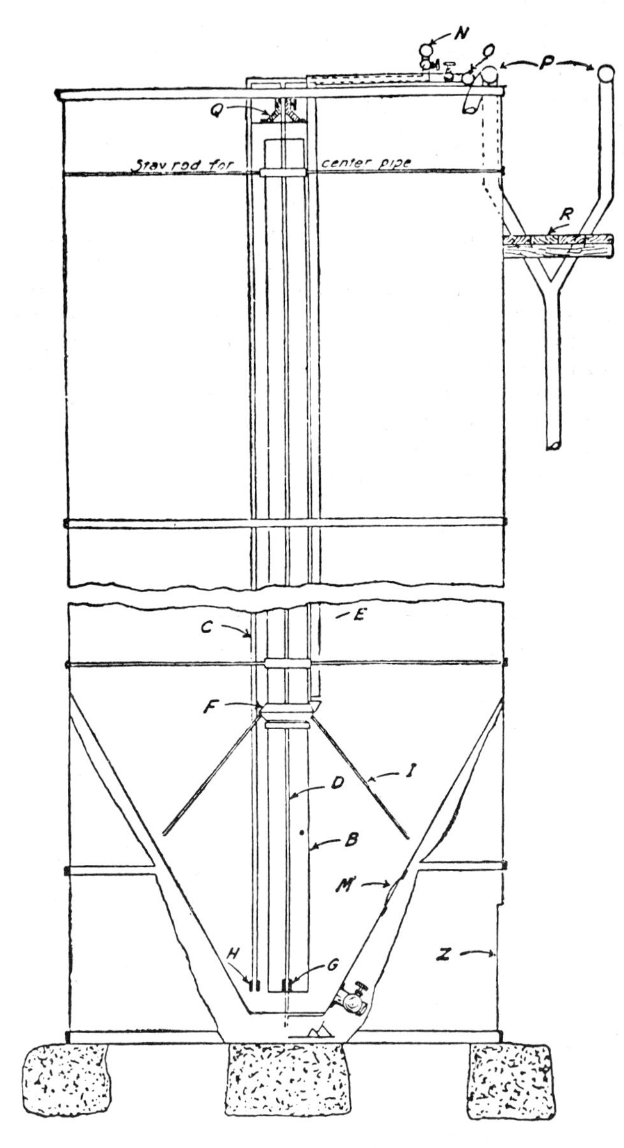

Fig. 199. Brown or Pachuca Washer

Air system, O, D. Outlet at G.
Auxiliary stirring air *via* C, E.
Water and solution system, N, E.
Splashplate, Q. Slimes feed, P.

soluble salts, or can be combined with pre-aeration to oxidise pyrrhotite, etc., and adjust the protective lime content of the pulp which may contain between 5 % and 45 % of *plus* 200 mesh material as it leaves the grinding circuit. Lime sufficient to maintain an available alkalinity of between 0·002 % and 0·025 % CaO is used as the "protective alkali". It also aids settlement of the arriving pulp by flocculating the fine particles. The pulp is next thickened to a slurry containing 35 %–45 % solids, usually in continuous thickeners. The clear overflow water is either re-used in the grinding plant or run to waste if foul with dissolved salts. (If the ore is ground in cyanide, this overflow is a valuable "preg" and goes to precipitation when a stage of thickening is used before agitation.) Part of this lime may be added during transit to the fine ore bin ahead of wet grinding if oxidation of pyrite is to be suppressed, as would be essential when grinding in cyanide. If, however, such oxidation aided the process by cleaning or stimulating the exposed gold surfaces mild pulp acidity might be encouraged ahead of thickening of an aqueous pulp.

Where discontinuous transfer is still practised the mill pulp is settled in a collecting tank from which as much as possible of the water is overflowed or decanted. The settled slurry is then diluted with barren cyanide solution as it

Fig. 200. *Dorr Agitator*

is being transferred into agitators. This method has largely been replaced by one in which thickeners receive pulp from the grinding circuit and send a thickened slurry to the agitators. Where space is limited, or winter conditions severe, tray thickeners are sometimes used. For agitation the pulp is thinned to a water-solid ratio of about 1·2:1, with sufficient dissolved cyanide to give a strength (in terms of KCN) of 0·01 % and protective lime 0·0005 % CaO.

Various types of agitator have been developed. The Brown or Pachuca is a tall cylinder with a conical bottom (Fig. 199). Air is blown in so as to cause the central pipe to act as an air-lift pump, lifting the pulp to the splashplate or baffle, whence it falls back into the tank and is duly recirculated with fresh aeration. These tanks are made of steel, and vary from 13′ to 33′ in diameter and 45′ to 55′ in height. They are either worked discontinuously or in series. If discontinously, the tank is filled with pulp and is then "blown" for the desired period of agitation (from 3 to 12 hours), periodic checks being made for strength of solution and protective alkali. If work is continuous, the required holding capacity of tank space is arranged in series. New pulp is fed continuously to the head tank and an equivalent quantity is displaced from the last tank. There appears to be little metallurgical difference between

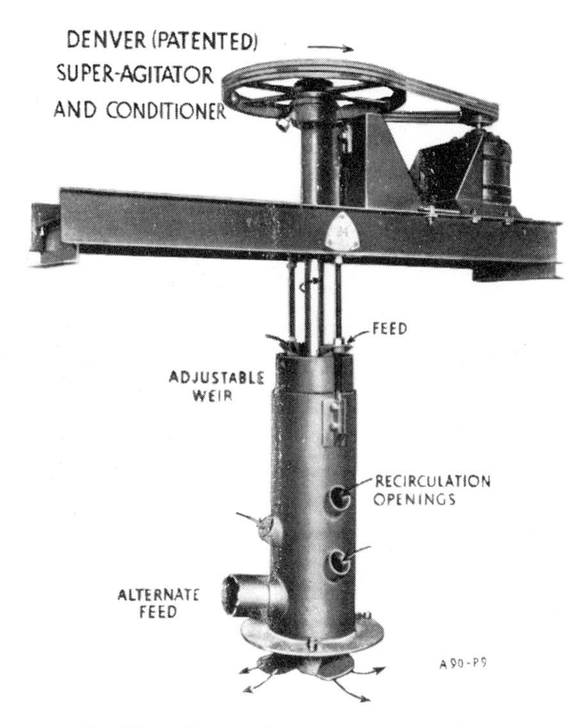

Fig. 201 Denver Super Agitator Mechanism

results obtained by batchwork and continuous working. When a tank becomes choked with settled slurry, it is restarted by introducing solution through the washring, or by working down a pipe ("lance") carrying compressed air, till it stirs the packed solids in the cone and frees the blockage. Series operation avoids loss of working time otherwise wasted during the filling and emptying of batch-operated tanks.

Intermediate in design is the Dorr agitator, which combines air-lift with gentle mechanical stirring of the contents of the tank (Fig. 200). The appliance resembles a Dorr thickener, but the rakes rotate from 1 to 4 times a minute, ploughing settled slimes to the central air-lift shaft. From this the pulp is delivered to a rotating launder and falls back into the tank, carrying with it some air. The scrapers are necessary because the flat-bottomed tank does not receive the sluicing possible with the cone of the Pachuca. The power used is low, but aeration is less copious than with the Pachuca, and is sometimes aided by the use of extra compressed air blown down into the tank.

The Wallace agitator (Fig. 201) is mainly mechanical in action. A 27″ impeller runs at 200 r.p.m., near the bottom of a 12′ × 12′ cylindrical tank, and above it a central standpipe draws air down as the impeller creates a vortex, agitation and entrainment of air thus being produced. It can also be used to give pre-aeration ahead of the Dorr agitator.

In some cases the central lift of the Brown agitator has been dispensed with, better dispersal of the air appearing the result: Clemes[7] finds that when continuously agitating 75% −200-mesh pulp in 33′ × 48′ Brown agitators with 60° cones, using ½ cu. ft. of 25–30 lb. pressure air per minute per ton of solids in a 1·40 specific gravity pulp, there is but little accumulation of sand when running with no central pipe. It is claimed that the longer column of pulp gives a better hydrostatic head than with the Dorr, and consequently stronger aeration and dissolution of oxygen.

Separation

After optimum solvation of the gold has been reached (or in some processes concurrently with this solvation), pregnant cyanide must be displaced. The first wash with cyanide solution, applied to the new feed, dissolves the bulk of the gold. This solution is sent direct to the precipitation plant. However separation has been effected, the solid fraction is now wet with pregnant solution which films each particle and is held by capillary action in the spaces between particles. This cyanide must be removed by washing with barren cyanide, low-value cyanide, or water. Cyanide solution in various stages of gold-enrichment is available for this work, but the amount of new wash water introduced must not exceed the foul cyanide liquor discarded, since solution storage facilities are limited.

The various washings produce cyanide solution which is either (*a*) rich enough to be sent to the precipitation section, (*b*) mildly auriferous and ready to pick up more gold before precipitation, or (*c*) foul enough to need regeneration or to be run to waste. In addition, barren solution is returning from the precipitation section. The two main methods of effecting primary separation are C.C.D. (counter-current decantation) and filtration. A combination of the two is also used. C.C.D. utilises the method in which upgrading or downgrading is applied by stages (Fig. 202). Any number of retreatments desired are used in line, the down-graded product from each appliance providing the head feed of the next succeeding appliance, and being joined there by the upgraded product from the next appliance lower down the line.

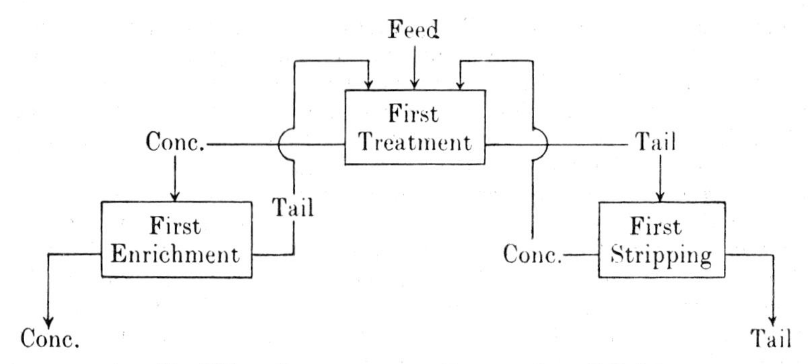

Fig. 202. *Counter-current Decantation (C.C.D.)*

This principle is applied in several ore-treatment processes. In gold cyanidation barren solution (or at the last stage wash water) is worked from the tail end of the process up toward the feed end, while the ore travels down by similar stages toward the tail end, losing part of its gold at each transfer (Fig. 203).

As shown in this diagram, the ore is being milled in weakly auriferous cyanide solution, and the overflow from the first thickener (T_1) goes to precipitation. Aeration (A) is applied to the slurry from this thickener. The barren solution leaving "precipitation" for thickener T_4 must be re-aerated before it can dissolve gold. Fresh water to T_5 balances the loss of weak cyanide in the slurry leaving T_5 as underflow. If reduction of the loss of cyanide, gold, or water justified the cost, this slurry could be filtered to recover most of the liquor. A desirable working detail is that the slurry entering each lower thickening stage should be turbulently mixed with the overflow cyanide

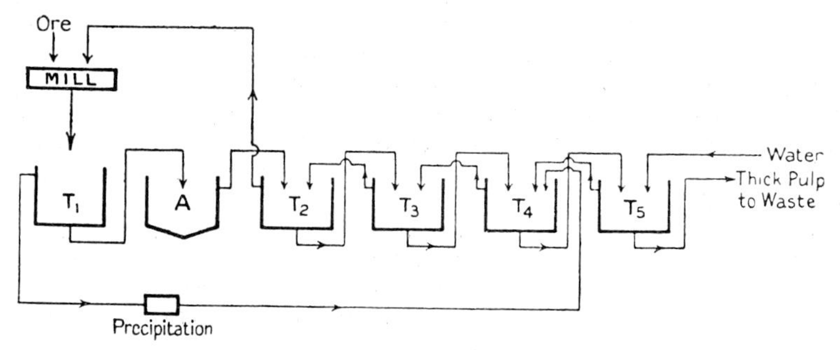

Fig. 203. *C.C.D. in Cyanidation*

coming upline from the thickener below. This turbulence entrains necessary oxygen and also helps to displace pregnant cyanide from the particles. C.C.D. is not used on the Rand, where intermittent decantation (similar in principle) was practised for many years until it was displaced by continuous or intermittent filtration. The latter methods do not lock up so much liquor and the problem of handling foul cyanide is simpler. Metallurgical results are considered better with Rand bankets when separation is by single-stage filtration.

C.C.D. may be applied through a series of thickeners, filters with intermediate re-pulping of the filter cake, or via classifiers. There are two basic steps. The first requires concentration of the solids into the smallest convenient bulk made possible by removal of the liquid, followed by re-mixing of these solids with more dilute liquid which is advancing counter-current to them. The second step is displacement of the liquid film surrounding each grain of solid.[8]

The operation of filtration, together with description of the machines used, is dealt with in a later chapter. Two main types of filter have been used, intermittent and continuous. The intermittent type of filter is disappearing because continuity of operation gives better control and more compact housing, and saves operating time. Filtration is usually applied to slurry delivered from thickening, the arrangement being on the lines shown in Fig. 204.

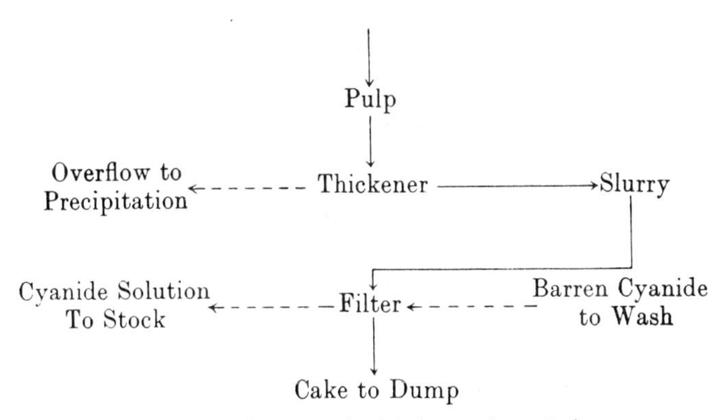

Fig. 204. Removal of Solution from Pulp

Instead of dumping the filter cake it may be repulped, either in water or barren cyanide, and then refiltered. It has been found with some ores containing readily oxidised material that the exposure of the filter cake to air drawn through to the interior of the drum by the internal vacuum provides an intensive re-aeration, and a useful final extraction of gold. This can also be the case when, for any reason, aeration during pulp agitation has been inadequate.

Since the force pulling the pulp on to the separating membrane of a continuous filter is produced by vacuum, a certain amount of de-aeration of the liquor occurs at this point. For precipitation of gold from solution this is very useful, as the next section shows. Unfortunately, however, carbon dioxide is also removed under vacuum, and this leads to precipitation of lime salts in the pores of the filter membrane. Periodic removal of this lime is therefore necessary in order to maintain a pervious septum.

In some plants which grind the ore in cyanide it is found that the moisture leaving the process (on the discarded filter cake) balances the new intake of water and provides a convenient discarding channel for foul mill solutions. Some gold is lost in the discarded filter cake. Re-pulping and re-washing are stopped when further recovery ceases to pay for the cost of extra treatment. The influences which bear on efficiency of solvation are considered in Chapter 15 under Three-phase Systems and shown in (15.6) as a rate-determining equation.

Clarification and Precipitation

The position at this stage is that cyanide solutions of various strengths and degrees of contamination, aeration, and gold content are circulating through the plant or held up in tank storage. The richest of these solutions is sent continuously to the precipitation section to be stripped of its gold. The operational sequence is of the nature shown in Fig. 205.

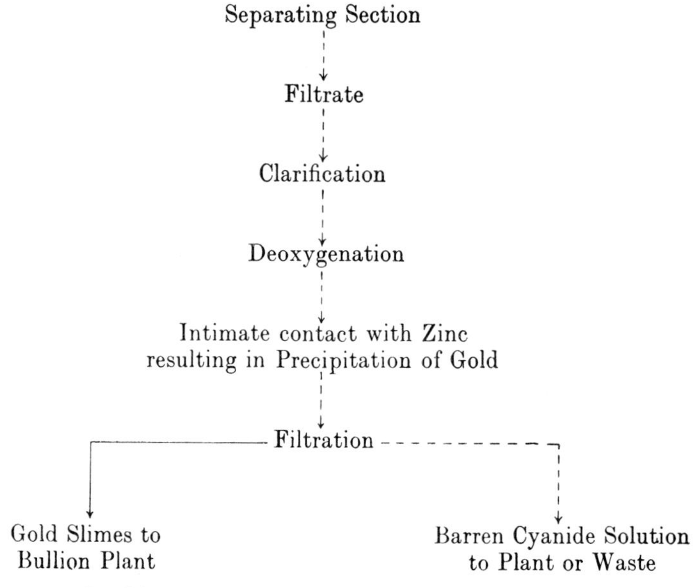

Fig. 205. *Precipitation of Gold from Pregnant Cyanide*

As the filtrate leaves the separating section it is more or less cloudy, owing to the presence of material too fine to be trapped on ordinary cloth filters. If zinc was introduced into solution in this condition the process of precipitation would be interfered with, partly by films of slime settling on the zinc and partly by adulteration of the gold precipitate with such slimes. This undesired slime would flocculate and choke the filters used to retain gold slime. It is therefore standard practice to clarify the pregnant filtrate from the separating section, the aim being to turn it into a limpid liquid. The oldest method is to pass the liquor through a tank containing a bed of sand. Colloids are arrested as a film at the top of the bed, which can be skimmed off periodically. A certain amount of de-oxygenation accompanies this clarification. At one time this was aided by the addition of iron filings to the sand, to take up oxygen by rusting. This practice has been abandoned since it introduces undesirable chemical reactions between the oxidising iron and the free cyanide. Small thickeners have been used for clarification, the rake mechanism being operated once daily and lowered a fraction of an inch, so as to plough a thin skimming of foul sand to the central well for discharge. In a variation of this the thickener works upon a bed of proprietary filtering material, which is similarly shaved down.

The method of precipitation worked out by Merrill and Crowe, which embraces clarification, is in widespread use today and is described below. A short description of precipitation in boxes filled with zinc shavings is desirable, as this gives some insight into the problems involved and provides a practical method of working the small, undercapitalised type of venture. Zinc shavings are swarfed on a lathe or spun from molten zinc just before use (to minimise surface oxidation). They are dipped into 10% lead nitrate (or acetate) solution, which precipitates on them a darkening coat of lead. Thus an electro-couple is formed which has a stronger precipitating action on the gold-bearing cyanide than has pure zinc. Lead nitrate may also be used in the agitation section to precipitate soluble sulphides, and any excess arriving at the zinc will help to maintain the PbZn coupling.

The zinc shavings (or extruded zinc wire) are packed into steel boxes, so made that the pregnant solution will flow upward through each box, over a weir, and down to the bottom of the next box in the series through which it is to rise. Shavings are contained in wire baskets, through which precipitates can fall to the bottom of the box for collection, *via* a drainage plug at one downsloped corner. In use, the zinc in the head box of the series (five to twelve boxes) disintegrates. New zinc is added to the final box, and is worked up toward the head as replacement is made. Minus 20-mesh shavings are periodically removed from the head boxes and sent to bullion treatment. With too weak cyanide, a white precipitate forms on the shavings of zinc. This masks the surface and inhibits precipitation of gold by preventing contact. By raising the cyanide strength at entry this precipitation is largely avoided. If the solution has been de-oxygenated before presentation to the zinc such trouble is minimised.

Care must be taken lest extremely fine gold, as slimes, becomes detached from the zinc and is flushed away. Sometimes the final box is left empty or is packed with coke in order to trap such slimes, which can be periodically reclaimed. No attempt should be made to dress the zinc boxes while cyanide is flowing. Up to 10 tons of solution can be precipitated daily for each cubic foot of carefully packed filiform zinc (weight about 5 lb.). During clean-up, all gold slime must be carefully collected from the boxes and transferred to the vessel in which cleaning up and/or acid treatment are to be performed.

In most modern plants zinc dust is used for precipitation, usually in a Merrill-Crowe (Merco) simultaneous clarification-precipitation unit. A typical layout is shown in Fig. 206. The first step is the clarification of the pregnant solution coming from the section where most of the solids have been removed. This solution may be slightly cloudy or fairly clear, but it carries minute particles such as the hydrates of aluminium, iron, and magnesium.

The pregnant liquor runs into the clarifying tank A, which is kept filled to a constant level by means of a float valve B. In this tank are hung filtering leaves. Solution drawn through these leaves by vacuum flows to a manifold and thence to the de-aeration tower D. One type of filter leaf has a light canvas cover which is coated with diatomaceous silica before being put into service, a special compartment and connexions in the clarifier tank being provided for the pre-coating and testing of new leaves before their clarified filtrate is permitted to join the main flow.

In the de-aeration tower the sparkingly clear liquor flows, under automatic control, over grids which expose it to vacuum. Here most of the dissolved oxygen is removed. In this tower carbon dioxide is also abstracted, thus tending to cause the dissolved lime (protective alkali) to precipitate out. This tendency is held in check by the fact that there are not many solid nuclei in the liquid on to which such precipitates can begin to form. In some plants sodium phosphate salts are added in minute quantities to retard this precipitation of lime.

De-aerated cyanide solution is drawn from the bottom of the tank by a special pump sealed against air leakage. On its way it receives a carefully measured addition of lead-activated zinc dust G which has been wetted with barren cyanide. The zinc presents a large active surface to the auriferous cyanide and precipitation occurs with great rapidity, possibly in accordance with equation 16.11 or 12.

This is one of several possible equations for the precipitating reaction. The almost instantaneous precipitation effected in the Merrill-Crowe process is important. If much calcium were present and sufficient time elapsed the zinc would be filmed over with calcium zincate:

$$Ca(OH)_2 + Zn = CaZnO_2 + 2H. \qquad (16.13)$$

Given time, calcium carbonate could precipitate on the zinc dust, thus insulating zinc and preventing it from playing its designated role in the reaction. From the special or "precipitation" pump the solution goes to the precipitation press, or filter E, where the excess zinc dust and gold slime are arrested and held until the next clean-up, while the now barren solution flows or is pumped to storage for re-use, or is discarded. Usually between 0·02 lb. and 0·06 lb. of zinc is used per ton of "pregs", but when silver is present the quantity needed may be increased as much as tenfold.

The use of a soluble lead salt (acetate or nitrate) together with the zinc stimulates gold precipitation. Since the addition of lead to the cyanide liquor might lead to the formation of insoluble lead salts (sulphide, carbonate, sulphate, or chloride), it should either be added ahead of clarification or completely precipitated on to the zinc before reaching the clarified "pregs". When the cyanide solution already carries appreciable silver or copper these will activate the se zinc, rendering the uof lead unnecessary. If much

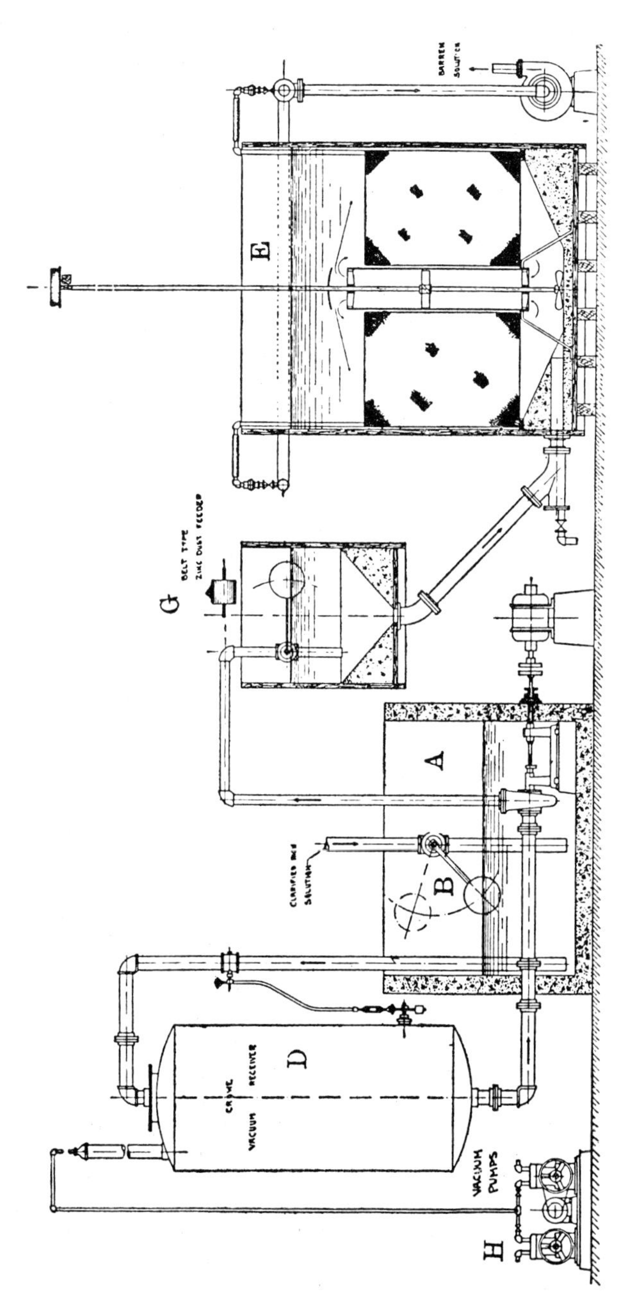

Fig. 206. Merrill-Crowe Precipitation Unit

copper is present, it may cause trouble by plating out on the zinc. This is not likely to occur with zinc dust, though it can happen with shavings where the total available zinc surface is much more limited. If the quantity of zinc present is insufficient for complete precipitation of the metal cyanides, reaction is preferentially directed to the gold. This tendency can be made use of by "starving" the zinc dust into the "pregs" in two successive additions, with filtration of slimes after each. The first slime will be gold-rich, and the second copper-rich.

Research has been applied for several years to the possible use of ion-exchange methods of stripping the gold from the pregnant solutions, and processes incorporating this method are working in some mills. The main advantages which have been claimed are the possible continuous solution and precipitation of gold in one operation of a resin-in-pulp (R.I.P.) type; the reduced need for clarification or even filtration of the pulp before precipitation; and the possibility of recovering other dissolved metals on the resin, thus making a by-product or, at least, regenerating foul cyanide. Little has yet been published.

Treating the Precipitated Gold

The production of bullion from the gold-bearing slimes is briefly outlined in order to complete the record, since this work is usually the responsibility of the mill superintendent. If thorough clarification and de-aeration have preceded precipitation upon clean-surfaced zinc dust a big proportion of the slime consists of gold and/or silver, with the zinc content below 10%. With precipitation of poorly clarified liquor upon zinc shavings only a minor fraction of the product consists of gold. Treatment with sulphuric acid dissolves the excess zinc, the resulting sulphate sometimes being used to pickle timber that is to be sent underground. In Australia the acid stage is usually omitted. The insoluble residue from the sulphuric bath is next calcined in an oxidising atmosphere, so that any base metal will be converted to oxide and enter the slag when the calcine is treated in the melting furnace.

When treating zinc shavings the sludge may be deposited on a 60-mesh screen resting a few inches below the top of a rectangular wooden tank, which may be lined with lead if no mercury is present. The tank is filled with water, and loose gold and lightly adherent deposit is flushed or gently scrubbed through. The residual oversize is returned to the boxes. Alternatively this material, which is rich in gold, may be acid treated and worked for its bullion. The excess water in the box is removed, care being taken that no slime is washed away. A little sodium aluminate can be added to flocculate and precipitate such slime, and the effluent water may be filtered through a double layer of calico, which is periodically burnt to ash and then cyanided.

Some liquid is left in the tank, and to this enough sulphuric acid is now gently added to form a dilute solution. Gas is given off as the acid attacks the slime, and this is sometimes very poisonous, carrying hydrocyanic fumes, arsenic, and antimony. The tank must therefore be connected with a good out-draught by means of a hood. Instead of this acid treatment, sodium acid sulphate is occasionally used:

$$2NaHSO_4 + Zn = Na_2SO_4 + ZnSO_4 + H_2. \quad (16.14)$$

The contents of the tank are stirred with a wooden paddle, more water and acid being added when the reaction slows down, until no more gas is evolved. The tank is now filled with water. The contents are next allowed to settle, and the liquid is decanted. The residue is rewashed several times, possibly using hot water to remove all the zinc sulphate. The sludge can then be transferred to a filter press for further washing. Moisture is removed by squeezing, followed by drying on iron trays which have been coated with whitening or bone ash. Calcining is done at 500° C. to 550° C. The calcine is then mixed with a suitable flux and transferred to a melting furnace. A typical charge is:

Dry calcine	100 parts
Silica sand	25–40 parts
Borax	40–60 parts
Soda ash	10 parts
Fluorspar	5 parts
Manganese dioxide	0–15 parts

When a high-grade precipitate is melted without previous calcining, the mixture can be:

Precipitate (at −20% moisture)	75 parts
Borax	50 parts
Manganese dioxide	45 parts
Silica sand	25 parts

Part of the borax may be replaced by soda ash, sodium carbonate or fluorspar, and part of the MnO_2 by sodium nitrate.

Oil-fired melting furnaces are in general use. Small plants use a tilting furnace in which a graphite crucible is placed. Medium-sized plants use a tilting reverberatory furnace. Large mills use stationary reverberatories either with the conventional firebrick hearth or, as is usual on the Rand, arranged as reverberatory pot fusion furnaces. After melting, the fused product may be poured into a conical mould, the gold later being remelted into a bullion bar, or the crucible may be tilted to decant the bulk of the slag, after which the last inch or so is thickened with dry sand, raked off and later remelted to recover its residual metal. After this raking, the clean bullion is poured direct. A sample may be taken before pouring, by dipping a small heated ladle into the molten gold. Furnace by-products—slag, ash, and linings—usually carry some values which are periodically reclaimed by grinding and retreatment, not necessarily with the main run of ore.

Testing Solution and Materials

Efficient working conditions in the cyanide plant must be maintained if adequate separation of the gold is to be achieved. The most important factors affecting gold solution from point to point are:

A
1. "Free" or "available" cyanide.
2. Total cyanide.
3. Hydrocyanic acid.
4. Protective alkali.
5. Precipitation efficiency (gold in "barrens").

These (with the possible exception of No. 3) are controlled by routine checks made at frequent intervals during the day.

B
6. Degree of fouling of circulating solutions by zinc, copper, iron as ferrocyanide and sulphocyanide.
7. Available oxygen.

These are periodically checked.

C
8. Quality of zinc dust.
9. Available CaO in lime.

These affect new supplies.

Hydrogen cyanide (hydrocyanic or prussic acid) does not dissolve gold and is therefore not "available", this term referring to that portion of the cyanide salt in a suitable condition ("free") for making such an attack. Not all of the "total cyanide" present is thus free, since substantial portions may have taken up copper, zinc, etc., and thus become unable to dissolve gold. Since the oxidation of various minerals in the ore releases acidic products into the solution, a slightly alkaline condition is maintained by adding calcium oxide. (Too high a content of lime slows down the cyanide process.) The solution is checked periodically for loss of alkalinity and an appropriate correcting addition is made. This lime combines with free hydrogen cyanide, restoring the availability. Finally, routine checks are made by the shiftsmen as to efficiency of precipitation. If insufficient fresh zinc surface is present, the gold will not be adequately removed from the "pregs", and loss may result. Simple rapid tests have been devised for the routine control of the foregoing conditions, and they are made by the mill operators. These tests are described below.

It is also necessary to decide when a cyanide solution should be discarded, either because it is too foul for further reclamation or because newly arriving water must be accommodated in the system. Control tests are made in the mill laboratory. This also applies to the available oxygen, though a colorimetric test can be made in the plant. Without adequate oxygen, as Elsner's Equation shows, cyanide does not dissolve gold.

Tests of new supplies are also made in the laboratory. The criteria of zinc are:

(*a*) Metal content and purity.
(*b*) Mesh size (available surface).
(*c*) Hydrogen emission (surface condition).

Lime, which may be kilned locally, is tested for its content of CaO in order that the quantity to be added to a given bulk of liquor can be correctly calculated.

Application of Cyanide Solution

Stock of cyanide solution can be built up, applied, circulated, and discarded in a variety of flow patterns. Barren cyanide leaving the precipitation section can:

(1) Return to the mill circuit.
(2) Be regenerated or purified, and returned.
(3) Be used as a weak displacing wash.
(4) Be discarded.

Regarding (1), usage varies. In North America wet-grinding is commonly performed in cyanide solution, but the practice is less common elsewhere. The Rand bankets were originally treated by amalgamation. With the introduction of the cyanide process the layout and tradition of grinding in water persisted, particularly since amalgamation is adversely affected by cyanide, which also attacks copper-containing plates. With the introduction of "all-sliming", most new Rand plants went over to grinding in cyanide. Stamp batteries, still in use on the Rand, require so much liquid that adequate storage of cyanide solution from this source might prove too costly. Ores containing clay which is not removed before treatment tend to have high values in the tailings when crushed in cyanide, possibly owing to sorption by the colloidal clay of auro-cyanides. Where soluble metal ions or cyanides must be dealt with by special treatment or pre-washing, grinding in cyanide is not practicable. Against these disadvantages can be set some important gains. "Float gold" is dissolved during milling in cyanide, theft is reduced, and since no new water is transferred from the grinding to the agitating section, the problem of discarding excess spent liquor is simplified.

In many plants which grind the ore in cyanide solution no problem of discarding foul cyanide exists, because the amount of liquid leaving the mill with the final filter cake sufficiently exceeds that arriving on the new ore to require a steady addition of make-up water.

Barren solution issuing from the precipitation section is sometimes too foul to be reused. Various methods of regenerating such liquor have been developed. At Flin Flon it is acidified and the gaseous hydrogen cyanide evolved is passed through a tunnel in which sprayed limewater traps it so that it can return to work as calcium cyanide. Before it can be used for the dissolution of gold the barren solution must be re-oxygenated. This process can be aided by using barren cyanide as a displacing wash for the removal of pregnant cyanide during filtration. The use of ion exchange resins for the regeneration of mill solutions and recovery of cyanide has been shown to be feasible by Goldblatt.[9]

Solution Control

At various points in the solution circuit it is essential to know how much cyanide is present in "available" form (capable of reacting with the gold in the ore). Since this cyanide must be slightly alkaline, the "protective alkali" must also be maintained. Not all the cyanide present is available. Efficiency of precipitation is also tested by a rapid method which shows the approximate amount of residual gold in the "barren solution". The final routine test is for efficiency of oxygenation, since the ability of the solution to dissolve gold depends on adequate content of air. These tests are made at regular intervals during each shift. Periodically the amount of zinc, copper, ferro-cyanide, and thio-cyanate in solution may be checked in the mill laboratory. The tests are made to determine the available CN– ion, and the results must be translated in terms of the compound used to provide these ions. Tests are reported in terms of potassium cyanide, but this is not used commercially, the chemicals most employed being sodium cyanide and calcium cyanide.

Free Cyanide

The reaction between silver nitrate and potassium (or sodium) cyanide is:

$$AgNO_3 + 2KCN = KCN \,.\, AgCN + KNO_3 \qquad (16.15)$$

$$KCN \,.\, AgCN + AgNO_3 = 2AgCN + KNO_3. \qquad (16.16)$$

The latter (AgCN) cyanide is insoluble, so that when this stage of reaction is reached a permanent white precipitate appears. This end-point is made yet more visible if one or two drops of a 10% solution of potassium iodide are added before commencing the titration. If 13·046 gm. of $AgNO_3$ are dissolved in distilled water and made up to one litre, then each c.c. of this solution, titrated into 100 c.c. of the cyanide solution (to the point when a precipitate forms), is equivalent to 0·01% of free KCN. To convert this to pounds per short ton, the volume in c.c. of $AgNO_3$ used to produce a precipitate is multiplied by 20.

Instead of titrating 100 c.c., the test may be run on 10 c.c. of solution. Each c.c. of silver nitrate used now represents 0·1% of free KCN or 2·0 lb./short ton of ore.

Total Cyanide

If 5 c.c. of 10% sodium hydroxide is now added to the sample which has been titrated, all the free cyanide, hydrocyanic acid, and zinc double cyanide is converted to sodium cyanide. Further titration with silver nitrate until an insoluble precipitate begins to form shows the total cyanide figure. This, however, takes no account of any sodium ferricyanide or ferrocyanide present.

A simple semi-automatic method for monitoring the actual dissolving strength of cyanide solution is reported[10]. Gold leaf is attached to a glass slide mounted on a slowly revolving wheel. At each revolution the slide passes through a light beam, and a photo-electric cell shows the rate of dissolution. The method takes care of several factors, and includes warning of inadequate oxygenation.

Protective Alkali

The 100-c.c. sample used to find the free cyanide can be used for this determination if no alkali is added after completing the first test. It is titrated with deci-normal oxalic acid (6·3 gm./litre of $C_2H_2O_4.2H_2O$), after adding 0·5 c.c. of a 10% solution of potassium ferrocyanide and a drop of phenolphthalein indicator. Titration is continued until the purple colour disappears.

One c.c. of oxalic acid is equivalent to 0·04% alkalinity in terms of sodium hydrate.

Efficiency of Gold Precipitation

To a litre of "barren" solution add enough sodium or potassium cyanide to bring the strength to about 0·1%. Next add a pinch of fresh zinc dust and two drops of lead acetate (10% solution). Agitate thoroughly in a stoppered vessel for a minute, and then allow the solids to settle. Decant off the clear liquid, transfer the residue to a porcelain dish and evaporate it to dryness after adding 10 c.c. of aqua regia. Take up with HCl into a small test-tube, cool, and then gently run in a few c.c. of fresh stannous chloride solution, slanting the tube so that the liquids do not mix. At the line of contact a brownish-pink stain develops if any gold is present. If this becomes strongly brown or purple, the solution carries appreciable gold. With experience the operator learns to estimate the gold content with sufficient accuracy to provide a running check of the efficiency of precipitation.

Other Tests

The available oxygen is tested by adding pyrogallic acid and making a colorimetric comparison, using standard colour tubes. Ferrocyanide, zinc, soluble sulphide, and copper are checked periodically by standard assay methods.

Cyaniding Difficult Ores

Some gold ores present special difficulties in cyanidation. Substances may be present which destroy the cyanide; gold may be in a finely disseminated state in the sulphide minerals; graphite may lead to premature precipitation of the gold from the cyanide; oxygen-consuming minerals, notably pyrrhotite, may compete for the available oxygen introduced during aeration of the pulp, thus slowing down the dissolution of the gold; oxidised copper minerals may react with cyanide to form stable complexes which cannot dissolve gold whether or no the pulp is alkaline; small amounts of nickel or chromium ion interfere with precipitation of gold by zinc from pregnant solution. An excellent brochure which deals with cyanidation chemistry[13] has been published by a trade organisation.

When the gold occurs in intimate association with tellurides or arsenopyrites, special methods are needed to ensure adequate extraction in a reasonable time. One of the older treatments involved extremely fine grinding followed by agitation in cyanogen bromide (BrCN)—cyanide solution to which cyanogen bromide was periodically added. The solution was held near neutral with the least possible margin of protective alkali.

Arsenic is not necessarily a cause of trouble in cyanidation, but it is frequently associated with antimony, or with partly oxidised iron minerals. These minerals may present problems and reduce efficiency of gold extraction.

Two important developments have modified the metallurgical difficulties with such ores. One is the use of froth-flotation to remove selected ore constituents such as auriferous sulphides from the main bulk of the run-of-mill ore. This yields a comparatively rich concentrate which can stand the cost of intensely fine grinding or other specialised treatment. The second is the steady improvement in roasting methods, which can today be applied in such a way as to ensure almost complete destruction of the sulphides and their associated trouble-making compounds at a controlled temperature. "Dead-roasting" of the whole of the ore to expel all tellurium, sulphur, and/or graphite is rarely practised, since it is expensive and difficult to control within desirable temperature limits.

One difficulty with antimonial ores is that although protective alkali is needed to protect the cyanide, antimony sulphide (stibnite) is soluble in alkalis and the compound thus formed decomposes cyanide to thio-antimonite. One possible reaction for either arsenic or antimony is

$$2\,Sb_2S_3 + 6\,Ca(OH)_2 = Ca_3\,(SbO_3)_2 + Ca_3\,(SbS_3)_2 + 6\,H_2O \quad (16.17)$$

$$Ca_3(Sb_2S_3)_2 + 6\,NaCN + 3O_2 = 6\,NaCNS + Ca(SbO_3)_2 \quad (16.18)$$

Pre-leaching of such ores with 2% to 4% of caustic alkali, oxidation roasting, and prolonged weathering are among the methods which have been used to mitigate this trouble.

Where carbon is present in such a form that it absorbs the gold from its solution in cyanide, it can sometimes be removed by pre-flotation. Special carbon-depressing reagents have been developed for removing such undesired material from concentrates before cyanidation. Graphite is sometimes present in such a form that it will rise as a scum on the classifier if a little paraffin or diesel oil is fed in with the ore, and this scum can then be removed from circuit. One mill takes advantage of the fact that the graphite in its ore preferentially sorbs mineral oils and causes such particles to become coated with an oily envelope, after which they can no longer sorb gold from the cyanide solution.

Since oxygen is essential if the cyanide is to dissolve gold, any mineral present in the ore which is oxygen-avid may reduce the efficiency of extraction. Pyrrhotite is the most active of these compounds. In some forms it is comparatively innocuous, while in others it is extremely unstable, taking up oxygen so eagerly as to make the task of providing sufficient excess air difficult, and adding to this the further undesirable effect of reacting with the cyanide to form thio-salts incapable of dissolving gold.

Among possible reactions between pyrrhotite and cyanide are

$$Fe_5S_6 + NaCN = NaCNS + 5FeS \quad (16.19)$$

$$FeS + 2O_2 = FeSO_4 \quad (16.20)$$

$$FeSO_4 + 6NaCN = Na_4Fe(CN)_6 + Na_2SO_4 \quad (16.21)$$

in which series both cyanide and oxygen are consumed. In alkaline cyanide the reaction may be

$$2FeS + 2NaCN + 3H_2O + 3O = 2NaCNS + 2Fe(OH)_3 \quad (16.22)$$

and in neutral solutions

$$FeS + 7NaCN + H_2O + O = NaCNS + Na_4Fe(CN)_6 + 2NaOH \quad (16.23)$$

Pyrrhotite may be regularly distributed through the ore body, or may vary in its concentration and/or instability from point to point along the lode. Pre-aeration at low alkalinity, followed by cyanidation, is used at Sub-Nigel to combat this reaction.[12]

If it is not convenient to remove pyrrhotite before treating the balance of the ore, or to segregate ore containing excessive amounts until it can weather down, then care must be taken to ensure ample aeration during agitation. It has been suggested that apart from trouble for which pyrrhotite is directly responsible, it is also able to act "as a sort of fulminator, and not improbably as a catalyst . . . (and) . . . quite small quantities (can) introduce serious amounts of cyanicides of many different kinds. . . ."

The gold-treatment plants in the North Ontario region commonly reinforce the aeration provided by the standard agitator of the Dorr type by introducing auxiliary pressure air deep into the tanks. In addition, the cyanide solution in which the ore is milled is kept down to a low strength so that milling heat and freshly sheared sulphide-mineral surfaces shall have as little opportunity as possible to decompose the cyanide.

Cyaniding Flotation Concentrates

Pre-concentration of auriferous sulphides by froth-flotation is sometimes employed in order to remove these minerals for special treatment before cyanidation. This is a process in which air-bubbles are used to remove selected minerals from an ore, in the form of a mineralised froth which rises from the pulp and is skimmed off. When applied to gold ores it lifts fine particles of metallic gold, metal sulphides which may be auriferous, such as pyrite, marcasite, pyrrhotite, chalcocite, bornite, chalcopyrite, galena, arsenopyrite and stibnite, and certain other minerals, including graphite. Such a concentrate, beside being greatly enriched in its gold content, carries most of the cyanicides and other minerals likely to cause trouble in cyanidation. The tailing from this flotation, if it still carries sufficient gold to repay straight cyanidation, gives but little trouble since most interfering minerals have been removed.

Not all the minerals in the flotation concentrates are gold-bearing. Graphite, which can cause premature removal of gold from the cyanide solution, is unlikely to carry values. If it can be removed before cyanidation, the process will benefit. Part or all the gold may be disseminated in minute specks in some of the base metal sulphides. Possibly the non-auriferous sulphides can be removed from the float so as to simplify the treatment of the residue. When a single float is made we speak of "bulk" flotation. When a series of minerals are floated separately, the term is "differential" flotation. A bulk float can be retreated by differential flotation.

The main characteristics of concern in cyaniding a flotation concentrate are:

(*a*) The high gold value of the material treated.
(*b*) The small bulk in which the value is now concentrated.
(*c*) The rapid oxidation of this sulphidic material during aeration and its effects on cyanidation.

(*d*) The need for adequate exposure of gold minutely disseminated in particles of pyrite.

The effect of (*a*) and (*b*) is to justify more costly methods of concentration than could be economically applied to run-of-mill ore. (*c*) means that sulphuric acid will be generated during aeration, and that the alkalinity of a flotation pulp must be watched very closely if the cyanide is not to be destroyed. If such a pulp is aerated in the presence of lime, hydroxyl ions react with the surface of the particles of pyrite and sulphide ions are liberated into the pulp. Here they react with oxygen from the aeration, and thiosulphate and sulphite ions are formed, also sulphur, all of these in time oxidising to form sulphate ions.

If cyanide is also present, the above reaction may be accompanied by the formation of a thio-salt, which is useless for the dissolution of gold. Pre-aeration in the absence of cyanide is therefore preferable, if possible followed by decanting away of the water in which the sulphur salts are now dissolved. For some ores pre-aeration without added lime has been recommended.

The best procedure is arrived at by means of empirical experiments in the ore-testing laboratory. The whole problem is frequently aggravated by the fact that the flotation concentrate must be reground to a fine slime in order to expose most of the locked-up gold to the cyanide solution. This greatly increases the new sulphide surface and, consequently, the reaction with oxygen. This reaction is not comparable in completeness with the roasting treatment described below, in which the oxygen penetrates the fairly solid lattice of the sulphide and destroys it. Broadly the difference is one of degree, as between singeing and burning to ash. Aeration of freshly ground pulp has a relatively gentle oxidising effect on the newly developed sulphide surfaces.

To sum up, the efficient cyanidation of flotation pulps may need adequate regrind, followed by aeration sufficient to stabilise the disturbed surfaces, possibly removal of the fouled water and, finally, cyanidation with copious aeration and vigilant watching of the protective alkalinity of the pulp. A secondary benefit, obtained when all these matters are controlled, is that the pregnant solution is not foul, and consequently precipitation is more efficient.

These controls are more easy to apply to batches of concentrate than continuously, particularly when the chemical reactions in the pulp are intricate. The operator can deal more specifically with the minor variations in a batch than with obscure changes in a continuous flow-line. In a small plant handling, say, ten tons daily, the flotation concentrate would be ground, collected into an agitating tank, aerated in water which would then be decanted, and finally cyanided in the same tank, each stage being carried to a satisfactory completion. One such plant uses five home-made Pachuca agitators to share the work, each holding up to five tons of dry solids. The flow-sheet is shown in Fig. 207.

A tank is charged for 12 hours, the solids being settled and the overflow water returned to the mill. The thickened pulp is then aerated without lime for 18 hours and allowed to settle. Clear foul water is syphoned to waste and 12 hours of agitation in weak cyanide, with lime, begins. The "pregs" are decanted and a 6-hour second agitation is given.

If copper is present in such a form as to be attacked by cyanide, it may preclude direct cyanidation of a flotation concentrate. Copper sulphides can usually be removed by differential flotation. Such cyanicides as the oxide minerals—malachite, azurite, cuprite, and tenorite—can perhaps be leached with sulphuric acid.

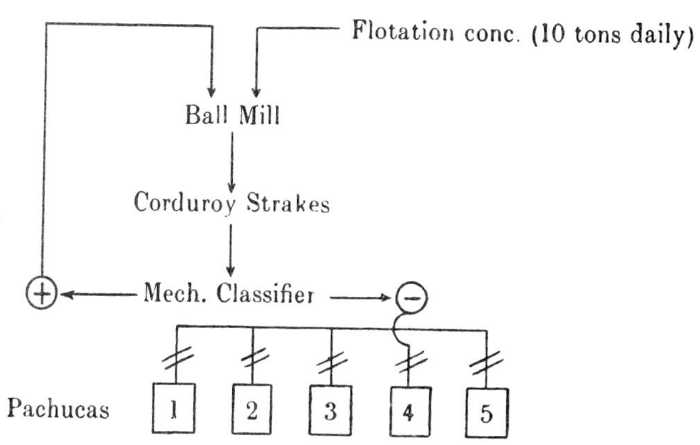

Fig. 207. Batch Cyanidation of Flotation Concentrate

Generally, cyanide should be used in weak concentration (0·02%–0·05%) to keep down consumption, or from 0·10%–0·8% with silver ores.

The very fine grinding used to liberate disseminated gold may lead to filtering trouble, and counter-current decantation may be used to avoid this, the final thickener underflow being discharged to waste. Pregnant solutions from the treatment of flotation concentrates are rich and may be somewhat recalcitrant; special care is needed in checking the efficiency of precipitation. If the effluent from precipitation is high in gold value and is returned to use or run to waste, serious loss may occur.

By the use of flotation in suitable cases, considerable grinding economies and improvements in recovery are possible. Only a fraction of the ore is subjected to costly intensive grinding and that fraction can be given special treatment which could not be afforded if applied to the whole of the ore.

Flotation can also be applied to the tailings from the cyanidation plant. In this case the ore may be ground in cyanide and given the normal extraction treatment. The final tailing is passed through the flotation process. The auriferous sulphide thus collected is then given appropriate further treatment.

Roasting

Since excessive consumption of cyanide arises from reaction with unstable sulphides in the ore, roasting methods are sometimes used to deal with these troubles. The change of state produced by pyrometallurgy is sometimes an essential step in the economics of chemical extraction. Other uses of roasting techniques are considered in later chapters. A flotation concentrate is usually sufficiently rich in sulphur to be self-roasting, or to require but little extraneous fuel. Pyrite, marcasite, and pyrrhotite react with oxygen to yield sulphates and oxides. The oxide-forming process follows the general line:

$$FeS_2 \rightarrow FeS — FeO \rightarrow Fe_3O_4 \rightarrow Fe_2O_3 \qquad (16.24)$$

as modified by roasting temperature and availability of oxygen. If salt is added in roasting, its general form of reaction is

$$2NaCl + RSO_4 = RCl_2 + Na_2SO_4, \qquad (16.25)$$

the metal (R) being first oxidised from its sulphide to the sulphate form. Salt is used in carefully guarded quantity in roasting at Lake Shore, the slight loss of gold chloride in the oxidised iron being compensated by better recovery when cyaniding the calcined product. Chalcopyrite in a roaster feed is converted to copper sulphate or oxide; the arsenic in arsenopyrite can be volatilised off as As_2O_3 and trapped in flue collectors. Gold may be lost (*a*) when furnace draught is so high as to carry fine particles to the stack; (*b*) when temperatures are excessive and the surface of the gold acquires a coating which insulates it from attack by cyanide; (*c*) when a sinter is produced in which the gold is trapped inside aggregates of incompletely calcined material.

Two main types of roaster in use at present are the Edwards and the Dorr FluoSolids (Fig. 208). At Raub in the F.M.S.[13] a 15-rabble Edwards furnace handled a flotation concentrate which was sometimes self-roasting (i.e. it used its own content of sulphur as fuel) and sometimes assisted by extra fuel. Recovery was best (about 91%) when the temperature was low and the calcine porous. At Que-Que in South Rhodesia gold concentrates from a number of small workings are treated in a customs plant which uses an Edwards roaster and wood fuel to produce a calcine from which 90% or more of the gold is recovered by cyanidation, and up to 85% of the arsenic by volatilisation. At Lake Shore[14] flotation concentrates carrying from 16% to 17% moisture are fed to Edwards roasters, at a sulphur content of 20% to 30% and self-roasted. These concentrates are produced by flotation of the cyanide tailings, which are exposed to sulphur dioxide gas generated in the roasting furnace as part of the treatment needed to render their sulphide fraction floatable. The rabbles which move the material through the furnace are air-cooled. When desired, this air can be used in its heated state instead of as cold air, thus raising the roasting temperature. In addition to this operating control the sulphur content of the incoming feed is steadied at between 26% and 27% by admixture of sulphur-barren material. Should the temperature in the furnace rise unduly, the moisture content of the entering feed is raised to 22% by reducing the thoroughness of water removal, and this immediately makes a difference at the feed end. Toward the discharge end of the roaster a cooling hearth is formed by incorporating a 10½″ drop in the bed which prevents material from being dragged back by the revolving rabbles. The gas still issuing from this "dead calcine" remains in the furnace and the residual heat is used in the combustion section. If the cooling hearth were open, a fume nuisance would arise and this heat would be lost. The concentrate (a filter cake) entering the furnace is very sticky, and the rabbles

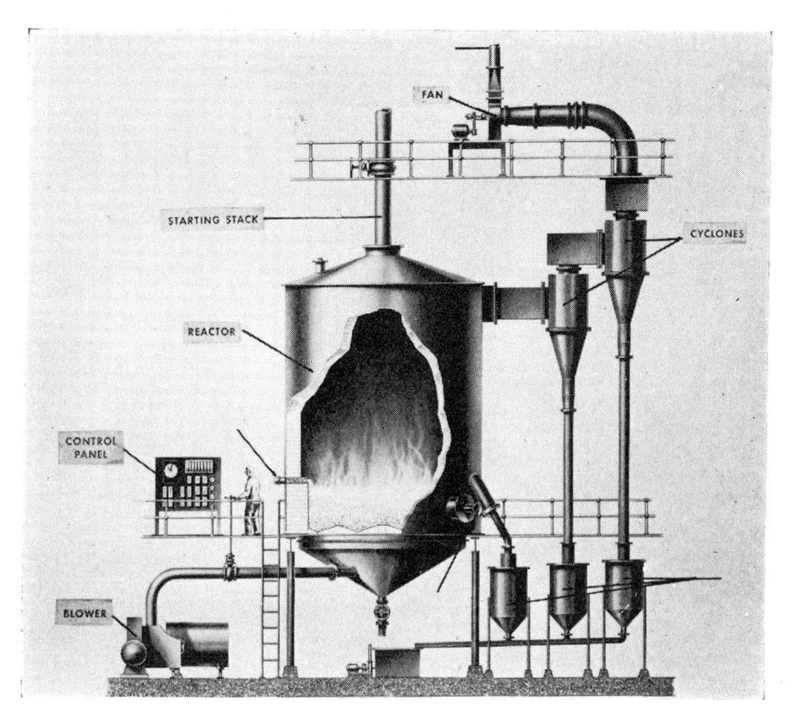

Fig. 208. *Dorr FluoSolids Roaster* (Dorr Co.)

must bring back enough dry, powdery charge in their circular motion to coat this putty-like material with a film of dust. The filter cake now shreds up into dust-coated lumps as the rabbles stir it. As the material is worked forward into the hotter section of the furnace the trapped water turns to steam and disintegrates the lumps. Thus the cost of an independent drying section is avoided and the dust set up is kept sufficiently low to avoid the need for costly trapping arrangements. In the roasting zone the charge has a flocculated appearance. Progress is regulated by the gradient and by drops between rabbles. These regulate the amount of backward travel. The calcine is drawn from the cool discharge end by a raking mechanism and agitated in water before being adjusted for alkalinity with lime. The main object of roasting at Lake Shore is not to eliminate sulphur but to open the material to cyanide attack by rendering it porous. The modifications used at Lake Shore reduce flue dust at feed and discharge ends to a minor nuisance of no great economic importance, the weekly clean-up of flue dust being some 0·4% of the feed, carrying only an ounce or so of gold per ton. It is trapped ahead of the stack.

In "FluoSolids roasting" as practised at one Canadian plant[15] a flotation concentrate directly produced from the ore is treated. Among the sulphides are arsenopyrite, carrying gold in solid solution, stibnite (which tends to make trouble by becoming pasty during roast), pyrite, and some sphalerite and chalcopyrite. When direct cyanidation was practised, less than 80% of the gold was recovered. Now, with jigs and strakes first taking out free gold, and cyanidation of a roasted flotation, concentrate applied to the gravity tailing, recovery is nearly 95%.

In fluidisation gas—in this case air—reacts as it is blown through finely ground solids, which form a fluid suspension as the air flows upward. Owing to the mobility of this fluidised bed the roasting temperature is fairly uniform throughout, with no local overheating. Self-roasting is probably feasible on a dry feed containing only 12% sulphur as sulphide. There must, however, be no fusion during the reaction, as nodules would fall and might choke the bed. Dust from the roaster is trapped in cyclones. Temperature is controlled by injecting water. The ideal working gas addition at Cochenour is indicated by a clean white plume from the smoke stack, a yellow colour showing the need for more air and red or black plume the loss of dust due to too much air. A black calcine is preferred to a red one, the colour produced depending on the quantity of air used.

Research by Parker[16] on auriferous pyrite which after "sweet roasting" carried a final cyanide residue of 6 dwt./ton showed that the loss in Kalgoorlie calcines was associated with pyrrhotite. This was formed during roasting of the pyrite, by fusion, and was associated with the rate at which the temperature in the material rose together with the rate of expulsion of sulphur from the pyrite. It was found that the best (porous) leaching structure was formed in a fluidised decomposition-bed held at 675° C. by means of air control, which regulated the rate of oxidation. Even then, with the complex Kalgoorlie concentrates, the tailing from cyanidation of the calcine assayed 4·9 dwt., a high figure by Canadian standards though low for some East African operations. A description of process development for a refractory gold ore has been given by Tait.[17]

Some Special Techniques

Carbon

Carbon, whether present as graphite in the natural ore body or added as activated charcoal, has the power of sorbing gold from its solution in cyanide. This can be a serious nuisance when treating graphitic ores, since gold-bearing graphite then goes to waste. From time to time charcoal has been used instead of zinc as the precipitating agent and it is thus employed in current practice to a limited extent.

When such carbon occurs in the ore it can be removed by roasting, as at Ashanti; floated out in the grinding circuit by the use of a little paraffin or diesel oil; depressed from the ore by special flotation agents; or oiled over by lubricating oil added to the grinding circuit and thus insulated from cyanide. Fortunately, such graphite is not always a menace to plant efficiency. Several theories to account for the sorbing action have been advanced, none of which will be discussed here.

Gold is deliberately sorbed from its cyanide solution in some plants where special circumstances would lead to difficulties with the use of zinc as a precipitator. A method is described by Edmonds[18] and other applications are described in the literature of the subject. The advantages claimed[5] are that better recovery is made from foul solutions; that cyanide loss is lower; and that refining is cheaper. In recent years methods have been developed for removing the charcoal from the pulp. One is flotation, in which cyanidation, aeration, and continuous removal of charcoal is used, thus permitting continuous dissolution and precipitation of the gold. Another is the use of charcoal which has been rendered magnetic and can be removed expeditiously by magnetic methods. In a third method a wire cage containing activated charcoal is immersed in the tank containing pregnant cyanide. As gold dissolves into the cyanide it is taken up by the charcoal, which is removed as requisite.

Lewis and Metzner[19] have described the activation of carbon derived from coconut, hardwoods, etc. The process "selectively removes the hydrogen or hydrogen-rich fractions from a carbonaceous raw material in such a manner as to produce an open porous residue". Charcoal can be exposed to CO_2, steam or a mixture of these gases at between 700 C.° and 1000° C. Successive use of CS_2 and H is still better, using fluidised beds.

The gold-bearing charcoal can be sent to a smelter, or calcined, its ash being re-cyanided.[20]

Chlorination

The use of chlorine to extract gold from its ores pre-dates cyanidation. The high cost of chlorine in the nineteenth century made it uneconomic (1881) but this objection no longer exists. Gold is dissolved fairly quickly in the presence of nascent chlorine, particularly when chlorinated brine is used,[21] and some of the difficulties presented by cyanicides are avoided. The method is not used today.

Losses

In theory the amount of gold present in the tonnage of ore entering the plant should all be accounted for in the equation for metallurgical balance.

$$\begin{aligned}\text{dwts. milled} = {} & \text{dwts. in bullion} + \\ & \text{dwts. in tailing} + \\ & \text{dwts. in solutions} + \\ & \text{dwts. locked up in plant.} \qquad (16.26)\end{aligned}$$

In practice it is not usually feasible to satisfy this equation from working data, though ore such as the Rand banket, with a fairly smooth distribution of value through a uniform type of trouble-free gangue, makes close approximation possible.

Discrepancies from true balance arise from several causes:

(*a*) Difficulties in computing head tonnage and value.
(*b*) Sampling errors at all stages.
(*c*) Spillage of "pregs", concentrates, and auriferous sands during treatment.
(*d*) Theft.
(*e*) Errors in sampling tail values.
(*f*) Variable "tie-up" of values in movement through the plant.

Losses from (*c*) and (*d*) always lower the totals on the right-hand side of the above equation. They can only be kept down by good supervision and a layout which makes theft difficult and spillage recoverable. (*f*) tends to compensate out over the working year, since gold lodging in inaccessible crannies comes back into circuit during repair and overhaul. (*a*), (*b*), and (*e*) present problems which vary with the nature of the ore and the sampling facilities available in the plant. Size reduction is essential if the sample is to be mixed and reduced in bulk, yet the size reduction is affecting the constituents, malleable and brittle, with their respective high and low densities, very differently. Fortunately, errors tend to balance out over a period. The tendency is, however, to rely on stock-solution and tailings assay, together with the returns of bullion production, to give the information needed for efficient control, and to place major reliance on good management and layout to keep theft and spillage in check.

Leaving aside (*c*) and (*d*), physical losses are associated either with the tailing sands or the discarded cyanide solution. Chromium in the ore can seriously reduce the efficiency of zinc precipitation.[22] Nickel can behave similarly. Losses of this sort are avoided either by selective mining, removal of the offending mineral before cyaniding, or a chemical treatment of the solution to precipitate out the Ni, Cr. etc., before precipitation.

Losses in the solid residue can arise from (*a*) inadequate washing, in which case a certain amount of pregnant cyanide is carried away with the tailings; (*b*) sorption by graphite from the ore of dissolved gold, and loss of that graphtie in the tailings; (*c*) insufficient cyanidation and/or aeration; (*d*) the presence of coarse metallics in the cyanidation section, where they cannot be completely dissolved in the allotted time; (*e*) inadequate unlocking of the gold and (*f*) tarnishing films upon the gold which prevent the liquids reaching the metal and attacking it.

Barsky[23] *et al.* have shown the maximum rate of dissolution of gold to be 3·25 mg./sq. cm./hour, which is equivalent to penetration of a flat gold surface at the speed of 1·68μ/hour. Thus a 325-mesh particle, being 44μ in thickness, needs 13 hours and a 100-mesh or 149μ particle 44 hours at least for complete dissolution. If all of the particle is not open to attack, these times must be increased. Hence, failure to arrest coarse metallics by gravity methods associated with the grinding may lead to otherwise avoidable losses in cyanidation.

A common cause of loss is tarnish—the formation upon the gold of coatings, usually of oxidised iron. Such "rusty" gold cannot be amalgamated or cyanided. This has been studied by Head[24] in an examination of gold particles in tailing dumps. These inhibiting films are difficult to detect and, in most cases, almost impossible to remove by commercial methods, particularly when they are associated with partially oxidised materials.

Other Metals and Minerals

The four process stages listed above for cyanide can be elaborated somewhat with respect to chemical treatment of base-metal ores. Welch[25] gives them as:

(*a*) *Preparation of run-of-mine ore for leaching;* this stage may include:
   (i) grinding;
   (ii) physical concentration of values, or removal of specific impurities by physical methods;
   (iii) roasting;
   (iv) special chemical treatments to render values soluble in the leaching operation, or to prevent consumption of leaching reagents by impurities.

(*b*) *Leaching,* in which values are selectively dissolved by an appropriate liquid reagent; this step requires study and control of leaching reagent composition and concentration, pulp density, temperature, pressure, and reaction time.

(*c*) *Separation of leach liquor and tailings* by settling, thickening, filtration, washing and clarification.

(*d*) *Recovery of values from clarified leach liquor* by one of the following processes:
   (i) precipitation;
   (ii) ion-exchange;
   (iii) solvent extraction.

(*e*) *Recycling of leaching reagent,* after adjustment of composition to stage (*b*).

Stage (*a*) uses mineral processing methods to pre-concentrate the minerals which are to be attacked, and/or to remove or passivate those which would weaken the chemical action. In this Symposium a number of methods of chemical treatment were considered at research and operating levels. The specialising reader should consult its 22 Papers and their bibliographies. Selected examples illustrative of industrial practice are given in Chapter 23 under the name of the principal metal or mineral extracted. The rest of this chapter is concerned with chemical reactions and techniques used in the leaching of other ores than those worked primarily for their gold content.

Pressure Leaching

In most chemical reactions the temperature and the intimacy of contact between the reacting phases are rate determining factors. Batchwise and continuous systems have been developed which process superheated pulp under pressure, thus making possible—or accelerating—reactions which would not be industrially economic in the leaching environments thus far considered. The autoclaves used in this work must have linings resistant to the chemical attack used, and corrosion-proof agitating mechanisms. These specialised methods are applied to concentrates rather than to run-of-mill pulps, and permit either the solvation of a mineral species or the direct precipitation from the solvated product of a metal[28] or a refined precipitate.[29]

Bacterial Leaching

Biological research has shown that selected strains of bacteria can be made to accelerate acid leaching. The micro-organisms found to render copper minerals more readily leachable[26] are *Ferrobacillus ferrooxidans, Thiobacillus concretiverous,* and *Thiobacillus ferrooxidans,* all of which occur in the effluent waters of some mines in the southwestern United States. Research has shown that some species of bacteria can feed on ore minerals in a highly acidic environment. Apparently the sulphide is not attacked directly, but ferrous iron dissolving as the result of chemical oxidation is converted by their action to acid ferric sulphate, the process chemistry being

$$2\,FeS_2 + 7O_2 + 2H_2O = 2\,FeSO_4 + 2\,H_2SO_4. \qquad (16.27)$$

$$4\,FeSO_4 + 2\,H_2SO_4 + O_2 + \text{bacteria} = 2\,Fe_2(SO_4)_3 + 2\,H_2O \qquad (16.28)$$

$$7\,Fe_2(SO_4)_3 + FeS_2 + 8\,H_2O = 15\,FeSO_4 + 8\,H_2SO_4 \qquad (16.29)$$

or, by reaction with any copper sulphide present

$$Cu_2S + 2\,Fe_2(SO_4)_3 = 2\,CuSO_4 + 4\,FeSO_4 + S \qquad (16.30)$$

The resulting ferrous sulphate is then re-oxidised by further bacterial action to ferric sulphate, together with the elemental sulphur, thus

$$2S + 3\,O_2 + 2\,H_2O + \text{bacteria} = 2\,H_2SO_4 \qquad (16.31)$$

This work confirms earlier research[27] in which the nutrient and environmental requirements of bacteria used to accelerate the leaching of metal sulphides were studied.

Use of Ion Exchange

The sudden emergence of uranium as a major strategic element during the war years of 1939–44 led to urgent development of new methods of extraction. One result has been the intensive study and worldwide use of ion-exchanger resins at an intermediate stage of chemical extraction and/or concentration. The general principles of IX were considered in Chapter 15 and are further discussed in Chapter 23 under Uranium. Since IX is proving applicable (mostly at the research level to date) in the processing of several other mineral species it seems probable that industrial use will in due course extend to the extraction of gold, silver, copper, nickel and other metals from their pregnant solutions.[30] An outline of the methods now in commercial use is therefore given at this point, together with some considerations affecting choice.

To the extent that mineral particles adsorb ions from the aqueous phase of their solid/liquid system such particles are solid ion exchangers, a reaction relied on in preparing (conditioning) an ore pulp for flotation. The present discussion is not concerned with operations where low-solubility species remain substantially unchanged throughout their concentration, but with ion exchangers which act as intermediaries in the transfer of ionised compounds, and which are not constituents of the original ore.

The solid ion exchangers used in mineral processing are resins chosen for their stability (resistance to swelling), toughness when exposed to abrasion or breakage, loading capacity and ionic sign. "Loading capacity" is definable as the number of iogenic groups per specific volume of weight or ion exchanger and apparent or effective capacity as the number of exchangeable counter-ions in that volume. The cation exchangers include sulphonated coals and sulphonated, carboxylic and phosphoric resins. Anion exchangers include a range from weakly to strongly basic resin matrices.

The operating cycle in IX commences with the loading of the exchange sites by ions drawn from the ambient solution, which must in its turn be capable of receiving the ions then displaced from these resin sites. Exchange continues till the Donnan equilibrium is reached in the semi-permeable resin bead, at which point the work required for further adsorption exceeds the residual electrostatic force available to procure more penetration. Choice between competing ions for a capturing resin site is determined by the relative sizes and valences of these ions, and by the relative concentration (availability) of the attracted species. The chemical nature of the ionized groups affects resin capacity. A weak acid such as carboxylic ($-COO^-$) is only ionised at a high *p*H, and at a low *p*H combines with hydrogen to form the $-COOH$ group. A strongly acidic group such as $-SO_2^-$ remains ionised at a low *p*H.

Several methods of loading the ion exchangers are in industrial use. In the oldest, the pregnant liquid ("royals") is separated from the sands in the pulp after leaching is complete by decantation or filtration. This may be followed by further clarification so as to reduce the entrainment of adulterating solids in the end product and to avoid masking the surface of the resin beads with slimes. The pregnant aqueous liquor with its load of B-type ions now percolates downward through a vessel such as the IX column shown in Fig. 209.[31] This column is packed with resin beads loaded with ions of type A. As the solution enters it deposits its B-ions in a narrow zone near the point of entry, and takes up the equivalent number of A-ions from the resin. As the solution continues to percolate downward the resin becomes progressively loaded and the zone of exchange travels through the column until the resin bed approaches saturation. This stage is termed the "break-through point", and B-ions now begin to appear in the effluent, though the lowest zone is not yet fully loaded. To avoid risk of loss, a series of three IX columns is usually kept on stream. The first is switched out when break-through point is reached. New pregnant solution is then led to the second column, the original third

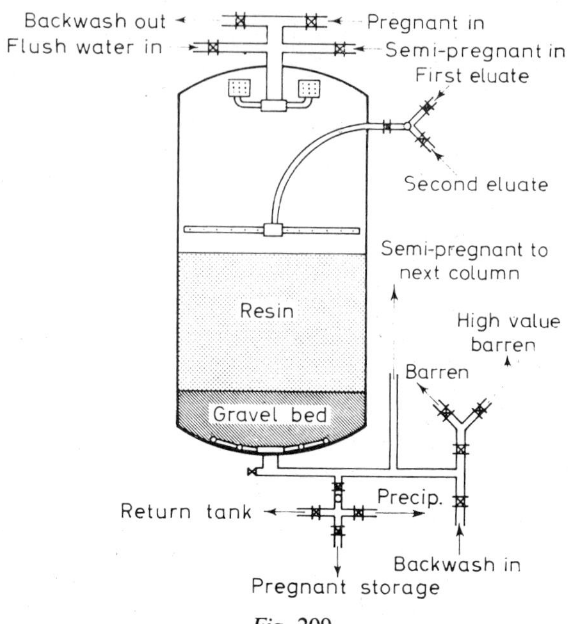

Fig. 209.

becomes the second, and a freshly prepared column is switched in behind it. This arrangement allows three in-line columns to be continuously loading while a fourth one is being regenerated (unloaded and re-activated).

When loading of the first column is stopped, its resin is not usually saturated. Some channelling of the bed may also have occurred, leading to incomplete contact between beads and liquor, and if clarification was incomplete some surfaces may have been masked by slime. The next stage is usually a stirring of the bed by rising water sufficient to scrub the beads without flushing them out. At the same time, deposited slimes are detached and flushed out. This operation, called "backwashing", which also removes any residual undrained pregnant solution, is sometimes omitted.

The next stage is elution, in which the captured ions are displaced from their sites when a suitable chemical solution, the eluant, is flushed through the column. On completion of this operation the resin sites are restored to their original state and the beads are said to be regenerated. The now pregnant eluate is run off either to storage or to precipitating treatment and the column is given a further backwash to remove its residual eluant. It is then ready to go on stream once more.

The chemical constitution of backwash water (which may be acidified) and of the various eluants used (re-cycling, new and special) are specific to the problem of displacing captured ions from the resin and re-activating its sites. During their working lifetime, the beads become progressively loaded or "poisoned" by undesired ions which have been picked up from the pregnant solution and which are too firmly held to respond to the eluant used for normal regeneration. This build-up reduces the loading capacity and is dealt with for most poisons by a periodic special regeneration. Some of these shorten the working life of the beads. The poisons are of two broad types, chemical and physical. Chemical poisons are ions which adsorb to resin sites too firmly to be removed by standard elution. Physical poisons (polymerising colloidal silica and organic, oily or fungoid and bacterial accretions) may inhibit the diffusion of pregnant solution into the beads. When this goes too far for efficient operation, the resin bed is replaced.

The final stage of IX, recovery in solid form of the values from the eluting solution, is performed by chemical methods which yield the desired precipitate. This is settled out, filtered and dried for despatch.

One variation on the use of static beds is in use in some American plants. The pregnant liquor is thoroughly clarified, as no slimes can be tolerated. It then is fed to adsorption. For this there are three groups of three columns, two of which are on stream while the third is used for elution. The tenth (backwash) column receives saturated resin from any one of the six columns of stream as it becomes saturated. It can also return resin to any of these columns. In addition, in one plant, there are two spare columns used for periodic regeneration of poisoned resin. In operation the pregnant solution is split between two parallel lines for adsorption. When the first column is saturated it is cut off and water-flushed. Its resin load is then pumped to the backwash column. Stripped resin is then pumped into the empty column from the first fully eluted column. Within sixteen minutes the regenerated and re-loaded column is back on stream, this time as the third in line of adsorption. The loaded resin is thoroughly washed in the backwash column and then transferred to the eluting series of columns to be stripped. Among the operating advantages claimed for this method are avoidance of flow of pregnant liquor, better resin loading, reduced bulk of eluant and a higher concentration ratio of "pregs" to eluate.

An entirely different approach is made in the resin-in-pulp method of IX (R.I.P.). In this, the coarser sand particles are removed by repeated classification before the pulp, diluted to some 10% solids, is allowed to make contact with the resin. It then flows through a series of rectangular cells, called "banks" in each of which a wire basket loaded with beads is jigged up and down with sufficient force to dilate and contract its load. With fourteen cells in line seven are adsorbing ions from the pregnant pulp, five are undergoing elution, one is receiving preliminary backwash and one post-eluting backwash.

Solvent Extraction

This term has, unfortunately, more than one technical use in mineral processing. Broadly, it applies to any process wherein chemicals in a liquid phase selectively dissolve a designated mineral from its ore. In current practice the usual meaning is "liquid-liquid extraction", a form of ion exchange. One definition reads:

"Solvent extraction operations are those in which the separation of mixtures of different substances is accomplished by treatment with a selective liquid solvent. At least one component of the mixture must be immiscible with the treating solvent so that at least two phases are formed over the entire range of operating conditions used[32]." Perry goes on to list the types as liquid-liquid, leaching, washing and precipitative extraction.

In liquid-liquid exchange, instead of a solid resin and an aqueous pregnant solution, the former is replaced by an immiscible organic liquid. Kerosene, itself inert but rendered active by dissolved chemicals of which it is the carrier, is in wide industrial use. The two liquids are mechanically stirred as they flow through a series of mixing vessels, under conditions which produce some temporary emulsification and are then separated. The value, now concentrated in the organic phase, is next retrieved, probably by chemical precipitation after washing with aqueous extractants.

Among the attractions of solvent extraction are its simplicity and continuity of operation. The immiscible phase (in mineral processing, an organic liquid) must be highly selective for the species of ionised compound which is to be transferred from the aqueous phase. Both phases must be quick to demulsify and separate when stirring ceases. The general lines of the flowsheet are shown in Fig. 210.

An excellent summary of solvent extraction has been issued by the U.S. Bureau of Mines.[33] The two essential stages are extraction and stripping. During extraction the required constituent of the aqueous solution ("pregs") is transferred to the immiscible organic liquid which acts as a carrier for reacting chemicals relatively insoluble in water. The liquids are then separated by stripping, after which the transferred values are recovered from the organic phase probably by washing with chemically treated water followed by precipitation and filtration. Effective and fast phase separation is affected by

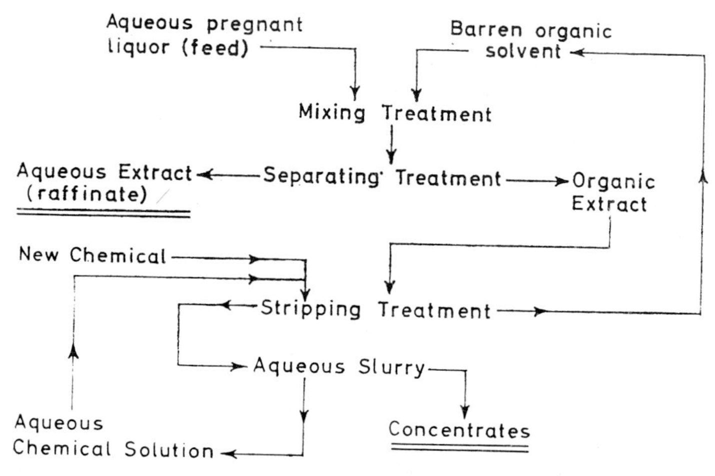

Fig. 210. *General Scheme of Liquid-liquid Extraction*

the relative densities of the liquids, the pH of the aqueous phase, temperature, viscosity, absence of colloidal silica or other slimes and surfactants, and interfacial tension. The organic phase must concentrate the values into a small bulk while leaving unwanted impurities in the aqueous phase. The chemical selectivity involved depends on the specific formation of extractable complex ions, e.g. $UO_2(SO_4)_2^{=}$, and on the insolubility of the reacting agent and the reaction product in the aqueous phase. Extraction of uranium by tertiary amine sulphate proceeds thus.[33]

$$(R_3NH)_2SO_4\,(org.) + UO_2(SO_4)_2^{=}\,(aq.) = (R_3NH)_2.UO_2(SO_4)_2\,(org.) + SO_4^{=}\,(aq.) \qquad (16.32)$$

Ion-pairs such as that in this equation are neutral complexes which react chemically as single units. Only oxonium systems are at present used in mineral processing. Stripping is performed by aqueous extraction with some pH modification. Selectivity between the phases may be reinforced by the use of salting-out electrolytes in the aqueous feed. These are salts which contain the anion of the extractable species and a non-extractable cation.

The organic solvent, or diluent, is not usually a reactant but rather a vehicle for the reacting chemical used to transfer the reaction product from the aqueous phase to the organic one in which it is more strongly soluble. In ion-pair extraction the whole organic phase may be employed in this way.

In addition to the reactions thus far considered a "modifier" is often used to produce synergism. Synergism has been defined[33] as the "co-operative effect of two or more extractants that exceeds the sum of the individual effects". This phenomenon, though widely exploited, is not yet fully clarified by research.

As in the case of solid IX resins, there is a maximum loading of the extracting solvent above which its reacting chemical/s cannot take up further ions. The proportioning of pregnant liquor and solvent during the mixing stage must be sufficient to avoid such saturation.

The term "non-aqueous leaching" has been tentatively suggested[34] to describe techniques in which direct attack by an immiscible liquid is made on solid mineral particles either dry or in aqueous pulp. This does not necessarily involve ion-pair exchange since the sought value is characteristically insoluble in water. This method of extraction is at present at the research stage, and poses such economic problems as reagent cost and specially resistant materials of construction of an industrial plant. Among proposed techniques are direct solvation of sulphur by dimethyl disulphide. Other extractants under test include liquid ammonia, liquid sulphur dioxide and liquid chlorine.

References

1. Lord Fleck. (1962). *Nature*, July
2. Thompson, P. F. "Dissolution of Gold in Cyanide Solutions." *Trans. Electrochem. Soc.*, 91, p. 222.
3. Barsky, G., Swainson, S. J., and Hedley, N. "Dissolution of Gold and Silver in Cyanide Solutions." *Trans. Amer. Inst. Min. Metall. Engrs.*, 122, p. 660.

References—continued

4. McLaurin, R. C. "Dissolution of Gold in a Solution of Potassium Cyanide." *J. Chem. Soc.,* 63.
5. Rose, T. K., and Newman, W.A.C. *The Metallurgy of Gold,* Griffin & Co.
6. Dorr, J. V. N., and Bosqui, F. L. (1950). *Cyanidation and Concentration of Gold and Silver ores,* McGraw-Hill.
7. Clemes, A. "Modern Metallurgical Practice on the Witwatersrand." *J. Chem. Soc. S. Africa,* 46.
8. Roberts, E. J. (1960). "Countercurrent Decantation, When and Why." *Trans. A.I.M.M.E.,* 217.
9. Goldblatt, E. (1956). "Recovery of Cyanide from Waste Cyanide Solutions by Ion Exchange." *Indust. Eng. Chem.,* 48.
10. Eicholz, G. C., and Josling, C. A. (1963). Dept. of Mines Tech. Bull., Canada, TB 43, March.
11. Hedley, N., and Tabachnick, H. (1958). "Chemistry of Cyanidation." Min. Dress Notes, American Cyanamid Co.
12. King, A., Clemes, A., and Cross, H. E. "Treatment of Gold Ore Containing Pyrrhotite at the Sub Nigel Ltd." Trans. I.M.M. (London), 56.
13. Bitzor, E. C., and Nines, C. B. "Some Milling Problems at the Raub Australian Plant." *Eng. Min. J.,* 141.
14. Lake Shore Staff. "Milling Investigations into the Ore as occurring at the Lake Shore Mines." *Trans. Can. I.M.M.,* 39.
15. Matthews, O. "FluoSolids Roasting of Arsenopyrite Concentrates of Cochenour Willans." *Trans. Can. I.M.M.,* 52.
16. Parker, O. J. (1957). *Proc. Aust. I.M.M.,* Part 1, June.
17. Tait, R. J. C. (1961). *Can. Min. & Met. Bull.,* April.
18. Edmonds, H. R. "Appl. of Charcoal to the Precipitation of Gold from its Solution in Cyanide." *Trans. I.M.M. (London),* 27.
19. Lewis, and Metzner. (1954). "Activation of Carbon." *Indust. Eng. Chem.,* May.
20. von Bernewitz, M. W. "Charcoal as a Gold Precipitant in Conjunction with Flotation." *Eng. Min. J.,* 141.
21. Putnam, G. L. "Chlorine as a Solvent in Gold Hydrometallurgy." *Eng. Min. J.,* 145.
22. Bell, H. D. "Chromium in Cyanide Solutions." *J. Chem. Soc. S. Afr.,* 36.
23. Barsky, G., Swainson, S. J., and Hedley, N. "Chemistry of Cyanidation." Am. Cyanamid Co. Tech., Paper 21.
24. Head, R. E. "Physical Characteristics of Gold Lost in Tailings." *Trans. A.I.M.M.E.,* 134.
25. Welch, A. J. E. (1957). "Extraction and Refining of the Rarer Metals." Symposium I.M.M. (London), March.
26. Sutton, J. A., and Corrick, J. D. (1963). *Min. Eng. (A.I.M.M.E.),* June.
27. Razzell, W. E. (1962). *Can. Min. & Met. Bull.,* March.
28. Mitchell, J. S. (1956). *Min. Eng.,* Nov.
29. Forward, F. A., and Vizsolvi, A. (1963). *Ethylene Clycol Leach Process, etc. 6th I.M.P.C. Congress,* Pergamon.
30. Everest, D. A., and Wells, R. A. (1963). *Undeveloped Potential Uses of Ion-Exchange in Hydrometallurgy. 6th I.M.P.C.,* Pergamon.
31. Ayres, D. E. R., and Westwood, R. J. Uranium in S. Africa 1946–56. Symposium, Vol. 2.
32. Perry, J. H. (1950). *Chemical Engineers' Handbook.* 3rd ed., McGraw-Hill.
33. Bridges, D. W., and Rosenbaum, J. B. (1962). *Metallurgical Applications of Solvent Extraction.,* U.S. Bureau of Mines, I.C.8139.
34. Lewis, C. J., and Drobuck, J. L. (1963). *Min. Eng. (A.I.M.M.E.),* Nov.

CHAPTER 17

PRINCIPLES OF FROTH FLOTATION

Introductory

To achieve selective flotation of mineral particles, whether by their removal in a froth or by the less-used method of agglomeration, specific characteristics of one or more of the mineral species present must be adequately developed. Provided these produce sufficiently marked differences of behaviour in the presence of air, they can then be exploited. If a particle is to be held in a mineralised froth, it must be ground to a fineness at which downward pull of gravity is insufficient to overcome its adhesion to an air-water interface. The usual commercial separation entails the lifting of a heavy metal sulphide away from a relatively light gangue by the agency of air bubbles rising through the pulp. This buoyancy results from adhesion of the particle to a comparatively large bubble. The adhesive force with which a particle clings to the air-water interface is opposed by the gravitational drag due to its mass. For successful exploitation of differences in surface properties most ore minerals must be ground finer than 48–65 mesh. A light mineral such as coal (density *circa* 1·4) can be floated at 10 mesh, provided the bubble system on which it is borne is developed as a quiet layer of froth. Random changes of direction, acceleration, and collision may tear too large a particle out of its bubble. At the other end of the flotation size-range, the surface characteristics of *all* particles in the pulp are *more* similar at very fine sizes. Somewhere below 10μ, and for most ores at about 3μ to 5μ, it becomes increasingly difficult to control and exploit differences in surface properties with the accuracy needed to depress gangue, and float concentrate. Typically, flotation is practised between the limits 60 mesh and 5μ. This is quite apart from any consideration of "break" or of liberation mesh.

Flotation is not concerned to alter the chemical nature of the particles involved, but to modify their surfaces. Most freshly exposed mineral compounds have characteristic levels of surface energy, though their natural force is reduced between mining and milling by some degree of oxidation, sorption, or contamination. It is usually possible, by the use of suitable chemicals and conditions, to maintain in the pulp a "*climate*" in which one specific mineral adsorbs a special reagent, while the others remain indifferent. This adsorbed chemical sets up a monomolecular aerophilic layer over part of the particle's surface. If this area is large enough, the whole particle is attracted to the air-water interface of a bubble, where it stays until it is removed with the layer formed when a multitude of these bubbles have survived conversion into a froth above the pulp.

If an ore contains two or more minerals which can be sufficiently liberated from one another in the flotation size range, flotation may be possible. Their surfaces must differ or be caused to differ in wettability. The whole surface of the particle is not necessarily involved. An incompletely liberated particle of middling may be caused to react provided sufficient energy of the required type is available to ensure that it is captured by a bubble. All metallic sulphides can be floated away from their siliceous gangues, and many non-silicates can be separated from silicates. Coal, gold, sulphur, phosphate, and fluorite are among the ores commercially treated. The most notable exception to industrial flotation is cassiterite. Here the problem is economic rather than technical.

Each problem involving the separation of constituent minerals in an ore contains two groups of factors:

(1) Factors fixed by the nature of the mineral species constituting the ore and their intergrowth.
(2) Factors variable as the result of mining and mineral processing.

The more important of these factors may be summarised, as regards their bearing on the flotation process. Those of group (1) include the bond-structure, chemical composition, crystal size and shape of each mineral in the ore; its specific liberation size and "break"; its surface chemistry and physics; rate of surface-energy change after shearing, in air and in water; its grindability and density; its solubility and reaction rate with associated ore-minerals in pulp ground in the available mill water.

Those of group (2) which can be brought under technical control so far as cost-factors warrant it, include some or all of the effects of variance on (*a*) each mineral and (*b*) the combined pulp system. Of the hundreds known to exist, the dominant controls are exercised on variation of grinding conditions and classification; surface modification by use of alkalis, acids, dispersants, wetting agents, resurfacing ions; pulp climate modification by removal or increase of potentially surface-active ions; specific coating of preferred surfaces after due modifying preparation to render them aerophilic; retention of aerophilic particles in a suitable mineral-attracting froth.

The general scheme for a one-concentrate treatment is shown in Fig. 211. The terms there used are defined in the Glossary.

History

The use of the conditioned surface for selective attraction of mineral particles goes back to the pre-Christian era. Herodotus records the stirring of silt in ponds by Ethiopian women, who used goose feathers covered with bitumen to pick out the specks of gold contained in that silt. From the thirteenth century onward, lapis lazuli has been concentrated by a flotative use of melted resin. In 1860 Haynes patented the use of selective clinging of particles at an oil-water interface. From 1877 the Bessels brothers were floating Bavarian graphite from an aqueous pulp to which small quantities of oil had been added. Bubbling was effected either by boiling or by chemical generation of carbon dioxide gas in the pulp. Their patents were registered in 1877 and 1887. Flotation of the same ores today uses air to render the

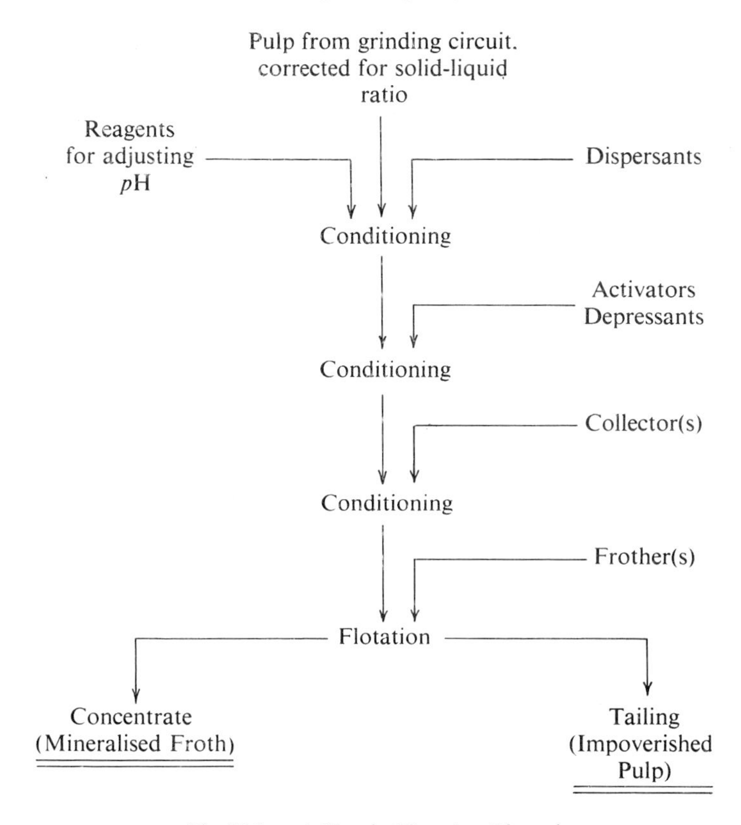

Fig. 211. *A Simple Flotation Flow-sheet*

selected particle buoyant. In 1886 Carrie Everson patented some reagents, following her observation of the flotation of galena during the laundering of textiles which had come in contact with ore. In 1902 Potter (Australia) and Froment (Italy) recognised the action of gas in lifting oiled particles of sulphide to the surface of a pulp. In 1904 Elmore patented methods in which gas was introduced either by electrolysis or by the use of vacuum. His cell still finds limited use in coal flotation. A great stride forward was made by Sulman and Picard in 1905, when they aerated the pulp by means of a submerged pipe. In the following year, when the true nature of surface conditioning was recognised, they patented the use of strong agitation of the pulp *plus* the use of very small quantities of oil. Hitherto anywhere up to 5% of the pulp volume was added as oil, and although the sludges from certain refineries were preferred, nothing of importance was known as to the chemistry of the flotation process, although the role of the "sulphur hook" in attaching aerophilic oil to the particle surface was being studied. In 1913 Bradford patented the use of $CuSO_4$ as an activator for sphalerite, but the significance was missed. The first collector not based on oil (*x*-naphthylamine) was patented in 1917 by Corliss. In 1922 the use of cyanide to depress sphalerite during the flotation of galena was patented by Sheridan and Griswold, and in the next year or two commercial flotation of these sulphides in distinct successive operations (differential flotation) was achieved by chance. Up to that time only bulk flotation of the sulphides was possible, and separation of the various metals in the float was made by expensive smelting methods. One account of the discovery of a differential method begins with the laboratory separation of lead and zinc sulphides, attributed by the workers to a change in reagent dosage. A plant was built as a result, and failed to reproduce these results. Close checking showed that the laboratory cell used a brass impeller to agitate its pulp. Similar impellers were fitted in the plant and the process became a success. The reasons are today fully understood and are discussed later. Other milestones in progress are the Sulman and Edser patent of fatty-acid soaps in 1924; Keller's discovery of xanthates in 1925; and Whitworth's development in 1926 of organic di-thiophosphates as collectors. Those early days of groping are vivid memories of many mill men still interested in the subject, but flotation is no longer the sport of empiricism. It has become a science applied industrially to processes as far apart as the preparation of special flour for invalid diet and the waterproofing of sparking-plug terminals. The early years, and the machines and struggles have been placed on record by Rickard, Pryor, Diamond, Hines, Crabtree, Vincent and others.

Flotation and Agglomeration

In the process of *froth-flotation* adhesion is obtained between mineral particles and air rising through a pulp. This temporarily buoyant combination rises, forms a froth, and is removed. The following steps are taken:

(*a*) The ore is ground in water to at least −48 mesh.
(*b*) The pulp thus formed is diluted to between 25% and 45% solids.
(*c*) Small quantities of reagents are added to *modify* the surfaces of specific minerals.
(*d*) A reagent specifically attracted to the desired float-mineral is added, which partly coats it with an aerophilic surface.
(*e*) A reagent is added which will help to establish a suitable froth at the surface of the pulp.
(*f*) The pulp is aerated in suitable vessels, usually in series.
(*g*) The mineral-bearing froth is removed from the impoverished pulp.

In agglomeration a similar conditioning treatment is applied, usually to a much thicker pulp containing classified feed from which all particles below the usual sizes treated on shaking tables have been removed. This slurry is then aerated. Aerophilic particles become agglomerated by edge-adhesion or by being stuck together by minute air-bubbles. These glomerules usually arrange themselves round a large air-bubble. On reaching a shaking table, such a light glomerule rises and runs straight down to the tailings discharge.

Alternatively, it can be washed over the sides of a conveyor belt adapted for this variation of the flotation process.

The Particle Surface

It is usual to regard surface modification in the pulp as a re-balancing between aquated ions and charged lattice points. No sweeping simplification is adequate in our present state of knowledge. If the polar strength of the water molecules, aided by a suitable *p*H, exceeds the attraction binding a mineral ion into its crystal lattice the ion will hydrate and move into the aqueous phase provided contact is made. The stronger the hydrating attack, the wetter will be the residual surface. Ions resisting aquation remain bound electrostatically in accord with the crystal geometry in the solid substrate.

The particle surface is not homogeneous, since it includes both hydration-prone and water-repellent charged points. The species represented by the particle is not necessarily pure or homogeneous and its surface may be screened to a varying extent from the water phase by contaminants or through deterioration. However small the particle may be, it bears the scars of battle set up by its explosive shattering during mining, followed by wrenching and shear during comminution. The weakest of its lattice planes may have suffered incipient displacement or weakening, and the apparently smooth surface is probably interrupted by sub-microscopic cracks. In these crevices and eroded pits dissolved gases from the aqueous phase may precipitate and cling, thus shielding the area they cover from chemical attack. (Later in this chapter the role of dissolved air and minute bubbles receives attention.) These, and other considerations discussed later show that the diagram of the electrical double layer shown in Fig. 187 is an over-simplification of a complex state. The truer picture is of a cratered, creviced and ionically irregular discontinuity layer extending a short distance into the solid and only partly available for reaction with the ambient phase. It is, perhaps, significant that research shows that only some 2% to 5% of the particle need accept a collector agent to ensure flotation in a number of cases.

Hypotheses must often take into account the surface area of the solids under examination. It should therefore be remembered that it is not possible to dehydrate a surface at normal temperatures and that the use of heat might introduce complications; that the adsorption methods used in accurate surface measurement measure *either* the dry *or* the hydrated area, but not both fully; that no reliable method of assessing the area of incipient cracking and cleavage which should be reckoned as part of the surface has yet been found; and that in consequence there is some discrepancy between various measuring methods.

Surface Changes

Most mineral surfaces, when freshly cleaved, are readily wetted and prefer water to air. After a period of exposure, however, some minerals sorb hydrocarbons from the air to a sufficient extent to reverse this preference. In cities where the atmosphere is heavily contaminated, this effect is strongly marked. In a series of tests made in London, freshly crushed emery was found to wet completely. On exposure to air this material rapidly lost its attraction for water. The progressive contamination of its surface (measured by capillary tests) began to show after a minute or so. Within an hour it had attracted from the atmosphere an organic coating which produced the smell of burnt lubricating oil when the emery was gently heated. Contaminated air is not the only source of random introduction of hydrocarbons. In mechanised mining approximately 0·2 lb. of lubricant is sent undergound for each ton of ore extracted, and part of this returns to the mill, where it is distributed through the pulp. Modern lubricants are chemically treated to improve their wetting adhesion to metal bearings, and to some extent they also react with mineral surfaces. Just as most minerals are readily wetted when their surfaces are clean, so do they become water-repellent when coated with such hydrocarbons as greases, oils, paraffins, and alcoholates. If the particles thus contaminated are sufficiently small, they then tend to adhere to an air-water interface, and to be floated.

A generalised coating such as this would be useless as a stage in selective separation, so flotation is preceded by *conditioning* treatment, during which only the specifically desired mineral is rendered aerophilic. In the course of this conditioning, any random contamination of the ore surface must be dealt with if it would undermine efficient exploitation of differences between particles of the various minerals. Specific reaction of the surfaces of the particles undergoing treatment must take place if the compound beneath each surface is to be collected correctly either as a concentrate or as a tailing. The quantity of chemical reagent needed to condition a clean surface is very small. In the case of chalcopyrite, 0·01 lb./ton of collector agent ensures the flotation of substantially all the −150-mesh particles of this mineral from a 2% assay value in the mined area, although this carries nearly one-tenth of an acre of surface.

The condition of the mineral surface is important. If it is to be brought into control, it must be sufficiently clean to react with the chemicals introduced into the pulp. Oil, fouling, tightly adherent slimes, iron stains, etc., would interfere with the treatment and would call for special remedial measures. During comminution very finely ground particles (slimes) of one mineral species sometimes attach themselves electrostatically as "slime coatings" to the surface of larger particles of a dissimilar species (del Guidice[4]). This tendency is countered by the use of *dispersing agents* and/or *p*H control. Some minerals exhibit inherent or "native" floatability, a phrase which indicates proneness to attach to an air-water interface without preliminary chemical treatment. Gaudin[5] considers this to be a fundamental characteristic of crystals which have no ions at their surface.

Most inorganic crystals are not molecular aggregates, but of ion-bonded structure. This distinguishes them from most organic substances. The crystal structure of a mineral affects its adsorptive properties and consequently its response to the heteropolar surfactants used in the flotation process. The lattice energy of the crystal is that needed to separate its ions to an "infinite" distance, and has chemical significance, e.g. in surface modification of the mineral particle before flotation. The greater the lattice energy, the more resistant is the particle to chemically induced change. Those of silica and alumina are about 3,700 kilocalories/mole rendering them inert and poorly responsive to surface-activation by flotation reagents. That of sphalerite is 800, and this mineral, though virtually insoluble in water, is moderately responsive to appropriate surfactants. The lattice energy of sodium chloride is only 180 kc/mole and it is readily ionised and highly soluble in water.

In practice native flotability and relative wettability are vulnerable to contamination between severance and treatment of the ore unless cleansing precedes chemical *conditioning*.

Surface change or "ageing" may occur between mining and conditioning through atmospheric attack in which there may be no contact with contaminants. This affects the subsequent treatment. In one instance[6] it was noted by Mellgren and Rau that "mild oxidation of a galena surface by exposure to moist air for six hours did not affect the adsorption density, but with prolonged exposure there was a large increase in the uptake of xanthate". The thermodynamics of floatable surfaces differ from those of the multiphase systems usually considered in chemical engineering, where the volume of the substrate phase is the dominant object of interest. In a readily blended system equilibrium is attained when free energy is at a minimum, so the driving force which leads to change is thermodynamically considered in terms of decrease in free energy. Chemical reaction is less directly concerned with handling, mixing and rate of change than is the flotation process. Here the phases in contact are virtually immiscible and mutually insoluble. The solid particle has been so reduced in mass that its gravitational drag is small and its surface-volume ratio high. Its behaviour (whether aerophilic or hydrophilic) depends critically on adsorption to this surface from the ambient aqueous phase. Whatever may be the substrate species, the surface of the particle determines whether it will adhere to a bubble or remain wetted. The force which determines its ability to attract external ions is its available surface energy. This is at its highest and most characteristic value at the moment when the particle is broken during comminution, with its crystal surface bonds newly sheared. From this moment the particle is a weak ion exchanger, capable of responding to environmental pressures.

Tests for Floatability

The surface energy of the phases in contact in the flotation system air-water-mineral determines the relative attraction between the various species of particle in the pulp toward *either* the air *or* the water phase, or in other words their degree of floatability. For a liquid the surface tension γ can be measured, but for solids its value can only be indirectly assessed. For a liquid it is the force which the surface on one side of a line 1 cm. long exerts on the other side of this line, expressed in dynes/cm. Surface energy is the product of γ and area, expressed in ergs.

Surface energy is also definable as the work required at constant temperature and composition of the liquid to increase the surface area by 1 cm^2. This represents the work required to bring additional molecules to the surface against the attraction of surrounding molecules in the liquid body. Since the kinetic energy of these molecules rises with temperature increase while inter-attraction is reduced $\gamma \propto \frac{1}{T}$ and vanishes at boiling point.

The pressure within a curved liquid surface is higher than the external pressure. A soap bubble of radius r with surface free energy $4\pi r^2 \gamma$ undergoes change when r is decreased by dr to $8\pi r \gamma \zeta r$. The tendency to shrink is balanced by the pressure difference ΔP across the film. For an N-bubble

$\Delta P = \frac{2\gamma}{r}$ and for a F-bubble $\frac{4\gamma}{r}$

Among the techniques used in the measurement of γ are capillary rise, bubble pressure, dropping weight, adherence measurement of a ring, or of a mica plate. In the two last-named the force required to detach a completely wetted wire loop or vertically suspended thin plate is measured. Unfortunately there is at present no direct means of measuring the tension at a solid surface. A surface atom is only partly bonded inward to the general lattice of a crystal, and for the rest is attracted outward to counter-forces in the surrounding phase or phases. It has higher potential energy than an atom in the substrate. The surface energy is not localised in a monolayer in the surface discontinuity but may be diffused to a depth of some 10^{-7} or 10^{-8} cm. Its strength is determined by the difference in polarity of the contacting phases—that is, the difference in intensity of the molecular forces at work in these phases. Indirect determination of γ for some minerals indicates the following approximate values.[29]

Gypsum (crystal face 010)		39 erg/cm^2
Calcite	100	78
Fluorite		146
Apatite (prism)		186
Feldspar		358
Quartz		780
Topaz (prism)		1080
Corundum		1550

That of water is 72·3.

In the research laboratory, surfaces can be tested and reactions studied by the method of *contact angle measurement*. The full experimental procedure is described elsewhere.[7] In brief, a clean smooth surface of mineral is placed in distilled water. A bubble is pressed down upon it. If after half an hour (induction time) no adhesion is visible, the surface is assumed to be clean (completely wet). A suitable chemical reagent is then added to the water, at a given pH, and after a short reacting period the air bubble is again pressed to the surface. If the latter has now become aerophilic, adherence follows.

The angle across the water phase is called the contact angle θ (Fig. 212). This angle is the resultant of three tensions (T). The contact angle (θ) is normally expressed in Young's equation as

$$\gamma S = \gamma SL + \gamma L \cos \theta \tag{17.1}$$

where γ is the free energy per unit area and S and L are the solid and liquid. This was originally stated by Young in words rather than symbols.[9] Since $\sqrt{S}$ is more or less unmeasurable the equation as stated above is of limited practical use. It can also be written in terms of work (W)

$$W_{SL} = \gamma S + \gamma L - \gamma SL \tag{17.2}$$

By combining 17·2 with Dupre's equation the expression is obtained

$$W_{SL} = \gamma L (1 + \cos \theta) \tag{17.3}$$

which shows the equlibrium for a liquid at rest on a solid. The righthand components of Eq. 17.3 arc measurable and can therefore be used to calculate the force of liquid adhesion to a solid surface. A fuller discussion of contact angle is to be found in research literature, (e.g. by Leja and Poling.[10]) The incidence of hysteresis on the measurement of θ has been studied by Gaudin *et al.*[11] This hysteresis is definable as the difference between the maximum and minimum values of θ under stated conditions of measurement. It expresses the resistance to movement of the wetting perimeter and depends on the direction of spread—gas displacing liquid at the solid surface or *vice versa.* Its magnitude depends partly on the rate of movement of the 3-phase perimeter and is up to 30% higher with static than with dynamic friction. It is also affected by roughness of the solid surface and the extent to which it has been modified by adsorption.

Contact-angle measurement, despite certain defects, not least of which is the difficulty of obtaining suitably large plane surfaces of mineral, provides a powerful investigating technique. For the given surface condition of the specimen θ is the index of surface energy. If θ is *nil,* the mineral will not float under test conditions. Even a contact angle of a few degrees indicates that there is some floatability. θ is not a characteristic of the mineral phase in a conditioned system, but of an aerophilic organic group in the molecules of the adsorbed chemical, which is called the collector agent. Floatability is determined by the surface coating, not by the substrate. The contact angle of all minerals conditioned by (i.e. carrying a monolayer of) potassium ethyl xanthate (a collector described later) is approximately 60°, regardless of whether these minerals are copper, lead, zinc, etc.

The contact angle method of establishing the conditions under which a given surface becomes aerophilic is open to certain criticisms and experimental difficulties. Chief among these are the problems of obtaining a representative crystal of the required mineral of a sufficiently large size (over 0·5 cm.^{2} in plane area) which is also truly representative. Next comes the question whether the finally prepared surface is still characteristic after its intensive polishing with abrasives under water. If the mechanical action used distorts the surface lattice or even produces a *Beilby layer,* the evidence as to induction and adhesion of air is suspect. A further disadvantage is the static nature of the test system produced, which differs vitally from the dynamic one in which the mineral particle must adhere to a *coursing bubble* (one rising toward the surface of the test fluid). Finally, the need to preserve complete cleanliness makes progress tedious if a new surface is prepared many times when such variables as pH and reagent concentration are being studied.

Two newer techniques have been developed, each of which overcomes most of these experimental difficulties while preserving the essentials of basic control. The first of these is the bubble pick-up method. In this small quantities of the mineral which is to be tested are crushed, sized to the required mesh range, and cleaned by shaking in distilled water. The first reference to its use is in a paper by Cooke and Digre.[12] An air bubble was

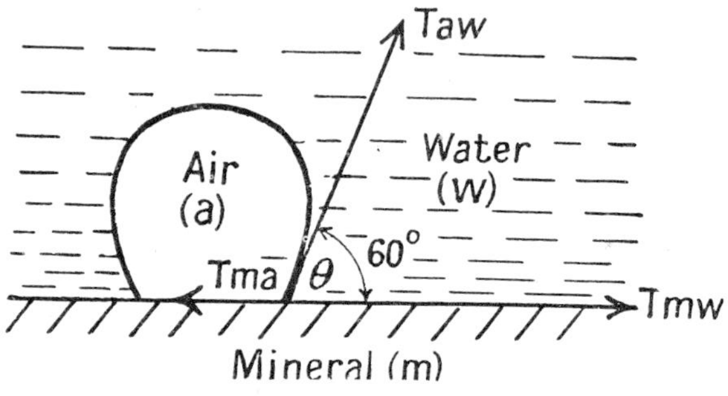

Fig. 212. *The Contact Angle*

pressed against particles and then lifted. The quantity of grains adhering was noted and it was found that correlation with contact angle tests on similar material was good, while the pick-up method was more sensitive to changes in reagent concentration than was measurement of θ. Improvements made by Sun and Troxell[13] start with the sizing, cleaning and placing in 200 ml. of distilled water of 0·5 g. of material. After adjustment of test chemicals and allowing the determined conditioning period for reaction to occur, the particles are swirled to the centre of the beaker, pressed on by an air bubble held in the concave end of a glass rod. This is then raised gently, moved to a clear part of the beaker and tapped. The particles which drop are then examined. Those so strongly aerophilic as to remain attached are usually visible. With the bubble pick-up technique it is possible to relate changes of reagent and pH to all stages between non-attachment and strong attachment of particle to air-water interphase, and to make a series of tests fairly rapidly on a small number of particles.

In the Hallimond tube, a modified form of which is shown in Fig. 213, the particles are held on a porous surface of sintered glass at the bottom of the tube containing distilled water and the testing chemicals. A stream of air bubbles is blown upward through the sinter, and any particles adhering to these coursing bubbles are floated. On emerging, each bubble explodes and its load of particles slides down to the receiving pocket from which they are retrieved as required. This apparatus marks an important further advance.

Levitation of particles is fully dynamic, and the conditions are to some extent similar to those in a commercial flotation cell. A small sample is treated, and the floating and non-floating fractions can be separately weighed and examined.

The only factor not brought into the testwork by the Hallimond cell is the influence of the frothing agent and its associated froth column. This can for many purposes be left out of the tests until optimum conditions for levitation have been established. Frothing can be studied under the same rigorous conditions of cleanliness in a simple apparatus devised by the author[14]. A sample of five grams or more can be treated, and given controlled flotation with any desired degree of thoroughness in removing a true mineralised froth. The wettability of small samples of powdered ore can also be assessed by the

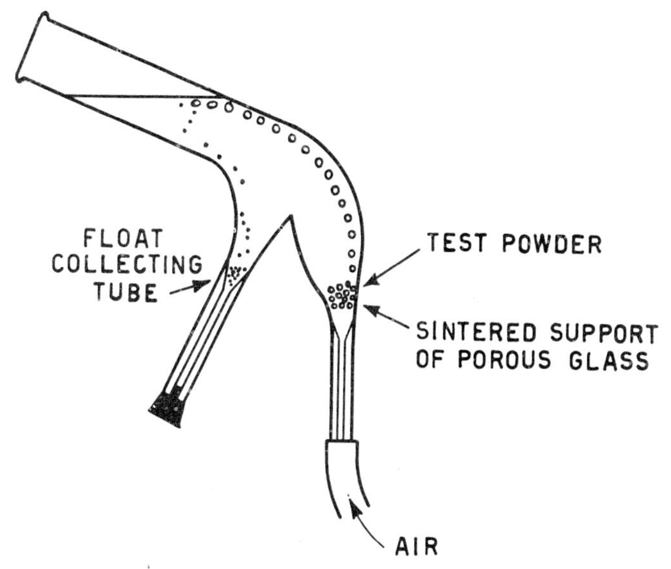

Fig. 213. *The Modified Hallimond Tube*

"phase inversion" method[11]. This relies on the mechanical stabilization of an emulsion by an insoluble interphase. Either a water-in-oil or an oil-in-water emulsion can be produced if the hydrophobicity of the intervening mineral powder is suitably modified by a surfactant. The method is simple, requiring only the screening of a small sample of mineral through 325 mesh, shaking of the phases, e.g. 2 ml. each of benzene and water, and observation of effects of change of collector or *p*H. Four degrees of wettability are shown: (*a*) benzene and water separated with mineral in water; (*b*) oil-in-water emulsion with contact angle below 90°; (*c*) water-in-oil emulsion with θ above 90°; and (*d*) separate liquid phases with the mineral in the benzene.

Many freshly cleaved surfaces exhibit a natural contact angle of a few degrees. Sulphur, ozokerite, graphite, and some coals and coal resins have a high enough θ to float without aid from collector agents. The greater the difference between the natural or induced contact angles of the minerals in the ore, the more controllable is their selective separation by flotation. Among those examining surface activity in its fundamental aspects are Moilliett[15], Adam,[16] Sutherland,[8] and Gaudin.[5]

Activation Theory

In flotation technology activation is the development on the surface of a designated mineral of selective reaction, usually attraction into the air-water interface. This renders the activated particle more eager to be drawn into a coursing bubble, rise and persist as part of a selectively mineralised froth. To induce such changes reagents called activators are used. Other chemicals, used to work to the opposite purpose are called wetting agents or depressants. Both types are surface-active agents or surfactants, capable of influencing conditions in the air-water-solid triphase. The reagents used in the conditioning of mineral surfaces are for the most part charged with inorganic ions (+ or −) chosen for their ability to increase the difference between the adsorptive powers of the various species in the pulp with respect to the collector agent used to gather the preferred particles into a mineralised froth.

In studies of the attachment of collector agents to minerals Plaksin[10] found the tenacity and density of fixation of xanthate to depend on the structure of the crystal lattice and the cleavage features developed during grinding. Sulphide minerals have heterogeneous surface properties which attract non-uniform distribution of xanthate and give irregular response to the stimulation coming from the aqueous phase. The electro-chemical potential of a species may vary widely (e.g. between 0·2 and 0·75 volts for galena). Heterogeneity in a continuous surface can lead to preferential "spotty" attachment to air, preferential adsorption of a surfactant or localised patches of oxidation. The same sulphide surface has been found to have both anodic and cathodic areas. In the case of non-sulphide minerals (e.g. fluorite, calcite) the crystal structure influences both the action of gases from the pulp and surface interaction with reagents. Even a pure mineral compound, prepared synthetically, exhibits a variety of responses to surface modification. Ore minerals treated industrially are not pure and the study of their reaction to treatment must be partly empiric. Plaksin's work has been summarised by Rogers[3] as showing (*a*) that in the absence of oxygen the fresh surface of a sulphide mineral is wettable to some extent; (*b*) oxygen is adsorbed from water in preference to other co-existing gases; (*c*) adsorption occurs in stages; (*d*) first oxygen is adsorbed, next xanthate is fixed. With non-sulphides such as metal oxides, silicates, carbonates, sulphates and halides much weaker adsorptive forces are at work than the chemi-sorptive bonds between surfactant and metal sulphide. Electrostatic and Van der Waals forces provide the weaker and less selective adsorbing mechanisms for carboxylic collectors.

Metal sulphides are the most commonly floated ore minerals. They are only slightly soluble in water, but most of them oxidise superficially in moist air or oxygenated water. Some are lustrous, some dull; some are good conductors and others poor; some are tough, some brittle, some soft. Despite this lack of common physical qualities, all float readily with a suitable collector. The reason is not found in their common factor, low solubility. When sufficiently oxidised to exhibit relatively high solubility, the metal ions then generated precipitate the collector, thus wasting this expensive chemical and diverting it from the work of surface modification. High solubility would therefore be a most undesirable quality. Indeed, when it is encountered (through partial formation of oxy-acids of sulphur at the surface by oxygen attack) it must be neutralised as a preliminary conditioning stage. In most minerals the metal-sulphide lattice is strongly bound together by internal ionic forces. It is then immune from attack across the interphase by solvating counter-ions at normal (near neutral) *p*H values in the pulp. A few unstable sulphides (notably pyrrhotite and marcasite) readily take up external oxygen, thus producing chemi-sorbed surface layers of sulphite and thiosulphite. This raises the acidity (reduces the *p*H) of the pulp to some extent, and robs it of oxygen. The role of dissolved oxygen in activation of a sulphide surface is controversial.

The argument which requires a chemical for flotation holds that collector-attachment is a stoichiometric reaction across the interphase. Wark, Cox, Sutherland, and others[8] postulate an ion-exchange process as the attracting mechanism. The chemical hypothesis requires incipient oxidation, and there is considerable test evidence for the necessity of oxygen in the pulp as a prerequisite of collector-coating. If oxidation is considered as de-electronation, with the removal of one or more electrons from the oxidising lattice point, a mechanism is provided for attracting a negative ion. The purpose of this attraction is considered later. There is at present insufficient evidence for Taggart's claim of chemical oxidation, but much that suggests the need for conditions which confer positive charge on a good percentage of discontinuity lattice points at the appropriate stage of conditioning. This suggests that the function of the hydrogen ion in the pulp is to promote an attracting potential at the surface of each specific mineral within a certain *p*H range or below a specific *p*H value. Activation is facilitated in accordance with the inverse solubilities of the metal sulphides. Those of mercury, silver and copper effectively replace zinc at the surface of sphalerite.

Study of the flotation of corundum by Modi and Fuerstenau[17] shows that it can be regulated by *p*H control of the mineral surface charge, suggesting a close connexion between ion association, electrokinetic behaviour and floatability. The adsorption of collector molecules is at its maximum at the iso-electric point of a mineral, and therefore its floatability with that collector. The electrical character of the surface is determined by two potentials, electrochemical and electrokinetic. With excessive transfer into solution of one type of ion (positive *or* negative) the electrical equilibrium is disturbed and the surface becomes charged. Ions of opposite charge now find it harder to escape from the crystal lattice into the liquid phase. Ions already in solution are drawn to the vicinity of the surface to balance its charge and an electrical double layer forms. Its inner component is due to the charged surface and is spread on, not into, the mineral while outer counter-ions are more mobile, being stirred by thermal vibration in the liquid. Counter-ions next the surface move with the mineral while those further out diffuse away, setting up the zeta potential (ζ). If a change in the magnitude of ζ is accompanied by a change in its sign, then the ions of reagents at work in the pulp are potential-determining and can penetrate to the inner layer of the zone of shear and thus influence adsorption to the mineral surface. Xanthates have been shown[29] to affect the electro-chemical potentials of conducting minerals.

The structure and stability of hydrated surface layers depends on the nature and interplay of the phases in contact. The process of hydration affects solubility, attachment of particles to bubbles and coagulation. There is a tendency in Russian research to analyse flotation in terms of surface hydration. Wettability and change in floatability due to adsorption of heteropolar substances can be investigated by contact angle tests. Attraction of dipoles of water to surface lattice points is due to free atomic or molecular force and surface hydration increases with the adsorption of hydrated (aquated) ions. If during comminution strong electrostatic forces are set up the mineral surface tends to be strongly hydrated because of ruptured bonds, and uncompensated atomic and ionic charges. Where such charges are small the hydration is slight (e.g. with graphite, sulphur, molybdenite or talc). Such minerals have weakly adherent bonds across the cleavage slip planes and are characterised by native floatability, air clinging to natural cleavage planes. Gas adsorption considerably lowers the hydration of a mineral surface.

Multivalent cations appear to be hydrated by from six to eight molecules of water, but anions attract a much weaker hydrating sheath. If dissolution of a mineral constituent is to occur the energy of hydration must exceed the lattice energy. In hydrated sheathing, the water molecules immediately in contact with the true ion appear strongly oriented, the effect decreasing outward. Polar molecules become considerably hydrated and non-polar ones only slightly. This hydration is exothermic.

Physical adsorption due to Van der Waals forces of some 3 to 4 calories/mole is unstable, not particularly selective and is easily reversed. Multimolecular sorbed layers are possible. In chemisorption, owing to interatomic action, the heat of adsorption is from 20 to 50 times higher, stable attachment greater and de-sorbing harder to accomplish. Chemisorption produces lower solubility of the resulting compounds at the crystal discontinuity lattice. It is aided when the sorbed ions have ionic radii similar to those in the lattice. Physical adsorption can overlie chemisorption.

It is easy to confuse electro-adsorption with the first stage of chemical action. Many physical chemists consider that adsorption must in any case precede full chemical action. While the possibility of chemical action in surface modification is clearly established as an essential energising activity for many minerals, ion exchange at a surface monolayer is a leading factor in the work of collecting and modifying flotation agents.

Flotation Reagents in General

The chemicals added in small and controlled quantities to the pulp before or during flotation are, broadly, of three types—collectors, modifiers and frothing agents. The division between these may be blurred, in which case control of selectivity suffers.

Collectors, sometimes called promoters, are surfactants which preferentially adsorb to specific types of particle surface, and by so doing render part of the surface sufficiently hydrophobic to cling tenaciously in any available air-water interface until finally removed as part of a mineralised froth when the pulp is aerated.

Modifiers (regulators, activators, depressants and dispersants) are reagents which intensify the selectivity of collector action. This they achieve in various ways discussed later, the general effect being *either* to increase *or* to decrease the hydrophobicity of a specific surface.

Frothers are reagents which tend to stabilise coursing bubbles as these emerge from the pulp during its aeration, and which facilitate the retention of specific minerals in a blanketing layer of mineralised bubbles. This layer is removed continuously from above the pulp, bearing with it the desired concentrated mineral/s.

Collectors act either by forming distinctive monolayers on part of the mineral surface or by forming transition phases at the surface of no specific composition or distinguishable nature. Modifiers act by increasing or diminishing the attraction for collector ions through the electrical double layer which surrounds each particle.

Collectors

A collector must sorb to the selected mineral surface so as to cover it partially with a film which is water-repellent. Success in achieving this specific sorption is essential if the particles thus modified are to be removed from the pulp as a mineralised froth. It is commonly necessary to develop the latent differences between the various mineral species before the collector is added, in order to help this culminating action. This is done in a series of conditioning steps, in which some of the minerals are caused to lose their attraction to the collector, while in others it is increased. The delicate reactions by which this differentiation is accomplished are caused by the addition of small quantities of surface-active chemicals, sometimes in quantities amounting to only a few grams per ton of ore. Technically, the mill operator is more likely to maintain efficiency with "starvation" quantities than with over-doses of these expensive chemicals, and economically he is concerned to keep down costs by using them thriftily.

In the early days of flotation, oils were the only collectors used, and little or no regard was had to their composition. Operations were in consequence erratic and inconsistent. Since oil is also a somewhat inefficient frothing agent, it was not possible to maintain separate control of concentration of collector and texture of froth, though this is important in plant practice. Oils are still used in flotation, and usually contain refinery sludges which remain after distillation and "cracking" with sulphuric acid. Such residues are cheap and contain alkyl sulphonic and sulphuric acids. Operators using these oils draw them from one source, thus assuring a fairly consistent composition.

The discovery of non-oily chemicals with higher selectivity—the xanthates and dithiophosphates, notably—changed the whole picture. The commercial use of "oily" collectors has shrunk considerably since 1925, their place being taken by "chemical" (non-oily) reagents. Their "smearing action" has not been clearly elucidated, but its generalised nature led to low-grade froths in which random flotation of oil-coated gangue occurred. With the chemical collectors now in universal use, reaction is specific to the required floating minerals.

The most widely used collectors in sulphide flotation are the sodium and potassium salts of certain acids containing a hydrocarbon group. These are anionic collectors and include the xanthates and dithiophosphates. Each collector molecule is heterogeneous. It contains a polar and a non-polar group, the latter being a hydrocarbon. The mechanism by which the polar end of the molecule becomes attached to a counter charge at a lattice point on the surface of the mineral is not completely agreed by research workers, but the effect is to anchor the molecule with its non-polar group pointing outward. When enough lattice sites have thus become occupied a hydrocarbon film is established over part of the surface of the particle. This film is aerophilic and is attracted to any available air-water interphase. If enough of the mineral surface has been thus coated the particle can be held in a flotation froth. If its hydrophilic area dominates, the particle will remain unfloated in the pulp. In this connexion the word "hydrophilic" does not necessarily imply positive wetting attraction to the water phase of the water-air system. A surface may be in this state, or may be indifferent to its surroundings. If a middling, there may be a tug-of-war between its aerophilic and aerophobic surface areas. The mass of the particle aids the wetting-down process. The collectors used for floating metal sulphides, which carry positive charges at their surface lattice points, are anionic. Both anionic and cationic reagents are used for non-sulphide minerals, where the surface charge is complex.

Xanthates

These are the salts of xanthic acid, a sulphydril compound, which in theory would be formed by the reaction

$$R\,.\,OH + CS_2 = R\,.\,O\underset{\underset{\displaystyle S}{\|}}{C}SH \tag{17.4}$$

Where R is a hydrocarbon group. Xanthic acid is the unstable and almost unknown reaction product. The name "sulphydril" or "mercapto" refers to the monovalent radical $-SH$. The general formula of xanthate is

$$RO\underset{\underset{\displaystyle S}{\|}}{C}-SNa$$

the H of xanthic acid being replaced by an alkali (Na, or K) to form a sulphydrate. The R group in the case of sodium ethyl xanthate is C_2H_5O, and the structural formula of the salt is

```
       H   H   :
       |   |   :
   H — C — C — O — C — S — Na
       |   |   :   ‖
       H   H   :   S
Non polar group : Polar group
```

The elongated xanthate molecule is from 5 to 7 Å long. The contact angle formed with potassium ethyl xanthate (KEX) is about 60° and with amyl xanthate (KAX) 85° to 90°. The solubility in water of the xanthate series decreases fourfold with each addition of one hydrocarbon group while the selectivity tends to increase with lengthening of the HC chain. The practical working limit is reached with hexyl xanthate. In water the xanthate molecule dissociates to form a heteropolar anion. This adsorbs to a charged lattice point on the mineral surface to form the compound with the bound metal point (Me)

```
R — O — C = S
        ‖
        S — Me
```

The xanthates are prepared by reacting an aliphatic alcohol (or in the manufacture of rayon, cellulose) with carbon disulphide and a caustic alkali.

$$C_2H_5OH + NaOH + CS_2 \rightarrow C_2H_5 - \underset{\underset{S}{\|}}{OC} - SNa + H_2O \qquad (17{\cdot}5)$$

If this somewhat unstable compound oxidises, it becomes dixanthogen

```
  C2H5        C2H5
   |           |
   O           O
   |           |
S — C — S — S — C — S
```

Dixanthogen has low solubility in water, and if used direct as a collector must be emulsified.

In an acid pulp the breakdown of xanthate is accelerated. Xanthates are most active in the *p*H7 to *p*H12 range. As pulp alkalinity increases the competition at the mineral surface between xanthate and hydroxyl ions results in decreased adsorption of the collector, a phenomenon particularly noticeable in the flotation of pyrite.

Xanthates are strong reducing agents.

$$2\,KEX + CuSO_4 \rightarrow Cu(EX)_2\ (\text{cupric xanthate}) + K_2SO_4 \qquad (17.6)$$

$$2\,Cu(EX)_2 \longrightarrow 2\,Cu(EX)\text{ (cuprous xanthate)} + (EX)_2\text{ dixanthogen} \quad (17.7)$$

Another mode of decomposition is

$$2\,KEX + \tfrac{1}{2}O_2 + CO_2 \longrightarrow (EX)_2 + K_2CO_3 \quad (17.8)$$

During decomposition depressing agents, such as thiosulphate or alkali, may form.

Dixanthogen has certain advantages when used as a virtually nonpolar collector. It is insensitive to *p*H and is not precipitated by heavy-metal ions in the pulp. Its relative insolubility can be overcome by feeding it into the ballmill.

Decomposition of xanthate in storage is usually due to entry of water.

$$6ROCSNa + 3H_2O \longrightarrow 6ROH + 2Na_2CS_3 + Na_2CO_3 + 3CS_2 \quad (17.9)$$

Drums should be stored lying on their sides to minimise this risk. Sodium and potassium xanthates ionise readily in aqueous solution, the metal ion dissociating and leaving the CS_2O group negatively charged. The higher homologues of the alcohol (non-polar) group of the xanthate molecule are formed by adding a CH_2.

The formulae of the lower alcohols from which xanthates are manufactured are:

Methyl alcohol	CH_3OH	
Ethyl alcohol	C_2H_5OH	(CH_2 added)
Propyl alcohol	C_3H_7OH	(CH_2 added)
Butyl alcohol	C_4H_9OH	(CH_2 added)
Amyl alcohol	$C_5H_{11}OH$	(CH_2 added)
Hexyl alcohol	$C_6H_{13}OH$	

The trade names under which the most-used xanthates are sold by two leading American producers are given in the Table below. These chemicals are also manufactured in Europe.

TABLE 32
Xanthates in wide use

Chemical Name	*Formula*	*Dow*	*Cyanamid*
Pot. Ethyl Xanthate .	$C_2H_5OCS_2K$	Z.3	Not numbered
Sod. Ethyl Xanthate .	$C_2H_5OCS_2Na$	Z.4	325
Pot. Isopropyl Xanthate .	$C_3H_7OCS_2K$	Z.9	322
Sod. Isopropyl Xanthate	$C_3H_7OCS_2Na$	Z.11	343
Pot. Sec.-butyl Xanthate	$C_4H_9OCS_2K$	Z.8	301
Sod. Sec.-butyl Xanthate	$C_4H_9OCS_2Na$	Z.12	—
Pot. Amyl Xanthate .	$C_5H_{11}OCS_2K$	Z.6	350
Pot. Sec. Amyl Xanthate	$C_5H_{11}OCS_2K$	Z.5	—
Pot. Hexyl Xanthate .	$C_6H_{13}OCS_2K$	Z.10	—

Considered in respect of their stoichiometric reaction, the metal xanthates have very low solubilities (below those of the metal sulphides) and therefore are formed preferentially. This favours sorption, quite apart from any question of stoichiometric reaction.

With regard to the carbon chain of the non-polar group, the increasing length of this decreases the solubility of the hydrocarbon and increases the avidity with which the group repels water and clings to the air phase of the aerated pulp. This change in attractability can be used as a controlling factor in flotation. Since the xanthates are ionised it would appear to be a matter of cost as between the use of the potassium and the sodium salt, but the former, though slightly dearer, is preferred by most operators. In the flotation of sulphides quantities varying between 0·02 lb. and 0·2 lb./ton of ore are used. The quantity needed depends on three factors:

1. The total surface to be conditioned with xanthate (function of assay value and fineness of grind).
2. The loss through direct precipitation by dissolved metal salts.
3. The oxidation of the surface.

If an excess of xanthate is used floatability is reduced, but the rate of flotation increases. Loss of concentrate may be partly due to flocculation of over-conditioned particles which then become too heavy to remain as aggregates in a bubble.

In the case of a deeply oxidised sulphide, anything up to several pounds of xanthate may be consumed per ton of ore treated, and this adds seriously to the operating expense. The loss is partly through direct chemical reaction between metal ions dissolved in the pulp and xanthate ions, which causes precipitation of metal xanthate. It is due partly to the porous nature of the oxidised surface. Alkali sulphides are sometimes added for the purpose of "surface closure"—re-sulphidising the oxidised surface—but the reactions involved are intricate and difficult to control. Broadly, the best treatment of ore prone to oxidise is to give it as little opportunity to do so as possible.

Thiocarbonates

These compounds are derived from the carbonate, and have collecting power for sulphide minerals. Their structural formulae are

Sodium ethyl monothiocarbonate $$O=C\begin{matrix}\diagup OC_2H_5\\ \diagdown S\,Na\end{matrix}$$

Dithiocarbonate $$S=C\begin{matrix}\diagup OC_2H_5\\ \diagdown SNa\end{matrix}$$

Trithiocarbonate $$S=C\begin{matrix}\diagup SC_2H_5\\ \diagdown SNa\end{matrix}$$

Methyl or butyl alcohol may replace the ethyl.

Thiocarbamates

In the xanthates the carbon atom connecting the hydrocarbon group to the polar group is oxygen. In the thiocarbamates the connecting link is nitrogen. The general formula for potassium diethyl dithiocarbamate is

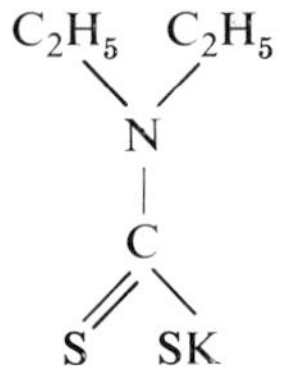

$$\begin{array}{c} C_2H_5 \quad C_2H_5 \\ \diagdown \;\; \diagup \\ N \\ | \\ C \\ /\!\!/ \;\; \diagdown \\ S \qquad SK \end{array}$$

These compounds are good collectors for some sulphide minerals.

Dithiophosphates

The general formula of these compounds is:

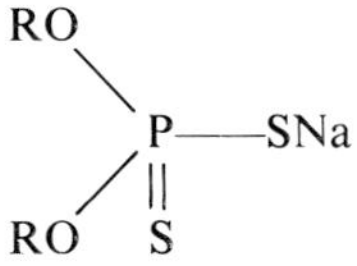

$$\begin{array}{l} RO \\ \quad \diagdown \\ \qquad P\text{——}SNa \\ \quad \diagup \;\; \| \\ RO \quad\; S \end{array} \quad \text{(sodium diethyl dithiophosphate)}$$

Here the connecting atom is phosphorus. They are marketed under the trade name of "aerofloats". They are weaker collectors than the xanthates, and are sometimes used to float the first mineral of a series, in order to avoid collecting the less eager sulphides during this first stage of treatment. Again, xanthate may be used after aerofloat, to bring up the weaker particles of the mineral sought. This is one form of "scavenging" treatment. The aerofloats are reaction products of phosphorus penta-sulphide with phenol, alcohol, a mercaptan, thio-alcohol, amines, and nitriles. A compound with phenol and alcohol is the aerofloat mostly used, and in the concentrator at Bingham, Utah, where the aerofloats were first developed, the reagent is made in batches by a mill hand as required. It is soluble in water and forms almost insoluble salts with heavy metals. It is more sensitive to depressing action than xanthate, and therefore has special value in differential work, particularly when iron sulphide must be kept down. Aerofloat 15 is a black reaction product of cresol with 15% of P_2S_5, the excess cresol acting as a frother. A-25 carries 25% of P_2S_5, and A-31 is similar with an added 6% of thio-carbanilid. A-239 is a di-amyl dithiophosphoric acid, neutralised with ammonia and thinned with 10% of ethyl or iso-propyl alcohol. The ammonium ion may account for the success of this reagent in floating copper in the presence of hydrated iron. Aerofloat 241 is a water-soluble form of A-25, used as a fast-acting collector for the sulphides of silver, copper, lead, and zinc in alkaline circuit, and as a non-selective collector and frother for sulphides in acid circuit. A-242, a water-soluble form of A-31, is similar to A-241 but has stronger collecting power. The dry aerofloats are thiophosphate compounds neutralised with solid carbonates (sodium or ammonium). Quantities used in sulphide flotation vary up to 0·25 lb/ton of ore.

A defect of the aerofloats, in conditions where very close circuit control is essential, is that they are both collectors and frothers. It is advisable to keep the two reagent-functions distinct.

Thiocarbanilide, etc.

The structural formula for this thiourea derivative is

$$H_5C_6HN-\underset{\underset{\displaystyle S}{\|}}{C}-NHC_6H_5.$$

It is only slightly soluble in water but disperses well, and where used is usually added with the feed to the ball mill. It has selective qualities which have some application in the flotation of copper and lead from complex sulphides.

The "minerec" reagents are derived from xanthates by oxidation, and have some application as collectors for such reducing minerals as sulphides. A third group of non-ionising reagents is the hydro-carbon oils, which confer aerofilic quality by smearing rather than ionisation at the mineral surface. Some of these are marketed under the name "Nujol". Another proprietary name "Flotagen" is given to mercaptobenzthiazole, the formula being

N
C—SH
S

It bases the American Cyanamid 400 series (also 125) and is used in alkaline circuit for oxidised lead ores. Non-polar oils are bonded by Van der Waals forces, by thio-compounds inherent or developed in the supplier's oil refinery, or possibly by long-chain dipolar molecules such as higher alcohols of low hydrating strength. The commercially supplied oily collectors vary both in composition and contained surfactants, but tend to be consistent when coming from some special source.

A completely saturated mineral oil or plant glyceride is not a collector *per se*. It depends for its polar groups on impurities, additives such as sulphated or sulphonated compounds picked up during commercial production, or on unsaturated organic fractions. Where oil refinery residues are used as collectors it is usually found desirable to rely on one source and its named by-product. Hydroxyls, carboxyls, methoxy, sulphur or nitrogen groups may be active. "Reconstructed" oils into which elemental sulphur has been incorporated by heating to some 250° C. or by reaction with concentrated sulphuric acid at about 50° C. (perhaps after the addition of alkali) have some use. Carbon disulphide or other sulphur additives have been employed.

Among the oil industry's by-products are the "green acids"—mixed sulphonation products from the cracking process. These, and the "mahogany soaps" similarly produced, are in industrial use in the flotation of some oxides and silicates, notably of iron minerals.

"Insoluble" collectors, when dispersed into the pulp as emulsions, form chemically unaltered films on mineral surfaces. Emulsification can be effected in a "*colloid mill*" where the reagent is sheared into droplets and blended with any desired additives. Selectivity of these reagents is sometimes improved by the blending in of a surfactant such as sodium oleate or a sulphonated fatty acid.

Fatty Acids and Soaps

Fatty-acid collectors have the general formula $CH_3.(CH_2)_n.X$, where X is either a carboxyl, sulphate or sulphonate group and n usually exceeds 9. The sulphate group is less readily adsorbed by calcium than is the carboxyl, and is sometimes therefore used in moderately hard process water to avoid the need for pre-softening and the risk of precipitation. It also aids the selective flotation of other minerals in the presence of unwanted calcium.

The most widely used fatty acid collectors are those linked to the carboxyl

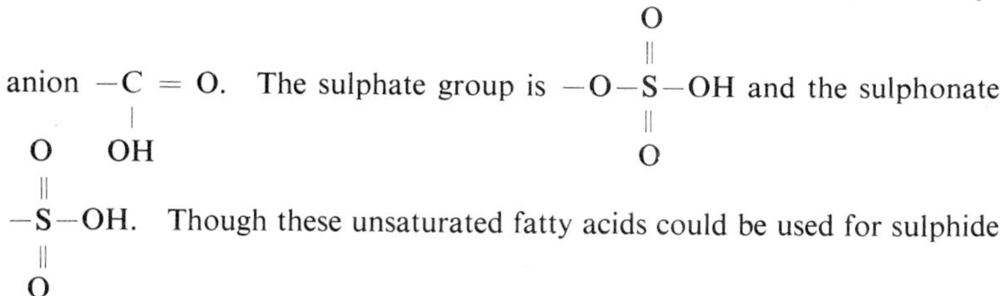

anion $-\underset{\underset{OH}{|}}{C}=O$. The sulphate group is $-O-\underset{\underset{O}{\|}}{\overset{\overset{O}{\|}}{S}}-OH$ and the sulphonate $-\underset{\underset{O}{\|}}{\overset{\overset{O}{\|}}{S}}-OH$. Though these unsaturated fatty acids could be used for sulphide flotation (as were some saturated soaps in the early days of flotation) it is simpler in practice to take advantage of the more specific action of the xanthates and to reserve the fatty acids for non-sulphide minerals. Treatment of mixed sulphidic and oxidic ore therefore proceeds in two steps (Fig. 214).

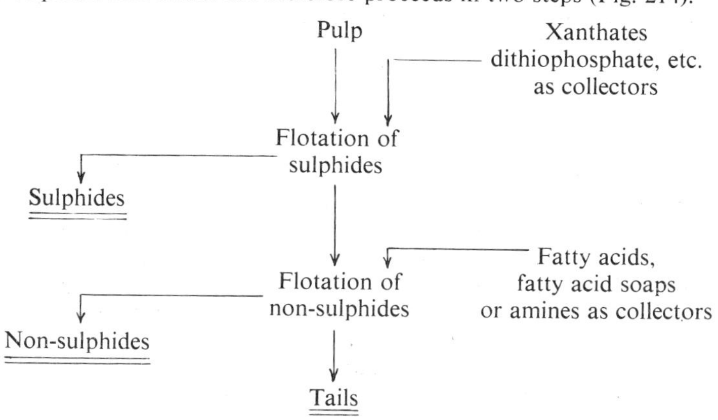

Fig. 214. *Staged Flotation*

The basis for collector action appears to be the abstraction of the carboxylate ion $(CO.OH)^-$ by the non-sulphide mineral. This reaction proceeds slowly, particularly with saturated fatty acids. Gaudin and Tournesac[19] found that 24 hours were needed for an approach to equilibrium in the sorption by barite of laurate. The carboxylic fatty acids are good collectors for polar minerals which contain alkali-earth metals (Ca, Ba, Mg, Sr), carbonates and non-ferrous sulphates. Adsorption by oxides and silicates is weaker. Process water containing a substantial trace of the above alkali-earth ions must be softened before carboxyls can be used. Sulphonated fatty acids are less susceptible to direct precipitation.

The saturated fatty acids are homologues of the basic structure C_nH_{2n+1}. COOH or $C_nH_{2n}O_2$. They include, in ascending order of carbon chain

Formic acid . .	H . COOH
Acetic acid . .	CH_3 . COOH
Lauric acid . .	$C_{11}H_{23}$. COOH
Palmitic acid . .	$C_{15}H_{31}$. COOH
Stearic acid . .	$C_{17}H_{35}$. COOH

The soap is produced by substituting a sodium ion for the hydrogen of the carboxyl $-CO_2Na$ instead of CO_2H.

The basic formula of the unsaturated fatty acid mainly used in flotation is $C_nH_{2n-5}COOH$. The formula of the widely used reagent, oleic acid, is $CH_3(CH_2)_7CH = CH(CH_2)_7COOH$ and of its sodium soap or oleate $C_{17}H_{33}COONa$.

Writing of the carboxylic collectors Taggart[20] says: "Fatty acids and soaps are collectors for all minerals which, in water, free an earth—or heavy—metal ion, or onto the surface of which such an ion can in any way be plated." The "plating" or re-surfacing ions available in most pulps include Pb, Fe″, Ba, Ca, Al. By their indiscriminate reaction with such common gangue minerals as quartz and calcite they lower the grade of non-sulphide concentrates unless they can be brought under control. ". . . it is for this reason that they have been supplanted for heavy-metal collection by the sulphydrate collectors, which do not form insoluble salts with the earth metals. But since no such substitute has yet been found for the earth-metal minerals, the fatty compounds constitute the principal collectors for many of these. The carbon content must be C_8 or greater, preferably greater than C_{12}, on account of the relatively high solubility of the earth-metal soaps of the lower acids, and it should not be greater than C_{18} or C_{20}, because of the low solubility of the higher alkali-metal soaps, in which form the collectors are usually dispersed. . . ." The random activating effect of Ca ions is commonly reduced by softening the mill water used in fatty-acid flotation. Close *p*H control is helpful, particularly where the more specific action of oleic acid (as compared to oleate) is desired.

A cheap source of fatty acid is tall oil or talloel (Am. Cyanamid 708). That used in flotation is usually a distillate consisting mainly of linolenic acid (an impure oleic acid), rosin acids, with some saturated fatty acids (palmitic and stearic). The composition of tall oil varies considerably. Its collecting properties for magnetite have been studied by Kivalo and Lehmusvaara.[21] The experimental flotation of hematite ores with an emulsion of tall oil and fuel oil has been reported by Kihlstedt[21] in connexion with progress in the separation of apatite from hematite. The resins in tall oil act as frothers, and this action may become excessive in alkaline pulp, particularly on slimy ores. The tannin compounds in crude sulphite liquor from the wood-processing industry (the main source of tall oil) are depressants. The reagent used in the commercial flotation of phosphate rock is prepared by dissolving tall oil in fairly strong caustic alkali (Cyanamid 708 of the 700 series), whereas the Kihlstedt reagent, named UMIX, is an emulsification of tall oil, No. 2 fuel oil and a water-soluble alkyl-aryl sulphonate, precisely prepared. This latter reagent, like one developed by the author and his colleagues on the basis of reaction between oleic acid and *inter alia* mahogany soap[22] is indifferent to slime and is not affected by ordinarily hard water.

The general formula of the alkyl sulphates is $CH_3(CH_2)_n\text{-}SO_3Na$ and of alkyl-aryl sulphonates $CH_3(CH_2)_n\text{-}C_6H_9\text{-}SO_3Na$. Most mineral and vegetable oils contain oleic acid or an associated compound. Among the commercially obtainable water-soluble collectors based on fatty acids are the following:

(*a*) Sulphuric esters of alcohols, in which R is a hydrocarbon having from 10 to 18 C's

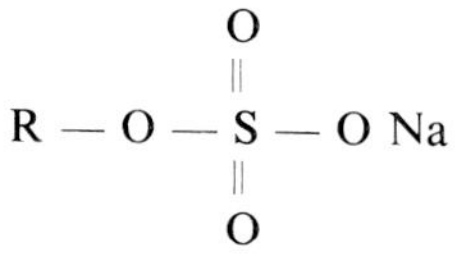

$$R-O-\overset{\overset{\displaystyle O}{\|}}{\underset{\underset{\displaystyle O}{\|}}{S}}-O\,Na$$

(*b*) Sulphuric esters of fatty acids (R_1 is a low-molecular alcohol radical and R a hydrocarbon with less than 10 C's)

$$COOR-O-\overset{\|}{\underset{\|}{C}}-O.Na \quad \text{or} \quad \begin{array}{l} CH_2\,OOCR \\ \quad| \\ CH_2OSO_3Na \end{array}$$

(*c*) Sulphated esters of amines $R-CONC_2H_4O.SO_3Na$

(*d*) Alkyl sulphonates, with R a 12–16 C hydrocarbon

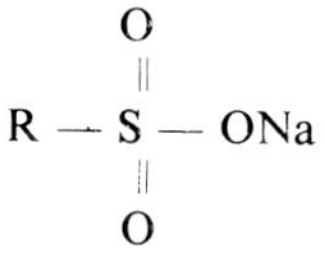

$$R-\overset{\overset{\displaystyle O}{\|}}{\underset{\underset{\displaystyle O}{\|}}{S}}-ONa$$

(*e*) Alkyl sulphonates with amide interlink (R has 15 to 18 CH)

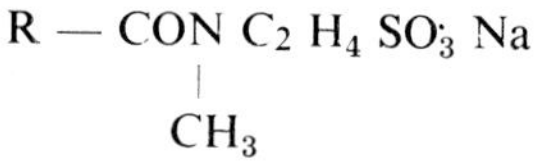

$$\begin{array}{l} R-CON\,C_2\,H_4\,SO_3\,Na \\ \qquad\quad | \\ \qquad\; CH_3 \end{array}$$

(*f*) Di-alkyl sulphonates (R has 4 to 8C8s in its alcohol)

$$\begin{array}{l} \text{R COOCH}_2 \\ \quad | \\ \text{R COOCHSO}_3\text{ Na} \end{array}$$

(*g*) Alkyl-aryl sulphonates with R of up to 12 C.

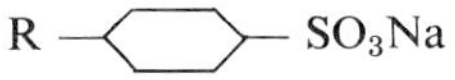

$$R-\langle\,\bigcirc\,\rangle-SO_3Na$$

An important series of reagents for the flotation of iron oxides, mica, chlorite, chromite and garnet, ilmenite and talc is marketed by American Cyanamid in its 800 series. These are anionic sulphonation products of complex structure, containing such ingredients as mahogany soap and "green-acids". They are water-soluble or water-dispersible and work best in an acid pulp (*p*H 2 to 4).

Oleic acid and its sodium or potassium soaps are used in the flotation of oxides, halides, silicates, and oxygen-salt minerals, e.g. fluorite, apatite, and garnet. The relation between the adsorbing mineral and the fatty-acid collector is not one of direct exchange reaction or simple monolayer formation. It appears to be influenced by the action of cations such as those of calcium, barium or iron at or near the surface of minerals such as silicates.

The adsorption of oleate ions on fluorite[18] is influenced by temperature, being 193 g/ton at 16° C. and 240 at 30° C. With temperature decrease soap and fatty acid molecules associate to form micelles, and activation suffers. Micelles also form when the collector concentration exceeds 250 mg./l., but this tendency is opposed if the oleate is emulsified with a very little pine oil.

Excessive use of fatty acid reduces its collecting power in other ways. It can be adsorbed into bubbles to a sufficient extent to impede the picking up of particles, or even to stabilise a more or less unmineralised froth. It can also compete with the molecules of collector at the mineral surface by masking it with hydrated micelles.

Amines

These are cationic collectors for non-sulphide minerals, the prefix "amino" or the suffix "amine" denoting a $-NH_2$ group. They are derived from ammonia by replacing its hydrogen with an organic radical. The nitrogen core can be depicted $\left[\begin{array}{c} | \\ -N- \\ | \end{array} \right]^+$ Each line is part of a hydrogen (H) or hydrocarbon (R) bond. A variety of structures is thus possible.

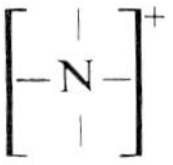

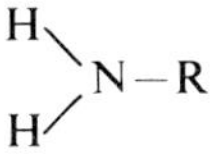

$$\begin{array}{l} H\diagdown \\ \quad N-R \\ H\diagup \end{array}$$ Primary amine

$$\begin{array}{l} H\diagdown \\ \quad N-R \\ R\diagup \end{array}$$ Secondary amine

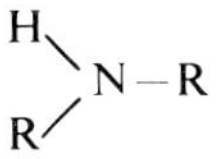

$$\begin{matrix} R \diagdown \\ \quad N-R \\ R \diagup \end{matrix}$$ Tertiary amine

$$\begin{matrix} R \diagdown \quad \diagup R \\ N \\ R \diagup \quad \diagdown R \\ OH \end{matrix}$$ Quarternary amine

A further classification of amines is into alkyls, aryls and alkyl-aryls, according to whether the nitrogen is attached to a carbon chain, to a ring, or to both. Amino salts include chlorides, phosphates, sulphites and sulphates.

These reagents are mainly used to float oxides, silicates and spars. Selectivity is poor and the reagents are sensitive to their concentration in the pulp, re-dissolving from the mineral surface if too diluted.

Quaternary ammonium salts are reaction products of tertiary amines treated with alkyl halides

$$\begin{matrix} R \diagdown \\ R - N \\ R \diagup \end{matrix} + Rcl \; — \; \begin{matrix} R \\ R \diagdown \, | \\ N - cl \\ R \diagup \, | \\ R \end{matrix} \tag{17.10}$$

the N–Cl bond being ionic. Nitrogen is also pentavalent in aqueous or acidic solution.

$$R-NH_2 + H^+ + OH^- \longrightarrow [RNH_3]^+ + OH^- \tag{17.11}$$

The solubility of amines (e.g. lauryl) is poor in an acidic pulp but increases sharply above pH 8 to a maximum ionization at pH 9.

The mechanism of collection has been investigated by several workers using contact-angle and bubble pick-up techniques.

Taggart and Arbiter[23] worked with dodecylammonium chloride on the minerals wollastonite ($CaOSiO_3$), calcite, barite and fluorite. Their conclusion was that metathesis occurred between metal and collecting cation.

$$BaSO_4 + 2Am^+ \longrightarrow Ba^{++} + Am_2SO_4 \tag{17.12}$$

Research at Mass. Inst. Technology includes an important study of the collection of silica by dodecylammonium acetate, by P. L. de Bruyn[24]. Adsorption is not strong, as the amine is displaced by one or two cold-water washes from various non-sulphides. Mono- and divalent cations do not affect the flotation of non-sulphides by amine collectors but such trivalent ions as those of aluminium or iron are strong depressants in the flotation of quartz with lauryl amine, apparently being preferentially adsorbed.

The special field of industrial use of amines is in the flotation of silicates, with pH and regulating agents closely controlled so as to make the best of the rather poor selectivity. They are useful in a secondary separation of silicate after bulk flotation by other collectors, then removed. Amine adsorption is in the electrical double layer rather than to the mineral surface. Floatability is sharply decreased in the presence of slimes. The reagents are not affected by hard water and have been used even with sea water. Among the commercial operations are the cleaning of silica from phosphate concentrates and the removal of micaceous contaminants from china clay.

Guanidine, a strongly basic water-soluble compound formulated $HN{-}C(NH_2)_2$ is used in one process for the flotation of copper from nickel in a matte consisting of their ground sulphides. Diphenyl or di-ortho tolyl guanidine are mentioned in British Patent Spec. 602028/1958 in this connexion. The diphenyl has also been suggested by Hines for sphalerite flotation.[25]

In general terms, good separation can usually be achieved between acidic (silicate) and basic (oxidic or carbonate) minerals. Silicates respond to such cationic collectors as the long-chained amines, while the basic minerals attract the anionic fatty acids.

In addition to the foregoing there are several oily liquids which act as vehicles for hydrocarbons insoluble in water, and which themselves are almost insoluble but can be emulsified. Their action is obscure. They include petroleum, tar and gas oils, tar acids, diesel oil and creosote. They aid in closing the surface of oxidised ore and in stiffening flotation froth.

Modifying Agents

The chemical and physical additives discussed in this section are variously described in flotation literature as regulators, conditioners or more specifically as depressants, activators, wetting agents, flocculants and dispersants. The practical requirement from the operator's viewpoint is the ability of the additive to sharpen selectivity as between the mineral species floated and that (or those) left in the pulp. Though a slight over-simplification, this can be considered in terms of a conflict between wetting and drying-out forces. Surface hydration favours retention in the pulp and air-avidity attachment of the particle to a coursing bubble followed by seizure in a mineralised froth. Reaction which increases the hydrated area on the particle favours wetting. Thermodynamically the spread of one fluid on another, or the replacement on a solid surface of gas by liquid (or *vice versa*) is due to a nett decrease in the total interfacial free energy of the system. Gravity, mechanical force and chemical change also affect spreading. The regulating chemicals used in flotation stimulate the desired surface activity of specific minerals in the pulp. Modifying agents therefore either reinforce or make possible the adsorption of collectors in the case of floating species, or aid hydration if they are active depressants. A reagent is not necessarily "active" in the sense of being measured into the pulp.

In the previous section collector agents were considered in connexion with a clean and attracting surface. In the laboratory such conditions are readily attained, but in plant practice this is not the case. Mill water, the dirt and oil contributed during mining, transport, and comminution, and the attempts of all the newly sheared particles to pick up from these additives anything which will reduce their surface potential, lowers the free energy of the surface lattices of the various minerals in the ore. This gives them some degree of superficial similarity, since they draw their contaminants from a common environment—the pulp. Even a surface which has escaped random sorption may require development of its ionic potential before it is adequately reactive. Conditioning agents are used to bring out latent aerophilic or hydrophobic qualities at appropriate stages of the treatment.

The first modifying agency to affect the particle is mechanical. Comminution detaches it from the parent mass, gives it form, surface, and energy due to lattice deformation. The next is physical. Classification determines the weight and mass of the particle on release from grinding. The third agency, simultaneously at work, is surface modification of the new surfaces developed by comminution. This change proceeds continuously from the birth of the new surface to the time when it is either in chemical balance with its surroundings or removed from them. These surroundings contribute oxygen from the air, ions from mill water and other minerals in the pulp, and surface-affecting reagents introduced during treatment. Study of the preceding section "Collectors" shows that, despite the thousands available, choice is usually narrowed down to a few xanthates, dithiophosphates and fatty acids. Before specific collector-action can be assured, there must therefore be equally specific preparation of the surfaces to which that collector is to be attracted. This is achieved by the use of modifying agents. One group of these is used to increase the attraction of the collector toward the surface to be floated: another group increases the wettability of the surfaces which are to remain unfloated. A third group removes interfering ions from the pulp water. A fourth disperses masking slimes from particle surfaces. A fifth plates, or re-surfaces, selected minerals. The main groups of modifying agents are:

(a) p*H modifiers*	Alkalis Lime, CaO, Soda ash Na_2CO_3 Sodium Hydroxide NaOH Acids Sulphuric acid H_2SO_4 Sulphurous acid H_2SO_3
(b) *Wetting agents*	2-ethyl-hexyl sodium sulphosuccinate (Aerosol) Sodium Silicate Na_2SiO_3 Organic colloids
(c) *Precipitating Agents*	Chemicals forming insoluble or non-ionised compounds with such ions as Ca, Ba(+); CN, CO_3, PO_4, SO_3(−)
(d) *Dispersing Agents*	Sodium silicate Na_2SiO_3 Polyphosphates
(e) *Re-surfacing Agents*	Ba, Ca, Cu, Pb, Zn, Ag(+) notably copper sulphate

Most of these have dual roles—e.g. a wetting agent (*b*) does its work by surface modification in which hydroxyls or other "wetting" ions increase their hydration of counter-ions bound in the discontinuity lattice. Conversely, re-surfacing agents (*e*) promote the adsorption of hydrophobic hydrocarbon groups to preferred lattice points and thus increase the aerophilic/hydrophobic ratio of the surface area, thus increasing floatability.

Parts of the surface not coated by collector may carry depressants. The degree of hydration, in addition to the foregoing causative influences, is influenced by adsorbed or occluded gas films which act as a barrier between the mineral and modifying agents. Such gas films, if not displaced during conditioning or agitation of the pulp, favour the flotation of the particle affected. Up to a point increased adsorption of collector improves floatability, but further coverage (if indeed this occurs) serves no useful purpose and may diminish selectivity and therefore grade of floated concentrate. Where the flotation of a sulphide mineral by a xanthate collector is concerned, optimum reagent dosage appears to depend on a balance between oxygenation at the surface and adsorption of xanthate. This further suggests that too much reagent can depress the float by overloading the electrical double layer at the expense of the oxygen dissolved in the pulp.

By far the most important conditioning factor is the *p*H of the pulp. *p*H value has two aspects, qualitative and quantitative. In that it measures the balance between hydroxyl and hydrogen ions it indicates the pressure-potential available at a reactive interphase. In that it is sustained by a concentration of acid or alkali which can be stated in terms of its titration value as a quantity per volume, it shows how much reagent is available for replenishing and sustaining the correct supply of ions. *p*H thus indicates intensity, and dosage with the *p*H regulating chemical one element in the rate of reaction. Most minerals have a range of *p*H within which they float best and above which it is almost impossible to secure attachment to an air-bubble. The *p*H of the pulp is maintained at the desired level by the addition of lime, sodium carbonate, or occasionally sodium hydroxide or ammonia to increase the alkalinity. Sulphuric or sulphurous acid is used when *p*H needs to be lowered. The substance of the chemical itself plays some part in the overall effect—lime, for example, sometimes has a depressing effect on galena and pyrite—but the primary purpose of *p*H control is to maintain the balance of hydrogen and hydroxyl or other counter-ions at a specific ratio. Different minerals float most readily at different alkalinities. The change in floatability from copper sulphide to iron sulphide illustrates this point. Using potassium ethyl xanthate as collector, the maximum *p*H at which the following minerals can be floated is shown in Table 33.

TABLE 33

(*After I. W. Wark,* "Principles of Flotation", (1938): Australasian Institute of Mining and Metallurgy.)

Mineral	*Max. flot.* p*H*	*Formula*
Chalcocite	$< 14{\cdot}0$	Cu_2S
Covellite	13·2	CuS
Bornite	13·8	Cu_3Fe3_3
Chalcopyrite	11·8	$Cu_2S \cdot Fe_2S_3$
Pyrite	10·5	FeS_2

No two specimens of mineral can be relied on to conform completely in lattice structure, and this is particularly the case where the specimens come from different mines or different depths of the same mine. Differences in conditions of geological deposition affect the internal lattice discontinuities, and the inclusions of minute traces of other elements which cause lattice distortion and variations in behaviour. It is possible that the ability of a mineral to float within the *p*H range determined by contact angle measurement is connected with its iso-electric point. When near its zero-potential a particle tends to flocculate. Flocculation in the sense of crowding to an air-water interphase, accompanies flotation. The role of the dielectric constant in flotation-conditioning might prove a fruitful subject of research, since it appears that the greater the conductivity of a metal compound, the more readily (as compared with other metals similarly compounded) does it accept a high *p*H and a xanthate-type collector.

It is usual to depict the role of the hydroxyl ion in the pulp as a screening one. If the metal sulphide has good positive ionisation at its surface lattice points, then hydroxyls are pictured as swimming in the zeta layer, so that they form a loosely attracted blanket of counter-ions. The thicker this blanket, the harder it is assumed to be for the collector anions to thrust them apart and anchor electro statically to the mineral. In the light of what is known of lattices, there are weaknesses in this type of argument. Radio-tracer studies suggest that only part of a clean lattice becomes collector-coated. Again, the negatively ionised heavy molecule of xanthate is cumbrous, compared to the equally charged hydroxyl ion. The only advantage the former would possess in competition for coupling to the lattice point would be the formation of a molecule of metal xanthate of lower solubility than that of the substrate of metal sulphide. In the conditioning tank in which this battle between the ions is fought, mechanical agitation is causing the particles to rub and scrub upon one another and there is little or no cessation of this between the conditioning period and aeration. Furthermore, the aqueous phase of the pulp is an electrolyte, although the only "electrode" external to the system is the grounding effect of the equipment and the piped water used in the process.

Turning to the role of the hydrogen ions, these are but lightly bound to their water molecules, and can easily pass from one to another, or free themselves to become protons. In either form they can set up cell action at the particle surface, either by electronation (oxidic reduction) or, since single stripped protons are the smallest atomic particles, by penetrating the lattice and attacking the sulphur points. It will be seen, then, that though in one litre of water only a minute percentage of the H_2O molecules dissociate into hydrogen and hydroxyl ions,

$$(H^+) \times (OH^-) = 10^{-14}, \qquad (17.13)$$

the balance between the two concentrations has a profound effect on the heterogeneous state referred to in this book as the "pulp *climate*". If the hydrogen and hydroxyl ions be pictured as a restless tide lapping at the shores of the surface lattices in the pulp, and if these surfaces be considered as having a greater or less electrolytic attraction for these ions, according to the type of lattice displayed by each particle, then some measure of intermediate agreement becomes possible between the exponents of chemical and of ionic activation. If the position revealed in the study of ion-exchange is valid for the capture by a positively charged lattice-point of an anion from the zeta-layer, then hydroxyls have the best chance, carbonate ions the next and xanthate ions the least, in order of ionic size. If chemical reaction enters, the xanthate, on reacting with a copper atom on the lattice, forms the least soluble product and, therefore, that with the lowest energy and greatest stability and irreversibility.

If the first stage of stoichiometric reaction is necessarily mono-molecular adsorption—the view of some physical chemists—then it is possible to render it reversible. Ions and molecules of all kinds are adsorbed to, and escape from, the surface of the particle continuously. While the entropy of the system is still increasing, this traffic is greater in the direction having the higher rate of stabilisation, whether it be the surface of the particle or the surrounding liquid. Transfer through the zeta zone surrounding the particle can therefore be stimulated by adding to the liquid phase reagents which can energise this cross-traffic.

The amount of each kind of ion present on the surface at a given instant depends on the ion's nature, on the character of the solid, and on the unsatisfied potential energy residual in the whole closed system. Once the free energy has been reduced to its minimum, the flow through the zeta zone is in balance in each direction. The only circumstance which can put an end to this interplay is the sealing off of the active areas. This could be accomplished by the occasional arrival of a xanthate ion, towing the non-polar portion of its hydrocarbon group. On anchoring, it seals the lattice point with an inert substance, not affected by electrical forces.

Hypothetical considerations such as these can only be tentatively put forward. Such practical tests as contact angle measurement show the empiric limits of H-ion concentration within which a mineral can be floated. Other methods, not involving such elaborate preparations, use a miniature flotation cell of some kind to study the effects of *p*H and other reactions. Even a simple shaking in a test-tube or stoppered cylinder can be made to give valuable indications as to floatability.

The *critical* p*H value* is one below which a given mineral will float and above which it will not, in the presence of controlled quantitities of other reagents held at constant temperature. Table 34 gives values for varied conditions, and the graphs in Figs. 215 and 216 show the *p*H line below which contact is made[8] and above which is it not. When differential flotation is required between two sulphides, both of which are floatable by the same collector, four controls based on collector and *p*H can aid selectivity. These are:

(*a*) Critical *p*H, at which mineral *A* will float and mineral *B* will not.
(*b*) Choice of collector which gives best gap in critical *p*H.
(*c*) Balance of concentration of ions of collector against hydroxyl anions.
(*d*) Developing (*c*), use of "starvation" quantities of collector, thus adding preferential sorption by the most strongly attracting mineral surface.
(*e*) Relative concentrations of hydroxyl and other competing anions.

TABLE 34

CRITICAL *p*H VALUES FOR VARIOUS COLLECTORS: ROOM TEMPERATURE

(*After Sutherland and Wark* "Principles of Flotation, p. 118 (1955): Australasian Institute of Mining and Metallurgy.)

Collector and Concentration	*Critical pH Value*			
	Sphalerite	*Galena*	*Pyrite*	*Chalcopyrite*
Sodium di-ethyl dithiophosphate 32·5 mg per litre	*x*	6·2	3·5	9·4
Potassium ethylxanthate 25 mg per litre	*x*	10·4	10·5	11·8
Sodium di-ethyl dithiocarbamate 26·7 mg per litre	6·2	> 13	10·5	> 13
Potassium iso-amyl xanthate 31·6 mg per litre	5·5	12·1	12·3	> 13
Potassium di-n-amyl dithiocarbamate 42·3 mg per litre	10·4	> 13	12·8	> 13

x Sphalerite does not respond to these collectors in the concentrations used.

Subsidiary effects of *p*H modification may include precipitation of ions which would otherwise react direct with a collector added later and thus render it unavailable for surface activation. Dissolved heavy metals react thus with xanthates, and earth metal ions in "hard water" with carboxyls. Flocculation can be lessened by *p*H regulation, thus avoiding the loss of mineral in floccules too heavy to be lifted by bubbles.

The two main types of *p*H regulator are (*a*) those which contain hydroxyl ions (e.g. NaOH, $Ca(OH)_2$; and (*b*) those which hydrolyse to yield hydroxyls, (e.g. Na_2S, Na_2CO_3, Na_2SiO_3). In the hydrolysis of sodium sulphide

$$Na_2S + H_2O \rightleftharpoons 2Na^+ + SH^- + OH^- \qquad (17.14)$$

the SH^- anion influences the metal-sulphide in the discontinuity lattice.

Three types of mineral are affected by hydrolysis: (*a*) Those which, when hydrolysed, yield equally strongly charged basic and acid ions, (*b*) Those whose cations can form strong bases and weak acids and (*c*) Those whose cations form strong acids and weak bases. The (*b*) and (*c*) types give, respectively, an alkaline or an acid pulp in mill water uncontrolled for its *p*H. Hence such minerals may affect the pulp climate by loading the process water with ions or by affecting the adsorption potentials of other mineral species present.

H^+ and OH^- ions modify the electrical double layer and the zeta-potential surrounding the particle, and hence affect the hydration of its surface. They can displace collector ions, affect the ionization of the collector molecule, modify the electrical properties of the mineral or act as potential-determining

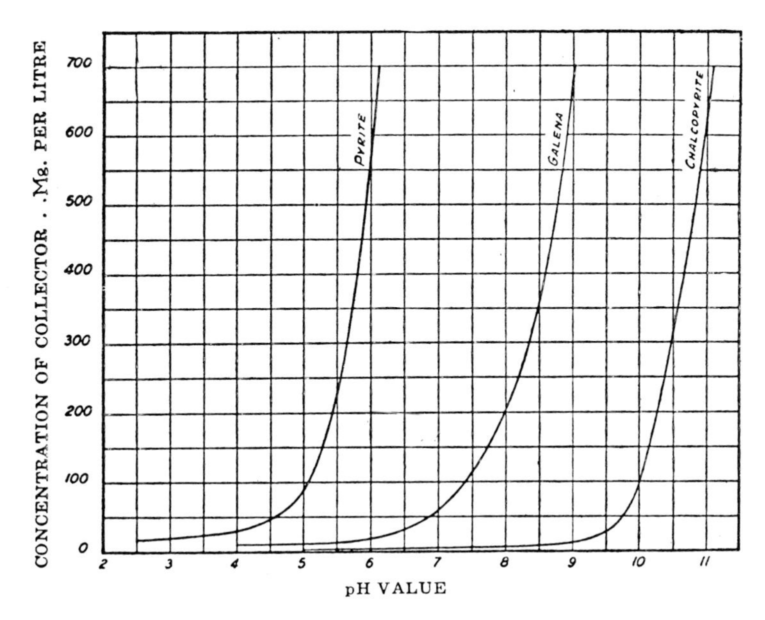

Fig. 215. *Critical* pH, *below which mineral floats, using sodium diethyl dithiophosphate collector*. (*After Sutherland and Wark, ibid., p*. 119)

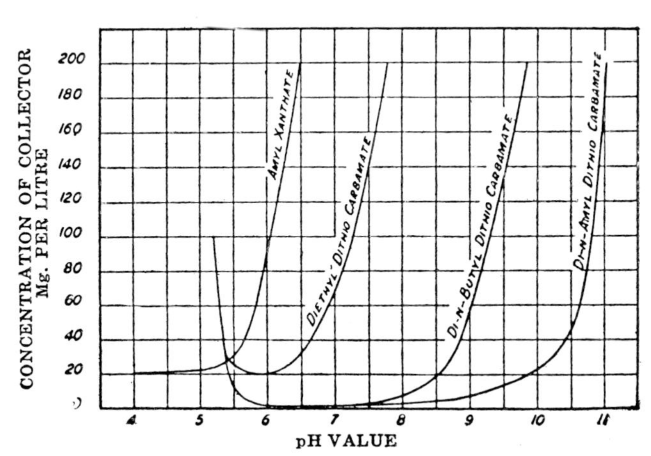

Fig. 216. *Critical* pH-*collector relationship in floating sphalerite* (*After Sutherland and Wark, ibid., p*. 119)

ions. The *p*H strongly affects the electro-kinetic potential for a specific mineral compound, which usually diminishes from *p*H 5 upward. If hydrogen and hydroxyl ions are potential-determining for a compound, it is very sensitive to *p*H change. In other cases these ions concentrate in the zeta-potential layer.

The depressing action of lime in pyrite flotation is due to desorption by hydroxyls of xanthate from its surface. Hydrogen and hydroxyl ions can also modify the situation in the zone of shear at the particle surface by barrier action or by competition with the collector at the discontinuity lattice.

The most widely used *p*H modifier is lime, which is used to render the pulp alkaline in the flotation of metal sulphides with sulphydril collectors. It is cheap, readily available in most localities, and with the exception of pyrite and to a lesser extent galena, does not compete with the collector at the surface of sulphide minerals. In a study of slime coating del Guidice[4] concluded that anions at the mineral surface react with cations in the sliming mineral to form a less soluble precipitate on the surface of the particle failing to float. Microphotographs showed galena heavily masked by calcite and consequently being lost as a tailing. On conditioning with sodium silicate, this slime-coating dispersed. Whether this is the reason rather than the still incompletely understood balance between hydroxyls and carbonate ions and their modifying effect on surface energy, it is common to avoid the use of lime as a *p*H control for the flotation of galena or pyrite. Light has been shed on the problem by Gaudin and Charles[13]—who found that oxygen in solution increases the adsorptive power of pyrite for calcium more than for sodium. In these cases soda-ash (commercial sodium carbonate) is the alternative. Caustic soda has limited use, mainly in the floatation of gold. Soda-ash has the advantage of precipitating calcium ions from hard mill water, and thus reducing their plating effect, which could otherwise lead to random activation of unwanted minerals. When fatty-acid soaps are used as collectors of non-sulphides, calcium must be kept out of the circuit if possible and soda-ash is the *p*H modifier used if an alkaline circuit is employed. A secondary useful result of treatment in alkaline pulps is the avoidance of corrosion in the plant. This saves wear and reduces the formation of hydrated iron salts, which have depressing and reagent-consuming effects.

Where an acid pulp is required sulphuric acid is normally used, though sulphurous acid has a moderate field of employment. In the flotation of cyanide tails for recovery of auriferous pyrite the intense depressing effect of the residual cyanide ion is thus removed as the first conditioning step. The acid has a dispersing effect which helps to produce clean sulphide surfaces for further conditioning.

Process Water

The chief chemical constituent in the pulp is the mill water, together with its content of salts and ions. If a pulp is being conditioned at 33% solids (by weight), the water-solid ratio is 2 to 1. If a reagent is added at a concentration of 0·06 lb/ton of ore, this is 0·02 lb/ton of pulp, a little under 10 g. This reagent, like the others added before and after it during a series of conditioning operations, must react with a large acreage of mineral surface. In a big plant many tons of pulp flow through the conditioning section every minute of the day, so timing and manipulation must be smooth and simple. The flotation process depends for its efficiency on accurate treatment during conditioning, so steady throughput rate in the grinding circuit is necessary even if no reagent is added there. It is, however, customary to bring the pulp to its correct *p*H value in the closed grinding circuit. If a series of minerals is to be floated, the *p*H is adjusted as part of the reconditioning process after the first mineral has been floated.

Mill water is usually drawn from the local rivers and lakes. The mineral content may vary between summer and winter in temperate climates or between the dry and wet season in the tropics. While rainy conditions prevail, the water is less likely to carry calcium, magnesium, carbonate, sulphate, and other ions than when it is being drawn from deep springs. In a sub-arctic winter water drawn from below the ice of lakes is de-oxygenated by algae, and has been found by some operators to produce too brittle a froth for good flotation unless it receives preliminary re-aeration. This can also apply to water recirculated from tropical tailings ponds, if vegetable growths have stripped the water of its air. If the water is pumped from underground, or even if the ore carries a substantial percentage of moisture from that source, it may carry metal ions or vegetable acids derived from the rotting timber of old stopes. These additives are fairly harmless as a rule. If the ore tends to oxidise and produce metal ions, these may precipitate some of the small quantity of collector with unfortunate effects on recovery, which depends in the last resort on *all* of the reagent being on the desired surface. Unstable sulphides—some pyrrhotites and marcasites, notably—may de-oxygenate the pulp, which requires oxygen for the effective sorption of xanthate to the minerals. The chemical role of the air carried in mill pulps at all stages in flotation is not entirely clear, but evidence has accumulated as to its importance. Unless threshold oxidation of a sulphide takes place, surface excitation may be insufficient for the collector reagent to sorb. If oxidation proceeds too far, direct chemical reaction between metal ions and collector may occur in the liquid phase, with consequent poor flotation of the inadequately conditioned particle. The careless use of oil underground can so contaminate the ore and foul the mill water as to upset process control in the mill.

Mill water may depress or activate the pulp constituents if over-charged with various ions. Those most liable to build up are Cl^-, SO_4^-, HCO_3^-, $CO_3^=$, Na^+, K^+, Ca^{++}, Mg^{++}, H^+, and Fe^{++}. When the intake of fresh water must be restricted the condition of recirculating process water must be watched and if necessary corrected by appropriate treatment. Salts which cause water hardness are particularly liable to react with fatty-acid reagents and form insoluble complexes.

Mill floor washings are liable to carry oil or grease, and must not be allowed to re-enter the circuit or to foul pumps or sumps connected with returned water. Frothing agents, which are not appreciably adsorbed by concentrates, overflow with these mineralised froths. If the water containing them is returned it should be added after conditioning, when it may usefully reduce consumption of this reagent. Where the presence of metal salts ionized into the mill water is possible, their effect must be considered before re-use. Earth salts, notably dissolved calcium, are liable to precipitate as scale in pipes, metering devices, filter cloths etc. and should not therefore be allowed to build to a dangerous level.

Conditioning

The collector action is partly controlled by the *p*H value, but further conditions may need to be satisfied before the collector is added, to heighten the difference between aerophilic and hydrophilic mineral surfaces. This is obvious in the case of differential flotation (Fig. 217) where the desired products are floated successively. When a series of values are floated, a further conditioning is necessary after each flotation.

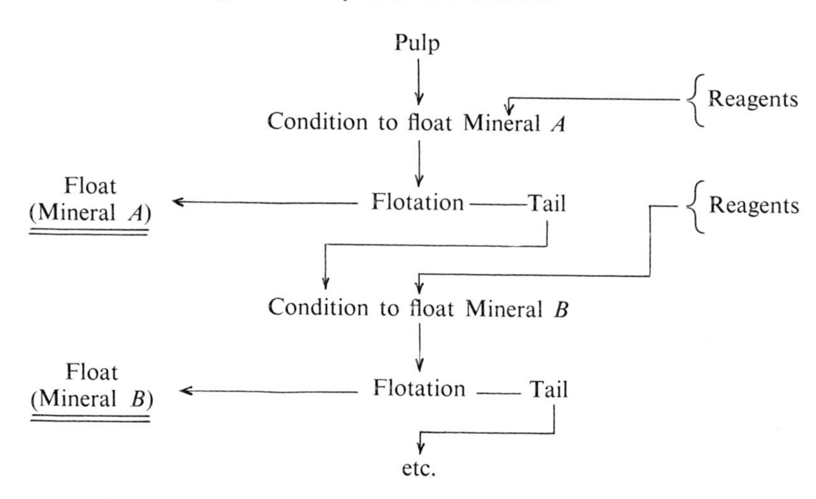

Fig. 217. *Differential Flotation*

In any conditioning treatment some of the following things must be done:

1. Obstructive compounds in the pulp must be rendered innocuous.
2. Ions released during grinding must be passivated if they would otherwise set up undesirable reactions.
2. Slime coatings sorbed to surfaces so as to interfere with necessary reactions must be dispersed.
4. Mineral in the pulp which is not wanted in the float must be rendered hydrophilic.
5. The desired forces of attraction must be developed specifically and exclusively on the particles which are to be coated with collector.
6. Porous surfaces must be closed to minimise consumption of collector.

7. Non-floating or weakly-floating minerals wanted in the froth must be resurfaced with ions which will attract the collector.

The reactions involved require only small amounts of chemical since aerophilic surface layers are for the most part involved, and then only over a fraction of the particle area. The quantities of chemicals must be added with fair precision, and the conditioning period must be sufficient for reaction to be reasonably complete. This may involve warming the pulp, but usually the mill temperature is adequate for reaction. Finally, the pulp must be agitated so that the added chemicals are thoroughly distributed.

Other Modifying Agents

Reagents used to hold down a mineral during flotation are called *depressants*. Those used to promote its acceptance of collector-agent or otherwise to aid its flotation are *activators*. Those used to clean slimed surfaces are *dispersants*. The same reagent may be a depressant at one H-ion concentration and an activator at another.

Cyanide. The anion CN^- is a powerful depressant for sphalerite and pyrite. It is widely used in alkaline pulps to keep down these sulphides while copper and lead are being floated with a xanthate collector. The pulp must not become acidic as HCN is a weakly dissociated acid. The available CN^- depends on the pH value, which must therefore be held within controlled limits in order to maintain the desired degree of depression. The CN^- ions thus made available, rather than the OH^- ions introduced by the regulating alkali, provide the controlling factor. The dissociation constant of hydrocyanic acid is

$$\frac{[H^+][CN^-]}{[HCN]} = 4 \cdot 7 \times 10^{-10} \qquad (17.15)$$

at 18°C. Research[8] shows that for a specific mineral and collector there is a critical value for CN^- concentration above which flotation is not possible. Gaudin[5] has tabulated the relationship between CN^- and HCN with change of pH.

TABLE 35

pH	*Millimoles/millimole of total alkali cyanide* as CN^-	as HCN
7·0	0·0005	0·9995
7·0	0·005	0·995
8·0	0·045	0·955
9·0	0·32	0·68
10·0	0·825	0·175
11·0	0·979	0·021
12·0	0·999	0·001

(To obtain CN^- in mg/mg of NaOH, multiply figures in second column by 0·531; (*after Gaudin*).

The critical CN^- concentration for various collectors rises from 3×10^{-8} millimoles/litre for dithiophosphate, $1{\cdot}6 \times 10^{-5}$ for ethyl xanthate, 5×10^{-5} for diethyl dithiocarbamate to 8×10^{-5} for amyl xanthate when the collector concentration is $0{\cdot}155 \times 10^{-3}$.[8]

In the flotation of chalcopyrite with cyanide in the pulp the ratio of xanthate ion to the square of the CN^- ion is a critical constant. On either side of this point the solubilisation of copper by cyanide determines the route taken by the mineral during its flotation. Equilibrium constants for some other metal-cyanide anions are

Copper	$2{\cdot}0 \times 10^{-24}$
Gold	$5{\cdot}0 \times 10^{-39}$
Silver	$1{\cdot}8 \times 10^{-19}$
Zinc	$1{\cdot}2 \times 10^{-18}$

CN^- reacts with metal cations in two stages.

$$Cu^+ + CN^- \rightleftharpoons CuCN(\text{insol.}) \tag{17.16}$$

$$CuCN + CN^- \rightleftharpoons Cu(CN)_2^- \tag{17.17}$$

The end product is a complex soluble salt. Gold and silver form the complex $Me(CN)_2^-$ and cadmium, mercury, nickel, and zinc $Me(CN)_4^{=}$.

Metals the xanthates of which are insoluble in cyanide are not depressed by this reagent. They include antimony, arsenic, bismuth, lead and tin. Those with moderately soluble xanthates can easily be depressed by a small concentration of CN^-.

The depressant action of lime or soda-ash (commercial sodium carbonate) serves to multiply the effect of the cyanide salt, by unlocking those CN^- ions.

TABLE 36

EFFECT OF CYANIDE AND LIME ON FLOTATION OF COPPER AND IRON MINERALS

(After Tucker, Gates, and Head)[26]

Alkalinity	*NaCN lb. per ton*	*Per cent. of Mineral floated*			
		Chalcopyrite	*Chalcocite*	*Bornite*	*Pyrite*
Neutral	Nil	92·0	92·0	91·7	70·0
	0·5	89·2	79·8	95·2	21·0
	1·0	85·4	87·3	90·5	19·0
0·0005% CaO	Nil	94·3	99·0	92·2	36·0
	0·5	91·4	87·3	95·0	14·0
	1·0	88·1	79·8	90·5	13·0
0·003% CaO	Nil	93·3	60·3	96·0	23·0
	0·5	92·1	97·0	95·0	11·0
	1·0	90·0	96·8	94·5	11·0
0·02% CaO	Nil	91·6	15·2	93·8	2·0
	0·5	91·2	23·7	62·4	1·0
	1·0	89·3	16·6	73·1	2·0

There are three possible explanations for the potency of cyanide. It may sorb to the mineral surface with greater tenacity than that of hydroxyls. It may remove Fe ions from the surface by complexing them, as it does those of copper and zinc. Copper is present on many sulphide surfaces, and as it has strong attraction for xanthates, it must be "complexed" to cupro- or cupri-cyanide if it is not to render floatable the mineral carrying it. The effect of cyanide on floatability (contact-angle criterion) is shown in Table 36 and Fig. 218. This graph shows the demarcation between clinging (*o*) and non-adhesion (*x*) in contact-angle tests of bornite, using potassium ethyl xanthate as 35 mg/l. as collector. There is a critical concentration of CN^- above which no flotation can occur.

Sulphides.—Soluble sulphides are often used as activators in the conditioning of oxidised sulphide surfaces. They can also act as depressants if added in excess. If $S^=$ ions are adsorbed to a metal-sulphide surface they increase the negative charge and also the adsorption of xanthate ions. Since this is a depressing action, true metal sulphides are floated after straight collector-conditioning before the residual pulp is further conditioned with a soluble sulphide. Hydrogen sulphide, like hydrocyanic acid, is a weak acid. The ionisation of its sodium salt Na_2S is largely governed by the *p*H of the pulp, as is the case with other weak acids. The ionisation product is

$$H_2S \rightarrow HS^- + H^+. \qquad (17.18)$$

In this form the sulphide ion competes with the collector for sorption. If it dissociates further, to

$$HS^- \rightarrow H^+ + S^=, \qquad (17.19)$$

its divalent ions favour combination with oxidised metal-sulphide surfaces to give an unstable metal-sulphide. The dissociation constants for the successive stages K_1 and K_2 are

$$K_1 = \frac{[H^+][HS^-]}{[H_2S]} = 10^{-7} \qquad (17.20)$$

$$K_2 = \frac{[H^+][S^=]}{[HS^-]} = 2 \times 10^{-15} \qquad (17.21)$$

The metal-sulphide layer at the oxidised mineral surface, so long as it can be sustained, is available for the attraction of xanthate and flotation of the particle bearing this sulphidised surface. The position, however, appears different from one where the external discontinuity lattice of a true metal-sulphide is concerned. Between the sulphidised monolayer and the true sulphide substrate lies a zone of change, in which the sulphur bond has chemisorbed oxygen to form sulphites, hydrosulphides, thiosulphates, hyposulphiates, and hyposulphates. These are trying to sorb oxygen and so increase their stability, so the fragile sulphidised layer is attacked on both sides and soon oxidises away. The durability of the sulphidization by the

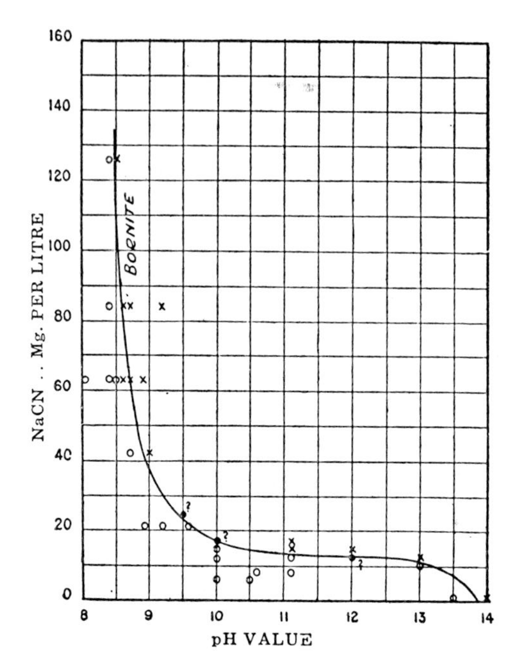

Fig. 218. *Effect of CN Ions on Floatability (after Sutherland and Wark, "Principles of Flotation", p.* 126. (1955): *Australasian Institute of Mining and Metallurgy.*)

added $S^{=}$ ions is determined by its depth, and therefore by the rate of diffusion into the surface during conditioning. This film is easily displaced by pulp agitation. Pre-sulphidization of some minerals (e.g. malachite) reduces the xanthate requirement considerably. Excessive residual sodium sulphide after formation of the surface film has ceased depresses flotation by xanthate. Floatability is, however, recovered after re-aeration. The time needed depends on the rate of oxidation of hydrosulphide and sulphide ions in the pulp. The period of agitation with air present is therefore important in sulphidising treatment. Oxidation of a surface monolayer of $S^{=}$ or HS^{-} takes only a fraction of a second. Sodium sulphide is therefore added in stages along the line of flotation cells, to suit the kinetics of the ore's reaction. Excess sodium sulphide desorbs the collector if the mineral surface is not masked by fine air bubbles. Soluble sulphides may also precipitate heavy metals and cause them to activate particles indiscriminately and so lower the concentrate grade.

Surface oxidants.—Some sulphide minerals show a marked deterioration in floatability when their freshly broken surface is given opportunity to oxidise beyond the threshold degree which facilitates adsorption of xanthate. Several sulphoxy compounds can be present in the zone of shear surrounding a particle, apart from those intentionally added to the pulp. There are six oxides of sulphur (27)—SO, SO_2, S_2O_3, SO_3, S_2O_7 and SO_4, and twelve oxy-acids.[5] Complex thio-compounds are formed by a number of metals. The factors which affect the degree of oxidation include pulp temperature, time, aeration, *p*H, extent of barrier films at the particle surface, and presence of oxidising agents. The systems formed in surface layers are complex and vary not only between species of sulphide, but also between minerals of the same species derived from different geological formations.

Depressants sometimes used include lime, alkali sulphites, permanganates, arsenates, chromates, ferricyanide, tartrate, acid-phosphates, citrates, oxalates, and tungstate. They are shown in descending order of effectiveness as depressant for pyrite in the presence of potassium ethyl xanthate. The general action of all the above depressants is to neutralise positive lattice ions.

Differential Depression.—When more than one mineral has been recovered simultaneously by *bulk flotation* it is often necessary to re-condition the product and to re-float one or more constituent species from the mixture. To achieve this the mechanism of depression may require selective desorption of collector, with destruction of the original collector followed by specific re-activation, or by specific weakening of aerophilic quality.

Specific surface oxidation or destruction of a xanthate film has been aided by the use of permanganate, dichromate or ferri-cyanide. These oxidise the xanthate to di-xanthogen. The alcohol group of the collector has been preferentially vapourised by controlled heating prior to re-floating, as in the separation of a molybdenite float from a copper sulphide sunk fraction. In the removal of entrained silica from a phosphate-silica float, the anionic fatty-acid collector is dissolved with caustic soda and the silica is then floated with a cationic collector. Sphalerite which has been activated with copper so as to float with galena can be depressed by re-floating the mixed concentrate after conditioning with cyanide to dissolve the copper.

Sodium silicate. This is much used as a dispersant, for removing slimes from sulphide surfaces to which they have sorbed. Up to 0·5 lb/ton is used for dispersing quartz. The reagent sold as "sodium silicate" can vary widely in its silica-sodium ratio, with corresponding changes in its value as a dispersant and gangue depressant (Sollengerger and Greenwalt). It also tends to embrittle flotation froths. The salts are the meta-silicate (Na_2SiO_3), disilicate ($Na_2Si_2O_5$) and orthosilicate (Na_4SiO_4). It may also be present as hydrated silica. The commercially supplied composition is $mNa_2O.nSiO_2$ and the n/m ratio or modulus mostly used varies between 2·2 and 3·0. With a lower modulus the pulp tends to alkalise and silicates are depressed. Above 3·0 a coarsely dispersed hydrated silica is produced.

The dissociation of sodium silicate is complex, and preparation of its solution for use must be standardised if the content of colloidal silica is not to vary unpredictably, with consequent random deposition of silica gel. Ions, molecules and micelles of silicic acid are strongly hydrated and if adsorbed may have depressant effects. Sodium silicate yields ions which are potential-determining for silicates and aluminosilicates, which are preferentially depressed.[29] A small addition activates the flotation of hematite by sharply decreasing the negative zeta potential, but a further concentration raises it again, and different dissociation products of the silicate then become adsorbed.

Sodium metaphosphate has limited use in calcite dispersion. In the sense that a dispersant restores the true surface to the mineral being activated it is an activator, but sodium silicate is regarded by most workers as a depressant. It is attracted by common-ion effect to silicate lattices, where it acts as a wetting agent.

One difficulty in the depression of slimes is the great surface area of these, when colloidal and clayey. They sorb expensive reagents, contaminate froths and stabilise them undesirably, and lead to difficulties in the settling of pulps after treatment. They are rarely troublesome in the treatment of unaltered ores but may be difficult with deposits which have been exposed to weather, such as leached and shallow ore bodies and dumps. The methods of handling them are preventative rather than chemical. If such slimes can be removed before fine-grinding without undue loss of values, the problem is simplified. Grinding should be kept to its minimum and the collector agent may be found to act best if added in the grinding circuit, so that it can sorb before the particles become slime-coated.

Apart from depressant action and the precipitation of fatty-acid collectors as insoluble soaps, lime is used in sulphide flotation, *inter alia* to depress pyrite. It appears to form a mixed surface film with ferrous and ferric hydroxides and insoluble lime salts, thus reducing xanthate adsorption. When it is desired to float pyrite despite interfering lime, ammonium ions can be added to reduce or prevent adsorption of calcium.

Starch is one of several water-soluble polymers which act as depressants when adsorbed. Some of the polar groups of its molecule orient toward the mineral while other remain free and thus increase the hydrophilic quality of the surface. Starch, and such other high molecular-weight colloids as albumen, caseinate, gelatine, saponin and quebracho have shown depressing effects in the flotation of several metal sulphides. The general formula of starch is $(C_6H_{10}O_5)n$, with traces of phosphorus and silica. On warming to dissolve it the giant molecules break down into polysaccharides or dextrins. With non-sulphide minerals the depressant action of starch is due to hydrated films on the minerals.

Quebracho is a wood extract containing tannic acid. It is an excellent depressant for calcium salts in the flotation of fluorite by fatty-acid collectors, but its concentration must be closely controlled if it is not to react with the calcium in the CaF_2. The chemistry of its reaction is obscure.

Regarded as hydrogen-ion phenomena, activation and depression are relative to the *p*H. There are two special aspects of activation not thus dealt with. In common-ion activation, an ion in solution is attracted to discontinuity lattices in which similar ions are bound, provided excess binding force is present at the surface of the particle. Thus, if the soluble sulphate of a metal is added to a pulp which contains particles of the same metal in sulphide form, the particle surface tends to become more basic, metal ions being drawn out of solution and captured at the surface by unsaturated lattice points. This depressing action can be used to keep down sphalerite when galena is floated with a xanthate or dithio-phosphate, by adding zinc sulphate to the pulp. The zinc cation is attracted into the hydrated sheath which surrounds the particle and aids depressing action by increasing the population of accompanying hydroxyls. Tests[29] show that zinc hydroxide begins to precipitate on a sphalerite surface at a *p*H of 5·2 and that at *p*H 7·4 flotation has fallen from $35\frac{1}{2}\%$ to 10%, even a copper-activated sphalerite being affected. The zinc-coated surface can be reactivated by using ammonia to dissolve the hydroxide, or sodium silicate to peptise it. Addition of common ions sometimes increases the insolubility of the sorbing collector, and in some cases allows economy in the use of this reagent. There is, however, danger of direct chemical reaction between sulphate and collector, resulting in precipitation of the latter.

Activation by Re-surfacing.—The modification of a mineral surface by adsorption of metal ions has been referred to already. As widely practised in the industrial flotation of sphalerite it is achieved at the surface of a weakly floatable sulphide by building on to it a more actively floated metal. The reagent most used for the work is copper sulphate, $CuSO_4.5H_2O$, and the mineral to which it is most applied is sphalerite. Zinc sulphide floats weakly or not at all with the xanthates. It is usually associated with galena and copper in ores, and traces of the latter are "killed" during the first stage of a differential float when the lead is being taken, by the uses of sodium cyanide. After the galena has been floated, the pulp is brought to a *p*H of 9 or more by adding lime. This lime is a depressant at such a *p*H for pyrite, which if present might become activated in the next conditioning stage. A suitable quantity of copper sulphate is now added, as a saturated solution. Most of it, after reacting with residual cyanide, precipitates out as hydrate or carbonate. These salts act as a reservoir and feed copper ions to the pulp to replace those which now plate to the zinc lattice-ions thus:

$$(ZnS)^{Zn^{++}} + Cu^{++} \rightarrow (ZnS)^{Cu^{++}} + Zn^{++}. \qquad (17.22)$$

If for any reason deactivation of the changed zinc particles is now needed, the reaction can be driven from right to left by the addition of solvating CN^- ions. In the short reacting period used in a plant, the copper layer is kept as tenuous as possible, as it is the zinc beneath, not the minute trace of surfacing copper, which is wanted.

The case may arise of copper ionising into the pulp during grinding of an ore containing zinc or other attracting sulphides. This would make differential flotation impossible and would also lead to heavy consumption of xanthate by direct precipitation. Here, the addition of sodium or calcium sulphide to the dry ore entering the grinding section can sometimes induce precipitation of the copper as an insoluble sulphide and thus prevent such general activation. Several metal ions can act in this way. If they then accept a collector, this may cause the mineral to float. Copper is the most positive in action of these, and in the fractions of a pound used per ton of ore is cheap.

Any metal in the electromotive series can displace from solution those below it. The series is K, Na, Li, Ba, Sr, Ca, Mg, Al, Mn, Zn, Cr, Cd, Fe, Co, Ni, Sn, Pb, H (hydrogen), Cu, As, Bi, Sb, Hg, Ag, Pd, Pt, Au. The

TABLE 37

RE-SURFACING AGENTS

After Dow, "Flotation Fundamentals" (1958), p. 21

Active Agent	*Added as*	*lb/ton of ore*	*Common use*	*Collector*
Cationic:				
Cu^{++}	$CuSO_4$	0·1–2·0	Activator, Zn, Fe, Co, Ni sulphides	Xanthate
Pb^{++}	Acetate	0·1–2·0	Activator, stibnite	Xanthate
Pb^{++}	Acetate	0·1–0·3	Activator, halite	Fatty acid
Ca^{++}	CaO or $(OH)_2$	0·5–10·0	Depressant for pyrite; activator for silica	Xanthate/Fatty acid
Zn^{++}	$ZnSO_4$	0,2–2·0	Depressant for sphalerite	Xanthate
Anionic:				
0^{--}	Air	—	Depressant, pyrrhotite	Xanthate
SO_3^{--}	Na_2SO_3	0·5–2·0	Depressant, sphalerite	Xanthate
S^{--}	Na_2S	0·5–20·0	Activator, Pb, Cu oxides; depressant for sulphides when excessive	Xanthate
CN^{--}	NaCN	0·5–1·0	Depressant, Cu, Zn, Fe sulphide	Xanthate
SiO_2	Na_2SiO_3	0·5–2·0	Depressant, gangue slime; activator for silicates	Xanthate/Cationic
CO_3^{--}	Na_2CO_3	0·5–5·0	Activator, Pb, Fe sulphides	Xanthate
		1·0–10·0	Depressant, gangue	Fatty acid
Organic colloids:				
Dextrin		0·1–1·0	Depressants for gangue slimes, especially carbonaceous slimes	Xanthate
Starch				Fatty acid
Lignin sulphonate		0·2–5·0		

electromotive series (displacement, Volta, constant series) is an "arrangement of the elements in order of their relative potentials. At the top of the list are the most negative elements, which displace all anions following, and are thus reduced. At the end of the list are the most positive elements which displace all preceding cations and thus become oxidised."[36] Thus, copper salts, being lower in the series, displace zinc at the sulphide surface. Broadly, to be an effective activator the surfacing metal must form a relatively insoluble sulphide. In our present state of knowledge of metal-sulphide solubility in an aerated pulp, specific tests on an ore are essential. A list of common re-surfacing agents is given in Table 37.

Depression by re-surfacing is sometimes advanced as the reason for the effect of sodium chromate or dichromate, salts used to depress galena from bulk floats of the sulphides of lead, copper or zinc.

Reducing agents are of practical value in flotation. In differentiating lead (floated) from zinc sulphide (unfloated) sodium sulphite (Na_2SO_3) or bisulphite ($NaHSO_3$) are often used, particularly in the presence of oxidisable copper minerals, which would tend to ionise during comminution and then to activate the zinc.

Whatever reagent plan is decided on in a particular case, the possibility of mutual interference must be guarded against. Where copper sulphate is used for re-surfacing it must not be allowed to plate out on the steel work of the machines and piping. Sodium silicate must not be used in a pulp made strongly alkaline with lime, which would react to precipitate out calcium as silicate.

In addition to specific reagents the quantity used per ton of ore, the reaction time given, and the pulp temperature can be used to aid in controlling surface changes during the conditioning stage. On completion of this stage the pulp is ready for aeration, having first been activated and next dosed with a collector specific to the activated mineral about to be floated.

Aeration and Frothers

The term "*N-bubble*" (an abbreviation of "neo-bubble") is used in this book to mark an important distinction in physical state. It is applied by the author to an immersed bubble of air (or any other gas) which has the essential characteristic that at the moment of its arrival in water the surface tension at the air-water interface is at its maximum for the system, being for pure water and air of the order of $\gamma = 72{\cdot}3$ dynes/cm. From that moment until its emergence from the water as an independent bubble the surface tension at the interface is progressively reduced by seizure from the aqueous phase of any molecules which will lower the surface tension. This change is important in several ways in connexion with the capture of floatable particles of mineral from the pulp, and in the establishment of the type of mineralised froth required in the flotation process.

The N-bubble, beside having an interfacial tension approaching that of air-water during much of its immersed period, has only half the surface area of a free bubble (*F-bubble*) which has risen clear of the aqueous phase and has achieved independent existence as a hollow liquid spheroid with two air-water faces. For the F-bubble to exist, even briefly, a surface tension substantially below that of water is essential. This is usually achieved in flotation by sorption to the interface of molecules containing hydrocarbon groups. Compounds containing suitable molecules for this purpose are selected and introduced with considerable accuracy as regards quantity and dispersion, so as to be available at the correct time and place in the treatment of the pulp.

There is also a dynamic difference between N and F bubbles. The N-bubble is an air-pocket pressed on by the surrounding pulp, through which it moves in accordance with its buoyancy. It is vibrating violently at the moment of entry deep in the pulp and is also distorted during its upward drift, particularly as it presses through any turbulent zones. Thus, since it is departing from and returning to a spherical shape without change of volume (other than a slight expansion as the hydrostatic pressure decreases upward) the area of the N-bubble is variable. The F-bubble is free from hydrostatic pressure and such movements as it makes after emergence are of a completely different nature

Air can be introduced into the flotation cell by several methods. These are considered in the next chapter. To be effective this air must search the whole body of pulp. Each N-bubble must be of a suitable size. Too small an N-bubble (under 0·5 mm in diameter) is indiscriminate in its adhering power and tends to promote too stable a froth. If extremely small, it remains in the pulp and cannot lift itself, let alone a particle. Power is used in aeration of the pulp to increase the surface energy by expanding the air-water interface. In a sense the air is ground into the pulp, an operation characterised by the same low efficiency in terms of end-product as is ore grinding. If too large an N-bubble is introduced into the pulp, its buoyancy is so high that it rushes to the surface. This reduces the time available for fixing particles in the N-bubble wall. It also sets up turbulence in the frothing zone which, as will be seen later, is thoroughly undesirable. Broadly, the smaller the diameter of the N-bubble, down to about 0·5 mm–1·0 mm, the tougher will be the emerging froth, while a useful upper limit of size is of the order of 4 mm. A further determinant of froth stability is the size-range of its constituent bubbles. Those produced in a mechanically stirred cell tend to cover a wide range of sizes and to mat together, while those formed when air is blown in are more closely sized and form a less stable foam.

The size of the N-bubble is influenced by these factors:

(*a*) Size of the aperture from which it emerges.
(*b*) Hydrostatic head against which its contents are compressed.
(*c*) Surface tension of the interface with pulp formed as it emerges.
(*d*) Speed of emergence, and the volume and pressure of gas behind it.
(*e*) Turbulence of the surrounding pulp.

Four methods of aeration are possible, of which two are widely used. In the so-called "mechanical" cell air is sheared into the pulp by a submerged impeller which is also receiving air from an external source. In the "pneumatic" cell air is blown in through diffusing mats or nozzles. These two methods may be mixed. In the third (little used) method a vacuum is set up and air precipitates from the aqueous phase on to the most hydrophobic particles with sufficient lifting power to buoy them to the surface. Some measure of precipitation is also thought to occur in the low-pressure zone swept out behind a fast-spinning impeller. This zone of cavitation increases toward its periphery. The fourth mode of aeration, rarely seen today, is that used in "cascade" flotation, where pulp falling turbulently over a weir entrains air.

Practical details of flotation machines are discussed in Chapter 18, but since most of those in use today incorporate mechanised aeration, a brief discussion of this vital link in the oxygenation, levitation and selective removal of particles from the pulp is appropriate at this point. A zone of low pressure is generated near the hub of the impeller. If this is in direct communication with the atmosphere or is fed by a controlled supply of low-pressure air this air blends with the pulp before leaving at the periphery of the spinning sytem. Air and pulp are discharged tangentially at differing pressures, and along their line of departure a cavitated low-pressure void is maintained behind the tips of the impeller blades. Thus, air is compressed into the pulp during its accelerating flow along the front of the blade, and re-precipitated to some extent where pulp is sucked back or slips back over the top. As the issuing stream quietens and changes to a rising movement some air precipitates preferentially on to aerophilic particles. This aids their incorporation into the larger (coursing) bubbles as they thrust toward the surface. It has been calculated[29] that in a pulp of S.G. 2,7 which contains 30% solids and is rising one metre from the impeller-swept zone to surface there are some 3 ml of precipitated air per litre. Much of this pulp is in circulation in the vertical plane. When a cell is started up the oxygen content soon shows a threefold increase.

When a frothing agent (discussed later) is added to the pulp the surface tension is lowered, power requirement for driving the impeller drops some 10% and both vacuum behind the blades and re-precipitation of gas are increased. Bubbles precipitated from an aqueous pulp are below 0·2 mm in diameter, and on addition of pine oil this is halved under running conditions in which the average coursing bubble is some 0·9 mm in diameter. The size range of bubbles generated in "mechanical" cells varies widely in accordance with settings, dimensions and speeds, but tends to lie between 0·05 and 1·5 mm diameter, the bulk being around 1 mm. Bubbles diffused from pneumatic sparging arrangements are much larger and more even in size, averaging between 2·5 and 3 mm diameter.

Despite contortions, vibrations and collisions bubbles do not noticeably coalesce while rising. Each has a hydrated enveloping barrier surrounded by a slip stream, and an air-water interface toughened by surfactant molecules. These fend off collisions and discourage size growth in the body of the pulp. The situation changes abruptly on emergence. Bubbles between 2·5 and 6 mm rising freely through water (i.e. unobstructed by mineral particles and unwarped by turbulent currents) steady down to the shape of a flattened oblate spheroid. Bubbles below this size change their shape rapidly, periodicity being about 1/1,000th of a second during ascent. Bubbles larger than 6 mm have widely irregular shapes, indented or bellying out.

The peripheral speed of the impeller and the geometry of its enclosing shrouding, is critically related to the intake and blending into the pulp of the indrawn air. The balance between air and pulp passing through the impeller is important, and too high an intake of pulp can prevent aeration. For a given pulp density and depth of impeller immersion there is an optimum point of balance between radial intake of pulp (as regards distance along blade from hub) and its peripheral discharge for a given speed.

The Bubble.—In order to simplify a discussion involving these, and other factors, it is helpful to begin with a system air-distilled water. At 4°C water achieves its tightest molecular packing. It expands below this temperature to the ice lattice structure at 0°C. Above 4°C it expands with the increase of molecular disturbance due to rising kinetic energy. At ordinary working temperatures water contains lightly associated polar molecules and fugitive lattice formations. The position of a molecule in water is considered to be constrained by these light binding forces, so that it is always in some degree of electrical association with the molecules surrounding it. The lower the temperature and the higher the ion-dissociation, the greater, therefore, is its binding. This is shown by the change in viscosity, which is 18 millipoises at freezing point, 10 at 20°C, and 4 at 70°C at normal barometric pressure. At boiling point the kinetic energy of the molecule transcends the restraint of the adhesive forces, and the system converts to the vapour phase. In Fig. 219 two molecules are depicted. That lying well within the water is in balance with the attractive forces on all sides of it (subject to slight unbalance due to its movement in the system), but the molecule at the surface is only half balanced. It, like the other surface molecules, is being drawn inward to the body of the liquid, and the force of this attraction is related to surface tension (γ). At 20°C, in air, γ is 72·8 dynes/cm. In order to minimise this potential energy the water reduces its area as much as possible. Since the sphere contains the maximum volume for the minimum area, a drop of water takes a spherical shape. Larger volumes conform to the overriding pull of the earth's

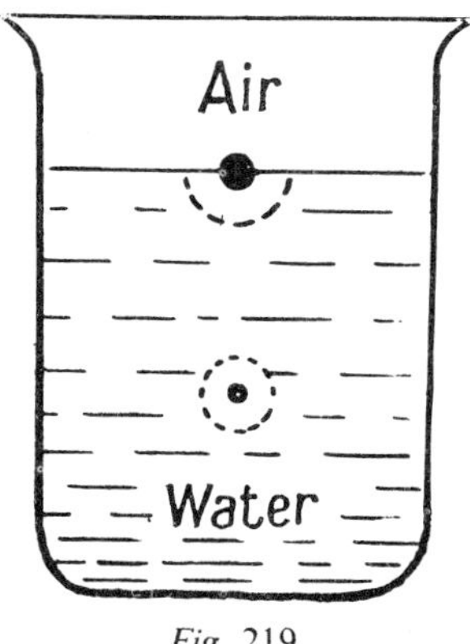

Fig. 219

gravitational field. When an N-bubble is forced into water it creates new air-water interface, the shape of which tends toward the spherical but is distorted since it lacks thermodynamic stability. This N-bubble ascends to the surface at a rate determined by its volume. It then explodes, since by doing so the total area is reduced, and the surface energy is again restored to minimum potential. This is why a pure liquid cannot form permanent froth.

Consider next a system in which heteropolar molecules of a liquid of low surface tension have been dispersed evenly through the distilled water. The most-used frothing reagent is a commercial distillate of wood called pine oil. Assume that a few drops of this low-solubility liquid, having a surface tension of the order of 27, have been dispersed through the water. All the molecules in the mixture are moving restlessly as before, so that occasionally one of pine oil reaches the air-water interface. The position is very different from that of an unbalanced molecule of water. The force attracting the pine oil downward is only one-third as strong as that acting on the water molecule, and in fact the main body of water is exerting an outward thrust against it. The tendency, therefore, is for molecules of pine oil to find their way to the interphase and stay there. This results in the lowering of the surface tension of the mixture, and gives it some tendency to form a foam. To set up a permanent foam it would be necessary for the added substance to form molecular linkages at the surface. This property is only weakly developed in pine oil, which can produce a fragile and short-lived froth when shaken with water. This will be shown later to be a necessary characteristic of a good frother in flotation. If more pine oil is added, so that the water is super-saturated, the undissolved excess will disperse as droplets, or float to the surface to form a non-frothing layer.

If air is introduced to water containing free surface-active agents (e.g. collector or frother molecules) these adsorb to the air-water interface with their polar groups oriented to the water. Thus, the N-bubble strips pine oil from water as it rises, and deposits it near the surface on bursting. If a column of such water is aerated, most of the oil is rapidly transferred to the surface layers. From then on, each arriving N-bubble has almost the full surface tension of water, which it momentarily reduced by loading with pine oil molecules as it emerges. This should be remembered when considering frother action in the flotation cell, in which the concentration of pine oil near the top must exceed that at the point of entry of the pulp, which is near the bottom.

For flotation to be fully efficient, the frother should not itself possess collecting qualities, nor should it be capable by itself of forming a stable froth, since this would reduce the attraction into the air-liquid interphase of mineral particles.

The Two-phase Froth.—An air-water froth is thermodynamically unstable owing to its considerable free surface energy. It can increase its stability by breaking down. The liquid film between the surfaces of a new F-bubble is about 1μ thick, and is attenuated as its enclosed water drains down, until the hydrated surface boundaries touch. After this, a shock can complete disruption. The main thinning influence is the force of gravity, but evaporation, capillary action and environmental pressure are also active. Inside a blanket of froth the bubbles tend to pack with planar interfaces, curving where they conform to adjoining bubbles.

If the optimum concentration of surfactant frother molecules is exceeded the stability of the froth decreases, since a hydrated layer between these molecules no longer exists. Highly stable froths are formed by evaporation of water charged with strongly soluble alkaline salts.

The Three-phase System.—The whole system—water, frother, and aerophilic particle—can now be considered. The newly born N-bubble is momentarily in rapid vibration[26] with maximum surface tension for the system. Pine oil molecules and mineral particles compete for positions in the interphase. The "values" have been selectively conditioned with a collector agent. The gangue minerals also have some aerophilic qualities, due to their sorption of lubricants and any other available hydrocarbons during their journey from mine to conditioner. All the particles in the pulp may have some degree of floatability as they arrive at the aerating stage of treatment. There is, moreover, a great excess of air-water interphase over packing bodies, so that all can be accommodated. This point is brought out by Taggart who shows the rise in assay value of the desired mineral to occur almost entirely in the froth column and not in the pulp. Particles which are completely wetted are unlikely to rise by air transport. They may, however, be carried over into the froth by agitation. Hundreds of mineral-bearing N-bubbles emerge from the pulp for every one finally removed as a concentrate-loaded bubble from the top of the froth. The great majority of N-bubbles fail to become bubbles. They burst, releasing air at the surface of the froth, and particles which start to slip back into the pulp. Of those which do develop into bubbles, only a minority stabilise sufficiently to be separated with their cargo of concentrate. The great excess of N-bubbles over bubbles finally needed to transport the concentrate out of the system is essential to commercial flotation. Oversized particles which are heavy enough to overcome the surface tension force holding them in the interphase may drop out during this period of violent adjustment. Particles which are lightly held may be shaken or torn away by the collisions and turbulences in the agitated pulp. A further important feature of the change from N to F bubble state is that the clinging grains of mineral now lose the aid of the pulp density, as they move from water-solid system of pulp to one which is largely gaseous—the froth column.

If a single N-bubble with its interfacial load is now considered, the chances of its survival in bubble form are remote. In Fig. 220 the N-bubble *a* is

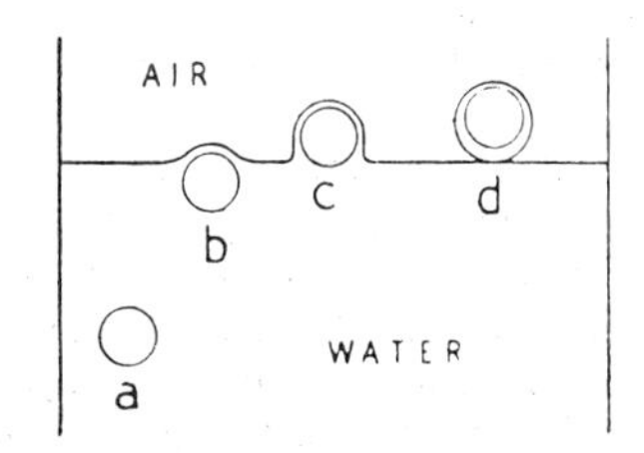

Fig. 220. Transition—N-Bubble to F-Bubble

drifting upward, carrying molecules of frother and particles of mineral with the bulk of their surface area in the liquid phase and part of it at the interface (since θ is normally well below 90°). Arriving at *b* the pulp surface blisters and a second surface is created, which proceeds *via* stage *c*, in which all water not held in the capillary system of the forming bubble drains back to the pulp, to the independent bubble form *d*, detached from the pulp on which it floats.

There must be a further addition if the bubble is to persist. Even if the N-bubble had an adequately filled mineral interphase, sufficient to ensure persistence at one surface, this would not suffice for two such surfaces, and the F-bubble has double the surface area of the N-bubble. In part, this deficiency might be corrected by additional frother molecules from the enriched surface layer of the pulp, but more than this is necessary if explosion is to be avoided, with return of the ensuing rain of particles to the pulp. Explosion does, so far as observation can establish the facts, supply the stabilising additive. Most of the bubbles explode, and their passengers start to drop back to the pulp. As they do so, however, they are picked up by emerging N-bubbles and stabilised into two-faced systems, the "armoured bubble" of flotation. It is therefore essential, in grasping the physics of bubble action, to think of abundant masses of loaded N-bubbles arriving continuously, of which only a fraction survive as removable froth. Water rich in surfactant molecules is sluicing the outsides of emerging bubbles. These molecules replace the more weakly adherent mineral particles, and also stabilise the inner face of the F-bubble as it forms. There is, indeed, some possibility of displacement of the desired adherent concentrate particles at this stage, if collector coating has been inadequate or its adsorption weak. The froth when formed should not be of a persistent character, not only because this would lead to difficulties in subsequent handling, but also for reasons of froth-column action discussed below. The stabilising ingredient in the froth is the mineral particle, which is built into the N-bubble (and after emergence, the bubble) rather as bricks are built into a wall. The mineral particles line the interphase, to which they are attracted by their ability to reduce the surface tension below that of water. The pine oil acts as the "mortar" between the particles at the interphase, the whole system rising to the surface of the pulp. Here it is converted into a froth persistent enough to be removed from the cell.

Next, the froth column itself may be considered. For simplification, the N-bubbles may be pictured as uniform in size, though in practice they are not. Inspection of a froth column (using a glass-walled cell) shows that the bubbles immediately above the pulp are the smallest, and that those at the top (the layer which is removed in treatment) are the largest. Further, if an ore such as galena is being floated from a siliceous gangue, only the uppermost layer of bubbles will be seen to be heavily loaded with submetallic and black-grey particles of PbS, while most of the froth column is buff or off-white in colour, being lightest nearest the pulp line. In the Taggart experiment referred to earlier in this section, a copper sulphide ore was being floated. The assay value rose from 0·74% Cu at the bottom of the pulp to 0·77% at the pulp surface, and climbed from that to 38·8% at the top of the froth (Fig. 221).

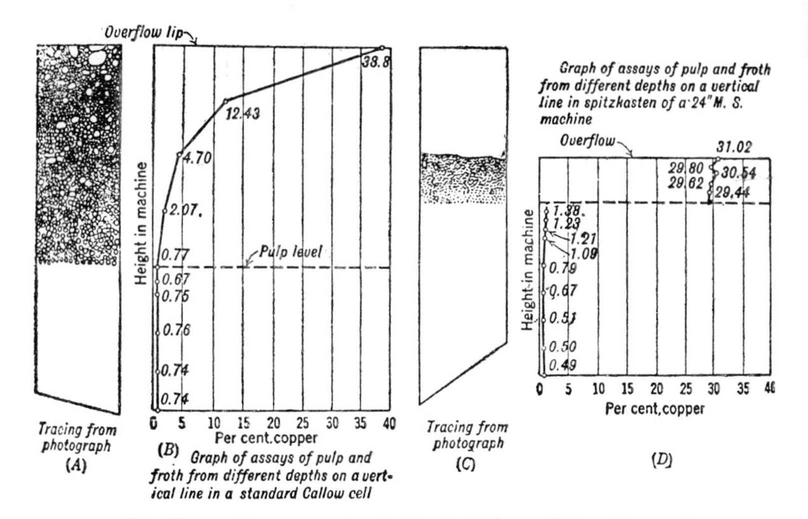

Fig. 221. *Concentration in Froth-Column* (after Taggart)

Taggart observes:

"All grains are falling relative to the bubble walls, streaming through the inter-bubble spaces. Both dark and light grains (Fig. 222) fall upon and slide along the upper surfaces of bubbles, but below the horizontal bubble equators only dark particles cling. Groups of dark particles collect at the lower poles of the bubbles, forming a pendant tip from which particles appear to string out and fall away. Cross observation of the column shows by colour that the concentration of dark particles in the inter-bubble spaces increases from bottom to top of the column. This is readily confirmed by sampling (Fig. 221). With all particles falling and the bubbles rising this change in concentration can only occur by reason of the fact that the average rising rate of the bubbles lies between the average falling rates of the light and dark minerals."

The picture obtainable by direct observation is one of a sudden change in ratio of gas to liquid from the predominantly liquid pulp to the predominantly gaseous froth. An equally sharp change is from the coursing bubble and free-swimming particle in the pulp to the packed N-bubbles at the bottom of the froth with the spaces between them full of particles in pulp, pressed on all sides by the air-pulp interphase. Here, if anywhere, the test conditions which apply to bubble pick-up attachment must be valid. The need for several types of mineral particle, present in abundance, can now be appreciated. Somewhere between the one-walled N-bubble at the bottom of the froth-column and the fully two-walled F-bubble at the top the unwanted particles must be squeezed out of the interphase and sluiced down into the pulp. There must be an excess of fully aerophilic particles at the top in order to ensure this. This is why a concentrate should not be removed prematurely (for example, in a closed grinding circuit). The valuable particles must help

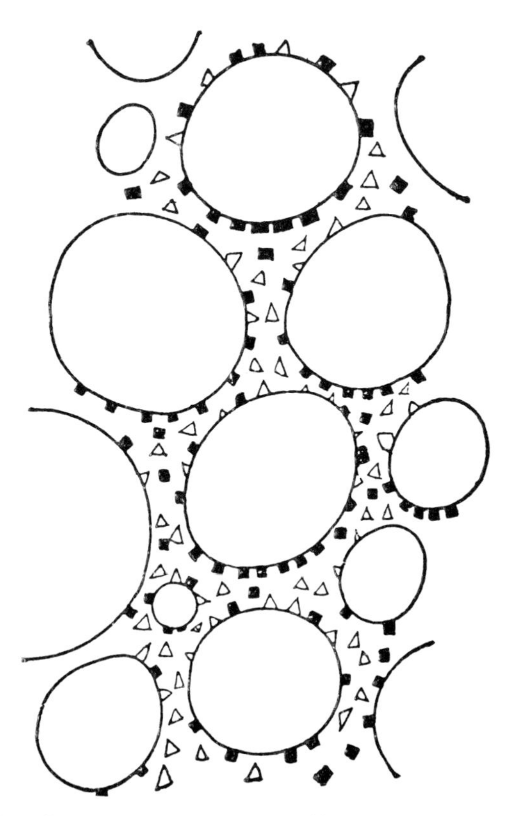

Fig. 222. Concentrating Action in Bubble Column (after Taggart)

to clean the froth. From above, down, aerophilic cling-weakness and response to the cumulative sluicing effect of the water draining down is increased. The porosity of the froth blanket is finer toward the bottom because the bubbles are smaller. The sluicing effect is, therefore, more vigorous in this area and the restraint on the material attempting to rise is greatest.

What cannot be observed in the froth column is the electrical modification. It is a matter of common plant observation that excessive lime in the conditioning (perhaps accompanied by excessive metal ionisation during surface modification using copper salts) reduces the liveliness of the froth. Without this liveliness the true concentrating action above described is impaired. If enough lime reaches the froth to induce flocculation of the gangue, true froth action is not obtained, since stability is increased. The degree to which the particle is submerged in the wall of a F-bubble depends on the contact angle. Strongly hydrophobic particles, notably heavy-oxide slimes, may form a "dry froth" in which these project into the air. Experienced plant workers rely on visual inspection of the froth to guide their controls, and association between a *p*H upset and a "sticky" or dirty froth is one element of control. A further activity probably proceeding in a froth column which contains incompletely passivated particles is renewed surface activity between the oxygen in the strongly aerated froth and the mildly excited mineral surfaces. Enough has been written to show that the cleaning action in the froth is complicated, and very important for good recovery. It should be studied and experimented with. The froth column can be standardised for height and condition to some extent as the result of specific tests on the mill pulp as it is produced in the local mill water.

The mechanism by which bubbles increase in size upward through the froth is mainly entropic. Although their surface tension has been appreciably lowered by armouring, it is still considerable. When several tiny bubbles merge, the total air volume remains the same but the total bubble surface shrinks considerably. Stated mathematically, n bubbles, each of the same radius and having a combined area A, will when merged to form one new spherical bubble of volume equal to their combined volume, and area A_1, have shrunk in area thus:

$$\frac{A_1}{A} = 3\sqrt{n}. \tag{17.23}$$

In a specific case 100 double-faced bubbles of 1 mm diameter would shrink to form one bubble of 4·6 mm diameter and the total surface area would drop from 6·283 cm^2 to 1·354 cm^2. Therefore, decrease of total surface—and consequently free energy—occurs with the merging of small bubbles to form larger ones. A constant supply of new small bubbles is arriving from below to replenish the leakage of skimmed bubbles. The air from these bursts through the uppermost layer. Since particles are showering down from burst bubbles, all those at the lower level are able to capture what they need to fill their surfaces. At the same time the total available area is always inadequate, since new particles are arriving on the merging bubbles, and only a minor percentage of those constituting the froth are being withdrawn at the top. Thus the upborne stream of particles finds a decreasing available area of interphase in which to cling, and at the same time it is in competition with descending particles for that space. Only those particle surfaces having the highest tenacity in terms of contact angle (or aerophilic attraction) will persist. The others will be crowded into the spaces between bubbles, and these spaces are sluiced by the water draining down to the pulp from burst bubbles. Under these circumstances the gangue drops back and the correctly conditioned particles, by virtue of their coating of collector, persist and are withdrawn as concentrate. The entropic change in the froth layer is aided in some flotation cells by a gentle rotating action of the pulp. The froth does not rotate so there is some mechanical dragging force between top of pulp and lower layers of F-bubbles. Its amount depends on the stilling effect of the baffles in the body of the cell.

An interesting minor corollary is that, during this laundering process in frother-rich bursting bubbles, the gangue particles appear to lose the organic contaminants which caused them to float. If one observes a series of cells through which a stream of pulp is flowing, the first to receive the pulp is found to hold a high froth column, while the last "scavenger-cells" have little or no froth head. It is often said that the assay value of a concentrate can be regulated by increasing or decreasing the height of the frothing column, but this is only partially true. Unless there are particles in the pulp which are prepared to dwell in the froth, its height cannot be raised. There is no difficulty in maintaining such a column when subjecting a removed froth to further cleaning, since the pulp now carries an abundance of collector-coated particles with which to stabilise emerged bubbles. Flotation circuits are arranged in cascade to take advantage of this fact.

It is sometimes necessary to assist the breakdown of too persistent a froth. Mechanical agitation usually suffices, but research suggests that some silicones (e.g. the I.C.I. "Silicolpase" antifoams) might have some value. Spraying with an oppositely charged surfactant has succeeded experimentally, but the nature of the stabilizing solids at the interface must be studied in each problem. Foams and froths can be stabilized by traces of starch, protein, detergent soaps, colloids or slimes as well as by collector-coated particles of concentrate.

Frothing Agents

A widely held concept of the frothing agent is of a surfactant of low solubility and surface tension, with no collecting power or reactance with collector agents. The exceptions to such generalisations are sufficient to enjoin some caution. The polypropylene glycol methyl ethers are completely water-soluble. Unaided pine oil has some small collecting power. These and other anomalies await further elucidation by research.

Frother molecules are heteropolar. Their hydrophilic groups are characteristically non-collectors (e.g. OH, COOH, C=O, NH_2, N). Hydroxyls are the commonest, in such reagents as terpineols ($C_{10}H_{17}OH$), cresols ($CH_3C_6H_4OH$) and alcohols ($C_5H_{11}OH$, etc). Froth stability increases with the length of the hydrocarbon chain.

Many other frothers beside those based on terpineol (pine oil, eucalyptus oil) are marketed. Some, such as the cresols, have collecting properties. Others are derived from alcohols. A frother must have a hydroxyl group attracting it to water, and a non-polar group sufficiently long in its carbon chain to keep solubility low. It should not ionise or possess collector qualities, since it is desirable to control frothing and collection separately, starvation of the former aiding with manipulation of the height of the bubble column, and of the latter sharpening the selectivity of sorption.

Pine oil is an aromatic alcohol product of the distillation of wood, the active principle of which is terpineol α, β or γ. These vary in the situation of their hydroxyl group. They are strong surfactants, sharply decreasing the surface tension of water. Where, as with pine oil, the hydroxyl group is directly connected to the hydrocarbon chain or ring structure, there is but little collecting attraction. save for such natural hydrophobes as coal and thocolite (a uranium-bearing hydrocarbon). The terpineol structure is

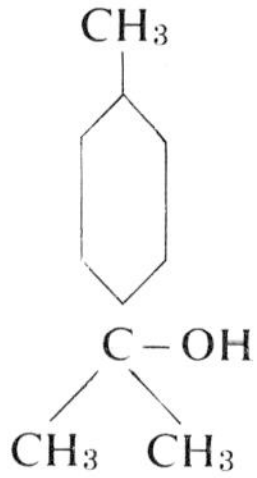

CH_3

$C-OH$

CH_3 CH_3

and that of commercial cresylic acid (a mixture of phenol homologues, cresols and xylenols).

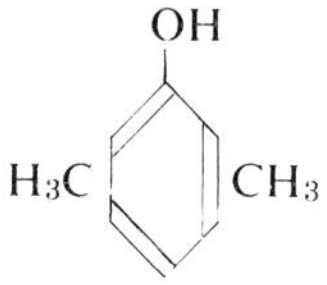

OH

H_3C CH_3

These carry the hydroxyl which links the low-solubility HC group to the water in the air-liquid interphase of the bubble. As the flotation process developed precision of control, practice moved away from the confused use of these compounds in the dual role of collector and frother. The modern method exploits specific chemical and independently controlled frothing action in separate reactions. Consideration of the possible significance of the alcohol group led to successful use of amyl and hexyl alcohols to improve selectivity. This in turn suggested that these reagents owed part of their value to their greater solubility. Today the largely empiric position has been reached that relative insolubility of the frother is by no means a *sine qua non*, and that selectivity may actually be improved if completely water-soluble frothers are used. Here research, though active, is not abreast of empirical application.

For sulphide flotation a series of water-soluble polypropylene glycol methyl esters—the Dowfroths (Nos. 200 and 250) is now well established. The formula is:

$$CH_3—(O—C_3H_6)_X—OH$$

and the number refers to the molecular weight. They produce a lively froth which breaks down well on leaving the flotation cell. The American Cyanamid water-soluble frother is Aerofroth 65. Others, based on the higher alcohols, are also marketed. Froths can be modified by the use of aerosols such as the latter Company's Aerosol OT. These disperse gangue in non-sulphide flotation.

The solubility[3] of some frothers is

Methyl amyl alcohol	17 g/l at 20°C
Capryl alcohol	1·28 g/l at 25°C
Pine oil	2·5 g/l at 25°C
α-terpineol	1·98 g/l at 15–20°C
Cresylic acid	1·66 g/l at 20°C

(*After Booth & Freyberger*)

In addition to the foregoing a number of compounds in use as collectors have frothing properties. These include long-chain carboxylates such as fatty acids, long-chain sulphates, sulphonates and amines.

The requirements for a selective type of froth are related to the texture produced by a small addition, since this affects its wetness, sluicing action and retention of the desired mineral species. The frother should not promote obduracy of texture in the absence of true concentrate particles. It should be insensitive to *p*H and to dissolved salts coming up from the pulp. The amounts added (in lb/ton of ore) vary between 0·01 and 0·25 for pine oil; 0·01 and 0·36 for cresylic acid; and 0·005 and 0·25 for methyl isobutyl carbinol, the average additions being 0·07, 0·09 and 0·07 respectively.

If frothing agents are present at the point of entry of air to the pulp they aid in the formation of smaller bubbles and, by thus slowing down the coursing speed, improve dispersion and contact time.

Particle Size

The flotation of fine particles often presents special problems, some of which are discussed in a later section. In the flotation of sulphide minerals the films adsorbed on very small particles are relatively unstable. Large particles are more liable to be lost by gravitational drag, especially in a dilute pulp or where collision or excessive turbulence tends to tear them out of the coursing bubble. Help given by the increased buoyancy of a denser pulp is somewhat offset by friction and turbulence.

Adsorbed collector may facilitate flocculation in which tiny bubbles provide the bonding force. If the resulting "aeroflocs" are too heavy to float and mechanical dispersion of these formations in the cell does not take place, loss results. Large particles are aided upward by copious aeration in which coursing bubbles give an extra lift, but such particles are unlikely to persist into the removed froth unless special facilities are provided in the overflow zone. The accompanying loss of sorting action also entails lowered concentrate grade. Sometimes, as in the flotation of metallic gold, the froth is stiffened with oil or oily collectors, again at the expense of concentrate grade.

In some cases it has been found helpful to deslime the feed, thus removing hydrated slime from the air-pulp interfaces and reducing its tendency to coat larger particles. This applies to *minus* 4 micron material. Where cationic collectors are used the pulp must usually be deslimed before effective collector-action is possible. It is sometimes feasible to float a highly slimed froth from the first cell of a series and to handle the desired larger particles lower down, the collector being added *en route*. A flocculated slime tends to float and a coagulated one to remain in the pulp. If the latter type masks larger particles it should yield to the dispersant action of sodium carbonate or silicate.

Fine slimes lead to high consumption of reagents, since they present a larger specific surface. They increase the pseudo-viscosity of the pulp, hinder bubble rise and tend to lower concentrate grade by over-stabilizing the froth. To the extent that specific surface is proportional to the transfer of ions into the pulp from the mineral the climate of flotation may be affected.

Since hydration occurs during wet grinding and is opposed by surface aeration there is room for research on the relative effects of dry grinding where slime-formation is a problem. Stable hydration of small particles is lessened in the presence of air and collector.[29]

Research has shown a relationship between collector concentration and maximum floatable size. A higher concentration of collector is therefore necessary in floating a long-ranged feed than is required if grinding reduces all desired constituents below the size readily floatable. As regards slimes and the problems they introduce, care in the grinding circuit should be concerned to keep these to a minimum. In running a plant, as in other humanly controlled activities, it is easier to keep out of trouble than to get out.

Mineralization of Bubbles

A particle which had its surface fully covered by collector would find it difficult, if not impossible, to cling to an air-water interface completely packed with frother molecules. The incompleteness of adsorption of the contacting phases is an important factor in attachment. Mineralizing contact may result from collision between bubble and particle, from precipitation of gas from water or through a mixed process in which precipitation initiates contact. Tests made by Whelan and Brown on coal showed that attachment on collision with a coursing bubble increases with the density of the particle concerned, which by its greater mass is better able to break through the slipstream or energy barrier surrounding the bubble. Taggart attaches importance to the precipitation of air behind the impeller and therefore to the need for dissolved gas in the process water. Other research[7] supports this by showing how a minute bubble, such as might result from precipitation, initiates bubble attachment or the build-up of a buoyant system. Field reports confirm the need for good aeration of process water if a sustaining froth is to be produced. The available gas also influences the kinetics of attachment.

Formation of the coursing bubble occurs in various ways in the flotation cell. It may be diffused by blown-in air, in which case it starts in a state of intensely rapid vibration, changing shape and area hundreds of times per second. It may be sheared into the pulp after induction *via* the low-pressure body of air near the hub of the impeller of a mechanical cell and further sheared by being milled with the solids in the turbulent zone swept by the tips of the impeller blades. Precipitated bubbles can be pressed against particles which settle below the impeller and are being drawn back toward the hub for re-circulation. Formation, coalescence and attachment of bubbles to solids are elements of flotation kinetics.

In agglomeration-flotation the deslimed and activated particles are tacked together by minute air bubbles and form a flexible envelope (glomerule) round an aqueous centre. This relatively light glomerule is removed in the tailings band of a shaking table or by such other means as treatment on a specially manipulated conveyor belt.

In froth flotation the energy barrier between the hydrated mineral surface and the bubble must be surmounted by force, either due to the adsorptive attraction of the collector or to collision resulting from mechanised energy. The acceptance of a particle by an air bubble can be complete in 0·005 second.

The point of impact between bubble and particle is inter-related with slippage along the surface. Whelan and Brown[38] found that the percentage attachment decreased as the point of contact moved away from the vertical bubble axis when coursing bubbles hit drifting particles.

Where the magnitudes are reversed, Dzieniewicz and Pryor found that a small bubble attaches more readily to a polished surface than a large one[7] and, once attached, improved the chance of attachment of larger ones which arrived later. Two stages at least operate—first, saturation of the pulp with small bubbles, which may be below visible size and second, attachment of the particle thus "primed" to a larger bubble. Without the existence of conditions in which the precipitation of gas from pulp can take place, flotation is reduced and may even cease.

In both mechanical and pneumatic cells air precipitation takes place in the body of the pulp. The special froth quality of the deep pneumatic cell (the Britannia) suggest that bubbles may emerge on the particles with pressure relief as the pulp streams upward. The amount and overall effect of precipitation, horizon by rising horizon, is qualified by the initial aeration and agitation.

Particle shape plays a part in mineralization. A sharp edge, beside being the most likely point for a bubble to form as it emerges from water, is better able to cleave the energy barrier round the coursing bubble. Roughness of the surface of a hydrophilic particle aids wetting by trapping water. Conversely, a surface which shields a pocket of air helps contact with a bubble.

The greater the hydration of the mineral surface, the harder it is for gas molecules in the surrounding aqueous phase to diffuse to the solid and initiate bubble formation. Where a surface is incompletely wetted, molecules of air should find it easier to displace water from the oriented sheath round the particle than to force apart water molecules in the fully aqueous phase. This aids coalescence and bubble formation at the surface rather than in the liquid phase.

Some freshly cleaved sulphide minerals hydrate so strongly as to have a natural zero contact angle and in this state to be indifferent to xanthate activation. Only when such a surface has had time to become oxygenated does some degree of hydrophobicity appear. Research with various gases dissolved in the pulp[29] shows oxygen to be the most actively adsorbed. This adsorption occurs in three stages (*a*) reversible adsorption (*b*) active attachment and (*c*) chemisorption with superficial oxidation. The initial surface layer of molecular oxygen dissociates to form a monatomic layer of atoms coupled to charged metal points in the lattice. Electron-transfer ensues and oxidation begins. An oxidic film then starts to form and at an early stage in this reaction the maximum adsorption of competing collector ions can be achieved. Excess oxidation reduces the adsorption of collector. It has been suggested that this is a sort of "softening-up" process in which the oxygen penetrates the surface layers and weakens the bonds between the ions.

This would facilitate ionization, ion mobility and either ion exchange with collectors or ion migration with solubility increase.

Theorising in this field of surface physics is necessarily tentative in our present state of knowledge, but the inferences drawn from research and confirmed by empiric industrial experience broadly confirm the suggestions in this section as to the vital interplay between aeration of the pulp and selective anchorage of the particle first in the coursing bubble and later in the overflowing mineralized froth.

Flotation of Sulphides and Non-sulphides

Unoxidised and uncontaminated sulphides are not floatable, though the role of oxygen in initiating xanthate acceptance is not fully agreed. Plaksin *et al*[39] observe that oxygen is preferentially adsorbed from water in successive stages and at rates depending on the oxygen-affinity of the specific sulphide. Adsorption of oxygen must precede fixation of xanthate and the amount required for maximum floatability with this collector rises for the sulphide minerals from a minimum with galena, *via* pyrite, sphalerite, chalcopyrite and pyrrhotite to a maximum for arsenopyrite. Oxygen control has been used in Russia to sharpen selectivity in floating copper, lead and zinc.

Where non-sulphide flotation is concerned these considerations no longer enter. Taggart[37] sums up an *ad hoc* approach thus. "Chance is as good as anything else to determine the order of such tests"—tests which embrace a variety of cationic and anionic reagents. The relation between crystal structure, surface properties and adsorption has been considered in detail by Aplan and Fuerstenau.[3]

Some Special Cases

For most minerals flotation works successfully in the *minus* 65 mesh *plus* 10μ range, with adequate control on the industrial scale. Below the 10μ size high recovery and concentrate grade becomes increasingly difficult to achieve economically, largely because of the smallness of the particle. Derjaguin and Dukhin[30] consider the approach of the particle to the bubble to take place in three stages as it passes through three zones. The outermost (zone 1) is that of the ambient liquid in which no surfactant forces operate. Zone 2 is the bubble's diffusional boundary layer and zone 3 a liquid wetting layer across which either an attractive or a repellent force operates between particle and bubble. Their theory is concerned with particles so small that inertial forces play little part. The particle moves by gravity in zone 1 and the balance between its inertia and the resistance of the medium determines its ability to enter zone 2. Here electrophoresis is active and may bring it to zone 3, where it can cling in the air-water interface.

For the flotation by standard methods of *minus* 5μ material a comparatively new technique called ultra-flotation has been developed[31]. In this, as commercially used to float contaminants away from kaolin, chosen "carrier minerals" are ground to a slime and conditioned in such a way as to cause them to attach to the unwanted contaminants and then float them off. The minerals so used include fluorite, barite, calcite and silica, and they slime-coat minerals themselves in the *minus* 1μ size range. The removal of anatase has been described by Greene and Duke[32].

Two other types of slime problem can arise. In one the sought concentrate tends to slime, as may happen with galena. In the other, a gangue mineral such as barite is the source. In either case slime coating leads to misplacement of the desired float and, because of the large specific surface of the slime, excessive consumption of reagents. De-sliming is one remedy, and the use of starch to adsorb on to the gangue may also have some success. The possibility of using ultra-flotation for slime flotation is at present in the laboratory stage. It has been discussed by Meloy.[3]

Another special application of flotation is the recovery of IX resins, described by Bhappu[2] and Singewold. It is suggested as a method of removing loaded resin during the resin-in-pulp extraction of uranium from its ores.

Ion Flotation

This, as described by Sebba[33,34,35] is a process which differs from the collection of already existing particles by flotation. The proposed concentrate is at first in aqueous solution, either as ions or as colloid. It must be converted into a product which has hydrophobic sites and then attached to a bubble which collapses on reaching the surface. This produces a floating scum rather than a froth of the normal flotation type, although at this stage of research a transient froth is required. The colligend ions (the aquated ones which are to be stripped from the aqueous phase and floated) are conditioned with a non-micellar collector and levitated in a fragile froth. The term "ion flotation" does not exclude recovery of sols, gels or polynuclear ions. The collector-attracting ion must become insoluble in water before capture by a coursing bubble. The collector itself must not be, or allowed to become micellar since single surfactant ions are required. Reagents such as sodium stearate are prepared by crystallisation from solution in anhydrous methanol, ethanol or acetone.

An ion in aqueous solution is not the simple structure written into formulae and equations. Water is a dipole molecular aggregate with a high dielectric constant. The ions it takes up are not "naked" but surrounded by a hydration sheath of water dipoles. If reasonably stable ion *plus* water would form a sort of hydrate with coordinate bonds and directional qualities. An alternative hypothesis pictures continuous exchange between the hydrating molecules and the ambient water. The degree, if any, of anion hydration is not at present known.

When the conditioned water is aerated, bubbles tend to concentrate the surfactant and carry it to form a surface layer sufficiently strong to re-dissolve the colligends and return them to the aqueous phase. The type of froth used in normal flotation is therefore unsuitable. One which is evanescent, breaking through the surface scum without building up, is required. The rate of air arrival should only just exceed that of foam collapse, and turbulence must be avoided.

If the potentialities of ion flotation suggested by current research can be translated into terms of process economics the handling of leach solutions, mineralised wastes, process water etc. may be affected, and the still visionary possibility of "mining" sea water for its dissolved mineral wealth brought a step nearer.

Flotation Kinetics

This subject has been defined by Arbiter and Harris[3] as "the study of the variation in amount of froth overflow product with flotation time, and the quantitative identification of all rate controlling variables". Operating variables, together with those of the pulp constituents, run into hundreds and their interaction extends the possible permutations into almost astronomical figures. Approach to this study is therefore empiric and confined to the control of selected key factors in a severely simplified system, which the author calls "the pulp climate".

Tomlinson and Fleming[40] concluded from studies mainly concerned with the behaviour of apatite and hematite that under precise laboratory control of size, mineral purity and aeration the rate of flotation for a single mineral of given size is first order and overall rate for a mixture of its sizes depends on the weight proportions and the rates of flotation of the individual sizes. With inhibited conditions flotation rates of fine sizes vary with the square of the particle radius for readily responsive materials. In intermediate or inhibited flotation dependence on particle size is always less than in free flotation. In flotation of a wide range of mixed minerals the concentrate grade increases with particle size.

During conditioning several activities can proceed simultaneously. Chi and Young[41] include the ionization of reagents, diffusion, adsorption, chemical surface reaction, desorption or diffusion of reaction products, sliming and flocculation. These authors and others[17] agree that recovery of floatable material from a pulp follows first-order law. This is defined by the equation

$$\frac{-\mathrm{d}(A}{\mathrm{d}t} = k_i(A) \qquad (17.24)$$

where reactant A changes at constant volume and k_i is the velocity constant. That put forward by Imaizumi and Inoue,[11] is

$$\frac{-\mathrm{d}Nm}{\mathrm{d}t} = k_1 Nm \qquad (17.25)$$

where t is flotation time, Nm the number of particles in the cell at the instant t, k_1 a rate constant applicable as a criterion of floatability.

References

1. Rickard, T. A. (1916). *The Flotation Process*, Mining & Scientific Press.
2. (1961). Quarterly of Colorado School of Mines Vol. 56, No. 3 (Vols. 1 and 2).
3. (162). *Froth Flotation, 50th Ann. Volume*, A.I.M.M.E.
4. del Guidice, G. R. M. (1934). *Trans. A.I.M.M.E.*
5. Gaudin, A. M. (1957). *Flotation*, McGraw-Hill.
6. Mellgren, O., and Subba Rao, M. G. (1963). *Trans. I.M.M. (London)*, 72.
7. Dzieniewicz, J., and Pryor, E. J. (1950). *Trans. I.M.M. (London)*, 59.
8. Sutherland, K. L., and Wark, I. W. (1955). *Principles of Flotation*, Aust. I.M.M.
9. Young, A. (1805). *Phil. Trans. Roy. Soc.*, 84.
10. Leja, J., and Poling, G. W. (1960). *Int. Min. Proc. Congress*, I.M.M. (London).
11. Gaudin, et al. (1963). *6th Int. Min. Proc. Congress (Cannes)*, Pergamon.
12. Cooke, S. R. B., and Digre, M. *Trans. A.I.M.M.E.*, 184.
13. Sun, S. C., and Troxell, R. C. *Trans. A.I.M.M.E.*, 196.
14. Pryor, E. J., and Liou, K. B. (1948). *Trans. I.M.M. (London)*' Oct.
15. Moilliett, J. L., Collie, B., and Black, W. (1961). *Surface Activity*, Spon.
16. Adam, N. K. (1941). *The Physics and Chemistry of Surfaces*, O.U.
17. Modi, H. J., and Furstenau, D. W. (1960). *Trans. A.I.M.M.E.*, 217.
18. Eigeles, M. A. (1950). *Metallurgizdat.*
19. Gaudin, A. M., and Tournesac. (1954). *First World Congress on Detergence*, Paris.
20. Taggart, A. F. (1945). *Handbook of Mineral Dressing*, Wiley.
21. Kivalo, P., and Lehmusvaara, E. (1957). *Int. Min. Proc. Congress*, Stockholm.
22. Br. Patent 708475; U.S. Patent 2698088.
23. Taggart, A. F., and Arbiter, N. (1946). *Trans. A.I.M.M.E.*, 169.
24. Bruyn, P. L. de. (1955). *Trans. A.I.M.M.E.*, 202.
25. Hines, P. R. (1959). *Trans. A.I.M.M.E.*, 214.
26. Tucker, et al. *Trans. A.I.M.M.E.*, 183.
27. Sidgwick, N. V. (1950). *The Chemical Elements and Their Compounds*, O.U.P.
28. Sollengerger, C., and Greenwatt, R. B. (1957). *Trans. I.M.M. (London)*, 65.
29. Klassen, V. I., and Mokrousov, V. A. *Introduction to the Theory of Flotation*, Butterworth.
30. Derjaguin, B. V., and Dukhin, S. S. *Trans. I.M.M. (London)*, 70.
31. U.S. Patent 2,990,(58.
32. Green, E. W., and Duke, J. B. (1962). *Trans. S.M.E.*, A.I.M.M.E., Dec.
33. Sebba, F. (1959). *Nature*, Oct., 184.
34. Sebba, F. (1963). *Royal School of Mines Jnl.*
35. Sebba, F. (1962). *Ion Flotation*, Elsevier.
36. Haeck. (1964). *Chemical Dictionary*, Churchill.
37. Taggart, A. F. (1951). *Elements of Ore Dressings*, Wiley.
38. Whelan, P. F., and Brown, D. J. (1956). *Trans. I.M.M. (London)*, 65.
39. Plaksin, I. N. et al. (1957/58). *Trans. I.M.M. (London)*, 67.
40. Tomlinson, H. S., and Fleming, M. G. (1963). *6th I.M.P.C. (Cannes)*, Pergamon.
41. Chi, J. W. H., and Young, E. F. (1962/63). *Trans. I.M.M. (London)*, 72.
42. Imaizumi, T., and Inoue, T. (1963). *6th I.M.P.C. (Cannes)*, Pergamon.

CHAPTER 18

FLOTATION PRACTICE

Introductory

Though a book cannot replace the need for practical experience it can give its reader some understanding of the principles which are utilised in good technical control. The presentation of these in Chapter 17 was inevitably somewhat repetitive and tentative. In plant operation, however, a clear-cut procedure must be agreed upon and adhered to, despite gaps in our fundamental knowledge and our conflicting hypotheses. The alert operator will keep in touch with research in his field and be prepared to adapt the milling technique whenever new discoveries, confirmed by local tests, justify change. He is concerned not only with technical efficiency but also with the pressures of management which take into account the overall economics of operation. Working conditions in an industrial plant must therefore be flexible enough to be adaptable to change in the business world. In what follows only the technical aspect of operation can be considered.

The main components affecting steady control (without which efficiency cannot be maintained) are

(1) Ore (texture, crystal interlock, variation in composition).
(2) Process water (seasonal variation, progressive fouling if re-circulated, waste disposal).
(3) Reagents (nature, reaction rate, correct dosage, effect).
(4) Machines (max. and min. capacity, adjustment, reliability).
(5) Pulp (particle size range, solid-liquid ratio, density change along circuit, *p*H stability, slime effects).
(6) Flow rates (aeration effects, operating sequences, time-temperature changes, steady flow, changes of ore grade).
(7) Product grades (concentrates, tailings, re-circulating middlings, partly processed stockpiles).
(8) Re-circulating build-ups (classifier returns, middlings re-treatment, return water).
(9) Adjustment lag (sampling accuracy and assay speed, monitoring, automatic control, surge suppression.

The interdependence of these items is not always realised. Any one factor in a chemical process which wanders from its norm of application changes the working conditions for which all the other factors depending on it have been fixed, and erratic running *must* follow. Though such factors as change in composition and texture of the ore are to some extent unforeseeable, they must be allowed for. If loss of control of one item is bad, multiple lack of control is a great deal worse. Systematic checking, a good standard of chemical cleanliness, and care on the part of the mill manager to interest the workmen in methods necessary for a good standard of technical efficiency are essential to success.

The Ore

It is important to observe any changes in the tenor of the incoming ore, preferably well ahead of treatment. Variation in the crystal structures and interlocks may have an important effect on liberation and optimum mesh-of-grind. Changes in the proportion of associated minerals may be important.

As the mine deepens and develops, new minerals may enter the feed and affect the concentration process. A patina or other indication of oxidation on a particle too small to have received much attrition grinding, points to the arrival of oxidised mineral which may either need special treatment or exclusion. The most feebly floated particles rise in the scavenger froth. They should be studied until the operator is thoroughly familiar with their normal appearance, shapes, sizes, and rate of production. Any marked change in the scavenger float will then give a clue to more subtle changes in the ore from which they have come.

Primary metal sulphides are usually stable and react with good predictability in the flotation circuit. Secondary metal sulphides are less reliable. They may occur as coatings on primaries, or as penetrations along cleavages in the altered ore in which they occur, as when covellite coats a pyrite crystal. Since flotation is selective to a surface, and not to a core, a completely or partially coated particle of alien mineral may float as though it were a true concentrate. When geological conditions have favoured migration and redeposition of the valuable ore-constituent, the problems of grinding and of producing a clean concentrate are complex. Oxidised ores, according to the extent of penetration and lattice "decay", are softened, mixed, and prone to send metal ions into solution. Increased consumption of reagent, over-grinding and the production of slimes are possible troubles under such circumstances.

Oxidised or altered ore may arise from geological changes, or from slowness in sending severed mineral from the mine to the mill. Hold-ups in stopes, ore passes, and ore bins must be minimised when oxidation-prone sulphides are being worked, particularly when unstable pyrite is present. Stope washing, periodical clean-up of trucks, excessive production of "fines" when blasting, and contamination of ore in transit by casual oil and foul water may lead to poor recovery. Sometimes ore can be drawn from the various stopes in such proportions as to produce a fairly steady blend. Sometimes markedly different drafts of ore can be kept separate and treated according to a plan. An ideal arrangement is for the mill to receive part of each stope sample during valuation and to keep its own stope plan. These samples are examined and tested, and the most suitable grinding and conditioning treatment can be worked out. Then, provided good liaison is maintained between mine and mill, the right handling is assured at the right time. The more prevision is exercised, the better the mill will do its work. Blending of mill feed from separate bins becomes increasingly desirable where the stopes are widely separated vertically or horizontally, or where one mill serves several mines.

When the feed is of high grade it is easy to produce a highly mineralised froth and top-grade concentrate. At such times the possibility of insufficient collector agent in the scavenging section must be guarded against. When feed is low-grade it is harder to stabilise a highly loaded froth, and one of the final cleaning cells may be switched to a lower-grade section if cells and launders have been flexibly coupled.

Pulp Preparation

The most important factor in preparing the flotation feed is correct wet grinding. Mineral particles as coarse as 28 mesh have been seen to float, but this is abnormal, and even a 48-mesh sulphide needs a thick pulp to help it to remain in the froth. Laboratory control of the milling operation includes assays and micro-observation of each screened fraction in a representative tailings sample, made in order to find the amount of loss and the reason for its occurrence. The coarsest particles are least likely to be adequately liberated. If grinding all the ore to a finer mesh improves recovery sufficiently to yield a profit after the extra work is paid for, it should be done. It may be that the trouble lies at the other end of the scale and that the losses are due to over-grinding. This kind of loss is not clearly brought out by tests of screen products and may pass unnoticed in the plant. The sizing method called beaker decantation[1] is easy to use and reveals the trouble. Tests are made on a composite sample cut from the routine daily tailing samples, in proportion to the tonnage represented by each of these samples over the period concerned. This may be a normal monthly check, or may be for a shorter period when adjustments to the grinding circuit are proceeding. Each fraction is assayed for its content of valuable metal. Results are worked up as in the following table:

TABLE 38

ATTRIBUTION OF LOSSES IN COMPOSITE TAILINGS SAMPLE

(1) *B.S. Mesh*	(2) *% Weight*	(3) *Assay Value*	(2) × (3)	(4) *% Value Lost*	
+100	2·2	0·5	110	3·7	
150	7·8	0·4	312	10·4	26·2% of loss at +200#
200	14·5	0·25	362	12·1	
240	6·3	0·2	126	4·2	
300	8·9	0·2	178	5·9	
40μ	11·3	0·25	283	9·4	
30μ	11·4	0·25	285	9·5	
20μ	12·8	0·3	384	12·8	
10μ	14·0	0·3	420	14·0	44·8% of loss at −30μ
−10μ	10·8	0·5	540	18·0	
	100·0	0·3	3000	100·0	

In this instance the heaviest section of the losses is in the -10μ band of the sizing band, while nearly half the total loss lies in the -30μ zone. This would not be clearly revealed by routine screening down to 200 mesh, which would only show 26·2% of the loss above that mesh, and would fail to bring out the significance of the much heavier loss in the fraction that had been ground extremely small. The remedy is better adjustment and control of ball mill and classifier. It may be necessary to "scalp" the closed grinding circuit and lift out heavy-metal sulphides which are retained there beyond optimum grind. This can be done by adding flotation agents to the mill discharge and placing a unit flotation cell between this and the classifier.

The maximum floatable size of a fully liberated particle depends mainly on its density, aerophilic attraction, and shape. The "toughness" or degree of permanence of the froth is a modifying factor. The passing N-bubbles do not affect the support given to the particle by pulp density, but this support virtually disappears in the froth, which consists mainly of air. Fall is resisted by the toughness of the froth and the strength with which the mineral surface is attracted into the air phase.

Coal has nearly the density of the supporting pulp and there is no difficulty in bringing 10–14 mesh material up to the frothing zone, though with the loss of pulp support from thence on it may need aid from the froth texture when a good deal of such coarse-meshed feed is being treated. For metal sulphides the upper limit is 48–65 mesh and for metallic gold 100–150 mesh. Metallic gold has a high density (19·2) and poor collection quality, so needs a tough froth if it is not to slip back into the pulp. Mineral oils are sometimes fed to the grinding circuit for this purpose, their passage through the mill ensuring emulsification. At the other end of the scale, selectivity diminishes when treating particles smaller than about 5μ and good recovery and grade of concentration become more difficult to obtain.

The gangue particle size is limited only by considerations of mesh-of-grind for liberation and smooth transport through the plant. The larger the particle, the greater its tendency to settle and cause trouble but if machines, pipes, and launders have been properly designed *and* installed, the risk is low. Large particles need more vigorous agitation, with use of more power. They may cause excessive abrasion of the flotation machine.

The reasons for the difficulty often experienced in the selective frothing of very fine particles—say *minus* 5μ—are complex, varying from ore to ore, and not fully understood. In the author's field experience it has been found that, provided the commercial value of the mineral raised justifies meticulous circuit control, particles down to 2μ can usually be floated at a good recovery-rate. The difficulties centre mainly on the disproportion between the mass and the surface of an extremely small particle, which causes it to behave as though its surface tension is unusually high. As a result it reacts prematurely, becoming oxidised, flocculated, or slimed before it reaches the conditioning section. It is then difficult to coat with collector. The physics of attachment to an N-bubble are also more delicate, since a particle of low mass is repelled from the slipstream which surrounds a fast-rising N-bubble. It must therefore be offered small, slow-moving bubbles to sorb into, or achieve its attachment in the mixing zone round the impeller. When a small particle (say

-4μ) reaches the froth column, it tends to overflow with the concentrate regardless of its composition. The down-drag of gravity hardly affects it when opposed by the prevailing upward drift of the bubbles. Its disproportionately high ratio of surface to volume (as compared with the average particle treated in the pulp) has given it undue selectivity of the energy-neutralising forces available during treatment, and thereby reduced its specific reaction. The relative scarcity of such small particles in the tailings can only be inferred in our present state of knowledge. Two indicators are observable in most plants. *First*, a miniature shaking table can often remove further value from the tailings band it is set to monitor, which would not be so easily done if the feed to it had not reclassified itself to some extent. *Second*, the last scavenger froth is white and foaming in many cases, which would not be the case if the fraction of very small particles was similar to that at the head end of the section. The conclusion drawn may at first sight seem to contradict the sort of evidence given in Table 38, but this indicates the need for research rather than objection, since the changes of size analysis, cell by cell, for feed and discharge (the latter of both frothed minerals and cell tailings) have not yet received the close attention than can best be given by a combined effort between the research laboratory and the operating plant.

One method of combining the economy of coarse grinding with the recovery and grade best ensured by fine grinding is a stage treatment, when the type of ore permits (Fig. 223).

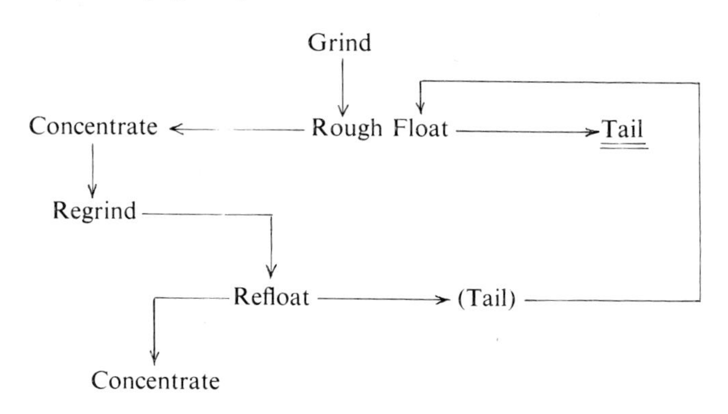

Fig. 223. Stage Grinding and Flotation

The regrinding and reconcentration are repeated as many times as necessary to bring less and less feed to a high stage of liberation. A good working example of this is the practice at Climax Molybdenum Inc. For such a method to succeed the sought mineral must be flotation-avid, so that by means of intense activation and hard pulling of the froth fairly coarse middlings can be recovered in the float, and a low-grade tailing discarded at a coarse mesh-of-grind. In the case of a graphitic mineral such as molybdenite associated with a hard and abrasive gangue, conditions for stage concentration combined with stage grinding are almost ideal. Molybdenite floats readily and strongly, and removal of a substantial portion of gangue at an early stage makes it possible to adjust the later grinding stages to the soft value rather than to the tough gangue. The first condition for success is that the mineral which is to float shall be so distributed through the ore as to produce a substantial amount of middling at flotation mesh. The second is that surface exposure of value in each such particle must suffice to ensure flotation despite the neutrality of the unactivated gangue also exposed on the surface. Finally, the aerophilic attraction must be unusually strong, since a limited area must overcome the gravitational drag of a relatively coarse particle incompletely adjusted to the air-water interphase. The possibilities of selective grinding are sometimes present, and even a simple scalping operation may be worth considering.

The rate of travel through the circuit of the properly liberated particle must be such that it has a correct dwelling time in each conditioning stage and in the line of flotation cells. The tonnage treated daily is fixed, and the plant capacity available for processing that tonnage is inelastic. Variations in dwelling time can be adjusted in most conditioning tanks by varying height of overflow and consequently the retained volume, also (providing some de-watering or economy in dilution is possible) by varying the solid-liquid ratio of the pulp. Up to a point, the higher the density of the pulp (the greater its percentage of solids), the less will be the requirement of reagent and stirring power to keep the particles suspended, and the shorter the conditioning time. In the cells, a thick pulp is better searched by the N-bubbles and it gives better support to the float. The working rule is to "rough" in a thick pulp and clean in a dilute one, in order to obtain a high-grade concentrate. The word "thick" depends on the slime content, and the slimier the feed the more dilute must be the pulp at its roughing stage. Dense pulps are unsuited to treatment in pneumatic cells.

Heavy sulphides are roughed at 30%–50% solids, the figure chosen being arrived at by laboratory testing. A disseminated gold pyrite would be roughed at any dilution down to 15% solids. Cleaning treatment is made at between 30% and 8%, in appropriate stages. Mechanical cells have a faster floating time than the equivalent volume of pneumatic cells. All cells have a maximum permissible rate of feed for efficient work but no minimum, provided that there is sufficient concentrate in the pulp to squeeze up an adequately rich froth in the head cells.

Although variation in the feed rate does not endanger flotation *per se*, it is thoroughly bad for conditioning, as a correct chemical concentration cannot be maintained unless the pulp is consistent both in texture and flow rate. Finally, temperature of the pulp, though relatively unimportant in a circuit using full addition of reagents from the start, becomes significant when starvation amounts of reagent are used to maintain critical control. A rise in temperature hastens reaction time, and aids in mixing by reducing pulp viscosity.

Wood chips broken down from mining operations must not get into the grinding circuit. If wood is ground, it turns to a sticky "wood flour", which adheres tenaciously wherever it can. Apart from obstruction, this material produces organic depressants during its decomposition. Wood should be removed on picking belts or ahead of crushers, and chips should be caught on trash screens wherever pulp is overflowing.

Conditioning

Conditioning must provide the continuously passing pulp with the optimum reaction time required, the appropriate reagents being added at each stage in a series of operations. The efficiency of the conditioning depends on:

(*a*) Thorough mixing and dispersion of each reagent through the pulp.
(*b*) Repeated contacts between molecules of reagent and *all* the particles concerned, at each conditioning stage.
(*c*) Development time for such contacts to produce the desired reactions, if necessary in successive stages.

The work to be done includes:

(*a*) the dispersal of slime coatings.
(*b*) the correction of H-ion concentration.
(*c*) any needed depression or activation, including resurfacing.
(*d*) possibly final adjustment of *p*H.
(*e*) selective coating by collector agent.
(*f*) addition of the frother.
(*g*) adequate stabilisation of correctly loaded F-bubbles at the top of the froth column.

Since the last additive (*f*) is not reacting chemically it needs dispersion only, not time. A certain amount of entrainment of air by splash, cascade, and vortex occurs during conditioning. If a frother was present a mineralised froth would be formed and would be a nuisance during the conditioning stage. Good operating control requires the accurate use of small quantities of reagents and their thorough distribution through the pulp. If the mill water is returning from dewatering end-products it may carry some frother, since this reagent is not permanently removed by reaction with any of the minerals. This might lead to the formation of froths or scums of the most air-avid minerals, which could thus withdraw an undue share of collector agent. It is therefore desirable that the mechanism of conditioning provides for the beating back of such froths into the body of the pulp. A particle which has once floated will float again whenever equally favourable opportunity is given, provided its aerophilic surface has not been destroyed, so there need be no fear in "drowning" such material.

The work of the reagent may be hampered if it is not mixed continuously into an equable stream or body of pulp. If the addition stage calls for 0·05 lb of reagent per ton of ore, harm will be done if all that 0·05 is surged in with a hundredweight of ore, or if the pulp itself is surging so that it alternates between double and half its correct rate of flow. Unless *all* the particles have a chance, a small proportion of them may take up the whole of the ration. · Reagents which are readily soluble in water disperse quickly into the pulp, but relatively insoluble oils must be thoroughly mixed, or a small amount of mineral will attract an undue proportion. It is often feasible to emulsify oils by mixing them with hot solutions of sodium carbonate. A means of emulsification is the "colloid mill", which mixes oil, water, and carbonate intimately as they pass through a pump-impeller rotating at high speed. The effluent is fed to the ball mill before it can reseparate. Occasionally machine oil or grease finds its way into the pulp—perhaps through leakage or carelessness in retrieving spilt ore from the mill floor. This leads to uncontrollable volumes of barren froth anywhere along the cells or the tailings line. Nothing much can be done to avoid the resulting loss of concentrate, but steps should be taken to ensure that the same thing does not happen again.

The conditioning series, reagent concentration, and reaction times are worked out in the laboratory and applied in the plant. Some of the machines

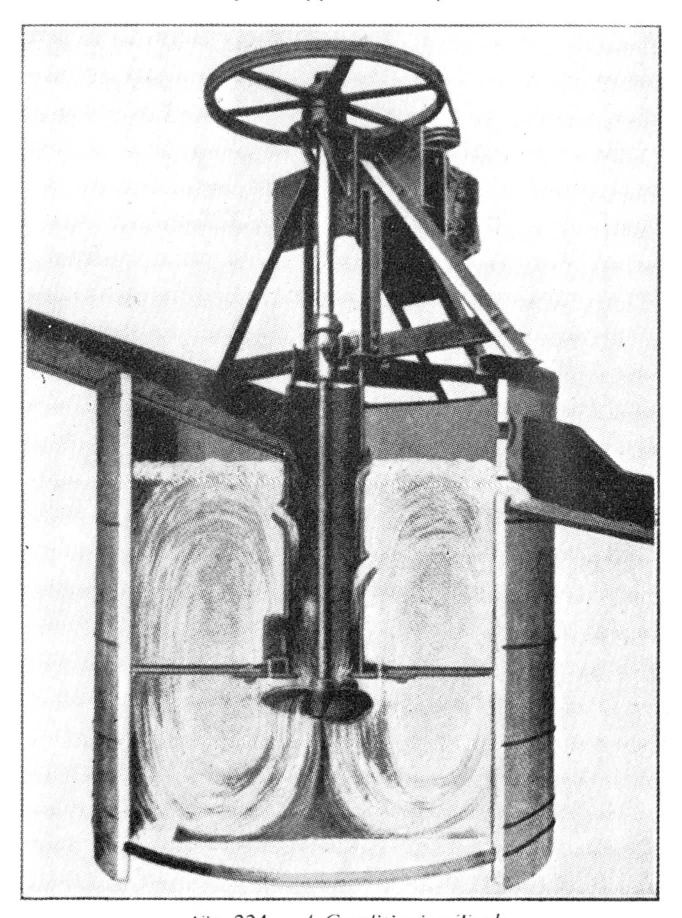

Fig. 224. A Conditioning Tank

in the flow-line can be used for conditioning. The ball mill is a good agitator and emulsifier. Occasionally a collector can be added there, to ensure its presence at the moment new surface is exposed, provided no edge-adherent scum of rich mineral then appears in the classifier. Removal of a paraffin-floated layer of graphite from the classifier has been made ahead of cyanidation.

The first stage of *p*H control may be an addition of lime in the fine-ore bin. This has the advantage of reducing deterioration of sulphides. Final *p*H adjustment would then be possible by feeding milk of lime (slurried $Ca(OH)_2$) to the classifier. Another favoured addition point is the surge tank between the classifier and the pumps which deliver to the flotation cells. Good mixing and agitation are assured, but the dwelling time may be erratic. Conditioning in the grinding circuit is, of course, only possible before taking the first float of a series, unless there is to be further grinding between floats.

Conditioning tanks (Fig. 224) are cylindrical, and up to 16 in. diameter by 16 in. high, though a dimension of 10 in. would be usual with a moderately coarse grind. Types such as the Denver and the Knapp & Bates are not sanded up in the event of power failure, and the latter's slatted weir facilitates return of prematurely floated mineral to the mixing zone. Capacity is adjustable by provision of discharge ports at different heights. Some conditioners have mild pumping ability, but in the event of breakdown this may lead to flooding of the cells from which they normally draw their pulp.

The criteria for scale-up from laboratory conditioning requirements to those in a full-sized operation have been mathematically considered by Chi and Young[2].

Reagents

Reagents are added to the pulp either as solids, immiscible liquids, emulsions, or solutions in water. Feeding arrangements are described later. There is nothing static in the supply trade, new reagents being constantly developed. Samples are usually sent by manufacturers on request, and are tested in the mill laboratory. There are no fixed rules as to type and qualities, though of the thousands which have been marketed relatively few are in universal use. The tendency is for each plant to find the specific combination, sequence, concentration, and staging time most suited to its own ore, and for this rearrangement to be peculiar to the plant concerned. Since the amount of reagent adsorbed is in ratio to the surface area entering reaction, the assay value of the reacting mineral and the fineness of the grind are the main determinants of optimum quantity added per ton of ore.

When the "fines" in the feed are excessive, and particularly if they derive from barren earths and heavily oxidised ore, a dry-looking froth may appear. Frothing agent should then be reduced. Little else can be done to help matters. The risk of high tailing loss and poor concentrate grade, with heavy consumption of reagent, is great when such froths appear, and the only fully satisfactory remedy is selective mining or removal of troublesome material during crushing. De-sliming as the first conditioning operation may be practicable.

When a change from one type of flotation cell to another, or variation in capacity or other adjustment is made, the reagent consumption may be affected. In flotation there is a correct dosage of each chemical. Nothing is gained, but efficiency may be reduced by exceeding it. The process is sufficiently elastic to tolerate some departure from optimum chemical dosage, but heavy over-addition defeats its own ends. Each improvement of acceptance by the air bubbles of correctly conditioned particles reduces the dosage needed. Trial-and-error (Am. "cut-and-try") work should be done in the plant itself, changing slightly one dosage or tempo only, and watching assay results for a long enough period to ensure that any alterations observed arise from the variation, and not from fluctuations in ore constitution or plant operation. In such experimental work sizing analyses of accurate samples are safe guides as to steadiness of grind.

The method used in adding a reagent must ensure thorough mixing in the pulp. Such solids as lime may be added in the fine-ore bin or to ore entering the ball mill, the latter point also serving for thiocarbanilid and soda-ash. Reduction of lime to a slurry before addition is best, since it promotes good mixing. Soda-ash, being readily soluble, can be added with the feed as a solid or dispersed in aqueous solution. When the conditioning plant is exposed to marked seasonal variations of temperature, conditioning times should be adjustable to suit changes in reaction rate. The frother is added last, usually to pulp leaving the conditioning section and near its point of entry to the line of flotation cells.

When lime is used in treating a stable sulphide ore, it can be added as a slurry to the grinding circuit to produce the required *p*H in the pulp. If, however, the ore tends to oxidise on its way from the mine, addition of some dry lime before it enters the bins may be helpful.

If there is any possibility of the reagents reacting with each other instead of upon the ore, they must be used in successive vessels, the collector coming last. If, for example, sphalerite is being activated by copper sulphate, this treatment must be completed before the pulp is transferred to the xanthate conditioning tank, or stoichiometric reaction will take place between copper and xanthate, the reagents will be "killed" and the sphalerite left unconditioned. Similarly, only a trace of copper ion should remain free in the pulp after completion of conditioning or again the collector will be attacked. Ideally, this trace should be less than the copper in a saturated solution of copper xanthate.

Stage addition of collector (starvation) is helpful when making a high-grade concentrate. Stage dosage improves the selectivity of cationic collectors, which are somewhat indiscriminate in their choice and must be helped toward the most eager attracting surfaces if clean separation is to result. Stage addition when the desired mineral is present both as its sulphide and oxide requires delicate control if soluble sulphide is being used in conditioning. A slight excess of this reagent depresses both minerals. It may prove economically sound to float the sulphide mineral before conditioning with soluble sulphide. The sulphidised surface formed on the oxide mineral is fugitive and speed in flotation is essential once it has been formed. Stage addition of soluble sulphide between cells is common practice in the flotation of metal oxides which have been surface-sulphidised.

Where fatty acids such as oleic acid are used, a cool pulp leads to freezing of the reagent. This can be mitigated either by warming the pulp or emulsifying the reagent. Most of these emulsions break down fairly rapidly, and if the pulp is cool enough in the cells an obdurate and "sticky" froth results. This is frozen, and holds all the particles entrained in it with such tenacity that it is difficult to break down or to clean. Special modifications of the oleic acid structure have been developed by means of which partial sulphonation is induced. The oleic acid can then be peptised in a fairly stable dispersed state, and a normal breaking froth in a cool pulp can be maintained.

Storage, Handling, and Feeding of Reagents

Lime is usually obtained locally, in truck or car lots. It should be bought in terms of its content of CaO, and checked for purity. Stones choke metering passages, and vegetable or carbonaceous debris can be a source of chemical trouble. Lime swells on slaking, so storage must allow for this, and also for keeping it dry. The general rule, as for all chemicals, is to use supplies in seniority of arrival. All reagents should be so arranged that the oldest is used first, with handling kept to a minimum. Suitable precautions against the effects of wetting, heat, cold, smelter fume, corrosion of containers, and dangerous leakage of poisons and acids must be taken. Reagents are packed in sacks, multiwalled paper bags, drums, and barrels, each presenting its own storage problems. Some types of reagent can, with bad storage and entry of rain, be ruined. The special problems of cyanide require their own security measures. Any dust escaping to the air during emptying of the container of this deadly poison *must* be drawn away or it may be breathed with fatal results. Similarly, precautions must be taken to ensure that water discarded as mill effluent does not carry poisons or noxious chemicals. Mercaptans are liable to offend in this respect, where aerofloats are used and mill wastes discharge into rivers. Pine oil is pervasive, though less unsavoury. If it runs out of stock of a reagent, the mill may be forced to stop, affecting the whole mining activity. It is essential that the mill manager makes periodic personal check of his stock, and of the condition of the less stable chemicals in store.

Reagents can be fed dry by spreading them upon a belt feeder, the cross-section of the layer being adjusted by means of skirt boards which regulate the width, and a horizontal adjustable scraper to maintain the correct height. The feeder should slide above a flat deck to complete accuracy of cross-section of the delivered layer. The belt is moved slowly by any suitable gear mechanism, and a rotating comb at the discharge end can be used to break up and deliver the lime, etc. A portable feeder (Fig. 225) consists of a reservoir delivering to a disc rotated at adjustable speed. From this the reagent is removed by an adjustable plough. A vertical scraper-shaft in the cone of the reservoir prevents arching of the material. In a third type, an electro-vibrator feeds reagent along a delivery launder from a hopper.

Part of the lime is usually fed wet. It can be drawn from storage to a

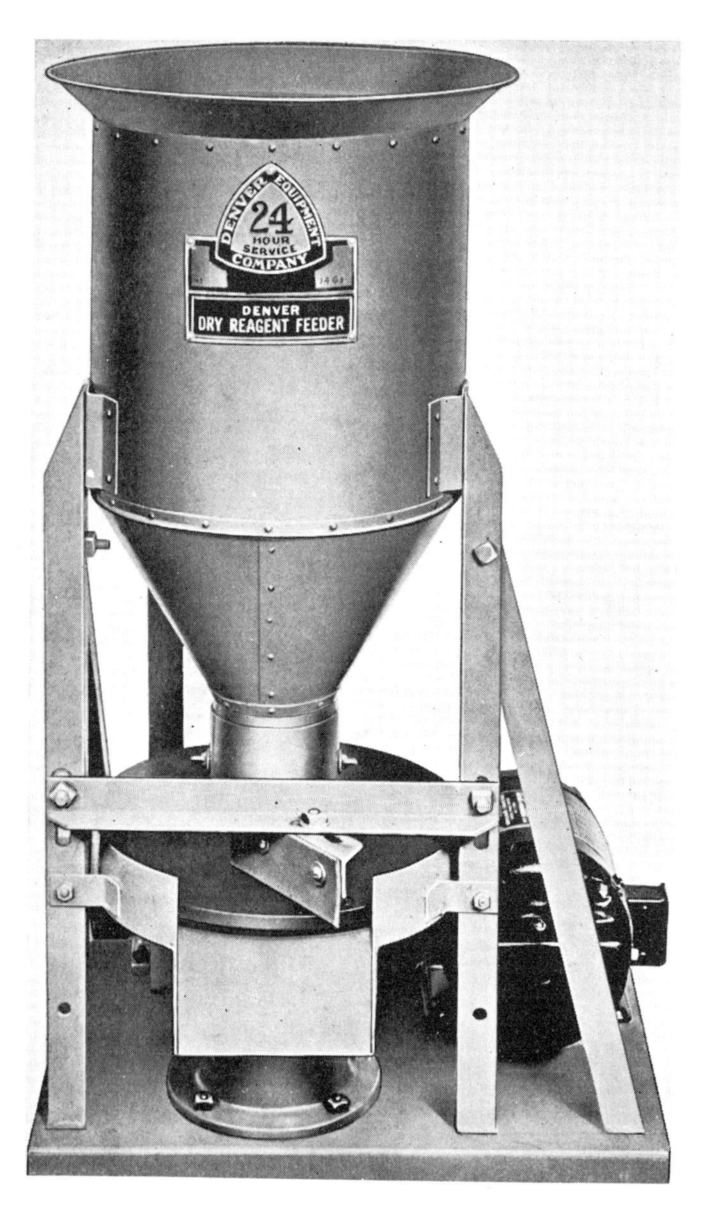

Fig. 225. The Denver Dry Feeder

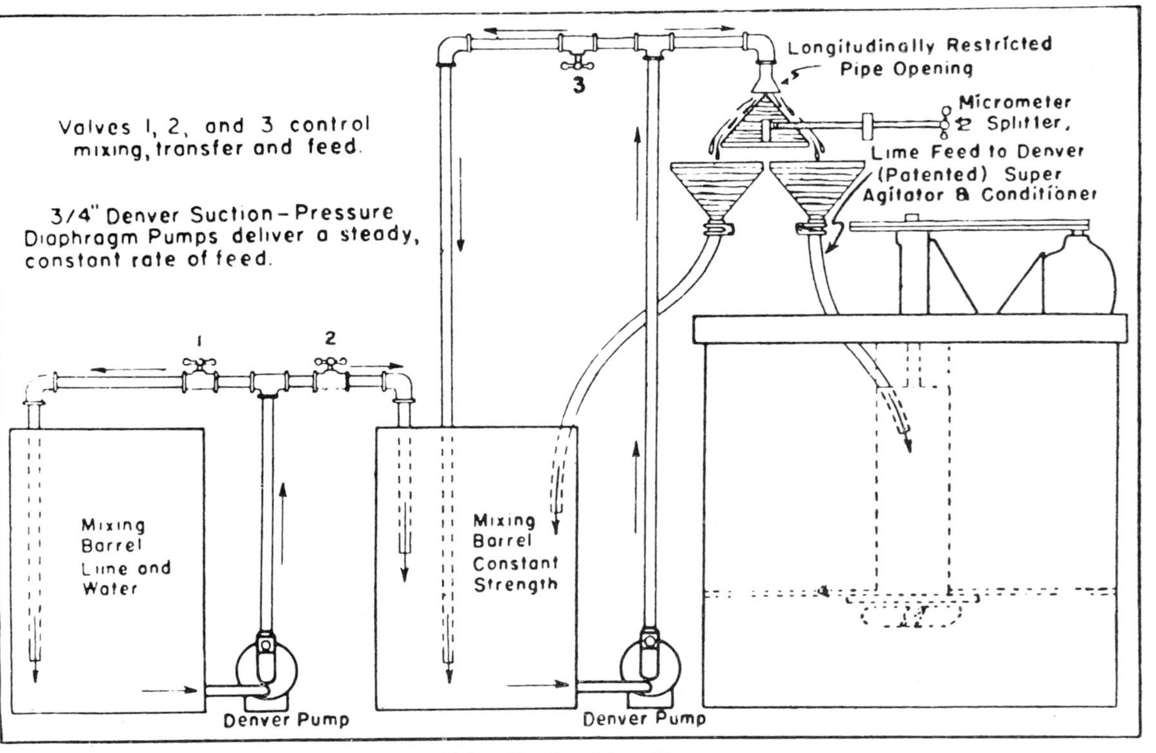

FIG. 226. *Feeding Lime Slurry*

small ball mill closed with a mechanical classifier, to a grinding pan, or to a machine resembling a coffee-mill, run wet or dry. Alternatively, a batch can be slurried as required. The slurry is held in a tank, and stirred by paddles to keep it from settling. In a large operation distribution may be made by tapping from ring mains, the final nozzles being so made as to be easily adjustable and cleared. The essential point is to keep the slurry in movement and to free any choked point expeditiously. A splitting arrangement, such as that shown in Fig. 226, can be used. Finely ground lime and water are measured into the mixing tank and kept agitated by circulation through the pump and valve 1, valve 2 being shut. When the consistency and strength of the slurry is satisfactory, it is drawn to the "constant strength" barrel through valve 2, valve 1 being closed. Here it remains in constant circulation *via* the pump, valve 3 and the adjustable splitter. The lime needed to adjust *p*H is drawn from the splitter to the conditioner. At Hayden, Arizona,[3] an automated plant calcines 200 tons of impure limestone daily, to produce 100 tons of burnt lime and deliver a slaked grit-free slurry to the concentrator.

A liquid such as pine oil, which is almost insoluble in water, is either fed direct at 100% strength, or (rarely) emulsified to a suitable dilution before use. Aqueous solutions such as the xanthates are made up in 5%, 10%, etc., solution strength. Usually the day shift makes a 24-hour supply. Most routine work involving stores issue is best done by day, when all the departments are open, visibility is best, and workmen most alert. The reagent stock is pumped to an overhead tank (Fig. 227), whence it is drawn down by gravity to the appliances which measure it into the pulp. Solutions which are corrosive (sulphuric acid, copper sulphate, etc.) are kept in lead-lined or wooden tanks. It is advisable to lay out the distribution scheme with care, not only to aid distribution of liquids but also to ensure that no reagent feeder will run empty or break down without warning, and that no delivery pipes become choked or displaced without the derangement being readily noticed. Alarm arrangements can be devised to show when the level of liquid is low in a dispensing tank. If the delivery pipes are put where the operators can see the outlets clearly, cessation of flow can easily be observed. With only one reagent missing from a conditioning series, the process is upset, and total failure to float may result until the trouble is rectified. The monetary risk involved justifies reasonable expenditure on the reagent feeding layout. The final metering devices should be readily accessible, as they need adjustment and calibration at regular intervals during each shift. If they can be grouped at a commanding point, the whole work connected with this vital part of process control is simplified. Nothing in a flotation plant more surely invites trouble than putting feeders in dark or awkward places. The author has seen a layout where the shiftsman had to climb a ladder at two-hour intervals and reach precariously to fill a reservoir. In addition, he had several long climbs to make when calibrating, because the only open point where a measuring vessel could be used was on the floor below. The loss in efficiency directly due to bad conditioning was high. Minor points in layout are the desirability of a good working light, an equable temperature so as to obviate readjustments due to changes in viscosity, and the choice of feeders which do

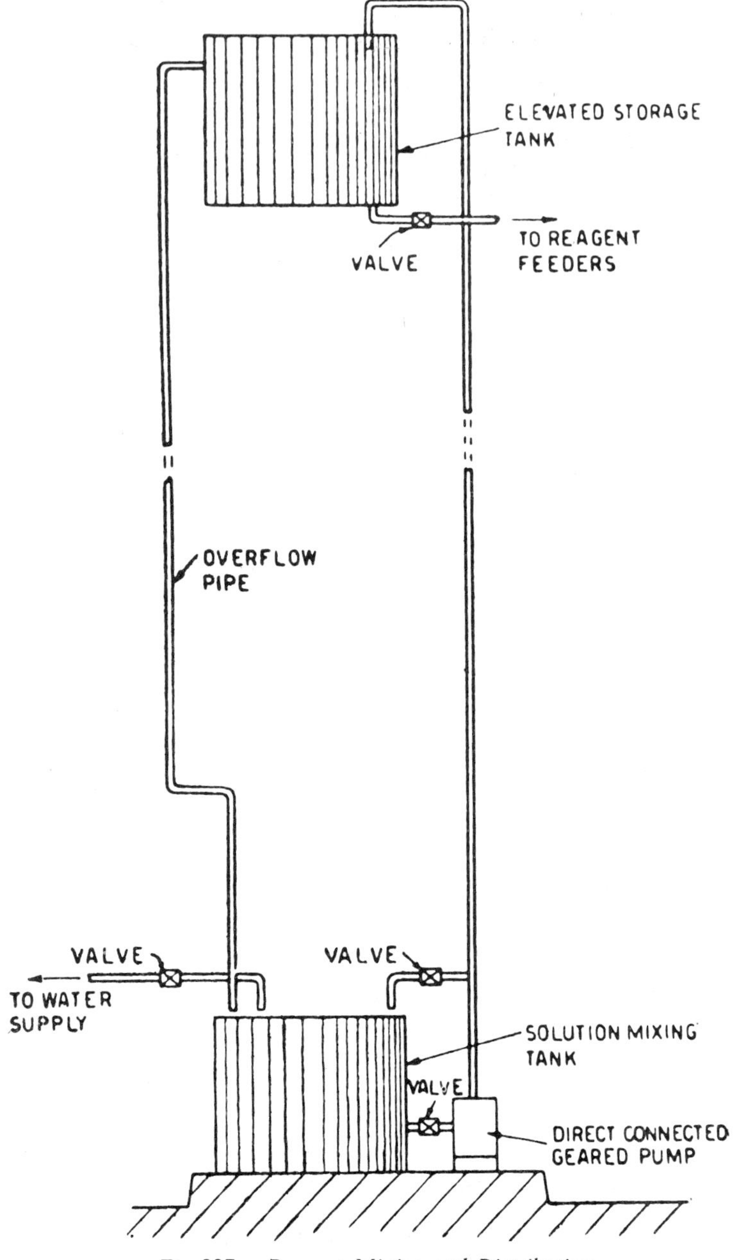

Fig. 227. Reagent Mixing and Distribution

not accumulate stale and decomposing chemicals. If highly corrosive reagents are in use, the materials of construction with which they make contact must be suitably chosen. This applies to tanks, pipes, or other channels, pumps and any supports, gears and bearings liable to be splashed.

An emergency feeder can be improvised from an empty oil drum fitted with a needle-valve tap. This is also useful when experimenting with variations in points of reagent addition, but needs constant care, as rate of delivery is uncertain. It is also handy during emergency working, or for testing new reagents which otherwise would contaminate the regular delivery lines.

Among the oldest mechanical feeders were flat-faced pulleys which dipped into the bath of reagent, usually pine oil. Any desired width of the film lifted on the pulley face was wiped off by an adjustable scraper and run to the process. Pulley speed was variable. A variation was the grooved pulley, with scraping fingers working in the grooves. Since the viscosity of oils varies inversely as temperature, heating arrangements in the bath are desirable where this type of feeder is still in use. Electro-magnetic pumps have been used, but are obsolete. A simple arrangement, used where close accuracy is less important than great simplicity, is a syphon tube mounted on a wooden float in the reagent tank (Fig. 228). It is made with 8-gallon and 50-gallon tanks, is non-mechanical, and can be stood anywhere. A heating lamp is desirable for viscous reagents. Since the syphon rides on the float, the head is constant. The micrometer screw adjusts the fall through the syphon and hence the delivery rate.

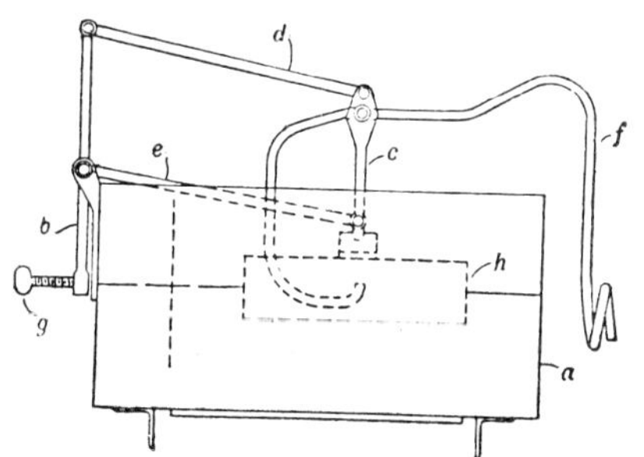

Fig. 228. *Syphon Reagent Feeder*
a, *Tank;* b, c, d, e, g, *syphon adjustment;* f, *syphon;* h, *float*

The most widely used reagent feeder (Fig. 229) has small buckets mounted at varying horizontal distances on a vertical revolving disc. A variable gear allows changes in rate of rotation, and a retractable tipping device governs the number of buckets tipped at each revolution. In a variation, buckets are added or removed from the disc as required. The tipping bar is raised or lowered to govern the amount spilt at each contact. Large and small buckets are available and for oils dipping nails can be substituted. The rotating buckets keep the solution stirred. Corrosive liquids are handled in fibre or alloy buckets. The arrangement is simple, cheap, sturdy, and precise.

Evaporation may cause buckets to clog, so periodic cleaning is desirable. A number of discs can be driven from one shaft, each dispensing a different reagent from a separate compartment in the tank. In feeding copper sulphate, which is very corrosive, buckets deliver water on to crystals in a wooden tank containing a saturated solution, displacing an equivalent quantity of saturated solution which overflows to the plant.

For large deliveries a circulating flow can be used, control being exercised by drawing off reagent as required *via* a V-notch weir. When oils are used late in the flow-line, so that they cannot be emulsified in the mill, various emulsifying reagents and mechanisms are employed. The mill hands are not chemical engineers, and must be taught the importance of careful reagent mixing and control. Locking covers can be put on reagent feeders to reduce unauthorised interference.

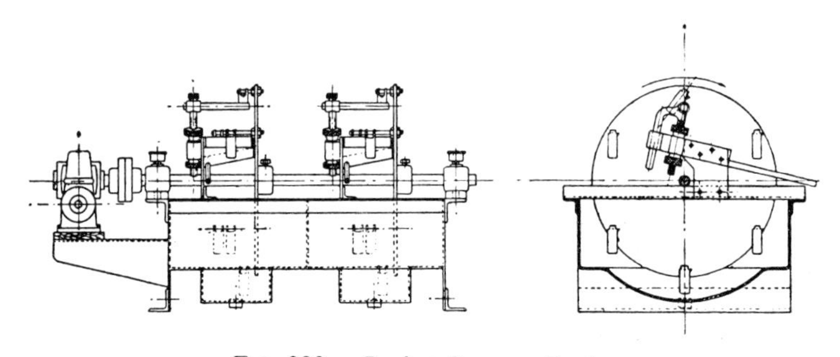

FIG. 229. *Bucket Reagent Feeder*

A table of instructions should be typed and kept at each calibration point, showing the mill hand what quantity in ml/min is to be delivered into a graduated cylinder kept for the purpose of checking. A robust seconds-hand clock in a glass-fronted case is a better investment than a fragile stop-watch for timing purposes during this check. The table should tie ml/min at normal reagent strength to lb/ton of ore, in terms of various settings of ore feed rate to the flow-line. Check should be made hourly. Use can be made of the tonnage rate given by weightometers, etc. One point which can be dealt with by an experienced shiftsman is change of collector rate with change of richness of ore. If, for example, the norm is 0·027 lb/ton K.E.X. for a 3% metal-sulphide, a rich patch of ore needs a higher rate of delivery to coat the extra surface offered, owing to the combined effect of richness and over-grinding. Ability to recognise changes in richness of feed is an important quality in a shiftsman, and is acquired by thinking of surface areas as well as of tonnage rates.

It is probably a lesser evil to overdose with collector than to starve, but reagent is wasted and concentrate grade lowered. One reagent which *must not* run wild is that responsible for maintaining the correct *p*H. If hydrogen-ion control is lost, the whole subsequent treatment may be upset. *p*H is checked at short intervals by colorimetric or electro-potential methods at the classifier overflows in a small plant, and continuously by automatic methods where it is of critical importance. Since the pulp is copiously aerated during flotation, *p*H often drops between head cell and final scavenger. This should be checked periodically by the mill laboratory. Any reagent can be fortified by further additions at suitable points along the flow-line. It is sometimes good practice, in differential sulphide flotation, to add a "starvation" dose of xanthate in the first conditioning stage and to reinforce this somewhere along the rougher section of cells. The overflow of pine oil with froth in the earlier sections may starve the scavenger cells of this reagent, and this should be checked from time to time.

Increasing attention is given in newer plants to automatic control of reagent addition, and to the checking of its concentration and action along the flow line.

Flotation Machines

The design of the cell in which the concentrate is floated must facilitate as many as possible of the following duties:

1. Reception and aeration of the pulp without allowing settlement of solids.
2. Discharge of impoverished tailings after aeration has removed a mineralised froth.
3. Avoidance of short circuiting of pulp from entry to discharge without being worked in the cell.
4. Search of the full pulp volume with N-bubbles of suitable sizes, thoroughly dispersed, and in adequate quantity for froth-column concentration.
5. Provision of a zone where a quiet blanket of mineralised froth can form and from which gangue can drop back into the pulp.
6. Discharge of "floats" and "sinks" by separate channels.
7. Controllability for pulp level and height of froth column.
8. Aeration without letting too large bubbles, or "bursts" of air create disorder.
9. Provision for easy re-start after mechanical failure, without "sanding up" of mechanical parts and discharge orifices.
10. Efficient use of power, mill space, and impellers.
11. Easy maintenance with no odd corners where wood, debris, or lime scale can accumulate and become a nuisance.
12. Provision for quick and easy changing of feed and intercirculation channels in the line of which the cell forms a unit.
13. Ability to cope with maximum-sized sinking particles in the feed without risk of accumulation and choking.
14. Working adjustment between new feed circulating past the entry-point of air and of pulp recirculating inside the cell.
15. Arrangement for periodic bottom discharge of accumulated sand too coarse to flow over the tailings weir.

Usually cells are arranged in series, each cell in a "bank" receiving the tailings from the one preceding it. The height of the froth column is determined for each cell by adjusting the height of the tailings overflow weir. Since the overflow lip for the froth is fixed, the difference between this and the weir level sets the height of the frothing column, subject to it being possible for the froth to reach that height. This is largely determined by the availability of aerophilic mineral particles. The new feed enters the rougher section of a typical bank. The froth column is kept high at the head of the section and lessened from cell to cell by progressive raising of weir height. The scavenger cells have but little mineral with which to stabilise a froth, so their weirs are raised till, by the end of the scavenger section, the pulp is almost spilling over to the froth launder. This end setting is part of the policy of "pulling the cells hard" so that every particle showing a tendency to float is removed. In the cleaner section the pulp is thinned and the weirs kept low, so as to maintain a thick layer of froth and obtain the maximum possible cleaning action.

The alternative to individual tanks is a series of long troughs, each containing an appropriate number of aerating and agitating devices. A single weir controls each trough. Current practice tends toward elimination of individual square cells, each with its separate impeller, feed entry and tailings weir, in favour of long tanks with unmechanised froth overflow. This trend is marked in roughing and scavenging circuits where large tonnages are handled. The Denver "free-flow" cell (Fig. 230) illustrates this simplification.

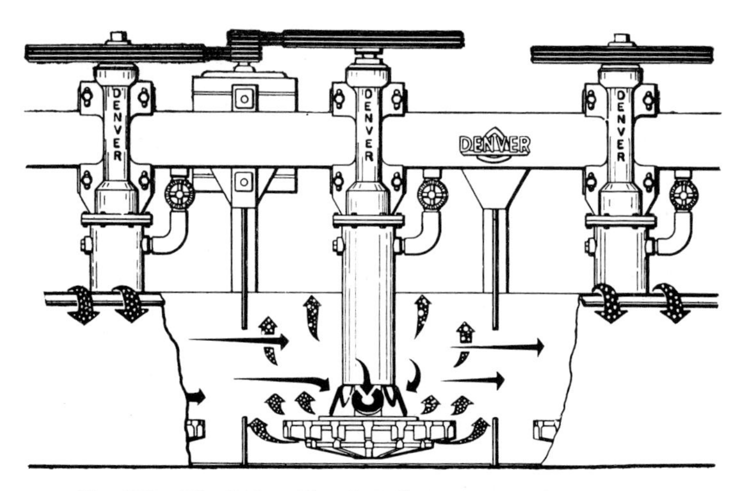

Fig. 230 *The "Free-Flow" Cell* (Denver Equipment Co.)

The four main types of cell are:

(*a*) Mechanical (self-aerating or supercharged, and mechanically operated).
(*b*) Mechanical (air blown in, agitation mechanical).
(c) Cascade (air entrained).
(d) Pneumatic (air blown in, air agitated).

Of the numerous cells manufactured, a few have been selected for description. Between them they incorporate the most important features of all cells. Developments in cell design are continuous, and the reader desiring detailed information concerning a particular type is better served by the manufacturers' bulletins than by a somewhat static text-book.

The history of cell design and development has been described by the author elsewhere[1]. The hydrodynamics of flotation cells are still inadequately understood despite much attention at research level[5]. Satisfactory adherence of the particle to the bubble depends not only on chemical preparation, but also on the geometry and dynamics of the cell. So many variables are perforce left uncontrolled in industrial flotation that a scientific basis for comparison of performance between various types of cell, with varied settings and working conditions, is apt to be misleading. Operating skill is by no means a negligible factor and this varies between shiftsmen.

Cells of type *a* include the Denver, Fagergren, Humboldt, Massco and Knapp and Bates, and several other proprietary makes which, while incorporating special variations, conform in the main to a few general principles. Tanks are square, round, or long (troughs), in the latter case having multiple impeller units.

Fig. 231. Denver Supercharged Flotation Cell

The standard Denver cell (Fig. 231) has a square tank. The illustration shows this cell with a bonnet above the stand pipe (A) through which air can be blown down to the vortex near the hub of the impeller (G). At this hub there is a partial vacuum which allows air to be drawn down to the impeller blades (B) where it mixes with the pulp entering from the previous cell in line *via* the feed pipe. This pulp has fallen over an adjustable weir (D_1) and, in due course the tailings from the cell will pass under a baffle and over (D_2) to the next cell in line. Sand which is too coarse to rise to this discharge weir is by-passed through the cell by means of the sand-relief ports (E). Weirs are raised and lowered by various mechanisms, operated in this instance from the rods (F). When the stand pipe is arranged to receive pressure air the cell is said to be supercharged. When the pipe is open the cell relies solely on air drawn in, which varies with speed and condition of the impeller. A cleaning rod can be used to clear the sand-relief ports (E). At the side of the cell is a pipe which can be connected to the middlings return system. Two of the stationary baffles (H) are shown. These receive pulp flowing from the tips of the impeller-blades and past the stationary diffuser (I). At (J) is a re-circulating port, called the "key stone". In some machines this can be closed by a sliding plate, thus controlling re-circulation in the body of the cell. In the author's experience better control is maintained when the keystone aperture is permanently closed, and recirculation is confined to pulp drawn to the zone between impeller tips and stationary baffles. A cross-section through a standard Denver cell is shown in Fig. 232. Here the stand pipe is open and the middlings pipe is shown connected to a return launder. A crowding baffle (K) is depicted. The purpose of this is to push the rising bubbles forward toward the skimming paddle (L) which removes a froth layer from the cell. These baffles can readily be improvised where it is desired to hasten removal of the risen material. They reduce the accumulation at the back of the cell of mineral which fails to move over to the skimming zone, and which may become over-oxidised by prolonged exposure, fall back into the pulp and then be lost owing to its changed surface condition. The pocket or spitzkasten, (M) below the skimming paddle is used in older Denver machines to provide a quiet zone where the froth crowds out unwanted material for return to the pulp. It is not much used in new installations, as better results can be obtained by use of a rougher-cleaner-scavenger arrangement of cells, in connexion with which a lively froth is removed as soon as possible for any required re-cycling. By taking advantage of the ease with which a froth from any section can be returned counter-current to join the feed with an earlier cell in the line, upgrading can be more accurately controlled than by the use of spitzkastens. Near the bottom of the cell is set a horizontal impeller which acts as a centrifugal pump. The impeller is driven by a vertical shaft from an overhead motor at a peripheral speed not less than 1350 ft/min and not usually more than 1800 ft/min, though up to 2000 is occasionally used. There is an important connexion between peripheral speed and efficient aeration. Several factors influence choice of speed. The feed to the cell is introduced near the centre of the impeller. This feed may have come from the conditioning section, from the previous cell in the bank, or it may be a middling product overflowing as froth and being returned from cells further down-line. In addition, any desired amount of pulp already inside the cell can be drawn down to the vortex at the centre of the impeller

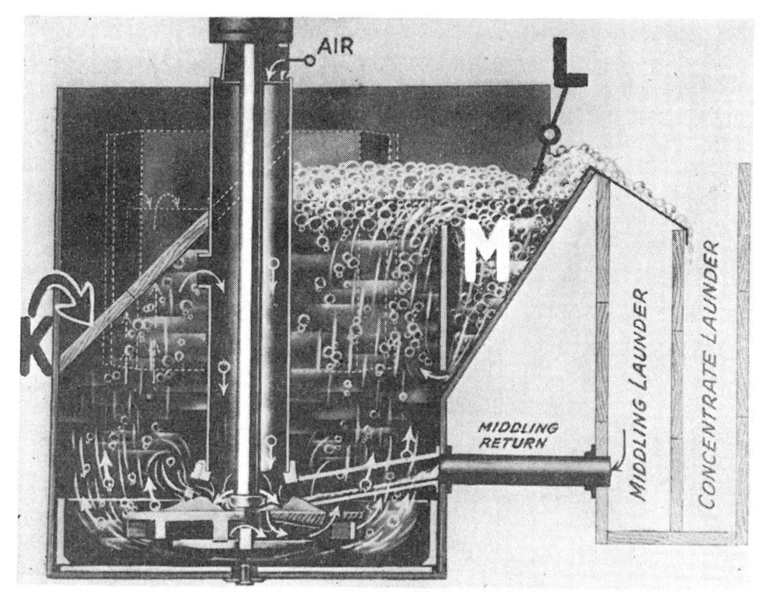

Fig. 232. Cross-section, Denver Flotation Cell

nd centrifuged outward. Stationary baffles check the swirl of the outflowing ulp. Some pulp may creep back along the bottom of the cell toward the entral vortex. All of the pulp in the swept area is caught up on the impeller lades and accelerated to its terminal velocity (the peripheral speed of the npeller less slip) before being abruptly arrested and turned over from orizontal to vertical flow. The amount of slip is partly governed by the learance between the impeller and the stationary hood above it, and to a ess extent by the shape and number of blades and their deformation with rear. The other slip-factor is the ratio of pulp to air being pumped through ne impeller. If all available pulp passages are open, there is less chance for ir to enter. The feed pipes are larger than is necessary for the pulp they arry (unless the cell is being grossly overfed). If the internal pulp-port is hut, the rest of the feed openings (new and return) and the standpipe sur- ounding the impeller shaft are available for drawing in air. The adjustments herefore centre on balancing the aeration by

(*a*) regulating clearance between impeller and hood;
(*b*) regulating internal circulation of pulp;
(*c*) regulating impeller speed.

practice the gate adjusting (*b*) is usually kept shut, and the makers' speed- tting is used. For a float requiring extra aeration, impeller speed may be creased above 1800 ft/min peripheral, but wear increases as the sixth power velocity and power consumption with it, so this is an expensive alteration.

Alternatively, special bonnets can be fitted to the standpipes. This allows extra air to be blown down into the pulp, the cell then being "super-charged" Two types of impeller (Fig. 233) are supplied for this machine, the "receded-disc" being preferred with finely ground ore. The hood clearance is adjusted by raising or lowering the impeller. Clearance must give ample space for the largest particle to pass without touching (say a minimum of $\frac{1}{10}$ in. and should be adjusted with the machine running and an ammeter in circuit A sudden rise in current as the gap is closed shows that it is too narrow Surging of the pulp and irregular arrival of quantities of big bubbles shows that the gap is too wide. The hood and leading upper edge of the impelle sustain heavy abrasive wear from the passing pulp and are often lined with rubber. Abrasion cleanses mineral surfaces and can be helpful in the flotation of particles inclined to become coated with slime. Since the impeller

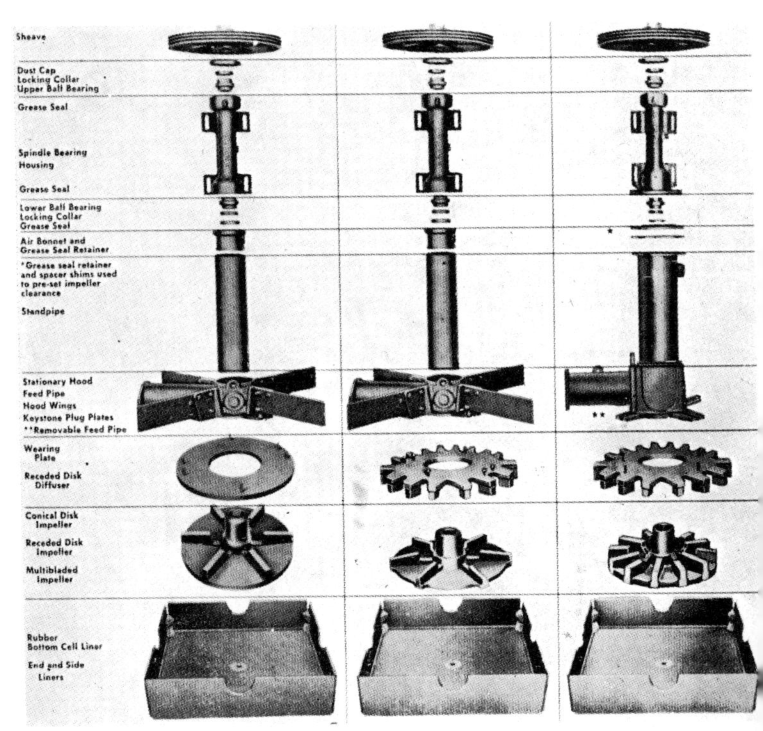

Fig. 233. Types of Impeller Assembly

wear down, adjustment should be a regular feature of the mill maintenan programme. It is good practice to reverse the rotation of impellers monthly intervals to equalise wear, and so prolong their life. Once t rubber wears down and the underlying metal is exposed, further wear very rapid. Figure 234 shows two receded-disc impellers, the rubb

Fig. 234. Denver Flotation Cell. Discarded Rubber Impellers

coverings of which have been worn to (or beyond) discard point. Poor flotation would be obtained with impellers in this condition.

In the event of power failure sand settles above the hood, so the cell can be restarted immediately power is restored. The height of the tailings weir is adjusted by slats or by a slide. The vertical difference between the tailings overflow and the bottom of the mechanical skimming paddles used to remove froth is the height of the frothing column. The greater this is made, the more severe is the cleansing and up-grading effect. Although the froth normally rises over the whole horizontal area of the cell, it can only be removed at the skimming side. This leads to the accumulation of ageing mineral in places not much affected by skimming and also, in the case of difficulty floated material, to the falling back of particles into the pulp. Where copious froth is needed to get the float out before the weakly aerophilic minerals drop back, special launders can be had which gather the froth more effectively. In the flotation of an abundant concentrate from a minor gangue in such cases as treatment of coal or phosphate, scrapers can bring out the froth from the front half of the cell. At the tailings end provision is made for the release of accumulated oversize too heavy to flow. This should periodically be discharged from the circuit before it begins to impede smooth working. One shiftsman controls up to 75 cells. He uses a vanning plaque, or inspects a miniature shaking table working on a cut from the bank tailings, to provide a visual check on performance.

Variations on the general Denver design include the Humboldt (Fig. 235) and the M.S.S.A. Fahrenwald, in which pulp recirculates through apertures in the hood. The size of the cell is determined by the economics of maintaining a given volume of pulp in suspension and of transporting it through the cell by means of the impeller, a centrifugal pump not necessarily being

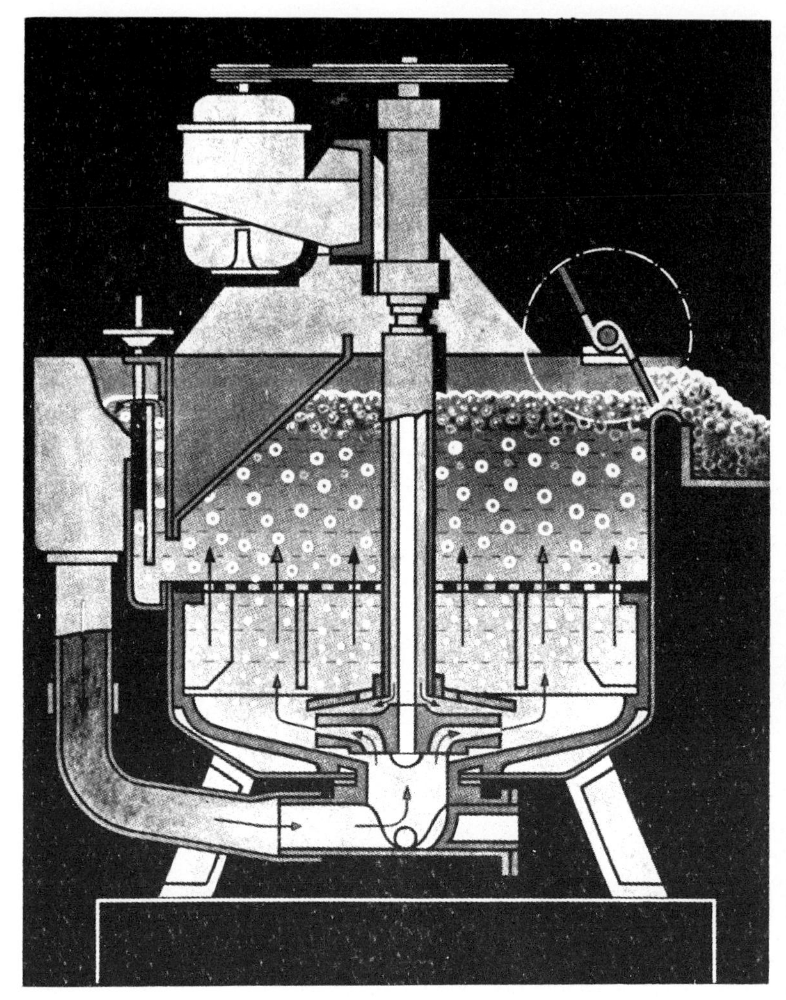

Fig. 235. Humboldt Flotation Machine

designed to be an efficient pulp-making device. The larger the individual cells, the easier a bank is to control. Against this must be set two weighty considerations. If one large cell breaks down, the process is far more upset than if one of a number of small cells is out of action. Secondly, the larger the cell the more force must be applied to keep its pulp from settling, and this is obtained at the cost of power and wear.

The Fagergren cell (Fig. 236) has a rotor and stator instead of impeller and hood plate. The power required to "grind" air into the cell is much higher than for a Denver of equal capacity, and the aeration is far more voluminous.

The Fagergren can therefore float reluctant and slightly aerophilic particles of the sort which slip back from most froths. The older Fagergrens have single assemblies in square-section cells, but the trend is toward a long rectangular trough, divided into sections each containing a rotor and stator. The feed enters below the first partition. Tails go over partitions from one compartment to the next and the froth level is adjusted at the end tailings weir, which has a sandgate below for removing accumulated coarse material.

Fig. 236. *Fagergren Cell* (Wemco)

The cell lacks one great advantage possessed by the Denver, positive pull from other cells to the impeller without a special pumping system. Any moving of pulp from one point to another therefore requires an independent pump, unless gravity flow is possible.

Cells of type *b* are typified by the M.S. Sub-A (Mineral Separations Sub-Aeration). Here pulp enters below the hub of the impeller, which is an agitating device only. Air is blown in at low pressure by some such device as a Rootes blower and is beaten up with the pulp. Froth is skimmed off by paddles, and tailing overflows a slat weir and goes to the next following cell. Again there is no positive suction for "shunting" pulp inside the bank of cells.

Fig. 237. *The Agitair Impeller*
(International Combustion)

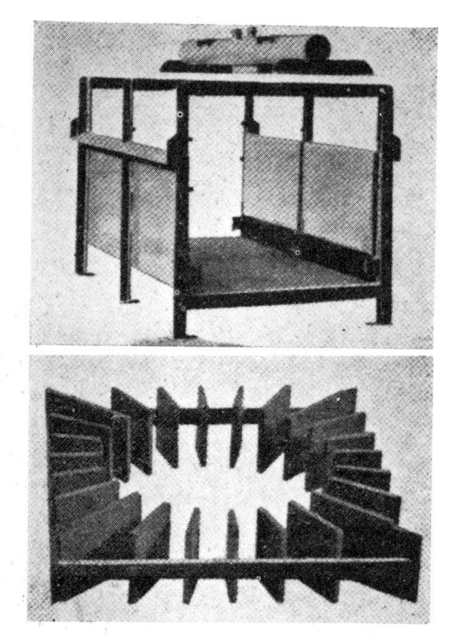

Fig. 238-9. *Agitair Tank Assembly and Baffles*
(International Combustion)

In the Agitair cell, a long trough divided into several compartments, the pulp moves from feed end to discharge. In each compartment an impeller (Fig. 237) rotates inside a baffle system (Fig. 238-9). Air is blown in at from $\frac{1}{4}$ p.s.i. to $1\frac{1}{2}$ p.s.i. through a hollow shaft (Fig. 240) and sheared into bubbles as it enters the pulp. The volume of entering air is controlled for each compartment separately. Impeller speed can be varied from 900 ft/min (peripheral) to 1750. Adjustment of speed and air entry permit operational control of the froth column, so that it flows counter-current to the pulp. The assembly is shown in Fig. 240. These cells are modest users of power

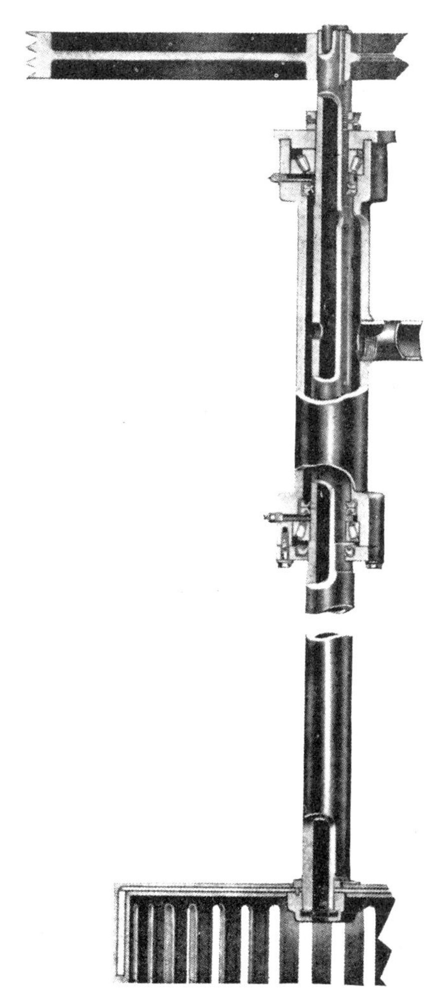

Fig. 240. The Agitair Impeller Assembly (International Combustion)

and produce a copious froth. This has won them preference in mills handling ores of poor floatability, which require a big volume of froth to help their mildly aerophilic particles to overflow. They have, indeed, found considerable application in flow-sheets once served by cells of type *a*. They lack the ability to effect positive displacement of slime which is a valuable feature of the violent impeller action in cells of type *a*. Against this must be set the fact that improvements in desliming during the conditioning period have today reduced dependence on this action.

Type *c* cells (cascades) use air entrained during the fall of splashing pulp into the cell to effect flotation. Their use is largely confined to simple home-made devices at the tailings end of a process where finished pulp is running down to waste and can be cascaded.

Type *d*, the completely pneumatic, is one of the original forms of flotation cell. In the Southwestern cell (Fig. 241) air is blown down vertical pipes from a distributing header, and agitates the pulp in a long trough. The froth rises to a quiet outside zone whence it overflows while the pulp, which has risen to a top baffle under the pumping influence of the air, drops back and is lifted again a little lower down the trough, thus slowly moving in spirals from feed to discharge. These cells have a more voluminous and fragile froth than those of type *b*. Maintenance is low, and the only power cost is for low-pressure air. If internal recycling is needed, an air-lift pump must be used. In the Britannia cell, a much tighter froth is produced, resembling that of cells of type *a*. This is achieved by making the cell several times as deep as the ordinarily shallow pneumatic cell (8 ft to 10 ft). The back pressure of the extra hydrostatic head causes smaller bubbles to be generated. An advantage is that for the same floor space several times the cell capacity is available. Another is that with the larger volume of pulp passing through the number of collector-coated particles on the bubbles is much higher. This aids the cleansing effect in the froth column by increasing the competition between particles.

A newcomer in the field of coal flotation is the cyclo-cell[6]. Agitation is effected in submerged vortex chambers in which a high-speed jet of water is formed into a hollow cone through which air is sheared into the cell in tiny bubbles.

Cell Economics and Design

Comparison between mechanical and pneumatic cells should be based on metallurgical performance when treating the same pulp in parallel streams. Several large operators have standardised their cell type after mill tests of this sort, but even here results are suspect, since much depends on the shiftsman's skill and on his prejudices. The author uses laboratory Fagergrens, K.B.'s, Denvers, and cells evolved in the Bessemer Laboratory for student instruction, and finds the first three about equal metallurgically, while those which he helped to design give the best results in the hands of learners. Perhaps if he had helped to evolve one of the other types it would show the best results in similar circumstances. An operator trained on one type of cell will prefer it to others. Any cell which sweeps the pulp with an adequate volume of N-bubbles of the right size, and provides a quiet place for forming and cleaning the froth column, is a good cell. The further criteria are that it shall be

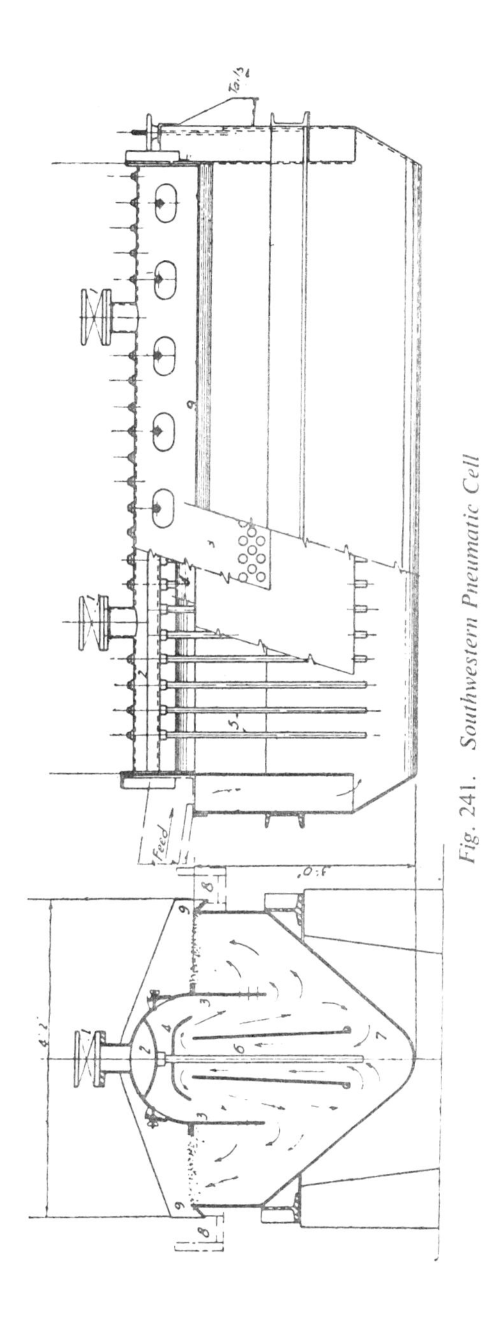

Fig. 241. Southwestern Pneumatic Cell

easily and cheaply maintained, thrifty in power consumption, and easily fitted into the circuit. On the criteria of power and reagent consumption, product grade and overall costs, pneumatic cells probably are best, but the tendency in older plants is for them to be replaced by mechanically agitated machines.

The pneumatic has not the transporting power of the mechanical cell, nor can it handle such a coarsely ground pulp. If sliming conditions are present, the positive agitation of an impeller has a scouring effect not possible with simple pneumatic agitation. For floating a single mineral from a simple ore, particularly where vigorous frothing is called for, pneumatic cells have their uses, especially when the plant must be inexpensive. Usually with expansion of treatment rate, the flotation section is changed to use motorised cells. When differential flotation is practised, the mechanical cell is dominant. With a complex ore requiring running changes in the number of cells roughing, cleaning, and scavenging, a machine which has positive pumping action (such as the Denver) is often favoured. It is only a minute's work to remove or insert a plug or two in a double launder, and the section's capacity can thus be trimmed to suit a change in the ore—a convenience in a medium-sized plant treating 500–1500 tons/day. A simple multiple-launder system is shown diagrammatically in Fig. 242. Nine cells are served, and new feed

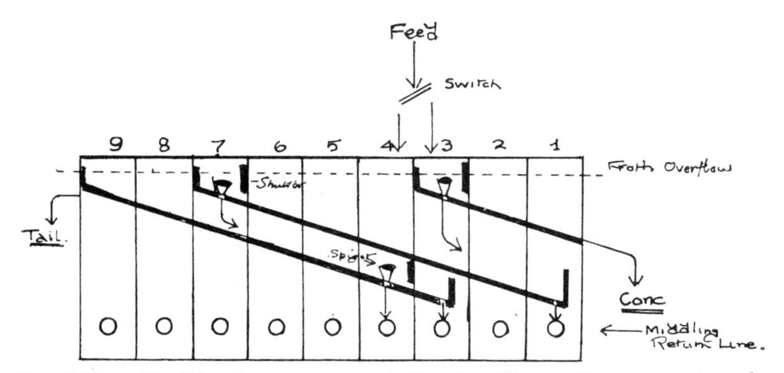

Fig. 242. *Flexible Arrangement of Rougher-Cleaner-Scavenger Launders*

can be brought to either No. 3 or No. 4, while the roughing section can contain up to five cells, ending at No. 7, or be shortened at either end. If the spigot at 7 is removed and the shutter on its right dropped, the froth from No. 7 cell falls with the scavenger froth from 8 and 9 to the middlings return pipe at either cell 3 or 4, according to whether the spigot and shutter for cell 4 are on or off. Thus, a lean ore which fails to produce sufficient crowding of the froth when the cleaning section is working on three cells can be strengthened by diverting new feed to No. 3 and the froth from No. 3 to the rougher concentrate launder. With larger operations, the maximum capacity for each section is usually built in and pulp-shunting is not necessary. The technical criterion is the metallurgical balance—the grade of concentrate as well as the percentage of value recovered being important. Factors in maintaining full running time and smooth working are simplicity of the machines and the ability of local labour to handle them. The corners of cells trap ground-up wood and other organic materials, which decompose and release depressants. Regular cleansing often improves recovery. In some plants a monorail is installed above the bank, by means of which a unit cell can be taken out of line for repair and a spare one can be dropped in. A circuit using substantial quantities of lime may be troubled by precipitated scale, which builds up and clogs pipes and other passages unless these are regularly serviced.

The size of the cell selected depends on the total flotation time required and the space and handling arrangements in the mill. In terms of capacity a cell has four dimensions—length, breadth, height, and dwelling time (rate of throughput). Part of the time is transferable to the conditioning section, and part can be varied by changing the number of cells in line. This latter element of capacity is based on the aeration requirements. The individual cell's length and width depend on the agitating and searching power generated by the impeller, which at n revolutions/minute is a function of the square of its diameter. Power consumption in unaerated pulp is a linear function of this diameter, but in practice this dimension is limited, owing to the high rate of wear should peripheral speed exceed some 1800 ft/min. The greater the swept area, the more slowly must the impeller turn. The cell's overall plan area can best be expanded by using troughs instead of square cells. Height can increase volume of pulp held, but the lifting must be done by the impeller normal to its plane of rotation. Since this is done across the baffles required to check surging, this is an inefficient mechanical design in terms of pumping action. Power tests favour the mechanical cell which works in the shallowest volume of pulp, and as most of the particles in that pulp are barren, the less power used to move them around, the better. Long troughs are suspect for similar reasons, save where they handle large tonnages. Most of the separating work is done in the first few feet, the balance of the distance travelled by the non-floated particle between entry and discharge being an insurance against bypassing of the impellers by values. Such bypassing can be more accurately controlled, as can the other operating conditions, in a bank of individual square cells than in a long tank. Here are the elements of compromise in design for a specific case of plant design.

Finally, there is the probability that new reagents may be developed which will shorten the flotation time. When this happens, provided the mine does not increase the tonnage sent for treatment, some of the flotation capacity is no longer needed. A single cell is more flexible than a trough, whether it is being tacked on to an under-aerated bank or removed to reduce floating time.

Agglomeration

This was earlier defined as the flotation of coarser particles on shaking tables and conveyor belts. Air is used chiefly in minute bubbles, which hold aerophilic particles loosely together. Thus, aided by edge-adhesion between hydrocarbon-coated particles, glomerules are formed, either around an entrained bubble of air or as a watery aggregate. This aggregate is in composition relatively large and light, so it works out to the top of a horizontally moving band of feed on the shaking table. It then rolls down over the riffles and is removed nearly opposite the feed entry.

Conditioning is usually carried out in a thick slurry of deslimed feed, fairly closely classified as is customary in table gravity working. Conditions are much the same as for flotation, *plus* the use of thickish oil to help to smear the solids, instead of the frothers used in cells. The slurry is aerated by being cascaded and tumbled in the air as it falls to the separating surface. The heavy-metal sulphides or other floatable minerals now form rounded, flexible glomerules which roll across the table to the gangue withdrawal launder, while the relatively light gangue works over to the heavy concentrate discharge end. Desliming is important as the slimes would collect in the oil interfaces

Fig. 243. Holman-Michell Flotation Table

and interfere with conditioning. The feed size range can be as high as $\frac{1}{8}$ in. and as low as 200 mesh, but the process is usually applied between the limits of 6 mesh and 65 mesh, over a three-to-one range between maximum and minimum particle in the feed. The use of a thick pulp saves reagent, and aids in the smearing of oil. Dilution is performed at delivery to the table, and cascade action is helpful in dragging in more air. The glomerules may burst against sharp riffle-edges but will then stream as a skin-float down to the tailings launder. The Holman-Michell flotation table is illustrated in Fig. 243. Air is blown into the pipes which cross the table deck, at a light pressure. In the operation shown, sulphides are being agglomerated and removed from a rough concentrate of cassiterite.

When separating a concentrate on conveyor belts, as practised in phosphate treatment, the thick conditioned pulp is delivered to the middle of the belt and is turned and ploughed as the belt passes under stationary fingers. Sprays are directed on to the pulp and agglomerated phosphate aggregates wash over the edge of the belt while the gangue remains wetted and travels to the discharge end.

Middlings

Good technical and economic control require intelligent treatment of the middlings. The grinding work should aim at producing just enough liberation to float the maximum economically recoverable tonnage of concentrate at the coarsest floatable mesh. This concentrate will then consist of fully, largely, and moderately liberated particles. The first-named can be cleaned up to a high grade of concentration but the latter, since they are still locked with gangue, must either be taken as they are or given further liberating treatment. In a small plant highly specialised treatment is unjustifiable. All that can be done in the absence of further grinding is to shake down the entrained gangue in one or more cleaning sections. This gangue owes its flotation in the scavenger cells to random contamination which is scrubbed away or dissolved during further recirculation of the froth. The middlings, however, accumulate until their final removal either in the concentrate or the tailing, according as the cells are being worked. The second alternative, possible in rather larger plants, is to pass the middlings from the scavenger cells to a special regrind circuit, and thence to return them to the conditioners. By this means further liberation is achieved and the work of upgrading can be done gently. The scavenging section which floated the original particles will trap their values more easily after regrind. Return of scavenger concentrate

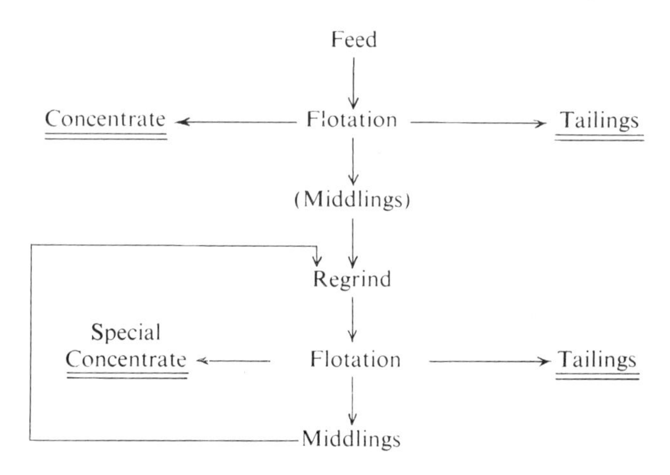

Fig. 244. Flow-sheet—Special Re-treatment of Middlings

to the main grinding circuit for further liberation is usually unsatisfactory, since this has already passed these particles out, and is therefore not adapted to the finer comminution they need.

A third alternative is rarely used except in a big plant. It is only necessary if, in addition to incomplete liberation, there is some special feature about the middlings which makes it desirable to remove them from the main circuit. In such a case the scavenger middlings might go to a regrind circuit and independent flotation (Fig. 244).

The terms "small", "medium", and "large" plants in the foregoing context are not defined. A 200-ton plant making uranium concentrates worth £2000 per ton would justify machinery that would be uneconomic in a 100-ton concentrator producing a £100 per ton concentrate.

Control and Modification

However well a plant is run, small changes in performance are continually occurring. It is part of the work of the shiftsman to detect such changes and to trim his section of the flow-line accordingly. The vanning plaque is commonly used to test the pulp. Continuous check on tailings can be given by diverting a "cut" over a miniature shaking table. The sense of touch sometimes serves to detect a change in coarseness of classifier overflow. The colour and appearance of the froth soon become familiar to a trained observer, who will see at a glance when things are beginning to go wrong. His work may be aided by the provision of fluorescent or monochromatic light, or the exclusion of daylight from the flotation section. It is often possible to devise a simple "spot-assay" method which the shiftsman can use to check the issuing products. A colorimetric assay for copper, accurate to within about 10% of the copper content, can be completed in a few minutes. In the case of a complex mixture, such as zinc and lead sulphide, an assay for associated iron is applied in one plant and gives rapid information. Oxidised copper is revealed by shaking a few grams of ore in a test-tube with ammonia, filtering, and noting whether the filtrate is colourless or blue. In a German plant which floats lead carbonate (cerussite), the tailings are panned and a few drops of sodium sulphide are added to the residual heavy mineral. If this is cerussite it turns brown. In all cases what is wanted is rapid warning of trouble-making conditions, so that a costly loss of values can be avoided or minimised. The assay office is several hours behind the operation with its information, and anything which can be done to reduce the time-lag is valuable. Microscopic inspection, particularly of tailings and scavenger froths, sometimes gives a means of control. The mill official concerned should give attention to the possibilities of rapid and simple tests capable of indicating changes in the flotative qualities of the ore. He should train the shiftsmen in the use of those found advantageous and instruct them as to any changes in reagent practice or machine setting needed when such changes occur. Check on *p*H is rapidly learned. The Lovibond comparator can be used with a suitable indicator, and is less likely to be damaged than is a portable *p*H meter.

The mill laboratory should be used not only in connexion with routine checks but also to try new reagents, treatment rates, and other variables. When tests have proved that an improvement can be made, the necessary changes must be initiated by the mill superintendent and closely watched by trained men in the early stages. The shiftsmen cannot be expected to understand and control a circuit alteration until they have acquired experience of it. If more than one factor is to be changed, the variations must be introduced gradually, giving the plant a day or more to settle down after each cautious alteration. Here lies the greatest danger since the day shift, during which changes are usually initiated, hands over to workers who have had no chance to get used to the new look of things. Alterations in a flow-sheet have a knack of leading to, or being accompanied by, sudden choke-ups in the mill pumps and piping, as most experienced engineers have found. Changes should therefore be so arranged that the responsible engineer is on hand for the following twenty-four hours or longer, while the shiftsmen are being trained to the new conditions. Do not alter more than one thing at a time, unless this is quite unavoidable. If the change leads to the re-circulation of a different type of middling, some days will elapse before the cumulative effect of this recycling can be assessed. It is essential that time be given for this to happen before proceeding with any further change, if confusion and misleading inferences are to be avoided. The difficulties of carrying through a change successfully are greater than would theoretically appear, but the pleasure of success is well worth the effort entailed.

The best concentrate is made in the first cell of a given section if conditioning has been correctly done. From this point on the froth mineralisation diminishes and the grade of float drops. There are several reasons for this. *First*, the head cell of a series contains the richest pulp, from which to draw true concentrate into its froth. *Second*, the later cells, which have less true concentrate, are less able to squeeze contaminated gangue back into the pulp from their lower-grade froth. *Third*, the height of the froth-column diminishes in the later cells, since there are fewer floatable particles to stabilise it. *Fourth*, the extremely fine particles entrained in the froth are removed near the head of the section, and the bubbles are more fragile after their departure. Thus for a given series of cells (whether set to rough, clean or scavenge) the falling-off in grade of skimmed froth is accompanied by a decrease in fully liberated value, and an increase in the middling and contaminated gangue which is then able to take its place. If a bank of cells is divided into rougher, cleaner, and scavenger sections, the correct length of each section varies with the assay grade of feed. It is sometimes helpful to have a flexible system of launders and returns so that these sections can be adjusted in accordance with changes in grade of feed. The height of the frothing head determines the amount of sorting work done on the floated particles. To measure this, a floating gauge can be made and marked in accordance with the depth of submergence in similar pulp ahead of the cells. The frothing head should be diminished slowly from head to tail, the last scavenger cell or so showing a barely mineralised fugitive froth. The weirs of these final cells are set high so that the pulp almost brims over into the froth launder.

In the conditioning stage, the higher the percentage of solids the less agitation needed to maintain all the particles in suspension, and the more concentrated will be the reagents in the pulp. This last point is not so critical as it sounds, since the aqueous phase of a 15% pulp is 94½% by volume as against 87·5% for 30% solids. Pulp density is usually dictated by that of the overflow from the grinding circuit. In the flotation cell, the higher the pulp density the easier it is for a particle to reach the frothing zone, but against this must be set the need for good mobility of the solids. The practical upper limit is 35%–40% solids, in the case of a slime-free feed. The presence of primary slimes and colloidal material might reduce the permissible percentage of solids, possibly even to 15%. In the scavenging circuit a high density aids the "reluctant" particles which are being forced to float. Most of these are incompletely liberated, surface-contaminated by lyophiles, largish, or of awkward shape for holding in a bubble, so they need help. With a thicker pulp there is more collision between bubble and particle and a longer dwelling time in the cell for a given weight of solids. Before the pulp reaches the scavenging section, most of the concentrate has been withdrawn, and thus it may have become unduly thin. At Rammelsberg it has been found worth while to switch pulp reduced below a critical operating density, thicken it, and then return it to the line of cells. Other advantages of working with thick pulp are the increased capacity thus conferred on the system and the better carriage of particles in the transporting stream.

Just as thickening aids in forcing up the last reluctant particles which would otherwise be lost as tailings, so does thinning of the pulp aid cleaning of the rough concentrate. This leaves the roughing section as a froth which breaks down to about a 2:1 liquid-solid ratio as it flows toward the cleaning section. Any desired dilution can be made by spraying this rough concentrate *en route*. Cleaning is usually practised on pulp containing about half the percentage of solids used in roughing. A limiting factor in cleaning is the dilution produced by the tailings from the cleaner cells, which are nearly always sent to join the rougher feed, the whole bank of cells being arranged as in Fig. 245.

This system can be extended as required in either direction, the number of component cells in each block being suited to the processing time required. The pulp is usually conditioned and floated at the prevailing mill temperature. Sometimes the movement of copper on to sphalerite is speeded by warming the conditioner, and where fatty acids are used in the flotation of non-metallic earths a temperature of 30 C may be justifiable to ensure that the oleic acid used is above its freezing point.

Fundamental improvements in process control have been made possible by such technical aids as continuous fluorescence analysis from samples drawn from appropriate points along the flow-line. These give rapid indication of change and, if coupled to correcting devices, make rapid corrections of drift or irregular running. Other monitoring and controlling arrangements deal with such matters as changes in pulp density, height of frothing blankets, reagent availability and building up of re-circulating sands or pulps. These are discussed in Chapter 21.

The largest number of cells is included in the rougher section and the smallest in the final cleaning, the scavengers usually being run together. Though the diagram shows the cells in separate blocks they would, if square units, be arranged in an unbroken line. The chief advantage in having two scavenging stages is that the first section can be "pulled" moderately hard, while the final "save-all" cells can be pulled very hard, their froth being withdrawn for special retreatment if unsuitable for return to the main circuit. Where a high-purity product (e.g. acid-grade fluorspar at 97·5%+ CaF_2) is required, several stages of recleaning may be employed. It is good practice to return

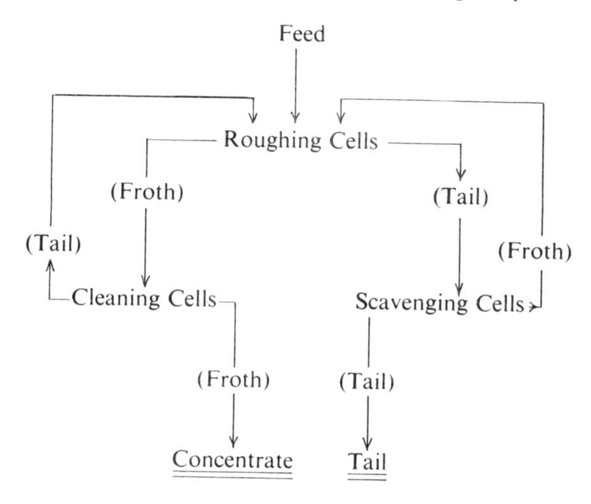

Fig. 245. A Rougher-Cleaner-Scavenger System

a froth from a lower-grade section to a point in the preceding part of the line of cells where the entering pulp has about the same assay value. Once a particle has responded to flotation, it will almost certainly float again if returned to the same cell, provided its surface has not deteriorated during recirculation. An exception to this general probability would be a particle of oxidised ore which had floated under the fugitive influence of conditioning with a soluble sulphide. Re-surfacing would be needed, perhaps by a small addition of the sulphide reagent to the scavenger float on its way back to the preceding series of cells.

The rougher froth usually overflows at between 35% and 45% solids. It may need the aid of water in the launder to break down the froth, to help it to run freely, and to dilute it into the cleaning cells. In the preceding section it was observed that a small plant could not normally justify a special re-treatment of scavenger froth, which must build up and overflow till its contained middlings either report as concentrate or tails. This leads to cyclic surging, the circulating load building up until some circumstance initiates a release. No amount of reagent or circuit manipulation can do much, since only further liberation can upgrade such particles. The concentrate may be deliberately downgraded by letting the higher-grade middlings work through.

It may be feasible to run an extra-high-grade concentrate with a favourable entry of ore, and to make a lower-grade product with poorer ore, thus preserving a rough average. If the scavenger float is non-oxidising, it may be possible to run it to temporary storage and to have an occasional regrind of this accumulation put through the main circuit, the grinding conditions being set specially for its treatment.

Since the scavenger froth may contain minerals other than those required, depressants may be used to decrease their floatability while on the way back to the previous section, provided they do not upset the legitimate float. Again, this froth might contain particles which for some reason had not taken their activation properly in the conditioning stage. Examination and test of samples would establish the reason for the failure of such mineral to float properly, and corrective action would suggest itself. The last scavenger cell should show a barren froth, ensuring that everything floatable had by this time been removed. This barren appearance should be checked by an occasional assay.

Where a regrind circuit receives the scavenger floats, the tailings from one of the cleaning stages may also be fed in at the same point. Just as the mineral in the scavenge gives a valuable picture of the escape to tailing of value, so does the tailing from a cleaning section indicate the nature of particles depressing the final assay-grade of the concentrate.

Practice—Some Single-Concentrate Treatments

Small differences in the crystal structure and chemical constitution of minerals of a given species may affect the qualities exploited in flotation. Plant practice is, therefore, adapted to the specific ore treated, not to the general type of mineral described in mineralogy as galena, bornite or whatever may be the sought value. The requirements for liberation, activation, gangue depression, optimum froth texture, etc., follow a general scheme for the separation of a designated metal-sulphide from its associates in the ore, but specific details are modified for physical, chemical, and economic reasons. The possible objectives are:

(*a*) To float one or more valuable minerals simultaneously.
(*b*) To depress one or more minerals from a bulk float.
(*c*) To float one or more minerals in order to concentrate a value intimately associated.
(*d*) To float one or more gangue minerals.

(*a*) is the normal procedure. It could be used to make a mixed bulk concentrate of, say, copper minerals which needed no further separation before shipment. Again, when two sulphides readily float together (e.g. the sulphides of copper and lead) they are sometimes bulk-floated, one being depressed in a further treatment.

Treatment (*b*) is used to clean a floated product by depressing undesired constituents (e.g. in removing the last of the lime, silica, and iron from a fluorspar required at "acid" grade) or to separate two values by specific

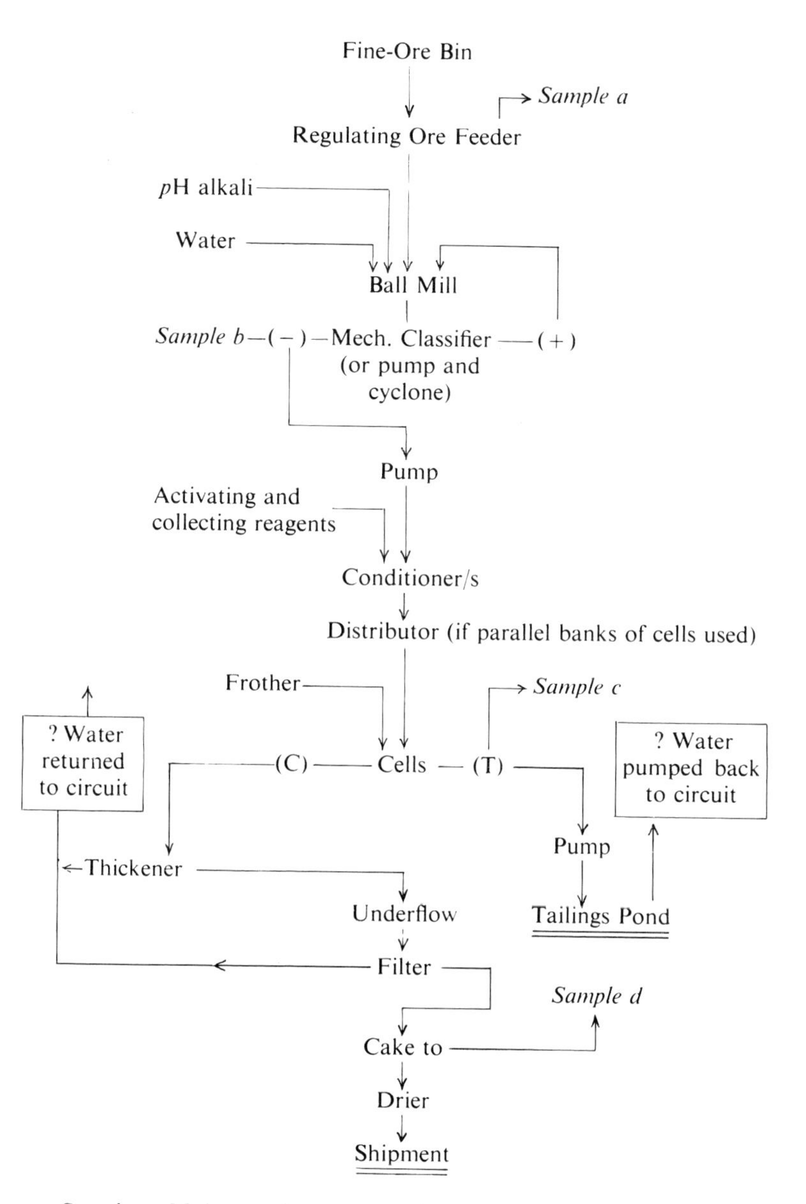

Sample a Moisture, size, assay grade.
Sample b pH, % solids, size, assay grade.
Sample c Mesh assay (monthly), assay grade (each shift).
Sample d Assay grade, moisture.

Fig. 246. *General Flow-sheet for Flotation of any Single Mineral Concentrate, with Discard of Tailings*

depression of one from the bulk float. Method (*c*) is indirect, and is used in such cases as the flotation of auriferous pyrite as a prelude to special treatment of the fraction thus selected, or for upgrading a fraction of an uranium-bearing ore by floating the minerals most closely associated with it. Method (*d*) is an alternative to (*b*) used in such cases as the final upgrading of a dirty phosphate float by first destroying its activation and then using amines to remove the silica. Fig. 246 shows a general flow-sheet for single-product flotation. Possible addition points for reagents are indicated, together with those at which control samples may be taken.

The conventional reagent plan for floating the single minerals named away from their gangues is shown in Table 39. The reagents prefixed R are Cyanamid reagents. In the column "Frothers", higher alcohols are often used to modify the frothing agent named. Since the minerals of any named type vary in their response to reagents, this Table can only indicate the best starting point for an investigation. Some typical industrial treatments of ores, most of which include flotation, are outlined in Chapter 23 under the name of the principal element in the concentrate.

Broadly, the collectors used to float metal sulphides are thiophosphoric (the aerofloats) and sulphydric (the xanthates). When such minerals are partially (or in some cases completely) oxidised soluble sulphide may be used ahead of, or concurrently with, the collector to restore for a short time an attracting sulphide-metal surface. With the commonly floated non-sulphides (excepting some copper, lead, and zinc minerals which accept resulphidisation) the surfaces of the particles are relatively inert. It is correspondingly more difficult to find a reagent combination which is strongly specific to one only of the minerals present. Fatty acids and their salts are the most used anionic collectors. Despite their higher cost, cationic amines are also in commercial use. The consuming industries often demand a high standard of purity in the concentrate they purchase. To attain this special care must be given not only to specific collection, but also to specific depression and the lively froth-column and multiple recleaning in which upgrading is achieved. The main groups of non-sulphide minerals treated by flotation are:

1. Oxidised heavy-metal minerals.
2. Non-silicates of the alkalis and alkali earths.
3. Silicate minerals.
4. Inert minerals.
5. Carbonates.

The usual approach aims at depression of the unwanted minerals in order to obtain a selective mineralisation of the froth. Depression achieves two objects. It makes the non-floating surfaces more wettable, and also reduces consumption of collectors by limiting the surface area which will attract them. A vital controlling factor is the ionisation maintained in the pulp liquid.

The most difficult operations are those which must separate minerals having closely similar surface properties, such as fluorite from calcite; apatite from calcite; fluorite from scheelite; beryl from felspar. The collectors used are strongly reactive and their selectivity is poor. The crux of control, therefore, lies in depression.

General Guide to Reagent Practice

(* Capitals are used to indicate common commercial ores.)

Element	Mineral*	% Principal Constituents	Average Sp. Gravity	Dispersants *p*H Regulators Selectivity-differentiating agents	Depressants	Activators	Collectors	Frothers	Remarks
Aluminium	Alunite $K.Al_3(SO_4)_2.(OH)_6$	K 9·4 Al 19·6	2·7	Na_2SiO_3	Excess Na_2SiO_3		Oleic acid R.708	Cresylic acid Higher Alcohols	Associated minerals include Kaolin, Pyrite
	BAUXITE Hydroxides $Al_2O_3.2H_2O$	Al 39·13	2·5	NaOH Polyphosphate Paraffin			Oleic acids R.801 R.825	Higher Alcohols	Associated minerals include Clay, Kaolinite limestone, Fe dolomite, Ti
	Corundum Al_2O_3	Al 53·4	4·0	NaOH	Excess Acid		Oleic acid and its salts	Pine Oil Higher Alcohols	
	CRYOLITE Na_3AlF_6	F 54·3 Al 12·9 Na 32·9	2·9	Orthotoluidine		$CuSO_4$	Oleic acid and its salts	Higher Alcohols	
	FELDSPARS $(AlSi_3O_8)$ {Na, K, Ca}	abt 18	2·5	HF		HF	Amines (Cationic reagents)	Pine Oil Higher Alcohols	Not concentrated for Al
	KYANITE Al_2SiO_5	Al 63	3·6	NaOH Scrub Aerosol		Deslime scrub	Fatty-acid Salts Oleate R.825	Pine oil Higher Alcohol	

TABLE 39—*continued*

Antimony	Stibnite Sb_2S_3	Sb 71·7 S 28·3	4·7	Na_2CO_3 Na_2SiO_3	Excess Alkali NaCN	$CuSO_4$	Xanthates (Ethyl sec. Butyl)	Pine Oil	Associated minerals include Quartz, Barite, Galena, Gold. May be floated for bullion content or depressed before cyanidation
Arsenic	ARSENOPYRITE FeAsS	As 46 Fe 34·3 S 19·7	6·0	Na_2CO_3 Na_2SiO_3	NaCN Lime	$CuSO_4$	Xanthates (Ethyl to amyl)	Pine Oil	Associated minerals include sulphides of copper, iron, lead, zinc, antimony. May be auriferous
Barium	BARYTES $BaSO_4$	Ba 58·8	4·5	Na_2SiO_3 Na_2CO_3 Citric Acid Aerosol	$AlCl_3$ $FeCl_2$	$Pb(NO_3)_2$	Oleic acid and fatty acid soaps. Higher-alcohol sulphates	Pine oil Cresylic acid	Associations— Iron, manganese, galena, sphalerite, chalcopyrite
Beryllium	Beryl $Be_3Al_2Si_6O_{18}$	Be 14·1 Si 66·8 Al 19·1	2·7	Washing, HF	H_2SO_4	$Pb(NO_3)_2$	Amines Oleic and fatty acid Salts. R.825		Associations— Granite, schist, mica, clay, slate
Boron	BORAX $Na_2B_4O_7 \cdot 10H_2O$	B 11·4	1·7	Aniline, Starch, Dextrin Quebracho		$BaCl_2$ $Pb(NO_3)_2$	Fatty Acids	Aniline Xylidine Pyridine	

Calcium	Calcite $CaCO_3$	Ca < 40		Na_2CO_3	Quebracho Na_2SiO_3 $K_2Cr_2O_7$	Oleic acid Sulphonated oleic Fatty acid salts	Pine Oil	
Calcium	Dolomite $CaMg(CO_3)_2$	Mg 13·2	2·9		Quebracho	Fatty acids and salts	Pine Oil	
Carbon	COAL	C < 95	1·4	Na_2SiO_3	Tannin Quebracho	Fuel Oil Paraffin, Cresyls	Pine Oil Cresylic acid	
Carbon	Diamond (boart CARBONADO)	C < 100	3·5			Natural grease, petroleum jelly, on Tables		Associations—Shale, clay, quartz, iron, Fe + Mg Silicates
Carbon	GRAPHITE	C < 99	2·0	Na_2CO_3 Na_2SiO_3	Starch 600 Series	Paraffin Fuel oil	Pine Oil Long-Chain alcohol	Associations—Spinel, shale, mica, silica, schist
Carbon	Shale	?			600 Series Cyanamid	Fuel oil Paraffin	Pine Oil Higher alcohols	Such shale may impede cyanidation unless removed
Cerium	MONAZITE (CeLaYt)PO_4 Formula doubtful		5·1	Na_2CO_3 Na_2SiO_3	Strong Acids	Oleic acid Fatty-acid salts	Pine oil	

TABLE 39—*continued*

Chromium	CHROMITE $FeOCr_2O_3$	Cr 46·5	4·5	NaOH Na_2CO_3 Fluosilicate		Phospho-molybic and p. tungstic acids	R.800 Series Fatty acids Alkylo-amines	Pine Oil Fuel Oil	Associations—Serpentine, black sands
Cobalt	COBALT Oxides CoO(OH)	Indeter-minate		Na_2SiO_3 Na_2CO_3		Poly-sul-phides Heat	Xanthate Fatty acids and salts	Pine Oil Higher alcohols	
Copper	AZURITE $2CuCO_3Cu(OH)_2$	Cu 55·3	3·8	Na_2SiO_3 Na_2CO_3 Na_2S	Quebracho	Poly-sulphides	Fatty acids and salts. Xanthates Aerofloat 25	Pine Oil Gas Oil Cresylic	
Copper	BORNITE Cu_5FeS_4	Cu 63·3 Fe 11·1 S 25·6	5·0	Na_2SiO_3 Lime	NaCN		Xanthates Aerofloats	Pine Oil	
Copper	CHALCOCITE Cu_2S	Cu 79·9 S 20·1	5·7	Na_2SiO_3 Lime	NaCN		Xanthates Aerofloats	Pine Oil	
Copper	CHALCOPYRITE $CuFeS_2$	Cu 34·6 Fe 30·4 S 35·0	4·2	Na_2SiO_3	NaCN		Xanthates Aerofloats	Pine Oil	
Copper	COPPER Cu	< 100	8·8	Na_2SiO_3 Lime	NaCN		Xanthates Aerofloats	Pine Oil	
Copper	CUPRITE Cu_2O	Cu 88·8	5·9	Na_2SiO_3 Na_2CO_3 Na_2S	Quebracho	Poly-sulphides	Fatty acids and salts Xanthates	Pine Oil Gas Oil Cresyl	

Copper	MALACHITE $CuCO_3Cu(OH)_2$	Cu 57	3·9	Na_2SiO_3 Na_2CO_3 Na_2S	Tannic acid	Poly-sulphides	Fatty acids and salts Xanthates	Pine Oil Gas Oil Cresyl	
Fluorine	FLUORITE CaF_2	F 48·7 Ca 51·3	3·2	Na_2SiO_3 H_2SO_4	Quebracho $K_2Cr_2O_7$		Oleic acid Fatty acids (sulphonated) R. 825	Pine Oil	
Gold	Gold Au	Au < 100 Ag	17·5	Na_2CO_3 if any	Lime NaCN	R.404	Xanthates Aerofloats	Cresylic Pine Oil	Pulp density 30% + solids
Gold	Petzite $AuAg_3Te_2$	Au 25·5 Ag 42 Te 32·5	9·1		NaCN		Xanthates	Pine Oil	
Gold	PYRITE Auriferous			See under Pyrite					
Gold	Sylvanite $(AuAg)Te_4$	Au 24·2 Ag 13·2 Te 62·6	8·1		NaCN		Xanthates	Pine Oil	
Iron	Hematite Fe_2O_3	Fe < 70	5·0	H_2SO_4	Tannin Phosphate	De-slime	R.801 R.825 Tall oil Oleic	Pine Oil Fuel Oil	
Iron	Limonite $2Fe_2O_3$ $3H_2O$	Fe 59·8	3·7	H_2SO_4	Tannin Phosphate	De-slime	R. 801, 825 Tall oil Oleic	Pine Oil Fuel Oil	
Iron	Marcasite FeS_2	Fe 46·5 S 53·4	4·7	H_2SO_4(?)	NaCN Lime	$CuSO_4$	Xanthate	Pine Oil	

Iron	Pyrrhotite Fe_7S_8	Fe 60·4	4·5	H_2SO_4(?)	NaCN Lime	$CuSO_4$	Xanthate	Pine Oil
Iron	PYRITE FeS_2	Fe 46·6 S 53·4	5·0	H_2SO_4	NaCN Lime	$CuSO_4$	Xanthate	Pine Oil
Lead	ANGLESITE $PbSO_4$	Pb 68·3	6·3	Na_2CO_3		Na_2S Phosphate $CuSO_4$	R.404 Xanthate Aerofloat	Pine Oil Cresylic
Lead	CERUSSITE $PbCO_3$	Pb 77·5	6·5	Na_2CO_3		Na_2S $CuSO_4$	Xanthate R.404 Aerofloat	Pine Oil Cresyl
Lead	GALENA PbS	Pb 86·6	7·5	Na_2CO_3 Na_2SiO_3	$K_2Cr_2O_7$		Aerofloat Xanthate	Pine oil
Lithium	Micas (lepidolite, etc.)			H_2SO_4	Glue, Starch HF	Lead Nitrate	Cationics (amines) R. 825 Resin Soaps	Pine oil Cresylic
Lithium	SPODUMENE $LiAl(SiO_3)_2$	Li 3·7 Al 14·5	3·1	Aerosol	Starch Dextrin		Oleic R.825	Pine oil
Magnesium	Brucite $Mg(OH)_2$	Mg 40·3	2·4	Polyphosphates			Xanthate	Pine oil
Magnesium	Magnesite $MgCO_3$	Mg 28·8	3·1		Tannin		Oleic and oleates	Pine oil
Magnesium	TALC $H_2Mg_3Si_4O_{12}$	Mg 19·2	2·8	Polyphosphates	Starch Glue		Short-chain Amines R.825	Pine oil Paraffin
Manganese	Pyrolusite MnO_2	Mn 63·2	4·7	Na_2SiO_3 Na_2CO_3	Phosphate Quebracho		Fatty acids and salts	Pine oil Paraffin

TABLE 39—*continued*

Manganese	Rhodocrosite $MnCO_3$	Mn 47·8	4·0	Na_2SiO_3 Na_2CO_3 NaOH			R.708 Hydrolysed veg. oils	Pine oil
Mercury	CINNABAR HgS	Hg 86·2	8·1		Na_2SiO_3		Amyl Xanthate	Pine oil
Molybdenum	MOLYBDENITE MoS_2		4·8				Min. oil, Xanthate Aerofloat	Pine oil
Nickel	PENTLANDITE (FeNi)S	?	4·8	Na_2SiO_4 Alkali (?)	Lime	$CuSO_4$	Pentasol Xanthate	Pine oil
Niobium	Pyrochlore $Na_1Ca_1Nb_2O_6.F$		4·3				Fatty acids and their salts	Pine oil
Phosphorus	APATITE $Ca_5(FCl)$ $(PO_4)_3$	P_2O_5 42	3·1	NaOH Na_2CO_3	HF		Fatty acids and salts	Pine oil
Phosphorus	Phosphate rock $Ca_3(PO_4)_2$		3·2	NaOH			Fatty acids	Pine oil
Tungsten	SCHEELITE $CaWO_4$	W 63·9	6·0	Na_2SiO_3 Na_2CO_3 Aerosol	Excess Na_2SiO_3 Quebracho		Oleic and fatty-acid salts	Pine oil

TABLE 39—*continued*

Tungsten	Wolframite $(FeMn)WO_4$	W 61	7·3				Oleic R.708, 710	Pine oil Fuel oil
Uranium	Carnotite $K_2O.2UO_3V_2O_5.8H_2O$	?	4·1	Na_2CO_3 Na_2SiO_3		$Pb(NO_3)_2$	Fatty acids	Pine oil Cresylic
Uranium	Pitchblende $xUO_2y.UO_3$	?	7·2			$FeCl_3$	Fatty acids Amines	Pine oil
Vanadium	DESCLOIZITE $4RO.V_2O_5.H_2O$	?	6·0	Na_2CO_3	Ca. ion	Softened water	Xanthates	Higher alcohols
Vanadium	VANADINITE $(PbCl)Pb_4V_3O_{12}$		7·0	Na_2SiO_3 Na_2CO_3 to pH_9	Hard water (lime salts)	Softened water	Pot. ethyl and amyl xanthate	Methyl isobutyl carbinol
Zinc	Smithsonite $ZnCO_3$	Zn 52·1	4·3	Na_2CO_3 Na_2SiO_3		Sod. sulphide	Fatty acids Amine acetate	Pine oil
Zinc	SPHALERITE ZnS	Zn 58·7	4·1	Lime	$ZnSO_4$ $NaHSO_3$ SO_2	$CuSO_4$	Xanthates	

Ejgeles[7] extends the list of depressive factors noted by earlier workers. These include:

(*a*) Decrease of amount of collector on the mineral surfaces.
(*b*) Direct influence of depressants on the mineral surface.
(*c*) Change in the process of froth formation.
(*d*) Change in the iono-molecular composition of the liquid phase of the pulp.

He also observes that selectivity of non-sulphides is achieved by depression rather than activation. The pulp "climate" is important in this respect. Research has shown dextrin and sodium silicate to be of value in the separation of fluorite from calcite; dichromate and dextrin for fluorite from barite; starch and lignin sulphonate for diaspore from pyrophyllite. All this work was done using softened water.

Experimental work tends to show that there is no direct relationship between quantity of collector sorbed and floatability. Depressants not only modify the general sorptive attraction, but can also reduce the area on a particle available for such sorption by occupying part of it or by replacing activating agents. Most mineral particles are capable of adhering to a bubble. Adhesion, if weak or reluctant, fails to persist in the mineralised froth in a turbulent flotation cell. When the collector is added, selective vigorous adhesion takes place in micro-seconds rather than after several minutes. Some depressants on the contrary, have a decelerating effect on adhesion. These include sodium silicate, potassium chromate, and dichromate, alkalis which modify the *p*H of surfaces activated by calcium or iron, sodium triphosphate, dextrin, tannin, sodium sulphide, and sodium oxalate.

The ionisation of the pulp's liquid phase is of importance in control. It is a common precaution to use softened water in connexion with such anionic reagents as oleic and other fatty-acid collectors. This is done (*a*) to reduce the influence of the cations in the liquid; (*b*) to avoid reaction of calcium with the fatty acid to form a soap, which contributes little or no selective activating value but can upset the froth; (*c*) to avoid having coagulating salts in the pulp. Softening is performed by the use of sodium hydroxide or carbonate, by lime treatment, or by ion-exchange. Another method, rarely applicable because of cost, is to complex the cations by using suitable reagents. Sodium triphosphate and metahexaphosphate are used to prevent the formation of hard crusts of lime in the filtration of cyanide solutions, but the use of these and similar compounds is rare in flotation.

The oxidised minerals include oxidised native metals, oxides of the heavy metals, and heavy-metal salts. Native metals can be floated readily with xanthate and a depressant for the gangue, provided particle size is not too large. Oxides are superficially "lazy" and resist the ionisation needed to induce collector coating. Non-silicate metal salts tend to sorb collectors too deeply, in which case a proper orientation of the collector molecule's hydrocarbon group is not procured. The general sequence of conditioning is surface-closure with a soluble sulphide, collection with one of the longer-chained xanthates, and use of creosote for a measure of "smear-collection" (collection by non-chemical surface adulteration), dispersion of gangue with sodium silicate, and flotation with a froth made fragile by use of such additives as iso-butyl-carbinol. Cerussite floats fairly well with a heavy dosage of amyl xanthate (up to several lb/ton), part of which can be omitted if sodium sulphide can be made to film the surface with lead sulphide. The copper ores, malachite and axurite respond to mild sulphidising treatment with xanthate and gas oil, or alternatively can be floated with fatty acids of the oleic type, partially sulphonated, and preferably well emulsified by grinding with hot water, sodium bicarbonate, and sodium silicate immediately before addition to the pulp. Again, gas oil helps with collection and frothing. Where siliceous gangue must be depressed, a moderately high *p*H aids by wetting the calcium sorbed to the surface. If a sodium oleate soap is the collector, any surface which carries calcium may sorb an oleate group and float, so *p*H up to $9\frac{1}{2}$–10, produced with soda-ash, is used to minimise this. Full depression after this calcium surface-activation requires a *p*H of 11.

Earths

In the flotation of earths low-solubility soaps are supposed to form between the Ca^+, Ba^+, etc., cations and the oleic anion in the discontinuity zone. Whatever the reason, the froth is heavy and tenacious, being so strongly flocculated in many cases as to earn the description "frozen". The author finds that modification of the oleic acid by mild sulphonation removes this difficulty in the case of fluorspar flotation[1], giving a lively, self-sorting froth. *p*H control gives some differentiation between the metal "earths". Tannic acid is a depressant for calcium, and the use of sodium silicate aids the selective wetting of the associated siliceous gangue.

Silicates

Silicates are somewhat similar in behaviour to the above minerals. The *p*H needs close control, and depression of gangue by a wetting agent is helpful. If cationic reagents are used, the water must be soft and the pulp fairly free of slime. Among the minerals floated by fatty acids or fatty acid amines are feldspar, kyanite, garnet, and mica.

Non-ionizing Minerals

The "inert" minerals include coal, graphite, and native sulphur. All these preferentially sorb hydrocarbon oils and float readily without a frother, or with a little pine oil. Sulphur acidifies its pulp, which usually must be neutralised with lime to prevent corrosion of metal surfaces. The main flotation difficulty is to disperse and wet down the gangue.

References

1. Pryor, Blyth and Eldridge. (1952). *Recent Dvelopments in Mineral Dressing*, I.M.M., London.
2. Chi, W. H., and Ypung, E. F. (1962). *Trans. I.M.M. (London)*, 72, p. 169.
3. *Ibid.* (1962). *Mining World*, March.
4. Pryor, E. J. (1961). "Flotation's Early Years" (*Qtly. of Colorado Sch. of Mines*), July.
5. Arbiter, N., and Steininger, J. (1963). *I.M.P.C. (Cannes)*, Pergamon.
6. Heal & Patterson Inc., Pittsburgh. Trade Bulletin.
7. Ejgeles, M. A. (1957). *I.M.P.C. (Stockholm)*, Almqvist & Wiksell.
8. Pryor, E. J. (1952). *Recent Developments in Mineral Dressing*, I.M.M., London.

CHAPTER 19

MAGNETIC AND ELECTRICAL SEPARATION

Introductory

The scientific understanding of the nature of magnetic force appears to have started early in this century with the work of Langevin, in France. He explained diamagnetism and paramagnetism in terms of orbiting electrons. Another French physicist, Pierre Weiss, followed closely with a qualitative hypothesis for ferro-magnetic behaviour. This was based on the assumption that suitable atoms behave as though they were tiny bar magnets which could be influenced so that they became aligned and acted as a single large magnet. He introduced the idea of the "domain" which today bases the theory of ferro-magnetism. In this, iron is considered to be composed of numerous small magnetized regions or domains, each containing millions of Fe atoms. If, as the result of external force these domains become aligned, the iron is magnetized.

Much later[1] quantum mechanics showed the nature of the aligning forces at work and provided experimental justification for the earlier hypotheses. Before 1939 only a few ferro-magnetic materials were in industrial use, but with post-war development in solid-state physics many new materials and outlets have been discovered. These include the ferrites[2] which allow alternating current to be used without the disabling effect of eddy currents that arise with rapid reversal of polarity.

With their present limited use in mineral processing, only brief reference to the ferri-magnetic oxides is justified. Their state is one in which elementary magnetic atoms do not line up in parallel as is the case in ferro-magnetism, but tend to oppose one another. This is due to their crystal structure. The oldest known magnetic material, lodestone, is a natural ferrite. Its formula, Fe_3O_4, can be written $Fe^{++}Fe_2^{+++}O_4^{--}$. All ferrites contain their trivalent Fe ions and their oxygen anions in a 2-to-4 ratio, while the divalent can be provided by any metallic atom small enough to fit into the crystal lattice. The general ferrite formula is $X^{++}Fe_2^{+++}O_4^{--}$. Resistance to flow of current is high because the crystals are ionic and the lower electron shells of their constituent atoms are filled. Magnetic strength, as always, is an effect produced by the spin of the electrons round their atomic nuclei. Pair electrons spin in opposite directions and therefore cancel out, but the atoms of such transition elements as manganese, iron, nickel and cobalt contain at least one unpaired electron, and thus are magnetic. The crystallographic structure has been discussed by Hogan[3].

Magnetic separation is applicable to materials in which a natural or induced degree of polarity can be sustained during passage through a field of magnetic flux. For ferro-magnetism to be usable this field must be steady, and is produced either by the use of permanent magnets or electro-magnets energised by D.C. In ferri-magnetism A.C. can be used as the problem of eddying does not arise.

For success the particle exposed to the magnetic flux must respond with sufficient strength to overcome inertial, gravitational and frictional constraint. It is aided in this by size control, controlled presentation of the feed, and by the use of appliances which take into consideration the effects of gravity and flow on the dry or pulped ore.

High-intensity or electrostatic separation is applicable where one particle species is relatively non-conducting and the feed to the system is sufficiently mobile and close-sized to allow the delicate electrical forces of repulsion and attraction to act on particles gathering charge as they move through a field of high electrical intensity and onward into one where insulation is suitably controlled and charge-dissipation becomes a discriminating force.

Separation by Magnetic Force

When a particle is placed in a magnetic field it either attracts or repels the flux flowing between the magnetic poles. A substance is paramagnetic when it concentrates lines of force to reach a flux density greater than unity and diamagnetic when it repels lines of force. Paramagnetic substances are attracted towards regions of greater flux-density while diamagnetics under the same circumstances are repelled. When paramagnetism is strong it is called ferromagnetism, and can sometimes be used for the concentration of ferromagnetic minerals. Diamagnetic forces are too feeble to be of use in mineral dressing, and minerals responding to these, together with many feebly paramagnetic ones, are termed "non-magnetic". There are degrees of susceptibility which are used to differentiate strongly and weakly magnetic minerals into separate products. The attracted particle must be deflected from the others in a moving stream of material. The size and S.G. of the particle, together with its freedom of movement relative to the moving stream in which it is being carried, are therefore factors influencing its separation. Magnetization per unit mass (σ) is:

$$\sigma = \frac{B-H}{4\pi\rho} \tag{19.1}$$

where B is the induced flux/unit area measured in *gauss*. One gauss equals one maxwell/sq. cm. H is the flux density in gauss of the exciting field. ρ is the specific gravity of the particle concerned. The susceptibility (ratio of intensity of magnetisation to magnetic field strength) may be stated as X

$$\frac{\sigma}{H} = X \tag{19.2}$$

and measured in electro-magnetic units (e.m.u.s.). The unit magnetic pole (m) is r Fμ, where r is its distance from a like pole, F the repulsion and μ the magnetic permeability compared with that of air (= 1). The unit H is F/m.

When the maximum magnetisation has been achieved a particle is saturated (σt). In magnetic separation at field strengths used in mineral dressing all ferro-magnetic minerals become saturated. They include magnetic pyrrhotite, magnetopyrite, and ferrosilicon. Magnetite has an X value of 93 e.m.u./g. Between this and non-magnetic minerals are a number of weakly magnetic and conditionally magnetic minerals. In addition, some iron minerals develop ferro-magnetic qualities after reduction roasting. Although at the lower end of the scale the susceptibilities of paramagnetic minerals vary between 10^{-4} and 10^{-6} e.m.u./g high intensity separators can in some cases treat such material. Strong fields (up to 20,000 gauss) are produced by powerful electromagnetics for this purpose and the pole pieces must be suitably designed to give the correct field gradient in the flux path. Permanent magnets are unsuitable for the separation of feebly-magnetic material since their fields do not exceed 7000 gauss. A ferro-magnetic particle orients itself in a field of uniform flux but does not move along it. In a non-uniform field it moves in the direction of greatest flux density, so as to concentrate in itself as much force as it can, provided the effect of this concentration is sufficient to overcome opposing friction, gravity, and (if in a pulp) viscosity and counter-flow. In order to produce a steep flux gradient the attracting pole is made wedge-shaped and the counter-pole flat (Fig. 247). The lines of force spring

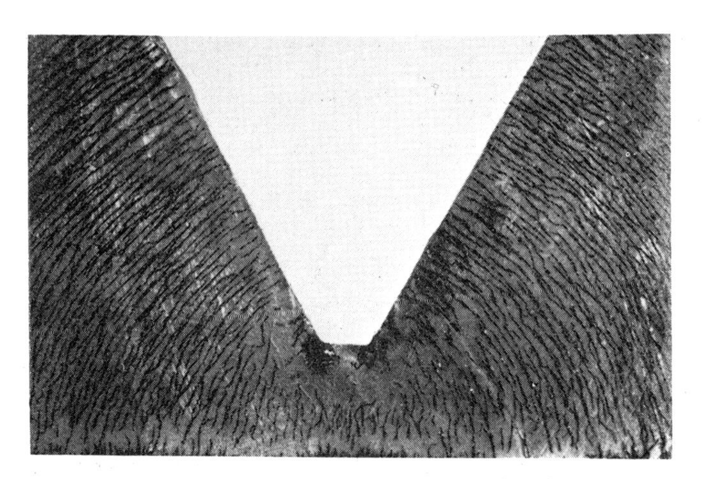

Fig. 247. Lines of Magnetic Flux (G.E.C.)

from the poles normal to the surface. Maximum attraction therefore occurs along the sharp edge of a wedge-shaped pole, while there is little or no tractive tendency immediately adjacent to a flat polar surface. The attractive effect is inversely proportional to the square of the distance between the attracting pole and the attracted particle. Magnetic alloys of high gauss are replacing electromagnets in a number of applications. They are described not only according to composition but also in terms of their coercivity (H_c), remanence (B_r) and maximum gauss (BH_{max}). Coercivity is the "magnetic force necessary to de-magnetize a substance which has been magnetized to saturation"[4]. Remanence (bid) is "the residual magnetism in a ferromagnetic substance (its hysteresis) after removal of the external magnetizing force". The properties of some magnetic alloys are shown in Table 40[5].

TABLE 40

*Alloy**	B_r (*gauss*)	H_c (*oersted*)	*BH max* (*megagauss-oersteds*)
35% Cobalt Steel	9000	250	0·95
Alni (Nial)	5600	580	1·25
Alnico, normal	7250	560	1·7
Alcomax II	13000	580	5·4
Hycomax I	9000	825	3·2
Hynico II	6000	900	1·8
Columax	13500	740	7·5
6% Tungsten steel	10500	65	0·3

* Figures for named Edgar Allan steels.

Electromagnets are used for heavy-duty work or where it is convenient to control the field flux by varying the D.C. current in the coils surrounding the high-permeability steels used. Specifications for such steels are given in B.S. 3100, which incorporates B.S. 1617[6]. The flux is controlled by varying the D.C. strength and the number of turns in the coil surrounding the iron core. A further running control is made by varying the air-gap between the passing ore and the attracting magnet.

Among the ores which may be directly concentrated by magnetism are those containing magnetite, franklinite, ilmenite, and wolframite. These and many others may respond, their treatment depending on the distinctiveness with which the magnetic mineral reacts at a suitable mesh. Such feebly magnetic minerals as biotite, garnet, basalt, zeolites, pyrochlore, muscovite and chlorite have been successfully concentrated on the laboratory scale and the industrial development of the sensitive systems needed is extending. A definitive list of magnetic minerals is not given, as the three main influences which must be exploitable in any separating method are specific to the occurrence of the given mineral and not general with respect to its type. These influences are:

(*a*) Impurities mechanically held in the grain of the particle.
(*b*) Dissolved impurities.
(*c*) Grain size of the crystal complex.

A tiny inclusion of magnetite in a non-magnetic particle could make it react as though feebly magnetic. Dissolved iron gives magnetic reaction to the otherwise unresponsive mineral marmatite.

The requirements for industrial concentration are that a thin layer of particles moves through consecutive magnetic fields, which are usually arranged with successive reversal of their polarity. This causes the ferro-magnetic particle to respond by turning through 180°, and thus frees any entrained gangue caught between it and the attracting pole. Successive magnetic fields may be increasingly strong, or the air-gaps progressively closer, in order to remove the most strongly magnetic particles first and the feebler ones later. The feed must be so presented that a particle of gangue will not pin down a ferromagnetic particle on which it may be lying. The machine must provide a suitably converging field at each point of separation, a means of regulating flux intensity, speed control through the field, scrubbing by pole-reversal to shake out entrained gangue, and separate discharge for concentrate, middlings (feebly magnetic), and tailings.

The type of machine used should suit the optimum liberation-mesh of the ore. In dry treatment moisture must be kept below 0·5% and the feed should be inside a specified size range. Its ferro-magnetic quality can be modified by pre-treatment. Leaching with an alkali has been described[7]. In the Murex process the ore is puddled with oil that has been loaded with magnetite. It is then pulped with water and sent to wet magnetic separation. Particles which have adsorbed this magnetised oil are thus separated.

The only method of pre-treatment in wide industrial use uses reduction roasting, in which the ore is heated to 550°C or more as it passes through an atmosphere containing free carbon monoxide, perhaps with some hydrogen if the source is producer gas. This reaction, which reduces part or all of the feebly magnetic iron minerals to Fe_3O_4, (a strongly magnetic oxide), is used in several systems—for example on hematitic and limonitic ores[8]. The feasibility of grain growth during roasting in a mildly reducing atmosphere (5% CO) has been investigated[9]. Non-magnetic and extremely fine taconites can "grow" magnetitic grains up to 40μ in diameter. Fluidized-bed reduction roasting of Lorraine ores at between 550° and 850°C takes place rapidly with *minus* 1 mm material[10]. Lump ore ground to 55% *minus* 300 mesh in a reducing atmosphere at 570°C gave a 90% recovery in pilot tests[11]. The Lurgi process[12] has proved successful with semi-taconites, minettes, siderite and hematite. In it the feed moves gently through a down-sloped tubular kiln and is reduced by gases fed in at desired points in its travel. These gases may be town, producer or blast, or may come from incomplete combustion of blown-in powdered anthracite or brown coal. Concentrates assaying up to 71% Fe and recoveries from 73% upward are commercially attainable after this reducing treatment. Other systems which follow reduction roasting with magnetic separation have been reported[13,14].

Applications

The separating principle used in magnetic separation has been developed in a variety of machines (Fig. 248). In some the magnetic material is lifted clear of the gangue, in others pinned while the gangue moves on. Air, water, and gravitational force are used to help or hinder movement of a selected fraction. Field intensity is varied by changing the windings and electric force in electromagnets, by alloy composition in permanent magnets, and by adjustment of air gap.

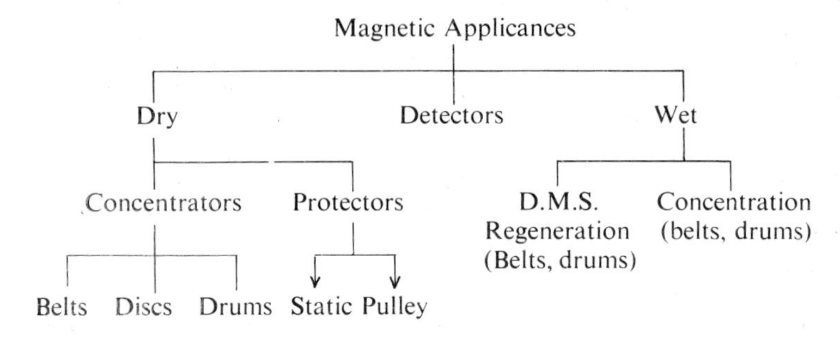

Fig. 248. *Types of Magnetic Separator*

The machines used can be broadly classified by function, thus:

(*a*) *Concentrators.* Those making bulk separation of magnetics from a stream of passing ore.
(*b*) *Purifiers.* Those used to remove small quantities of deleterious magnetic material from a product (e.g. from china clay).
(*c*) *Reclaimers.* Those used to return magnetic material to the circuit (e.g. cleaning magnets in D.M.S. process).
(*d*) *Guard magnets.* Those which safeguard machines or processes by removing tramp iron or detecting dangerous magnetic objects and initiating protective action.

Dry Magnetic Separation

This is practised at two levels low- and high-intensity. Low-intensity machines are in wide use in the concentration of lump ores and coarse sands[15], particularly in "cobbing" operations (a term for the picking over of coarse ore for removal of a selected constituent, in this case strongly magnetic particles). The rough concentrate thus obtained probably needs further comminution and magnetic treatment. An early type of machine, still used in various forms, is the low-intensity drum separator (Fig. 249). Its renewable non-magnetic surface rotates above a series of oppositely charged magnets. Most of the flux short-circuits from pole to pole beneath the drum, but enough leaks through to grip ferromagnetics, which roll with each change in polarity, till they are pulled clear of the free-falling gangue. This movement frees gangue which might otherwise remain sandwiched between the magnetic particles and the drum. This design does not give room for strong electromagnets with well separated alternate poles. With the modern development of powerful magnetic alloys this type of separator is obsolescent. Variations in which magnetic particles are lifted from a belt conveyor to the underside of a parallel short conveyor set immediately above, and similarly magnetized, are industrially used. A surface flux of 1000–1200 gauss is possible giving a field strength of some 600 gauss 2 in. from the upper belt, maintained either by a gang of four or six electromagnets or by permanent magnets.

The main loss with these machines is of middlings not sufficiently gangue-free to be picked up, of small magnetic particles adhering to or pinned under waste rock, and of feebly magnetic iron minerals. Low-intensity separation

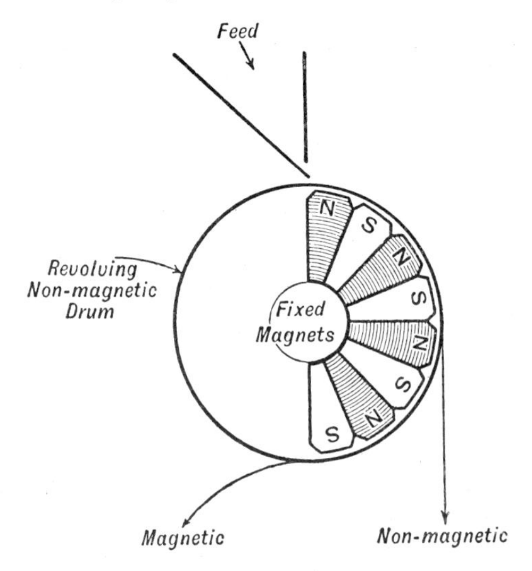

Fig. 249. Ball-Norton Drum Separator

tends to be replaced in the *minus* $\frac{1}{4}$ in. size range by wet methods. It remain s attractive where autogenous dry grinding is practised and in climates subjcct to freezing of ore products. The fecd must be reasonably dry, screened into suitable size ranges, de-dusted and spread into a layer only a particle deep, thus limiting capacity.

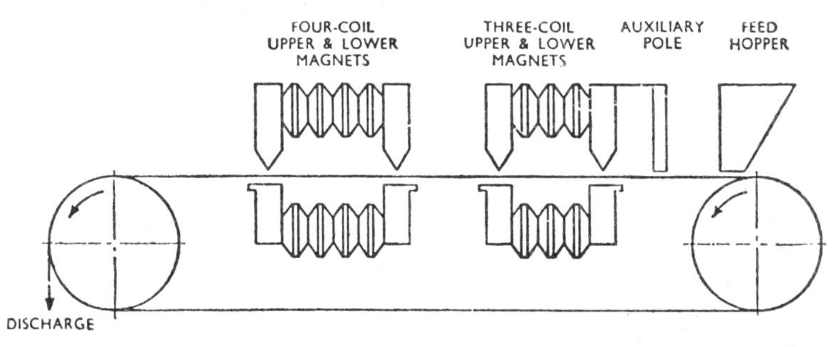

Fig. 250. Arrangement of Wetherill Type Separator with Coils in Upper and Lower Magnets for Fields of High Intensity

Before 1912 or thereabouts dry treatment dominated industrial practice despite the dusty conditions then accepted as inevitable. Only since the mid-fifties has a large-scale change to wet processing occurred and this is still in its emergent phase.

One of the oldest belt pick-up separators is the Wetherill (Fig. 250). Cross-belts (not shown) move between feed belt and wedge poles, and convey magnetic material out of the system. In the Rapid (Fig. 251) and Rapidity machines the same idea is exploited in a different manner, using tilted keeper discs to convey the rising particles away. The Rapid works across the feed belt and the Rapidity along two bridged belts, the latter arrangement giving higher capacity.

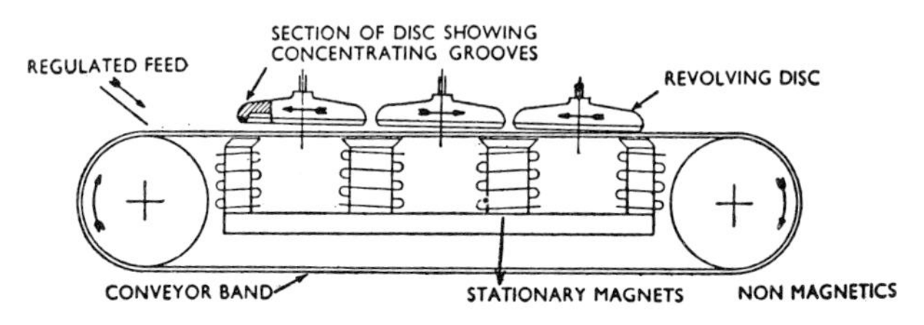

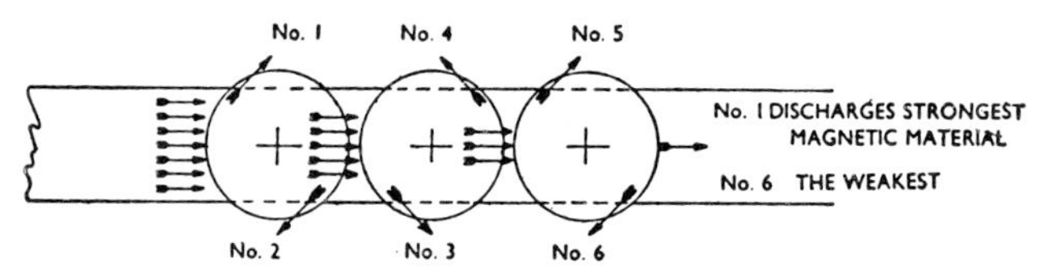

Fig. 251. *Rapid Magnetic Separator*

Before turning to high-intensity separating systems the use of protective magnets in the mill may be considered.

Most heavy crushing machines are guarded against the entry of uncrushable iron or steel by some magnetic device. The simplest is a suspended electro-magnet (Fig. 252), hung above the conveyor belt or non-magnetic

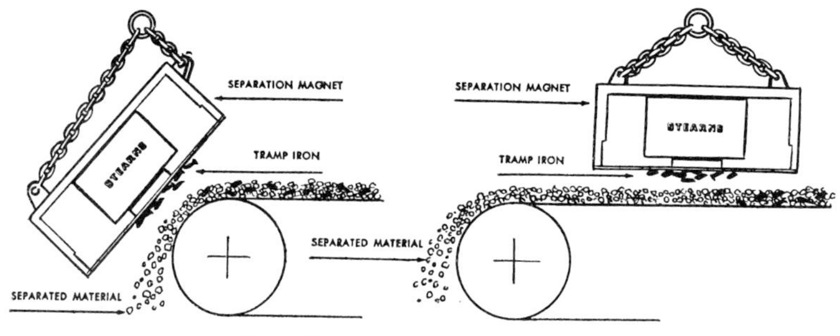

Fig. 252. *Guard Magnet* (Stearns)

feed chute. An objection to this arrangement is that it can seriously damage the belt, if an object such as a length of drill steel comes along, is picked up at one end, and then gouges the belt with its free end. This danger is minimised by suspending a rectangular magnet at a 45° angle across the discharge end of the belt. If possible, a short feed belt ahead of the main conveyor should be used, so that damage of this sort is confined to a small length of belting. The electro-magnet is periodically pulled clear and de-energised, upon which it drops its load. For small pieces of iron, a magnetic head pulley (Fig. 253) may be used at the discharge end of the conveyor belt. This

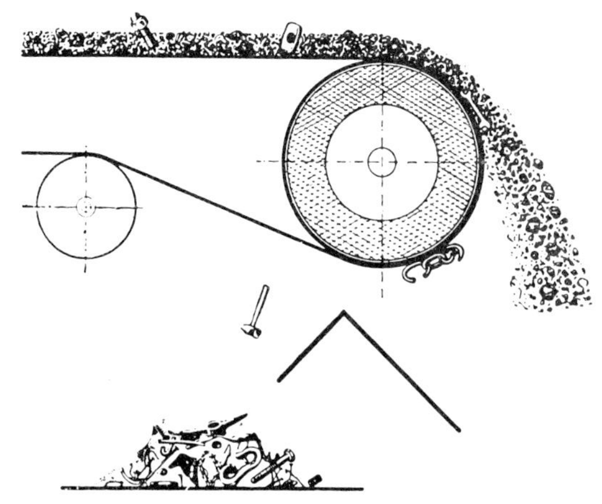

FIG. 253. *Separating Principle of Electro-Magnetic Pulley Separator* (Humboldt)

has the advantage that it is more likely to remove iron pinned down by a piece of rock than is the suspended magnet. It may have Alnico permanent magnets or be energised by D.C. In the latter case it must be inspected regularly to ensure that current is flowing, that cooling air-gaps are open, and that the pulley is functioning efficiently. To carry light pieces of iron clear of the underside of the pulley a light strip of steel belt lacing can be tacked in at intervals on the belt.

Special problems arise with non-magnetic steel. An excavator's digging tooth would smash something if it ran through to the crushers. Several sensitive methods of dealing with this type of problem have been worked out. Small inserts of magnetic steel can be drilled into all parts liable to break off in the mine and find their way up with the ore. These can sound an alarm and shut off the belt motor, if necessary marking the spot by dropping whitewash or chalk, since a loaded belt travels a little way before stopping. If there are two or more such search points, the risk of the attendant's failing to find all the signalling tramp metal at the first stop is removed. Mine-detector types of coil are used in some arrangements. Extremely sensitive arrangements can be used, but it is not necessarily desirable to stop the transport of ore for a tintack.

High-intensity dry magnetic treatment handles a considerable tonnage today. One economic advantage over wet concentration is that the latter must provide from 500 to 1,000 gallons of water per ton of concentrate made[16]. One type, the 30 in. Erzbergau dry concentrator, working with a maximum field of 23,000 gauss, has a throughput of 5 to 6 tons/hour for a power consumption of 1 kWh/ton. Feed is in the *minus* 300 ♯ + 30μ range, all *minus* 20μ material being removed by cyclone classification followed by filtration and drying[17]. The ore assays 25·2% Fe and 82% of the iron is recovered at a grade of $32\frac{1}{2}$% Fe.

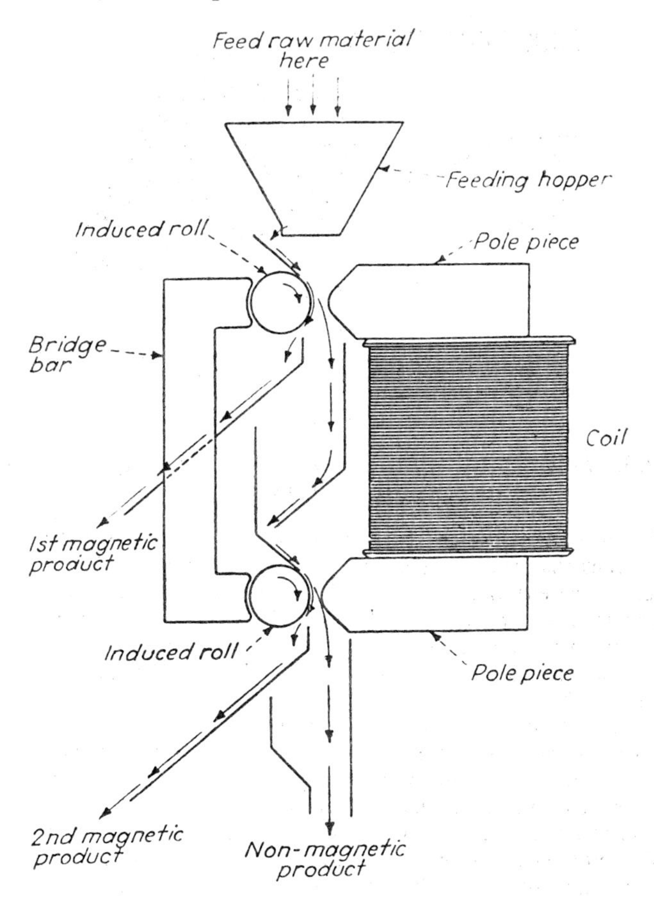

Fig. 254. Dings Induced-Roll Separator

Fig. 254 shows arrangement for weakly magnetic sands (8–200 mesh) in which the magnetic particle is deflected when falling through air. Up to seven laminated rolls, with progressively smaller air-gaps, are used.

Much of the concentration required today is concerned with sands and fine powders. In order to shake out trapped gangue particles by breaking up magnetized aggregates the modern separator develops a rotating field. This is produced by permanent magnets set with a pole distance of only a few centimetres. The design recommended by Runolinna[18] has one or more rotating drums, whose speed is so balanced as to use centrifugal force to discharge the more weakly attached and feebly magnetic particles. The magnets used in these drums can be closely packed. Cavanagh and Williams[19] have recorded studies made by the Ontario Research Foundation on sands varying from 6♯ down to −325♯. The basic developments in the work were guided by these considerations:

(1) For dry work a single layer is ideal.
(2) Economically, high operating speeds are needed.
(3) The shape of the magnetic field can be developed to be circular, spiral, or helical.
(4) No mechanical feeding and discharge devices should be used.
(5) Gravity should aid separation.
(6) An opposing air stream must replace the water-wash used in wet magnetic concentration.

(1) above can be achieved by using small permanent magnets mounted on a drum which rotates at a different speed inside an independent stainless steel drum. The magnetic particles are thus constrained in movement. Four types of machines have been developed from a "rougher" working on −6♯ +100♯ material down to one handling −325♯ dusts. The essential difference is between the appropriate dynamics for easily settled material and that easily airborne. In a specific case the relationship between average particle size, magnet design, drum speeds, and differentials is considered in setting the machine. Two types of rougher are used. In one the feed travels on a belt at 400 ft/min and passes over the double drum, of which the inner (magnet) drum is rotating in a reverse direction at 400 ft/min. The non-magnetics are thrown clear as the belt turns over the drum, while the magnetite clings as it is accelerated by the influence of the field flux, until it is separately discharged as the belt leaves the drum.

In the other "rougher" type the magnet drum is mounted in a second unit above the feed belt, and travels in the opposite direction. Gaps and flux intensity are so adjusted that the magnetic particles are picked up more than once and dropped, before finally adhering to the upper belt. As they climb they are again accelerated by the differential effect. The upper belt moves at twice the speed of the feed belt and the magnet drum rotates in the opposite direction, as before. An air blast is directed upon the climbing particles to aid separation.

−60 +200♯ material is treated in a different way. Dust is fed from a hopper on to a rotating stainless steel drum which forms its bottom. The inner magnetic drum rotates inside this drum, but at a different speed. Its magnetic particles thus aid with the feeding by stirring the material as they feel the influence of the reversing flux. As feed leaves the hopper it is stratified with non-magnetic gangue on top. This material leaves first, magnetic middlings later, and concentrate on the rising side of the drum. Detachment is produced by mounting the drums eccentrically, so that the gap widens sufficiently for centrifugal force to overcome magnetic attraction in the concentrate discharge zone. Air elutriation aids the separation. With this appliance magnetic flocculation may occur if the drums run too slowly, the proportion of magnetic material is high, or the population in the air stream too dense. This can be overcome by an arrangement of magnets which induces a helical field on the drum surface. Usually the magnets are mounted herring-bone fashion on the inner drum. Magnetic particles fed toward the centre of the drum migrate outward and are discharged at the sides, while the gangue falls off centrally. Several drums are normally used in series to produce a good concentrate grade. The machine makes a finished grade at −150# from ore containing not more than 30% of −325# material.

Very fine dust must be treated in an airborne stream. A device called the magnetic precipitator has been developed. The entering air travels in a helical path, dust being pressed against the inner rotating wall of a cylinder, while an outer system of stationary or rotating magnets induces a helical magnetic field. The magnetic particles move upward against a draught of air and at the top enter a weaker magnetic field and are thrown off, while tailings are blown down to a central discharge. All these machines have proved successful in pilot-scale runs, an Aerofall mill doing the dry grinding. Runolinna[20] has described development work in Finland by Professor Laurila on a drum type of machine. High field strength improves separation (in the tests 900 gauss 2 mm from the drum surface (Fig. 255) at a flux density gradient

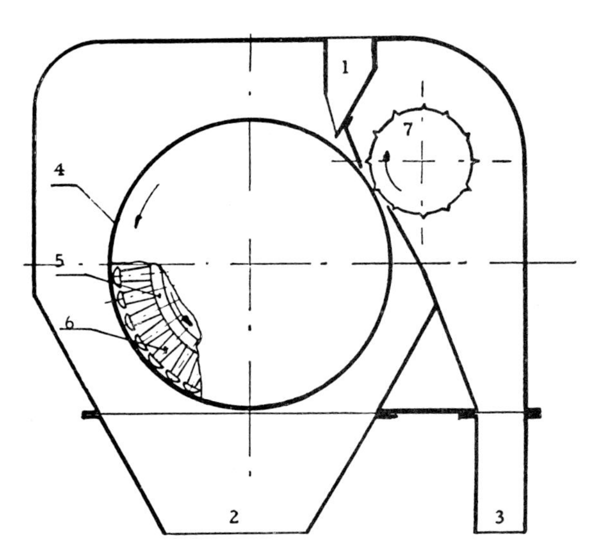

Fig. 255. The Laurila Separator (after Runolinna) 1. *Feed entry.* 2. *Non-magnetic discharge.* 3. *Concentrate discharge.* 4. *Drum.* 5. *Wheel carrying magnets.* 6. *Magnets.* 7. *Induction roller.*

of 50 gauss^2/mm was used). The drum is run at a speed somewhat below "critical", which is defined as the point where centrifugal force equals magnetic attractive force. A peripheral speed of 4·75 metres/second gives good practical results.

In an alternative arrangement[21], the Mortsell separator, the feed falls over three drums in series. The two upper ones rotate slowly, and centrifugal force is low. The top drum picks up most of the magnetic particles and drops its tailings to the second one for scavenging treatment and discard. The concentrates from both drums are delivered to the third, which is smaller and rotates at a regulated speed. This is so adjusted as to remove by centrifugal force any middlings which would lower the grade of the retained concentrate.

A pick-up type of drum separator (Fig. 256) is designed for use where a

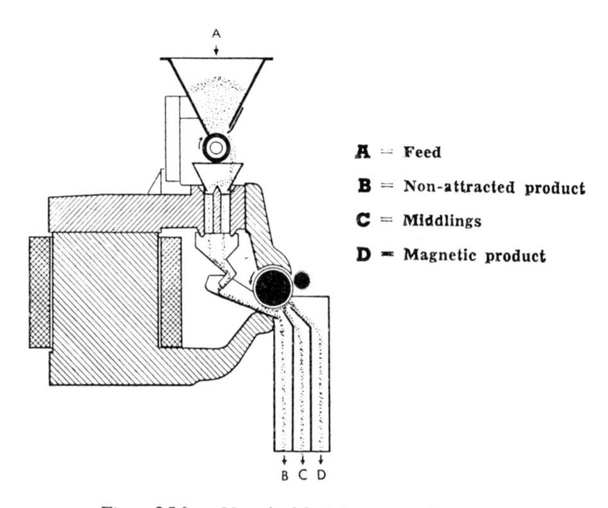

Fig. 256. *Humboldt Magnetic Separator*

large proportion of magnetic matter is contained in the feed and a high grade concentrate is required. The magnetic material is lifted from the passing stream and carried forward proportionally to its tenacity, to separate middlings and fully magnetic drop-off points. Where the magnetic permeability of the feed is low, or its percentage small, the feed comes directly on to the drum, from which non-magnetics are thrown while seized particles are carried down till they leave the field and fall clear. This is a roughing treatment.

High-intensity separation in which D.C. energised magnets are used has been discussed by Palasvirta[22]. He notes that the rebounding of tailings particles from the pole face back to the induced roll (see Fig. 254) increases with rotor speed thereby reducing the grade of the magnetic product. This can be suppressed by choke feeding, when the stream of particles blankets the rebounding gangue. Within the sorting capacity limits of the machine feed rate then becomes an operating control with a fast-run rotor.

Wet Magnet Separation

Wet magnetic separators are widely used in the concentration of ferro-magnetic sands and for purifying the magnetic media used in dense-media separation. The magnetic log washer is a spiral classifier with magnets below its trough. These cause the ferro-magnetic material to flocculate, drop, and be dragged up, while barren pulp overflows at the weir. In the Crockett machine (Fig. 257) a belt (4) dips into a trough (2) through which the pulp is

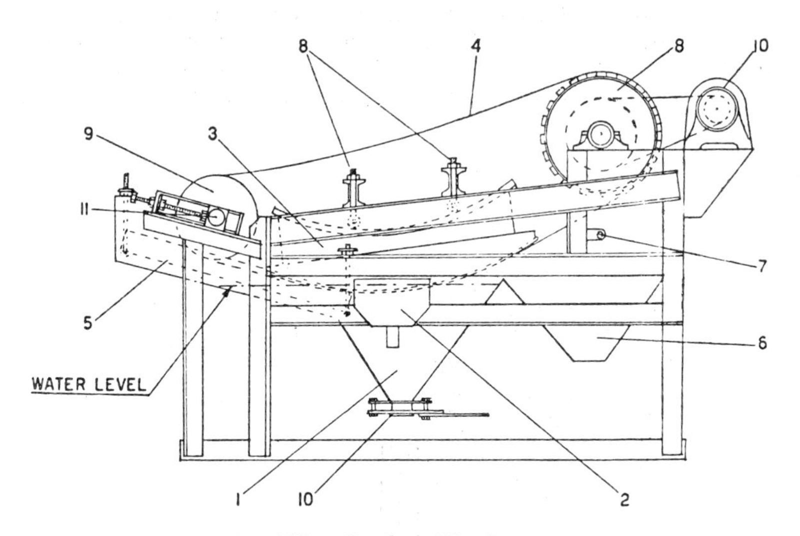

Fig. 257. Crockett Wet Separator

fed. Magnets (3) above the belt lift and convey the flocculated ferro-magnetics (6) forward and out while the barrens (1) run on to tailing. The Linney develops the same idea differently. When these machines are used for dense-media work, the residual magnetism must be destroyed before returning the ferro-silicon to the separating bath. This deflocculation is done by passing the material through an A.C. coil.

In northern Europe drum-type wet separators are much used for concentrating finely ground iron ore. The "pick-up" principle is used (Fig. 258), magnetic particles being lifted from the pulp while the gangue streams through. Hydraulic currents keep the pulp in suspension, and the picked-up fraction passes through a series of N and S fields before it is removed by sprays.

The need for efficient methods of cleaning foul ferro-magnetic media from D.M.S. baths stimulated the development of machines such as the Crockett. Later, these were largely replaced by drum-type machines containing electro-magnets, which were more compact and efficient. With the improved permanently magnetic alloys now available, these have in turn given way to drum separators of this type. In a review of current practice. Bronkala[23] gives as reasons for this change weight reduction, elimination of the need for

D.C. energising current, greater reliability and much lower cost of maintenance. The problem of insulation of the electrical wiring also disappears.

The drum separator has from four to seven poles alternating north and south, at the back of a non-magnetic drum which dips into a receiving tank and transports the magnetic fraction of the feed to the discharge point. Feed

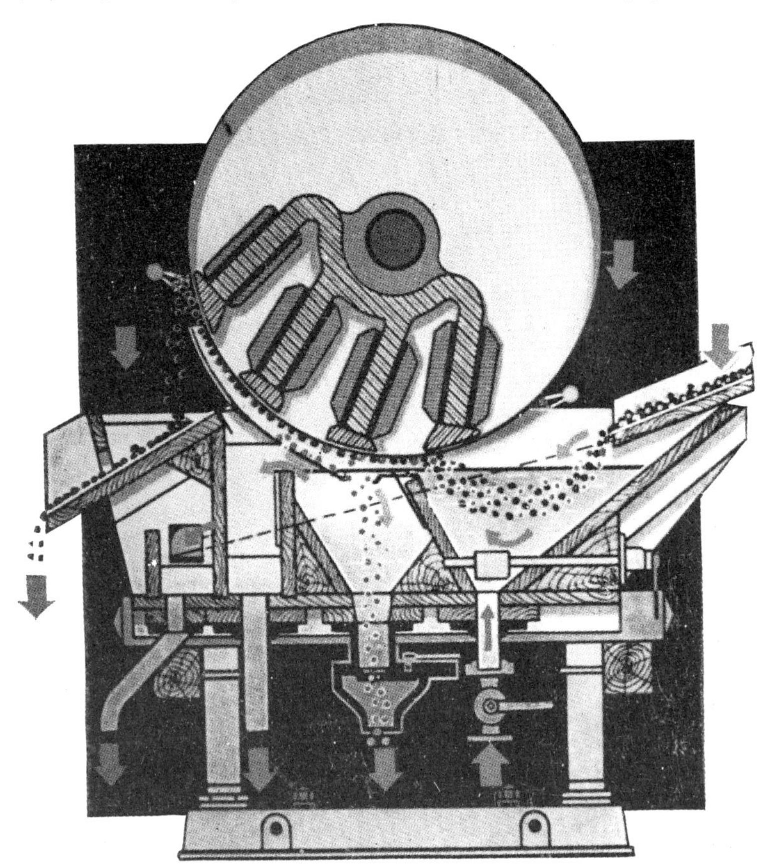

Fig. 258. *Principle of Separation in Wet Separator* (Humboldt)

slurry is spread the width of the receiving tank as it leaves the feed box. Fig. 259 shows a two-drum system in which the tails from the first drum are scavenged in the second. With the most recently developed ceramic magnets the second stage can often be omitted. A spigot usually delivers part of the tailings discharge, thus avoiding retention of particles too coarse to overflow. The rest of the tailings leave by weir overflow, thus ensuring a steady height through which final washing of the magnetic concentrates is performed.

Choice of drum diameter and width swept by the magnets depends on volume of slurry passing, its solid content, the percentage of magnetics, the desired operating efficiency and concentrate grade. Feed in single-drum work should be below 80 gal/min and 25% solids. With a double drum up to 125 gal/min and 50% solids can be handled. A thicker pulp should be diluted. The standard drum widths are 30 in. and 36 in., which allow a discharge rate of magnetic product of 3 and 5 tons/hr respectively with a single drum, and double this with two-stage work. The effluent from a well-

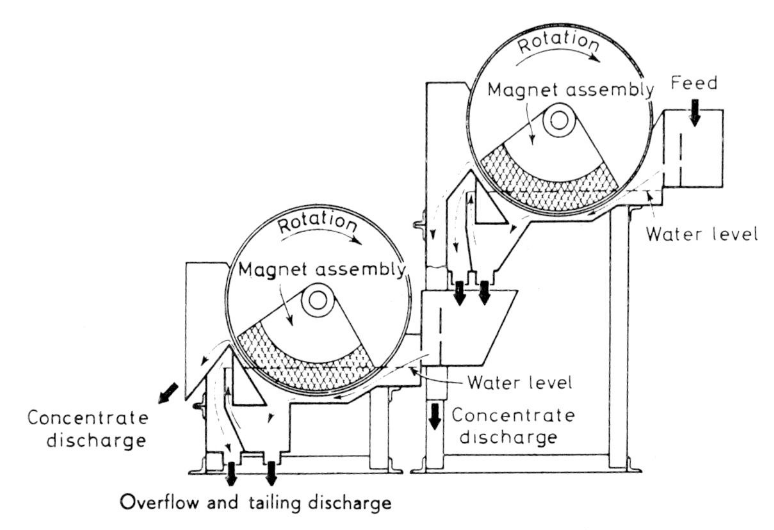

Fig. 259. *Double-Drum Wet Separator*

run primary drum contains from 1% to 5% magnetics in the tailings, up to 80% of which can usually be recovered in a second stage. In treating the effluent from a cyclone D.M.S. plant, where the ferro-magnetic media is finer, losses may be up to ten times as high ($\frac{1}{2}$ lb/ton or so with ferro-silicon and 1 lb with magnetite for ordinary pulps).

The Jones magnetic separator was described by its inventor[24] in 1960, and is now in commercial use. Detailed tests made in Canada between 1959 and 1961 have been reported[25]. The machine (Fig. 260) operates over cycles of approximately four seconds, the effect being continuous. A high-intensity electro-magnetic field is applied by the magnet 13, through its poles 12, to the plates 11. These are grooved—17, 18—so as to concentrate magnetic force at their peaks and provide valleys through which the non-magnetic solids pass to the discharge funnel in its 16 position. The feed is kept stirred in its hopper 1 and admitted through the valve 2 for two seconds. It is then cut off and low-velocity water is admitted *via* 9 and 3. As this flows two rams 5 initiate short high-pressure pulses which spread the retained magnetics and stir them on the plates while flushing out middlings, the funnel having moved to delivery position 15. After about $1\frac{1}{2}$ seconds of washing the magnetizing current is cut off and a short surge of high-pressure water enters *via* valves 6 and 3, the funnel being at the same time in position 14. The cycle is then repeated.

Tests on some 34 minerals were made and good concentration of such moderately magnetic minerals as hematite, ilmenite, garnet and biotite were made, down to sizes as small as one micron. Efficient separation depends on adequate liberation, distinctive discrimination in magnetic quality of the species treated and a well-dispersed pulp. The operating economics thus far

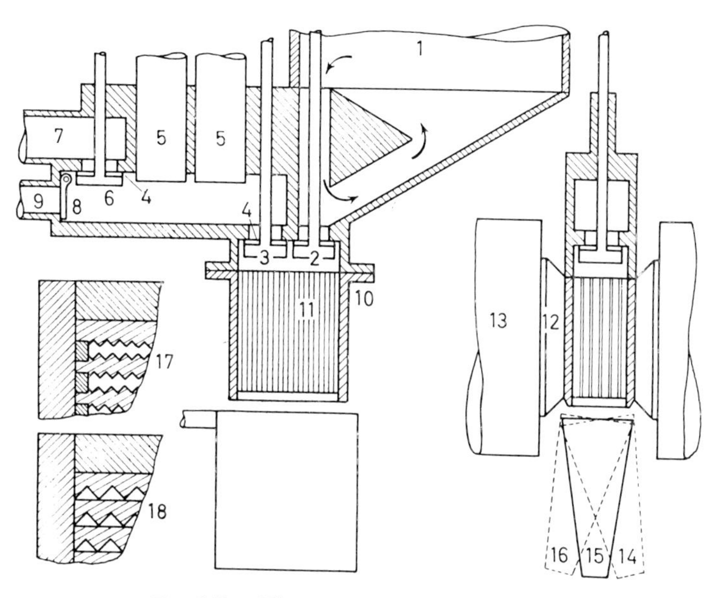

Fig. 260. *The Jones Magnetic Separator*

reported are attractive, less than 3kWh/ton of feed being needed. This machine should fill a special gap in the work of purifying high-grade clays, talc, glass sands and ceramics where the presence of iron is undesirable.

Another cyclic arrangement the Ferro Filter, has been used at Kipushi in the Congo to remove renierite (a germanium mineral) from a copper-sulphide flotation concentrate. The filter, a horizontal disc, works in a 135-second cycle of three stages—loading, flushing and discharge after cutting off the electro-magnetic flux. The disc is fed from above and discharges into a three-compartment tub.

High Tension Separation (H.T.S.)

This, which includes electrostatic separation, is an operation applicable to the dry separation of small particles. It is based on their relative ability to acquire and to retain electric charge applied at a high voltage. Suitably exposed minerals can capture (or alternatively lose) electrons and then be attracted to, repelled from or neutralised by other bodies which are either grounded or charged. Physical contact between the particles and these latter bodies is usual, but not always an essential step.

Static electricity is that in which the electrical charge is temporarily fixed on the charged body. With more than one region of fixed charge in a system there is regional interaction, modified by charge magnitude, polarity, distance between the charged regions, and di-electric constants of the particles concerned. These forces are exploited in H.T.S., which is basically dependent on electrostatic differentiation.

The main avenues along which charge can be acquired by the particle are:

(*a*) By conductance.
(*b*) By ion bombardment (gaseous).
(*c*) By friction.
(*d*) By thermal strains.
(*e*) By light or radiation conductivity.

If two types of particle are placed between a positive and a negative electrode, so as to rest on the former, the non-conducting particle receives equal charges from both sources and so remains neutral, whereas the conducting particle has its negative acquired charge neutralised, and hence ends with a total positive charge in line with its capacity. This is case (*a*)—charge by conductance.

In case (*b*) the particles can be freely moving through ionized gas from which one of their ionic species captures counter-ions and thus acquires charge. If the particles now make contact with an electrode or a grounded surface, only non-conductors retain their charge, and that in the area not in physical contact with the surface. In case (*c*) when two dissimilar compounds rub together electrons are transferred and a frictional charge develops. In (*d*) thermal strains in crystals may set up local areas of opposed charge—called the pyro-electric effect. Effect (*e*) is a photo-electric effect in which where incident light or x-rays causes electrons to be emitted so that the particle acquires positive charge.

The exploitable forces are delicate and are typically used to treat particles of low S.G. or high ratio of surface to mass. The limiting particle size is about 3 mm diameter, but shape affects the practical limit. Electrostatic processes have been broadly classified by Fraas[28] thus:

(*a*) Those which depend on difference in contact electrification either between substances which compose particles or between these and the separating surfaces of the appliance used.
(*b*) Those exploiting differences in conductance.
(*c*) Pyro-electric polarization processes.
(*d*) Methods depending on differential di-electric constant.
(*e*) Photo-electric or photo-conductive processes.

An important factor in modern industrial usage is charging by ion bombardment. The ions concerned are producible between an electrically charged wire and a grounded or charged conducting body separated from it by an air gap. Air round the wire becomes ionized and is attracted toward the grounded body, where it discharges its ions. If the voltage of the wire is sufficiently high the ionized corona is visible as a luminous discharge. Mineral particles which enter the electrified field are bombarded with gaseous ions from which they acquire negative charge.

Two types of charge are possible[26]. If the particle is falling freely gaseous anions tend to attach electrostatically to positively charged points in its surface lattice. It then acquires overall negative charge as part of this positively charged discontinuity lattice becomes neutralised and/or the particle as a whole becomes a charged condenser. After partial saturation the nett negative charge cannot increase, since further arriving gas-ions are repelled from its now negative field.

If instead of falling freely between the corona-producing wire electrode and the grounded surface the particle travels on the latter surface as it moves past the corona discharge, it receives the same ion bombardment as in the case above, but a new condition is added. If the particle is a good conductor or if its surface is conductive, the arriving anions pass from it to ground. If it is a good insulator, this leakage will be slower, less or even absent. A conducting particle cannot, therefore, retain all or much of the charge induced by ion bombardment but a non-conductor can. Build-up of retained charge to the final strength (the steady-state) is practically instantaneous and the polarity of this charge is opposite to that of the plate on which the particle rests.

If the grounded surface is moving (e.g. as a drum or roll in rotation) the particle is carried out of the ion-bombarding field into one where its electrostatic charge attenuates to its steady-state strength. A non-conductor retains full charge, an intermediate one suffers some loss, but a conductor's charge first decays to neutrality and then reverses its polarity until it has acquired the charge density of the roll. Thus, the non-conductor remains electrostatically coupled to the roll while the conductor either loses its grip or is repelled.

With a two-component mixture of particles of which some are negatively and some positively charged, separation can be effected by exploiting the attraction of each species toward an oppositely charged electrode, and its repulsion from one of like charge. Figure 261 shows separating conditions for (*a*) free fall, (*b*) fall along an inclined plate, (*c*) fall between rotating rolls instead of plates and (*d*) the additional use of an ionizing electrode. In (*b*) the particles fall on the positive electrode. If non-conducting but initially charged they take a steeper path on leaving the plate while the conductors acquire positive charge and are repelled into a flatter trajectory. The roll-type separator (*c*) aids removal of strongly adherent particles. In (*d*) the ionizing electrode is either a highly charged wire or one with a series of needle points (the spray electrode). It can either take the place of the negatively charged roll or reinforce its action.

The complexities of the surface physics of conditioned minerals discussed in earlier chapters are also encountered in H.T.S., since the ability of a particle to acquire or retain charge depends in part on its crystalline nature and in part on the state of its discontinuity lattice, particularly with respect to adsorbed moisture. It has been suggested[27] that the treatment of minerals with organic reagents can stimulate their electrical response by giving them differing affinities for adsorbed water, thus varying their surface conductivity. A hydrophobic surface has a sharply decreased conductivity, and it is possible that the conditioning of minerals before H.T.S. may have some analogy to that preceding the flotation process. Pre-treatment, however, does not markedly affect the electrical response of minerals which have high conductivity.

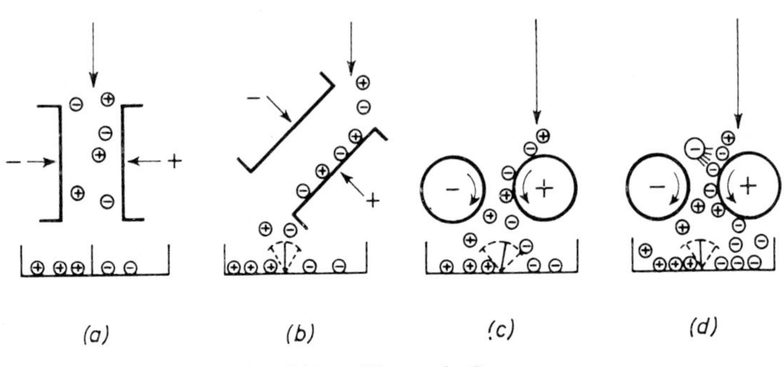

Fig. 261. *Electrode Systems*

Some Practical Considerations

Three types of particulate reaction are used in practice, and are defined by Hudson[33]:

(*a*) Conductance charging, in which particles passed over a metal surface charged by induction are themselves charged by conduction from it.
(*b*) Convective charging, in which particles are "sprayed" with a discharge from a sharp edge, points or fine wire.
(*c*) Contact-potential or frictional charging in which particles acquire a charge while passing over a surface and maintaining frequent intimate contact with it.

On completion of the charging process the particles pass through a high-intensity electric field and are deflected in accordance with the sign and magnitude of their charge. In view of the non-static nature of the convective process, in which there is some current flow, and of the differential time-element in discharge rates of various types of particle, the term "electrostatic" is not entirely satisfactory. For the (*b*) type defined above the term "high-tension" separation is perhaps preferable. Modern methods of rectifying alternating current to yield steady D.C. at high voltage have broadened the application of these methods, which are used in the benefication of various minerals not readily treated by the flotation process.

H.T.S. is sometimes applicable to material too coarse for froth flotation, or where dryness is important. It is also useful where small quantities of relatively valuable concentrates are to be removed from a gravity concentrate (e.g. in recovery of diamonds) and in arid country.

The surface potential of a substance (its contact potential) is a measurable fundamental property important in assessing its amenability to separation. The potential between the electrodes of the separating appliance is limited by corona or by spark-over. The working limit of the corona charge is 100 statvolts/cm. Spark-over instead of corona discharge occurs when the distance between spherical electrodes is less than their diameter. Spark-over voltage is proportional to the density of the ambient gas, and hence to its pressure and temperature. It is highest in a nitrogen atmosphere and diminishes *via* air, CO_2, O_2 to H_2. The ionization discharge varies with humidity. Fine wire, bars with projecting points, and knife-edged electrodes are in use. The Linari-Linholm electrode[29] is a rod 1 in. in diameter carrying a uniform row of sharp metal needles set at 0·125 in. centres, with a gap between electrodes of $1\frac{1}{2}$ in. Ionizing electrodes excited by A.C. have been experimentally used[30].

Complete separation is not likely to be obtained in a single pass (Fig. 262).

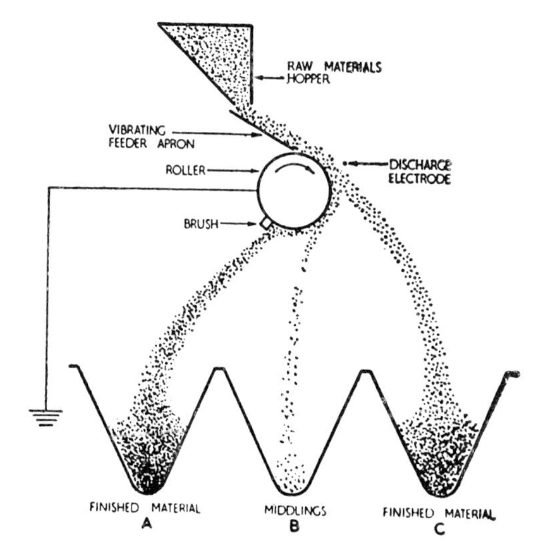

Fig. 262. A = *Non-conductive material.* C = *Conductive fraction.* B *may be returned for further treatment*

The usual procedure is to re-pass *either* the residual fraction at lower potential of ionizing electrode *or* the deflected fraction with a higher potential, or to re-treat both fractions as shown in Fig. 263.

Material is heated while passing in a single-layer stream from the feed hopper to the grounded separating roll. The needle-point electrode charges the passing stream. Good conductors discharge rapidly to the grounded roll and thus become electrically inert. As they descend further on the rotating roll they are strongly attracted toward the gas tube electrode and are deflected

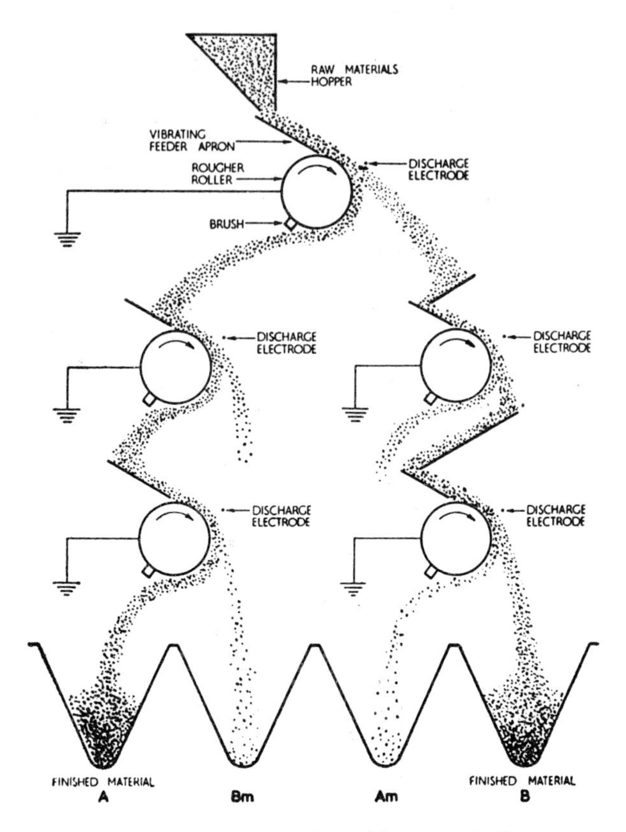

Fig. 263. *Operating Principle in Electrostatic Separation*
(Sturtevant Engineering Co.)

into products and middlings hoppers, by an adjustable splitter. The strongly adherent particles are swept off by a revolving nylon brush on the rising side of the roll. 50,000 volts at 1500 micro-amps are used.

A detailed diagram of one unit in a commercial machine is given in Fig. 264.

Instead of di-electric separation through air, separation in such liquids as kerosene and nitro-benzene have been practised on the laboratory scale[31,32]. Particles with di-electric constants higher than that of the medium tend to adhere to point electrodes while those with a lower D fall through. No important industrial application appears to have emerg ed but the method has had some use in sorting particles in the laboratory, prior to micro-observation.

Where the mineral surface is contaminated wet attrition, chemical solution or de-dusting may improve separation, provided the cost of treatment and re-drying is justified. A superficial film of dust can impair efficiency. Linari-Linholm[28,29] found that in H.T.S. diamondiferous salt-water gravels reacted differently from fresh, and used conditioning with a dilute solution of

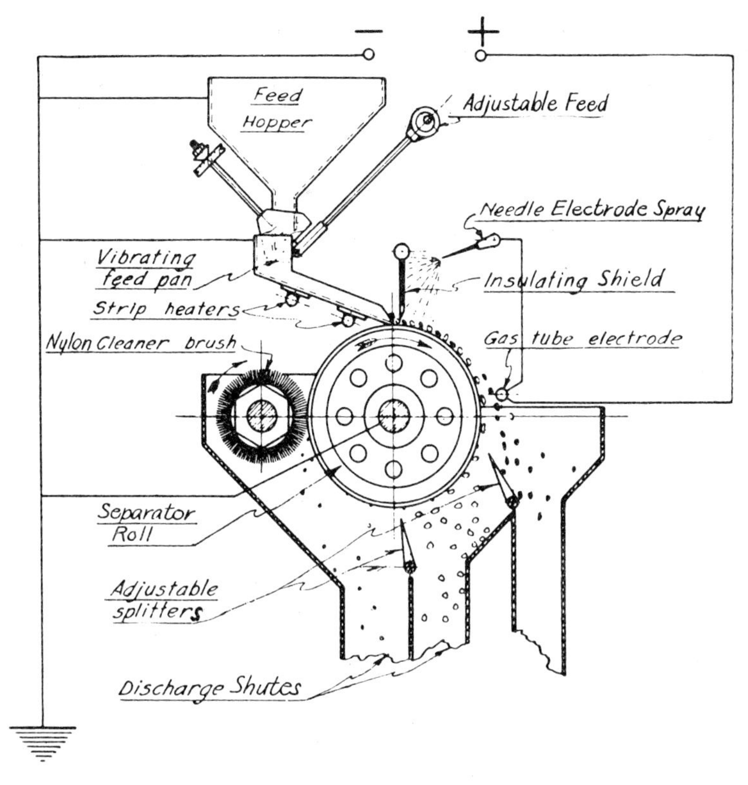

Fig. 264. *Single Unit in Electrostatic Separator* (Kipp-Kelly)

salt in acid to render the fresh-water gravel conductive. Amines, barium chloride, hydrofluoric acid, sodium cyanide and copper sulphate have produced a selective difference in conductivity, as has controlled humidity. For most minerals differentiation can be improved by raising the feed temperature, heating as high as 600°C being quoted by Fraas.

Conductive solids are of three types—metallic, ionic and semi-conductors. The last-named are highly sensitive to impurities a small change of which may produce a large change in conductivity. Semi-conductors with excess anions are "deficit" or "positive-hole" types, while those with stoichiometric excess of cations are termed "excess" or "electronic" types. An excess cation acts as a donor and an excess anion as an acceptor, the donor giving up electrons more readily. Changes in stoichiometric proportions under reducing or oxidising conditions while heated can be retained on cooling, but without such change the mineral reverts to its original state. Two possible aids to improved separation are therefore possible. One is to heat the feed to a prescribed temperature and to separate products at that temperature. The other is to heat and after re-cooling separate at a lower temperature.

In a specific case the maximum particle size which can be treated depends on shape and S.G. Heavy sulphides up to 6 mesh may be amenable, and up to 1 in. coke. The lower end of the size scale is about 325 mesh, smaller particles no longer flowing with sufficient freedom through the separating field.

In research on the basic principles which underlie mineral behaviour in corona-discharge roll-type separators the Warren Spring Laboratory has developed two novel laboratory appliances. One, christened the insiloscope, has stainless steel electrodes working in a humidity-controlled atmosphere in a brass box. Measurement of rate of charge leakage from mineral grains has found this to vary only with respect to surface conductivity with changes of humidity. The second instrument is a variation on the high-tension separator in which a flat disc spinning at 80 r.p.m. is used instead of the conventional roll. Mineral grains are fed radially so as to pass under a stationary line filament. As they lose charge they are thrown off centrifugally into one of the compartments surrounding the disc.

Some Industrial Applications

Among a growing list of minerals now treated in part or wholly by high-tension separation are Florida beach sands containing separable ilmenite, zircon and rutile and Indian and Australian rutiles and ilmenite. Other reported commercial separations include cassiterite from scheelite, scheelite from arsenopyrite, wolframite from quartz, monazite from rutile, columbite from euxenite, cassiterite from columbite, phosphate from quartz, feldspar from quartz and from mica, and diamond from heavy associated minerals.

The following short notes on the largely trial-and-error methods at present in use are mainly concerned with advanced research and pilot-scale tests. In work on a somewhat coarse-grained specularite a small-diameter grounded rotor has been used with a double electrode (a small static cylinder and a movable thin tungsten wire) at a potential of about 40,000 v (D.C.). Surface conductivity was found to be the discriminating factor[15], quartz minerals adhering to the rotor while such conductors as hematite were repelled. Feed must be dry, separately sized as granular and fine, de-dusted and clean as regards particle surface.

In all roll-type machines the centrifugal force is several times that of gravity. In the absence of an electrical field the thin layer of ore (ideally spread across the rotor one particle deep) is classified by air as it leaves the roll in accordance with the kinetic energy of each free-moving particle (K.E. $= \frac{1}{2}\, mv^2$), and receives this energy in accordance with the firmness of its contact with the roll. Once an electrical field operates a more complex situation arises. Non-conductors are seized rapidly and firmly and acquire maximum peripheral speed. They adhere until brushed off after physical contact with a rotating conducting brush on the rising side of the rotor. Moderately non-conducting particles may slip a little, and leave when the balance between gravitation, centrifugal acceleration and electrical attraction toward the grounded roll swings enough for them to be flung off. Conductors are the least likely particles to be gripped, and therefore acquire the least kinetic energy from the rotor. As they are in any case being electrically repelled they are the earliest to be thrown outward and the flatness of their trajectory can be augmented by attraction toward the positive electrode system (Fig. 261*d*). Between these are the swept-off non-conductors falling from the separating zone are two types of middlings—gravitational, and ionically charged[34]. Ideally four kinds of product can be gathered from separate trajectories. Furthest away from the grounded rotor and following the flattest path are the electrically repelled good conductors. Next come middlings which have mainly responded to gravity and air-classification, but which have some electrical charge. Third are the ionically charged middlings detached mainly by centrifugal force but mildly charged when leaving the rotor. Last are the non-conductors too firmly pinned to the surface of the roll to be removed by centrifugal force alone.

Conductance, charge etc. are relative to each mineral in the feed and to the purity and surface cleanliness of each of these species. Centrifugal speed and charge intensity are variables affecting the sorting, as is the positioning of the charging electrodes. The charging electrode has two parts, one ionizing by its corona field and the other not. The angular and radial positions of these electrodes relative to each other and to the feed roll affects the response of the passing particles. Roll speed and electrode voltage must be correlated. The electrode system should be high and fairly close to the feed roll, with the ionizing portion closer than, but not beneath, the non-ionizing electrode.

Titanium-bearing beach sands which contain relatively heavy minerals such as ilmenite, rutile, zircon and monazite together with lighter ones can be pre-concentrated by gravity treatment so as to separate the bulk of the silica, garnet, epidote etc. The heavy-mineral concentrate thus produced can be treated on the lines indicated in Fig. 265, 266.[36] Recovery of values by H.T.S. from such concentrates is usually high.[34] The method is particularly interesting where high-grade products must be made and the densities of the mixed minerals lie too close for effective gravitational separation to be used. Good response has been obtained from chromite-bearing sands, tungsten mixed minerals, microlite and tantalite, and cassiterite and columbite. Advanced laboratory tests on potash-bearing ores have been described[37]. Pilot-scale work on coarse pebble phosphate[38] and feldspars[39] has been encouraging.

One important industrial application is the concentration of diamond from associated gravity concentrates produced by D.M.S., jigging etc. The diamond is distinguished from the accompanying minerals by its very low conductivity[40]. In earlier appliances trouble was experienced when the gravels harboured such metallic objects as wire and nails, which gave trouble by short-circuiting the pin-point electrodes used. This has been remedied by shielding these points with a strip of phenolic laminated material (Celeron). This has the effect of intensifying deflection of the gravels. The strip consists of layers of materials with different insulating qualities which cause it to act as a condenser. This leads to crowding of the emerging field lines and thus increases the intensity of ion bombardment in the air gap. The commercial separator replaces the pin-pointed electrode

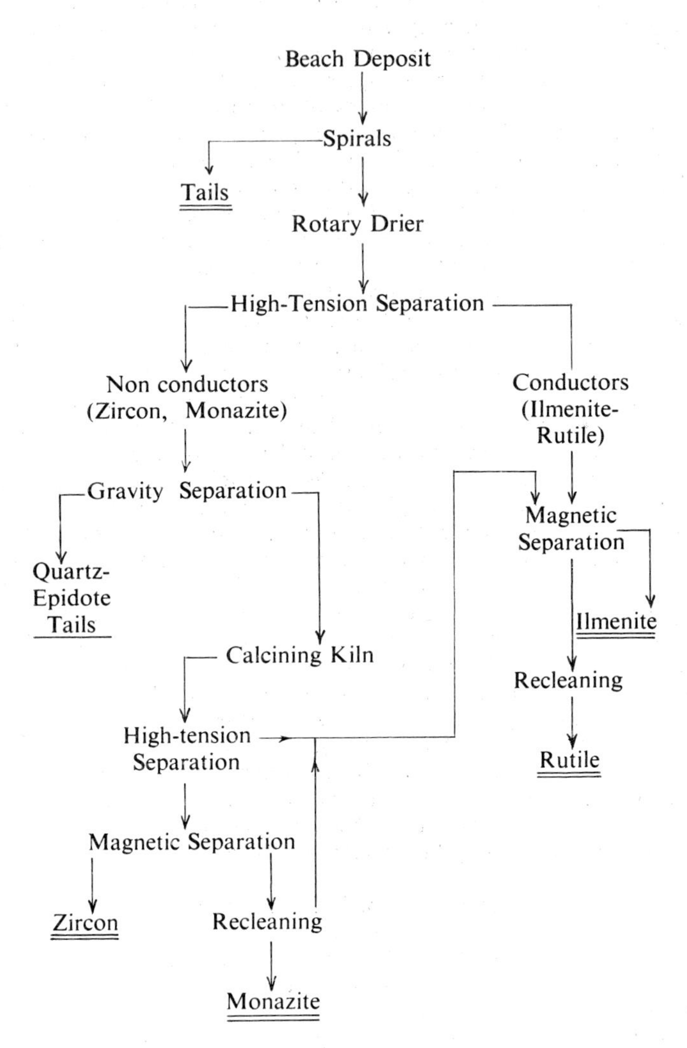

Fig. 265. *Treatment of Beach Deposit* (after Carpenter and Griffith, American Institute of Mining and Metallurgical Engineers, Feb. 1957)

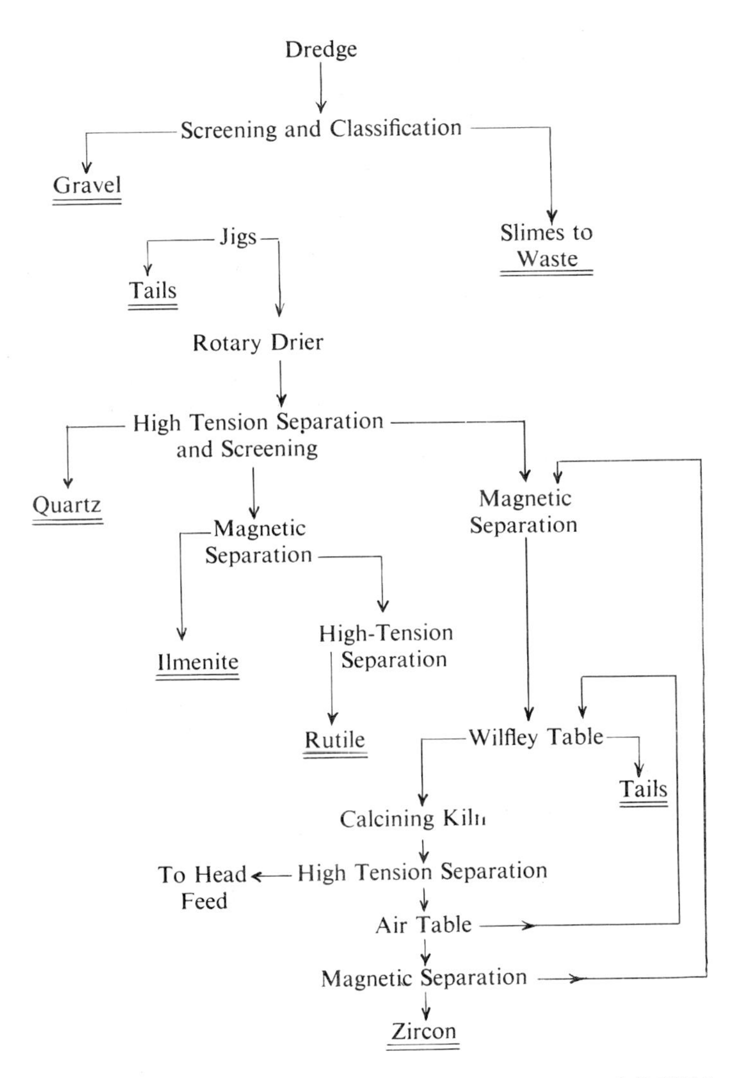

Fig. 266. *Treatment of Marine Alluvial Deposit* (after Carpenter and Griffith)

by a suspended insulating strip of Celeron which hangs from a charged high-potential metal rod when used to separate diamonds larger than 3 mm from gravels. For finer feeds only roll electrodes are needed.

At one plant which has now been producing for a decade *minus* 6 mm D.M.S. concentrate is attrition-cleaned in a tumbling mill, washed on a vibrating screen, dried and de-dusted. Warmed gravel (-6 mm $+1{\cdot}9$ mm at about 30°C) goes through six stages of separation. Each assembly has 5 ft long electrodes with a positive polarity of 22,000 v and the concentration ratio with feed at $\frac{1}{2}$-ton/hour is 200 to 1, for 99% recovery. In north-east Angola a diamondiferous concentrate not amenable to grease tabling contains staurolite, tourmaline, limonite, ilmenite, magnetite, zircon and quartz. Attrition-tumbling removes most of the limonite and electrostatic treatment on the lines mentioned above the ilmenite and magnetite. When negative polarity was used instead of positive the staurolite and tourmaline remained non-conductors while the diamonds were strongly deflected.

H.T.S. is in wide industrial use in electrostatic precipitators which are used to remove fine dust and fume from contaminated air or gas. The air passes between negatively charged wire electrodes and earthed or positively charged plates (the collecting electrodes). Particles are ionized, picked up by the plates, and periodically shaken down into hoppers by rapping. The subject has been discussed by Platksin and Olofinsky[41].

References

1. Goldman, J. E. *Magnetism's Role in the New Physics*, Ford Motor Co.
2. Willis Jackson. (1956). "The Ferrites". *Proc. Inst. El. Eng.*, Oct., 104B.
3. Hogan, C. L. (1960). *Scientific American*, June.
4. Pryor, E. J. (1963). *Dictionary of Mineral Technology*, Mining Publications.
5. Edgar Allen, Ltd. *Magnetic Materials*, Pub. 153.
6. B.S. 3100, Brit. Standards Inst.
7. Herzog, M. E., and Backer, M. L. (1963). *I.M.P.C. (Cannes)*, Pergamon.
8. Eketrop, S. (1957). *I.M.P.C. (Stockholm)*, Almqvist & Wiksell.
   Ibid. (1960). *I.M.P.C. (London)*, I.M.M.
9. Dahlem, D. H., and Sollenberger, C. L. (1963). *I.M.P.C. (Cannes)*, Pergamon.
10. Boucraw, M., Koskas, R., and Michard, J. *Ibid.*
11. Leon, J. B., and Worner, H. K. *Ibid.*
12. Mayer, K. *Ibid.*
13. Gagyi-Palffy, H., Palfi, G., and Halasz, H. *Ibid.*
14. Hencl, V. *Ibid.*
15. de Vaney, F. D. (1960). *I.M.P.C. (London)*, I.M.M.
16. Roe, L. A. (1958). *Trans. A.I.M.E. (I.M.M.)*, 211.
17. Goltz, A., and Neumann, K. (1955). *Erzaufbereitung-sanlagen in West Deutschland*, Springer-Verlag.
18. Runolinna, U. (1961). *Acta Polytechnica, Scandinavia (Chem. Ind. Mer.)*, 16 (303).
19. Cavanagh, P. E., and Williams, E. W. (1957). *Trans. Can. I.M.M.*, Sept.
20. Runolinna, U. (1957). *I.M.P.C. (Stockholm)*, Almqvist & Wiksell.
21. Kiblstedt, P. G., and Skold, B. (1960). *I.M.P.C. (London)*, I.M.M.
22. Palasvirta, O. E. (1959). *Trans. S.M.E.*, 214.
23. Bronkala, W. J. (1959). *Eng. and Min. J.*, 164.
24. Jones, G. H. (1960). *I.M.P.C. (London)*, I.M.M.
25. Wyman, R. A., Stone, W. J. D., and Hartmann, F. H. (1962). Dept. of Mines, Ottawa, T.B.36, June.
26. Barthelemy, R. E., and Mora, R. G. (1960). *I.M.P.C. (London)*, I.M.M.

References—continued

27. Kakovsky, I. A., and Revnivtzer, V. I. (1960). *I.M.P.C. (London)*, I.M.M.
28. Fraas, F. (1962). *Electrostatic Separation of Granular Material*, U.S. Dept. of Interior B. of Mines Bull., 603.
29. Linari-Linholm, A. A. (1950). *J. Chem. Met. and Min. S. Africa*, Oct.
30. Fraas, F., and Ralston, O.C. (1948). *A.I.M.E.*, T.P. 2408, July.
31. Hatfield, H. S. (1924). *Trans. I.M.M. (London)*, 33.
32. Holman, B. W., and Shepherd, St. J. R. C. (1924). *Trans. I.M.M. (London)*, 33.
33. Hudson, S. H. (1953). *Recent Developments in Mineral Dressing*, I.M.M. (London).
34. Barthelemy, R. E., and Mora, R. G. (1960). *I.M.P.C. (London)*, I.M.M.
35. Nilkuha, C., and Hudson, S. H. (1962). *Aust. I.M.M.*, Dec.
36. Carpenter, J. H., and Griffith, R. F. (1957). *A.I.M.E.*, New Orleans Meeting, Feb.
37. Le Baron, I. M., and Knopf, W. C. (1958). *Trans. A.I.M.E.*, I.M.C. Cpn., Oct., Chicago.
38. Oberg, F. N., and Northcott, E. (1958). I.M.C. Cpn. Chicago.
39. Northcott, E., and Le Baron, I. M. (1958). I.M.C. Cpn., Chicago.
40. Linari-Linholm, A. A. (1962/3). *J. S. Af.*, I.M.M., 63.
41. Plaksin, I. N., and Olofinsky, N. F. (1959/60). *Trans. I.M.M.*, 69.

CHAPTER 20

TESTING AND RESEARCH

Introductory

To obtain maximum recovery in any operation the material which is to be treated must first be brought into the condition which ensures the most effective response to the process used. Since the term "maximum"—whether used in connexion with efficiency or recovery—is not necessarily synonymous with "maximum profit" (the ultimate goal), treatment in a commercial plant is concerned to bring the feed at each stage to a condition which ensures optimum working efficiency in terms of maximum profit. This, ideally, involves the following stages of process-development.

(*a*) Fundamental research.
(*b*) Interpretation of (*a*) in terms of technology.
(*c*) Specific laboratory application of (*b*) to an ore (batch testing).
(*d*) Pilot-plant application, a continuous operation with suitable tolerances, of the laboratory method chosen as most suitable.
(*e*) Development of the pilot-plant method to the requirements of large-scale commercial practice.
(*f*) Cost analysis, comparison and assessment at each appropriate step.

These stages are not always followed. New methods are often worked out before the research scientist becomes aware of the problem and of the empiric solution which has been arrived at in the plant. Hence, in practice methods are at times used which are either harmful, needless, or only required because the whole problem has not been correctly analysed. In mineral processing, as elsewhere, it is sometimes possible to maintain good practice despite ignorance of basic principles, but the cost is apt to be excessive. Improvisation, occasional helplessness and frustration are the price paid for lack of fundamental knowledge. Logical activity, a keen and abiding interest, and accurate control are among the pleasures which result from correct application of scientific principles.

The orderly development of an ore body from original discovery to—and through—its processing treatment usually proceeds along the following lines (Table 41):

TABLE 41

Stage	*Place, method*
(*a*) Preliminary identification	Location, central lab
(*b*) Amenability tests	Central lab., miniature methods
(*c*) Drill cores available	Central lab., micro-tests

TABLE 41—*continued*

Stage	*Place, method*
(*d*) Tonnages opened out	Central lab., batch tests
(*e*) Alternative treatments known	Central lab., miniature pilot tests
(*f*) Flow-sheet decided on	Full pilot-scale tests on mill site
(*g*) Concentrator built and operating	Mill lab., routine tests
(*h*) Variations from original treatment plan	Mill lab., aided by central lab. at need

Stage (*a*) is concerned with simple and portable equipment controlled by the geological or prospecting units concerned with the search for promising ore deposits and their preliminary probing. The objective is the location of an ore body large enough to warrant economic exploitation. Samples are tested either in the field or at an existing central laboratory equipped for this work. If these exploratory tests are sufficiently encouraging stage (*b*) follows. Here again crude field tests may be made, or selected samples sent for examination. These are not probably fully representative, and will include hand-picked specimens in which ores thus far identified are well displayed, in order to aid the mineralogical evaluation with respect to amenability. As the project develops deep probes are made by core drilling or the sinking of exploratory shafts and tunnels, and samples characteristic of large tonnages of unaltered ore become available. These warrant closely detailed examination ((*c*) merging to (*d*)) by the central laboratory. In this each type of mineral present is identified, established as regards its bonded structure and purity, and assayed for grade as a percentage of the deposit represented by a bulk sample. These tests are made on one or more bulk samples of the ore, accurately representative, and shipped unaltered to the laboratory. If the deposit changes in character from place to place, more than one such sample of a few tons is needed. A few gallons of the proposed mill water (both dry and wet season) will also be wanted should the test work show that the quality of this water is liable to affect process efficiency.

Part of the bulk sample material will later be used in miniature pilot-scale tests (*e*) in which the various alternatives disclosed by batch treatment are compared in continuous "runs" designed to simulate operating conditions. As these may require expert control and product testing, they should be conducted in the central laboratory unless it is considered that the ore has changed by ageing during transit and storage, or that the process water to be used in the final concentrating plant may introduce special problems which cannot be exposed by using the laboratory's water supply. Otherwise the convenience of a wealth of specialised equipment under experienced technical control outweighs any small difficulties imposed by distance.

Stage (*f*) has now been reached, and construction of the mine and mill buildings is at design level or actively proceeding. The finally agreed flow-sheet has emerged and a proving run may be considered wise on the proposed site, using local water and freshly mined ore. The mill laboratory is brought into full use during this testing, both for assaying products and process evaluation. In due course stage (*g*) is reached, when the mill laboratory settles down to routine tests which will continue through the life of the mine.

Since the quality of the concentrating work is to depend largely on information gained from examination of shift-by-shift samples every precaution should be taken to aid quick and accurate handling of each manipulation which is to be be made during the coming years. Bad planning usually leads to short cuts, risky procedures and deterioration of control as novelty is succeeded by routine.

As the mineral deposit is opened up it may happen that adjacent ore bodies are discovered or that the deposit changes somewhat at depth. Newer methods of treatment may be found applicable and economically possible, or by-product minerals not originally allowed for may become worth recovering. Any of these happenings may call for tests, pilot check and appropriate modifications of the original flow-sheet (*h*).

Physical and Chemical Examination

On the whole, a specific mineral compound is broadly consistent as regards its reactions during mineral processing. This statement is, however, subject to some qualification. No mineral species occurs naturally in a pure state, and such modifying agencies as lattice distortion, solid solution, surface deterioration, and cross-reactions with other constituents of the ore or the conditioned pulp treated in chemical and flotation extraction must be revealed during test-work, if an efficiently controlled process is to result when large-scale operations are started.

Impurities diffused into the particle or pentrating *via* its visible or incipient cleavages affect its chemical constitution and stability as well as its physical strength and conductivity. Even a trace too minute to be readily detected may have an important bearing either on practicable treatment or the commercial value of the end product. For example, a little antimonite in an auriferous sulphide complicates the extraction of the gold and may lead to loss; a little chromium in the cyanide solution may upset recovery by precipitation with zinc. Evolutionary testing must therefore be concerned with both technical and economic viability of the process which is being developed.

Something of the lattice structure of each mineral species in the ore should be known. An assay of a copper deposit which simply established the percentage of that element without specifying the amount and compounded form of each copper mineral present might be dangerously misleading, since any bornite, chalcopyrite, malachite and chrysocolla present would present its special problems, have its own cost of recovery and its own efficiency of extraction. The economic purpose of testing is to arrive at maximum overall profit in operation. It must therefore be able to indicate both cost and probable percentage recovery with increasing clarity and reliability as the investigation proceeds.

The "advanced" alloys used in modern engineering demand call for high-grade concentrates able to meet tight specifications, and mineral processing is expected to play an important part in meeting this demand. In the by-product phase of economics, yesterday's impurities in an ore may become tomorrow's special values. Uranium, rhenium and germanium are cases in point. The main considerations in this section can be summarised thus:

(*a*) Recognition of each mineral species present.
(*b*) Determination of the pattern of its structure and of its purity.
(*c*) Assessment of the nature and possible influence of these impurities both as regards the mineral and in the pulp in which all impurities are free to react.
(*d*) Solubility product effects.
(*e*) Liberation-mesh of each species desired in its separate concentrate.
(*f*) Specific radio-activity and its exploitation.
(*g*) Detection and study of trace elements of potential economic interest.

Thorough examination on the foregoing lines is not possible until a representative bulk sample is available. It is preceded by amenability tests made on small samples of the more readily exposed ore deposit during preliminary geological and prospecting work, discussed later. When such a bulk sample has been obtained a "complete assay" is desirable. The term has been defined[1] as "one complete within the requirements of a special investigation, as regards identification of each mineral species in a sample of ore; establishment of its formula and possibly its stereo-structure; and correct quantification of all elements likely to enter the problem of devising a suitable method of treatment".

The term "complete assay" when used in mineral dressing refers rather to the acquisition of complete structural information bearing on the problem than to 100% identification of the components of the ore. The soluble salts and ions liberated in grinding are found by analysis of filtrate. If they are of a nature liable to affect flotation, the sooner this information is obtained the better.

Occasionally, with dumps and with shallow and secondary deposits, organic compounds of presumably vegetable origin may be found in the filtrate. These may be depressive, particularly when fatty-acid collectors are used.

Ordinary methods of chemical analysis are used to detect and estimate quantities of the elements, and when possible of their compounded forms. The total sulphur should be attributed to the metal bases, and important metals must be identified with reasonable precision in their proportionate occurrence as sulphide, carbonate, oxide, silicate, etc., since the acid radical is a prime determinant of the flotative behaviour of most metal compounds. Where the same elements vary in proportion, as in an ore containing a mixture of copper, sulphur, and iron varying from chalcocite to chalcopyrite, the combined efforts of the analyst, structural chemist, and crystallographer may be unable to unravel the tangle. Problems of this kind occur repeatedly, and should be sorted out as far as possible. The siting of the various elements in their lattices is sometimes important, in which case specialised investigation is needed. Several minerals called cassiterite cannot be brought within 20% of clean concentrate form, because other elements are bound into the lattice structure and nothing that mineral processing can do will remove these adulterants.

The most difficult problems are set by the "trace elements". These may be present in indefinitely small quantities in one of the ore minerals, and ordinary chemical analysis cannot detect them. Their presence may be established by some special behaviour. Fluorspar, for example, fluoresces in ultra-violet light if its lattice contains copper, but not if it is pure. The obvious line of attack in identifying the elements present is to make a complete spectrographic investigation. This is frequently disappointing, and misleading, because the attack is usually made on too broad a front. The ore should be broken down into constituents, either by selecting liberated particles under a microscope or by use of some concentrating method to divide a sample into its main constituents. For rough work these are next analysed spectrographically, but a complex compound may require a further preliminary analysis. This is done by taking any suspected fraction into solution and precipitating out its constituents separately with chemicals of known spectral purity. These precipitates are in turn examined in the spectrograph.

Chromatography can be used at suitable stages in work of this sort. Chromatographic analysis has been defined[1] as "separation of components of a mixture into zones, one or more of which can be identified by colour, etc. (*a*) by adsorption column, adsorbing from solute in a tube packed with cellulose, alumina, lime etc. (*b*) by electrochromatography, passage of electricity across a column or a paper strip along which a solvent mixture is flowing, causing migration of specific ions to one side of the flow-line (*c*) by electrophoresis—using electric current to aid migration (*d*) by paper partition, with separation into bands as a suitable solvent flows past a drop of solution which contains the compounds undergoing identification or quantification." Elements closely allied in periodic groups which change their valence properties from element to element by varying the electrons in the inner orbits are particularly difficult to separate when only small traces are present. If they can be got into solution, and then separated by chromatography, the spectrograph has its best chance to complete identification of a given product. Investigations at this level are unsuited to the mill laboratory.

Radio-activity checks should be used on unknown ores. A radio-active constituent may be a useful guide in test-work, since its travels are easily followed by means of a Geiger-Müller counter.

Among the potentially exploitable characteristics of the sufficiently liberated particle are:

(1) Specific gravity (S.G.)
(2) Shape and cleavage
(3) Refractive index
(4) Hardness and toughness
(5) Streak and colour
(6) Ferro-magnetism
(7) Radio-activity
(8) Di-electric constant (permittivity)
(9) Solubility product
(10) Liberation mesh and "*break*".

Specific gravity can be determined for a single particle by immersing it in a series of progressively heavy liquids, up to the maximum available with Clerici solution (S.G. 4·2). A small grain can be placed in a drop of solution under the microscope, and check on the thoroughness with which it has been wetted is then possible. Porous minerals may give misleading indications. The particle shape resulting from comminution is a possible factor in treatment since it affects both specific surface and rubbing contact. Cleavage, liable to be associated with structural weakness or obdurate deep sorptive contamination, can be observed under the microscope.

The refractive index aids in the identification of transparent minerals, which merge with the ambient liquid when immersed in one having the same index value (Table 42[2].)

TABLE 42

REFRACTIVE INDEXES OF COMMON MINERALS (*after Winchell*)

Liquid	*Refractive Index*	*Mineral*	*Refractive Index*
Ethyl alcohol	1·36	Fluorite	1·434
Chloroform	1·44	Orthoclase	1·523
Glycerine	1·47	Gypsum	1·524
Castor oil	1·48	Quartz	1·547
Xylol	1·49	Muscovite	1·582
Benzene	1·50	Beryl	1·582
Cedar oil	1·51	Calcite	1·601
Clove oil, or monochlorbenzol	1·53	Topaz	1·622
Fennel oil	1·54	Tremolite	1·622
Nitrobenzene, bitter almond oil, or anise oil	1·55	Dolomite	1·626
Monobromobenzene	1·56	Aragonite	1·633
Bromoform	1·59	Apatite	1·633
Cassia oil	1·60	Barite	1·640
Monoidobenzol	1·62	Diopside	1·685
a-Monochlornaphthalene	1·64	Kyanite	1·723
a-Monobromnaphthalene	1·66	Epidote	1·750
Cadmium borotungstate	1·70	Corundum	1·765

A Russian technique for identification, in which solid reagents are ground together with specimen crystals, has been described by Hosking[3]. Characteristic colours are produced for a wide range of elements and their compounds.

The hardness on Moh's or Knoop's scale is assessable by drawing a grain of the specimen, under controlled pressure, across a smooth face of the mineral surface used in these scales, and noting whether abrasion or scratching results. One difficulty here is that brittleness and toughness vary according to which of the three crystal axes is being put in tension. Grindability tests on a bulk sample are more informative. The colour of the powdered mineral, or that of the line drawn by it on an unglazed ceramic surface (its streak) is a valuable indicator. The colour of the crystal itself can vary considerably, and is to that extent misleading.

Ferro-magnetism is assessed by the use of a laboratory electromagnet in which the current can be regulated, or by means of a magnetic alloy. Radioactivity, if specific to one of the minerals comprised in the ore, is exploited in mechanized sorting and is easy to check. The di-electric constant affects amenability to high-tension separation.

The solubility product is sometimes a useful guide both to the original condition of the freshly severed ore and to its ageing during transport and storage. Two weighed samples are finely ground in distilled water. One is immediately filtered and the other after standing for a defined period, preferably after settlement and exposure of the moist solids to air before remixing with the decanted water and filtering. Both filtrates are evaporated to dryness, weighed and analysed. Original compounds which might influence the pulp climate, and those which may develop as the result of ageing before treatment, are thus discovered, and the final flow-sheet can be designed to handle any problem thus exposed.

Liberation mesh gives a clue to optimum mesh of grind (*m.o.g.*), which is further considered below. In this connexion the microscope yields valuable information. It can reveal the mode of interlock of the ore minerals and the type of association in particles of middlings, thus indicating the efficiency of liberation. When it is possible to locate particles of valuable mineral which have reported with the tailings, microscopic inspection may suggest, or even expose, the reason for their failure to respond to treatment and thus open the way to a remedy. Surface contamination, inadequate liberation, and partial oxidation are usually revealed by thorough micro-inspection.

In order to examine a specimen of ore a properly polished surface must be prepared. Harsh pressure and the use of high-speed abrading grind lead to deformation, formation of a Beilby layer, a matted surface in which undue removal of the softer constituents has occurred, or to deep scratches. Hallimond[4] recommends the use of a slow-running cool surface such as can be mounted on a horizontal turntable, run at 50 to 100 r.p.m. With this a series of sized quick-cutting abrasives is used, thorough removal of each from both lap and specimen preceding the use of a finer size. The specimen can be trimmed to preliminary shape by a rock-slitting saw (a thin disc armed with diamond-dust cutting edge and rotated at a high speed). Quite good work can be done by hand trimming to about a 1 in. cube and producing the surface by light hand pressure of the specimen on a piece of wetted plate glass sprinkled with abrasive. Those in common use include jeweller's rouge, silicon carbide, magnesia alumina and diamond dust. Aluminium foil, paper-backed and not surface-varnished, makes an excellent medium-holding surface if trimmed and attached to a small turntable. To examine aggregates of mineral sands or crushed products the sample can be set in lucite and then polished in the same way. A bibliography of microscopy as applied to mineral processing accompanies a paper by Amstutz[5]. His classification of intergrowth patterns is shown in Fig. 267.

Opaque specimens are studied in the metallurgical microscope by means of a vertical reflector, used either with monochromatic or polarized light. Another technique now coming into use employs the x-ray micro-analyser (Fig. 268)[6]. Electrons are focussed on the specimen and the resultant x-rays are viewed by reflection. A method of identification which mixes electrochemistry with microscopy is chromography[7]. In one development[8] the smooth surface of the specimen is pressed against cartridge paper soaked in an appropriate attacking reagent while current is passed for about a minute. The ions then transferred from the minerals to the paper are rendered visible in characteristic differentiated colours by development and the distribution of

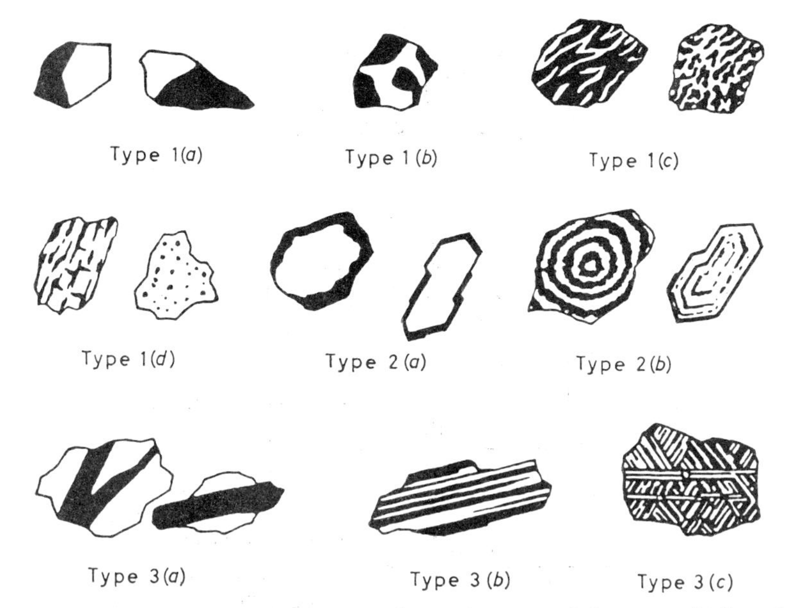

Fig. 267. *Geometric Classification of Basic Intergrowth Patterns of Minerals*

Type 1a: *Simple intergrowth or locking type; rectilinear or gently curved boundaries; most common type, many examples.* *Type* 1b: *Mottled, spotty, or amoeba-type locking or intergrowth; simple, common pattern; many examples. Type* 1c: *Graphic, myrmekitic, or "eutectic" type; common; examples: chalcopyrite and stannite; quartz and feldspars; etc.* *Type* 1d: *Disseminated, emulsion-like, drop-like, buckshot or peppered type; common; examples: chalcopyrite in sphalerite or stannite; sericite, etc. in feldspars; tetrahedrite in galena etc.* *Type* 2a: *Coated, mantled, enveloped, corona-, rim-, ring-, shell-, or atoll-like; common; examples: chalcocite or covellite around pyrite, sphalerite, galena, etc.; kelyphite rim, and other rims.* *Type* 2b: *Concentric-spherulitic, or multiple shell-type; fairly common; examples: uraninite with galena, chalcopyrite, bornite; cerussite-limonite; Mn- and Fe-oxides, etc.* *Type* 3a: *Vein-like, stringer-like, or sandwich-type; common; examples: molybdenite-pyrite; silicates; carbonates; phosphates, etc.* *Type* 3b: *Lamellae, layered, or polysynthetic type; less common; examples: pyrrhotite-pentlandite; chlorite-clays, etc.* *Type* 3c: *Network, boxwork, or Widmanstatter-type; less common; examples: hematite-ilmenite-magnetite; bornite or cubanite in chalcopyrite; millerite-linneite; metals, etc.*

Between most of these nine common locking types there are naturally gradational transitions with regard to both, pattern and size. Particle or grain size data are a prerequisite of any accurate study of rocks and mineral deposits and enhance the value of this chart (after Amstutz).

these minerals is readily scanned through a low-powered microscope.

In addition to the study of prepared samples, the microscope is used in the processing plant to examine granulated minerals at various stages along the

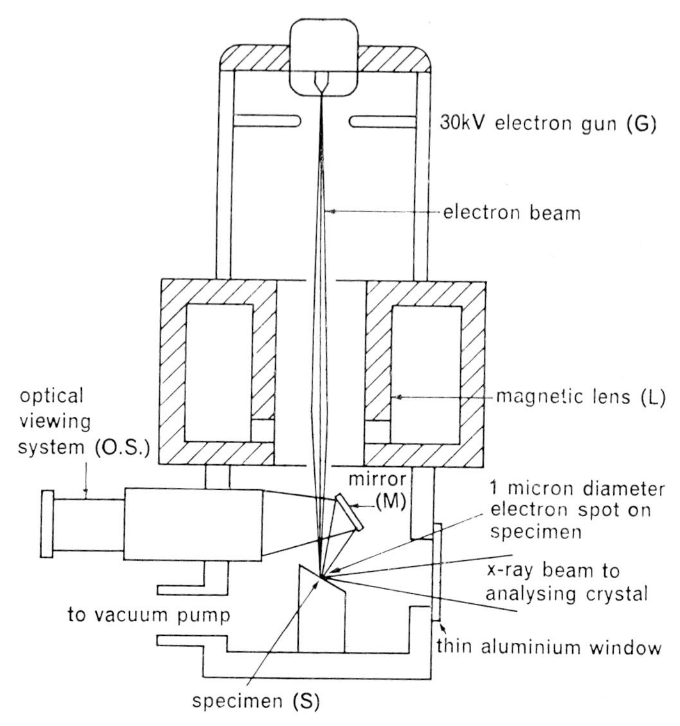

Fig. 268. *Schematic Diagram showing the general features of an x-ray micro-analyser*

flow line. Immediate information is thereby given on matters concerned with working controls, such as misplacement of product, surface patinas on particles, grain size and concentration. A further field of use is in sub-sieve sizing, with the aid of either a stage micrometer or an eyepiece graticule. An invaluable accessory to mill control is a binocular microscope, through which wet material placed on a microscope glass slip and perhaps dispersed by the use of a drop of Calgon, or under water in a Petri dish (thus avoiding surface reflection), can be seen in three dimensions.

Various micro-manipulators can be obtained, though for most purposes a cork-mounted steel needle permits the operator to pick out a grain. Apparatus exists which permits micro-analysis of such selected material under micro-observation.

In appropriate cases selective illumination by fluorescent light can aid micro-identification. An ordinary microscope can be used, with a high-pressure mercury lamp and an isolating filter. In one method a drop of water is placed on a slide, and a small quantity of powdered ore is added. The slide is dried and examined for natural fluorescence. A drop of fluorescin dissolved in alcohol is then introduced. Quartz, which originally exhibited red fluorescence, is now a bright golden yellow; corundum changes from dark to light green. Fine scratches on an apparently polished surface can be shown up by rubbing with a smear of fluorescent oil or grease and then examining by incident illumination.

Amenability Tests

When prospecting has indicated the possible existence of a substantial ore deposit preliminary tests may be called for. Their main purpose is to identify the contained minerals and to check whether they should respond to standard methods of treatment. There is at this stage no substantial exposure of representative ore, so elaborate tests and large samples are not required. Some rough tests can be made in the field, using simple equipment such as the miner's pan or a blowpipe kit. It is, however, possible to make mistakes in relatively crude local tests, so a few small specimens are often chosen, which exhibit as clearly as possible the known minerals. These are sent to the testing laboratory and given miniature tests sufficient to show the type of each species and its reaction to gravity, chemical, magnetic or flotation action. Tests on these lines, some of which are described below, are called amenability tests.

Bulk Samples

The field preparation of a large sample designed for exhaustive test work is described in Chapter 21. It may be five or more tons in weight, and represents a large and partially proved body of similar ore. The testing laboratory may be called on to do one or more of three things: (*a*) it may investigate in detail the response of the ore to various methods of treatment, reporting in terms of percentage recovery and indicated process cost; (*b*) and arising out of (*a*) it may suggest the basis for a flow-sheet which will guide mill design, full-scale equipment and commercial processing, (*c*) it may compare current mill practice in the case of an already working mine with possibilities for improvement revealed by the tests.

Work on the sample will usually take some months to complete, starting with batch tests on small quantities (a kilogramme or less) and ending with small-scale continuous runs, called pilot tests. The first step is to cut a reliable head sample which will be truly representative of the bulk. This is subjected to a "complete analysis" and must be sufficiently trustworthy to be considered the representative feed in all further tests on the raw material. Some hundreds of small batch samples will be tested during the work, and part of the ore may be needed for dense-media tests and pilot runs on freshly ground feed. The whole sample is therefore gently but thoroughly dried. This should arrest any further ageing or acidic deterioration that may have started in transit. The pile is then mixed and an adequate large sample is withdrawn after coning and quartering. This is sealed up, while still dry and warm, to preserve it until needed for D.M.S. or pilot-scale testing. The rest of the ore is screened until it passes the coarsest mesh consistent with the need for thorough blending and with due consideration of the liberation mesh of the largest mineral grains observed in preliminary inspection. This would not include barren gangue, but should be in line with the requirements of good mixing discussed under sampling in Chapter 21. Probably the material is now below coarse sand size. It is now given a final mixing and sealed in small lots (say 7 lb. packets) in polythene bags, care being taken to check that it is dry. If there appears to be a danger of oxidation, some of the packets may be given the further protection of a small addition of dry calcium hydroxide to maintain alkalinity, but the fact must be recorded and the weight noted.

The final crushing size is chosen with regard to treatment in preparing these packets. If jigging is contemplated as part of the flow-sheet the ore is not brought down to sand size, for example. Crushing machines must avoid local overheating of the ore, particularly at the later stages, where a metal sulphide could easily be semi-melted in an appliance such as a laboratory disc mill.

Sizing and Sorting

Laboratory screens, elutriators, infrasizers and microscopes are among the instruments used in the routine check of mill samples. The main purposes served are:

(*a*) Check on particle sizes released from comminution to treatment.
(*b*) Check on sizes of mill feed, mill discharge and classifier on cyclone returned products.
(*c*) "Losses" attribution in mill products, notably tailings.
(*d*) Examination by size of re-circulating middlings.

In non-repetitive process testing and research two additions may be made to this list

(*e*) Elimination of the size variable in gravity test-work.
(*f*) Regulation of specific surface variations by close sizing.

Sizing and sorting methods and purposes were discussed in Chapter 8 in their general application. Subsieve sedimentation by beaker decantation is described in Appendix B[9]. Standardised methods of sizing with the optical microscope are set out in a B.S.I. brochure[10].

Item (*c*) in the above list can provide valuable information regarding the efficiency of comminution in large-scale treatment. In one application, suited to a small mill laboratory, a small weight of mill tailings is accumulated from each shift analysis and periodically used. First a cumulative head sample is taken, to check the routine shift tailings samples. The rest of the sample is sized and sub-sieved and each fraction is weighed and assayed. From the results the proportion of the total loss of value in the combined tailings is attributable to each sized fraction, thus directing attention to any remedial condition desirable in the grinding circuit. Regarding (*d*) above, sizing aids in the micro-examination needed to throw light on the reason for the existence of the middling.

Screen sizing can be used down to 200 mesh with good accuracy. Below this mesh the difficulty of weaving screens with precision increases, and the problem of ensuring that fine particles find their way through the meshes also becomes more severe. Screening can be taken finer, but is usually stopped at 400-mesh Tyler (nominally 38μ). A mixture of sizes tends to aggregate and thus falsify the work, whereas particles covering only a short range of sizes move freely with respect to each other and therefore report more accurately to their correct size-grade. Screening should therefore be commenced by washing a charge of up to 100 gm. (less for closely accurate work) with a gentle stream of water to which a dispersant such as sodium silicate has been added. This operation is performed with the material on a 200-mesh phosphor-bronze laboratory screen (or any non-corroding material). The screen is placed in a vessel such as a white enamel basin in which undersize is caught. After this washing the undersize is settled, collected, dried, and added to any further -200-mesh material produced in the rest of the operation. The oversize is also collected, dried and dry-screened through any desired series of screens, using a suitable vibrating apparatus.

The -200-mesh product, or better, enough -100-mesh or -150-mesh particles from a number of screenings to provide an adequate head sample is next treated by subsieve sizing. Of the three principal methods (infrasizing, elutriation, and beaker decantation), the latter is preferred by the author since it requires no apparatus which does not exist in a mill laboratory.

Comminution

If the testing laboratory is suitably equipped the coarse crushing tests can be made there during the reduction of the bulk sample considered above. Alternatively, a large sample may be sent to the manufacturer's more specialised testing plant. The test sample should be screened through mesh *a* and on mesh *b*, to accord with the usual by-passing of undersize round the crusher. The sizes are selected in accordance with the desired gape and set—for example 3 in. and $\frac{3}{4}$ in.—and between 1–2 cwt. of ore are needed. The product is in due course returned to the bulk sample and mixed with it.

The crusher setting and throw are accurately measured, and its power consumption is checked while idling after warming up. The sample is weighed and fed through the crusher fast enough to maintain a full load, the time taken and the average power readings being noted. If the crusher is of the cone type, its setting can be checked by passing a short length of lead piping (larger than the set) through and measuring the thickness on emergence. From the figures obtained for various feed and discharge settings and nett power consumed per ton, the coarse crushing behaviour of the ore is observable.

Grindability tests are made in free-path crushing appliances. Their general lines have been described by Bond *et al*[11, 12]. From these the "standard work index" (W) can be calculated.[13] This is the kWh/ton applied during the reduction of indefinitely larger material to a particle size at which 80% passes a hundred micron screen. The key formulae (Eq. 3.5, 3.6 and 3.8) were discussed in Chapter 3. If W_i and the comminution exponent r of Eq. 3.8 are known for the ore under test, this general equation shows the horsepower needed to produce the required size of product.

The grindability tests used to establish W_i and r can be made under dry or wet conditions, the latter starting with dry grinding. One recommended procedure[14] begins with a dried sample weighing 50 lbs. and screened through 6 mesh with precautions during crushing to avoid over-crush. A screen analysis cut from this head sample establishes the particle size which constitutes the 80% passing mesh of the feed. A bulk charge for testing is then prepared by consolidation in a 1,000 ml graduated cylinder by vibrating it on a laboratory screen shaker for half a minute so that 700 ml are prepared. This is weighed and charged into a smooth cylindrical ball mill 12 in. long and 12 in. in diameter. Two hundred and eighty-five steel balls, weighing 20·13 kg are also charged in. Bond recommends the following size distribution.

Nominal ball size	*Average ball size*	*Number of balls*	*Weight (grammes)*
1·50″	1·45″	43	8730
1·25″	1·17″	67	7197
1·00″	1·02″	10	705
0·75″	0·76″	71	2058
0·50″	0·61″	94	1441
		285	20131

This dry charge is given 50 revolutions at 70 r.p.m., and then removed and screened, all material of the desired finishing undersize (P_1) being removed. The rest is returned to the mill together with a weight of new feed equal to P_1. From the original screen analysis and the weight of feed the weight of P_1 producible by a given period of grinding can be calculated. This in grammes divided by the number of revolutions used (g./rev.) is the grindability G. From g./rev. for the first grinding period the number of revolutions required in subsequent test periods is calculated. The cycle is repeated until equilibrium of P_1 is reached in at least three successive grinding cycles.

If, as is customary, wet grinding will be used in the industrial plant, a circulating load is first brought into equilibrium by the above method. The charge is then diluted to, say, 65% ore and 35% water (by weight) and tests are continued in the same way until a new equilibrium is established. In precise work the oversize from each cycle is screened off, dried and weighed before the charge weight is made up with new material. By making an allowance for moisture after the first of these cycles the drying step can be dispensed with in making an approximately correct addition to the charge. Tests can be conducted with varied solid-liquid ratios between 50% and 80% solids by weight. Other variations are mentioned below.

Scale-up from laboratory tests to calculation of the size, capacity and power required for a given tonnage to be milled is considerably aided by these tests. A small sample cannot give assured predictability with regard to large tonnages of ore liable to change in its constitution, so comparison, empiricism and the intangible criticisms of experience may justifiably modify the findings. Considerable information has been acquired by manufacturers of milling plant, and much has been published. Using Eq. 3.8 as regards work required (W) and knowing the tonnage to be treated per 24 hours, the applied horse power (H) is H = WT/18 to which a 10% minimum overload allowance is usually added when selecting the drive motor. A second scaling-up method of selection is by "base mill" comparison with performances by other mills working on similar ore. Any variation in nature or in feed or product sizes could prove misleading with these methods.

In the author's view it is not wise to calculate from a laboratory rod mill to a large ball mill in one jump. The pulp's solid-liquid ratio is lower in the small machine than the large one, and abrasive grinding rather than impact is at work. Final decisions are reached either at the pilot stage or on the basis of comparison with ores of similar grinding characteristics. It will be observed that empiric tests are relied on in this class of testing, while strictly adhering to the basic rule of good research—altering one factor only at one time.

These factors may include variation of the charge; of the make-up of balls; of mill speed; of liner shape (not tested for on laboratory scale); of size analysis of new feed; of pulp chemistry (notably its *p*H and content of sodium silicate); and of solid-liquid ratio. Such changes as those from sliding motion of the crop load toward cataracting, and of the percentage of the volume occupied respectively by grinding bodies, ore and water can be studied as part of these tests.

Since the final objective is optimum profitability of an operation in which concentration is affected by the quality of the grinding, the effect of *m.o.g.* on recovery and concentrate grade must also be assessed. This probably involves a tie-in between grinding tests and those concerned with efficiency of treatment in the concentrating processes. Fine grinding adds to cost both as regards wear and power consumed. Current practice is turning somewhat from the use of steel balls to autogenous media. The grinding of run-of-mine ore by selected large pieces of ore or by added pebbles cannot at present be investigated by small-scale laboratory tests. For this work facilities are maintained by manufacturers.

A single short grinding run of, say, a 700 ml charge is a batch test. As soon as part of the product is discarded and part recirculated this returned portion constitutes a kind of middling and the operation is termed a locked cyclic test. Hitherto in this book middlings have been considered as incompletely liberated values but this concept can usefully be broadened to include gangue middlings of greater or less grindability. The object in grinding is not so much to remove as finished product all material below a specified screen size chosen in accordance with the liberation mesh of the valuable ore constituent as to release from the grinding circuit this valuable mineral together with all gangue particles, whatever their size, which are wasting mill capacity and power by being uselessly retained in the system.

In some cases there is sufficient difference between the SG of the value and that of the usually lighter gangue mineral for the classifier to be so controlled that relatively coarse gangue and finer value overflow together. Micro-examination of the returned grains, coupled with sizing analysis and check with a suitable heavy liquid, will throw light on such a situation and suggest useful economies in the final flow-sheet. These arise partly from increased capacity and partly from reduced overgrinding of value while a lighter tough gangue is being reduced in size. The final specification for overflow product in the full-scale plant may include some such instruction as "80% *minus* 100 mesh and not more than 7% plus 65 mesh", based on the feasibility of controlling classification so as to overflow up to 7% of barren gangue at the latter size.

Where facilities exist useful information as to the balance between grinding and classification can be obtained in a miniature continuous operation. The requirements are an accurate delivery of new feed, a small ball mill with the necessary facilities for sampling, speed variation and rapid opening up, and either a classifier or preferably a hydrocyclone closing the circuit. The relation between feed and overflow size, feed rate and circulating load, crop load and power consumption can be checked in such a system.

Gravity Separation

Since in many circumstances gravity separation compares favourably with other methods its applicability should be considered. Indeed, there is at present no widely used alternative in the treatment of such minerals as cassiterite, coarse metallics (notably gold), chromite, and beach sands containing ilmenite, rutile, monazite, etc. The determining criteria are the relative density of value and gangue; the "break" at which the minerals can be effectively separated by comminution; the size-range of the resulting values; and the specific surface (as pseudo-viscosity) of each readily classified fraction of the feed pulp. The concentration criterion (C) is

$$C = \frac{(S_h - S_m)}{(S_l - S_m)}$$

where S_h and S_l are the specific gravities of fractions which are to be separated and S_m is that of the separating fluid in which they are borne and through which they must move to separate exits. Where water is the medium, as in jigging and tabling, S_m is 1·0, but when an element of dense-media density enters (as in D.M.S. and the use of spirals) the contributory effects of viscosity and particle size appear. In tabling the shape factor modifies the separating effect. Broadly, when $C > 2{\cdot}5$, all sizes down to fine sands can be treated. Thus cassiterite (S.G. *circa* 7·0) is readily separated from silica (S.G. 2·7), the C value being of the order of 3·5. The C concept takes no account of the time factor which enters increasingly as the particle size diminishes. In the case of cassiterite, where direct commercial flotation does not take over at any stage, treatment of the subsieve sizes becomes gentler and less efficient down to 30μ or less, when even a widespread settling area fails to handle a limited tonnage economically. Thus, in test work for gravity concentration strong emphasis should be placed on feed preparation so as to maintain sharply classified material in the coarsest practicable liberation-meshes.

When C is $-2{\cdot}5+1{\cdot}75$ jigging should be possible down to some 60#. Below 1·5 jigging is still possible with moderately coarse sands, but below 1·25 direct separation under turbulent conditions is usually uneconomic. If a measure of centrifugal force is used, C is raised accordingly. This suggests tests with spirals in the treatment of sands which are just below a C value for good tabling. D.M.S., which "weighs" its feed under relatively non-turbulent conditions, is efficient at lower C values than is jigging.

Test work should commence with measurement of the specific gravity of heads, liberated minerals and middlings in a series of sizes corresponding to those which would be produced by hydraulic and/or free-settling classification of typical sands. If substantial liberation occurs at +20# the same measurements should be used on coarse screened fractions. A crushed sample might then be sized and each size could be tested with a suitable heavy liquid. Assay, or micro-inspection of the sinks and floats, would show whether dense-media separation held any possibilities. The density of the fluid would be adjusted until the sinking fraction reached an assay value high enough to bear the cost of upgrading into concentrate. If a substantial quantity of gangue could be rejected at this density and at a moderately coarse mesh (say + 10 mesh), a few hundred weights of the ore would be tested in a small dense-media plant at various densities of media close to those indicated by the heavy-liquid tests.

The heavy liquids mostly used are tetra-bromo-ethane or TBE (S.G. 2·96), methylene iodide (3·31) bromoform (2·89) and carbon tetrachloride (1·5). A simple testing apparatus can be improvised from two laboratory glass funnels set in vertical series. The upper one is closed by a rubber tube and screw clamp and filled with heavy liquid. The dried sample is stirred in, given time to separate into floats and sinks, and the latter are then drawn down on to a filter paper in the lower funnel. After removal the floats are similarly flushed down to a second filter paper.

If the valuable mineral has a density below 3·5 or thereabouts, dense-media separation might be used for production of concentrates. It is cheaper than other forms of gravity concentration, and far more precise, provided the work does not require too high a bath density. Since, however, most of the commonly treated minerals have densities inconveniently high for direct differentiation in a heavy-media bath, jigging has a field of application outside the range of this method.

There is a trend toward the use of higher bath densities, and the possibilities of such methods as the Stripa process may widen the applicability of D.M.S.

Heavy-liquid tests would also indicate whether concentration by jigging and/or tabling appeared promising. In these operations the object would not be to reject ore too low in grade to warrant the cost of treatment, but to make a concentrate or high-grade middling. A miniature jig might be used to give preliminary information, after which a larger machine could produce a few kilograms of concentrate. Considerable adjustment is needed if the full possibilities of jigging are to be explored. The density of the bed is often a critical factor, also the size of the bedding. The bed consists of solids +

liquid, and the average voidage is therefore a significant factor in its composition. If a jig bed were composed of material of s.g. 4·0 and voidage such that the solid-liquid ratio was either 60–40 or 70–30, the weight of a unit volume of bed would vary by 10% and its mobility and discriminating effect of passage would vary correspondingly. Tests should therefore take into account the shape as well as the size and weight of the bedding material. At the same time, the temptation to control this and any other factor beyond the condition which could be easily maintained in plant practice must be resisted. Wear of the bedding material must also be reckoned with. The variables in a batch jig test, using a jig of the Denver type, include:

(1) Mesh of supporting screen.
(2) Mesh range of feed.
(3) Mesh of bedding material.
(4) Shape of bedding material.
(5) Thickness of bed.
(6) Density of bed, including void water.
(7) Rate of stroke.
(8) Length of jigging stroke.
(9) Rate of feed.
(10) Setting of hydraulic water valve.
(11) Percentage water in feed.

In view of the considerable number of variables involved—and this is usually the case in good work, whatever concentration process is being applied—a severely limited number of the possible permutations must be explored. Fortunately, most of the above items will be found not to have critical importance in a specific case, but 2, 3, 4, 5, 6 and particularly 10 need close attention. If possible, a rapid means of assessment of results should be worked out before starting, as it is important that the effects of each alteration be studied before proceeding with the next variation to be explored.

Tabling tests can begin with individual treatment of each size from about 50 mesh down to, say, 20μ on a Haultain superpanner. If material can be concentrated by gravity methods (tabling, vanning, and spiralling included), it can be concentrated on a superpanner. If this instrument fails, no commercial method will succeed on fine sizes. The superpanner is a miniature shaking table into which are built most, if not all, of the motions used in panning. A sample of a few grams can be tested, or some hundreds of grams can be worked in successive additions and particle removals. There is, in fact, no great virtue in using more than a light charge when testing for a gravity process, but bigger quantities are treated when search is being made for particles of heavy minerals only present in very small amounts. The advantage to the microscopist of being able to examine a sample of, say, half a gram which should contain all the heavier minerals from a total of half a kilogram are obvious. Any desired texture of decking can be used on the superpanner, linoleum and stainless steel being sufficient for most work. Adjustments are for side and end shake, slope, bump, stroke length and speed, and sluicing water. A variation on the superpanner has been described[15]. It incorporates an improved shaking mechanism and a porous deck through which a cylic pulse of vertically jigged water is applied, thus stratifying the test material.

If superpanner tests are favourable, a suitably classified feed is run over a miniature shaking table. The usual settings for a commercial machine are applied, attention being given to rate of feed and to pulp consistency.

Spiral tests are made in a laboratory five-turn Humphreys spiral. The action in the spiral has an element of dense-media separation, mild centrifugal action which magnifies the differences between settling rates of similar-sized particles of different density, and sluicing action controlled by the introduction of wash water at various points down the spiral. In order to exploit these forces fully, a separating band of middlings is desirable, and the operator usually continues to add ore, removing a small band of finished gangue and concentrate, until a sufficiently wide middling band has been built up. At this stage the testing, and taking of samples, can be made to duplicate commercial practice.

In all gravity tests the same feed material can be remixed and recirculated as long as desired. A certain amount of degradation of soft minerals may occur. It should be kept low by gentle handling in the transport system.

Coal preparation is concerned mainly with the ash content of the product, which rises with the S.G. of the raw coal.

Preliminary studies are made before deciding whether cleaning is justified in a given case, and if so, the optimum rejection of ash. The standard test procedure commences with the cutting of a representative sample crushed to $-\frac{3}{4}$ in. + 10♯, weighing several kg.[16] From this a 400-g sample of dry coal is immersed successively in a mixture of organic liquids rising in specific gravity from 1·25 to 1·50 by 0·05 stages. The liquid can be prepared by blending carbon tetrachloride with toluene, the density being checked by hydrometer. After each separation the floating fraction is removed by skimming with a gauze ladle and dried. A sample of 10 g (or all if the float is less than this) is finely ground and mixed. From it a 2-g sample is taken, and ignited to ash. The ash is weighed. The residual (sinking) fraction from each immersion is roughly dried and tested in the next-heavy solution. High-grade coal floats in a liquid of density 1·12–1·35. Middlings of decreasingly good fuel value float between 1·35 and 1·6. Carbonaceous shale floats from 2·2–2·6, and clay between 1·8–2·2. The American term "bone" applies to their shale-coal, and intercalation of shale and vitrain. The waste removed as the result of cleaning the feed is a charge on the cost of production of the final saleable coal or "vend". It is therefore important to use as much combustible middlings as is economically justifiable. There are three main products from the washery:

(1) Good grade coal for sale.
(2) Low-grade coal for on-the-spot consumption in special boilers.
(3) Valueless waste.

Any separating process may be so controlled as to produce these three divisions, provided there is an outlet for the power raised by burning the low-grade fuel. The laboratory sink-float tests establish the correct density for each type, and the separating devices are controlled accordingly. To check the results obtained with the small sample, a bulk sample may be run through a washery test, the separating devices being adjusted to conform to the laboratory findings. These tests have the further value of showing what amount of particle disintegration must be expected in full-scale treatment of the coal, and what problems are likely to arise in connection with the clarification, re-use and change of the large quantities of water used. Some coals are found to be very simple in their constitution and their breaking liberation, so that a clean-cut density separation will drop the shale and lift the coal. Others have a wide middling band, which may need further liberation and separation. Laboratory batch and pilot tests can thus show the ash content for a given range and density. They can also be used to show how much re-crushing of various middling sizes is needed to liberate them to treatment point.

From the ash analysis of each weighed fraction floated in the heavy-liquid tests five graphs can be constructed. Cumulative yield percentage at each S.G. of the parting liquid is plotted against the mean percentage ash of the total floated at that S.G. to give the cumulative float curve. The cumulative yield percentage of sinks at each S.G. is plotted against the mean percentage ash to give the cumulative sink curve. The characteristic ash curve is obtained by plotting percentage ash of successive fractions against cumulative percentage yield of floats at the mid-point between the S.G. limits of each fraction. The S.G. curve plots the upper value of the S.G. of successive floated fractions against the cumulative percentage weight. This shows the percentage of clean coal yielded by perfect separation at any S.G. The Tromp, or distribution-partition curve relates the percentage floating in a heavy liquid to the S.G. of that liquid.

Gold Amalgamation

The use of mercury to entrap gold goes very far back into history, mention being made of its use by Vitruvius (13 B.C.). Its use became important from the sixteenth century onward, and with sluicing and panning was virtually the only gold-catching process in use before the arrival of cyanidation.

In mineral dressing, amalgamation is the process of separating gold and silver from their associated minerals by binding them into a mixture with mercury. Many other metals can be amalgamated, the list including zinc, tin, copper, cadmium, lead, bismuth, and sodium. The word "amalgam" is used to cover a variety of states, from those in which the solid metal is wetted down into the mercury in a putty-like mixture to those in which the metal dissolves in the quicksilver.

Gold is slightly soluble in mercury (0·06% at 20°C.), and two compounds ($Au_{19}Hg_4$ and $Au_{19}Hg_3$) have been isolated. When nitric acid is used to dissolve away the mercury from amalgam, the gold particles are recovered apparently in their original form. The wetting down of gold into mercury is not alloying, but a phenomenon of moderately deep sorption, involving a limited degree of interpenetration of the two elements (solid gold and liquid mercury).

In all wetting phenomena the surface tensions of the substances involved influence the nature of the reaction. All naturally occurring systems strive to arrange themselves in a state of maximum order, at a minimum potential energy. When water stands in a burette, a meniscus is formed with an up-turned rim because the surface tension between glass and air is higher than that of water and air, so that water is drawn up against the force of gravity until the glass/water/air system has arranged itself at the maximum possible stability. The surface tension of water is about 73 dynes/cm. That of mercury is 375, and its meniscus turns down at the rim, showing that the glass/air tension is sufficiently below that of mercury/air for repulsion to take place. Gold is readily wetted by mercury because its surface tension is still higher, and it is therefore absorbed into the quicksilver. There is also a third factor at work—the dense-media effect. The density of quartz is 2·7, of mercury 13·5, and of gold about 19·2. With mercury as the separating bath, quartz floats and gold sinks. Gravitational force acts to drown the gold in the mercury, apart from any lowering of surface tension which might result from sorption. Indeed, gravity may well be the most important force at work since extremely small particles of gold cannot be amalgamated. Surface contamination, however, may be partly responsible for this failure. If a gold particle has reduced its surface tension sufficiently—e.g. by becoming coated with an oily film—attraction into the body of the mercury may be replaced by repulsion for reasons similar to those which affected the meniscus in the mercury/glass/air system.

One vitally important condition for efficient amalgamation is that the surface of both gold and mercury must be clean. "Clean" in this connection implies purity at the surface, of a type which is studied more closely in the chapters dealing with the flotation process. There must be no substantial lowering of surface tension through the presence of impurities adsorbed to either surface. Another is that the mercury must offer an adequate receiving surface to the particle of gold. If mercury is divided into minute droplets—"floured"—it cannot open its surface to gold, nor do these droplets readily reunite. A serious loss of mercury occurs when it is handled so roughly that it becomes floured.

Mercury can be contaminated or "sickened" by several substances liable to be present in an ore pulp. Fatty-acid oils and their salts (greases) finding their way to the mill from oil leakages are a hazard with mechanised mining. Such oils have a low surface tension—30 dynes/cm or less—and if they make contact with mercury they coat its surface. This "sick" mercury no longer attempts to minimise its total surface tension by gathering itself into a sphere but rolls sluggishly as a tear-shaped globule. The attractive force across the interface is now that of the contaminating oil, and amalgamation does not result from contact of this sick mercury with clean gold. A further difficulty is that an oil film attracts graphite, talc, calcium earths, and metal sulphides, which tend to form an impermeable film even if the mercury below has not been sickened. Mercury is blackened by sulphur and by some sulphides, particularly those of antimony, arsenic, and bismuth. This tendency is minimised if the pulp is alkaline.

Amalgamation can be applied to run-of-mine ore or (more usually) to a gravity concentrate, provided the metal is *free-milling*.

The surface appearance of the gold should be examined microscopically. If clean and bright, there will be no difficulty in reconcentrating a strake concentrate with mercury in a tumbling mill. If the gold is rusty or stained, or the strake concentrate carries minerals likely to contaminate ("sicken") mercury, the appropriate measure to deal with these conditions must be worked out.

An amalgamation test can be made by grinding a suitably sized pulp with mercury, either in a laboratory mill run slowly with one heavy rod, or in an iron mortar. Sodium amalgam may be used instead of mercury, or a slightly alkaline pulp may be used to aid coalescence of any globules of mercury which have become separated, to facilitate their recovery by panning. Products are assayed, and tailings may be superpanned in the hope of collecting and examining particles of gold which have failed to amalgamate.

Chemical Extraction

Laboratory tests are particularly concerned with the natural solubility of the sought ore constituent and with its specific stimulation by economic chemical treatment. In each case the influence of associated minerals must also be considered, to minimise contamination of the dissolved product, to avoid consumption by them of costly reagents or to prevent interference with the desired reactions.

Natural solubility may be affected by ageing during transport (oxygen attack), and this, if it appears helpful, may suggest cheap methods of acceleration. These could include pre-treatment to remove an unwanted gangue mineral. An example would be the removal of dolomite before acid leaching. Roasting might be used to break down an insoluble sulphide to a quasi-sulphated state. If direct leaching with acids or alkalis is suggested, tests are required to show the *p*H effect, the concentration and combination of chemicals to be used, their consumption per ton treated under varied test conditions, the effects of process temperature control, intensity of agitation and aeration, and the redox reactions.

The kinetics of reaction and the rate-controlling steps are of vital importance in this testing, whether it is directed to the development of a new flow-sheet or to better process control in a plant already at work. The possible speed of reaction must be balanced against its cost in each method investigated, together with their effect on the efficiency of extraction and the grade of the product. A formidable amount of testing is usually involved, so wise planning of the handling of the thousands of samples which must be reduced and analysed is important. Many short cuts are available today—spectrography, chromatography, x-ray fluorescence etc—and these can shorten the time required by the older routines of the analytical chemist. Most workers still consider it essential to be able to weigh and examine products from time to time where such indirect methods are used.

The leach treatments most used in mineral processing deal with gold and uranium. Cyanidation tests follow the general lines of plant practice. Leaching or agitating conditions are stimulated and all such variables as particle size, pulp consistency, protective alkali, sodium cyanide strength, aeration time, gold precipitation, and wash-displacement are closely controlled and individually varied. The first thing to settle is usually the mesh-of-grind most likely to prove economic. This fixed, the most favourable variables listed above can be worked out. When the tests are being made on successive batches of ore (100–500 g each) the final "barren" cyanides from a group of tests should be retained and later used to treat a few successive batches of new feed. This reveals any fouling conditions, and is shown by a falling off in efficiency of extraction. Chemical analysis of the fouled solution will suggest the appropriate further action.

Routine checks of cyanide solution in the plant is discussed in Chapter 16 under Solution Control. To these tests may be added a quick method (Chiddey's) of analysis of the gold and silver content of pregnant solution. A gramme of sodium cyanide is added to an adequate measured volume which is then heated nearly to boiling. Twenty ml. of saturated solution of lead acetate are then added, after which the solution is cautiously acidified with hydrochloric acid, precautions being taken with regard to any hydrocyanic fume evolved. A piece of aluminium sheet is then introduced and the solution is boiled for 30 minutes. The aluminium is removed, any adherent lead being washed back into the solution. The lead sponge which has formed is collected, wrapped in a 2 in. square of lead foil, vented and cupelled at about 850°C. to yield a prill containing the gold and silver.

Kakovsky [17] has described an apparatus used in the study of convective diffusion, which has been used to study the leaching rate of gold, silver, copper and some sulphide minerals. Since chemicals freshly presented to the metal surface normally have a relatively fast rate of reaction, the speed of this diffusion is usually the rate-controlling step. His apparatus, an improved form of the Levitch rotary disc, can give reproducible information on cyanide concentration, oxygen addition and pressure, alkalinity of solution, temperature and the interplay of these variants. Discs 20 mm in diameter and 0·5 mm thick were rotated in his tests at regulated rates in the cyanide bath. They gave both absolute and recurrent rate-constants even when chemical reaction between the metallic disc and the solution was slow.

The conditions in which roasting of auriferous sulphides is performed prior to their cyanidation affect the character of the calcine. If partial fusion occurs the loss of gold in the residue may be as high as 6 dwt/ton. An investigation reported by Parker[18] shows the importance of temperature control.

Among the research Papers on the natural and experimentally stimulated leaching of uranium minerals is a group of three[19] which connect bacterial action with bacterial nutrients, temperature effects, pyrite breakdown and leaching kinetics.

This brief discussion of chemical testing for extraction is rounded off with a reminder of the extensive and growing literature concerned with IX resins and their testing in both fixed and moving beds. Further reading is available describing tests in such beds,[20] for the development of moving-bed IX to the commercial scale[21] and dealing with the potential uses of the newer resins.[22]

Flotation

Among the objects of these tests are:

(*a*) Research on unusual ores.
(*b*) Control of mill operations.
(*c*) Testing of new reagents.
(*d*) Trial of new methods.
(*e*) Improvement of concentrate grades.
(*f*) Check on new developments in the ore body.
(*g*) Amenability tests on drill cores and other samples.
(*h*) Development of a flotation flow-sheet.

Quite useful rough tests can be made with no better apparatus than a stoppered test-tube, cylinder, or separating flask in which the pulp and reagents can be shaken and from which a froth can be removed. The last stage of grinding for any such test must be under water, using a pestle and mortar if no better apparatus is available.

Scientific testing for flotation is based on these main principles:

(1) N-bubbles do not adhere strongly to most clean mineral surfaces, but to suitable absorbed films at such surfaces.
(2) Particles borne up in a flotation froth are either aerophilic owing to a hydrocarbon film, or are temporarily entrained.
(3) In the latter case, they can be got back into the pulp by simple cleaning of a correctly textured froth.
(4) If mineral particles should have become froth-borne after conditioning treatment and are not, there is a physical reason for the failure.
(5) This reason is probably:
   (*a*) Inadequate surface liberation.
   (*b*) Incorrect conditioning.
   (*c*) Slime masking.
   (*d*) Too small N-bubbles.
   (*e*) Too large particles.
   (*f*) Too violent aeration.
   (*g*) Short-circuit through the aerating zone of the flotation cell.
(6) An adherent aerophilic film on a mineral surface lattice is developed by conditioning with suitably ionised collector.
(7) The conditions under which such film is developed can be made specific for floatable minerals by methods broadly described in previous chapters.
(8) Conditioning treatment directs suitable collector agents on to specific mineral surfaces, after the latter have been prepared for surface activation.
(9) Most of such useful surface reactions are ionic. They lead to partial coverage of the mineral's discontinuity lattices with a layer of polarised collector.
(10) In addition to specific collection, the possibility of increasing the wetability of minerals not required in a froth may exist. This calls for the use of depressing agents.

The facts listed above, logically employed to meet the case of a specific ore, make it possible to apply scientific method to the problem of testing a mineral for the purpose of floating it. Trial-and-error methods are not appropriate in the light of modern knowledge of surface physics.

A common criterion of good technology is reproducibility of results when a method which has proved successful is repeated. This though good as far as it goes, would leave flotation test-work in an unsatisfactory and uninspired state. If a scientific approach is made and the tests are all reasoned out and recorded, whether or no they have fulfilled expectation, the work ceases to be a "cook-book" routine and is set on a proper scientific basis.

Observation, analysis, and step-by-step variations of the conditions under which the ore is subjected to flotation require skill, experience, and patience. The list of factors affecting efficiency of concentration given below, though long, is far from complete, and each important factor (for a given ore) must be varied in isolation during the tests, the remainder being held unchanged as far as possible. Assays are required on test products and take time to perform, even if a short method can be relied on. The ultimate success of the commercial operation reflects the care with which these tests are made, and the cost of making them is a mere fraction of the money which will accrue from improved efficiency of extraction if they have been correctly performed and accurately established in the plant. The factors include:

(*a*) Mineralogical character of the feed.
(*b*) Size distributions of minerals in feed.
(*c*) Grinding methods.
(*d*) Grinding times.
(*e*) Travel history of ore.
(*f*) Nature of mill water.
(*g*) Pulp dilution and temperature.
(*h*) Amount and nature of soluble substances released during grinding.
(*i*) Amount of character of each reagent used.
(*j*) Order of addition of reagents.
(*k*) Conditioning time for each reagent.
(*l*) Nature of series-reaction through the conditioning treatment.
(*m*) Kind of flotation machine.
(*n*) Time and degree of agitation in cells.
(*o*) Quantity and size of N-bubbles.
(*p*) Bubble-column height.
(*q*) Froth texture and tenacity with which each mineral species clings.
(*r*) Temperature.
(*s*) Slime effects.
(*t*) Flocculation in the pulp.
(*u*) *p*H changes during frothing period.
(*v*) Floating times (optimum) of various products.
(*w*) Rate of change of concentration of various pulp constituents as products are removed.
(*x*) Mutual interference of minerals and reagents during differential flotation.

(*y*) Interplay between particle mesh and recovery.
(*z*) Optimum mesh-of-grind.

An elementary testing scheme for a single-product flotation could follow a simple plan. Grinding mesh for optimum recovery can be established by fixing conditions of wet grinding in a laboratory mill, giving the product a standardised conditioning and flotation treatment, and examining various sized fractions of tailings for their residual content of value. If the grinding conditions are varied in strictly controlled stages and the flotation results for these variations are compared, the optimum treatment is revealed. This is adhered to in the next series of tests in which reagents, dosages, and conditioning time are similarly varied until optimum preparation of the pulp is established. Next, flotation period is worked out by removing from a properly ground and conditioned sample the froth produced in separate, but consecutive, two-minute skimming periods. This is done under varying conditions of initial pulp density and aeration. Analysis of the products give a preliminary indication of cell-capacity required and of the type of concentrate to be expected at various stages along a line of cells. This information can then be worked up by using "locked" or "cyclic" batch tests. In these, the two-minute concentrates from each test of a series are suitably pooled and retreated to simulate feeds (*a*) of the first concentrate (rougher) to a cleaning stage and (*b*) of mixing scavenged rougher froth (last stages) with a new batch of pulp. The effect of blending can thus be studied in miniature. Instead of this, it is possible to do this part of the test work in a continuous miniature flotation circuit. The importance of the locked test is that it shows what happens when border-line middlings particles are retained, while the bulk of the fed ore leaves the treatment as either concentrate or tailings. Ultimately this will be happening in the industrial plant and the problems which will then arise must be clarified on the laboratory scale as far as possible.

Laboratory cells for these tests are used to perform the conditioning and then the flotation of each test sample. They should have as many as possible of the following features:

(*a*) Ability to run for some hours continuously at easily varied speeds.
(*b*) Thorough aeration and agitation with minimum degradation of particles in the pulp.
(*c*) Accessibility for thorough cleaning.
(*d*) Control of aeration.
(*e*) Consistent skimming action.
(*f*) Good visibility of pulp and froth.
(*g*) Clean removal, from top of froth column only, of floated mineral.
(*h*) Quiet froth blanket.
(*i*) Reproducibility on the commercial scale.
(*j*) Arrangements permitting skimming at all heights of froth column.

Not all of these conditions are achieved in the manufacturer's laboratory cells, but they can be satisfied if the precision of the resulting work warrants it, and this also applies to machines large enough for use in the plant. Following on research into cell aeration, part of which has been published[23,24], cells were designed which embodied some of these requirements.

Reagents can conveniently be added from a dropping tube which has been calibrated. With a charge of one kilogram of solids, one millilitre of a 5% solution of reagent is equivalent to 0·1 lb./short ton. The tube for introducing 100% frothing agents can also be calibrated, though there is some inaccuracy due to change of viscosity (this varies inversely as the laboratory temperature, and specifically with different sources of pine oil). Pine oil, like other reagents of low solubility, can be thinned down for sensitive addition by emulsifying, say, five drops with 50 ml of water in an agate mortar or by solution in ethanol.

Flotation cells can be cleaned between tests by scrubbing with a solution of soda and then running them with clean sand and water. Tests should be made in distilled water to begin with, mill water only being used after a satisfactory treatment scheme has been evolved. A laboratory rod mill loaded with −10-mesh ore roughly simulates a standard mill receiving $-\frac{1}{4}$ in. feed. Aeration of the cell can be checked by filling a graduated cylinder with water, inverting it under water in the cell, and timing the rate of fall of the water-level in the cylinder. This should be about the same in all parts of the cell at a cross-section a little below the surface.

When hand skimming, a rougher froth should be scraped somewhat deeper than a cleaner froth. A final scavenger froth should be removed as completely as possible, even if some pulp is thereby included. To aid this scavenging "pull", the pulp in the laboratory cell must be diluted till it nearly brims over. In the specially designed cell mentioned above this is avoided by use of a tilting basin, and loss of pulp density and particle support is thereby avoided. In a laboratory cell designed by the author the froth-removal is performed by suction through an array of small tubes supported on the housing of the impeller shaft. This array can be slid so as to maintain the desired height of froth blanket. Vacuum for operating the suction can be tapped from the centre of the zone swept by the cell's impeller, using a receiving vessel to trap the removed froth. Concentrate grade can be improved by restricting aeration, diluting the pulp, or using a weak frother, apart from manipulating conditioning treatment. Recovery can be increased by reversing these methods and by stiffening the froth.

When non-sulphide ores are being tested, particularly in the presence of slimes, collectors containing fatty acids or amines should be added in successive "starvation" quantities with skimming between additions. Tests in which oxidised metal surfaces are treated by sulphidisation are best performed after straight sulphide flotation has been completed. Skimming should be completed quickly as the superficial sulphidisation is fleeting and aeration of the pulp aids reversion to the oxide.

Table agglomeration tests on −10-mesh+100-mesh mineral can be performed in a large watch-glass. The feed is deslimed, stirred with reagent and frother into a thick slurry, and shaken vigorously in a test-tube to introduce air. The slurry is then poured into the watch-glass using sufficient pulp to ensure quiet peripheral overflow. If glomerules form they will overflow into a surrounding vessel, the tailing remaining in the watch-glass.

The induction time for adhesion of air bubbles to mineral particles is one determinant of flotation time, tenacity of mineral retention in the froth blanket and general efficiency of levitation. Using a specially developed technique Eigeles *et al*[25,26] find induction time to be affected by such time factors as conditioning period and age of air bubbles. In advanced test work the optimum period for adsorption of reagents, also the pulp climate and aeration during this period should be examined. Ionic activation and depression introduced with the mill water must also be checked at a suitable stage, to ensure that laboratory tests conducted in distilled water are reproducible on the plant scale with local water. Bicarbonates, sulphates, chlorides, calcium and magnesium ions and their possible build-up in recirculating mill water should be observed in this connexion.[27]

Electro-Magnetism

Here tests are made to show the relative reponse of the ore's minerals to the influence of fields of magnetic or electrical force. The chief matters investigated may be summarised thus:

(1) *General Tests*
- (*a*) Particle size effect
- (*b*) Particle density effect
- (*c*) Degree of liberation

(2) *Wet and Dry Magnetic Tests*
- (*d*) Magnetic response and remanence
- (*e*) Response to feed rate, display and lifting gap
- (*f*) Effects of pulp density and hydraulic action
- (*g*) Flocculation and product grade

(3) *Electrostatic and Electrokinetic (Dry) Tests*
- (*h*) Particle conductivity
- (*i*) Particle permittivity
- (*j*) Natural and induced electrical leakage
- (*k*) Effects of ambient humidity and temperature
- (*l*) Feed, display and post-charging systems

The laboratory equipment used in magnetic testing includes variable flux coils by means of which the degree of lift of closely sized dry material can be measured. For wet tests the Davis tube, in which a column of pulp is shaken between the poles of an electro-magnet is widely used. The seized product is held while the relatively non-magnetic fraction is discharged. A more rapid apparatus is the Krupp magnetic balance, which rapidly shows the ferro-magnetic intensity of the constituent minerals in a pulp. In high-intensity tests the size variable should be tightly controlled. Remanence affects hysteresis, magnetic flocculation and hence purity of product. It is countered by the required series of pole reversals to which the feed must be subjected. In wet magnetic treatment the hydraulic force acting against gravity on the particles suspended in the pulp can be used to discriminate between feebly and strongly magnetic particles in suitably contrived continuous flow test apparatus. This determines the capacity required and the number of operations in series where progressively feeble minerals are to be recovered separately.

In high-tension separation three types of field can be set up[28]—the electrostatic, the convective and a mixed electrostatic convective system. Particles acquire charge in four ways: (*a*) by heat and friction; (*b*) by rubbing contact with a charged body or bodies; (*c*) by induction and (*d*) by bombardment with mobile ions. Of these the first three are electrostatic in that no external flux is moving. With (*d*) (high-tension separation proper) which is added to these, there is some current flow. The condition of the particle's surface affects its rate of loss of charge. If hydrophobic, as when it has been conditioned with a suitable reagent, it tends to reject moisture otherwise adsorbed from the surrounding air. Such moisture increases leakage from the surface, and thus varies conductivity.

De-Watering

Tests may be called for in connexion with the following main operations:

(*a*) Control of pulp density.
(*b*) Thickening before filtration.
(*c*) Filtration.
(*d*) Flocculation.
(*e*) Counter-current decantation.
(*f*) Replacement of fouled water.
(*g*) Maximum extraction of pregnant solution.

Settling rate is studied under various conditions.

The chief test is for settling rate in connection with the calculation of thickener capacity.[31] A measured quantity of representative pulp, which should be at the *p*H and percentage solids expected to be delivered to the thickener under working conditions, is placed in a large glass cylinder down the outside of which a paper strip has been pasted. The height of the pulp is marked 0 minutes, and the clear level is marked for five minutes, at one minute intervals. The pulp is then left to settle some distance. After this, enough clear water is decanted to leave an intermediate density when the pulp is remixed. The same procedure is repeated, aiming to have a series of initial water-solid ratios of the rough order 5:1, 4:1, 3:1, and 2:1. After four decantations, the pulp is left for nineteen hours and the dividing line is marked. It is then left a further five hours to find whether any further settlement occurs. The clear water is now removed and the percentage of moisture in the residual slurry is determined. If A is the area of the thickener (sq. ft. short ton of dry solids/24 hours), F is the initial and D the final pulp dilution $\left(\frac{\text{Water}}{\text{wt. of solid}}\right)$, and R the settling rate (ft/hr), then

$$A = \frac{1{\cdot}333\,(F - D)}{R}.$$

A 25% safety factor is allowed on the area/ton. If settlement is not complete in nineteen hours, the thickener must be made deeper.

In most cases natural or induced flocculation is at work. By decanting and re-mixing the thicker pulp the opportunity for collision and floccule-building increases. Slow stirring may be used to simulate the action of thickener rakes in cutting channels through the flocculated slurry, through which clear water can squeeze up. In one apparatus used for this purpose[29] a transparent-sided tank represents a segment through a thickener, and provision is made for feed and withdrawal as in full-scale working. Cycle chains are used to rake down settled material settling on the sloping bottom of the rectangular tank, with cross-bars at intervals. The chain descends on the side corresponding to the outer wall of a thickener, turns downslope to the sludge withdrawal point and then rises vertically. Fresh pulp is fed in continuously through a test lasting up to nine hours.

The natural flocculation and sedimentation of the pulp depends on its mineralogical composition. Clays impart pseudo-viscosity and are hardest to settle. Other factors include specific surface, an increase in which reduces settling rate; *p*H (flocculation being fastest at and near the iso-electric point); pulp temperature; dissolved salts and initial solid/liquid ratio. Tests are sometimes affected by ageing of the pulp and where possible should be made on fresh material. The first step in batch-testing a 1,000 ml sample is standardized mixing. A perforated disc should be vigorously plunged down the cylinder six times in as many seconds. Next, the descent of the "mud-line" of the settling solids is timed, a slowly revolving stirrer simulating rake action. In a series of these tests the effects of change in *p*H and of the addition of flocculants at various dosages is observed, without decantation. One method is detailed in a trade brochure.[30] The effect of centrifugal thickening through a hydrocyclone may merit examination in the case of very dilute or slimy pulps. The problem of dispensing flocculants into the pulp both in batch testing and on the plant scale requires special precautions if the whole reagent addition is not to be taken up immediately by a fraction of the feed. Gieseke[33] has described large-scale testing of coal concentrates produced by froth flotation, which are difficult to settle or filter without special conditioning.

Only a few of the many matters which may require filtration tests can be mentioned here. These, selected as of major importance, include:

(*a*) Formation rate of filter cake.
(*b*) De-watering rate of cake.
(*c*) Method of forming the cake.
(*d*) Effect of solid-liquid ratio.
(*e*) In rotary drum filtration, percentage of each revolution allotted to formation, displacement washing and drying the cake.

Dahlstrom[32] deals with the complex problem of testing at some length. Among his twenty-five bench-tested factors are (*a*) and (*b*) above; effect of pressure drop across the filtering media during these; temperature; filter media and filter aids; and, residual moisture. Tests are made using a filter leaf 0·1 ft^2, having a standard grooving. The filter media is clamped to this surface and provision is made for use of vacuum or pressure, with appropriate equipment for trapping filtrate, measuring flow rate, and for recording the volume and rate at which air is drawn through during the drying test. The slurry is adjusted as to flocculation, solid/liquid ratio and temperature and kept mixed. The test leaf is gently immersed, face down, just below the surface of the sample under recorded conditions of vacuum.

Tests in the Pilot Plant

When batch tests have completed the closely controlled examination of the ore constituents, and locked cyclic tests have shown as much as possible of their inter-dependence, development of a new flow-sheet is carried on in a continuously operating system. Here the optimum working conditions are more clearly displayed and a compromise is reached between the technically ideal treatment and one economically viable. Pilot tests in miniature mark the first stage of scale-up in which this working adjustment begins. They are best made in the central laboratory on two or three tons of the original bulk sample. By starting on a small scale and with a feed rate well below 100 lb/hour close scanning of the flowline is possible. The problem of re-circulating products can be met by the use of holding tanks if full continuity cannot be achieved on this small scale. Each variation of the flow-line suggested by batch tests should be examined. Well-planned tests will then confirm the general lines of the process and disclose potential difficulties not shown in batch testing. When this work has been digested, together with any further batch tests thus shown to be desirable, pilot testing should be continued on a larger scale on location. This move away from the highly specialised central facilities should not be made until the emphasis of the test-work has shifted from basic research to that mainly concerned with process economics and plant design. From now on tests are made on fresh ore and use the proposed mill water. The data recorded during the tests will deal not only with such things as particle sizes, recovery distribution and grade, but will also record reagent consumption, power required and any other matters of economic importance. A start may now be made with the training of technical workers later to be employed in the concentrator. If the ore changes from horizon to horizon the effects of blending and of treatment of varying types can be compared. Change in grindability, liberation mesh and their effect on products are now scrutinised in their bearing on circulating load, plant capacity at bottle-neck points and optimum rate of feed. This last is of direct concern to the mine's rate of ore extraction.

These continuous tests must show whether there is a difficult build-up of re-circulated material anywhere, whether in near-liberation mesh tough grains of ore or in the return treatment circuits (middlings and scavenger). These matters affect the provision of re-grinding facilities and the decision on optimum grade of concentrate and percentage recovery. If the concentrates are being sold, large samples may be called for testing by the proposed market and the specifications drawn as the result of their tests may require alterations in the treatment plan.

Even a very large pilot plant cannot provide complete answers to questions which depend mainly on changes in the ore as the mine develops. Too large a plant is costly, and can defeat its own object by making the findings of a trial run complex and obscure. Possible effects on flotation of seasonal change in the water supply should be assessed where possible. Tests should be run long enough to ensure that anything liable to upset the process by building up to unmanageable proportions is disclosed, so that it can be traced back to its cause and properly dealt with in the final operating plan and engineering design. Surging and an erratic flow-rate are wasteful and should be dealt with once for all in the pilot plant, where they can be exposed by suitable instrumentation and measurement of flow-rates.

An automated pilot-scale flotation plant has been described,[34] with a feed rate between 33 and 66 lb/hour. It handles a *minus* 10-mesh feed and includes automatic sampling and pulp level control and has proved reliable in the testing of a variety of ores. One rule-of-thumb suggestion regarding scale-up[35] is that pilot tests should be performed on at least fifty times the scale of bench and unit tests, and that the full-scale plant should have a capacity of not more than fifty times that of the pilot plant. A final test-production scale of between 10 and 25 tons per day is often found satisfactory. One variation is to test only a novel unit operation on the pilot scale and to rely on experience for the more common sections of the flow-sheet. Where mechanised sorting or dense-media separation are proposed, portable plants are available in many mining centres for such specialised unit tests, which should be made before the overall plant capacity is estimated.

As the pilot plant is used intermittently special care is needed with regard to cleanliness and the avoidance of corrosion. Plastic piping is non-corrosive and easily re-coupled when testing flow-line variations or flushing down between tests. It should be installed self-draining with no horizontal runs or crests and bends where solids can lodge. Reliable flow-control valves are needed, though in the final plant only on-off ones are usual. Accessibility and safety in the lay-out are important, with good lighting *plus* plug points for inspection lamps. Among the matters recorded during the tests are temperatures, changes of *p*H along the flow-line and pressure fluctuations. The sequence of sampling along the flow-line should be planned to allow for time-lags in the course of the pulp down-line, if the bearing of one sample on those following is to be interpreted correctly. Periods between sampling must be sufficiently small to expose fluctuations during treatment, but must not overwhelm the laboratory which handles them. Two types of analysis are required. Control assays feel the pulse of the current operation, while special and detailed ones track down the chemical and physical causes of variations in operating efficiency.

A second type of pilot testing is that designed to improve an existing operation. It may be feasible to retain the original pilot plant on care-and-maintenance for this purpose, or to design a large mill so that one treatment line can at need be used for pilot tests. In chemical engineering the technique called "evolutionary operation" is well established. It has been defined[36] as a method of process operation which has a "built-in" procedure to increase productivity. To put a designed experiment right into a main production line involves undesirable risks and may lead to premature abandonment of a valuable idea. The basic elements of evolutionary operation leading to improved efficiency are (*a*) variation and (*b*) selection. Small changes are made on two or three control variables. Their effects may be masked by the usual swing in production figures, but since this is not significantly affected constant repetition and improvement of these cautious and limited changes can show in improvement of the general process. In the chemical engineering industry the second component (selection) is achieved by the use of an "information board", which keeps the operators in touch with the work. A committee may be helpful in selecting and appraising the effects of the variables thus tried out. It includes the process supervisor, a research specialist familiar with the work and the plant manager. The arrangement ensures that all ideas worth following up are given a fair trial.

Summary and Application

No hard-and-fast rules have been suggested in the preceding sections of this chapter. The precise lines an investigation is to take should be dictated by the nature of the ore, not of preconceived opinions. Preliminary batch tests may use 50 to 100 lb. of material; combined, locked, and cyclic tests up to 500 lb.; miniature pilot runs up to 5 tons; and full-scale tests duplicating mill practice any quantity per day commensurate with the importance of the detailed economic picture. Batch tests, properly conducted, give the essential technical information, and do so under controls far more rigid than can be imposed on a continuous test. In the case of a difficult ore, batch testing is far more likely to succeed in producing a method than is pilot-scale work, but the latter is invaluable in training personnel, in evaluating costs, and in uncovering weaknesses not disclosed in the small batch tests. Such weaknesses are likely to arise from retention in a closed circuit. Some trouble-making ore constituents are not disclosed even by locked cyclic tests, or in the fouling of mill water and chemical solutions when recycled.

The consulting engineer will be called on to approve the final treatment scheme. This will require consideration of the following matters:

(*a*) Haulage costs from mine to rail.
(*b*) Freight schedules.
(*c*) Smelter penalties and contracts.
(*d*) Local cost of power, labour and materials of construction.
(*e*) Local fuel.
(*f*) Water position.
(*g*) Product disposal.
(*h*) Size of operation.

The tests will show all the alternative methods of treatment and will point to the one likely to yield the highest profit. This is not necessarily synonymous with maximum recovery or highest grade of concentrate. The possibility of changes in the ore body must be considered and the treatment designed to be reasonably adaptable if such changes occur. The location of the mill affects the capital locked up in stores, suitable protection against climatic conditions, and such matters as tailings disposal. Ample room for expansion should be reserved in site-planning. When the other buildings come into being expansion is restricted, should a big mine develop. The character of the process may have to be modified to suit the labour force. A few highly skilled men can work a difficult process, but too many undertakings are wedded to a "cheap native labour" policy which is incapable of running a complex ore-dressing operation smoothly.

Mineral-dressing costs fall into two divisions, capital and operating. The former must be charged out as depreciation over the working life of the mine, *plus* an obsolescence reserve, in case some new process is invented which must replace part of the flow-sheet. If the mill is near a smelter, high maintenance costs may be needed to offset the corrosive effect of the sulphur-acid fumes. Flotation plants are more costly to install than cyanide and gravity mills.

Working costs include wages, stores, power, and overheads. Breakdown allocations of cost should be made monthly, so that the working cost of each operation in the mill is kept in sight. These include crushing, conveying, grinding, pumping, classifying, lighting, heating, concentrating, sampling, thickening, filtering, tailings disposal, and shipping, with detailed analysis of the components of each cost. These figures disclose the insidious entry of wasteful practices and stress the importance of full loading of the plant, since labour is one of the main items of costs and is the same whether work proceeds at full or part capacity. The same applies to power and overheads. The mill should therefore never be so run as to risk breakdown leading to serious stoppage.

References

1. Pryor, E. J. (1963). *Dictionary of Mineral Technology,* Mining Publications.
2. Winchell, A. N. (1951). *Elements of Optical Mineralogy*, Chapman and Hall.
3. Hosking, K. F. G. (1960). *Mining Magazine,* May.
4. Hallimond, A. F. (1963). *Mining Magazine*, April.
5. Amstutz, G. C. (1961). 50*th Anniv. of Froth Flotation*, Vol 2, July, Colorado Sch. of Mines.
6. Allenden, D., and Mulvey, T. (1962). A.E.I. Eng., May/June.
7. Williams, D., and Nakhla, F. M. (1950/1). *Trans. I.M.M.* (*London*), 60.
8. Hosking, K. F. G. (1963). *Mining Magazine,* July.
9. Pryor, E. J., Blyth, H. N., and Eldridge, A. (1953). *Recent Developments in Mineral Dressing*, I.M.M. (London).
10. (1963). B.S. 3406 Part 4, Br. Standards Inst.
11. Maxson, W. L., Cadena, F., and Bond, F. C. (1934). *Trans. A.I.M.M.E.,* 112.
12. Bond, F. C. (1949). *Trans. A.I.M.M.E.,* 183.
13. Bond, F. C. (1952). *Min. Eng.,* 4.
14. Warren Spring Laboratory. (1962). Min. Proc. Information Note, No. 3.
15. Muller, L. D., and Pownall, J. H. (1962). *Trans. I.M.M.* (*London*), 71.
16. (1942). Rev. B.S. 1017. Sampling of Coal and Coke, Br. Standards Instn.
17. Kakovsky, I. A. (1963). *I.M.P.C.* (*Cannes*), Pergamon.
18. Parker, O. J. (1957). *Aust. I.M.M.,* Part I, June.
19. Miller, R. P., Napier, E., Wells, R. A., Audsley, A., Daborn, G. R. (1963). *Trans. I.M.M.* (*London*), 72.
20. Hancher, C..W. (1959). *Eng. and Min. J.,* March.
21. Maltby, P. D. R. (1959/60). *Trans. I.M.M.* (*London*), 69.

References—continued

22. Everest, D. A., and Wells, R. A. (1963). *I.M.P.C. (Cannes)*, Pergamon.
23. Pryor, E. J., and Liou, K. (1950). *Trans. I.M.M. (London)*, 58.
24. Pryor, E. J., and Dzieniewicz, J. (1950). *Trans. I.M.M. (London)*, 59.
25. Eigeles, M. A., and Volova, M. L. (1960). *I.M.P.C. (London)*, I.M.M. (London).
26. Eigeles, M. A., and Volvenkova, V. S. (1963). *I.M.P.C. (Cannes)*, Pergamon.
27. Agey, W. W., Salisbury, H. B., and Placek, P. L. U.S. Bureau of Mines Research Inst. 6189.
28. Vijaykar, S. V., and Majumdar, K. K. (1962). *Mining Magazine,* July.
29. Cross, H. E. (1963). *Jnl. S. Af. I.M.M.,* Feb.
30. (1959). *Cyanamid Flocculants,* American Cyanamid Co.
31. Dorr, J. V. N., and Bosqui, F. L. (1950). *Cyanidation and Concentration of Gold and Silver Ores,* McGraw-Hill.
32. Dahlstrom, D. A. (1959). *Trans. Can. I.M.M.,* Feb.
33. Gieseke, E. W. (1962). *Trans. Am. S.M.E.,* Dec.
34. Raffinot, P., and Formanek, V. (1963). *I.M.P.C. (Cannes)*, Pergamon.
35. MacDonald, R. D., and Stevens, F. M. (1959). *Min. Eng., May.*
36. Box, G. E. P. (1958). 43rd Ann. Meeting Tech. Assn. of Paper Industry, New York.

CHAPTER 21

SAMPLING AND CONTROLS

Purposes in Sampling

The first and most obvious reason for sampling is to acquire information about the ore entering the plant for treatment. The second is to inspect its condition at selected points during its progress through the plant so that comparison can be made between the optimum requirements for efficient treatment and those actually existing, should these not coincide. The third is to disclose recovery and losses, and to learn how to improve the former and reduce the latter. Just as in military circles reconnaissance before the fight, information during the fight, and a conference on "the lessons of the scheme" are essential to intelligent control, so, translated to the civil field, does the mill superintendent look ahead, keep his finger on the pulse of his charge, and hold "post-mortems" on such failures as high tailing loss and poor concentrate. "What's gone is past", but knowledge of the reason for failure is the first step toward preventing its repetition.

No readily visible financial profit results from sampling, which is a rather dull and costly routine activity. Without its methodical and careful fact-finding, however, the losses during treatment would be unassessed, unchallenged, and largely unrecognised, and would be out of proportion to the sampling cost. The quantity finally tested is a small fraction of the tonnage from which it has been reduced. Chances of error in a single sample exist. Careful use of a sound method, coupled with the taking of an adequate number of samples, keeps the overall error within tolerable working limits. It sometimes happens that the least dependable labour is assigned to the collection and reduction of samples, and that an unnecessary amount of dull manual work is called for in their crushing, grinding, and bulk reduction. It is true that a good shiftsman will get more out of his section than a bad one. It is equally true that a bad sampler will so distort the contents of the fraction of the original sample he finally takes to the assayer that the whole purpose of the work is weakened and the returns rendered deceptive. There is a tendency to rely on returns without checking the validity of basic data from which they are compiled. Mill return sheets can degenerate unless an efficient routine is maintained by sample-runners who understand the importance of their work.

In a Symposium on "Statistical Methods in the Chemical Industry" a Paper[1] compares a variety of appliances and methods used in automatic sampling. It defines sampling as " . . . the operation of removing a part, convenient in quantity for analysis, from a whole which is of much greater . . . in such a way that the proportion and distribution of the quality to be tested are the same in both the sample and the whole". Sampling is a statistical technique based on the theory of probability. In the plant it must minimise the errors arising from such variables as surging, settlement from a pulp and segregation of sizes in an ore bin. With moving material the best place to cut the sample is at a point where there is free fall, so that the stream can be traversed at right angles. The intervals between cuts across this stream must compromise between possible fluctuations in the quality being checked and the problem of reducing a large sample to the dimensions required by the analyst cheaply, speedily and with sufficient accuracy. An automatic cutter is usual, since it eliminates the personal variation of hand-sampling. Size of sample depends on aperture of the collecting device, speed and frequency of traverse. Start and stop of each cut should build to travel speed before the stream is entered and the traversing rate must be steady. A large sample may be reduced by a secondary cutter, perhaps with intermediate size reduction.

Sampling errors are definable as differences between the observed or calculated values and the true ones. They fall into several classifications. Accidental errors are unpredictable and arise from special circumstances, which can only be dealt with if they are seen to be occurring. The average error is the arithmetic mean of a series (plus or minus). A biassed, constant or systematic error is that in a series which is always wrong in the same way, thus producing a cumulative distortion. An error of observation is one due to misreading a signal or measuring index, or to faulty recording. A personal error, though biassed, may be random. The random error is not biassed and may therefore cancel out (as a compensating error) in a carefully observed series. A sampling error is the divergence of the sampling information from the true value, and arises from defects in taking, handling, reducing or treating the sample.

An error band is, statistically, the range of determined values inside which the correct value is presumed to lie. All sample values inside this band are taken to be experimentally valid. A commonly used formula for adjustment of error where Σd is the sum of the deviations of observed values from the mean value and n is the number of observations, is given by the root-mean-square formula

$$V_m = \frac{\pm \sqrt{\Sigma d^2}}{(n-1)} \tag{21.1}$$

A term used in statistical methodology is the "confidence interval". This has been defined[3] as "... the limits of error of quantity calculable from given data when allowance has been made for known chance variance in the collection of such data. The space between these confidence limits is the confidence interval within which lies the true value, shown with sufficient precision for the required purpose of the work in hand."

For a bulk sample the three main considerations are (*a*) the representative weight, (*b*) the size to which the bulk sample must be reduced before reduction for a given purpose can begin and (*c*) the magnitude of the tolerable sampling error. A nomogram developed by Gy[2] gives rapid and reasonably accurate solutions of these three basic problems, where (as is usually the case) the percentage error in assay grade shown by the sample is distributed about the true assay grade in a normal way. To use the nomogram the liberation mesh of the value, its assay grade when pure, the density of value and gangue, and the maximum size of the material being sampled must be known.

Sampling and process control are intimately associated. Here the dominant need is the most rapid possible approximate information on any essential quality in the material at the point sampled, and the rate at which it is passing that point. Accuracy of control and advance in permissible complexity of the flow-sheet have today led to a high efficiency and smoothness of operation. Research has now made possible continuous check on some vitally important processing factors, and the use of automation helps to maintain smooth overall operation.

Sampling Solids

Occasionally the mill engineer is asked to prepare a representative sample of ore either for a trial run, for dispatch to a distant ore-testing station, or in connection with the valuation of a property. This is one of the less satisfactory assignments, particularly if precautions against surreptitious tampering must be added to the natural problem. A proper proportion of freshly severed ore should be cut from each exposed rock face that can be safely reached. If the face has been exposed long enough to become chemically altered a new surface should be exposed by blasting. All the material, large and small pieces alike, should be gathered. If the face represents one-tenth of the presumed ore body, and a ten-ton sample is wanted, then a ton should be taken from a fresh, clean surface. If the ore is being shipped, it may well be sealed into clean, empty oil-drums, to minimise oxidation *en route*. If the ore varies from exposure to exposure, it may be wise to keep the sample from each such point separate, but usually the whole amount is mixed as well as possible before dispatch. Picked hand specimens which exhibit the constituent minerals can usefully be sent separately, to aid the distant consulting engineer with the microscopic work and contact-angle investigation. A few large crystals of the valuable mineral are most helpful for this purpose, if they can be found.

If the large heap of ore gathered is to be sampled on the spot, it must be reduced to a manageable quantity. The distribution, grain size and frequency, and the assay grade of the valuable minerals influence the accuracy with which size reduction can be performed. The most difficult type of problem is that of a very small quantity of value segregated in a large volume of gangue, such as that presented by metallic gold in quartz. Each pennyweight is 1/600,000th of a ton of ore by weight and 1/4,000,000th by volume. The quartz is brittle and the gold malleable so that crushing which will reduce the former to sand will only distort particles of the latter. The sample finally treated by the assayer is from one to four assay-tons. If the gold existed as a single particle weighing one pennyweight and the ton of rock were reduced in bulk to one assay-ton (29·1667 g in the short-ton system) the odds against any gold being found are over 30,000 to 1, while if it were in the portion finally tested the assay would indicate richness. The example chosen, though possible, is unlikely to be met in practice, but a low-grade gold ore illustrates the great difficulty in sampling. If instead of gold (S.G.19·2) in quartz (S.G. 2·7) a 10% sulphide of density 7·0 in a gangue of 3·5 is considered, and all particles are taken as of equal size, then the odds of one particle being misplaced so as to enrich or denude the fraction being prepared for assay are less than twenty to one, and the millions of particles under treatment reduce the chance of error to trivial dimensions.

Hence the relation between the scarce and abundant minerals is important, as also is their relative crushability. It would not be possible, of course, to reduce the ton of sulphide to particles of equal size, and it is commonly found that such a valuable fraction as a metal sulphide breaks into much smaller sizes than the gangue mineral. The first rule therefore is to decide on a proportion between the weight (or, if preferred, volume) of the largest sized particle in the bulk sample and the whole bulk. Much has been written on the subject, which is a happy hunting-ground for the statistical mathematician. In practice, some rule-of-thumb decision is usual for a given ore—say a weight ratio of 1:20,000. This would mean that the largest lump in the ten-ton sample must not weigh more than a pound. If the work were being done by hand, as might well be the case at the early development stage, this requirement would be onerous, but with crushing plant available it could easily be satisfied. For size reduction the coning-and-quartering method is widely used, and requires only a level working space. The material, crushed to size, is built into a cone, each shovelful being deposited at the apex so that coarse and fine particles distribute themselves fairly evenly all down the sides. Owing to the tendency for the rich sulphide to be more friable and to segregate, care is essential in mixing the sample accurately while coning it. The cone is next flattened and marked into four equal segments or quarters. Even shovelfuls are taken successively from each of the four equal segments and deposited so as to form four truly mixed cones. Meticulous care is needed in this separation to avoid undue inclusion of fines in any quarter. The material from opposite quarters is now rejected, so far as the sampling is concerned, though it can be sent for treatment. The retained cones, if the particle weight relationship is still suitable, are united in a new cone, which is flattened and requartered. The original sample is thus reduced, with occasional further crushing of the retained fraction, to a weight of 100 lb. or less, at a maximum size of, say, 5 or 10 mesh. This is further reduced to an assay size, first in one or more Jones Rifflers (Fig. 269) and finally at perhaps −60 mesh, by rolling a few hundred grammes on glazed cloth, forming a cone and flattening it before taking a final sample.

This laborious and lengthy procedure, though essential in an ore-testing station, would be out of place in day-by-day plant operation. Taking the case of a plant receiving about 1,000 tons of ore daily, and sending a sample of 500 grammes to assay, a first cut of about a hundredweight per shift might be made (1/20,000 of the original head). This could be taken by an automatic sampling arm set to traverse the delivery belt at suitable intervals or to intercept a falling stream of ore. Since the incoming tonnage must be accounted for, the sample would need protection against change in its moisture content, as the metallurgical balance sheet is calculated on dry tonnage received. In the sample reduction room a small jaw crusher and a disc grinding mill would handle the comminution. Reduction would probably use riffling followed by either rolling on glazed fabric or the use of miniature riffles. In addition to the sample sent to assay, a reference sample might be kept, and a quantity accumulated toward a monthly composite sample.

Where crushing machines are used for sampling, a pile of barren rock or sand should be kept, and some should be run through to clean each machine after a sample has been crushed. A little of the new sample should then be run through and discarded before crushing the rest.

A description of advanced design in sample reduction has been given in connexion with the iron ore mined at Kiruna in Sweden[4]. Here the ore

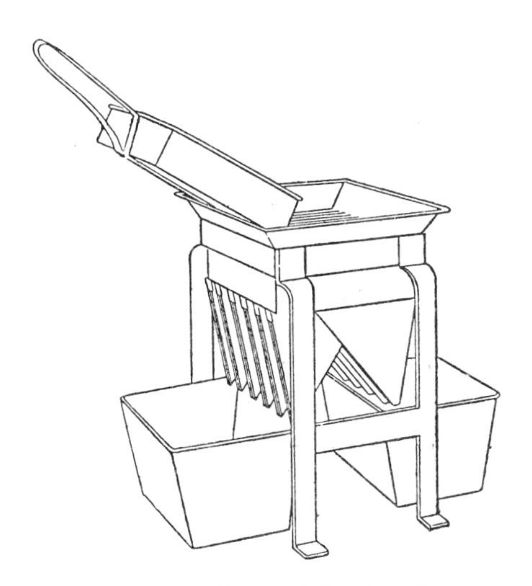

Fig. 269. *Jones Riffle Sampler*

varies in its phosphorus content between 0·017% and over 2%, and control of this element in the finished product is essential. About 1,000 samples must be assayed daily. Following statistical studies it was decided to take a weight of 1 kilogram as representative of a carload or of the ore from one round of shots during mining. For some grades of ore five samples are combined, so the plant can handle either 1 kg or 5 kg lots. Twenty processing lines are in use, through one of which each sample gravitates from a drum drier to a gyratory crusher and automatic splitter and thence to a mortar-type of grinder and a sample box. The entering sample is poured into a drum, its tag number dropping into a holder below to which the final sample will come. The sequence is automatic and between samples the drum is cleaned by a strong air blast; 90% of the sample is rejected after passing through the miniature gyratory crusher, and the rest of the work takes some three minutes.

Control by automatic devices of the full-scale crushing plant is discussed in a later section (Instrumentation and Automation).

Sampling Pulps

Process control depends partly on testing of samples cut from the passing stream of pulp. The information obtainable from these samples is discussed later. Provided the load of solids is not distorted by irregular surges of tramp oversize, abrupt changes in solid-liquid ratio, or uneven rate of passage, a pulp sample can be made with good accuracy. If any of the above conditions are present, they must be dealt with in any case in order to restore operating efficiency, and it would be a waste of time to test a sample representative of such poor operation.

The best place to take the sample is at a launder overflow where the pulp stream is falling quietly and with a shallow flattish cross-section. Such points exist where classifier overflow launders deliver to pump sumps or conditioners. They can be contrived (head room permitting) by interrupting the run of a launder so that it delivers to a second section. The sample cut can be made either by a periodic diversion of the whole pulp stream, or by means of a cutter (a rectangular bottomless box with its length normal to the flow). The cutter traverses the stream at a steady rate, delivering the material it gathers to a sample bucket by means of a flexible hose. The cutter is moved by a reversible switch. (Fig. 5, p. 18). Several types are made. The sample buckets are changed periodically by the sample runner and can be kept locked between collections if desired. In this connection, the author's experience is that if shiftsmen want to tamper with samples nothing can prevent it, but that proper relationship between management and labour should remove this temptation. When assay returns show that a shift has had a bad run, there is probably a reason for the trouble which can best be discovered by co-operative investigation. Hectoring only confuses the shiftsman and tempts him to take covert measures to ensure that future samples shall be perfect, regardless of their validity.

Sample cutters which take part of the stream all the time are dangerous, since there is usually segregation of the solids in various parts of the cross-section. The amount gathered at each cut, and the period between cuts, must be decided in each specific case. Some operators favour the use of a master-switch which sets all the automatic samplers going at once. The general purpose of the sampling is to provide information which will help with the preparation of a metallurgical balance sheet, which accounts for the units of value fed into the mill and those leaving it as concentrate and tailings. Specific information is also gained at each sampling point which checks the state of the pulp and its fitness for transfer to the next stage of treatment. This will control technical performance and give data for possible improvement. The general assay figures will show, subject to a fairly consistent amount of value temporarily locked up in the various treatment stages, what amount of saleable product has been made for the period and, when conjoined with the cost accountants' figures, what the cost of production has been.

Where the use of power-activated sampling devices is not warranted, various simple expedients are possible. A stream of water can be arranged to fill a small balancing tank and cause deflection of the whole pulp stream, the same movement causing the tank to empty and return the deflecting baffle.

When hand methods are used, care is needed to ensure both regularity of cut and precision of time during which it is made. Failure in these matters biasses the sample.

Material not flowing in a launder with adequate fall is more difficult to sample. Ball mill discharge arrangements vary, and here hand sampling is usual. Classifier returns can be picked up in a cutter made from a half-section of piping, and this applies to table discharge of wet solids. Filter-cake is moist and can be cut by hand.

Reduction of the sample of pulp must be performed accurately and methodically. The whole sample should be got into suspension, either by hand stirring or mechanical paddling, and then poured through a vessel which splits it into equal halves, the accuracy of division being checked by inspection. One half is rejected and the other split again, down to a convenient size for drying and treating. Carelessness can seriously "weight" the reduction of this sample. One common fault is to stir vigorously and then simply to pour away part of the pulp. The centrifugal action and faster settling rate of the heavier minerals under this treatment causes preferential concentration at sides and bottom of the container, so that the discard is not representative of the original bulk.

The liquid-solid ratio, when integrated with rate of flow, shows the through-put rate. Pulp density and flow-rate are both control factors of considerable importance, and can be recorded continuously. For an ore reasonably consistent in its S.G. the absorption of gamma rays by the stream of pulp indicates this within 2%. A capsule of cobalt 60 is placed on one side of the flow channel and scintillation counter on the other. Variations occur in accordance with the amount of radiation absorbed by the solids in the passing pulp. The rate of flow can be measured by a magnetic flowmeter spinning in the pulp at a point surrounded by a magnetic field. The induced voltage is registered by immersed probes.

The Head Feed

This sample is usually cut from the feed coming to the fine-ore bin. It is tested for moisture, assay grade, and size. The moisture content shows the correction needed to give the figure for dry tonnage received. The assay provides a rough check on the stope assays, and if sorting by hand or by dense-media separation has preceded entry to the fine-ore bin, it gives a checking figure for the value and tonnage removed in that treatment. The size analysis checks the reduction performed by the crushing section and helps to ensure that the wet-grinding section does not have to handle oversize. Since assay grade multiplied by dry tonnage gives the figure for units of value received by the mill, it is important to check the entering tonnage.

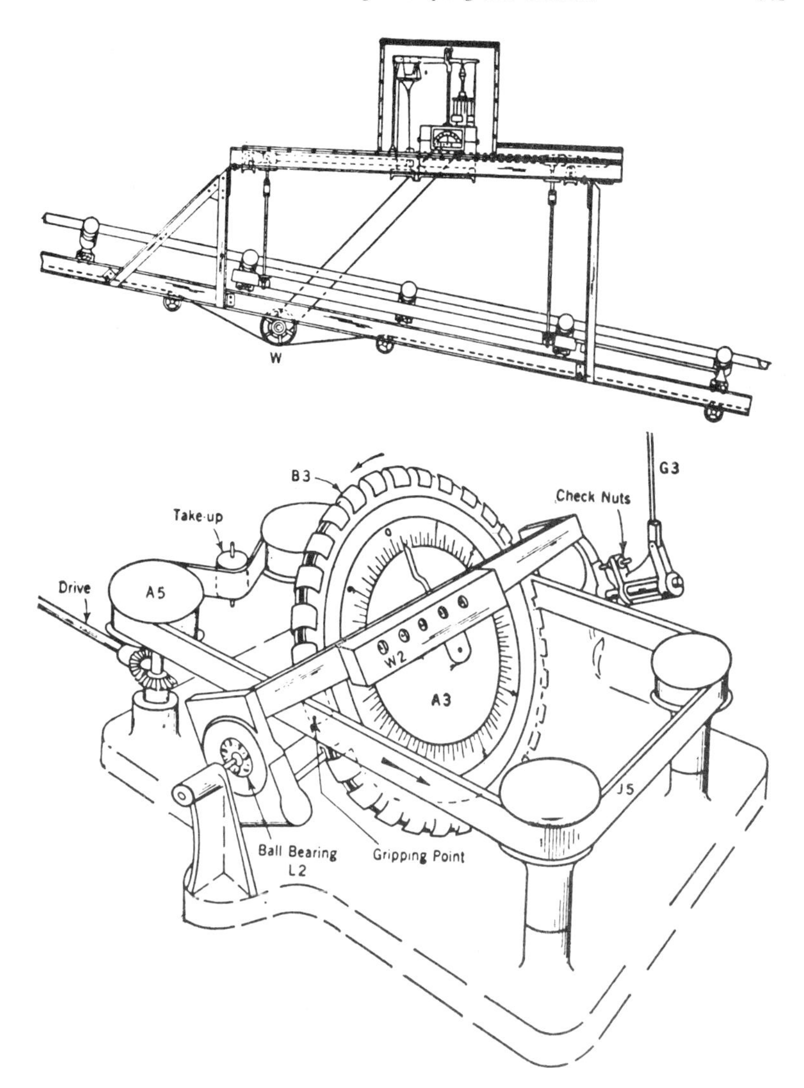

Fig. 270. *Operating Mechanism and Assembly of Merrick Weightometer* (after Hudson)[5]

Several types of machine have been developed for the purpose of doing this while the ore is in transit. One widely used appliance is the Merrick weightometer (Fig. 270).[5] Troughing idlers are mounted on a table which hangs from scale beams that register the weight of the passing load. This weight is integrated with the speed of the conveyor belt by the disc A3, which is driven from the underside of the returning conveyor belt through A5 and the crowned rollers B3. The disc shaft is carried in ball bearings L2 and is free to tilt under the pull of the weighing beam, transferred through G3. The travel of the endless belt J5 is in fixed ratio to that of the conveyor belt, and it is pressed against the disc rollers by guide rollers (not shown). While two rollers diametrically opposite are rotating the disc does not turn. This is zero position for an empty belt. When a load on the belt causes the disc to tilt, B3's are skewed across J5 and the disc rotates in proportion to that tilt. This rotation, recorded on the counter W2, shows the weight conveyed. Remote counters can be added, for purposes of distant control. To obviate belt slip and inaccuracy there should be a gravity take-up on the return side.

The "Adequate" Conveyor weigher picks up the belt on a single troughing idler, and can readily be installed on existing systems without structural alteration. The underside of the weighing component provides a compensating return idler when sticky material is handled. Any weightometer must be regularly calibrated, and this can be done by passing special weights over it or by anchoring a length of heavy chain so that it lies on the empty moving belt. If the entering ore changes seasonally from dry to moist, a layer of fines may adhere to the belt and pass over repeatedly, thus showing a delivery in excess of what the mill really receives. Recorded weight must therefore be corrected when necessary by making suitable allowance for variation in the weight of the conveyor belt, due to the clinging to it of an adhering layer of ore or to any other cause. This is done by adjusting the check nuts (Fig. 270).

Various methods of rate-of-feed recording are possible on the discharge side of the ore bin. The rate of travel of a feeder apron, the weight of ore per foot run, and a record of the running time give a rough check. Less satisfactory methods are the counting of trucks delivered, or use of the figure for ore sent from the mine. The Humboldt feeder is a duplex automatic hopper, one hopper filling while the other delivers its charge into the ball mill. Each delivery is automatically weighed and recorded.

The Hardinge constant-weight feeder (Fig. 271) automatically adjusts the

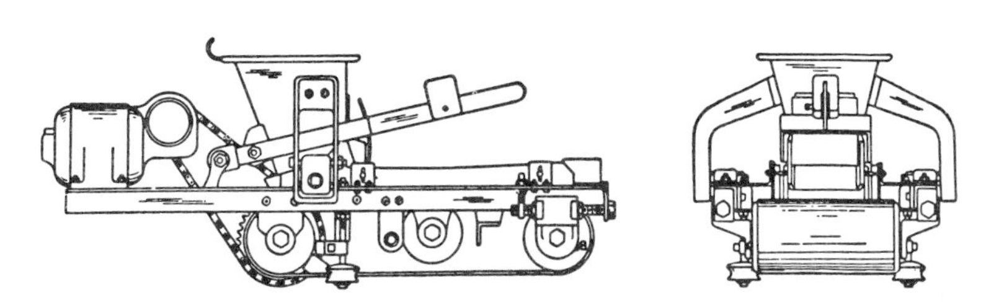

Fig. 271. Hardinge Feeder-Weigher (after Hudson)

weight fed, instead of the volume. It consists of an endless belt on a pivoted frame. Feed from the hopper above is received continuously. If the weight on the belt varies the frame tilts and the ribbon of ore drawn on to the belt is adjusted accordingly, because the clearance between it and the ore bin has altered. Rate of feed can be adjusted by means of a variable-speed drive and by optimum tilt produced by adjusting the counterweight. Belt travel is recorded, and since rate of discharge at constant weight is known, tonnage can be checked. It includes that of the moisture on the ore.

Rate of new feed can be controlled by signals which sense changes in the wet grinding load. One system measures variation in the vorticial vacuum of the hydrocyclone in a closed ball-mill circuit. This indicates change in the circulating load and initiates a corrective alteration in the rate of new feed.[6] Other methods include observation of the depth to which classifier rakes are buried and of load readings on the classifier motor; check of the pump load between mill discharge and bowl classifier; weighing of the underflow from the hydrocyclone; use of bubble pipes to gauge the density of the classifier pool or overflow; and measurement of sound level in the ball mill. One arrangement[8] measures the density of the mill discharge and maintains it at the set value by varying the input of feed water. This variation is used to signal alteration in feed rate of new ore. Another device uses very small changes in the fluid temperature through the mill (due to changes in ore grindability) to vary new feed rate.

When pebbles are used as grinding media the optimum loading can be maintained via signalled changes in the power draft, used to govern the addition of new supplies.[7]

In these forms of control, as with many others which call for periodic removal of pulp samples at key points along the flow line, the current trend in control practice is toward continuous monitoring which feeds appropriate signals to automated devices. These may be coordinated by computer control. Some systems are discussed later under Instrumentation and Automation.

Grinding Circuit Samples

Ball Mill Discharge

The most important control at this point is the shiftman's check of solid-liquid ratio. This is made at regular intervals, by collecting a canful of mill-discharge pulp and weighing it. If the pulp density is too high, the mill feed water must be increased, and *vice versa*. Sometimes a sample is accumulated during the shift and given a sizing analysis, and occasionally the *p*H of the mill pulp is checked, but this information is more readily obtained by sampling the products (+ and −) leaving the classifier. Mill-discharge density is a very important control since it affects the coating of balls and liners, and hence the grinding action in the mill.

If $p = \%$ by weight of solids in a pulp,

$q = \%$ by volume of solids in a pulp,

$Sp =$ specific gravity of the pulp,

$S_s =$ specific gravity of the solid,

$D =$ water-solid ratio by weight,

then[9]

$$Sp = p \cdot Sp + 1 - \frac{p \cdot Sp}{S_s} = \frac{S_s}{p + S_s(1-p)} = \frac{D+1}{D + \frac{1}{S_s}} \tag{21.2}$$

$$p = \frac{S_s(Sp-1)}{Sp(S_s-1)} = \frac{1}{D+1} \tag{21.3}$$

$$S_s = \frac{p \cdot Sp}{1 - Sp(1-p)} = \frac{Sp}{1 - D(Sp-1)} \tag{21.4}$$

$$D = \frac{1-p}{p} = \frac{S_s - Sp}{S_s(Sp-1)} = \frac{1}{p} - 1 \tag{21.5}$$

$$q = \frac{Sp-1}{S_s-1} = \frac{p}{S_s - p(S_s-1)} \cdot \tag{21.6}$$

The relations are shown in nomogram form in Fig. 102.

Classifier Returns

Hand samples are collected periodically, and a sizing analysis is made. This checks the efficiency with which the oversize is being returned and the undersize released from the closed circuit. It is also possible to make a check on the quantity of circulating load.

Classifier Overflow

This extremely important sample should be cut automatically. It is assayed for head value, which should check roughly with that of the solid head sample unless a concentrate is being withdrawn from the closed grinding circuit. Sizing analysis is also made on this fraction, to check efficiency of grinding. Since efficient concentration depends on optimum liberation this checking-point is essential. In addition to this sample, which is taken for laboratory use and under laboratory control, the shiftsman usually takes dip samples regularly at the weir and tests them for density and *p*H value. His running controls are used to adjust water and alkali, and so maintain the correct mesh-of-grind and alkalinity of the pulp.

Pulp density can be roughly assessed by the use of a hydrometer, though the solids may settle so rapidly as to upset the reading. The usual method is to weigh a known volume of pulp. Balances are made which give a direct reading of specific gravity.

When concentration of metal sulphides is practised in the closed grinding circuit, it is usually in order to avoid retention and overgrinding of a heavy mineral. The concentrate thus made joins the classifier overflow. In this case, provided they mix ahead of the automatic sample cutter, no special precautions are needed. If, however, a special concentrate is collected and permanently removed it must be brought into account. This would happen if strakes or jigs were used to trap gold and auriferous pyrite anywhere in the closed circuit. In South African practice the strakes are protected by security guards and pulp sampling does not usually begin until the ore has left this section and is ready for cyanidation. In Canadian practice, too, the physical difficulty of getting a reliable daily head sample places the emphasis rather on security measures than head sampling. All work of this sort should be mechanised as far as possible, only a few trustworthy men being employed.

Flotation Products

Concentrates

In a flow-line along which several successive products are removed, samples of each concentrate, rougher froth, and scavenger froth are needed. Pulp flowing through intermediate stages may also require sampling. The mos important of these is each final concentrate, which is sampled periodically during the shift. The "cuts" thus taken over a period are mixed and assayed for grade of end-product. This assay may include total sulphur, the sought metal, and any other metals. It must be sufficiently comprehensive to provide a basis for valuation if the product is being sold. Usually a simple assay for metal grade is sufficient, a little concentrate being retained from each shift assay to form a composite sample which is mixed and given the necessary comprehensive assay at monthly intervals. At that time, a portion may be sized, each screen and subsieve size being assayed for grade. The information given by these last assays as to tonnage and grade of each size-fraction is used to assess the work being done in the grinding circuit and the degree of liberation of each value.

The assays of the partially cleaned concentrates are only made periodically on accumulated bulk samples, except when alterations in the cell circuits are being planned. The important day-by-day value of these samples is their sizing analysis and microscopic appearance. These give direct information which can be used to maintain efficiency despite changes in ore mineralisation, or to improve any weak detail of the treatment.

Tailings

These are assayed from each shift's cumulative sample for grade. The *p*H may be checked as each cut is taken, to provide a cross-check on conditioning. The density of the pulp at overflow from the cells may also be checked by the shiftsman, who should log his figures here, as elsewhere in the plant when running check is made. This density check is only needed when over-dilution of the pulp might lower flotation efficiency. This could arise where a heavy draft of concentrate was made, or where tails from a series of cleaning sections diluted the new feed immoderately.

A suitable weight of each sample should be retained and mixed to form a special weekly or monthly sample. This is fractionated by screening and sedimentation down to convenient subsieve sizes—say to 15μ—and each size is assayed for percentage of the product which should have been floated. Heavy and light fractions of each size are obtained (say, by panning) and examined under the microscope in the expectation of finding the reason for their failure to float. If the lost value is heavily or completely oxidised, or is incompletely liberated, appropriate steps can then be taken to vary the flotation or grinding treatment, provided the extra recovery thus obtained justifies the cost of the change involved. The evidence of the shift log may be helpful, for example, by showing the drop in *p*H from grinding to tailing. Enough material should be kept in this composite sample to permit laboratory grinding and flotation tests to be made.

In most mills standard practice varies the reagent dosage in accordance with the operating results being observed. In a large and complex plant there are multiple addition points and human observation at one point does not necessarily fit happily with that at another. Methods of analysis now include colorimetry, polarography and spectroscopy as useful reinforcements for *p*H measurement and chemical spot-checks in the smaller and older mills. The problem of coordination and continuous assay and check is increasingly handled in newer plants by x-ray fluorescent analysis applied semi-continuously to the passing pulp, coupled with automated corrective measures.

Cyanidation Products

Samples of pulp are collected by cutting through the whole stream at an appropriate overflow point. Solutions may be sampled continuously by diversion of a steady drip or by periodic draw-off. Examination of the pulp samples is mainly concerned with losses in the tailings. Gold can be lost by incomplete washing of the discard, allowing pregnant cyanide to run to waste. This is checked by washing filter-cake and examining the filtrate. Another cause is coating of the gold by iron or aluminium oxide, preventing solution. A third is incomplete liberation, particularly of the gold carried in auriferous pyrite. A fourth is too short a time in the agitators. This is usually due to the failure of the gravity and amalgamation sections to trap and hold coarse metallics. If graphite is present in the ore, its effect on the operation must be watched.

The testing of samples of cyanide solution is concerned with the oxygen content, freedom from fouling salts, strength of the "available" cyanide, adequacy of its protective alkali, and the efficiency of precipitation. The quantities of pulp, pregnant cyanide, stock solution, etc., held up in the plant must be known if the day's recovery is to be correctly assessed. This involves recording of flow rates.

Dense-Medium Fluids

These are usually regenerated continuously. Sampling methods may be used to control any or all of the following:

(*a*) Optimum viscosity.
(*b*) Dilution by ore slimes.
(*c*) Rate of settlement.
(*d*) Bath density.

The last is a function of fluid density, of viscosity, and of the size-blend of the particles used to constitute the medium. With wear these become less angular and the loss of weight is therefore to some extent compensated by decrease in rubbing surface.

Mill Control

Routine sampling should give all the information needed for technical control and economic appraisal. In the foregoing sampling list discussion was limited to single-mineral concentration, but checking would have to cover the needs of a more complex operation by the addition of any further control points needed. A metallurgical balance sheet can look complicated, but when all the information from assays has been collated, its basic form reads:

Units received = Units in concentrate + Units in transit + Units in tailings + Units unaccounted for.

The word "units" is applied to each valuable mineral undergoing concentration. The tests on the various samples give, in addition to this profit-and-loss statement, technical information as to the moisture, crushed state, oxidation, etc., of the arriving ore, and as to the way in which the various processes have been conducted. The bulk samples direct attention to changes in the composition of the ore over a period of weeks and to the particle sizes and types chiefly responsible for losses. They give the mill superintendent the information he must have if his work is to reach and maintain high efficiency. They forestall trouble by showing its first appearance in any section of the flow-line, and by thus indicating the need for special attention and adjustment.

There is a basic difference in approach between the work of controlling a plant and that of testing ore in a laboratory. The two types of activity must not become confused. In a working plant all the conditions have been fitted together, though with a certain amount of flexibility, and all the operators should be concerned to maintain working conditions with minimum deviation. In the laboratory experiments can be made and new methods tried out with respect to any desired detail, but this could be fatal in the plant. Every definite change in the flow-line affects the whole of the work beyond the point where it is made, and there must be no change at any point until and unless all the consequences which will result have been brought into reckoning. It is for this reason, above all, that the sampling work must assist the operators to preserve the *status quo*.

The previous paragraph would need some qualification in the case of a system designed for treating ore which, despite correct use, was showing unaccountably poor or erratic results. The probable cause would be a change in the ore mineralisation not taken into account in the original test-work, and the place where preliminary investigation could best be initiated would be the mill laboratory. The monthly tailings sample would be a good starting point, and analytical methods might be worked out which would show whether the lost value was of the true mineral that should have been recovered, or whether there had been some change in composition. The mill laboratory would try to produce isolated samples of the suspected materials for x-ray, spectrographic, or other specialised tests only possible in a central laboratory. Special care is needed in these not uncommon circumstances, to ensure that the probable nature of the change has been discovered, and that the minerals suspected are the ones calling for tests beyond the resources of the mill laboratory. Such clarification as can be performed on the site aid greatly in the work of a distant specialising organisation and may make the difference between success and failure.

Use of X-Ray Fluorescence

The basic elements of the analytical systems increasingly in use for rapid or continuous analysis linked with plant control depend on a high-intensity source of x-ray radiation or the use of radio-active isotopes. In one simple system an x-ray tube directs radiation on the pulp sample, which may be a powder or a pulp, static or flowing in the latter case. Atoms in the sample are excited by the primary x-rays and emit secondary rays (fluorescent radia-ation) at wave-lengths characteristic of the elements thus excited. This fluorescent beam then passes through a slit (the collimator) so adjusted that a selected band of radiation is directed upon an analysing, or sorting, crystal such as lithium fluoride. From this it is diffracted through a second colli-mator, and is scanned by a detecting device such as a Geiger counter. The fluorescent energy is thus measurable and can, in the case of a continuously flowing pulp, be recorded on a strip chart. The signal strength is in pro-portion to the quantity of the selected excitation (and therefore of the selected atoms) irradiated by the primary x-rays. In the case of a streaming pulp this is the surface layer of solids in the sample. The "read-out" of radio-activity must therefore be corrected in accordance with the density of the thoroughly well mixed sample of flowing pulp. A description[10] of pioneer development work at Anaconda shows how such initial difficulties as the presence in a flotation sample, of bubbles, floccules, wood pulp and upsetting chemical reagents were overcome.

The detector can be set so as to move through a quadrant at twice the angular speed of the analysing crystal (Fig. 272). Since the wavelength of the fluorescent radiation is a function of the atomic radiation of the excited element, the position of the detector along an arc centred on the analyser identifies this element in accordance with Bragg's law

$$n\lambda = 2\alpha \text{ sine } \theta \tag{21.7}$$

where n is an integer for the specific diffraction, λ is the wavelength of the radiation, θ the angle of incidence and the interplanar spacing in the analysing crystal[11]. This arrangement allows qualitative to be added to quantitative analysis.

At Anaconda the original system[10] has been developed to give a continuous on-stream assay at thirteen points along the flow line[12]. This monitored information on feed, concentrates and tails is relayed to control points within 250 ft. In a system now under development on the Copper Belt, as part of the proposed use of fully automated control at Bancroft[13] part of the x-ray beam impinging on the continuous pulp sample is used to measure the density of the pulp and thus to correct the read-out.

Determination of the ash in coal has been usefully accelerated by means of either gamma or x-rays. Coal and ash differ in their reaction to short-wave radiation. As the ash content of a coal sample increases the scatter diminishes sufficiently to permit direct measurement of the ash content, provided moisture is controlled, the chemical constitution of the ash remains

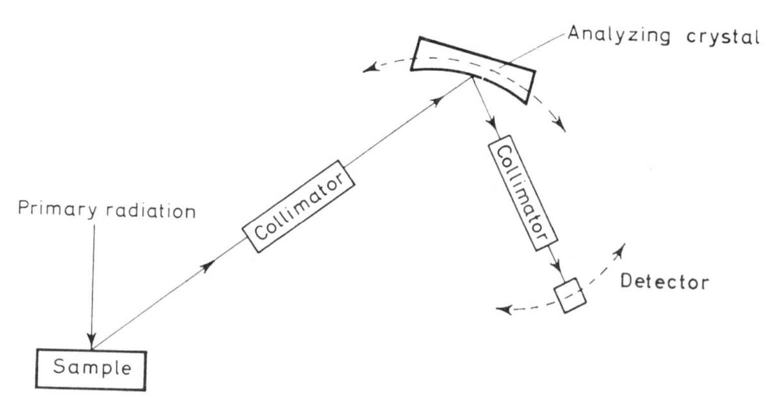

Fig. 272.

reasonably constant and an allowance is made for particle size. In the Simcar apparatus a radio-isotope (thulium 170) is used as the radio-active exciting source. The sample, at minus 1 in., weighs 30 lb and is collected in a steel box with a perspex window through which radiation is beamed. Moisture can be controlled by flooding the box with water and up to thirty samples can be analysed hourly, ash content varying between 2% and 30% with a maximum error of 2%.

A more elaborate system for continuous control and sampling has been put into profitable use by Dutch State Mines[15]. Samples of coal are cut at half-minute intervals, dried and ground in a hammer mill. They then pass through an x-ray beam and the reflected rays are analysed. The operation per sample takes some three minutes and its findings can be used as part of the automatic control in the washery.

The use of a radio-active source in connexion with coal was mentioned above. A Paper[16] on the use of isotopes describes their experimental employment in the rapid analysis of lead and tungsten ores.

Automatic Control

With the growing complexity of ore treatment, the importance of maintaining optimum conditions at key process stages has led to an increased use of automatic control. This is still in its infancy, but most mills have devices at critically important points which either regulate a process or give automatic warning when something is going wrong. Against this, it must be recognised that many mills have also put expensive equipment of this kind on their junk heap. The reasons are simple. While labour is untrained, a good automatic device, fresh from the manufacturer, is invaluable. As the shiftsmen gain experience and the automatic appliance becomes clogged by pulp and sand, the reliability of the man comes to exceed that of the appliance. Thus many excellent devices fall by the way, though several remain. It is therefore important not only to have the right appliance but also to be able to keep it in going order, and this is made easier if it is simple and reasonably robust.

If the whole chain of operations is divided into sections, and the minimum number of control factors essential to the efficient working of each section is known, then planning for automatic control can usefully begin. No ore should leave a section for the next in line until it is in the correct condition to serve as feed to that section. If any suitable automatic control can be used to check part of the condition at the point of transfer, it is worth consideration. Again, if any of the control factors in a given section can be made automatic, the effect on the men in charge of that section should be studied, to see whether the lessening of their duties will increase efficiency. There are human problems as well as technical ones in these matters.

The elements of an automatic control system include:

(1) Detection of change.
(2) Transmission of warning signal.
(3) Effective indication of variation.
(4) Actuation of correcting mechanisms.

These operations are performed in the order given above. There may be a time-lag network coupled with stage correction, to compensate feed-back or to prevent over-correction. If, for instance, *p*H is being measured in a conditioner fifteen minutes beyond the point where the addition of lime is controlled and the rate of addition of lime is automatically adjusted with each measurement, then there must be a time-lag between successive measurements of at least fifteen minutes so that the changed *p*H of the pulp will be measured at the next measurement.

Automatic devices can monitor a detail in a process and can either give appropriate warning of variation, keep a record, or take steps to correct a drift from normal working. In the last case, the correction can be complete at each resetting, or can be partial. The latter is to be preferred as a restoration to normal, if made abruptly, leads to over-correction, whereas step-by-step partial adjustment, with a suitable time-lag between steps, brings the condition to normal more satisfactorily and quickly.

Instrumentation and Automatic Control

Computers. These are increasingly used to receive, coordinate and integrate monitoring signals initiated at key points in the processing line. They then act on the combined information and thus ensure that any resulting correction is made in rhythm with the overall operation and not simply in response to a local disturbance. Computer-integrated corrections can be signalled to manual control points or applied direct in the form of automated adjustment. The current trend is toward increased operating control by computer, designed to maintain optimum conditions in routine running.

The three main activities suited to computer control are (*a*) simple recurrent problems, (*b*) complex development problems and (*c*) process control. Type (*c*) is discussed in this section. Computers are of two types. The digital computer is a high-speed adding machine which uses numbers rather than physical quantities when processing data. Its control unit determines the operating sequence in which calculations shall be made by the calculating unit. Its internal storage (as distinct from external filed information on magnetic tape, punched cards etc.) is called the memory unit. This holds the information required for the job in hand, in the form of machine language (symbols, signs, processing rules). The analog(ue) computer works by setting up a mathematical analogy of the problem in hand. A glossary of computer terms has been made by Drevdahl[17]. Computer control has been successfully superimposed on many automatically controlled processes such as those outlined below.

The simplest application of automation corrects one variant. Where several conditions react on the quality of the work and their automated sensing devices are coordinated, computer control is exercised.[22] Process function has been defined as "the sum of the individual operations . . . making up the process". In the functional, or black-box approach the system is not concerned with the mechanism of operation but with what is being done. The operational approach is concerned with the way in which things are done. The "language" of automation speaks in the form of electrical signals, shaft rotation, variation of pneumatic or hydraulic pressure, change of voltage, current, magnetic flux, temperature change, alteration of noise levels, density, *p*H flow rate, metering of light beams, change in resistivity of an electrified wire and so forth. Two main types of cognitive controls are used. In the open loop or open-end method there is no feedback, and constant check on performance is not made. In the feedback or closed loop system used in mill control, the monitored detail responds to a difference between the operating norm represented by the correct signal and a working change which has varied the signal. Appropriate correction (automatic reset) is then initiated. Digital or numerical control deals logically with data processing, while analog control is proportional to the dimension or function monitored. It uses signalled information as noted above (the "language" of automation) to represent the numerical variables in a computation, whereas digital control is only concerned with a step in decrease or increase of signal. Thus, analog control can compare the *p*H of the pulp at selected points with the operating norm required, and initiate the multiple adjustment needed to restore drift smoothly. The digital control would only recognise two states *plus* or *minus*, and would not work continuously to re-balance the flowline. The operational-digital technique of computer control combines the two methods of operational analog control and programmed digital control. The process controller gives feedback control based on input signals, coordination and adjusting signals.

Feeding and Blending

Control of the rate and quality of process feed is highly important to smooth and efficient running. In addition to the weightometers described earlier some newer machines use electronic sensing devices to improve accuracy. In one blending arrangement material A is continuously weighed and sampled as it passes from its feed belt to the blending belt. The weight is signalled to the sampling device and to a memory unit. The latter sends a delayed signal to the feeder which delivers material B on to the blending belt and feed rate is adjusted in accordance with the sampling information[18].

In cement practice one operator[19] stock-piles four heaps of raw material in accordance with the chemical content. This, involving quarry control and complex mathematical resolution to determine the lowest-cost blend for a specific kiln feed, is controlled by linear programming with a computer. Six ingredients must be proportioned within close limits. In another instance the automation of the delivery belt tripper has been described[20]. Kennecott's blending trippers traverse the system continuously, the automatic discharge limits being adjusted by hand so that each type of ore is routed to its correct storage. In the filling system servo-devices are used to sense the delivery position and the height of ore. When a bin is running low the tripper leaves its continuous traverse and deals with it. It also signals the main control centre if a bin is overfilled, shut down or wrongly working, sounding an alarm if there is trouble with choked discharge.

Comminution and classification. In the crushing plant automatic control can deal with such matters as sequential starting and stopping, temperature rise in bearings and loss of oil pressure. Transfer points can be scanned either by closed-circuit television or with a photo-electric cell. This normally receives a ray of light, which can be arranged so as to be cut off when ore piles up or conversely to emit a signal when the interruption of a properly loaded conveyor belt is stopped by its running empty. Such cells monitor a number of plant operations elsewhere, checking turbidity in bearing oil or in clarified solutions, and loading levels in ore bins, steady-head tanks etc.

Automatic control in wet grinding can use several types of signal. These include changes in:

(*a*) Grinding noise level in the ball mill.
(*b*) Power draft to mill.
(*c*) Circulating load in mill-classifier system.
(*d*) Pump loading (power draft).
(*e*) Pulp density (mill discharge or classifier overflow).
(*f*) Temperature rise through ball mill.
(*g*) Feed water to mill or classifier pool.

Noise level is controllable by means of a microphone set close to the mill. The signal is used to vary the rate of new feed when the sound intensity drifts from the pre-set level. Power draft to the mill should be held at its maximum. Decrease is due to underload of ore; overload of ore; wrong charge of grinding media; liner wear or incorrect solid-liquid ratio. The main use of power monitoring is in correction of ore loading. Where pebbles are the media and wear is fairly rapid it can be made the signal for addition. Williamson[23] has described the use of a sensitive power-watching arrangement which is used to control the periodic input of a fresh supply. Drift in solid-liquid ratio in the mill crop can be monitored by means of a radio-active source beamed on the mill discharge, or by semi-continuous weighing devices which check its density. The signal initiates any required change in the volume of head water added, and a flow meter which monitors this water then signals any further adjustment to the feeder supplying new ore. This ore may either enter the closed circuit or feed the preceding open-circuited rod mill[24]. The Paper describing this also mentions a control based on temperature change between feed water and mill discharge pulp, used to indicate grindability variations in the ore.

Closed circuit control in the mechanical classifier usually depends on maintained pulp density in the overflow. This can be assessed either by a radio-active gauge system or by bubble pipe. The latter consists of one or two vertically placed pipes at different levels in the pool. Compressed air escapes from the open lower end and the pressure, or alternatively the pressure difference between the two levels, measures the pulp density. Change in this is signalled to a motorised water valve which alters the diluting water as required. To avoid the turbulence of a raked pool the single-pipe system may work in a gathering funnel which receives the weir discharge. Other automated devices measure the return load from the classifier or cyclone and correct drift by regulating the new feed. The vacuum in the vortex of the hydrocyclone can be probed and monitored for the same purpose[25,26].

Gravity separation. In DMS the first step away from periodic hand-check and manual adjustment was continuous recording of the density of the bath media. This is now used in various systems to control changes in the use of diluting water and the rate of return of cleaned ferro-silicon from the densifier[27]. Control of the water gives an immediate response while trimming of the densifier rakes to increase or decrease the circulating load of solids makes its adjustment more slowly. In one system four zones in the separating bath are monitored by density gauges. Another (Fig. 273) has pneumatic controls worked by signals originating with the returning dense media. This is monitored both for S.G. and for pressure, thus affording a control level for height of bath. In a system developed for the concentration of diamonds[28] bath density is controlled by the differential pressures between two bubble pipes, and the viscosity is continuously monitored.

Improved de-sliming of a finely ground pulp in a hydro-separator was found by a magnetite producer to depend mainly on the hydraulic current introduced by four jets[21]. The volume of water thus admitted is controlled by means of a magnetic coil which senses the interface between a heavier magnetite-rich pulp and the supernatant siliceous middlings which are to be overflowed to waste. Signals from the coil to a pulsed timer adjust the main water valve and hold the interface at the required height in the hydro-sizer. Should an abrupt change in the ore cause the valve to remain at its open or closed limit too long, the speed of the pump removing the underflow from the separator is automatically varied until the situation returns to normal as signalled by the return of the control water valve within its working limits.

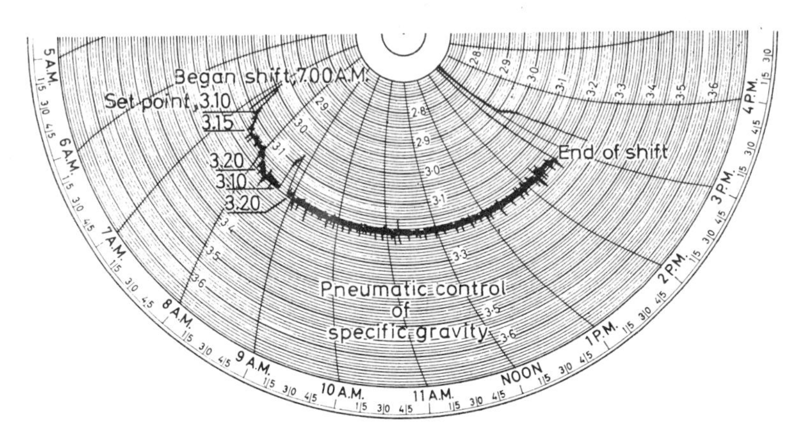

Fig. 273. Automatic Control of D.M.S.
(after Oss and Erickson)[27]

Chemical extraction. Automation of leaching circuits takes many shapes. pH control in connexion with maintenance of protective alkali in a cyanidation plant has been successfully demonstrated, using a conductivity probe to control the addition of lime[29]. A large-scale automation system, used in the leaching of flotation concentrates to produce cobalt and copper[30] at the respective rates of 4,000 and 100,000 metric tons annually centralises the indicators, regulators and remote control apparatus in a control room supervised by one or two operators. Control checks corrosion, thickening, blockage and sequence manipulations in two main circuits with an assembly of closely connected flowlines. So thorough is the system that interference by workers in the plant, save on special instructions from the remote control, is not allowed.

Froth flotation. Closed-loop control is used to anticipate, prevent or correct rapid change in one or more process variables. Process lag (delay of response of the controlled variable to a change in the manipulated variable[31]) should be as short as possible. Distance lag (delay in transmission of signal and response) should likewise be kept to a minimum by good planning and installation. Load change in mill instrumentation refers rather to that of an added reagent than to passing tonnage. In two-step control as defined by A.S.M.E.[32] the manipulated variable alternates between two limiting values.

In floating control the rate of change is a continuous function of the signal. In proportional control there is continuous relation between input and output. The variables monitored include flow rate, height of pulp in tanks or sumps, density and pressure changes, pulp conductivity and *p*H, pulp and/or froth level in flotation cells, thickener rake load, concentration of available reagent/s, change in circulating load, change in assay value along flow-line. In addition to methods of monitoring and signalling already mentioned is the use of radio-active detectors to trace the passage of specially activated additive minerals or reagents.

Developments in the automatic control of flotation are increasingly reported in the technical press[33,34,37]. Fluorescent check on the assay changes in mill pulp has proved valuable technically and economically. At Bancroft in Zambia[35] automatic continuous sampling and fluorescent assay of pulp from six points along the processing stream has been in use since early 1962 and is to be coupled with control by instrumentation activated by the assay signals in due course. Each sample is piped to its own irradiation cell mounted on a turntable and scanned in turn for sixty seconds, the cobalt and copper content being assayed.

Electro-magnetism. One example of magnetic control was given under *Gravity Separation*, above[21]. Another[36] controls the permissible magnetite in an ilmenite concentrate produced by flotation. Success depends on close watch on the iron in the tailings from each primary magnetic separator. The sample flows through a primary A.C. field and the magnitude of the induced secondary current in the pulp is measured. If it rises above the permissible 0·1% corrective action is taken. There is a lag of up to 2 hours between this point and the flotation from which this escaped magnetite would be concentrated sufficiently to spoil the ilmenite product.

Miscellaneous notes. In conclusion, a few of the other numerous reported uses of instrumentation are noted. Papers read at Cannes[38,39] refer to the automated preparation, use and tracking of reagents in which titration, absorptiometry, dosage, flow check, polarography and colorimetry can be instrumented. The processing of abrasive borax ores with the attendant problems of crystallised build-up and blocking of lines makes contact control difficult. Large-scale instrumented operation has been achieved[40]. In cement blending a digital computer "holds" a mathematical model of the process in which temperature and chemical composition of the slurry is detailed and "optimised", even the effect of changes in the weather being monitored, and allowed for in the automated adjustments[41]. In the filtration of cyanide residues feed level in the filter tank and that of the filtrate in the vacuum discharge is described[42].

References

1. Cook, P. E. (1959). Stevens Inst. of Technology Symposium. N.J., Jan.
2. Cy, P. ( ). "Sampling Nomogram". Minerals et Metaux, Paris.
3. Pryor, E. J. (1963). *Dictionary of Mineral Technology,* Mining Publications.
4. Eriksson, S., and Sundstrom, Y. (1961). *World Mining*, Aug.

References—continued

5. Hudson, W. G. (1954). *Conveyors and Related Equipment,* Wiley.
6. Anon. (1961). "Automatic Grinding Control System". *Can. Min. Jnl., May*
7. Williamson, J. E. (1960). *S. Af. I.M.M.,* Feb.
8. Weiss, N. (1960). *World Mining*, June.
9. Taggart, A. F. (1945). *Handbook of Mineral Dressing,* Wiley.
10. Bogert, J. R. (1961). *Mining World,* March.
11. Wood, R. E. (1959). *Min. Eng.* (*A.I.M.M.E.*), 214.
12. Lucy, W., Fulmer, T. G., and Holderreed, F. L. (1963). *I.M.P.C.* (*Cannes*), Pergamon.
13. Carson, R. (1963). *Rhod. Ch. of Mines Jnl.,* Sept.
14. Anon. (1964). *Simon Eng. Review.*
15. Balkestein, J. G., and Baerts, J. W. R. (1962). *Trans. S.M.E.,* 223, Dec.
16. Kalmakov, A. A., Polkine, S. I., Khan, G. A., and Smirnov, V. V. (1963). *I.M.P.C.* (*Cannes*), Pergamon.
17. Drevdahl, E. R. (1961). *World Mining*, June.
18. *Electro-Weighers*, Trade Publication (Birmingham) Ltd.
19. Nalle, P. B. and Weeks L. W. (1960). *Min. Eng.*, A.I.M.M.E., Sept.
20. West, H. H. (1963). *S.M.E.,* A.I.M.M.E., March.
21. Conroy, K. D., and Kachel, G. C. (1963). *Can. Min. and Met. Bull.,* Aug.
22. Amber, G. H., and P. S. *Anatomy of Automation*, Prentice-Hall.
23. Williamson, J. E. (1960). *S. Af., I.M.M.,* Feb.
24. Weiss, N. (1960). *World Mining*, June.
25. Roe, L. A. (1961). Min. Eng. (A.I.M.M.E.), Feb.
26. Anon. (1961). Automatic Grinding Control System. Can. Min. Jnl. May
27. Oss, D. G., and Erickson, S. E. (1962). *Trans. A.I.M.E.,* May.
28. Nesbitt, A. C., and Weavind, R. G. (1960). *I.M.P.C.* (*London*), I.M.M. (London).
29. Kelly, F. J., and Stevens, C. S. (1964). *Can. Min. J.,* Jan.
30. Piedboeuf, C., and Suys, F. (1963). *I.M.P.C.* (*Cannes*), Pergamon.
31. Lawver, J. E., and Barbarowicz, W. (1962). *Froth Flotation, 50th* Ann. Vol., A.I.M.E.
32. (1961). Amer. Soc. Mech. Eng. A.S.A.C.—85, N.Y.
33. Bredberg, J. H., and Carroll, J. T. (1963). *Eng. and Min. J.,* 164, Oct.
34. Anon. (1963). "Exper. at the Parc Mine". *Mining Magazine,* Jan.
35. Carson, R. (1963). *Rhodesian Chamber of Mines Jnl.,* Sept.
36. Runnolinna, U., and Heikkinen, M. (1962). *Mining World,* Aug.
37. Anon. (1963). "The Parc Mine Project". *Mine and Quarry Eng.,* Jan.
38. Blumkine, G. V., et al. (1963). *I.M.P.C.* (*Cannes*), Pergamon.
39. Lyfield, A. J. (1963). *I.M.P.C.* (*Cannes*), Pergamon.
40. Hegarty, A. (1960). *Mining Magazine,* Aug.
41. Anon. (1962). Japan's Most Modern Cement Plant, etc. Int. Systems Control Ltd., Wembley, April.
42. Mokkan, A. H. *Jnl. S. Af. I.M.M.*

CHAPTER 22

UNIT PROCESSES AND MACHINES

Introductory

In this chapter a somewhat incongruous mixture is presented. Each section deals in general terms with an essential feature of ore-dressing practice. Local practice and the reader's growing experience will best fill in details for some of the paragraphs, and the rapidly increasing volume of technical information accumulated in the research laboratories of specialising manufacturers should be studied when new plant is being considered.

Transport of Dry Solids

The movement of ore from the stopes to the mill is not usually the concern of mineral processing, but the manner in which this task is performed can affect subsequent treatment. If the blasting is done badly, too much fine material is produced in the wrong place, and in the wrong way. Leaking lubricants, if they fall on ore, become collector and frothing agents. Long delay between stope-face and ore bin initiates undesirable chemical changes. The *tempo* of transport, together with fore-knowledge of the flotative qualities of ore from different stopes, is sometimes of direct importance in process control. At the ore bins dry transport usually ends, and the solids are moved in water from thence to final discharge. In the older mill designs, great importance was attached to gravity fall. Today belt conveyors handle dry solids and pumps move pulps smoothly, cheaply, and efficiently, and it is no longer important to choose a hill-slope down which flow can proceed by gravity.

Dry ore can be moved through chutes, provided slope and cross-section are adequate, and awkward turns are avoided. Clean solids slide readily on a 15°–25° slope faced with bright steel, and on 40°–45° wooden surfaces. For most ores a working slope of between 45° and 55° is used. Too steep a slope is bad. Power is used in elevation of the feed, and fast-sliding rock is hard to control. At the delivery end some cheaply replaceable restraining device may be used, to brake the descending ore and deliver it gently. Iron sheeting, a curtain of old drill shanks, or heavy chain are suitable.

Bucket elevators are used for the elevation of dry or pulped ore. They consist of an endless chain or belt (Fig. 274) carrying the buckets vertically or at a steep angle. On reaching the top pulley they should turn at sufficient speed to fling their contents out with a measure of centrifugal force. Feed is directed into a gathering boot or to buckets rising from a boot. Wet ore can be drained through perforations in the buckets. A rod can be so suspended as to rap the discharging bucket and so help to shake out packing sludge.

Scraper elevators work by means of flights or rakes attached to a moving endless belt. The material is trapped against the underside of the housing and dragged up to the discharge point.

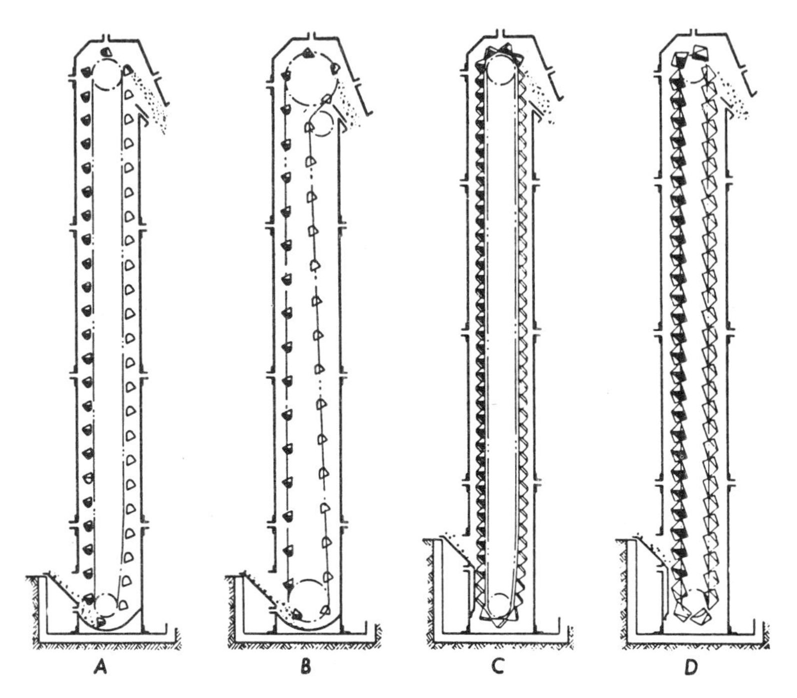

Fig. 274. *Four Types of Bucket Elevator* (after W. G. Hudson)

Belt conveyors handle the bulk of the dry ore moved through the modern plant. A system including a tripper is shown in Fig. 275. They include the *belt,* the carrying and return *idlers* (B and E), the take-up (F), the tripper and shuttle belt/s, if used, the drive (C), the belt cleaner (near D).

The standard "rubber" belt has a cotton foundation which must be strong enough to withstand the driving pull and the loading strains. The "weight" of duck in ounces, refers to a single thickness or "ply" 36″ × 42″. Plies are bonded together with rubber, and the pull needed to separate two plies in a strip one inch thick, is termed the "friction". It is between 12 lb. and 24 lb. and chosen for required duty and flexibility. The plies must not separate under the worst conditions (i.e. when rounding the end pulleys). This carcass (plies) is protected above and below by rubber, of specified tensile strength (800–4000 p.s.i.). Covers vary in thickness from $\frac{1}{16}$″ to $\frac{1}{2}$″. For severe service, a "breaker strip" of open mesh fabric may bind cover to carcass and confine any tear arising in use. Cord belts have additional longitudinal cords embedded in rubber. Steel wire cables are also available for use under

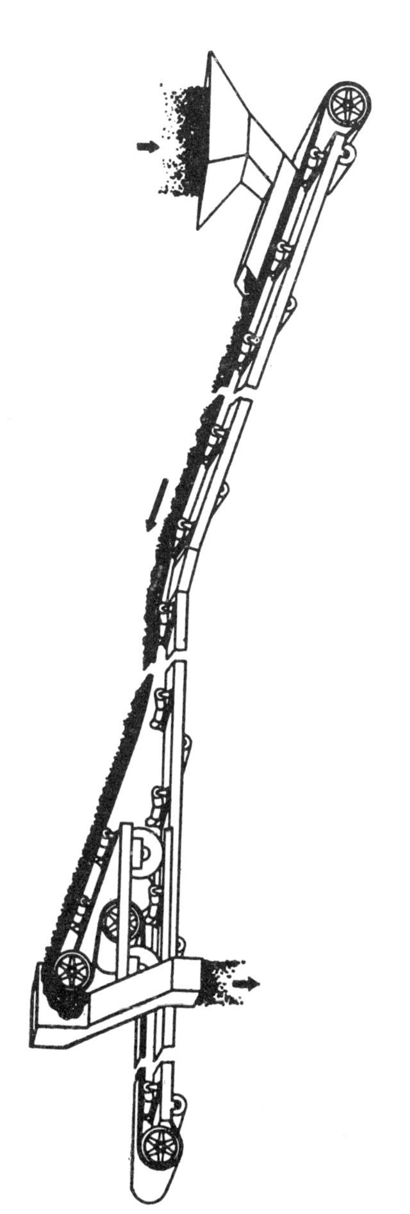

Fig. 275. *Inclined and Level Conveyor with Movable Tripper* (G.E.C.)

severe tension. Special splicing methods are necessary. Where oil-contaminated coal is moved, synthetic rubber coatings may be used.

Grierson[2] notes the increased use of nylon belting, which has thrice the strength of cotton, absorbs shock-energy resiliently and withstands up to 17% elongation before breaking. With these advantages go resistance to fungal attack and to abrasion. Nylon can be wefted with a cotton warp, mixed with cotton in both warp and weft or of all-nylon weave. The last material is suited to long-distance conveyors with high capacities and up to 1,000 H.P. pull. The safe working strength for most synthetic fibres is 10% of the ultimate tensile strength. Terylene and rayon have limited use. To minimise fire risk rubber and balata are often replaced by non-flammable polyvinyl chloride, despite its lower impact resistance, poorer tracking and shorter life.

Belts arranged in series are today used over great distances, but in ordinary milling operations a thousand or two feet suffice for most work. For very severe work (massive, jagged, or hot material) pans are used. These are steel sheets formed into an articulated system. The ordinary endless belt rides on troughing pulleys (idlers) on its loaded side, and driving power is applied shortly after it begins to return after passing over the head pulley, except in a small system when the head pulley may itself do the driving. To avoid the abrasive effect of passing the tensioned surface of the belt over the pulleys when it is gritty, it is sometimes washed or brushed at the beginning of the return run. In one system it receives two successive 180° turns on the return side, so that all contacts are between the pulley and the underside of the belt. The importance of this care for the surface is that the chief part of the cost of conveying is the replacement of an outworn belt, which can soon be ruined by bad treatment.

Troughing idlers (Fig. 276) are so spaced that there is but little sag on the loaded belt. If this became excessive, particularly on a steeply inclined belt, the slip and wear would seriously shorten the belt life. Wear is worst at the

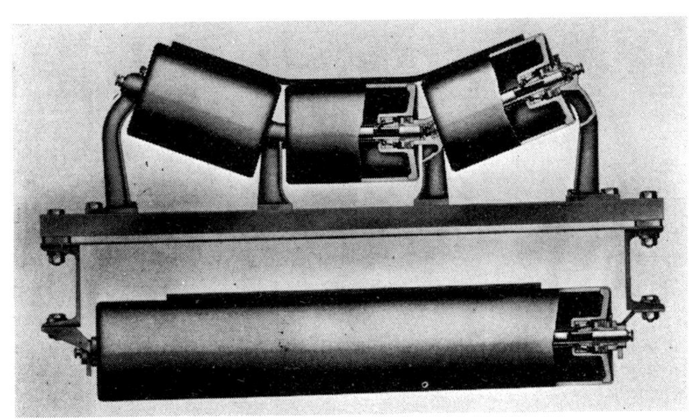

Fig. 276. *Troughing and Return Idler Unit* (G.E.C.)

feed end, and the shock of falling rock is avoided partly by delivering between idlers, rather than directly above them. Rubber-covered idlers are often used at such points. Transfer is kept to a minimum, to avoid this wear, and extra troughing idlers are used at the loading end.

Return idlers are flat and more widely spaced. If abrasive material is allowed to build up on the surface, undue belt wear will result, as there is appreciable cling due to tension, even though the belt is not loaded. This applies even more to the surface of the troughing idlers, which support the underside of the fully loaded and tensioned belt. One essential condition assisted by the idlers is the "training" of the belt. Misaligned structure, build-up on idlers, bad loading, wind pressure, unequal stretch of the belt, insufficiently heavy take-up, may cause the belt to climb off centre for part of its run. In old installations guide idlers, which bore on the edge of the belt, were used to prevent serious trouble, but they are obsolete. Troughing idlers with tilting blocks or self-aligning arrangements (training idlers) are used instead, combined with improved design. All idlers should run freely, and be checked regularly to ensure that they are doing their work properly, and not damaging the belt (for example, by being seized, or by having sharp stones stuck to them which would punish the belt each time they turned). The troughing pulleys are spaced at 3′–5′ intervals and the return idlers at about 10′.

By using wefted nylon it has become practicable to increase the troughing angle of the idlers to 45°, thereby raising the load-carrying capacity some 25% for a comparable width of belt. The more severe flexing stress along the "hinge lines" is met by superior elasticity and the need for fewer plies than with cotton duck. The deep trough reduces the major causes of wear—abrasion at the loading point and slipping on upgrades where a sagging belt rises over an idler. Better training and centring are also claimed. The belt speed must be kept above 400 ft./min. in order to ensure a good trajectory as the ore discharges from the belt. Idlers as steep as 53° have been used in one coking plant, where as a result spillage has virtually disappeared.

The take-up tensioning device (Fig. 277) acts on the return side. It should be kept at the lowest practicable level so as to minimise belt stress. Automated adjustment is possible.

The take-up adjusts the belt for stretch or shrinkage, prevents undue sag between idlers, and prevents slip at the drive-pulley. Provision for 1½% take-up is wise. Screw take-ups at the tail end have some use in short runs. Gravity operated arrangements (Fig. 277) which adjust the tension continuously to varying conditions, are usual in mill installations, together with a screw take-up. With a belt inclined at 13° or more, the return side provides sufficient weight for most, if not all, the take-up.

The tripper is used to distribute the discharge over a required length of ore bin or dump. Instead of discharging over a drive pulley at the head end, the belt is led back over a pair of bend pulleys. The load falls to a diverting chute (Fig. 275). The tripper may be fixed or movable. In the latter case, the bend (or snub) pulleys are on a travelling frame movable by hand, by power from the main system, or by independent motor. It can be automatically self-reversed.

The drive is at, or immediately following, the head end, and is supplied by

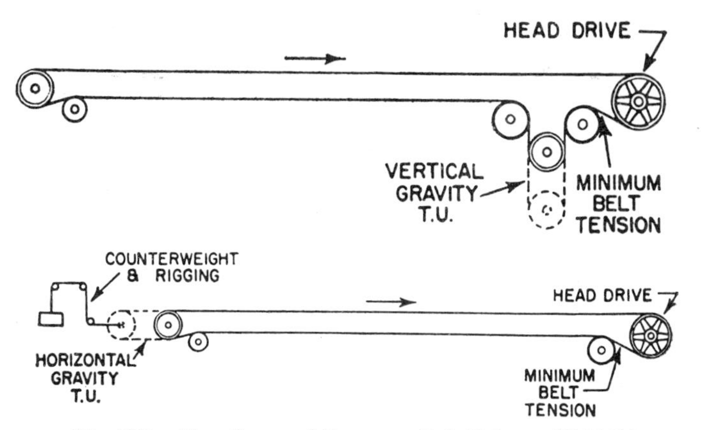

Fig. 277. Two forms of Conveyor Belt Take-up (G.E.C.)

one or two motors (the latter for heavy-duty work). With dual drive, the first motor is synchronous and the second slip-ring, with automatically timed acceleration and distribution of torque. Electrical interlock is usual in modern systems. The final unit must be well under way before the next in front can be started, and so through to the head feed. At stopping, the head machine stops first. If a magnetic head pulley is in use, the belt cleaner must be set well back, since magnetic material carries on round for a few feet.

The effective driving power depends on the difference between carrying tension (T_1) and return side tension (T_2).

$$\text{Effective belt HP} = \frac{(T_1 - T_2)S}{33{,}000}$$

where S is the speed (ft./min.) and T is expressed in lb./foot. Transmission of driving grip depends on the area wrapped round the driving pulley (the angle of wrap) and the coefficient of friction between this pulley and the underside of the belt, Tension on the loaded side is produced by weight of belt, load, inclination and mechanical resistance of the system.

Heavy-duty conveyors with high lifts and/or long flights are widely used. In the "Cable Belt Conveyor" two wire ropes at the edges of the belting take the driving tension. In the Horstermann conveyor, used in less heavily stressed work, the loaded belt is carried on 4′ square steel plates attached to a driving chain. In America the rigid structure is increasingly replaced by a system in which the idling pulleys are supported at 10′ to 20′ intervals on ropes in a light steel framework. Another radical departure from conventional design is the German "Serpentix" conveyor. This breaks away from the limiting requirement of straight-line movement from feed to discharge and can even turn spirally on a radius of less than 10′ while climbing, or wriggle snakewise through obstacles. The carrying belt is in short sections each forming a trough normal to direction of travel, and vulcanised to underlying steel pans which are bolted to a driving chain. Side discharge is used at the delivery end after which the troughs are righted, and thus are available for a return load. Another German development is the sandwich conveyor.[3] In this, a weighted belt over-runs the ore being lifted at an abnormally high angle.

The maximum recommended belt speed for granular material coarser than $\frac{1}{8}''$ is 800 ft./min. and for heavy, large or abrasive ore 400′. Light and fine sands should not be carried faster than 250 ft./min., and if discharge ploughs are in use the speed should be still lower. Troughing rolls may be spaced at from $3\frac{1}{2}$ ft. centres upward and return idlers from 8 ft. The T_1/T_2 tension ratio for a plain steel pulley with an angle of contact of 180° should be 2·19 and when snubbed to 240° 2·85, the coefficient of friction being 0·25. With a rubber-lagged driving pulley having a coefficient of 0·35 the figures become 3·0 and 4·33 respectively. With tandem drive and bare steel pulleys at 360° total contact the ratio becomes 4·8 and at 500° contact angle 8·86, Lagged tandem pulleys would increase the safe working ratios to 9 and 21·2 respectively. Corresponding belt tension T_1 would lie between 1·85 and 1·13.

Since the cost of a belt may equal that of the drive and framework, care in its maintenance is essential. Choice of the right belt is important. Starting and loading strains and stresses must be controlled. Tension should not be in excess of that required to avoid slip and sag. The belt must not bear on anything except its properly maintained drums and idlers. Joints must be square. Good design and maintenance at loading and discharge points are necessary. Belts can convey their load at any slope up to 15°, or if the ore is dry and not slippery, up to 20° or exceptionally 25°. On the Rand a certain amount of washing of the banket ore is performed by spraying water onto the rising belts. This assists the men picking off waste rock by displaying the true surface, but does not, of course, remove any gold-rich slime adhering to the underside of such waste. When the belt delivers downhill, the system must be suitably braked. When it elevates its load, provision is necessary to prevent run-back in the event of power failure. The tensile stress set up by a load too heavy for a single belt to handle can be reduced by using a series of belts each transferring to the next. Since the chief wear is at loading points, such transfer is kept to the minimum.

Several methods are used to minimise the shock of loading. In one common arrangement (Fig. 278) the fine fraction of the feed falls to the belt first and cushions the larger pieces which arrive later. The feed must not fall directly above a roller, unless this is well padded with soft rubber. A good arrangement is to bring the feed to a short unsupported length between two such rollers. Skirting-boards, used to prevent side spill, must not be so arranged that a stone could lodge between them and the belt, an occurrence which could cut and ruin the belt in a short time. The surface of the belt should be systematically inspected and all cuts, cracks, and damage sufficient to let water into the fabric should be dealt with expeditiously. Portable vulcanising units can repair long cuts quickly. If acidic water is allowed to penetrate to the textile, or mildew begins, the belt life is shortened. On some conveyors idlers are used to bear against the edges of the belt if it should ride up on its troughing idlers. If these seize, the edge of the belt saws into them, and its protective rubber coating is removed.

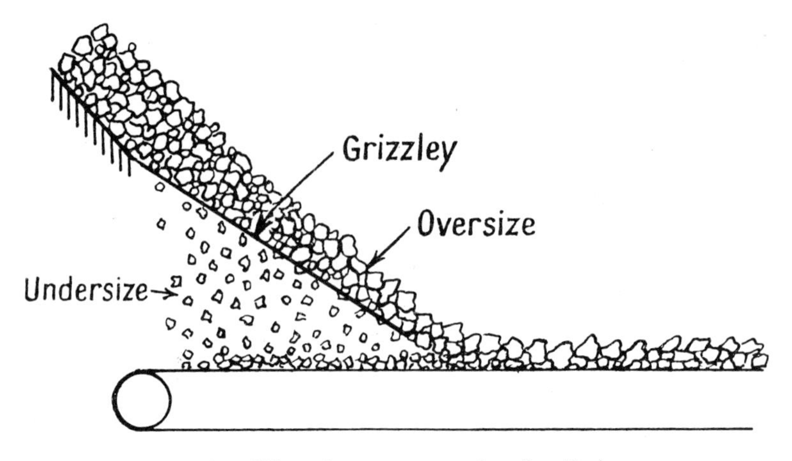

Fig. 278. *Arrangement at Loading End*

If such idlers *must* be used, they must be kept well lubricated and turning. Another cause of wear is the build-up of spilt ore on the return track, so that the belt surface rubs against it. Ore can also get on the inside of the belt and weaken it by pressing up as the loaded fabric goes over the troughing pulleys.

The usual working speed of a belt is of the order of 600 ft./min., and this is only suitable if the ore is free from dust. Each single section of a belt series must go straight from feed to discharge point, as bends in a normal belt line are not possible. The installation must be such that failure at any point will automatically stop delivery in front of that point. A choked discharge chute at the end of a series of belts would lead to chaos if the preceding sections did not stop at once. Interlock with devices being fed by the belts is also important for the same reason. It should not be possible to shut down any machine on the working line without arresting the feed at the same time, and similarly, motor failure should lead to the automatic tripping of all preceding machines.

Hydraulic Transport

This may include delivery of ore to the mill, handling of pulp through grinding and treatment, and/or product delivery such as that to tailings ponds or underground backfill points.

The pulp varies in corrosiveness, abrasiveness, and particle size of its solid content, and the channels of flow must be able to deal with these changes. At some points flushing water may be needed, and at others it must be withdrawn. At points of severe abrasive wear duplicate channels can simplify the task of maintenance. Buried piping should be accessible, if necessary by manholes at convenient distances apart for the use of drain-clearing rods and "go-devils". Pulp transport is the arterial and venous system feeding and relieving the organs of the plant, and if it fails the mill must stop. All the pulp—the solid, the liquid, and the deceptively small quantity of sticky, corrosive or oversize material which is most prone to settle and form the nucleus of blockage in a pipe—must move continuously.

Pulp may be made to flow through open launders. The correct slope is established by tests which disclose the settling characteristics of the material concerned—pulp and launder surface. The latter may be wood, metal, or rubber.

In plants of any size, the pulp is moved through piping, which should be kept travelling in as straight a line as possible to prevent abrasion at bends. The use of oversize pipe is dangerous whenever slow motion might give the solids a chance to settle and choke the pipe. A self-draining arrangement with a straight up-slope followed by a straight down-slope is less liable to give trouble than a straight discharge from a pump, unless the shiftsman can be relied on always to flush water through before shutting down the pumping line. Pipes carrying pulps containing lime may become scaled, and periodic treatment is helpful. Pipes of small bore can be scraped through if a rope can be passed, while larger ones can be worked through with rods or go-devils. Pipes should be accessible and identified as to function, to aid the work of the maintenance gang. Materials of construction include wood stave (for large diameter piping carrying tailings), iron and steel (which may be rubber lined for handling corrosive pulp), plastics, e.g. Saran, concrete and asbestos cement. The convenience of access provided by laundering can be achieved by cutting slots along the top of fairly flat runs of piping at defined points. The launder section may be rectangular (usual), V, or trough-shaped. Solids move by flowing in suspension when small and near the density of the pulp. Larger particles roll if equi-axial or slide if tabular, and there is some jumping due to eddies in the pulp stream. When pulp consistency changes, or rate of flow is low, solids may settle on the bottom of the launder and form a layer, which only moves when it has built up sufficiently. If the launder is too shallow for this "bursting head" to form, it will choke and overflow. The slope required varies in direct ratio to particle size, particle density, and percentage of solids. Fines are best moved in a deep flow, and coarse particles in a shallow one, down which they can slide. The water should more than cover the largest particle.

The principles which govern hydraulic transportation of solids involve a multiplicity of physical characteristics and design factors. Settling mixtures, as distinguished from non-settling, have been arbitrarily defined as those in which settlement of the particles in a stationary pulp is faster than 0·002 to 0·005 ft./sec.[4] There are four modes of flow at a given velocity of slurry movement. In increasing rate these are (*a*) a stationary bed with saltation above it; (*b*) a sliding bed with suspension; (*c*) heterogeneous flow; and (*d*) pseudo-viscous flow. Condition (*a*) exists when a deep settled bed has formed[5] but where the velocity is slightly above that at which the channel would be blocked by deposited material. In condition (*b*) the bed thins and begins to slide.

When the channel is a pipeline any partial blockage due to settlement restricts the cross-sectional area of free flow, and thus increases the pressure ahead of this settlement. The factors involved in pipeline design and installation[6] include:

(1) Solid-liquid ratio
(2) Average S.G. of pulp.
(3) S.G. of solid constituent/s.
(4) Size analysis and particle shape.
(5) Viscosity of fluid.

Among other factors are materials of pipeline construction, abrasive or corrosive effects of the pulp, pump characteristics, required location, distances, grading of line, steadiness of operation and maximum load. Here the Durand-Condolios sizing classification[7] is helpful. These authors have developed a relationship between pipeline diameter and S.G. for "Limit Deposit Velocity (LDV)" definable as the absolute minimum rate of flow to be allowed for in designing a hydraulic transporting installation. Their equation is

$$LDV = K\sqrt{\frac{2g\,D\,(r_1 - r)}{r}} \qquad (22.1)$$

where D is the diameter of the pipeline, r_1 the mass per unit volume of solids, r that of water and K a constant depending on the concentration and sizing of the solid particles. For a specified pulp density and mineral content

$$LDV = K_1\sqrt{D} \qquad (22.2)$$

and for variation in S.G. of solids

$$LDV = K_2\sqrt{\text{S.G. Solids} - 1} \qquad (22.3)$$

Design factors for a slurry pipeline must allow for change in frictional resistance where the walls of the piping are expected to roughen with use. For a long line, boosting stations may be needed, their distance apart depending on the maximum permissible pressure to be developed at each pump discharge. A typical phosphate slurry moved at 600 tons hourly might use 7,000 gal./min. of water at 15 ft./sec. and 40% solids. In Florida nearly 15 million tons of phosphate rock are pumped annually[8] and automatic control of series pumping is practised.[9]

Storage and Blending

The following are the main purposes of ore storage:

(*a*) To receive ore intermittently and deliver it smoothly and cheaply.
(*b*) To accumulate mill products for intermittent disposal.
(*c*) To maintain an adequate tonnage of ore for treatment.
(*d*) To smooth out irregularities in working (surge storage).
(*e*) To provide emergency catchment space.
(*f*) To facilitate balanced blending of dissimilar minerals.

The considerations which modify stockpiling policy include the chemical stability of the ore; the ease of reclamation (cost of double-handling involved) and icing-up.

The capacity for type (*a*) (mill head storage) depends on the hoisting cycle and days worked underground, and on the nature of the mill treatment. The more complex the flow-sheet and the more costly the plant used, the stronger becomes the argument for continuous milling. Not only must twice the plant be provided if the mill is only to be run half the day, but recovery is erratic during start-up and shut-down. An extremely simple operation, particularly one closely linked with the entry of raw ore, can easily be started and stopped, but such work as differential flotation tied to stage grinding should be run continuously once started. Dense-media plants usually run in phase with the delivery of ore to the secondary crushing plant, though there is an increasing tendency to work the crushing plant most of the twenty-four hours. If the mine is shut over the week-end, the mill must either have fine-ore storage to tide it over or be shut down for the same period. A plant requiring frequent overhaul might be run with a weekly shut-down, but a better arrangement would be to shut down one section at a time for maintenance, provided the tonnage treated justified the laying out of treatment in parallel.

A further condition to be studied in arranging ore storage is the effect of exposure. Unstable sulphides must be treated with minimum delay. Wet ore cannot be left exposed to extreme cold, or it will freeze and be difficult to move. Ore in transit represents a tie-up of working capital which cannot be released until the concentration is finished and the product shipped. The extra handling involved adds to the total cost.

Where development ore must be stockpiled until the mill has been constructed, there is little choice as to storage arrangements. Again, political considerations may call for special accumulation. A notable example was the formation of special dumps of open-cast coal in Britain during World War II, when the interruption of normal transport by damage through enemy action had to be provided against. Here, the added problem of spontaneous combustion had also to be met, by making the dumps of such dimensions that internal heat did not rise to danger point.

The angle of repose of broken ore varies between 37° and 45°, and the sliding angle, once movement has commenced, from 30° to 45°. An extra 5° to 10° must be allowed for, to ensure smooth withdrawal. To some extent this depends on the surface condition of the rock, a clean pile of close-ranging sizes being easy to control, while one of random sizes with deteriorated surfaces and half-consolidated fines can be difficult to start moving and to keep running smoothly. A good large-scale method of coping with seasonal surges, due to shortage of underground labour at sowing and harvesting times followed by plenty at mid-season (a common condition where "cheap native labour" is recruited from tribal sources), is shown in Fig. 279. The undersize needed to keep the plant running is screened off and treated, while oversize is formed in a long dump, above a tunnel equipped with a belt conveyor. Such ore, provided it is non-caking, slides easily and drawing rate is regulated by a curtain of chains. The dump can be formed by tripping from an overhead conveyor belt.

When mixed ore is blended, it can be reclaimed by an arrangement such as that shown in Fig. 280. The dump is formed by adding successive thin

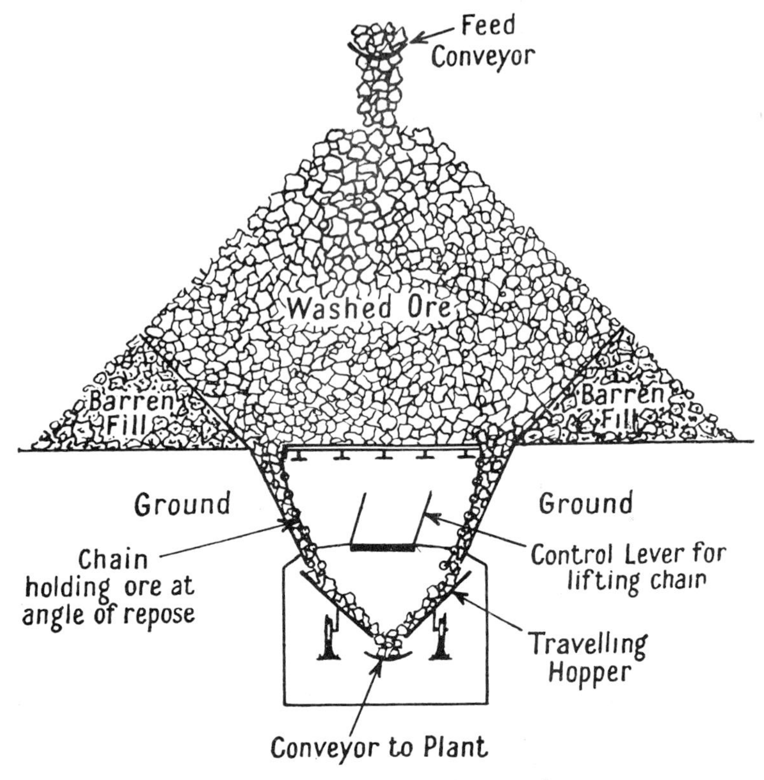

Fig. 279. *Stockpiling of Coarse Ore*

layers by conveyor and then reclaiming by loosening the pile by harrowing action and delivering the loosened ore to withdrawal conveyors. In a typical system[10] where iron ore from chemically different sources must be evenly blended the feed, crushed to *minus* $2\frac{1}{2}''$, is distributed from wing conveyors or "stackers" to the ridges of prism-shaped beds, as the longitudinal feed belt travels backwards and forwards to distribute a thin layer as the prism is built. Each bed holds 15,000 tons and is 56 feet wide at the base and $21\frac{1}{2}$ feet high. The bridge of the reclaiming machine spans this width and is carried on tracks. It is pressed face-on into the end cross-section of the prismatic pile. It supports a triangular harrow armed with steel teeth which oscillate at 26 strokes a minute as they "comb" their way into the pile. Loosened ore falls to the base of the bridge, which carries a plough conveyor. This delivers the now thoroughly blended material to a longitudinal conveyor running along the flank of the pile. The forward speed of the bridge is adjustable between $\frac{1}{2}''$ and $5''$ per minute, and the retracting speed is $30'$ per minute.

"Blending" is not necessarily synonymous with "mixing". A blend combines varieties of raw material into a consistent head feed in which the consistent grains are mixed in the required proportions so as to become physically and chemically consistent. Mixing, by Larsen's definition, does not imply the loss of identity of the mixed components since it does not overcome segregation into sizes. Mixing in bins may suffice for some purposes but the withdrawn material will be layered if segregation has occurred. In the Robins-Messiter system the finer material tends to collect at the centre and the coarser at the outer edges. Further segregating effects may put the finer ore in the upper part of the pile and the coarser at its toe. Voids are highest away from the stacker and lowest below the delivery point. As the pile builds, several hundred layers are thus spread, each containing an equal amount of ore. The harrowing action during retrieval repairs this segregation. Provided the pile has been correctly proportioned with its various feeds when building, a consistent plant feed is ensured. Excessively wet or dry ore, coarser and finer fractions, "highs" and "lows" in trouble-making contaminants have been blended into a composite feed for treatment.

These large-scale tonnages could not conveniently be scaled down to meet the needs of the ordinary operation without a considerable tie-up of ore in transit. The mill's needs can usually be met by other methods. Size segregation in the ore bin can be minimised by good design and planned distribution of the delivered ore. Where ores from various sources must be binned separately, or in separate sections of a long bin provided with an adequate number of independently controlled draw-off gates, the discharge from these points can be independently delivered at a prescribed rate on to a common conveyor. From this, these layers will be thoroughly blended during grinding.

Cruder storage arrangements, which are cheaper to start but dearer to reclaim, are long dumps of mixed size, formed by dumping from trams or trucks. Their ore is withdrawn by mechanical shovelling, bulldozing, or glory-hole. Where a discharge gate is used it should be at least three times as wide as the maximum lump of ore in the dump, to avoid the formation of an arch of jammed rocks. A running chute should be six times the width of the largest lump, to prevent side friction from arresting the flow of rock.

Tailings, when delivered dry for dumping, are usually fairly fine in size. If they are suitable for use underground, the only problem is one of transport. If dumped, they should go to waste ground well clear of present or future underground workings, and be downwind (prevailing) from the township.

Fine-ore bins are not intended for prolonged storage, but to provide continuous feed of crushed ore to the grinding section. They are made of wood, concrete, or steel. Wood, though most easily procurable in newly opened country, is the least desirable material of construction, because of the fire risk. Concrete is widely obtainable, very adaptable and easy to form to the required shape. The bin must be easy to fill, and must allow steady fall of the ore through to the discharge gates with no "hanging up" of material or opportunity for it to segregate into coarse and fine fractions. Sudden changes in size of feed reduce efficiency in both grinding and concentration. The discharge must be adequate, and drawn from several alternative points if the bin is large. The withdrawing mechanisms must be accessible for maintenance and clearing in the event of a choke. A flat-bottomed bin is easy to construct

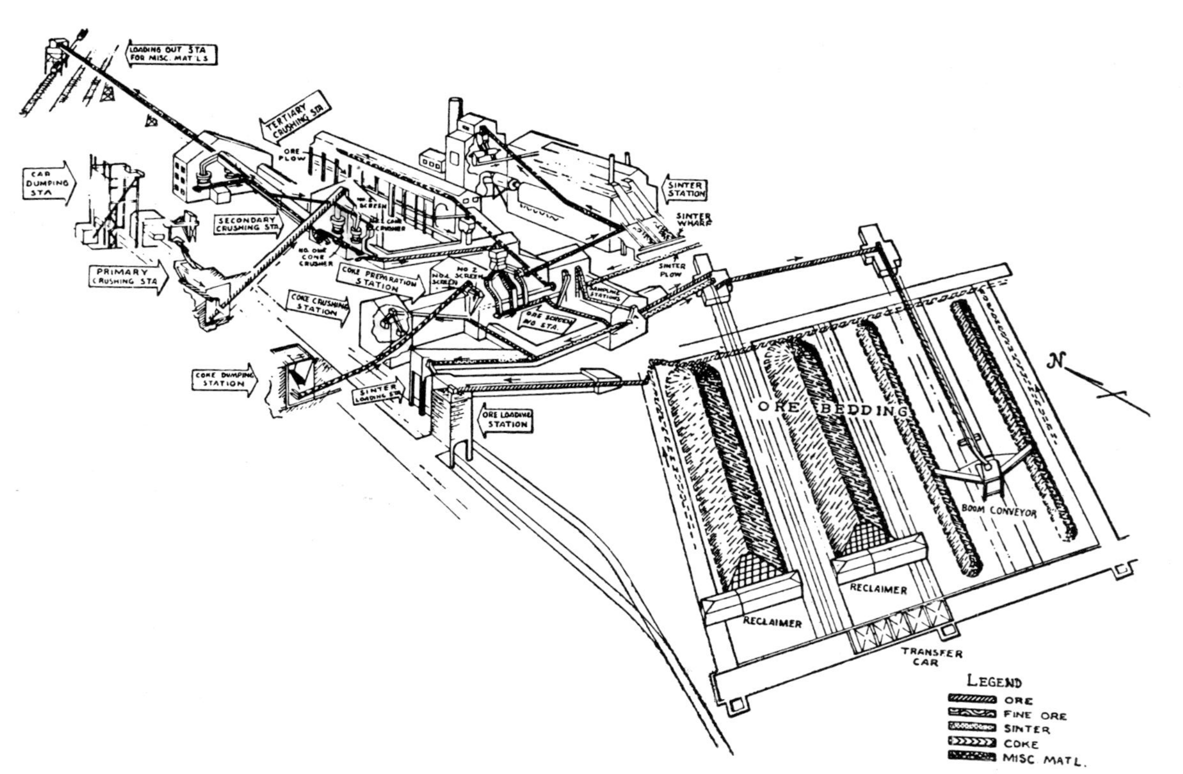

Fig. 280. *Large-scale Blending and Stockpiling*

and provides a cushion of rock to protect it against wear. Ore builds up to the angle of repose, the stationary volume providing an emergency reserve. This construction would be bad for holding easily oxidised ore, which might age dangerously and then mix with the fresh supply. Bins with sloping bottoms can discharge centrally or to one side. A fine example of binning with protection against both silicosis and freezing of wet ore in winter is the Hollinger fine-ore bin (Fig. 281). In calculating the structural strength of an ore bin, allowance must be made for "swell" of ore, wind stresses, and any subsidence due to underground work. Where the ore varies from stope to stope, separate bins may be needed, delivering to a common feeder, so that an even blend can be sent to the mill. In regular working the ore in the bins should be drawn so that the latter are nearly empty by the time the hoisted ore for the new week commences to arrive. The bins should be full at the end of the mining week. If the tonnage in the bins is an element in calculation of tonnage milled, the height in each bin must be logged daily at the time the

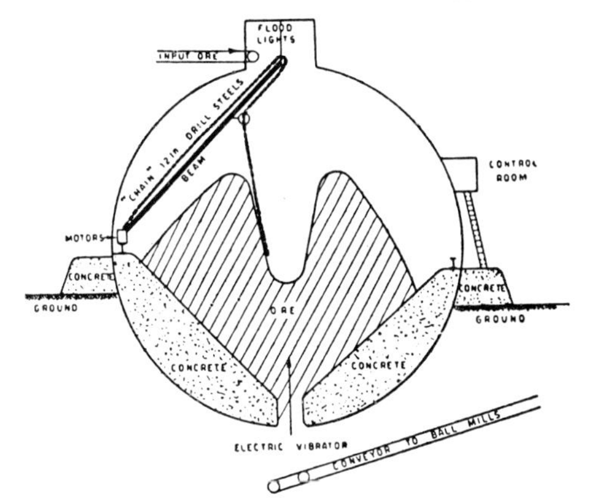

Fig. 281. *Hollinger Fine-ore Bin*

used tonnage is calculated, and any change in bin content must be allowed for to give true tonnage milled. If recording of milled tonnage is made on the withdrawal side of the bins, this does not matter.

Where the risk of freezing is negligible, an alternative to binning is formation of a dump above a tunnel which houses a withdrawing conveyor belt with regulating gates. The dump may be enclosed by a curtain wall. Gate regulation is by rack-and-pinion, adjusted so that while the conveyor belt moves, a layer of defined height is withdrawn quickly. The belt must be of heavy duty type. It does not undergo shock loading but there is the risk of angular material wedging and damaging it, unless it is arranged so as to be "flood-fed" with ore only when the angle of repose in the throat of the discharge chute is upset as material moves away. The freezing up of stock-piles, ore in trucks, etc., in severe wintry conditions can be mitigated by sprinkling it with salt or calcium chloride, provided the introduction of these chemicals does not upset further treatment or product specification.

Where the demand is seasonal (e.g. for agricultural phosphates) or periodic (as with shipments by sea) it is necessary to stockpile concentrates. One system, planned to prevent wetting of a sticky magnetite by the heavy local rainfall has been described.[12] Further requirements were rapid turn-round of ships taking up to 45,000 tons in a tidal basin having a variation in loading height of over 40′, and the need to accommodate delivery to ships of varying beam. At the land end a cone 100′ high is formed with an angle of repose of 45°, sheltered by iron sheeting hung tentwise from cables. A retractable wharfside loader with a 75′ boom swings vertically through an arc of 25°, and makes the final delivery into the ship's holds.

Fluidised mixing has been developed for the dry blending of talc concentrates, less than five microns in particle size.[13]

Pulp storage on an important scale is not practicable. The conditioner tanks used to provide reaction time in flotation must be agitated continuously, not only to provide mixing but also to prevent settlement and choking up. Dense-media liquids must usually circulate whether or no ore is being fed for the same reason, though many modern plants are able to reagitate settled media. Heating of pulp, where necessary, is done by means of steam coils. Wood, lead, and special alloys are used where corrosive pulps (such as sulphur flotation concentrates) are handled. Surge tanks are placed in the pulp flow-line when it is necessary to smooth out small operating variations of feed rate. Their contents can be agitated by stirring, by blowing in air, or by circulation through a pump (Fig. 282). At points in the plant where pulp must be run off in an emergency, special spill-tanks should be contructed. Consider, for example, a long rising pipeline used to transport tailings fed into it by pumping. If the pump broke down, this pulp would settle and block the mechanism, so a spill-pit capable of holding the contents of the delivery pipe is desirable. This should be cleaned out as soon as possible after normal working has been resumed. Similarly, spill-pits can receive the contents of classifiers when there has been power failure, and can hold the spillage of material washed down when floors are hosed. Diaphragm pumps, air-lifts, and air-turbine pumps such as are used for shaft-sinking are excellent for returning such spillage to the flow-line. A thickener should have some spare capacity, so that it can continue to store slurry while the filter which it serves is being overhauled.

Couche[14] has considered the complication introduced into slurry mixing by "non-Newtonian" fluid behaviour in which the shear rate of a pulp varies with the shape and dimensions of the containing system as well as with the type of agitation and the speed of the agitated slurry. Mixing has three main objectives—to produce and maintain homogeneity, to prevent settlement or flocculation and to accelerate reaction. The criteria differ in each case. In chemical solvation the essential rate-determining step may well be movement of the attacking chemical to the mineral surface and of spent chemical with its dissolved load from that surface back into the body of the pulp. Here the dominant interest is in stirring, not in mixing. If reaction does not depend on rate of presentation of attacking ions neither stirring nor mixing has a prime effect. If oxidation by the agency of introduced air is involved as in the cyanidation of ore pulps containing gold, distribution of this oxygen in a suitably fine aeration becomes part of the mechanism of agitation.

Rate of shear varies in different parts of the mixing vessel. Slurries usually have pseudo-plastic flow characteristics in which retardation increases with the distance from the moving impeller or other stirring agent. Design must therefore not consider conditions in the quietest part of the mixing tank so much as the type, size and speed of impeller which gives optimum mixing and extraction for a given rate of throughput. For homogeneity, the upward flow in the tank must at least be equal to the settling rate of the largest particle, though the power needed to meet this condition would not suffice to ensure homogeneity.

With so-called Newtonian fluid systems the rate of strain in shear ("rate of shear") is proportional to the shear stress. This rate is the velocity gradient across the direction of flow, and the coefficient of viscosity is the ratio of shear stress to rate of shear (a constant). With "non-Newtonian" systems such as a typical mineral pulp there are two main types of variation from this. Couche defines these as:

(*a*) Consistency deviations in which the rate of shear is not proportional to the applied shear stress.
(*b*) Variations with time of the ratio of the shear stream to rate of shear.

With such variations there is a yield point above which flow begins. Most mineral slurries do not exhibit such yield stresses.

A turbine mixer projects a high-speed fluid disc radially while a propeller projects a large cylindrical turbulent form at a lower velocity. For mixing the propeller is probably best, but where reactions are part of the process the rate-determining step requires displacement of liquid from the particle surface and the question of viscosity arises. This is best met by high turbulence such as is given by turbo-action.

One proprietary mixer[15] can be installed in a pipe line. It has a rotor-stator combination on the lines of the Fagergren flotation cell unit.

The mill's steady-head water tanks are placed at suitable high points and must not be allowed to freeze. Gauges show their state of load. A float-operated gauge is usual, or a simple overflow can show that the tank is full. Another arrangement is a scanning "electric eye". Tanks should be inspected regularly to check that they are not accumulating a deposit of silt. There should be at least several hours of reserve water in all important storages, in addition to the main supply.

This problem of water storage for the general working of the mill should be studied when choosing the plant site in order to avoid trouble and unnecessary cost. In countries with alternating monsoon and dry-season conditions, there is at times an embarrassing excess of water, though sometimes this can be utilised for flushing away an accumulation of tailings. Later comes a shortage requiring recirculation from impounding dams. In severe climates the availability of water during winter must be studied. Flotation uses from three to five tons of water per ton of ore treated, and gravity treatment from ten to twenty. Much of the water can be used repeatedly, but provision must be made for replacing spillage, drag-out, evaporation loss and fouled water.

Since the water in the mill is used for a multiplicity of jobs, it is essential to conserve adequate supplies which must allow for unfavourable rainfalls. Water is a transporting agent, a separating medium in gravity work, the main classifying control reagent, and may be the determining factor in selecting a method of treatment.

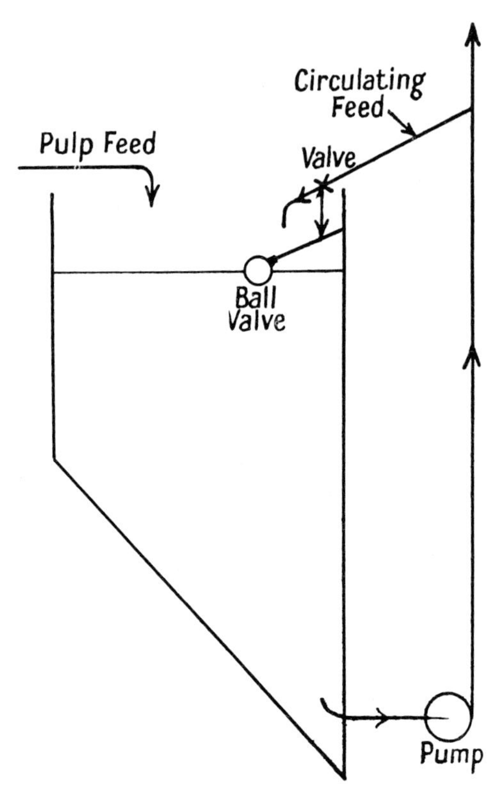

Fig. 282. *Surge Equalisation of Pulp*

Pumping

Pumps are used in ore dressing plants for a variety of purposes, to meet which special designs and modifications have been developed. The main types and duties are:

Centrifugal single lift pumps

Lifting and circulating water.
Moving acid solutions (special alloys, perhaps with rubber linings).

Moving cyanide solutions (no cyanide-soluble parts, possibly seals to exclude air).
Moving coarse and abrasive pulps (special alloys in wearing parts).
Moving finely ground pulps (perhaps rubber linings).

Pressure pumps for slimes

Screw-type impellers.
Plunger pumps (reciprocating).

Other pumps for thick slimes

Diaphragm pumps.

Miscellaneous

Frenier (ribbon-impeller).
Air lifts.
Hydraulic injection.
Pulsometers.

The centrifugal pump (Fig. 283) consists of an impeller which revolves in a casing. As the pulp flows on to the hub of the impeller it is picked up by the vanes of the impeller and accelerated to their outer tips. Pulp passage through a pump of the horizontal impeller type (Fig. 284) is indicated by arrows. Its speed at leaving is that of the circumference of the impeller, less losses due to slip. The velocity of the issuing stream is an important factor in determining the vertical height to which the liquid can be raised. It is modified by frictional loss in the delivery system. In ore dressing, single-stage

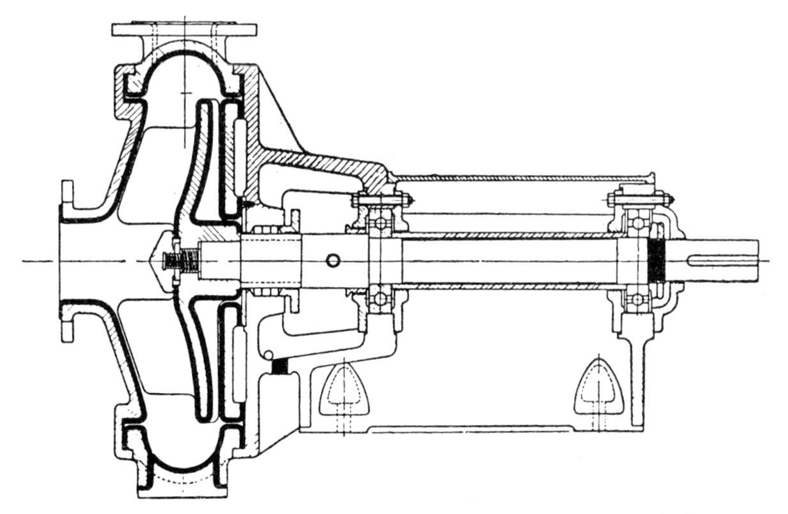

Fig. 283. *The Vacseal Pump* (International Combustion)

pumps are used, and the practical maximum lift is about fifty feet. If more is needed, a second pump is put in series with the first. The efficiency, in terms of volume swept, is between 50% and 75% when the pump is working at its

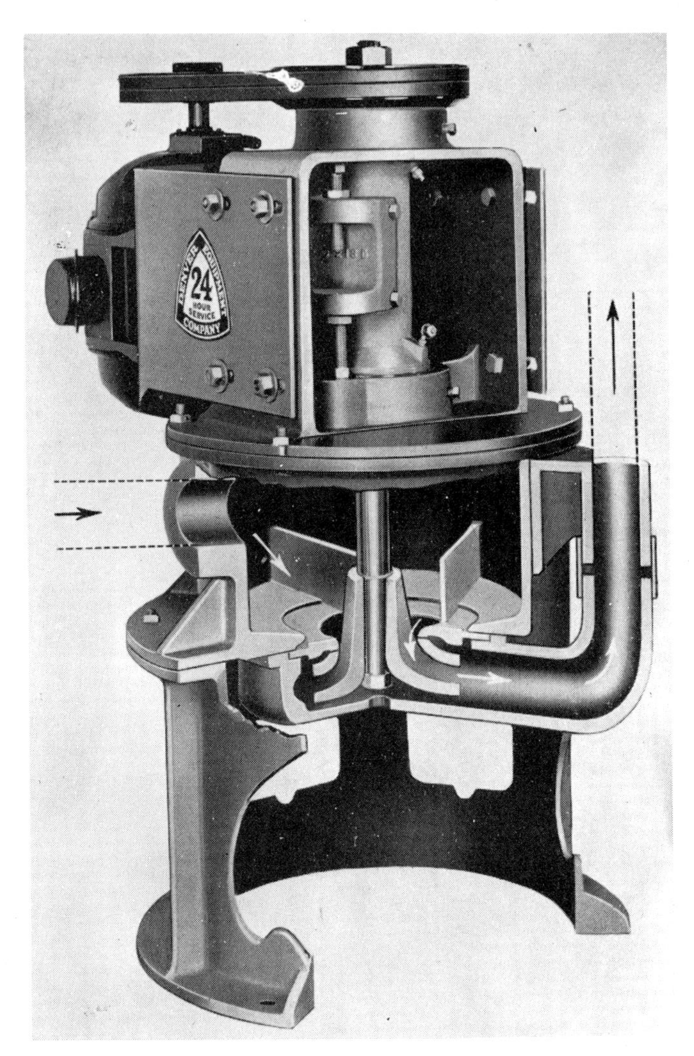

Fig. 284. *Denver Vertical Centrifugal Sand Pump*

full load, and drops when it is only partly loaded (drawing air). The pump casing must be protected against wear by means of renewable liners. These and the impellers, are cast of abrasion-resisting alloys, or are faced with rubber, the latter being preferred for finer sands and the former for coarser grit and gravel. The chief costs are for power and replacement of worn parts. There are two main modifications of design—glandless pumps, and those with a water-seal. The purpose, in each case, is to protect the shaft on which the impeller is mounted from being damaged by abrasive material from the pulp. This is done in the latter case by the introduction of pressure water to a lantern ring, so that it maintains a small positive flow through the stuffing box into the casing. Pumps of the self-protecting type are more widely used, since they are simpler to install and less liable to shaft damage such as would occur in the event of gland-water failure.

The glandless pump has for some purposes the advantage of not introducing water into the pulp.

A second differentiation is between pumps which are designed to lift by suction to the impeller and those which are meant to work on feed coming in under slight pressure. In theory all pumps can lift the feed a short distance, but in practice this is not the case. The Wilfley pump should be installed to draw down from a surge tank so that there is at least a 4′ head. The Vacseal, if given a priming arrangement, can lift a moderate distance by suction.

The pumping system must be in balance with the load handled by the plant. It must be able to deal with the maximum overload liable to enter the flow-line and yet be able to work at reduced capacity when for any reason the flow is restricted. Several methods of achieving this are possible. One scheme for a small operation employing only one pump is shown in Fig. 282. The ball valve controls recirculation of enough pulp to maintain a reserve and an inlet head. This arrangement is wasteful of power, and sets up undue wear. Where tonnage justifies it, the more elaborate system in Fig. 285 is good. A two-pump scheme for ensuring completely steady delivery to a classifier or a cyclone is shown in Fig. 286. Coupling units are marketed,[19,20] which adjust the pump speed to maintain a balance between feed and discharge rate when the inflow varies. The present range of "Inpower" couplings lies between $12\frac{1}{2}$ h.p. and 400 h.p. When the running speed of the pump is varied without other change, the capacity is directly in proportion, the delivery head proportional to the speed squared, and the power consumption to the speed cubed. Within the limits of variation of impeller diameter (usually an effect of wear rather than of design) the capacity varies directly as the diameter, the head as its square, and power used as its cube. If a pump runs badly, the impeller may be too much to one or the other side of the casing.

Periodically the seal must be renewed. After placing the seal, the packing should be left loose and tightened down gently when the pump is working, so that it beds down gradually. If properly packed, the shaft will turn easily by hand. The maximum desirable rate of flow is 8 ft./sec. on the inlet side and 12 ft./sec. on the discharge side of a centrifugal pump. Sharp bends are to be avoided, and change of section of piping should be made gently through tapering reducing joints. The piping should be self-draining on shut-down, and needs a reasonable velocity of flow through it to ensure against blockage. A rate of 5 ft./sec. is a good minimum for ordinary pulp, or 4 ft./sec. for fine material in reasonably dense pulp. The power consumption is proportional to the density of the pulp.

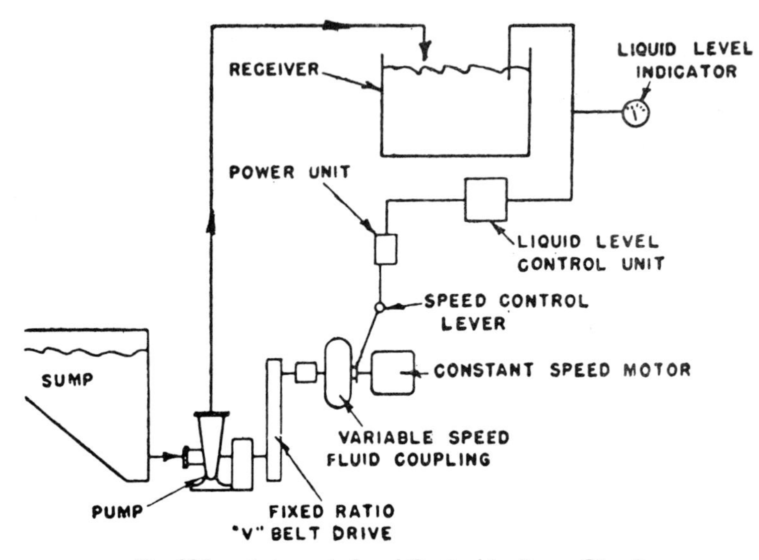

Fig. 285. *Automatic Load Control in Pump Circuit*

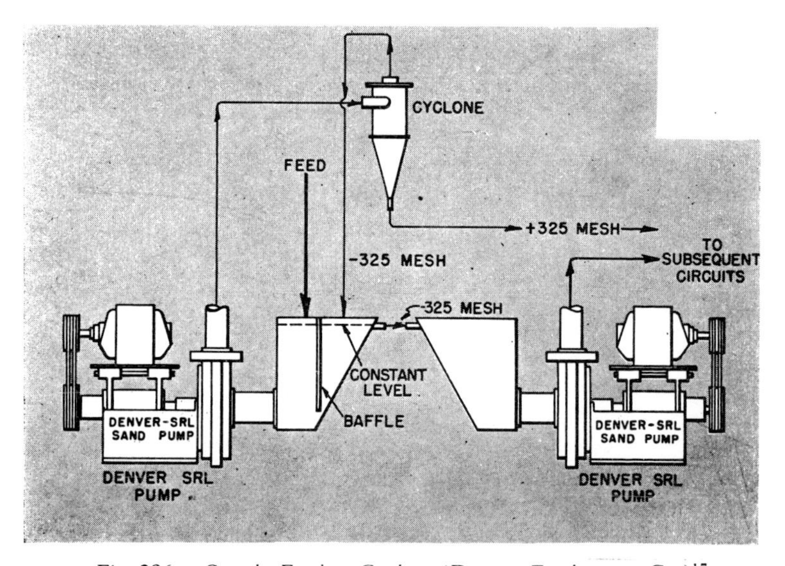

Fig. 286. *Steady Feed to Cyclone* (Denver Equipment Co.)[17]

Transportation by a combination of pumps and pipeline is economically competitive in many fields, notably where abrasion is moderate. The horizontal movement of *plus* 2 mm. gravel at the rate of 100 tons per hour requires some 6 h.p. per ton-mile, sand averaging 0·4 mm. about 2·2 h.p. and raw cement slurry at 40 – 50μ size only 0·12 h.p.[18] Vertical lift is simpler than horizontal transport where there may be settlement on route, in which case the velocity must be twice the settling rate of the largest particles in free fall.

Transport interruption arises either from jamming or instability of flow. The conduit must be at least three times the diameter of the largest particles and where there is a pump in the flow line the maximum particle diameter

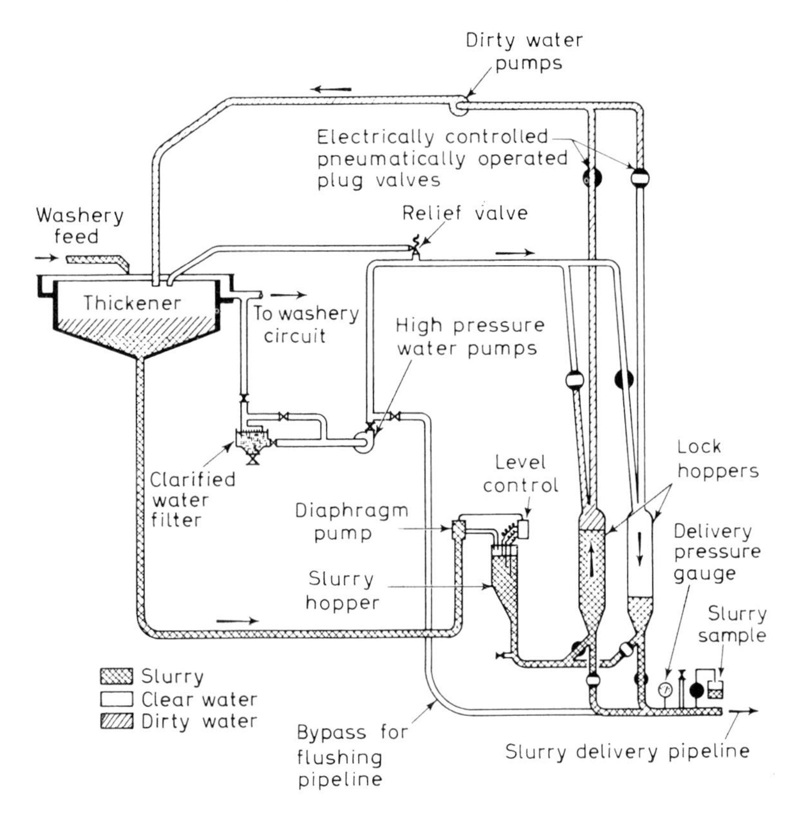

Fig. 287. *Lock Hopper System*

must not exceed one half of the smallest pumping cross section. If the delivery grade rises at all steeply means of emptying the load must be provided, operable swiftly in the case of a sudden shut-down.

The main maintenance costs arise from pump and pipeline wear and vary from below 0·03d. per ton-mile for slurries to over 3d. for gravels.

The diaphragm pump is used with most thickeners to remove the underflow. It consists of a flexible diaphragm with a flap valve, which is reciprocated vertically by an eccentric. Pulp is drawn up into the space swept by the diaphragm through a lower flap valve and overflows on the pressure stroke. The throw of the eccentric is usually adjustable without stopping the pump, so that the rate of withdrawal of settled slurry can be regulated. Since this slurry is of high density, the piping from the thickener is of small size, and the pulsing action set up by the pumping action helps to keep the material in suspension. In one arrangement the density of the thickener underflow is monitored by a gamma ray gauge. The resulting signals actuate an automatic control on the diaphragm pump which varies the stroke amplitude. This controls the rate of slurry withdrawal at the desired density.

Air-lifts are handy and easily improvised. They can be used to transfer pulp from section to section of a flotation bank where the cells do not provide suction, and for such jobs as collecting spillage and returning it to circuit. An air-lift has the advantage, sometimes useful in cyanidation or in oxygenating flotation water, of re-aerating the pulp. The inlet of the piping must be at least one-third of the elevated distance below the pulp.

The Mono pump consists essentially of a rubber stator shaped as a double internal helix, inside which a single helical rotor runs slowly with a slight eccentric motion. It is self priming and delivers to a height of 100′ or more.

In the Lock Hopper pump (Fig. 287) clean water at a high pressure is used to evacuate slurry from a filled hopper into the delivery system. Operation is continuous, one of two hoppers in parallel being filled while the other empties, the control valves being automatically worked. The Westfield installation in the figure delivers 28 tons of dry solids hourly some 3,600 feet at an 80′ lift, with a flow speed of 8·2 ft./sec.

Process Water

Among the factors which affect the use of water in ore treatment are:

(*a*) Availability of fresh supplies.
(*b*) Seasonal variation in quantity and quality.
(*c*) Disposal of radio-active, poisonous, tainted or acidic effluents.
(*d*) Softening or de-salting before use.
(*e*) Storage, conservation and re-circulation.
(*f*) Effect of build-up of dissolved salts when re-circulating.

Fatty acids are particularly sensitive to calcium ions, and water used for flotation may require pre-treatment to reduce hardness. Where rainfall is seasonal the concentration of such dissolved salts varies according to whether run-off or deep spring water is being drawn to the plant. Aeration affects both the effect of oxygen on the chemical action of flotation reagents and the stability of the froth, so water drawn from beneath a wintry ice cover or reclaimed from tailings dams in which algae have stripped the oxygen may need some re-aeration before use.

When newly arriving water carries deleterious ions the re-circulated water in a flotation circuit may have been improved during treatment by loss of ions to the solid products. Again, it may have built up fouling ions to a dangerous level by dissolution from the new ore. It is sometimes found advisable to dewater the pulp between comminution and conditioning ahead of cyanidation or flotation. Economy in consumption of flotation reagents is sometimes reported where a substantial re-circulation of process water is used. This could arise from retention of frothers not removed with the products; from over-supply of other reagents in the first place; from reagent change during recirculation (e.g. oxidation of residual xanthate to dixanthogen); or from the neutralising effect of process chemicals on harmful ions in the new mill water. If the tailings pond is distant from the concentrator, thickening or cycloning on the mill site may be used to facilitate the return of process water without exposure to evaporation or de-oxygenation. Only the thickened underflow is sent to the disposal area.

Scarcity often dictates conservation practice. In western America the copper producers suffer from a low rainfall and high evaporation[21]. Evaporation in storage is mitigated by filming the impounding area with cetyl alcohol. In 1960 the reclamation for re-use in thirteen copper plants in the western U.S.A. was about 50% of the total in use in most cases and exceptionally rose to a 4:1 ratio. One method of combating evaporation is to send tailings underground where their solids become backfill while the water is pumped back to the plant. Of the 330 million gallons used daily in 1961 by the U.S. copper industry[22] 46% is saline with over 1,000 parts/million of dissolved solids.

Removal of brine from brackish water by electro-dialysis is under active development and research.[23] The cleaning of fouled process water is a routine operation where scarcity or local requirement warrants the cost. In one plant water used in the carbonate-leach of uranium carried 628 parts/million of calcium carbonate. This gave scaling trouble in piping and on filter cloths. It was overcome by softening in which the residual sodium carbonate in the spent filter cake was used. At Butte, Anaconda, an effluent laden with ferrous sulphate is treated by mixing with a heavily limed effluent from other sources, or by direct precipitation with lime. Another Anaconda operation has the problem of disposal of radio-active effluent water.[25] This has been met by injection into porous sandstones underlying a fault-free impervious stratum, after partial evaporation in tailings ponds.

Washery Water

The problem of keeping the large quantities of water in circulation in a non-corrosive and reasonably clean condition is important in the design of coal-cleaning plants. Extremely fine colloidal clays, soluble salts, fine coal, shale particles, and acidic substances can be settled as a sludge or slurry. This must be treated to recover its content of useful fuel and then disposed of without undue detriment to the surrounding countryside. Periodically, the washery water is changed.

Prevention being better than cure, the first line of treatment is to de-dust the dry raw coal. This is only possible when the coal is mined under dry conditions. This not only removes some of the contaminants and the bulk of the fine fusain, but it may even yield a product suitable for use as pulverised fuel. Such a fuel can be used even with an ash content exceeding 40%, provided the flues of the furnaces are able to deal with the resulting fly-ash. In typical de-dusting systems (Fig. 288) the raw coal drops through

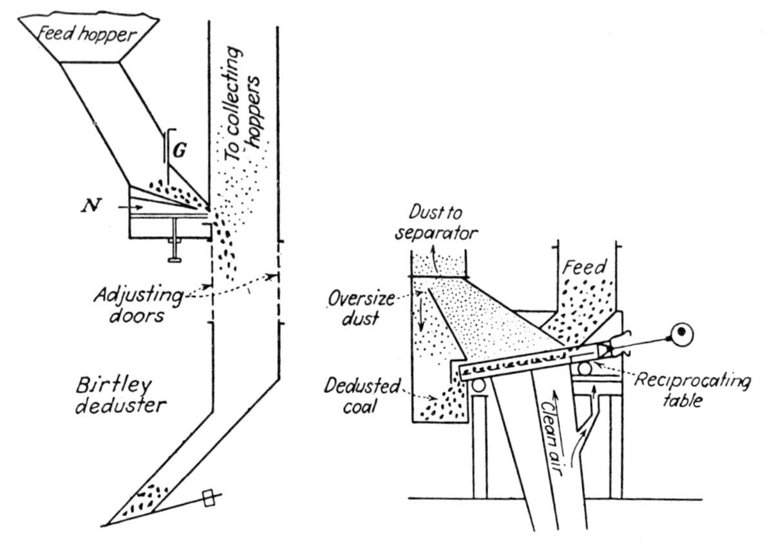

Fig. 288. *De-dusting Methods*

an aspirating column to a bottom discharge, the dust being trapped above.

A coal-washing plant uses large volumes of water. In a Baum washery treating 150 tons/hour the circulating water might be of the order of 250,000 gallons hourly. Despite continuous settlement of clay, etc., this water becomes too foul and acid for further use after a period, and must be replaced. Acidity must not proceed to the point where corrosion commences. Suspended near-colloidal solids and dissolved salts must not be allowed either to coat drying washed coal, upset cleaning operations, or pollute the local drainage area. The term "coal slurry" may be roughly defined as a suspension of fine coal and associated minerals, from colloidal and suspenoid sizes upward, in too dilute a state for direct filtration. A *coal slime* is such a suspension with more than 50% –200 mesh. A *coal sludge* is a partially thickened slurry, sufficiently dewatered to be filterable. These dispersions do not settle readily until the concentration of solids exceeds 10%. A typical water circuit is shown in Fig. 289, and a settling cone (thickener) in Fig. 290. Fine screens are used to treat the thickener discharge, from which they remove the coarser particles of coal.

Two main conditions must be satisfied in a close-circuited water scheme. *First,* there must be a sufficiently rapid rate of settlement, and *second,* it must be reasonably easy to handle the final silt (using scrapers, grabs, shovels, etc.). This second condition requires that the percentage of water be low enough to

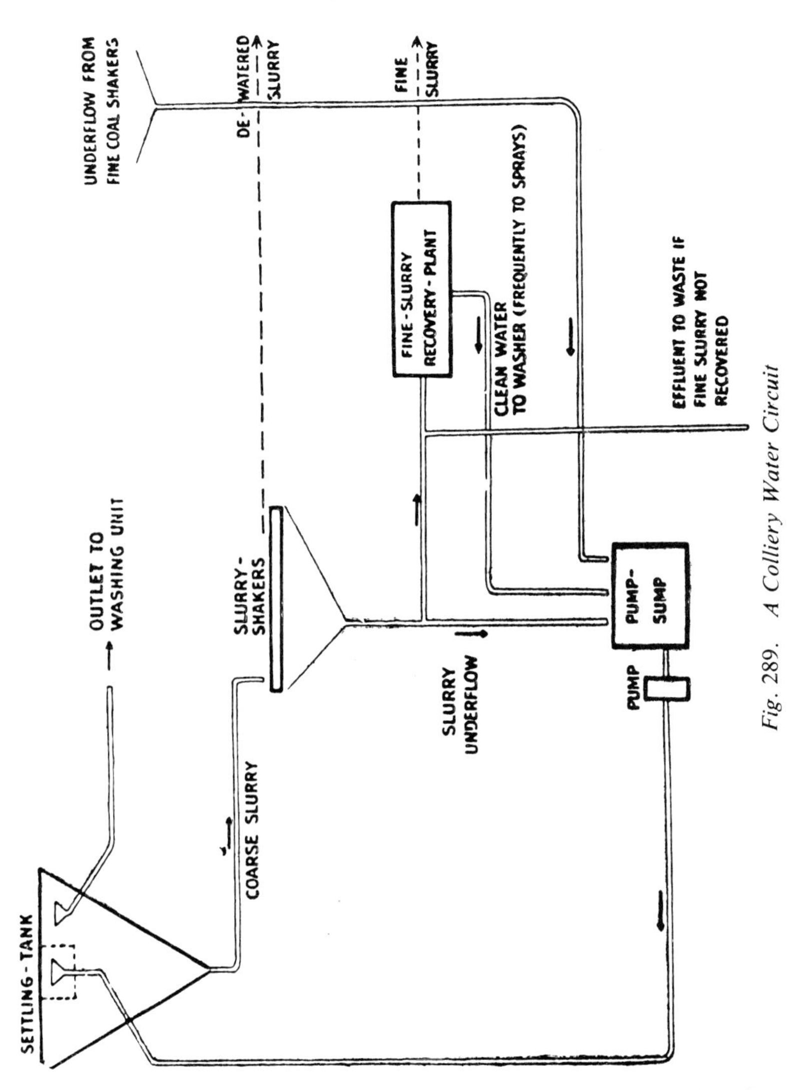

Fig. 289. A Colliery Water Circuit

allow solids to be handled, while still keeping the material sufficiently soft to be worked. Flocculation is usually necessary to hasten the settlement of clay. Starch, lime, and various proprietary compounds based on high-valency salts are used. The three stages are:

(1) Addition of flocculating reagents.
(2) Mixing till these are dispersed.
(3) Gentle movement to help collision of dispersed particles without breaking up formed floccules.

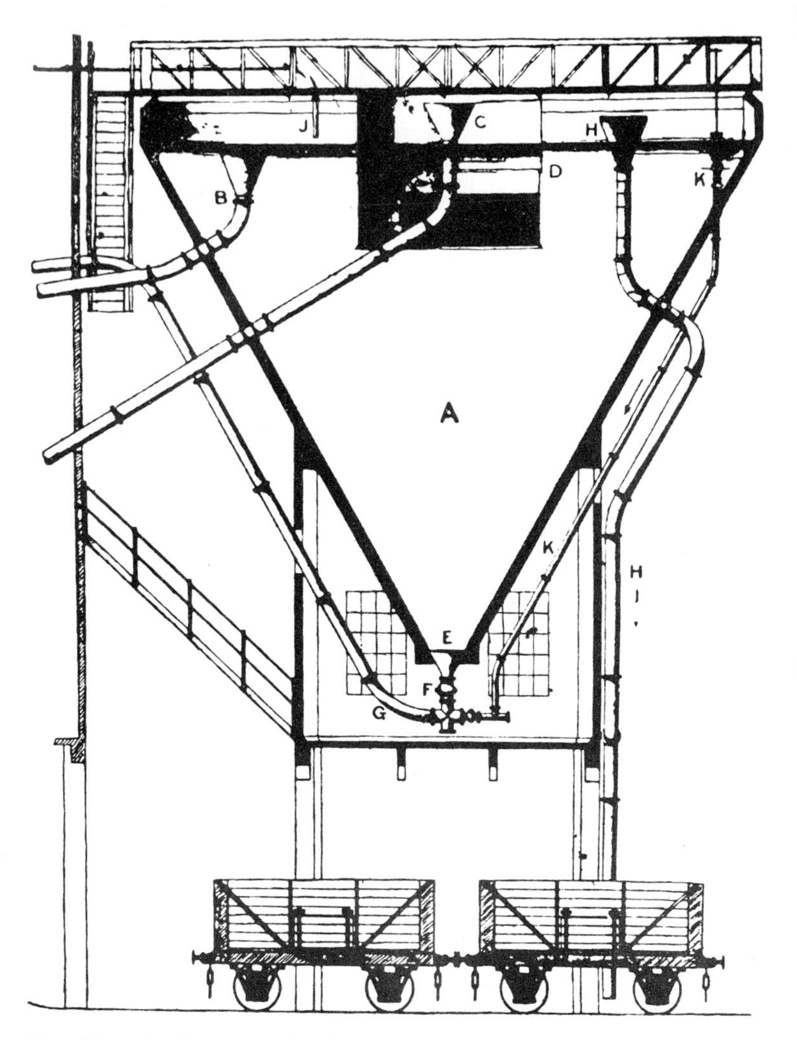

Fig. 290. Settling Cone for Washery Water. A = Thickener; B = Clear overflow; C = Entering foul water; E, F = Underflow discharge; G, K = Make-up water; H = Overflow to waste

The mechanism of flocculation by means of starch is complex and appears to comprise two stages. First there is charge neutralisation in the slime colloids with formation of incipient floccules. The starch then attaches these into large stable groups. The form in which the starch is dispersed is important. Alkaline starch is made by adding a potato-starch paste to hot sodium hydrate at a 5% concentration, making sure that the sol is thoroughly dispersed. Zinc chloride and calcium chloride are among the proprietary additives.

Two main techniques of clarification are possible. In one the whole of the water in the thickener is dosed. This gives a good settlement and a coarse slurry, but much underflow must be removed and filtered. In the other a moderately thickened natural sediment is underflowed and treated. The thickener underflow is filtered, while clear water is drawn from the top of the tank to the cleaning plant. An alternative to vacuum filtration is the running of the slurry to a rectangular concrete tank, with a sand layer in the bottom communicating to rows of well-points (draining pumping pipes). One tank is filled while others are drained down and dug out.

Flocculation

This subject was discussed in connexion with thickening (Chapter 9) and the physics of flocculation (Chapter 15). The condition of the solids fed to filtration affects residual moisture, filtering rate and general plant performance. Thickening, which normally precedes filtration, is usually accelerated by the use of lime or other flocculants. The latter may include trivalent salts, blended starch, metal salts, alkalis or organic compounds. When the overflow water from the thickener/s is re-circulated, a prolonged test on the full plant scale is desirable to ensure that any new flocculant does not upset recovery. Floccules which lock up an undue proportion of liquid must not be allowed to increase loss of pregnant liquor with the tailings.

When finely ground ore is chemically treated the final tailings may be highly dispersed, slow to settle and hard to arrest on a filter membrane. Several proprietary flocculants are today available which allow such a problem to be met without undue increase of thickening area. These flocculants have a strong affinity for minerals and must be carefully introduced in stages into the pulp stream, or the whole addition may be taken up by a small fraction of the solids. The available compounds include glues, starch admixed with metal salts, and such compounds as Separan[26], the Cyanamid series of aeroflocs and superflocs[27] and the Nalcolytes.[28]

The use of a mobile testing unit directly on the main pulp stream has been described.[29] In one test coal filtration was in trouble, the tailings filters being overloaded and recovery poor with an ordinary starch flocculant. A hydrolysed poly-acrylonitrile was substituted with satisfactory results. In another a *minus* 325 mesh coal froth was too wet at 40% moisture. Aerosol OT—75% brought this down to 23% and gave a fast-filtered and crumbly cake.

Filtration

A filter is a permeable membrane, septum or bed, so arranged that the pulp to be treated is presented to one side at a higher pressure than that on the other side of the septum. The bulk of the solids are held back. Material so fine that it can pass the porous barrier is part of the filtrate. The pressure drop may be due to simple gravity, to pressure on the pulp being fed, to a vacuum on the filtrate side, or to the effect of centrifugal force. Filtration is used to remove pregnant cyanide after leaching, and to clarify cloudy solutions.

If the viscosity in a pore is insufficient to neutralise the pressure differential between its intake and discharge, a liquid can flow through it. Poiseuille's formula for dynamic viscosity, v, as determined by use of the capillary tube, is

$$v = \pi p r^4 t / 8 l V$$

in poises, where p is the pressure difference, r the radius of the tube, l its length, V the volume of liquid delivered and t the time. Experimental verification has been made of its implications, but the practical difficulty in evaluating filter performance is that the porosity decreases rapidly as mineral particles arrive and modify the overall structure of the system—septum *plus* arrested layer. A further decrease in porosity is due to precipitation of calcium salts. As the particles are arrested, the larger ones bridge the orifices. Smaller particles next lodge in the interstices, so that after a very short run the filtrate no longer contains material such as passed through initially. The effect of this increase in l and the decrease in r is to decrease the dynamic viscosity considerably. In practice, the rate of filtration and the ability of the barrier to arrest solids depends on several main factors:

1. Available filtering area (direct).
2. Pressure difference across cake and membrane (direct).
3. Thickness of cake (inverse).
4. Pulp temperature (direct).
5. Specific surface of solids (inverse).
6. Size-range of particles (inverse).
7. Degree of flocculation (direct).

In *gravity* filtration the main function of the septum is to act as a support for the bed of material. In one sense, leaching tanks used in cyanidation are filters. Their thick beds are made possible by exclusion of most of the fine sand which would clog the interstices between coarse particles and prevent drainage. When the object is clarification, a thinner sand bed is used, the fine material being allowed to obstruct the interstices until rate of filtration and degree of solid arrest are adequately adjusted. Control can be maintained by such means as a skimming cut to remove some of the uppermost sand. Thickeners can be adapted for this purpose.

Pressure filters are robust vessels which contain porous filtering sections. Feed is charged in and air pressure is turned on. In the developed form, plate-and-frame filters, they are intermittent in action. The septum is stretched over a frame which has channels for pulp feed, wash liquor, and filtrate discharge. A number of these frames are assembled in a press, and pulp is pumped in. When pressure reaches a determined intensity, delivery of pulp through its force pump is stopped. The solid cake now filling each frame is washed by liquid under pressure, which can be discharged separately. The press is next opened and the cake is removed. These devices are much used in the chemical industry, but have declined in importance in milling.

Vacuum filters are of two types, intermittent and continuous. The most common intermittent type found in mineral dressing is used in clarifying pregnant cyanide or in catching precipitated gold slimes. A number of frames, each covered by a canvas envelope, are suspended in the tank which receives the feed. Piping incorporated in the frame connects it with a common "leader" served by a vacuum pump.

Intermediate between this and the truly continuous filter is the "Genter".

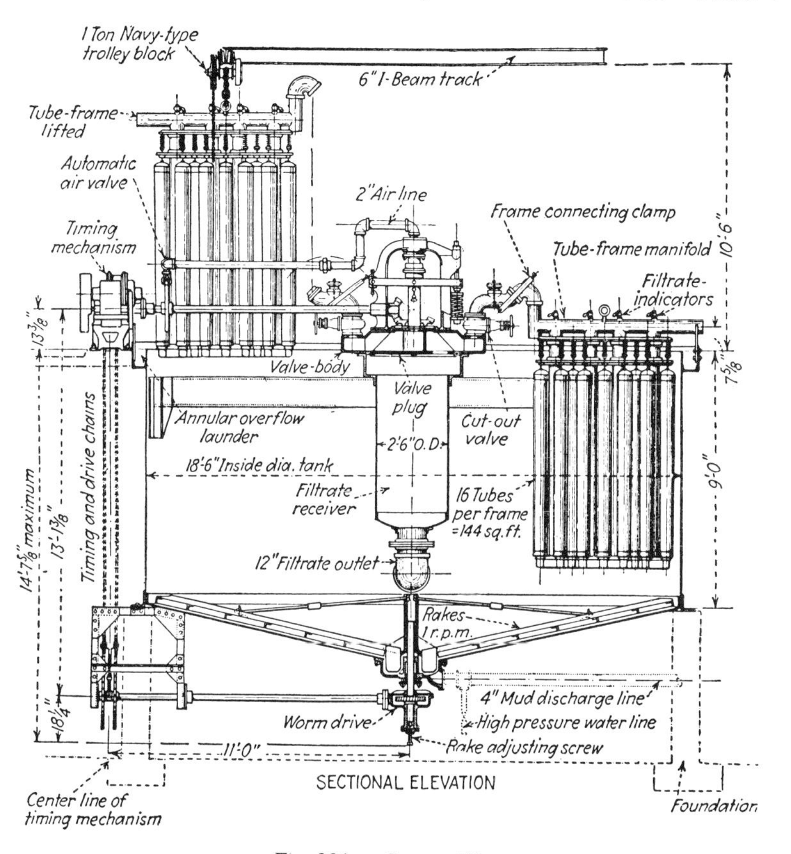

Fig. 291. Genter Filter

This type is semi-continuous and is a vacuum filter intermediate between a thickener and a completely specialised filter. The Genter is little used today, separate thickening and filtration being preferred. The cylindrical tank (Fig. 291) receives the pulp, and liquid is withdrawn by vacuum through the canvas "socks" hanging from each arm and submerged in the tank. These are kept distended by wooden balls. An automatically controlled valve cuts off the vacuum at regular intervals and reverses the liquid flow for a few seconds.

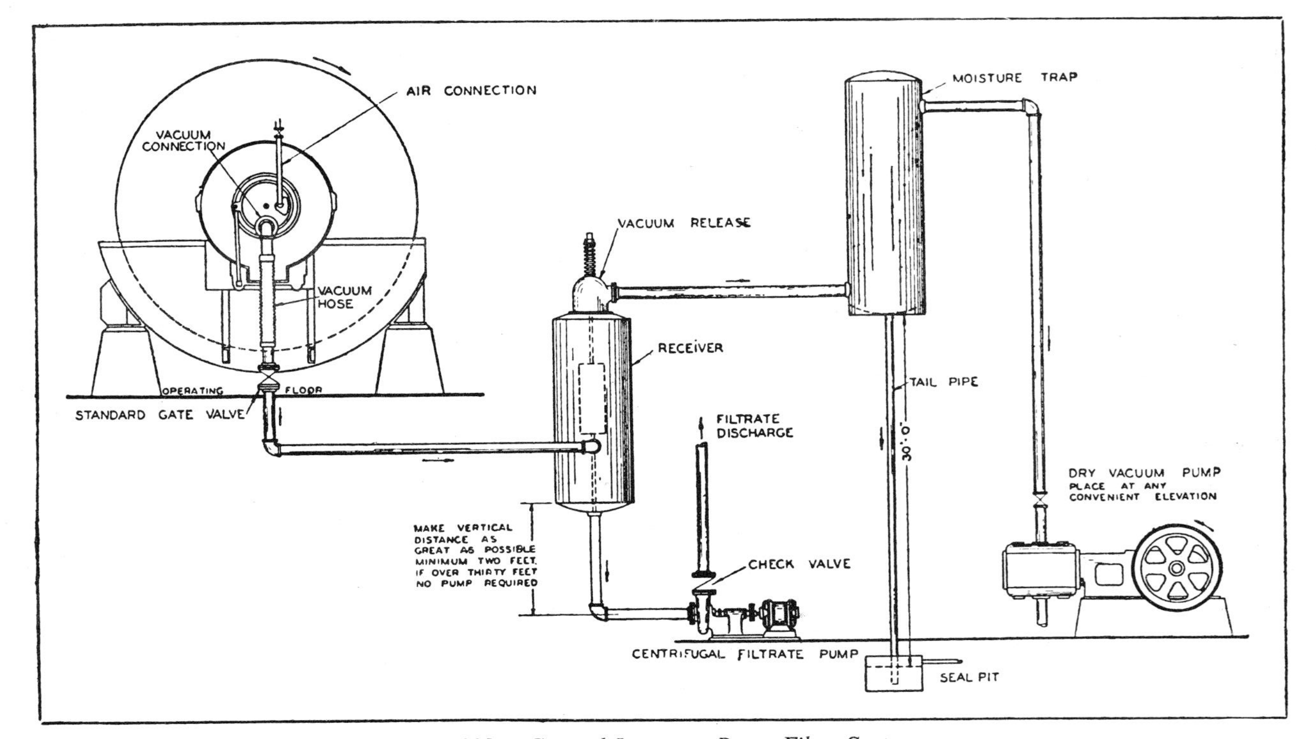

Fig. 292. *General Layout. Drum Filter System*

The sludge which has collected on the outside of each sock now slides down to the bottom of the tank and is raked to a centre discharge. Low-pressure air can be used to aid the displacement of the mud. This mud is intermediate in consistency between the flowing slurry of a thickener discharge and the solid cake of a drum filter.

Continuous vacuum filtration is today the standard practice in most mills. The main types of machine are the drum and the disc filter. These have several variants designed to meet special needs. Development research is also active in the creation of new types. The drum filter is normally fed with slurry underflowing from a thickener. If a dilute pulp, or an acidic one in a highly dispersed state (such as that from an acid uranium leach), is fed direct to the filter its solids are too mobile, and clog the system before a substantial cake has formed. A similar feed if first flocculated and thickened produces a wetter cake, but the rate of filtration is increased a hundredfold.

The general layout of a single-filtrate filtering system is shown in Fig. 292. The cake detached by the scraper knife usually falls to a conveyor belt (not shown). The cross-section of a drum filter is shown in Fig. 293. The machine consists of the drum, of metal construction. On its circumference are the filter plates, of wood or metal. These are grooved to permit flow of filtrate drawn through the *cloth*, a membrane wrapped round the drum and bound on by wire, so as to rest on the plates. The drum rotates at between two and twelve minutes per revolution, in accordance with the length of gathering time of cake required. This gathering is performed while the surface of the drum is immersed in pulp in the trough. The trough, of metal structure, receives pulp and keeps it from settling by means of the reciprocating arms of the agitator. Height of slurry is controlled between maximum and minimum level by manual or automatic methods. Vacuum during submergence is produced by connexion between the plate and the vacuum pump through a valve head and distributing ports. Each port serves its own plate. As a plate leaves the pick-up zone (Fig. 293) it passes progressively through the drainage, washing, drying, and "blow-back" zones, finally reaching the scraper-knife, where the cake and cloth, under the influence of compressed air, bellies outward.

This assists the removal of the cake. The valve-ports can be connected in various ways through the valve head in accordance with the desired series of operations. Vacuum, first filtrate, second filtrate (if required), and blow-back air are controlled by these settings. The scraper knife bears lightly on the spirally wound wire which holds the cloth on the drum, and aids the work of detachment mainly performed by the compressed air, which is either blown steadily or pulsed so as to cause the cloth to flutter. The grooves in the filter plates should be diagonal, to facilitate drainage of filtrate and minimise the possibility of a wet blow-back. The scraper is usually edged with rubber or belting to save wear on the cloth. In the Feinc system a string discharge is used, cake being discharged at the sharp bend. To minimise the formation of cracks which would lower the vacuum, Sullivan Consolidated Mines has fast-rotating rubber strips beating on the emerging cake. Deutsch Barit uses a perforated steel cylinder to bear against the cake and squeeze out some residual moisture. Vacuum is normally applied from the entry of the segment into the pulp to a point on the rising side where sufficient extraction of residual filtrate has been performed. The later this is left once the drum has emerged, the more vacuum capacity must be provided.

Variations on the standard drum include a downward converging two-drum

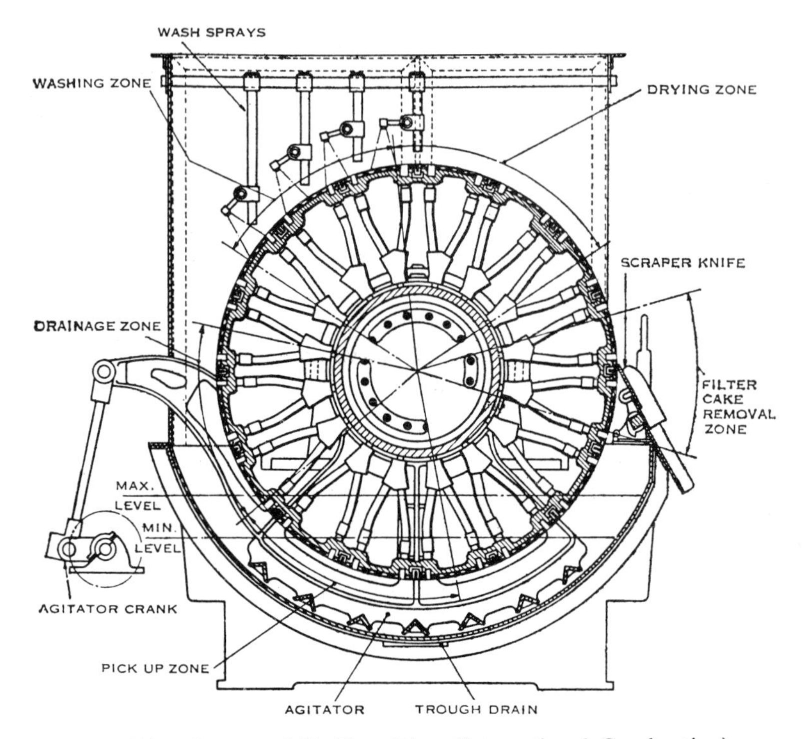

Fig. 293. *Section of Endflow Filter* (International Combustion)

system for fast-settling material which could not conveniently be held in suspension in the tank, and one in which feed first enters a small tank which distributes its underflow on to the top of the drum (Fig. 294).

In another variation the pulp floods down the rising side of the drum, thus sealing cracks in the cake. Where pregnant solution must be displaced before discarding the filter cake (e.g. in the cyanide process), the vacuum must be strongly maintained in most of the arc through which the cake moves after emerging from the slurry tank, so that abundant wash water can be used to displace and recover solution which would otherwise be lost. Blow-back must be set well clear of this washing. Cracks in the cake would endanger the efficiency of washing, and might overload the vacuum system. Regular attention to the vacuum pumps should include check on leakage of air past the valve seatings and correct tensioning of inlet valvesprings. Leaks in the

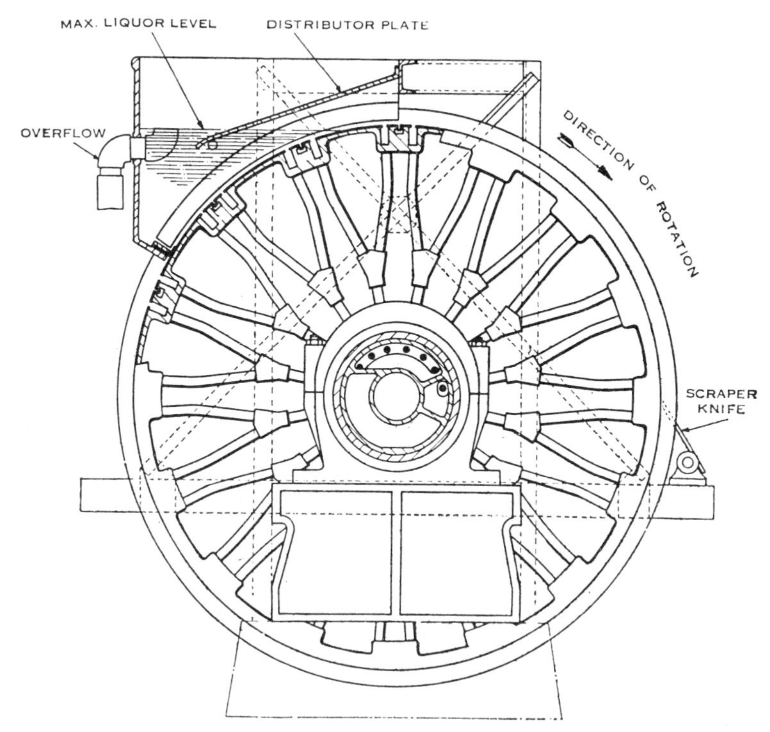

Fig. 294. Section through Endflow Top Feed Filter (International Combustion)

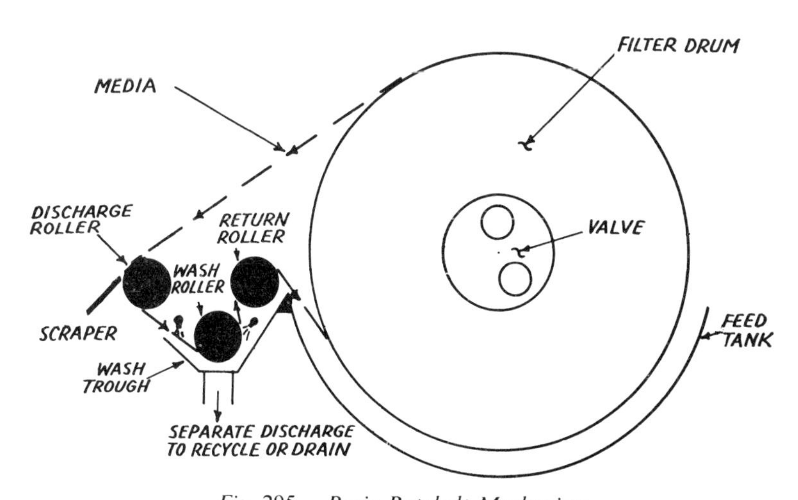

Fig. 295. Basic Rotobelt Mechanism

piping system can be detected by filling it with low-pressure water during a shutdown.

To overcome some of the difficulties in filtering colloidal slimes, a filter has been developed in which the whole cloth leaves the drum and can be washed from its inner side (Fig. 295). In a German system[31] clogging of the septum is reduced by the use of a belt made of sponge rubber on to which the pulp is poured. Apertures at the surface control passing particle size. These apertures open out inside the belt. Removal of filtrate is aided by passing the loaded belt through a pair of rolls which squeeze out residual liquor. Solids are lifted from the belt by the upper roll and brushed off.

Periodically the cloth of the standard drum filter must be changed. This takes several hours, during which feed must accumulate in the thickener serving the filter which is being reclothed. The wire retaining the cloth is loosed and reeled off, a special gear being used to speed rotation of the filter during the process. The new cloth is then fitted and wired on.

The *disc* filter, also called the American filter, is widely used. Instead of a drum this has a number of independent discs (Fig. 296) each composed of detachable segments and each running in its own compartment of the filter tank. Valving permits suction and air blow as with the drum filter. Two specially attractive features for the smaller plant are the ease with which cloths can be replaced—a skilled shiftsman being able to change a segment without stopping the machine—and the fact that more than one product can be handled in the one filter by using an appropriate number of compartments for each type of concentrate coming from the flotation circuit.

Canvas, cotton cloth, duck, nylon, and woollen weaves are mostly used as filter cloths. Glass fibre and stainless-steel meshed cloth have special use in chemical engineering for corrosive pulps, but are rarely found in mineral dressing. Porous rubber and other plastics have a limited field of use. Nylon is widely employed as it has superior resistance to abrasion and clogging, and cake moves better from its smooth surface. Its longer working life makes it cheaper than cotton. Terylene has high resistance to chemical attack and to temperatures up to 150° C. Particles are not necessarily individually smaller than the pores in the membrane, it being usual for several to stick together and bridge an orifice. Diatom earth may be mixed into the feed when extra clarity in the filtrate is sought. The coarser the overall porosity the drier will be the cake, but the lower will be the clearness of the filtrate.

The most suitable cloth for a given job is usually arrived at by experiment. In operation, particularly where lime is precipitated from the pulp-water during exposure to vacuum, the cloth and piping of the filter are apt to become choked. The usual remedy is a periodic scrubbing and washing with very dilute acid, together with a thorough rinsing and perhaps the use of an alkalising flush to neutralise any residual acid. The scrubbing can be mechanised. At Van Dyk (Rand) this time-consuming task, together with descaling right through to precipitation of pregnant cyanide, has now been replaced by the use of "inhibited" acid. Acid inhibitors are long chain complex organic compounds of high molecular weight. They form a film at the surface of the metal which retards diffusion across the metal-liquid interface but leaves limy deposits open to attack since these remain unfilmed.

A fairly high vacuum (say 26″ of mercury at sea-level) is necessary to maintain an adequate pressure-drop across the cake. Not only must the exhausting pumps have adequate sweeping capacity but the system must be free of leaks and give unobstructed passage from drum to pump. A perfect vacuum at the latter in a leak-free system is not alone enough, since vacuum is the result of general diffusion of gas molecules from a more densely occupied volume to one more empty. Hence narrow pipes, awkward bends, and rusty,

Fig. 296. *Denver Disc Filter*

rough conducting channels, by obstructing gas diffusion, reduce the vacuum available at the filter, however good the pump may be. Mere increase of pump speed cannot increase the vacuum if air flow is obstructed by bad piping, since diffusion depends on the freedom of molecular movement. The filter layout (Fig. 292) requires protection of the pumps against a carry-over of filtrate, which is being drawn from the collecting tank by its own pump. This may be provided by means of a so-called "barometric leg" if the vertical distance between moisture trap and seal pit exceeds 30′. A float in the receiver operates a regular check valve. At Van Dyk, an electrode signals any rise in filtrate which might bring it above the incoming pipe and thus cut off the vacuum in the drum.

Centrifugal filters have a limited application in mineral dressing, outside their use in the drying of coal. The pulp enters a horizontal or vertical drum which rotates at high speed. A cake of solids separates to the sides, whence it is scraped out and a relatively clear liquor flows through.

In electrophoretic filtering use is made of the fact that clays, when suspended in a suitable electrolyte, can be made to migrate under the influence of an electrical current. Pilot tests on unfilterable phosphatic material have indicated some commercial possibilities for this application of zeta potential.

Drying and Roasting

Drying

Minerals nearly always carry inherent or acquired moisture. This is undesirable at times for several reasons:

(*a*) It increases transport cost.
(*b*) Water may be a nuisance or hazard in ensuing treatment of a mineral product.
(*c*) Water introduces or increases humidity, corrosion and/or uncontrolled chemical action.
(*d*) Heat of reaction due to water may lead to a fire risk.
(*e*) Water may lead to difficulties in handling ore—e.g. frozen stockpiles or unstable cargoes.

Against these possibilities is the probability that dusting loss in transit is reduced when finely ground concentrates are exposed to wind *en route* or in exposed stockpiles.

Moisture may be deeply held in cavities or cracks, carried as a surface film, adsorbed or held by capillary attraction, or trapped by poor drainage in a badly designed bunker. As distinct from good drainage, drying is definable as removal of liquid (usually water) by evaporation. This may be effected by exposure to sun and air or accelerated by heating. In the latter case convection of hot air over and through the body of moist material is necessary, together with removal of the resultant humid air. Alternatively, radiant heat may be used. For efficiency there must be adequate contact between the drying agency and the wet surface. Air may be pre-warmed or may gather its heat from flues or plates over which the ore is passing. In rotary kilns a long cylindrical shell slopes gently downward from feed to discharge. Hot gases ascend counter to the stream of ore being tumbled as the shell slowly rotates. This arrangement gives exposure of the mineral, while the upflow of hot air from its entry, dry, to its moisture-laden discharge can be controlled by gates or fans. A similar arrangement is used in roasting or in magnetic roasting, a hot reducing gas being used in the latter case, to promote the chemical reaction.

In the Lowden drier reciprocating rakes move the ore over heated plates which constitute the roof of a furnace. Stationary kilns of various types receive the wet feed at one end (above or horizontally) and either let it gravitate or be rabbled to a discharge gate, moving counter to hot gases regulated to produce the desired drying or pyro-chemical effect. Roasting action was briefly described in Chapter 16 in connexion with the FluoSolids and Edwards driers, used in recovery of gold from auriferous sulphides. Further discussion is presented below. One horizontal stationary drier (the Hazemag) uses screw paddle action to work up to 100 tons/hour of asbestos through a drying process. The machine is compact. The Raymond flash drier (Fig. 297) is used for coal, lime or other finely divided material. The wet feed is mixed with dry material in the mixer B and circulated in a turbulent stream of hot air round the drying circuit.

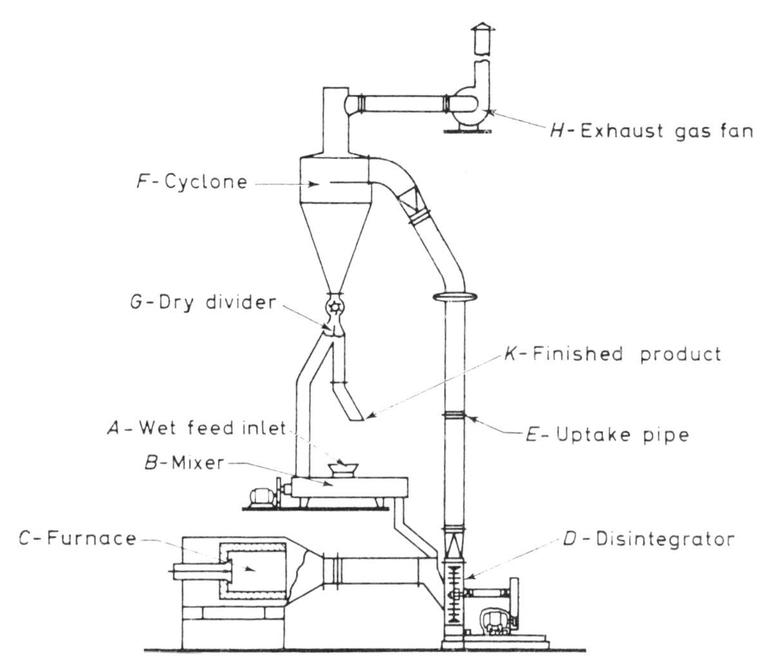

Fig. 297. Raymond Flash Drier (International Combustion Ltd.)

Roasting

Sulphidic concentrates are frequently roasted either to produce SO_2 as a stage in sulphuric acid manufacture or in connexion with further treatment (e.g. copper sulphate leaching). Provided a sufficient percentage of sulphur is present the process is exothermic, the driving force being oxidation of the contained sulphur to SO_2 with, at lower temperatures, some SO_3. Control may be used to avoid undue dilution of the sulphur-bearing exit gases. One of the oldest arrangements, the multiple-hearth roaster, has been developed into the flash roaster in which finely ground dry material is blown downward through rising hot air. In fluidised beds the feed is introduced into a chamber in which hot and partly roasted solids are teetering in a rising column of air. When this rising air balances the weight of the bed there is generalised stability. If the air current is increased beyond this expansion, mixing and agitation increase as in elutriation. The interplay between the size and density of the mass of particles, the size range of the load, together with the velocity and reaction characteristics of the gas can be held in control by choice of a suitable shape of container and by regulation of reaction rate, temperature and gas velocity. Schytil[32] has described two Dorr systems, the BASF and the INCO. Cooling control in the fluidised roaster is achieved in various ways. Water may be directly injected; the solid-liquid ratio of the newly-entering pulp may be varied; waste-heat boiler elements can be built into the bed; cold-roasting gas can be partly de-oxygenated and re-cycled or cold re-cycled material from previous roasting can be added with the feed. In the Dorr FluoSolids plant used typically to treat flotation slurries at *minus* 150 mesh low gas velocities are used. Where coarser grains are to be roasted cooling is controlled by cold re-cycled gas injected into the upper part of the bed. Schytil describes research on temperature effects in which a dispersion of pyrite in sand was used. The threshold roasting temperature is about 400° C., and results in copious sulphatisation. As a grain reacts the reaction product migrates into its substrate and this diffusion governs the situation. Pyrite converting to Fe_2O_3 is reduced 37% in volume and pores are formed. When sulphate is formed, however, the mol volume increases by 160% so that the pores are closed and reaction stops prematurely owing to the formation of a non-porous sulphate skin. The rate-determining step is controlled by the speed of this diffusion. At 500° C. retardation ceases and roasting proceeds with the availability of oxygen. At still higher temperatures reaction stops before complete oxidation, due to distillation of sulphur from the pyrite particles under the influence of heat at a rate exceeding the availability of oxygen for reaction.

Tait[33] has described research which led to improved recovery of a refractory gold ore carrying pyrite, arsenopyrite, stibnite, sphalerite, pyrrhotite and galena. In order to produce a good porous structure for subsequent cyanidation the arsenic eliminating temperature requires strict control of available oxygen in roasting. After this, further roasting is less critical as regards temperature save that with stibnite present the extraction is badly interfered with unless the control continues. At 480° C. gold extraction from the roasted calcine exceeds 90% even when 2% of antimony is present. During oxidation the temperature must not rise above this point until the charge has been visibly dead-roasted. Arsenic must be eliminated at *minus* 496° C. before this further oxidation at a somewhat lower temperature, starvation of oxygen being used in the first stage. These research findings were successfully transferred to plant scale in a two-stage fluid roaster.

Magnetic roasting of iron ores in a reducing atmosphere (Fe_2O_3 to Fe_3O_4 to FeO to Fe, or intermediately) was referred to in Chapter 19.

Tailings Disposal

At completion of treatment the tailings may be in the form of filter cake or, more usually, a pulp. A filter cake could be handled by any convenient form of solid transport, but a pulp must be gravitated or pumped to the impounding dam. One general pulp-handling scheme, used with various modifications by many operators, is now briefly described. A peripheral ditch is made round the impounding area, the spoil being thrown outward. Pulp is led into this ditch at a rate sufficiently slow to permit settlement of the coarser material, the ditch being blocked some distance down so that the slimes overflow inward. The coarse fraction is worked back into the mound on the outer side of the ditch till it has been formed into a raised wall, a foot or so high, with slime flowing inward. The pipe bringing the tailings is now extended, and the process is repeated until the whole site is thus enclosed. If water is to be returned to the plant, a penstock and withdrawal line is placed at a convenient low point in the impounding area and water from the periphery, having settled the slimes, is allowed to run back. This arrangement must rise as the impounding area rises. Piping is now placed on trestles so as to deliver into the dam from the low peripheral wall, and the coarser solids, which are the first to settle, are raked back to form a new wall a foot or more in from the first, and of the same height. Instead of using manual labour, this work can be done by tapping underflow from the feed pipes or by using classifiers and dropping the spigot product to form the wall, while the overflow runs into the enclosed area. This work, with systematic shifting of piping, and possibly, addition to the pumping facilities, is continued until the dam has risen to the desired height. The rate of progress on the walls must allow ample time for consolidation before the next step upward is commenced.

Earth dams depend on their dead weight to anchor them against pressure exerted by the impounded tailings. They must have low slopes, perhaps with a stepped structure, so that they will not spread and collapse under their own weight. The ground on which they rest is not laterally stressed to any extent, and therefore need not be strong. If a high tailings dam is needed (e.g. across a valley or gorge), and has a steep slope, the foundations are stressed and must be strong. While mill tailings, not too slimy to dry out and consolidate, suffice for a low earth dam, broken rock is needed in a high one, with an impervious core carried down to the solid rock. Sludge which has drained and settled is impervious but time must be allowed for this consolidation. Fluent slurry has lubricating powers which might cause the dam to give way if leakage occurred.

Windolph[34] has described the building of a 495′ dam designed to impound some 300 million tons of tailings. The construction of the earth toe was the key to stability. It was set on stable soil, peat bog and surface silts having first been stripped off. A rock filter 8′ wide and 3′ thick was placed at the bottom on the downstream side, to control seepage. Above this rose a 35′ dam 450′ long, the berm of which sloped at $2\frac{1}{4}$:1. A more usual batter is $1\frac{1}{2}$:1. This toe structure was underrun by six lines of perforated piping ending some 500′ upstream, and bedded in small gravel.

Given[35], discussing the relative merits of single large dams as against several small ones, notes that though initial construction of one dam is cheaper, flexible use and isolation in the cycle of filling, drainage, consolidation and raising is easier when small ones are operated in succession. Time is necessary for drying out and raising the berms. A minimum of 11 and a maximum of 18 acres per thousand tons daily should be allowed. The dam crest must be wide enough to carry piping and strong enough to take a dragline or other dam traffic. Feed lines from mill to dam, if gravitational, should be grade gently so as to minimise abrasion. Slope adjustment can be made by the us of drop boxes.

In one method of build-up tailings at around 46% solids are delivered t cyclones slanted inward 15° along the berm. The coarser and thicker unde flow falls nearest the inside of the wall and the overflowing slimes are laundere well into the impounding area. When the berm is sufficiently consolidated dragline picks up this underflow material. Each dam is rested for from 4 t 10 weeks between fillings and after rising from 30′ to 35′ a new dragline cre road is built. The use of mobile cyclones has been described,[36] a telescop arrangement of the feedpipe allowing up to 37′ of travel without having to sh off delivery.

The stability of slimes dams has been discussed by Donaldson,[37] who lis three causes of failure:

(*a*) Surface erosion by rain and wind.
(*b*) Seepage through the toe end.
(*c*) Shear failure due to low strength of incorporated slimes or c foundation soil.

Vegetative cover helps to prevent erosion and is easy to establish on flat su faces. On berms Chenik[38] suggests planting suitable grasses into holes fille with soil, or the use of 30″ lengths of reed inserted 16″ into the sand. Gras must be given a start by the use of a fertiliser and perhaps helped until a natu ral humus has been formed. Creeper grasses can be used initially, giving wa to tufted ones when a growing compost has formed. Marram and couc grass have been successful, and have also combated the nuisance of dust blow from the dam.

As an alternative to impoundment, mill tailings may be returned unde ground for hydraulic stowing in exhausted stopes or as backfill. The materi must be sufficiently deslimed (e.g. by removal of minus 15μ particles wit cycloning) to ensure that it will drain and consolidate stably *in situ*.

Soil mechanics has been defined as "that discipline of engineering science which studies theoretically and practically soils, by means of which and upo which engineers build their structures".[39] These include the foundations c heavy dynamic loads such as reciprocating crushers as well as the silts an slimes liable to enter the walls of tailings dams. The moisture content of soil is related to the surface available for wetting and specific surface is pro bably more important than solid-liquid ratio when considering the effect c this moisture. The colloidal fraction of the pulp sent to a settling pond nee not be great. If it segregates and finds its way through crevices it can influenc the creation of a slipping plane by virtue of its concentration locally rathe than its size percentage. Tailings may carry chemicals which are probabl flocculative in tendency and therefore strengthening. If the tendency reversed—for instance, through subsequent oxidation of contained sulphide —entrained slimes would be a source of weakness. Compressive action a the dam rose could produce a shear zone, and streaming potentials migh aid flow of water and further liquidation. If the retaining walls are reasonabl coarse there should be no risk. It is, however, well to choose a site wher failure would not endanger life or property. Streaming potentials are more important where the soil structure must carry heavy engineering loads. Thixotropic clays and thermo-osmotic migration of water in the subsoil, particularly of varved clays and silts subject to permafrost conditions, call for experienced judgement.

Maintenance

This section can only touch briefly on the routine plant activity required to ensure smoothness and continuity of operation, without which good effciency can neither be reached nor maintained. Among the elements of a well-conceived routine maintenance schedule are:

(*a*) Regular inspection of wearing parts, pipes and launders.
(*b*) Filed records for each machine.
(*c*) Relation of replacement of worn parts to tonnage treated.
(*d*) Regular inspection for leakage of solution piping and vacuum and compressed air networks.
(*e*) Routine de-scaling of pipes and monitoring orifices.
(*f*) Clean-up of corners of casings, tanks, reagent dispensers.
(*g*) Service check on automated controls and telltales.
(*h*) Check on use of correct brands and quantities of lubricants.
(*i*) Alignment checks on shafts, link motions, gears.
(*j*) Safety check on plant lighting, travelling ways, guards.
(*k*) Dust removal from electric motors, lodging points, etc., and check on dust handling system including particle counts, for preventing air pollution.
(*l*) Insulation tests, inspection for chafing of wiring, hot running, overload indications, excessive wear.
(*m*) Check on stores for quantity and quality (e.g. of chemicals liable to deteriorate).
(*n*) Spot-checks on performance of selected items, made non-recurrently.

De-scaling can be done by the use of suitable solvent acids, with due precaution against corrosive attack or retention in the working system after use. Descaling of pipes is done with rods armed with scraper heads, or by pulling spring-expanded scrapers through. Large pipes can be cleaned by means of an inflated rubber ball enclosed in a chain net, flushed through the system. If it sticks, it can be punctured and washed along. Rough-surfaced metal balls, with a radio-active cell inside, have been used. The point of blockage is readily detected by the use of a Geiger counter.

Corrosion through abrasion is minimised by use of resistant wearing surfaces, by limiting grain size and by gentle flow of pulp. Chemical corrosion calls for use of special alloys or plastic-clad surfaces. Local corrosion due to electro-chemical action can be met by cathodic protection or by special insulation of the piping.

Lubrication is a highly specialised matter, in which the manufacturer's recommendation, backed by a specialised survey, should be followed. Choice of oils and greases is made according to load, working temperature, liability to entry of water or chemicals, and the surface physics of the metals in contact. It takes three basic forms. True lubrication requires the interposition of a laminar, viscous or fluid film between the moving surfaces which is thick enough to prevent high spots on the two surfaces from making contact. Parting lubrication is that in which this layer is held under sufficient pressure to fend off the moving surfaces from each other. Special lubricants are those in which the oil adsorbs to one or both surfaces and maintains a wetting film which neither ruptures under load nor deteriorates at working temperatures.

Wrong loading conditions in heavily stressed components lead to metal-to-metal contact, shown by scuffed, ridged, galled or scored gears. Incipient welding occurs if the lubricant fails to maintain a barrier and to remove the intense local heat generated during sliding contact. Pitting is due to surface fatigue of the affected metal and is not appreciably ameliorated by lubrication. Abrasive wear, due to the entrainment of mineral dust, shows as a smooth matte surface. Here a more viscous oil may help things, but the best remedy is to suppress the dust at source.

Filed records are helpful in building up information as to the probable working life of parts of a machine not readily seen without shutting down. If it is known that a submerged impeller has an average life of so many hours inspection can await approach to that limit. Clean-up of odd corners often improves metallurgical recovery by removing depressive wood pulp from the corners of such places as flotation cells. Safety check should be made as though one is a stranger to the plant, since habit leads the workers to ignore danger points. Spot checks can be made outside routine checking, and deal closely with quality control. Quiet deterioration in performance is apt to go unregarded and an examination in which localised laboratory sampling aids inspection is sometimes illuminating.

References

1. *A Handbook on Belt Conveyor Design,* General Electric Co. Ltd.
2. Grierson, A. (1963). *Trans. I.M.M. (London),* 73, Dec.
3. Rasper, E. H. and P. (1960). *Eng. and Min. J.,* Nov.
4. Govier, G. W., and Charles, M. E. (1961). *Eng. Jnl.,* 44, No. 8.
5. Ellis, H. S., and Round, G. E. (1963). *Bull. Can. I.M.M.,* Oct.
6. Roberts, C. L. (1959). *Proc. Aust. I.M.M.,* Dec.
7. Durand, R., and Condolios, E. (1952). *Proc. Nat. Coal Board.*
8. Bowen, F. B. (1963). *Bull. Can. I.M.M.,* Oct.
9. Hardy, H. S., and Canaris, S. A. (1961). *Min. Eng. (A.I.M.M.E.),* March.
10. (1960). *G.E.C. Journal,* Vol. 27, 3.
11. Larsen, E. P. (1962). *Min. Eng. (A.I.M.M.E.),* Jan.
12. Jensen, V. J. E., and MacDonald, J. V. (1962). *Western Miner and Oil Review,* June.
13. Mclellan, R. S. (1961). *Min. Eng. (A.I.M.M.E.),* March.
14. Couche, R. A. (1961). *Aust. I.M.M.,* June.
15. Western Machinery Co., San Francisco, U.S.A., Trade Bulletin.
16. Stephenson, W. B. Trans. A.I.M.M.E., 187.
17. Denver Equipment Co. (1958). *Bulletin,* Jan–Feb.
18. Condolios, E., Couratin and Pariset. (1961). *Eng. Jnl. Canada,* June.
19. Townend, D. S. (1957). *Petroleum Times,* Sept.
20. Anon. (1964). *Mining Journal,* 3rd Jan., p. 8.
21. Michaelson, Ensign and Hubbard. (1960). *Trans. A.I.M.E.,* July.
22. Mussey, O. D. U.S. Geological Survey.

References—continued

23. Rapier, P. M. (1960). *Eng. and Min. Jnl.*, Dec.
24. Garman, C. N. (1962). *Trans. A.I.M.E.*, 223, June.
25. Lynn, R. D., and Arlin, Z. E. (1962). *Trans. S.M.E.*, Sept.
26. McCarty, M. R., and Olson, R. S. "Separan 2610", Dow Chemical Co.
27. Amer. Cyanamid Co. (1959). *Cyanamid Flocculants.*
28. Nalco Chem. Co., Chicago, Trade Bulletin.
29. Gieseke, E. W. (1962). *Trans. Am. S.M.E.*, Dec.
30. Cornell, Emmett and Dahlstrom. (1957). New Orleans *A.I.M.E.*, Feb.
31. Ritter, J. (1962). *New Scientist*, Jan.
32. Schytel, F. (1959). *Metallgesellschaft A. G.*, No. 1.
33. Tait, R. J. C. (1961). *Can. Min. and Met. Bull.*, April.
34. Windolph, F. (1961). *Mining Eng.* (*S.M.E.*), Nov.
35. Given, E. V. (1959). *Mining Eng.* (*A.I.M.E.*), July.
36. Holiday, J., and Wilks, R. (1959). *Eng. and Min. J.*, Oct.
37. Donaldson, G. W. (1960). *Jnl. S. Af. I.M.M.*, Oct.
38. Chenik, D. (1961). *Jnl. S. Af., I.M.M.*, 65, Nov.
39. Jumikis, A. R. (1962). *Soil Mechanics*, van Nostrand.

CHAPTER 23

SELECTED ORE TREATMENTS

In this chapter brief descriptions are given to illustrate representativ industrial treatment of a number of common ores. The arrangement is i alphabetical order under the name of the principal element or compoun recovered. Since more than one mineral is usually concentrated this selectio is arbitrary (e.g. lead-zinc sulphidic ores are listed under "lead"). Som reference is made to new or emerging methods of treatment not yet in com mercial use.

Aluminium

Bauxite

The formula is Al_2O_3. $2H_2O$, but this clayey or lateritic mineral include among its impurities iron oxide, phosphate and titania, together with silic Preliminary treatment often consists of a wash-scrubbing which leads to th discard of a substantial amount of alumina-rich tailing. This can be treate after a comparatively coarse grind and de-sliming with an 800 series collect in acid (sulphuric) circuit, with fuel oil to aid froth stabilisation. Where th concentrate thus produced carries iron or titanium minerals, tabling or hig intensity magnetic treatment has been used for final cleaning.

At Weipa, N. Australia, *minus* 3″ material is wet-screened at 250 ton/hou *Plus* $\frac{3}{4}$″ and *minus* 10 mesh sizes are rejected. The intermediate size, som 156 tons/hour, is sent without further treatment to the refinery. This screeni removes the bulk of the silica. The Jamaican bauxite treated by Reynol Metal Co. consists mainly of gibbsite and boehmite. It is low in silica b high in iron and is treated by a modified Bayer process.[1] The essenti reaction is

$$Al_2O_3 . 3\,H_2O + 2\,NaOH = 2\,NaAlO_2 + 4\,H_2O + \text{red mu}$$

In simplified form the flowsheet is shown in Fig. 299. Ore assaying 40 59% Al_2O_3 with 20 to 24% Fe_2O_3 and from 2 to 3% each of SiO_2 and Ti is fed at minus 2″ to rod mills where it is ground with spent liquor. T discharge slurry, at minus 20 mesh and 45% solids, is digested under pressu at 200° C. without mechanical agitation. After half an hour pressure released by stages and the hot slurry is settled in four-tray thickeners whi discharge red mud to be washed by countercurrent decantation and discarde The thickener overflow is cooled in heat exchangers to 65° C. and sent Pachuca tanks, where it is precipitated with "seeded" alumina from stora

The solution is agitated by compressed air for 65 hours during which alumina precipitates on the "seeds". These are then settled, washed and filtered before kilning to yield 99% Al_2O_3, anhydrous. The aluminous liquor remaining after precipitation passes *via* heat exchange to evaporating units and is then topped up with sodium hydroxide and re-cycled. Some 50% of the alumina is thus re-cycled, the balance leaving as precipitate.

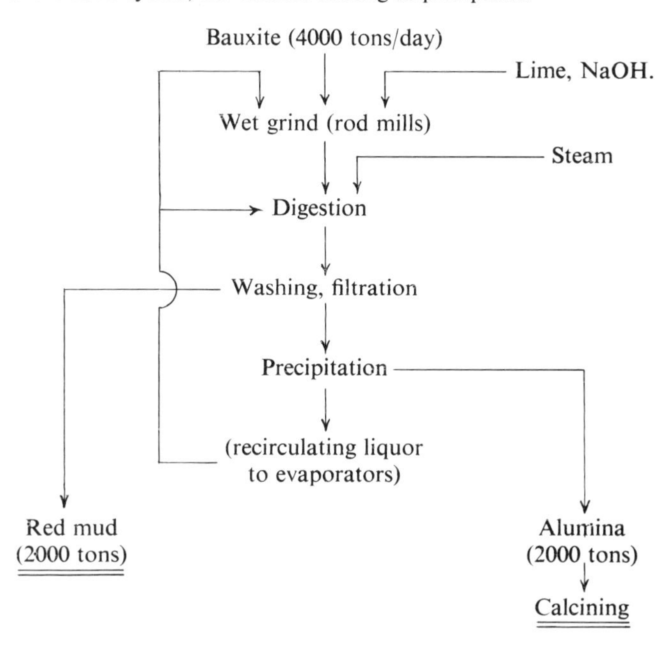

Fig. 299. Treatment of Bauxite

The usual specification for processed alumina allows up to 0·02% Fe_2O_3 and a little silica or titania.

In the normal Bayer process the ferrous oxide is roasted to ferric before leaching with sodium hydroxide. The ratio of seeding alumina to alkali is about 1 to 6 in the precipitation stage. An alternative treatment has been proposed by CSIRO, Australia,[2] for treatment of low grade clays not amenable to Bayer treatment. The key variation is reduction of the pregnant leach liquor with SO_2 followed by hydrolyzation in the absence of air and under pressure at 220° C., till basic aluminium sulphate ($3\ Al_2O_3 . 4\ SO_3 . 7\ H_2O$) is precipitated. This is filtered off and calcined to produce fairly pure Al_2O_3 and the exit gases are returned to the first stage of digestion.

Corundum (Al_2O_3)

*p*H is regulated by sodium hydroxide, the float being depressed by acid. The collector favoured is a fatty acid, such as oleic acid, or its soluble soaps. The frother is pine oil or a higher alcohol.

Cryolite (Na_3AlF_6)

Activated by copper sulphate, collected by fatty acid salts or oleic acid, aided by ortholuidine. Higher alcohols for frothers.

Feldspars (Na, K, Ca) $AlSi_3O_8$)

*p*H is regulated with hydrofluoric acid, which also serves as an activator and silica depressant. Flotation is promoted by the use of amines, and frothers include pine oil, fuel oil, and higher alcohols. The U.S. specification for feldspar calls for 19% + Al_2O_3, total alkali 11–12%, lime—2·5%, iron—0·1%, and low silica. The ceramics industry requires a slime-free granular product sized between 20 and 200 mesh. Associated ore minerals which should be removed include quartz, mica, garnet and particularly ferro-magnesians—the dark silicates which contain iron liable to affect the colour of fired ceramics. Treatment can be on the general lines of Fig. 300.

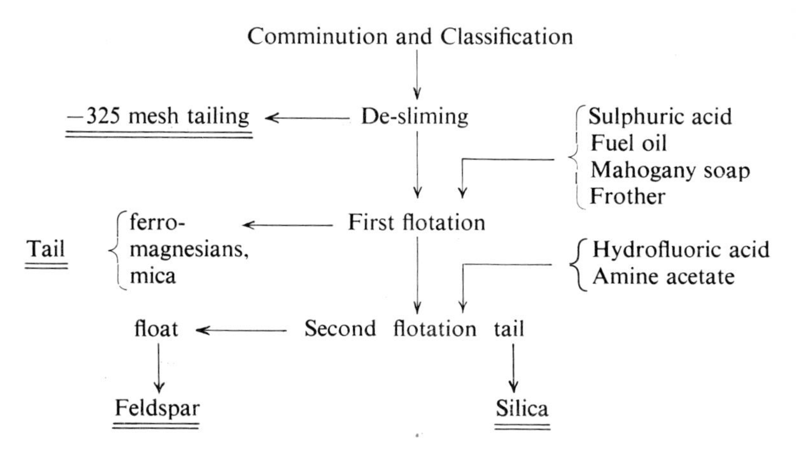

Fig. 300. *General Flow-sheet for Feldspar Treatment*

In one British feldspar-mica separation[8] coarse mica is removed as soon as liberated. This treatment is on shaking tables, and also in two further stages with intermediate grind. The final feldspar concentrate carries some 0·5% mica, representing 0·1% iron. Conditioning for this table flotation is at a *p*H of 4·6, and aliphatic amine acetate is the collector. American specifications of end product require a definite ratio between the potassium and sodium oxides and the silica. After first stage flotation to remove mica the pulp may be re-conditioned (water first being changed) with petroleum sulphate, Armac T and acid.[9] A final stage of flotation depresses the residual silica with hydrofluoric acid.

The economics of production must allow for costly reagents, abrasive solids, corrosion in the acid circuits, and a low selling-price. In such circumstances high recovery is less important than low operating cost. As an example, an American operator (United Lawson Feldspar Co.) treats an or carrying 60–70% feldspar, 20–30% quartz, 5% mica and some slate, garnet and iron to obtain a 50% recovery, much of the loss being in a −200# fraction. which is run to waste. After starvation conditioning with amine, fuel oil pine oil, and sulphuric acid, the mica is floated off at 30–40% solids. The tailing is thickened to 70–80% solids in a cyclone, further conditioned with a very little hydrofluoric acid, R.825, and pine oil, and garnet and iron are floated. Finally the feldspar is floated after depression of silica with further hydrofluoric acid and the use of more amine, fuel oil and pine oil.

Kyanite (Al_2SiO_5)

Treatment may start by scrubbing with a solution of sodium hydroxide followed by desliming. The collectors are oleic acid and fatty-acid salts, and perhaps R.825, aided by aerosol. Pine oil and higher alcohols are used as frothers, with sulphuric acid for cleaning. Pilot-plant tests are reported by Browning.[13] Laboratory and pilot-scale work is also reported by R. A. Wyman.[152] The flow-sheet of a small American producer (Fig. 301) shows several stages of flotation.[3]

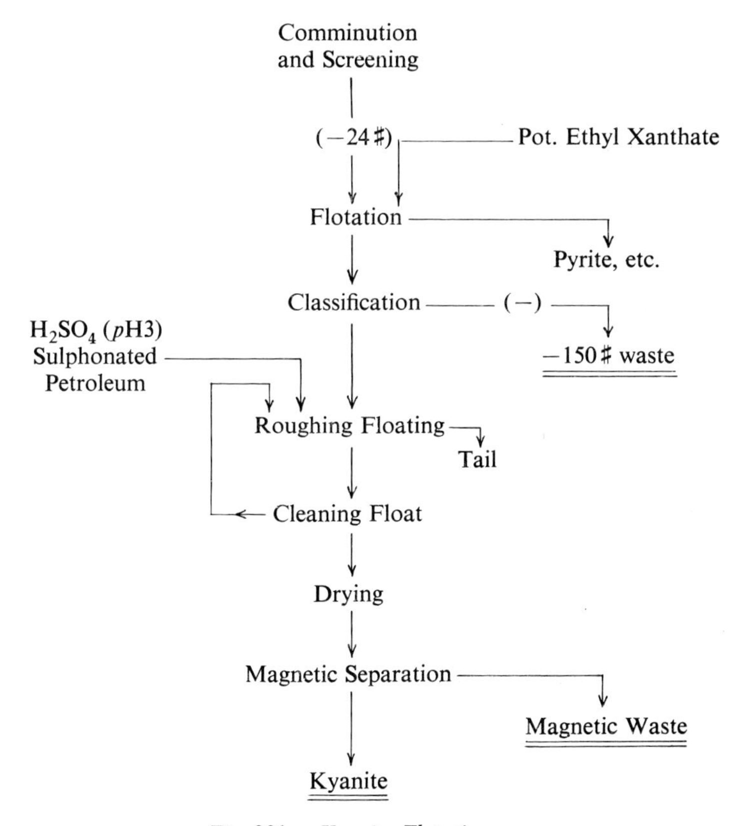

Fig. 301. *Kyanite Flotation*

The U.S. specification for kyanite and sillimanite ($Al_2O_3 . SiO_2$) calls for at least 59% Al_2O_3 and less than 39% SiO_2, with under 0·75% Fe_2O_3.

Antimony, Arsenic

Antimony sulphide (stibnite) and arsenopyrite are among the metal sulphides liable to contain sufficient gold to justify flotation as a stage in its recovery. The response of these minerals to a xanthate collector is much improved by pre-activation with copper sulphate in a pulp rendered alkaline with soda-ash. The specific dosage, reaction time, *p*H, etc., must be worked out in each case. If the concentrate is to be roasted and treated for its gold, it may be necessary to depress the antimony minerals or to remove them from the roasted product by alkaline leach before cyanidation.

At Turhal, in Turkey, a readily slimed soft stibnite with some mispickel and pyrite in a quartz gangue is first hand-picked to remove barren rock. At the same time small lump stibnite assaying up to 55% Sb is selected. The ore is next impact-crushed through 8 mm., screened into three fractions and jigged and tabled.[4] Middlings and slimes are floated after activation with copper sulphate, using locally produced reagents.

A standard plan calls for a large circulating load in the grinding section in order to minimise overgrind, since stibnite is very friable. Pulp is adjusted with lime to a *p*H around 7·6 and activated either with copper sulphate or lead acetate, up to 6½ lbs. of the latter being required in the conditioning stage. If arsenopyrite or pyrite are present they are depressed in the flotation cleaning stage by the use of 0·04 lb/ton of sodium cyanide. Where talc is present it is also depressed in this stage, using yellow dextrin at 0·04 lb./ton. Sodium isopropyl xanthate, also at about 0·04 lb./ton is the collector. Smelter prices are based on a 65% content of antimony and a grade below 50% is rarely profitable.

Asbestos

The objectives in milling asbestos are to free the potential fibre from the slabby gangue in which it is sandwiched; to maintain the maximum length of fibre in the end product; and to remove dust and grit at each stage of release and progressive opening out so that as little as possible is trapped in the final asbestos. Up to 10 tons of air are drawn through the passing material for each ton of fibre finally produced and special problems of dust catching and disposal arise. Work may start with impact crushing, or the use of Aerofall mills, Symons cones, impactors, ball mills or hammer mills as feed size decreases and fluffiness increases. After rough drying, the broken ore is repeatedly screened and re-crushed. It stratifies during passage along the screens, the undersize being waste or a middlings of gangue and unopened short fibre worth further treatment. The screen discharge goes on to finer comminution and fiberisation while freed fibre is lifted by air elutriation, using aspirators placed over the screens and trapping the lifted fibre in cyclones.

Good storage and blending may be needed to maintain an equable feed. The flowlines are flexible, since fibre varies in its length, strength and milling response. Since long fibre fetches the best price, gradual opening out (fiberisation) is practised, with repeated aspirations to remove finished material at the earliest liberation and to shake down entrained grit.

The Voorhees deposit in California[6] is milled as it gravitates through an eight-storey building. At the quarry it is broken by impact crushing to *minus* 6″. After rough drying it is hammer-milled to produce three sizes $-+\frac{3}{4}''$, $-\frac{3}{4}''+\frac{3}{8}''$ and $-\frac{3}{8}''$. From storage the rock is delivered to the fifth floor on to double-deck screens which are kept open by rubber balls vibrating in pockets between the decks. The feed stratifies with rock down and partly opened fibre above. Sufficiently fiberised asbestos is picked up by aspirators exerting an elutriating suction above the screen and trapped in air cyclones. On the sixth floor the material is cleaned and graded. Trommels with centrally mounted fans let short fibre through and blow entrained dust away. Final fiberisation of oversize is performed in buhr mills between carborundum discs before a final cleaning stage.

The test criteria on which sales are agreed are rather unsatisfactory.[7] A screen test is based on the use of a stack of three Hatschek square screens ($\frac{1}{2}''$, 4 and 10 mesh), operated as specified in the "1932 Quebec Standard". The modern trend is toward closer control, some specification of fibre strength and length, and of the degree to which it has been fiberised. Colour and behaviour when used as a wet filter may also be important. Purity as affecting di-electric quality and stability when used in brake linings are also possible matters of customer interest in the complex business of preparing a marketing grade.

Barite

Treatment aims at producing barium sulphate suited to several main requirements, in the paint, paper, pharmaceutical and other industries. Where colour is important it must meet rigid specifications, while lower grade concentrates can tolerate discoloration. These latter are used to blend drilling muds in oil wells and must have a density of at least 4·25, be 98% *minus* 200 mesh, and have a viscosity of less than 60 centipoise. The ores carry iron carbonates, quartz, iron oxides, calcite and sometimes fluorite. A generalised flow-sheet is given in Fig. 302, in which gravity concentration at the coarser liberation sizes is followed by froth flotation. In this sodium silicate is used as a dispersant in a circuit brought to a *p*H of between 8 and 10 with soda-ash and/or hydroxide. The collectors for the iron minerals may include R 824 or R 825, and tall oil or sodium oleate, with pine oil as frother.

At Rammelsberg the tailings from sulphide flotation (lead, zinc and pyrite) are first cycloned at a dividing point of 40μ[10] to remove floating pyrite and thicken the feed to between 45 and 40% solids. It is roughed with Resanol (a mixture of fatty acids) which has been fluidised with tall oil or oleic acid, *p*H being adjusted to 6·5.

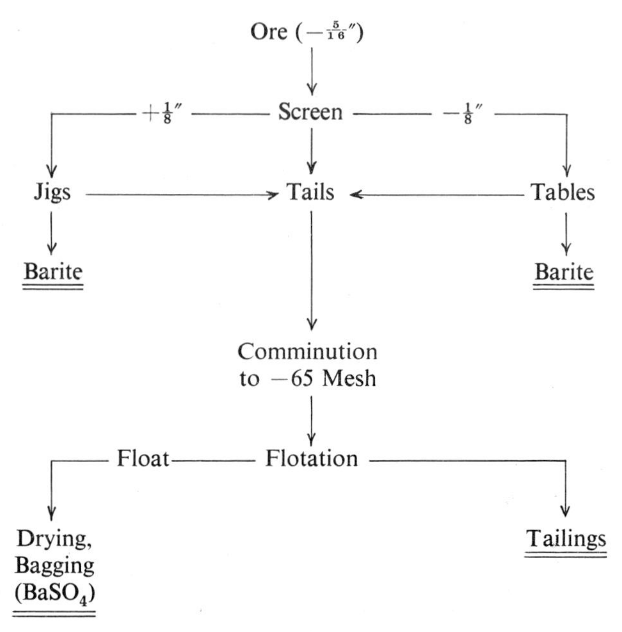

Fig. 302. *General Barite Flow-sheet*

If the product is to be sold as drilling mud, heating during the drying stage may be utilised to remove reagents undesirable for that purpose.

Beryllium

The pure mineral beryl (3 BeO . Al_2O_3 . 6 SiO_2) contains 14% BeO, and the market calls for 10% BeO minimum. The pegmatites in which this mineral is found contain as major gangue minerals quartz, albite and muscovite, while associated "heavy minerals" include cassiterite, celumbite and tantalite, rarely in economic quantities. Spodumene, a lithium mica, may occur in viable quantity. Hand-picking at the supply source is increasingly giving place to flotation. After floating off muscovite the pulp is conditioned at *p*H 5 with a tallow amine acetate and its beryl and feldspar are bulk-floated, using an alcohol frother.[9] These minerals are separated by conditioning with calcium hypochlorite followed by flotation of the beryl with a petroleum sulphonate.

Chemical attack at the research level has been described[11] in connexion with the upgrading of rough concentrates. For successful flotation pilot-plant studies[12] show that the basic requirements are (*a*) recognition of the effect of the heavy associated minerals on product grade; (*b*) determination of the flotation characteristics of these minerals and of beryl both in the natural and the acid-washed state and (*c*) selective removal of heavy minerals ahead of beryl flotation. By close attention to these matters even low-grade ores have yielded over 10% BeO in pilot operations. Beryl can be separated from feldspar and quartz with an anionic collector because it has zero charge a *p*H 3·7–4·6, when the other minerals are negative.

Spodumene may be recovered with beryl in a bulk float. Tests on a large pegmatite deposit containing these economic minerals have been described.[13]

Calcite

Limestone is the general term for rocks rich in calcium carbonate. Treatment may be directed to upgrading this either by comminution and production of an enriched slime fraction by classification, or by flotation of the calcite (and if present dolomite) away from silica and sericite. Sodium silicate is then used as a dispersant and oleic acid as collector agent, in a pulp made alkaline with soda-ash. Sulphonated fatty acids may also be used. A secondary flotation process is used when calcite must be reduced or removed before leaching an uranium ore with acid.

Carbon

This is dealt with under the headings "Coal"; "Diamond"; and "Graphite".

Chromium

The industrial mineral chromite ($FeO \cdot Cr_2O_3$) is marketed either in terms of its chromium content or of its Cr/Fe ratio where metallurgical grade is concerned. Chromite brick is assessed on the refractory character of the raw material. Beneficiation by gravity methods, which include DMS, tabling and the use of spirals, is concerned to raise the Cr_2O_3 content well above 40%. Iron may be partly replaced in the crystal lattice by other elements.

The gangue minerals are chiefly serpentine and olivine. At the I.M.D.C. Congress at Goslar, 1955, L. Wenz reported studies on a Turkish ore containing 30% Cr_2O_3, associated with oxides and sulphides. When ground to -100μ the serpentine dominated the fines, chromite the medium particles and olivine the coarse. A strongly alkaline pulp (*p*H + 10) was used, with sodium oleate and oleic acid (1lb./ton) to float the serpentine. The pulp was then rendered acid (*p*H 3) and chromite was floated with $\frac{3}{4}$ lb./ton of a fatty-amine collector. Olivine, which only floats in a neutral solution under these conditions, remained in the final tailing, the concentrate grade being 45% CrO_3.

Mixed pyro-metallurgical, leaching and electro-chemical methods are used to raise the Cr/Fe ratio to the degree required for ferro-chrome. In one approach applied to a table concentrate[14] the chromite is ground to *minus* 52 mesh and mixed with coke ground to *minus* 200 mesh before reduction roasting at 1,250° C. The product is acid leached in 10% boiling sulphuric acid to yield a residual ferro-chrome of 11/1 ratio.

Clay

Kaolinite or "china clay" is a hydrous aluminium silicate, of formula $Al_2O_3 . SiO_2 . n H_2O$. Its crystalline mode is plated hexagonal. Its main uses are as inert fillers, as a constituent of ceramics and as coating in the paper industry. If the end-product of processing is mainly below two microns it commands the best price provided it can be sufficiently bleached to be used as paper coating. Coarser particles, consisting of stacks of these plates, are saleable as fillers up to 325 mesh. More exacting specifications are satisfied by the Cornish china clays, which meet the demand for a non-staining clay slip low in fluorine, these requirements being basic to production of chinaware not discoloured by kilning and not liable to contaminate the atmosphere or corrode the brickwork during that process. China stone, an undecomposed granite used in connexion with the manufacture of clay slip, can be stripped of its fluorine by flotation of part of its mica.

China clay is either dug out or slurried from the working faces of quarries, by high-pressure jets and washed through sand traps, sluices and settlement pits to remove most of the unwanted sand. The balance is separated either by mechanical classifiers, hydroseparators or hydrocyclones, with some use of centrifugal classifiers. A typical American treatment[15] first pulps the pit clay by blunging, and then adds a dispersant such as sodium tetra-pyrophosphate before pumping to the treatment plant. Overflow from the de-gritting classification is screened to remove any mica. Crude kaolin is sometimes classified into two grades, coating (fine) and filler (coarse). Bleaching with zinc or sodium hydrosulphite is performed at a *p*H between 3·5 and 4, sulphuric acid being added to the naturally acid slurry. De-watering of the fine grade may start with thickening through a highspeed centrifuge and be followed by filtration, either on roll-discharge or string-discharge drums. Heating may be needed to reduce the viscosity and speed up filtration. The cake is then dried and bagged for shipment.

In pilot-scale tests anastase has been partially removed from a *minus* 2μ kaolin pulp together with some staining limonite.[16] The collecting agent was *minus* 325 mesh magnetite conditioned with alkali, talloel and fuel oil. This was added to a conditioned pulp and removed either by magnet or by froth flotation. In the former case some 26% was picked up in 5·4% of the feed, and in the latter 53% in 28% of the feed. This led to the development of a better carrier reagent, calcite at minus 325 mesh being chosen. This floats well with fatty acids and removes almost 90% of the titania.

Coal

The objectives in coal preparation include:

(1) Improvement of technical performance.
(2) Grading into sizes for sale.
(3) Removal of undesirable constituents.

Coal, particularly when produced by mechanised mining, is sent to surface in a mixture of sizes from dust upward, with random inclusions of chalk, included clay, wall rock, ankerite ((CaMgFeMn)CO_3), and pyrite. Its main uses are the production of heat, light and power in the electrical and gas industries; domestic consumption of solid fuel; metallurgical industries (coke); bunkering of ships (declining); steam-raising, as lump or as pulverised fuel.

Instead of leaving the ash to be heated and withdrawn during consumption (at a cost for labour and lost calories), it may be removed by wet-cleaning methods.

Some 30% is processed chemically to yield napthalene benzene, phenols, toluene, etc.[17] With the increasing use of uranium as a source of industrial power this trend should increase, and coal preparation may be expected to go far beyond its present duty of simply preparing raw coal for low-ash consumption, with which this section is concerned.

If coal with a high ash-content were used, this waste material would have to be transported from mine to consumer and removed from his grates at a cost for labour, manipulation, and loss of the heat uselessly withdrawn with the ash. Industrial plants are designed to work efficiently when supplied with raw materials of specified character, and one of these is the coal used for raising power. Such specifications may be concerned with size, carbon content, volatiles, ash (both "fly" and clinkering), sulphur, coking "swell", moisture, etc. It is the function of the cleaning plant to receive the raw coal, deliver a specified "vend" to the market, and as far as possible to produce a residue which can be profitably used.

Mined coal contains two kinds of ash-forming dirt—"fixed" and "free". The fixed ash is derived from the inorganic matter which grew in the tissues of the original coal-forming plants, together with fine silt entrained during deposition of the seams. There is also pyrite formed during the obscure processes of pressure, ferment, and bacterial action which changed the original peat, cryptogams, and lush sub-tropical swamp growths into the bedded deposits of the coal measures. These growths were complex and contained cellulose, resins, waxes and lignin. Their metamorphosed reaction products influence the special qualities of individual seams. The free dirt is extraneous to the true seam, coming from roof, floor, or "seat" measures. It includes clay, black shales intercalated with the coal, and coal locked to calcareous and other mined dirt. Fixed ash is not removable by standard cleaning methods, but free dirt can usually be reduced by suitable treatment. It includes massive pyrite deposited in the fissures or "cleats" of the coal seam, and salines and sedimentary intercalations brought in with percolating water.

The four main operations in coal preparation are:

(1) Screening or sizing.
(2) Mixing, blending.
(3) Cleaning (removal of incombustibles).
(4) De-dusting or de-watering.

For most purposes the calorific value of the coal is the quality required. The ash content of this "packaged power" is worse than useless. It must be paid for as coal, the freight charge of transport must be met, and the residual ash or clinker must be withdrawn from the furnace. Disposal of this waste product costs up to several shillings per ton. Its removal at source is justified if the cost is absorbed by a higher market value.

As in the case of ore dressing, the best preparation is the one showing the highest profit. The highest part of the cost has irrevocably been paid in winning the coal and delivering it to the plant, so it is important to use as much of this tonnage as possible. Seams and coals vary considerably. With some it is possible to send out a high percentage of cleaned coal carrying little ash and to reject a low percentage of unburnable rubbish, with but little middling. In other cases it is profitable to send out a lower percentage of the raw feed as cleaned coal and to use most of the balance as a low-grade fuel, suitable for power raising on the property in specially constructed boilers.

The capital cost of a plant bears less relation to capacity than is the case with hard-rock mills. Coal is easily crushed, and must be handled more gently than by methods using skips and bins (though these are taking over to some extent from trucking in "tubs" from underground to plant). Storage capacity for feed and product are therefore elaborate. The elements in construction comprise:

(*a*) Conveying.
(*b*) Storage.
(*c*) Switching.
(*d*) Cleaning.
(*e*) Grading.
(*f*) Dust collection.
(*g*) Drainage.
(*h*) Slurry treatment.
(*i*) Structural work.
(*j*) Water handling.
(*k*) Product disposal.

The oldest classification of coal in this country was by "rank"—the degree to which original coal-forming material had been changed by metamorphism from peat up toward anthracite—and was based on the fixed carbon calculated on a dry, impurity-free basis. The new classification is given in Table 43.

The organic constituents of coal, in Dr. Stopes's classification, include four main types. *Vitrain* consists of the broader bright bands, derived from wood and bark. *Clarain* is more finely laminated and of a less brilliant lustre. It is usually well cleaved, the coal breaking into small rhombs. *Durain* is dull, hard, and of somewhat higher density, with no visible lamination save when it carries a banding of vitrain. It is derived from finely comminuted plants and is tougher than "brights". *Fusain* is a soft, powdery, mineral charcoal derived from the cellular tissues of plants, and is the soiling constituent of coal.

The normally predominant types are clarain (with lower ash, brittle strength, and higher coking properties) and durain, with tough non-brittle strength. The difference in crushing resistance tends to segregate durain and clarain in the coarser grades, with vitrain and fusain in the smaller sizes.

The lower the rank of a coal, the greater is its tendency to absorb and retain moisture when stored in the open. This "slacking" results in excessive fines and a tendency toward spontaneous combustion when the coal surface is dry enough to deep-sorb oxygen.

Cleaning Treatment

The purpose of this is to remove "free" dirt. Coal varies widely through

TABLE 43

NATIONAL COAL BOARD CLASSIFICATION OF COALS

(Revision of 1956)

Coals with ash of over 10% must be cleaned before analysis for classification to give a maximum yield of coal with ash of 10% or less.

Group(s)	*Class*	*Sub-Class*	*Volatile Matter (d.m.m.f.) %*	*Gray-King Coke Type*	*General Description*
100*			Under 9·1	A	Anthracites
	101†		Under 6·1	A	
	102†		6·1–9·0		
200*			9·1–19·5	A–G8	Low-volatile steam coals
	201		9·1–13·5	A–C	Dry steam coals
		201*a*	9·1–11·5	A–B	
		201*b*	11·6–13·5	B–C	
	202		13·6–15·0	B–G	Coking steam coals
	203		15·1–17·0	E–G4	
	204		17·1–19·5	G1–G8	
	206		9·1–19·5	A–B for V.M. 9·1–15·0 A–D for V.M. 15·1–19·5	Heat-altered low-volatile steam coals
300			19·6–32·0	A–G9 and over	Medium-volatile coals
	301		19·6–32·0	G4 and over	Prime coking coals
		301*a*	19·6–27·5	G4 and over	
		301*b*	27·6–32·0		
	305		19·6–32·0	G–G3	(Mainly) heat-altered medium-volatile coals
	206		19·6–32·0	A–F	
400 to 900 :—			Over 32·0	A–G9 and over	High-volatile coals
400			Over 32·0	G9 and over	Very strongly caking coals
	401		32·1–36·0	G9 and over	
	402		Over 36·0		
500			Over 32·0	G5–G8	Strongly caking coals
	501		32·1–36·0	G5–G8	
	502		Over 36·0		
600			Over 32·0	G1–G4	Medium caking coals
	601		32·1–36·0	G1–G4	
	602		Over 36·0		
700			Over 32·0	E–G	Weakly caking coals
	701		32·1–36·0	E–G	
	702		Over 36·0		
800			Over 32·0	C–D	Very weakly caking coals
	801		32·1–36·0	C–D	
	802		Over 36·0		
900			Over 32·0	A–B	Non-caking coals
	901		32·1–36·0	A–B	
	902		Over 36·0		

* Coals of groups 100 and 200 are classified by using the parameter of volatile matter alone; the Gray-King coke types quoted for these coals indicate the ranges found in practice, and are not criteria for classification.

† In order to divide anthracite into two classes, it is sometimes convenient to use a hydrogen content of 3·35% (d.m.m.f.) instead of a volatile matter of 6·0% as the limiting criterion. In the original Coal Survey rank coding system the anthracites were divided into four classes then designated 101, 102, 103, and 104. Although the present division into two classes satisfies most requirements, it may sometimes be necessary to recognise four or even five classes.

Reference.—"The Coal Classification System used by the National Coal Board" (National Coal Board, Scientific Department, Coal Survey).

d.m.m.f.—dry mineral matter free.

the seam, and even more markedly between the various superposed seams the colliery may be working. The washing plant must accordingly be adaptable. The roof and floor of a seam (composed of sandstone, fireclay, shale) and the inter-laminated shale "partings" are often carbonaceous and easily mistaken for true coal, and are mixed with it during mining operations. Pure coal varies in density. Typical British figures are S.G. 1·35 at 5% ash, and S.G. 2·3 for completely free associated dirt (the equivalent of ore-dressing "gangue") with carbonaceous shales intermediate, in accordance with the dominance of coal or impurity in the locked middling. This gradation from light coal to heavy shale is exploited in gravity separation down to about 10 mesh, below which flotation is used. Sulphur may be controlled by hand-picking or by wet methods of removal. A fully developed flow-sheet using all main types of wet cleaning in shown in Fig. 303.

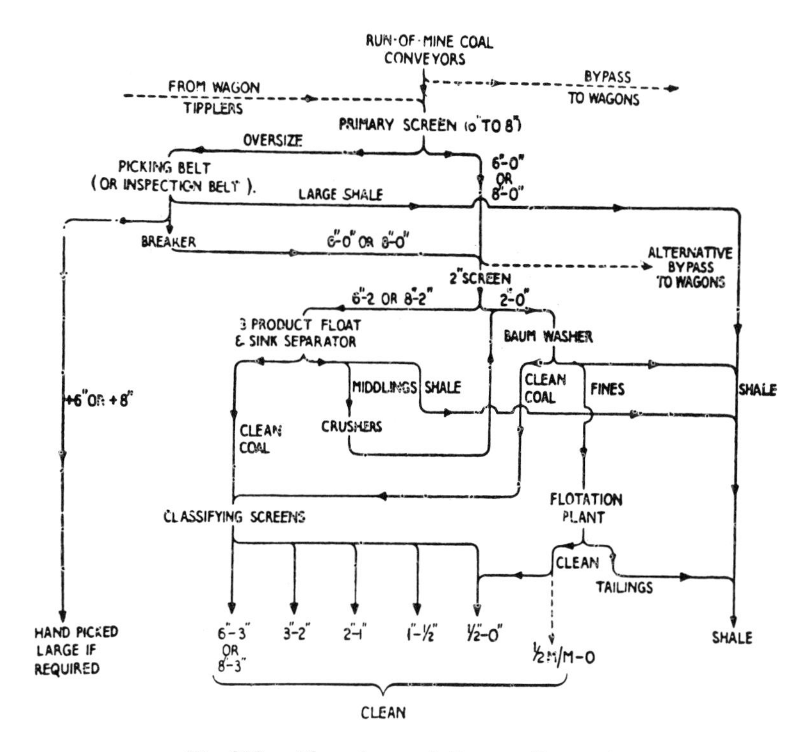

Fig. 303. *Flow-sheet.* 3-*Process Separation*

Finely disseminated impurities (fixed dirt) cannot be removed by washing treatment. One or two specially clean seams are treated by flotation, followed by chemical digestion, to produce carbon electrodes. Flotation is also used to upgrade washing fines and slurries. The free dirt removed in normal treatment must be sufficiently liberated to be separable at a grade size acceptable to the purchaser. It is separated by hand picking, or by gravity treatment based on density differences. Large lumps may still be hand-picked, but the tendency to market large coal is disappearing. Dense-media separation now does most of the work once performed by hand picking.

In collieries where the dust is not allayed by the use of water underground the feed may need de-dusting before wet cleaning, in order to minimise fouling of the washery water. Large coal and easily recognisable impurities may be removed by hand picking. The main cleaning methods are jigging, horizontal classification, dense-media or *sink-float* separation, and froth-flotation. Dry tabling is also practised, but its use is decreasing with the spread of dust-preventive wetting of the coal face.

Screening may precede treatment in order to facilitate sorting or transport, but is usually applied after washing, as a later stage in preparation for sale.

Upgrading by gravity methods was discussed in earlier chapters (under "Jigging" in Chapter 13, by Rheolaveur in Chapter 14 and by DMS or sink-float in Chapter 12). The efficiency of operation (E) is checked by laboratory sink-and-float tests, and is expressed in the equation[17]

$$E = \frac{\text{Yield of washed coal} \times 100}{\text{Yield of float coal of same ash content}}$$

The U.S. Bureau of Mines circular[18] defines some internationally used terms and criteria employed in the evaluation of coal washing performance.

Dry-Cleaning Methods

Part of the value of coal cleaning is lost if as the result of removing dirt a substantial amount of residual water must be shipped. At coarse sizes washed coal drains readily to a fairly low moisture, but with the smaller sizes capillary action in the storage bunkers and on the dewatering screens becomes a problem. Washed "slack" at $-\frac{3}{4}''$ carries over 10% of water, and smaller sizes are much more retentive, even when centrifugal force or vacuum filtration is applied. A 10% charge of water on ten tons of moist coal entering a coke oven requires nearly $2\frac{1}{2}$ million B.T.U.'s for its evaporation (over 27% of the heat applied to the charge). This would take more than a quarter of a 32-hour coking cycle, and if this feed moisture were reduced to 3%, six hours would be saved. Slurry, even after forced drainage, carries up to 25% of residual water and presents problems which still elude satisfactory solution. No dry cleaning method yet evolved gives a complete solution of the commercial problem. Those used include:

(1) Stationary devices worked with continuous or interrupted air streams.
(2) Dry reciprocating tables with air classification *via* porous beds.
(3) Appliances exploiting frictional and shape differences between coal and dirt.
(4) Methods using centrifugal force.

The machines include air jigs, pneumatic tables ,and other oscillating devices. With the increased use of water for suppressing dust in British coal mines dry cleaning methods have suffered a set-back.

Flotation

Compared with hard-rock treatment, coal flotation (as practised) is extremely simple. It is applicable to a fairly coarse mesh-of-grind, considerations of liberation permitting, and gives a docile float at *minus* 48 mesh. Current practice is extending froth flotation to coarser sizes, up to 2 mm. being reported.[17] In 1960 four million tons of *minus* 0·5 mm. concentrate were produced in Britain. In that year 92% of the United Kingdom coal output was mechanically mined and in many collieries the *minus* 12 mesh fraction was around 30%, and was relatively high in ash. In the U.S.A.[18] an average top size of 28 mesh is floated, alcohol possibly aided by kerosene and/or fuel oil being the reagent. Germany floats a *minus* 3 mesh feed, most of which is below 150 mesh, and brings its ash content down from 20% to below 8% in the product.

The fine coal thus produced is either dried for use as a powdered fuel, or blended with coarser coals. Low rank brown coal and lignite is hydrophilic, but when the carbon is above 80% the chemical structure changes and by about 89% carbon maximum hydrophobic quality is reached.

The flotation reagents most used are locally produced by-products rather than those of specialised manufacture. They include cresyls, pine oil, kerosene, creosote, fuel oil, gas oil and aliphatic alcohols. Among the reagent combinations favoured in current British practice are creosote and "cresylic acid" in 2:1 ratio, and medium fuel oil and "cresylic acid" in 5:1 ratio, the latter being used for lower-rank coals. These so-called cresylic acids are often mixed with phenolic compounds distilled well above the normal boiling range for commercial cresylic acid. The bulk of the feed floats, so withdrawal arrangements for the froth must be able to handle a large volume. The value of the concentrate is far lower than in ore treatment, so high percentage recovery is less important than a good grade. Coal, being light, can be worked in a dilute pulp. This aids the process by reducing trouble with slimes from associated clay.

Owing to the high residual water in the final product, the direct use of floated concentrate presents difficulties. It is increasingly mixed as a newly produced filter cake with washed smalls, in which condition it is easy to move through bunkers to automatically fired boilers, provided the blending has been properly done. Owing to this moisture limitation, flotation was once looked on rather as a purification process for colliery wash-water and sink-float liquids than as a direct beneficiating process applicable to fines.

The ash content of coal in washery water rises as the particle size drops. In a typical test the +14-mesh material has an ash content of $12\frac{1}{2}$% and the −200-mesh $43\frac{1}{4}$%. Flotation aims to reduce this. In a pilot test on feed carrying the ash percentages just quoted, the washery slurry was thickened in a hydro-separator with rising water at 12 ft./hour. This gave an overflow containing much of the − 200 mesh material, and a reasonably thickened underflow from an entering slurry which carried $90 < 95$% water. Roughing flotation of the underflow with a light fuel oil as collector and pine oil, the pulp containing $17 < 20$% solids, gave a good coal recovery with $12\frac{3}{4}$% ash at negligible cost. Despite centrifuging, the moisture of the product exceeded 25%. Provided the total output of flotation fines does not exceed some 10% of the available "smalls" suitable for blending, its use in this way is economically interesting.

In Britain the laws affecting discharge of washery water into rivers are stringent. Effluents from flotation plants must not contain more than 2 p.p.m. of phenols or 100 p.p.m. of suspended solids Flotation tailings are not now impounded, but flocculated and filtered.

Apart from the importance of flotation in cleaning contaminated dense media and in upgrading the high-ash coals in washery slurries, the possibilities of flotation in the direct treatment of low-rank coal is being studied. In Europe flotation is a useful stage in the further upgrading of already high-purity coals, as part of the treatment to remove fixed ash before chemical methods are used. The end-product is used in the production of carbon electrodes.

Dewatering

Residual moisture is of five kinds:

(*a*) That inherent in the pores of the original plants.
(*b*) Surface moisture filming the particle of coal.
(*c*) Moisture in particle cracks.
(*d*) Adsorbed moisture clinging by capillary attraction.
(*e*) Moisture trapped in storage bunkers owing to poor or hurried drainage.

Down to $\frac{5}{8}''$ (*nut*) coal readily drains, but below this size (*slack*) the water is increasingly retained. Below 5 mesh it may rise to 30% moisture even after dewatering on jigging screens. Down to about 40 mesh, centrifuges reduce the residual moisture to some 10 < 15%. The residue on 5 < 10-mesh coal in a typical test was 22% after five minutes and $19\frac{1}{2}$% after twenty hours, while for 150 < 200-mesh material the corresponding figures were 46% and $44\frac{1}{2}$%. All sizes drain rapidly to a natural limit, after which action is very slow.

The time taken for drainage depends partly on the bulk in the hopper. Since coal is fragile, it degrades easily under the pressure and loading shock of the overriding load. The inter-spaces between lumps then become choked, as do the draining orifices. Perforated pipes projecting up into the hoppers are useful, provided they do not become "blinded". Drainage perforations on elevator buckets, which can be helped to keep open by smart rapping with a knocker bar as they discharge, are also used. Dewatering screens are the most effective draining devices.

The following note on slurry handling is abstracted from *Information Bulletin* No. 55/1/56 N.C.B. Coal slurries do not handle well on continuous devices, and are usually pumped into settling ponds, drained, and lifted by grabs. If loaded to wagons, end-tipping is necessary with this sticky material or the use of special containers. If loaded to bunkers the sides must be well spaced and steep, with at least three or four feet of bottom opening. Continental practice with < 25% ash and < 45% −200# material of < 25% moisture uses two types of bunker. In one a row of six screw feeders under-runs the bunker and breaks the slurry into small lumps for delivery to mechanical stokers. In another a layer of slurry is deposited on one of fuel passing under the bunker on its conveyor belt. Here control of blend is made by setting a guillotine gate and any required blending and delivery rate can be obtained.

Dewatering of flotation concentrates has been cheapened and accelerated[19] by the use of flocculants before filtration. Two main types of centrifuge are also in use—basket and solid-bowl. The former are of three sub-types[20]—those which have scrapers to effect discharge, those with vibrating baskets to move the solids and those with "pushers" or "peelers" to remove these. The solid bowl types used in Britain are the Bird and Dynacone centrifuges.

The free liquid associated with a mass of small particles is either interstitial or held as a surface film. Voidage depends mainly on particle shape and size, and surface liquid on surface area. Displacement by unaided drainage can only use small forces to oppose the capillary hold, but a centrifuge of the type used in industrial plants develops up to several hundred times the gravitational effect. Experience with Durham coals indicates that *minus* ½″ feed to centrifuges and filters should be split at about ½ mm. (32 mesh). The coarser fraction, after centrifuging had a moisture of 4½%. In the solid bowl type of machine a cylindrical bowl, or one shaped like a truncated cone, revolves on a horizontal axis. Its settled contents are moved from the wall by a screw conveyor moving at a slightly different speed. Feed enters centrally and solids and liquids are discharged from opposite ends. Effective classification, in terms of residual moisture and of solids left in the liquid effluent, is related to the depth of liquid being centrifuged (the "pond" depth.) The shallower this is, the longer is the "beach" over which solids must move to the discharge and the lower the discharge moisture. Solid bowls are used in the *minus* ⅜″ size range and for de-watering flotation froth and tailings.

Automated thermal drying by use of a rotary kiln in one Canadian operation uses a drum 6′ in diameter and 40′ long, turned at 8 r.p.m.[21] Heating is by a mixture of oil and coal dust, the latter being regulated by the temperature of the exhaust gases and the oil acting mainly as a pilot flame. The system described has a throughput of 120 tons/hour of *minus* ¼″ coal fed at between 10% and 15% moisture and discharged at 4% moisture. About ⅜ gallons of oil and up to 40 lb of coal dust are consumed per ton dried.

Grading

The oldest method of grading made use of convalescent mining labour fit for light surface duty. The coal was displayed on picking belts and worked over for sizes, dirt, and pyritics. Below a handsorting size screening was used to produce the various sizes (variously named in different coalfields) such as "treble nuts" (3″–1½″), "double nuts" (1½″–1″), or "peas", beans, pearls, nutty slack, and fine slack (duff or dart). A quiet revolution in practice has been under way since the nationalisation of Britain's coal industry. New coal-preparation plants have better feed arrangements and more mechanisation, and a measure of standardisation of screen sizes has replaced the traditional confusion.

The nomenclature used by the National Coal Board is based on that set out in a B.C.O.R.A. and B.C.U.R.A. publication. Seven sizings are recognised, on round hole screens, the permitted variation and percentage undersize being stated. The groups are:

	Typical Screen Size (inches)
Large Cobbles	6 × 3
Cobbles	4 × 2
Trebles	3 × 2
Doubles	2 × 1
Singles	1 × $\frac{1}{2}$
Peas	$\frac{1}{2}$ × $\frac{1}{4}$
Grains	$\frac{1}{4}$ × $\frac{1}{8}$

Cobalt

Most of the world's supply is a by-product from the concentration of copper or nickel ores. At Chibuluma in Zambia[22] a 3% concentrate containing some 36% of iron, 44% sulphur and 10% silica is upgraded to about 20% each of cobalt and copper by mixed smelting and leaching. This starts with production of a low-grade matte in which re-cycled slag containing some cobalt forms part of the charge. The product assays 4½% each Co and Cu and is crushed, ground and calcined. Soluble sulphates are thus produced in fluidised roasting, and these are dissolved and upgraded first by reaction in a cobalt-copper-iron sulphate solution which removes a barren liquor and second by further leaching which upgrades the insoluble portion to shipping assay, the liquor becoming the leach solution for the next batch of fresh matte.

The pressure leaching of a concentrate assaying 17½% Co, 24% As, 20% Fe with some Ni and Cu has been described by J. S. Mitchell.[24]

After cleaning to an As/Fe ratio of 1·2:1 it is pulped to 26% solids and oxidised in a carbon steel autoclave cladded with stainless steel, with an additional lead lining and acid-resistant brick. The pulp is weired through six successive compartments of the 40′ long vessel. Each has its own agitator and aeration, and high pressure pumps force the feed in. Reaction is exothermic at an operating pressure of 500 p.s.i. and 190° C. On discharge, 90% of the cobalt is in solution as sulphate. The residual acid in the discharge is nearly neutralised with lime. Pregnant solution is filtered off and stripped of its gypsum and iron salts by precipitation methods. Copper is next removed by precipitation as a cement. Finally, metallic cobalt is produced by batch treatment in vertical autoclaves which hold 450 gallons. Anhydrous ammonia is added, the autoclave is heated to 190° C. and hydrogen is blown in, the pressure being 750–800 p.s.i. The cobalt amine now seeds down as metal.

The use of Flotagen (mercaptobenzothiazole *plus* sodium carbonate) is reported as a specific collector for cobaltite, smaltite, and erythrite from a diabase gangue. Diphenyl thiocarbazid $(C_6H_5NHNH)_2C = S$ has been used for the selective flotation of cobalt and nickel from copper and iron minerals. A mixture of roasting and flotation has been developed experimentally for the upgrading of Co-Ni stockpiles.[25] Complex Co, Ni, Cu, Pb, Fe sulphides containing about 3% each of Cu and Ni were roasted at 400° C. and floated with conventional sulphydric reagents to give high recoveries.

The beneficiation of a wide range of minerals including the arsenide of cobalt by gravity and flotation has been described with respect to Moroccan ores.[23] The flotation section activates with copper sulphate and uses sodium sulphide ahead of an amyl xanthate collector in cells specially designed to handle a tenuous froth.

Copper

The copper-bearing ores include so wide a range that industrial treatment uses nearly every processing technique, sometimes in highly specialised ways. Copper minerals range from straight sulphides *via* copper-iron sulphides to oxides, carbonates, silicates and chlorides. Often associated minerals must be recovered separately, including gold, silver, pyrites, cobalt, molybdenum, germanium, lead and zinc. The practical examples outlined in the following sub-sections illustrate the use of chemical extraction (leaching) and its modifications in the L.P.F. process (leach-precipitation-float); straight flotation of typical copper ores; differential flotation of mixed ores; and pyro-metallurgy in the segregation process.

Leaching

The chemical reactions used in leaching the more common ores are:

Azurite: $Cu_3(OH)_2 . (CO_3)_2 + 3\,H_2SO_4 — 3\,CuSO_4 + 2\,CO_2 + 4\,H_2O$

Brochanite: $Cu_4(OH)_6 . SO_4 + 3\,H_2SO_4 — 4\,CuSO_4 + 6\,H_2O$

Chalcocite: $Cu_2S + Fe_2(SO_4)_3 — CuS + CuSO_4 + 2\,FeSO_4$
(Covellite)

Chalcopyrite: $CuFeS_2 + 2\,Fe_2(SO_4)_3 + 2\,H_2O + 3\,O_2 — CuSO_4 + 5\,H_2O + 2\,H_2SO_4$

Chrysocolla: $Cu . SiO_3 . 2H_2O + H_2SO_4 — CuSO_4 + SiO_2 + 3H_2O$

Copper, native: $Cu + Fe_2(SO_4)_3 — CuSO_4 + 2\,FeSO_4$

Covellite: $CuS + Fe_2(SO_4)_3 — CuSO_4 + 2\,FeSO_4 + S$

Cuprite: $Cu_2O + H_2SO_4 + Fe_2(SO_4)_3 — 2\,CuSO_4 + 2\,FeSO_4 + H_2O$

Malachite: $Cu_2(OH)_2CO_3 + 2\,H_2SO_4 — 2\,CuSO_4 + CO_2 + 3\,H_2O$

Pyrite: $4\,FeS_2 + 15\,O_2 — 2\,Fe_2(SO_4)_3 + 2\,H_2SO_4$

Tenorite: $CuO + H_2SO_4 — CuSO_4 + H_2O$

Heap roasting and leaching have been practised in Spain and Germany since the sixteenth century. Dump leaching is currently an important source of copper.

Pit leaching has been practised for many years, but its methods have today been mechanised and accelerated. Weed[26] has described the Cananea method used since 1946. The material includes pit dumps (0·2% or so Cu) and caved stopes. The chalcocite this carries is readily leached in weak sulphuric acid carrying ferric iron. Water is led in, percolated, and pumped up to a precipitation plant built of wooden grids in wood cells. Shredded iron from old cans catches the contained copper. The pregnant solution carried 3·3 g./l. of Cu, 3 g./l, of ferrous, and 7·4 of ferric iron and has a *p*H of 2·3. It emerges from the precipitation cells with 0·25 g./l. of Cu and a *p*H of 2·9.

At Bingham, Utah, dump leaching, which produced 16,678 tons in 1962, is planned for expansion to 72,000 tons.[27] Several factors have led to this revival of an ancient technique. Leaching is largely self-sustaining, provided handling arrangements permit adequate exposure of the copper mineral to air and water. The role of bacteria in chemical attack is now understood and exploited. Huge tonnages can be moved by mechanised handling. Protective materials minimise corrosion of exposed installations. The process is cheap and requires but little manpower. For success the dump must not contain much acid-consuming limestone and must be underlain by an impervious stratum. Natural oxidation may contribute to the supply of sulphuric acid and the iron sulphate essential to leaching action. Any pyrite present must make contact with the leach liquors if it is to help the replenishment of acid and ferrous sulphate, but must not build up excessively re-circulating liquor or the retrieval of the copper from the pregnant solution will run into difficulties.

Treatment suited to dump leaching is also applicable to caved copper-bearing rock or ore in situ, as has been proved at Ray and Miami, in the American southwest mines. Here the ore is already broken or is pervious and the leaching liquor is sprayed over the deposit, whence it percolates through and drains to old workings. It is then pumped to surface and stripped of its copper. At Kennecott research has shown the conditions under which bacterial activity thrives include humidity, warmth and darkness. These bacilli oxidise pyrite to sulphuric acid and ferrous sulphate and convert ferrous to ferric iron. The acid takes up ferric sulphate which during attack on the sulphide copper is reduced to ferrous, thus forming food for bacterial sustenance. Leaching solutions are controlled to a *p*H of 2·3, at which hydrolysis of the copper is avoided.

Stripping of the pregnant liquor is by reaction with iron. Crumpled cans from which lacquer or tin has been removed are favoured, the reaction being

$$Fe + CuSO_4 = Cu + FeSO_4 \quad \text{This is followed by}$$
$$Fe + H_2SO_4 = H_2 + FeSO_4 \text{ and finally}$$
$$Fe + Fe_2(SO_4) = 3FeSO_4.$$

Scrap iron and sponge iron are also used. Stoicphiometrically, one unit of iron precipitates nearly 1·14 of copper but in practice the Fe/Cu ratio is usually nearer 3 to 1. The resulting low-grade cement copper assays from 50% to 90% Cu, and recovery is high.

Acid strength is usually below 7% and where necessary is maintained by the "autoxidation process".[28] In this ferrous sulphate catalyses the leach liquor in the presence of sulphur dioxide gas and oxygen and produces the needed strength of sulphuric acid and ferric sulphate. To do this barren solution from copper precipitation is diluted to some 5% strength of ferrous sulphate and reacted in aerating cells with gas produced by burning elemental sulphur.

At Bagdad, in Arizona, a leaching plant produces some 20 tons of cement copper daily from low-grade oxidic stockpiles and oxide ore overburden. Each of two stockpiles, in adjacent canyons, has its surface divided into 100′ square pools separated by embankments and delivering to a common overflow. Acidic water is sent from the precipitation plant to the leaching area and run on to one of two settling ponds. From the pond pregnant liquor containing 1 g./l. of copper passes to ten 2-compartment stripping cells in series. Each has a steel grid on which rests a load of de-tinned cans, the depth to this grid being 13′ and the cross-section 10′ by 9½′. Cement copper falls through and the ratio of iron to copper is 1·6 to 1.[29] N'Changa, on the copperbelt of Zambia, has the world's largest plant engaged in the electro-winning of copper. In the concentrator three products are made—a sulphide which is smelted, an oxide float containing from 11% to 14% acid-soluble copper and a low-grade oxide with about 2% copper. These are leached with spent electrolyte from the "tank house" where cathodes are made electro-chemically. The higher grade oxides are treated in agitators and the lower ones in pachucas. Pregnant liquor is filtered off and sent to the tank house.[30]

A useful recovery of copper from old mine workings, and from drainage areas in copper-bearing rocks, can be made by simple precipitation on scrap iron if the iron in the pregnant water has been reduced to the ferrous state. In one area the mine water is left in contact with pyrrhotite for this purpose,[31] the reaction being

$$Fe_7S_8 + 32\,H_2O + 31\,Fe_2(SO_4)_3 \longrightarrow 69\,FeSO_4 + 32\,H_2SO_4$$

and the effect that pyrrhotite replaces a substantial part of the scrap iron.

The leaching of atacamite, a copper oxychloride with the formula $CuCl_2$ $3Cu(OH)_2$ presents special problems. The method developed at Mantos Blancos, in Chile[32,33], uses percolation leaching with sulphuric acid followed by precipitation of the copper with sulphurous acid as cuprous chloride. The presence of chlorine precludes electro-precipitation and no local supplies of scrap iron are available. Precipitation is effected in spray towers through which gaseous SO_2 moves counter-current to the pregnant solution, reaction being:

$$(a)\quad 2\,CuCl_2 + SO_2 + 2\,H_2O \longrightarrow 2\,CuCl + H_2SO_4 + 2\,HCl$$
$$(b)\quad 2\,CuSO_4 + 2\,NaCl + SO_2 + 2\,H_2O \longrightarrow 2\,CuCl + 2\,H_2SO_4 + Na_2SO_4$$

At Bagdad Copper Cpn., a 5 ton per day pilot plant is working out a novel method. Sulphide concentrates from copper flotation are roasted by the Fluo-Solids method. The calcine is then leached and the copper is recovered by electro-deposition. Sulphuric acid generated during this work is thus available for heap leaching oxidic copper ore too poor to warrant normal treatment. Part of the expected saving in this process will be the cutting of transport costs by local production of 99·9% grade Cu instead of sulphide shipment to a smelter 425 miles distant.

Leach-precipitation-float (*L.P.F.*)

A typical flow-sheet for this process is that of the Hayden plant (Fig. 304).

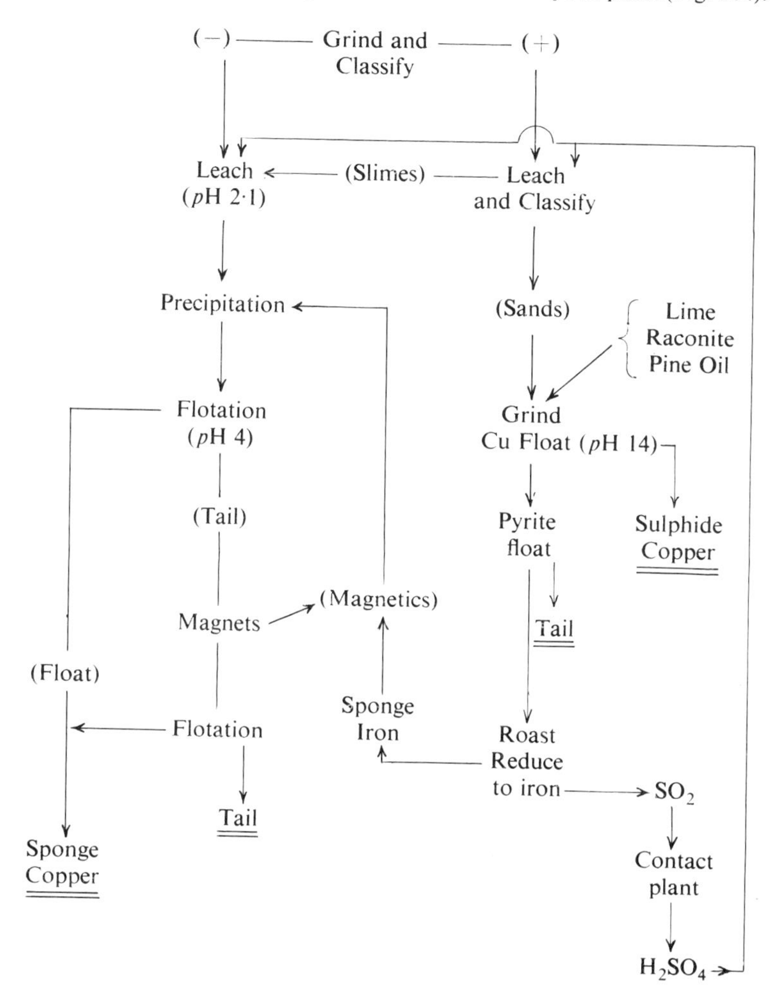

Fig. 304. *L.P.F. Process at Hayden*

In other variations of this scheme sulphuric acid is pumped over old dumps, ore piles, etc., *in situ*. The resulting pregnant solution is stripped of its copper, reconstituted and recirculated.

The use of sulphuric acid in the L.P.F. process has the further effect of cleaning partially oxidised sulphide surfaces during treatment with sulphuric acid, following which flotation with xanthate may be improved.

At Inspiration, Arizona, the old process called for leaching followed by electrolysis. With impoverishment of the mine at depth together with an increase in the proportion of sulphidic copper the old acid ferric sulphate process which took nine days has been dropped. Instead, leach contact with straight acid for four days now deals with the oxidic copper, leaving clean sulphides to be recovered by flotation. Ray (Kennecott) has developed a L.P.F. flow-sheet for recovering oxidic copper. The values include chrysocolla, cuprite, malachite, tenorite, and native copper. The carbonate, silicate, and sulphate are the main objects of attack. Copper is got into solution by standard leaching methods. The residue is reground, given froth flotation to recover the sulphide copper, and then refloated for its pyrite. This pyrite is calcined, producing SO_2, which after reaction to H_2SO_4 is used for leaching. The ash is reduced with coal to yield finely divided iron. This iron is circulated through the precipitation plant and is then removed by flotation as a copper-iron froth, any iron not caught at that stage being picked up, if needed, by magnetic separation of the tailing.

L.P.F. as a method of treating mixed sulphide-oxidic copper ores was developed from tests started in Arizona in 1929. "Oxidic" includes copper oxides, carbonates and silicate. The original full-scale operation (1934–1943) and a revived form (1957–1959) have been reviewed by Bean[34,35]. In the finalised process the sulphide minerals were first floated and the tailings, in which chrysocolla was the dominant copper value, were thickened and leached with sulphuric acid in air-lift agitators. Copper was then precipitated during passage of the "pregs" over de-tinned and shredded cans agitated in wooden drums, and floated. At Hayden it was found that there should be an excess of *minus* 35 mesh powdered iron in the secondary (acid) stage of flotation. Any unconsumed iron was removed by magnetic separation applied to the scavenger tailings. At Rosita, Nicaragua, highly refractory ores are treated by L.P.F.[36] The copper is precipitated on shredded iron, screened to pass 6 mesh and floated with Minerec A and Aerofloat 25. Three types of precipitate must be allowed for—a readily floated fluffy cement copper, a reluctant hard and granular cement, and copper-plated iron particles giving a low-grade but payable concentrate.

Flotation, sulphide copper

The copper-containing ores normally treated by flotation may be divided into two classes with respect to reagent response. The sulphides, with which are included native copper and such ores as selenides, tellurides, arsenides, and antimonides, respond readily to xanthate collectors, notably potassium ethyl xanthate. Selectivity is improved by care in grinding, by avoidance of undue oxidation through delay in treatment after severance, and by judicious use of polysulphides where such oxidation cannot be avoided. The main sulphides are chalcocite (Cu_2S), covellite (CuS), chalcopyrite ($CuFeS_2$), and bornite (Cu_2FeS_4). Minor sulphides include chalmosite ($CuFe_2S_3$) and cubanite ($CuFe_2S_4$). Broadly, the greater the copper-iron ratio the higher is the upper limit of pH and the lower the risk of depression through the use of lime, which is used not only to control pH but also to depress pyrite. This depression may be increased by pre-aeration to decrease the attraction of xanthate to the altered iron surface, and by the use of sodium cyanide in very small quantities. If copper is only one of the minerals to be floated, pre-aeration must be considered in relation to its possible effects on the later work. Copper, being the most readily floatable of the major sulphides barring molybdenite, is taken first in a differential treatment. The use of cyanide might be precluded in the case of an ore carrying appreciable gold which could best be recovered from a floated product.

At Noranda copper occurs mainly as chalcopyrite with associated pyrite-pyrrhotite, all carrying finely disseminated gold. The pyrrhotite is unstable and care is essential in the transport of the severed ore, since delay produces oxidation which has an adverse effect on subsequent treatment. Fine secondary grinding is necessary in order to free gold before rejecting pyrrhotite which, by reason of its instability, is a stong cyanicide. It is also necessary to float substantially all the copper before cyanidation of the auriferous pyrite. The flow-sheet (Fig. 305) of the heavily pyrrhotitic section of the

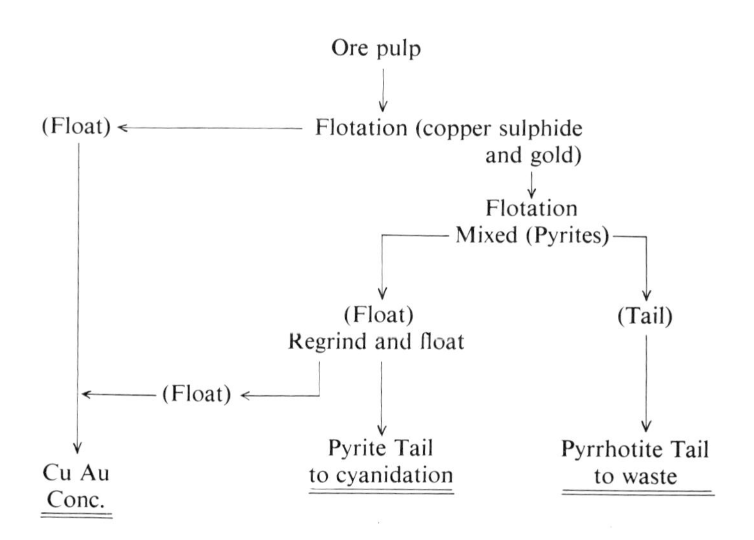

Fig. 305. *The Noranda Flow-sheet*

ore body shows in condensed form the method used. A fuller description is given by Ames.[38] At Adak (Sweden)[39] ores from two adjacent mines are treated, the principal minerals being copper sulphide and pyrrhotite. The pulp pH is under automatic control, lime being the alkali used. Potassium amyl xanthate is the collector. The flow-sheet (Fig. 306) is unusual in the number of re-cleaning stages it incorporates.

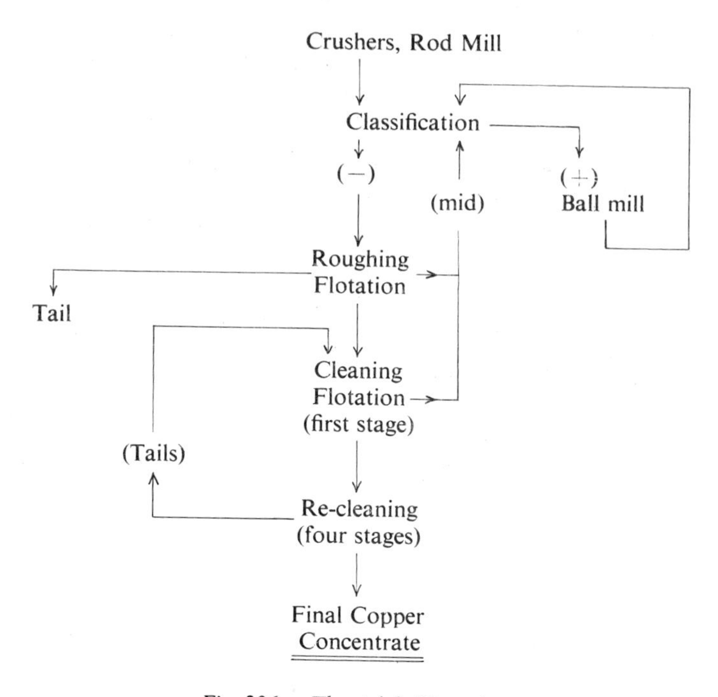

Fig. 306. *The Adak Flow-sheet*

The change from the pure sulphide Cu_2S which floats at *p*H up to 14, with increasing structural iron down to chalcopyrite (maximum tolerable *p*H 11.8), shows the coincidence of structural change and electro-chemical response. If it were necessary to float one copper-bearing sulphide away from others, this could be utilised. Normally, the purpose of the work is to float as much of the copper as possible for a specified grade of concentrate, and a suitable *p*H for this purpose is maintained. Pyrite is usually present in the ore, and a lime-controlled *p*H above 11 is used to aid its depression. Iron sulphide floats with increasing reluctance as the *p*H rises above 6·5 and lime has a depressing effect. When copper minerals are treated, there is some ionisation during grind and conditioning, and this results in a certain amount of activation of any non-copper sulphides present, as well as of selenides, tellurides, arsenides, or antimonides which may be in the ore. The soluble copper causing this can be complexed by the addition of a little sodium cyanide, but this may depress bornite and chalcopyrite which carry enough structural iron to be easily depressed by a conditioning treatment intended to keep down iron. Bornite is also prone to oxidise, and the pre-flotation processes (mining, transport, storage, and grinding) must deal with this. The reagent plan for the copper sulphides is dispersion of masking slimes (if needed) by the use of a specified sodium silicate (the trade terminology is loose as regards this reagent); *p*H control by calcium hydroxide; possible use of cyanide to de-activate copper-surfaced iron or other alien sulphides; and the most suitable xanthate (found by test). Usually potassium ethyl xanthate is preferred, in quantities varying from 0·01–lb./ton up to 0·1. The pulp is conditioned for *p*H and dispersion in the grinding circuit, and requires only a few minutes with the xanthate, at normal milling temperatures and from 20% to 30% solids. As with all floats, the higher the temperature (and therefore the lower the pulp viscosity) the shorter can be the final conditioning period. Also, the coarser the grind of the ore being floated, the higher should be its percentage of pulp solids in the first roughing stage. The frother is usually pine oil, added immediately ahead of the first cell. *p*H drop through the cell series is rarely serious enough to require attention, but occasionally a make-up addition of pine oil or xanthate in the scavenger section may be economically justified. Native copper floats strongly with a xanthate.

A special case is the reflotation of copper sulphide from a copper-nickel matte[40] as practised at Sudbury, N. Ontario. The *p*H is held between 10 and 12·4, a high OH^- concentration being needed to depress nickel sulphide. Xanthate may be used, but the preferred collector is diphenyl guanidine (0·5 lb./ton) or di-ortho tolyl guanidine, added during grinding. Pine oil may be added as frother, and lime is used to control alkalinity. From three to six rougher-cleaner stages are used with intermediate regrind and some 7%+ 325-mesh at completion of separation. The guanidine provides sufficient frothing action as a rule.

The chief non-sulphide ores to be floated are azurite ($2CuCO_3Cu(OH)_2$), cuprite (Cu_2O), and malachite ($CuCO_3Cu(OH)_2$). No satisfactory method of floating chrysocolla ($CuSiO_32 \,.\, H_2O$) has yet been developed. The formulae given should be accepted with reserve. X-ray and chemical analysis of malachite[41] proved that in the case of a typical ore no clear-cut formulation was justified. The general treatment plan for ores found in the Congo and Rhodesian copper belt commences with a xanthate float for sulphide copper. The tailings are conditioned with collectors based on fatty acids derived mainly from palm kernels and cottonseed oil. Sodium sulphide is added in the conditioners and along the flotation line. The prevailing mill temperature is subtropical, which aids the rate of reaction. Research[41] has shown that given close control of the hydroxyl ion balance against the carbonate ion, straight xanthate flotation is possible.

In the old N'Changa treatment, as described by Talbot,[42] an important consideration affected flow-sheet design. Coal must be brought over a congested railway system at considerable transport cost whereas electricity, now being developed from hydro-electric resources, was temporarily in short supply. Thus, both from cost and availability considerations, it was decided to leach as much concentrate as possible and to smelt only the residue. The flow-sheet was, therefore, developed as in Fig. 307. The main copper values in the ore are sulphides, native copper, azurite, and malachite.

At Kolwezi in the Belgian Congo copper and cobalt are bulk floated. The upper mining levels carry malachite, chrysocolla, heteroganite ($CoO \,.\, 2\,Co_2O_3 \,.\, 6H_2O$) and cobaltiferous wad, which are floated with hydrolysed palm oil, dispersion being aided by sodium silicate and carbonate. A mill-head of 5 to 6% copper and 0·2% cobalt yields a 26% copper concentrate with up to

85% recovery. In the deeper sections of the mine the ore is relatively unaltered. The main sulphide copper mineral is chalcocite, with some bornite and chalcopyrite in a silica-dolomite gangue and the cobalt occurs as carrolite. Simple xanthate collector gives over 80% recovery and high-grade concentration. Lime is used to give mild cobalt depression. Selective mining aids the mill, but the mixed ores of the transition zone must be helped by sulphidisation all along the flotation line, amyl xanthate as collector, and critical additions of gas oil and palm oil. The predominant oxide copper mineral is malachite. Associated cobalt oxides float feebly and their recovery from a 0·5% head is some 55%. Care must be taken to avoid entry of dolomite to the flotation circuits since as little as 0·2% can upset the work, calcium oleate soaps being formed. Further, this dolomite would float with the malachite and consume acid in the later leaching stages.

The open pit (oxide) ores from Musorori and Kamoto are floated at 40% + 65#. Tall oil aids frothing, and soft water at 20° C. pulp temperature is used. The collector is palm oil emulsified with hot solution of sodium carbonate in a colloid mill.

Ores mined within a 75-mile radius are treated in a central plant at Flin Flon in western Canada. Most of their gold and silver are recovered with the copper sulphides floated from the sphalerite at 55% *minus* 325 mesh.

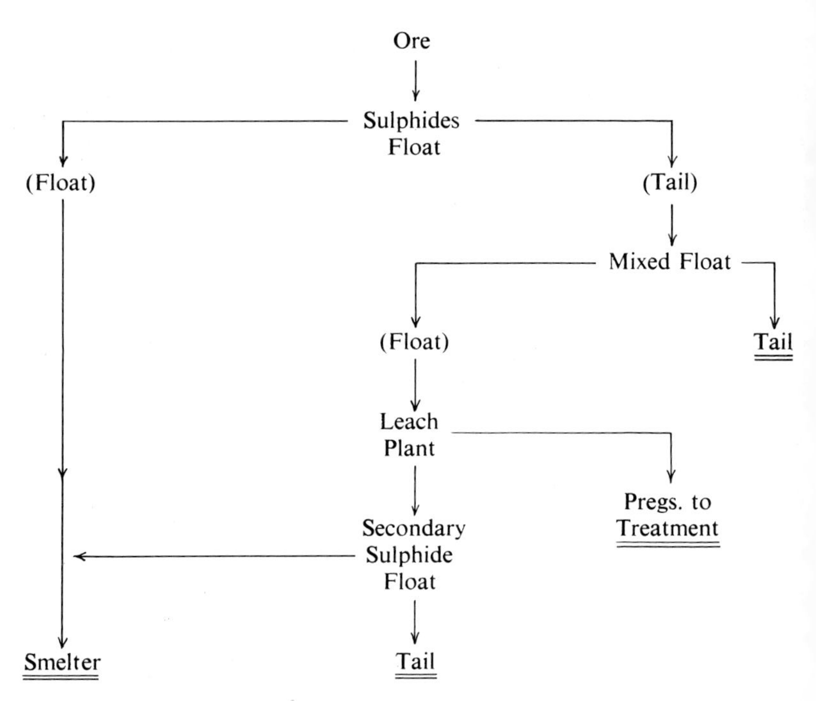

Fig. 307. *The N'Changa Flow-sheet*

In the subsequent zinc flotation live steam is used to keep the cleaning section of conditioning up to 30° C. Flotation tails are cyanided to recover the rest of the gold.[43]

Separate circuits are used for the sulphide and talcose ores (Fig. 308). Talc is roughed out, cleaned twice, and discarded, the tail from this float joining the feed to the secondary sulphide float. Copper is floated in two grind-and-float stages, the concentrates being combined and recleaned. Zinc is taken from the copper flotation tail, cyanidation being a final scavenging operation used to recover residual gold and silver after this operation. Special care is needed in the talc flotation to limit the loss of valuable sulphides. Some of the floatable talc mineral is allowed to escape to the sulphide circuit. These "insolubles" are depressed by adding sulphurous acid and dextrin at the head of the copper cleaner circuit.

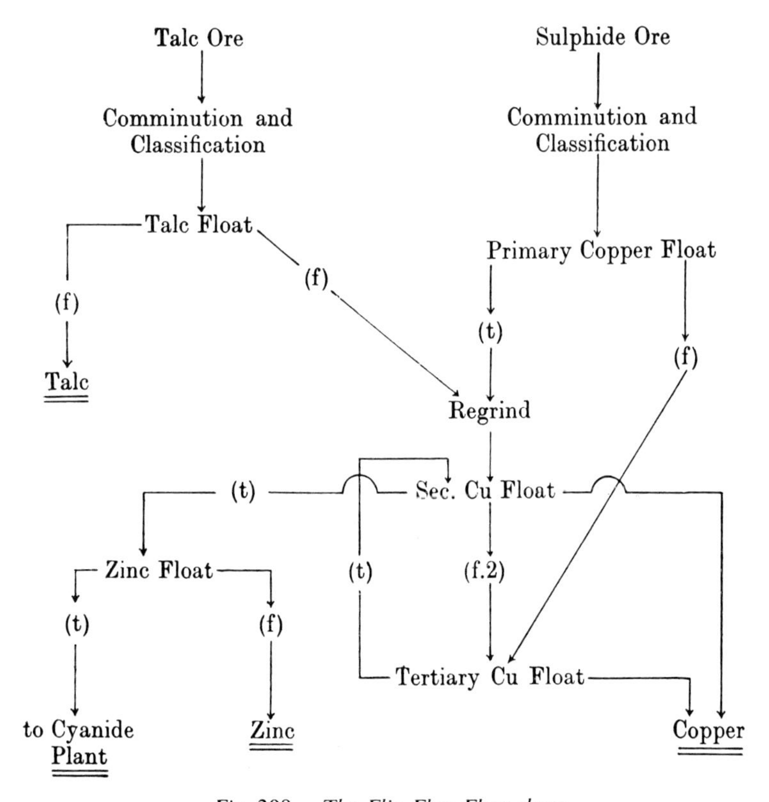

Fig. 308. *The Flin Flon Flow-sheet*

In the talc circuit zinc is depressed with 0·03 lb./ton of lime at the ball mill, together with 0·034 of zinc sulphate. The only flotation agent is pine oil (0·007). The copper reagents are lime (1·78), soda-ash (·55), zinc sulphate

(·63), Aerofloat 25 (·041), Minerec T-27 (0·076) in the primary and secondary stages. Sodium ethyl xanthate (·022), dextrins 615–620 (·41), and sulphur dioxide (S = 0·25) are used in the late stages. The zinc is floated with lime as iron depressant (2·57), copper sulphate (Cu = ·15), minerec 748 (·03), and Aerofloat 25 (·056) for roughing, with further copper (0·033) and ethyl xanthate (·029) in the scavenge-float.

A substantial amount of molybdenite is bulk-floated with sulphide copper, notably in America's western porphyrites. A typical concentrate assays 30% Cu and from 0·5 to 0·9% Mo.[44] Five main methods are used to separate these minerals from the float. Steam or dry-heating may be used to partially remove or destroy the collector agent, which adheres preferentially to molybdenite. Temperatures from 90° to 140° C. are used over periods from 1 to 4 hours (prior to re-flotation of the MoS_2 away from the copper using only a frothing agent). In a second method the molybdenite is depressed by organic colloids such as starch and glue. One plant applies this treatment to a float originally made with a dithiophosphate collector and re-floats the copper with dithiocarbonates after depression with dextrin. A third method uses sodium ferrocyanide in a weakly alkaline pulp to depress the copper, with fuel oil and an alcohol frother to lift the molybdenite. A fourth treatment depresses the copper with sodium hypochlorite in an alkaline pulp, in combination with the third of these methods. In a fifth method, copper minerals are depressed by use of phosphorus, arsenic or antimony salts. The products from these various methods of differentiation assay from 3% to 30% MoS_2 and to bring them to shipping grade repeated cleaning, re-grinding, roasting, leaching and the use of special reagents may be needed.

A summary of Russian practice in copper-molybdenum flotation has been made by Crabtree.[45]

The Lynn Lake deposits mined by Sherritt Gordon in North Manitoba contain copper as chalcopyrite, pyrrhotite, pentlandite and some cobalt, zinc and gold. Talc and actinolite are troublesome gangue minerals dealt with by organic depressants. Ore is ground to 50% *minus* 100 mesh and separated into magnetic and non-magnetic fractions. The former is re-ground to 90% *minus* 200 mesh, re-magnetised and floated to yield a 3% nickel concentrate. Non-magnetics are ground to 50% *minus* 200 mesh and the copper *plus* the rest of the nickel is bulk-floated. The concentrate is then acidified to *p*H 5·6 and re-floated in a 16-cell cleaning bank. Float from the first four cells, is alkalised with lime, and the nickel is depressed with cyanide, the resultant tailing joining the magnetic concentrate and bringing it up to an assay value between 10 and 12% Ni, while the float assays 31% Cu. The balance of the cleaner section is scavenged before discard.[46]

The Segregation Process

Details of a small commercial operation have been reported.[47] Crushed ore is calcined in a rotary kiln with sodium chloride and coke at between 740 and 810° C. The product contains metallic copper in a fine metallic form, and this floats readily with xanthate. The kilned ore is ground through 65 mesh, pulp alkalinity being controlled with lime. The final float assays up to 50% metallic copper.

Research and large-scale pilot tests show good possibilities for silicates and oxidic ores not readily amenable to straight leaching and/or flotation. Sutulov[48] estimates Chile's reserves of "difficult" mixed copper ores to be some 100 million tons assaying from $1\frac{1}{2}$% to 5% Cu. The general pattern of treatment starts with crushing to *minus* 10 mesh. The product is thoroughly mixed with common salt of *minus* 20 mesh size and *minus* 40 mesh coke. Up to 2% salt and 1% coke may be needed. Excess of these leads to embrittlement of the froth at the flotation stage and too much salt may react unfavourably during pre-flotation grinding even when the pulp is raised to a *p*H of 10·5 with lime.

Kilning is performed in a rotary calcining furnace indirectly heated, the main factor here being the furnace temperature. In tests in a 175-ton/day pilot operation on southwestern American oxide-sulphide copper ores the reaction temperature aimed at was 760° C., reached 22′ along a 48′ kiln 54″ in internal diameter and having a surface temperature of about 870° C. The copper is volatilised from its minerals as cuprous chloride and then reduced to metal flakes up to $\frac{3}{4}$″ by $\frac{1}{2}$″ which are deposited on the coke particles.[49] These flakes separate on agitation. The optimum temperature quoted in the Chilean tests is 750° C., volatilisation leading to loss beginning at about 800° C., but here pre-heating and slower calcining times were used. Ninety per cent recoveries were obtained save where there was calcium in the gangue when yield dropped below 86%. Specular hematite made its ores refractory, recovery falling below 70%. The collector agent found best with the Chilean ores was Z–200 with a cresylic acid frother. American work used amyl xanthate and methyl-isobutyl-carbinol at a *p*H of 11·5.

Water

Figures published by the U.S. Geological Survey show that about 100,000 gallons of water are used per ton of copper produced from U.S. domestic ores of which 70% is consumed in mining and milling. Of this 46% is saline with 1,000 p.p.m. of dissolved solids and 54% is fresh. In arid country considerable economy is practised by recirculation of mill water, coupled with precautions against excessive evaporation.

Diamond

The dominant facts in diamond concentration are that the concentration ratio is of the order of 20 millions to 1, that the stones must be recovered intact and undamaged and that, despite the security risks attendant on the handling of such values a high percentage recovery is required. Three main types of deposit are exploited—kimberlite ("blue ground"), marine terraces and alluvial gravels, the latter including offshore deposits. Kimberlite, the ore of volcanic pipes and fissures, has a blue-grey matrix which consists mostly of serpentine, with such subsidiary minerals as calcite, diopside, enstatite, ilmenite, phlogopite, pyrite, pyrope, and—rarely—a diamond.[50] Marine terraces and alluvial gravels need only screening before treatment, but kimberlite must be disintegrated. At Kimberley the stages of concentration today start with washing pans worked at a 32/1 ratio, followed by DMS at a 6/1 ratio. The sinking fraction is then re-concentrated on greasy surfaces at a 50,000/1 ratio and hand sorting at 2/1 completes the work. At Premier Mines on the Rand, grease vibrating tables take over from a 25/1 concentration by DMS and work at a ratio of 80,000/1.

Despite experiments with alternative methods, Kimberley retains pan washing for its clayey blue ground.[51] This clay acts as a natural dense media in a slurry blended to a density of 1.26 to 1.28. Mine ore at *minus* $1\frac{1}{4}''$ is fed to primary concentrating pans where 80% of the diamonds are held. The overflow from the inner concentric weir of the pan is screened through $\frac{3}{8}''$ and oversize is crushed to minus $\frac{3}{8}''$ and returned. Undersize is given a secondary panning, and thence goes to 6-mesh screens from which oversize is discarded. The underflow is cycloned to retain *minus* 6 *plus* 28 mesh and the cyclone overflow recirculates as "puddle" through the plant.

A washing pan has two concentric walls at 7′ and 3′ radii, the outer of which is 20″ high and the inner 12″. A sliding bottom door facilitates clean-up. A vertical shaft at the axis has ten stirring arms furnished with small triangular teeth which thrust the settled material outward against the inward flow of the slurry. A single circular tooth at the end of the arm moves settled material to a discharge port in the outer wall. Arms revolve at 8 r.p.m. The settling minerals are diamond, ilmenite, garnet and diopside. The rakes keep the bottom slurry in semi-suspension at a S.G. between 1·5 and 1·6 with a 50% slip past the stirring teeth. The concentrate taken at this stage should be 0·25% but for safety's sake a settled cut of 2% is removed. This is cleaned up by grease tabling after DMS. Since 1958 the *plus* 10 mesh fraction has been sent to DMS and the *minus* 10 *plus* 28 and *minus* 28 mesh sizes to separate grease table operations. The grease used is a refined petrolatum having four grades of penetration point, the one selected having the best hardness of arresting surface for the grain size of the diamondiferous feed. Diamonds caught on the grease are scraped off periodically by hand, tables being in pairs so that one is being fed while the other is dressed. About an equal quantity of adherent gangue is removed at the same time. The grease is melted away with boiling water and the diamonds are dried and sieved on 10-mesh. Undersize is wet-milled, de-slimed and again degreased in chromic acid, and finally concentrated either by flotation or electrostatic separation. From a daily input of 13,500 tons of kimberlite the final concentrate of diamond weighs 2,800 carats (1·25 lbs.). Until 1948 the grease tables used at Kimberley had a side-shake imparted by a vibrator but this was then superseded by high-frequency vibrations of the table deck.

Although diamonds from freshly mined kimberlite respond to grease tabling, those from alluvial and marine deposits do not. The surface is restored to its water-repelling state by gentle scrubbing of the DMS product and conditioned with a fatty acid if grease tabling is to follow. For electrostatic separation, one mine in S.W. Africa[52] sends the DMS concentrate to screens after drying at 130° C. Dust and oversize are removed and the *minus* 6 *plus* 1·9 mm. fraction is separated at 22,000 volts.

Checking of separating densities is performed either by use of a variant of the coal washery sink-float curve or by feeding radio-activated isotopes through the recovery plant and tracing their progress with Geiger-Muller counters. The S.G. of diamond is 3·5 but 90% of the associated mineral is 2·8.[53] Control of the viscosity of the separating media is important.

Fluorine

The commercially floated ore, fluorite, occurs in association with silica, calcite, barite and sometimes such sulphides as lead, zinc or iron. These last are floated or removed by gravity concentration before dealing with the fluorite. The market for acid-grade CaF_2 demands a high degree of purity, and the treatment usually calls for thorough recleaning to remove the last of the calcite and silica. The specification for acid-grade fluorite requires at least 97·5% CaF_2, and less than 1·5% SiO_2, with iron (as Fe_2O_3) below 0·5%. The conventional reagent plan controls *p*H by the use of soda-ash or sodium hydroxide. The depressants for gangue minerals, notably calcite, are quebracho (a tannin extract from the bark of a tree of that name), dextrin and sodium silicate. The collector agent is oleic acid or one of its modifications. In some plants a mildly acid pulp is preferred. The frother is pine oil. Hard water increases the difficulty of obtaining a freely breaking froth, without which it is not possible to depress the last of thc calcite. A further cause of trouble is the freezing range of oleic acid, varying with commercial reagents between 21° C. and 28° C. Unless the pulp is warmed well above this temperature, a "frozen" froth tends to form, and selecting action in the froth column is impaired. Reagents have been developed which modify the oleic structure without loss of collecting power. With these mildly sulphonated reagents, ordinary temperatures and untreated mill water can be used. The cool pulp also aids the specific depressing of calcite, and in a working plant control of the depressing quebracho reagent can be closely made by maintaining the frosty blue-white froth of high grade fluorite until the scavenging section, when a faint pinkness warns that excess is being approached. Potassium dichromate may be used to depress any sulphides not previously removed by sulphydril collectors. The all-important factor for high-grade concentration is a "lively" froth, in which the particles have not become stuck together by frozen oleic acid or calcium oleate. Sodium silicate is used as a dispersant in fluorite flotation.

Germanium

A method of recovery of renierite $(Cu, Ge, Fe, Zn, Ga)_4(S,As)_4$ from its flotation with copper sulphide has been developed at Kipushi.[54] It occurs finely disseminated in both copper and zinc sulphides in this deposit in a concentration of 220 p.p.m., which become 600 p.p.m. in the copper roughing float. This is passed through a magnetic filtering arrangement—the Ferro filter—on which a fraction assaying 0·919% Ge is held. The germanium is later abstracted by volatisation as fume during smelting in an electric furnace, the final product assaying from 4% to 9%.

Gold

General

In its ores gold is usually present as the metal, alloyed with metallic silver and perhaps copper. Its high specific gravity (19·3) causes particles of gold, even when of subsieve size, to settle readily from pulps in which the chief gangue mineral is silica. Gold is malleable, and a grinding treatment which breaks the gangue minerals may only flatten the metal without substantially reducing the size of its liberated particles. This differential grinding effect, by developing size differences, may assist gravity separation once the pulp is clear of the grinding section. Against this, the weight and malleability of gold particles causes them to be retained in a closed grinding circuit even after they have been adequately liberated.

In addition to "native" gold, the element may occur in forms of associations unfavourable to direct recovery by gravity methods. Metal sulphides such as pyrite, pyrrhotite, stibnite, arsenopyrite, and galena frequently contain inclusions of gold, perhaps as nodules only a few microns in diameter. Such sulphides are called "auriferous". It is not always practicable to grind these sulphides to the fineness required to liberate this finely disseminated gold, nor would it always be possible to trap it efficiently once it had been freed. The usual practice in such cases is to concentrate the gold-bearing sulphides at a relatively coarse m.o.g., regrind them, and then extract the gold by chemical attack.

A third class of gold ore has its values combined in the form of telluride or sulpho-telluride. These compounds are not malleable. They slime readily and their gold content is extracted by chemical methods.

The process selected depends, therefore, on whether the gold can be freed from its gangue at a sufficiently coarse mesh, or whether it is carried in a heavy sulphide which can be similarly freed. If the ore contains other valuable minerals, it may also be necessary to provide for their recovery. A notable example is the flow-sheet used in a number of mills on the Rand, where half the gold is recovered from strakes (a form of sluice) and the balance by the chemical process of cyanidation, after which the tailings are re-treated to recover uranium.

Another important factor must also be considered in commercial treatment. Whereas in separating base-metal values the concentrate forms a significant percentage of the ore, only a few pennyweights of gold per ton exist in the average run-of-mine ore. Thus a large quantity must be handled in order to extract a very small amount of extremely valuable concentrate. If all the gold is carried in a small percentage of the ore, as is the case with auriferous pyrite, considerable handling economies may be made possible by using froth-flotation to separate this pyrite from the barren gangue at an early stage. If part or all of the gold is disseminated through a siliceous gangue, all the ore must be treated.

Gravity separation of gold is practised on strakes, shaking tables, and in sluices and jigs. Amalgamation with mercury is used in connexion with this work. Froth-flotation can be employed to remove gold and sulphide minerals from a finely ground pulp.

The hydraulic trap can be used to catch a spigot product. It is a special form of hydraulic classifier, usually placed between the mill discharge and the mechanical classifier. It removes metallic particles of gold from the closed circuit. Jigs, unit flotation cells, and tables are similarly employed. Even when the whole of the mill feed is to be "cyanided"—a common term for chemical separation using a cyanide salt as the gold solvent—it is important to trap large particles of gold by gravity methods, since solution proceeds slowly and would be incomplete for such particles even after many hours of attack.

The cyanide process is discussed in Chapter 16 and the special problems connected with the use of froth-flotation for gold and auriferous sulphide are also dealt with later. The form which the flow-sheet takes is largely determined by the liberation-mesh of the gold and its association with the other minerals in the ore. The chief methods of treatment and their various combinations are summarised in Fig. 309.

These "mixed methods" do not exhaust the possibilities. The ore could, for example, be ground to say 90% −65 mesh, with gravity devices (1, 2, 3, or 4 alone or in combination) in the grinding circuit, and this could be followed by (5), the tailings then proceeding to (6) while the flotation concentrate passed to (8) and thence to (6).

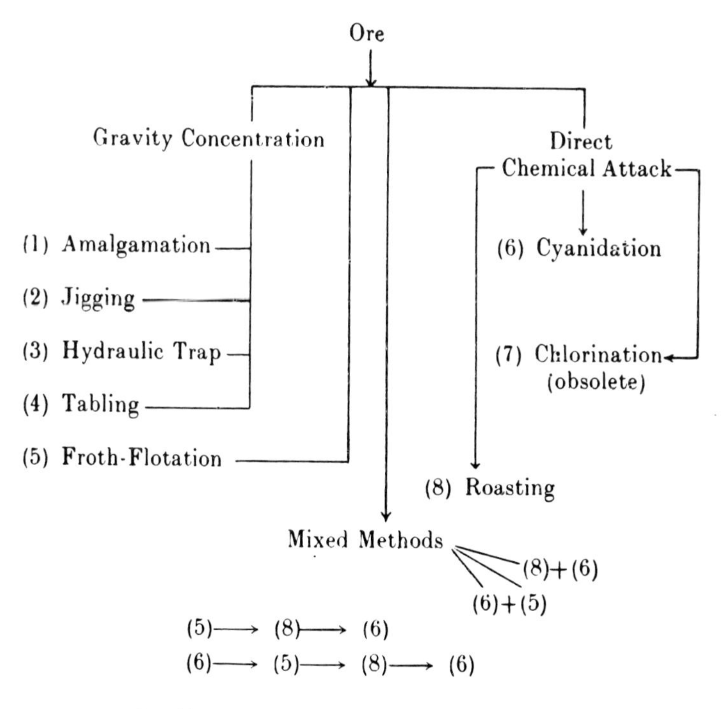

Fig. 309. *Methods used in Gold Concentration*

Modes of Occurrence

The process selected for separating any mineral from its associated gangue must take into account its condition after comminution. Gold, silver, and copper occur in nature in the free metallic state and also in chemical combination. The modes of occurrence of gold are shown in Fig. 310.

(1) Much of what follows applies both to silver and gold. Usually the native metal is a gold-silver alloy. Other elements and minerals may be present.

Free gold occurs in alluvial and eluvial deposits as the result of the action of weather upon the original ore, during long geological periods. Perhaps, as with the Alaska beaches, a mere trace of the precious metal existed in the original rocks, from which frost and tidal action have removed the lighter sands, leaving enriched patches or beds of pay-dirt. Under these circumstances only minerals which are heavy, tough, and chemically inert remain when the bulk of the gangue is weathered down and washed away. Concentration is therefore fairly simple, requiring only the use of gravity treatment to exploit differences between the specific gravity of the gold (19·0+) and the sands (2·7+). If the surface of the gold particle is bright and clean it can also be trapped by clean mercury, with which it combines physically to form an amalgam. Amalgamable gold is said to be "free milling". It is possible to make a gravity concentrate with no use of mercury, and to apply smelting methods direct to this concentrate in order to produce bullion. When the gold is not free milling, this is sometimes done. If the gold occurs in an auriferous sulphide (e.g. galena) and the latter is worth recovering, the gold may be produced as a smelted by-product.

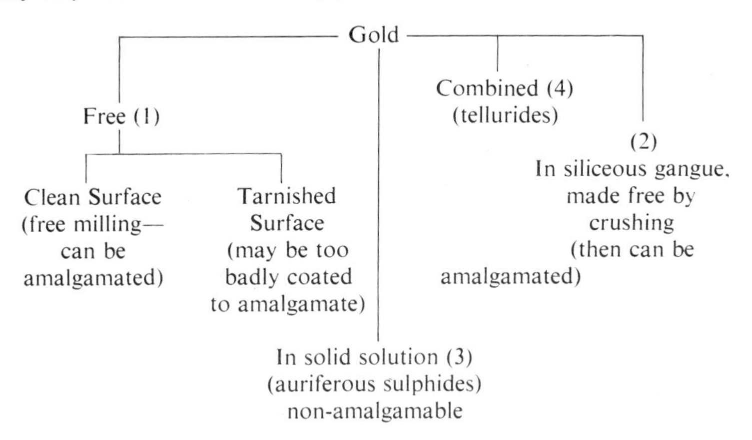

Fig. 310. *Modes of Occurrence of Gold*

During the geological ages of exposure to wind, rain, sea action and chemical attack from salts of associated minerals, the particle of gold may have acquired a patina or coating of some kind, such as oxidised iron. It retains the high density of clean gold but can only be amalgamated if the tarnish is thin and discontinuous, or if scouring treatment has removed this film. Such particles are sometimes called "refractory", though the term should be reserved for occurrences of types (3) and (4) in Fig. 310. These films probably originate with the solution of iron from sulphides in the original ore body from which the alluvium has been derived, and its redeposition as a hydrated iron oxide around the gold. Particles of alluvial gold occur in many forms, varying from large (and rare) nuggets down to specks only a few microns in size. The shapes include threads, wires, plates, foils, and spongy gold, and specific gravity varies between 12 and 19 according to the associated alloying elements.

Much of the world's lode gold occurs in a simple siliceous gangue, from which it must be partially or completely unlocked by crushing and wet grinding before separation. If at the end of the grinding process the gold is free fron gangue, clean and bright, it can readily be coated by mercury (amalgamated). Because of its simplicity and efficiency, amalgamation was the main process used for concentrating lode gold before the cyanide process came into use early in this century. If during comminution fragments of quartz are driven into the malleable gold, amalgamation is more difficult, and losses may occur. If the associated minerals reduce the chemical or physical cleanliness of the mercury, the amalgamation process may be unsuitable for direct application to run-of-mill ore. If the gold is so fine in grain that its complete liberation entails very fine grinding, amalgamation may again be an unsuitable method, partly because of the cost of such fine grinding and partly because very fine gold does not readily amalgamate. There are therefore several circumstances in which "straight" amalgamation is not the most efficient process.

The truly "refractory" or non-amalgamable ores are of two types. In (3) above (Fig. 310) the gold exists as metal which during the formation of the mineral deposit has precipitated from its solution in molten metal-sulphides, and is now disseminated through these sulphides as minute specks. Ordinary grinding of the ore might liberate these sulphides and might partially expose the gold locked inside them. More frequently it is necessary to separate auriferous sulphides from the siliceous gangue and subject these concentrates to intensive grinding before the contained gold can be exposed. Ores of type (4) are completely unamalgamable, as regards their tellurides and selenides. Direct chemical attack upon the gold is the only known method for its recovery.

Concentration before Amalgamation

It was stated above that gold could exist in the ore either in a "free-milling" or a refractory state. Both types can co-exist. It is, however, essential to ensure clean surface contact (clean gold *and* clean mercury) if good amalgamation is to be assured. Even if the gold can be superficially cleaned and scoured by grinding, any associated minerals liable to upset the purity of the mercury still remain. The selected treatment must avoid such contamination. Three approaches are possible:

(*a*) Direct amalgamation of all the ore.
(*b*) Amalgamation of a selected concentrate, after bringing deleterious material under control.
(*c*) Use of methods not involving amalgamation.

Method (*a*) can be used where a substantial amount of free-milling gold occurs and there are no risks of excessive loss of "floured" or "sickened" mercury. It has the merit of simplicity, since it is only necessary to provide contact between pulp and mercury, and to remove the resulting amalgam by simple separating appliances.

It is usually found best to take advantage of the high density of the gold and to concentrate it before amalgamation. This is helpful in several ways. First, it permits the production of a concentrate containing all the heavier gold-bearing minerals, including auriferous pyrite and partially unlocked gold, whether clean or "rusty". Second, it may enable the mill to discard an impoverished gangue from a coarsely ground pulp, thus making an important saving in grinding cost. Third, it removes a highly valuable product into safe control before it can accumulate in quiet crevices and joints in the plant. Fourth, the relatively small bulk of auriferous concentrate can easily be transferred to a room where it can be handled under conditions providing security. Fifth, this concentrate can be given an intensive grinding and surface scouring treatment which would be uneconomical if applied to the run-of-mill ore. Sixth, any contaminants liable to interfere with amalgamation can better be removed or chemically neutralised when such a small bulk is handled. Seventh, the presentation of the concentrate to mercury can be made under far more favourable conditions for successful amalgamation than when the same gold or silver is thinly disseminated through a great volume of gangue. Eighth and last, any gold still remaining in the concentrate after amalgamation is probably in a better condition for treating by the cyanide process.

Consider the hypothetical case of a gold ore containing 10 dwt./ton, about one half of which is free milling after adequate grinding. Tests have shown that cyanidation must be applied but that it is easy to remove the free gold by amalgamation. Two possibilities finally arise, (*a*) amalgamation of all the ground pulp, and (*b*) amalgamation of a concentrate. Tests now indicate that, when allowance has been made for loss of mercury, cost of plant, etc., the costs and results will be of this order.

Case (*a*)

Grinding 1000 tons at 2s. per ton	£100
Bulk amalgamation at 1s. per ton	50
	£150

Case (*b*)

Grinding 1000 tons at 2s. per ton	£100
Concentration to 0·1% at 6d. per ton	25
Grinding and amalgamation at 40s. of the ton of concentrate	2
	£127

Process (*b*) shows a definite saving. Suppose further that it is found that this intensive grinding of all the auriferous concentrate improves recovery, which with process (*a*) is 40% and with process (*b*) 50%. The real comparison is between 4000 dwt. recovered for £150 (= 9d./dwt.) and 5000 dwt. recovered for £127 (= 6d./dwt.). If further savings resulted by lowering the grade fed to the cyanide plant, the advantages in this case of preconcentration would be even more striking. Broadly, an operation of good magnitude and continuity will usually benefit by some measure of preconcentration.

Preconcentration Methods

Whether or no preconcentration is intentionally practised, it will occur in the grinding circuit. An appreciable amount of metal may be locked up temporarily in such places as the mill feed box, the liner joints, launders, and beds of mechanical classifiers. This represents at best a tie-up of values and at worst an opportunity for theft. Metallic gold should be removed from the grinding circuit as soon as possible after liberation. The chief preconcentrating devices used in the grinding circuit are:

(*a*) Jigs.
(*b*) Hydraulic traps.
(*c*) Shaking tables.
(*d*) Blanket, corduroy or rubber strakes.
(*e*) Mechanised strakes.
(*f*) Froth-flotation cells.

Two types of jig are favoured—the Pan-American, and the Denver. In each case the valuable concentrate can be delivered to a padlocked hutch. The jig receives the mill discharge, removes an auriferous concentrate and delivers the impoverished overflow to the next appliance in the flow-line.

Shaking tables can be used either to treat the whole of the mill discharge or to reconcentrate a rough concentrate produced by other methods, such as straking. In the latter case it is possible to improve the rough concentrate *en route* to the shaking table by giving it a further liberating grind or by removing abraded iron with an electro-magnet.

Froth flotation introduces special problems, which are dealt with in Chapter 17. It can remove auriferous sulphides as well as gold, even when they are too finely ground to be trapped efficiently by the straight use of gravity methods. Flotation concentrates are not always suitable for amalgamation. Among these there may be compounds which react with mercury in such a way as to reduce its selectivity for gold.

Strakes

The gold strake or corduroy table was described in Chapter 14. Strake concentrates are high in value, and strict precautions are taken in handling and transferring them. Spare cloths are not left lying around. Despite this, the open-strake system, beside being monotonous, unpleasant, and laborious to operate, is a security risk. There is the further objection to its use, that trapping of values from the flowing stream is only efficient during the early stages of deposition after clean up. As the sand beds down tightly, it becomes harder to trap the desired particles, channels begin to form, and concentrate is scoured away. Against these defects must be set the fact that a simple non-warping surface is easy to make, a distributing head can be constructed by a carpenter or formed in concrete, and that strake concentration needs no power, provided a few feet of fall are available. A further virtue of all straking systems is that they expose the sands to aeration and oxidation. This may be helpful in preparing for cyanidation, especially when the ore contains oxygen-avid sulphides such as pyrrhotite.

Several types of mechanised strakes are available. One consists of a continuous upward inclined belt of corduroy or other gold-catching material. The feed is distributed over its width. Heavy particles are entangled and in due course washed off after the belt has turned over the upper pulley, while the gangue washes down and discharges continuously over the tail end. A second type is the tilting tables, made in various forms. A third type of mechanised strake is the Johnson concentrator, which in addition to giving continuity of operation can be made thief-proof. It consists of a sloping steel cylinder 12′ long and 3′ in diameter, rotated seven times a minute at a 6° slope by a ½-h.p. motor. It is lined with ribbed rubber in which the concentrate collects, and from which it is washed by sprays on the rising side, to fall into a discharge launder set near the axis.

Another appliance is the "automatic" strake. In this an oblong plane supports a continuous band of corduroy cloth which is mounted so that its corrugations move parallel with the long axis, as it creeps along in response to a ratchet mechanism that actuates one of the rollers which move it round. The plane is tilted and the ore fed to the upper long side. Concentrate clings to the corduroy and is washed off after the cloth has left the supporting deck and is turning round the discharge-end roller. A continuous blanket table (Fig. 311) has also been developed.

Fig. 311. *Continuous Blanket Table* (G.E.C.)

In addition to the milling treatment of pulverised lode gold, a substantial quantity of concentrate is recovered from alluvial operations such as dredging. Here the problems are somewhat different, being mainly concerned with moving huge quantities of sand very cheaply, and during that movement to extract a very small amount of gold-rich heavy sand. Riffled sluices and jigs are the chief appliances used in this work.

Amalgamation Practice

The methods used to bring the liberated gold particles into contact with mercury are:

(*a*) Plate amalgamation, in which pulp flows over a viscous film of mercury anchored to a metal plate (surface contact).
(*b*) Presentation of pulp to a pool of mercury (immersion contact).
(*c*) Dispersion of globules of mercury through the pulp followed by collection of the amalgam (grinding contact).

Some operators consider that a little sodium amalgamated into the mercury improves recovery. The pool (*b*) has been made one terminal of an electric circuit. Pressure has been used to force the gold into the mercury.

From 1860 to 1925 much of the world's gold was concentrated by surface-contact methods. Plates were hung in the mortar boxes of stamp batteries, placed so as to receive the issuing pulp, and set in launders. Mortar plates were found unsatisfactory, as the splashing pulp scoured away both mercury and amalgam. The addition of mercury inside the mortar box was also discontinued owing to excessive flouring. Plates, where still used, may be of copper, silver-plated copper, or muntz metal. They are usually $\frac{1}{8}''$ thick, 4′–5′ wide, and 4′–12′ long, the size varying with tonnage and fineness of grind. If the liberated particles of gold are very small, the pulp must be dilute and must flow gently in order to ensure seizure by the mercury. The area of plate per ton per 24 hours varies between 0·2 sq. ft. and 3 sq. ft., chiefly in accordance with the thoroughness used to search the pulp. The plates are inclined between $1\frac{1}{2}''$ and 3″ per foot run, 2″ being common. They are mounted in such a way as to protect them from the jarring vibration of the stamps, which would cause the mercury to pack hard and to be less receptive to gold. A new plate may receive its first dressing with mercury into which a little silver has been amalgamated. When correctly prepared or "dressed" it is clean and bright, with its adhering film of mercury soft and not quite fluid enough to start "weeping" down-slope. All discoloration is removed from the plate metal, stains being rubbed with a solution of ammonium chloride, sodium hydroxide, or cyanide. Mercury is then sprinkled on to the oil-free, clean plate. A "bottle" made from a short length of iron pipe, plugged at one end and having a piece of cloth tied over the other, forms a good shaker. Riffles can be scribed across the soft finished surface with a whisk broom to aid arrest of gold. Pulp is allowed to flow at 10%–20% solids, just fast enough to ripple gently and to move evenly. As gold is trapped, the amalgam hardens and must be freshened and kept soft by the addition of more mercury, care being taken neither to allow hard spots to develop nor on the other hand an excessive fluidity. A mercury trap may be used at the tailings end of the plate to catch any detached amalgam (Fig. 312).

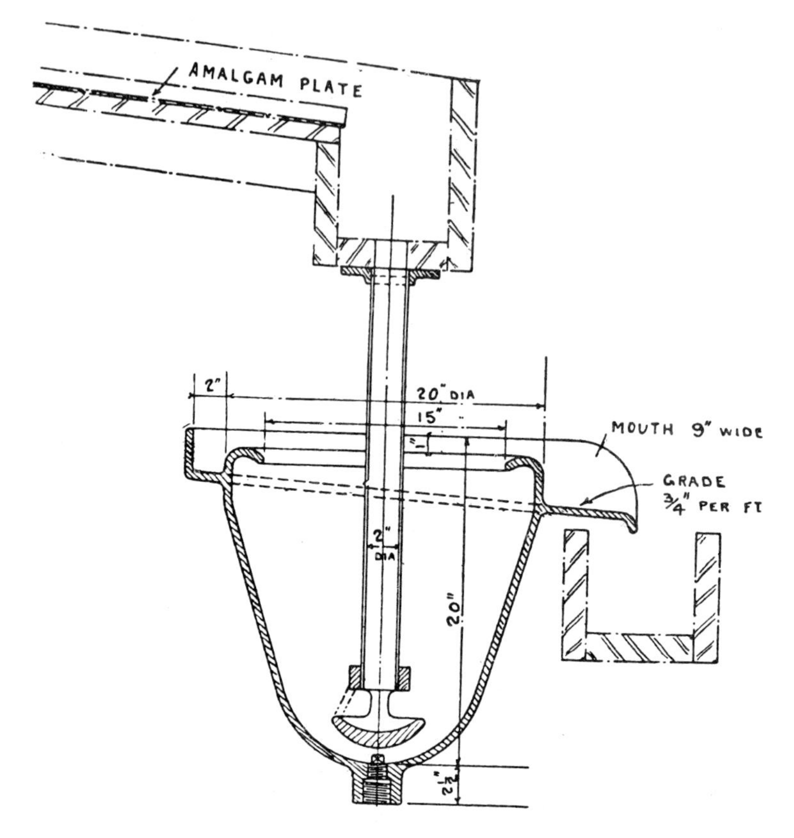

Fig. 312. *A Mercury Trap*

From time to time the plates are dressed. Pulp flow is stopped and the plates are washed clean. The amalgam is then worked upward, new mercury being added if necessary to soften it. A rubber or hardwood scraper is used to remove amalgam. Stains are dealt with, new mercury is put on, and work restarted. The time interval between dressings depends on the richness of the ore and on the rate at which the plates begin to show staining or "sickening" of the mercury.

Immersion contact, in which either an aqueous pulp or the dry ore is forced through a bath of mercury, has been developed in several ways, but its use has not become widespread. Gold is more readily wetted into a soft layer of amalgam than into a pool of mercury, and contact between the two metals is more thorough when the particles roll over plates than when a separating film of air or dust obstructs the union. An ingenious use of pressure amalgamation was at one time made on alluvial black sands.[55] A cast-iron bowl flaring in diameter from 17″ at the bottom to 23¾″ at the top rotated at 150 r.p.m., under which conditions a $\frac{1}{4}''$ thickness of mercury built up on its walls, forming a layer of $8\frac{1}{2}$ sq. ft., weighing 100 lb. The speed was adjusted so that even rusty and greasy gold stayed in the machine while the associated heavy black sands worked up and out.

The most common method of amalgamation in use today depends on grinding contact. It is applied intensively to concentrate produced by gravity methods. Its use in an amalgamation barrel is described in the next section. The older methods, in which mercury was ground together with the whole of the ore, sometimes led to serious working losses. These are largely avoidable when the concentrates are gathered into a small bulk and treated for removal or neutralisation of fouling substances before any mercury is added. The Patio Process has been described by Rickard.[56] Pan amalgamation in the clean-up pan (Fig. 313) is usually applied to a concentrate.

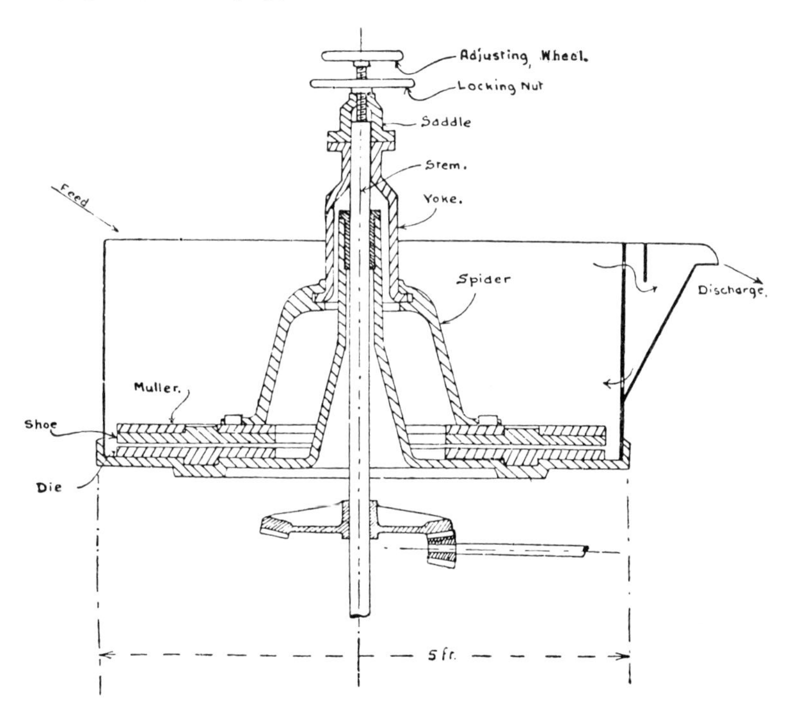

Fig. 313. *The Grinding Pan*

Arrastres were introduced into California from Mexico in 1850. They consisted of a low curtain wall of stone, surrounding a fairly level circular track paved with stone. Over this a beam attached to a central post was dragged by a mule. To it was attached a "drag stone" weighing some 150 lbs. This bore on *minus* $\frac{3}{4}''$ ore until it was reduced to a sand. A 10-ft. diameter arrastre held a quarter of a ton of ore and was ground for 4 hours.

Toward the end mercury was added and during the last few minutes water was flushed through. Further batches of ore were added for some days, after which the amalgam and any sand not displaced during flushing was retrieved and panned. The track was then repacked with clay between the stones and the work re-started.[57]

The Amalgamation Barrel

The "clean-up barrel" is a small batch-type ball mill (Fig. 314) and owes its wide use to the following advantages:

(*a*) It can grind the concentrate intensively, cheaply, and unsupervised.
(*b*) Any desliming or chemical treatment needed to improve amalgamation can be conveniently carried out as part of the preparation.
(*c*) Gold surfaces receive thorough cleaning treatment.
(*d*) Security working is simplified.
(*e*) Flouring of the mercury is avoided, and all values can be accounted for.
(*f*) The risk of mercury poisoning is minimised.

In operation, a charge of concentrate is weighed into the barrel, through an aperture in the cylindrical shell. The amount of water required to make a correct pulp density is added. Grinding media may be balls or old lengths of shafting. If the latter are used, a door in the end plate of the mill may be needed, so that the crushing load can be reduced after completing the grind. Lime or sodium hydroxide may be added to adjust the alkalinity, together with any chemicals required to react with trouble-making ore constituents during the grind. Grinding can conveniently be started during the afternoon shift, the mill being automatically stopped by means of a time switch when the desired period of grinding is complete. The morning shift removes most of the rods and if necessary gives the contents of the barrel a series of decantations, so as to remove as a slime any finely ground barren minerals liable to foul the mercury.

The mill is then set to run at half-speed. The mercury is introduced, the mill closed and started up. With the reduced rod-charge and the gentle rate of revolution, the risk of floured mercury is minimised. After the prescribed period of amalgamation, a vent in the shell is opened and the mill is restarted. At each turn pulp now leaves the mill and is guided into any suitable device for trapping the amalgam and letting the tailings overflow, either back to the grinding circuit or onward to the cyanide plant. The last of the pulp is washed out. In returning the tailings from the barrel to the flow-line care must be taken to adjust the sampling plan, or this material will be brought into account twice.

Practice, times, and barrel dimensions vary from mine to mine. At Dome (Ontario) a 3′ × 5′ barrel grinds the strake concentrate 16 hours with 150 lb. of mercury, 200 lb. of balls, 75 lb. of lime, and 4 to 5 tons of concentrate in a thick pulp. Another mine which had trouble through fouling of its mercury with soluble sulphides in the ore stopped it by using a small quantity of sodium plumbite. At Seal Harbour a ton of jig concentrate is charged daily to a 3′ × 4′ barrel, with 300 lb. of rods, 2 lb. of cyanide, 20 lb. of lime, and the lead carried in 30 used cupels. A 12-hour grind at 25 r.p.m. in a heavy pulp is followed by a short further grind with added mill cyanide solution. Slimes are then decanted off, and mercury *plus* ammonium chloride added for a 30-minute period of amalgamation. In another instance poor amalgamation of filmed particles of gold was rectified by grinding in dilute cyanide made alkaline with lime,[58] up to 3 lb. being consumed per ton of auriferous concentrate treated.

Unsatisfactory results from gold amalgamation may result from:

(*a*) Lack of suitable contact between gold and mercury.
(*b*) Gold in unsuitable condition,
 (i) too fine, or "floating" as flat buoyant particles,
 (ii) present as telluride,
 (iii) locked in sulphides,
 (iv) tarnished by surface films.

Fig. 314. *Denver Amalgamation Barrel*

(*c*) Gold lost as amalgam.
(*d*) Impure mercury.
(*e*) "Floured" mercury.
(*f*) Contaminants in the pulp, such as oil, grease, talc, sulphur.

Treating the Amalgam

From the plates, mercury traps, riffles, etc., a more or less fluid amalgam is obtained. This may carry sand, iron minerals, ore fragments and skimmings of fouled mercury. In a small plant this mixture can be purified by thinning it down with fresh mercury and agitating the mixture under hot wash-water, skimming off the rising impurities and giving them a special clean-up with pestle-and-mortar washing and grinding.

Whether the amalgam comes from these sources or from barrel amalgamation, it is next squeezed through chamois leather or canvas in an amalgam press. Nearly all the gold remains as a putty-like grey amalgam containing from 20% to 45% of bullion, though a little of the metal may be squeezed through. The mercury "filtrate" should be kept for clean-up work so as to avoid losing such gold.

The amalgam is now placed in a mercury retort and heated until the mercury has been distilled off. The residue consists of more or less clean bullion which is removed and melted with a flux of silica, soda-ash, and borax. These remove the remaining impurities as a slag. This, with discarded crucibles, furnace linings and other auriferous material is periodically ground and retreated.

Handling Mercury

The quantities used, and the intermittent way in which they are handled, make it unlikely that the workers in an amalgamation plant will be exposed to serious health hazards. The subject is covered in various official publications which deal with the large-scale use of mercury.[59] Chronic poisoning develops by insidious stages. Poisoning can result from exposure to mercury vapour in the atmosphere or from mercury spillage finding its way into the worker's system. Mercury should be stored in tight flasks and not left where food is eaten or stored. If spilt, it should be covered with water and immediately picked up. It should never be removed by blowing with pressure air since this might cause small globules to become airborne. Workers should wash their hands thoroughly after handling any mercury-containing materials. They should rinse their mouths before eating or drinking, and should wear special clothing in the work sheds. This clothing should be washed weekly.

Process Control

Process control in the ore-treatment plant has two aspects, financial and technical. In the case of gold both are complicated by the high value of the concentrate and the heavy density of gold, which hangs up in any quiet, rough place in the flow-line. Ideally, a known tonnage of ore would be sampled and assayed with good accuracy, and the various products would be similarly checked in order to produce a metallurgical balance sheet. This might be depicted as in Fig. 315.

Feed (1)

Straking (2)
Conc. (3) ———————— Amalgamation ———————— Tails (4)
etc.

Fig. 315. *Data for Metallurgical Balance*

Suppose the feed (1) to be 1000 tons assaying 5 dwt., then concentrate (3) might be 1 ton assaying 2000 dwt. The equation for metallurgical balance ($Ff = Cc + Tt$) would read:

1000 tons at 5 dwt.	=	1 ton at 2000 dwt.	+	999 tons at 3 dwt.
= 5000 units		= 2000 units		= 3000 units
(1)		(3)		(4)

In the rough and tumble of milling it is not possible to obtain a neat equation of this kind. Consider a developed flow-sheet (Fig. 314) for a plant using two stages of wet grinding, each with gravity extraction of its "metallics", rusty gold, and auriferous pyrite. Items (3) and (8) may consist of jigs, flotation cells, shaking tables, strakes, etc. The concentrates they produce are closely guarded and handled under conditions which provide a high degree of security. In order to bring this flow-sheet into metallurgical balance we must be able to equate the total pennyweights (or other units) of gold at certain key points, thus,

$$\text{Dwts. (1)} = \text{Dwts. (6(}-\text{))} + \text{Dwts. (15)}$$

The difficulties in obtaining a balanced equation of this sort are almost insuperable. Since accurate knowledge is the foundation of good control, this unfortunate fact is a challenge to ingenuity, not an inevitable disability. Although perfection in this important matter is not yet possible, a great deal can be done to obtain valuable information at a reasonable cost. If the flow-sheet is now considered in detail a good idea of some of the practical difficulties confronting the mill superintendent emerges, and the flesh of day-by-day operation begins to clothe the skeleton of basic principles thus far discussed. At the same time the matters requiring technical control are noted.

New Feed.—For accuracy, the following information is needed:

(*a*) Tonnage (dry).
(*b*) Assay value.
(*c*) Size analysis.

(*a*) and (*b*) show the total value delivered by the mine, for which the mill accepts responsibility. (*a*) can be known with fair precision, if the feeding arrangements into the scoop box (2) are properly calibrated and the moisture of the ore is checked. Alternatively, the tonnage drawn daily can be arrived at by checking the weightometer reading at a fixed hour, and compensating for tonnage rise or fall in the ore bin. With item (*b*) the difficulty begins. Gold (S.G. $< 19{\cdot}3$) is distributed through gangue (S.G. $> 2{\cdot}65$). The gold is

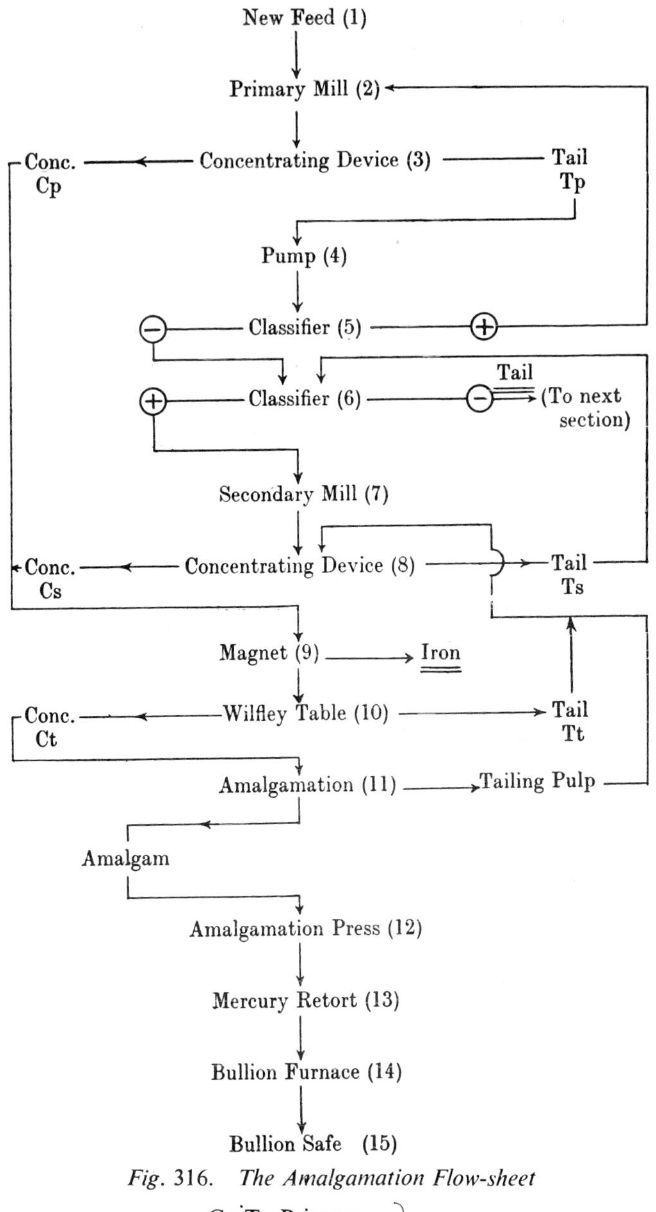

Fig. 316. *The Amalgamation Flow-sheet*

Cp Tp Primary
Cs Ts Secondary } Concentrate or Tail
Ct Tt Table

finely divided and locked into pieces of ore varying in size from, say, $\frac{3}{4}''$ down to a few microns. If the ore is "spotty"—i.e. if it has its gold irregularly distributed (like a very few plums in a very large cake)—a big sample must be taken in order to achieve some approximation of accuracy. If a 10% sample were used on a 1000 tons/day feed, its reduction to a bulk suitable for the assay office would be too costly. Yet, as mathematical analysis shows, only a large sample, ground fairly fine, can be made to yield a representative assay sample of a few pounds weight. Some psychological resistances to accuracy complicate an already formidable technical problem. The mill is judged by the percentage recovery, and that judgment ultimately resides with a partially informed and distant Board of Directors. It is tempting to give the mill the benefit of every doubt, real or imaginary. The stope assay plans, since they suffer from most of the disabilities which affect the mill head sample, are sometimes adjusted by a "milling factor". Again, moisture figures might be too low after a head sample had dried between collection and test. A wrong assay return can arise from difficulty in reducing the size of the original sample. However much care is taken, the head value as calculated from sample is unusually open to error in the treatment of gold ores. Better reliance can be placed on the figures obtained toward the end of the flow-line, and the head assay usually provides only a somewhat unreliable secondary check. As technical control improves in the mill, this state of affairs will gradually disappear. Its existence will be indignantly denied by some metallurgists, but the author has been at pains to verify the facts. In the case of finely disseminated and evenly distributed gold values such as are characteristic on the Rand, the head sample is closely representative of the grade. It then forms a reliable integer in the equation of metallurgical balance. Here, as in some other respects, the Rand "bankets" provide a model of milling docility rarely met with in the ordinary lodes.

Primary Mill (2).—This provides numerous traps in which metallic gold can lodge. Control is based on the concept of continuous flow accompanied by continuous separation into a poorer and a richer fraction, thus:

$$\text{Entering Pulp} = \text{Richer Product} + \text{Poorer Product},$$

all in terms of rate of flow and pulp consistencies. Should any constituent of the ore become static, the rate of flow ceases to be representative. If, for example, a serious amount of gold hung back in the mill and was released during occasional surges, any pulp sample would be unrepresentative. If cut during such a surge it would be too rich, and if between surges too poor. Over a long term the control data would be reasonably accurate, provided the trapped gold were all recovered and brought safely to the bullion safe. Since the feed box to the mill and the joints between the liners provide lodging points at which gold is retained, safety precautions must be taken to prevent theft when the mill is stopped, or is being relined.

Concentrating Device (3) and (8).—This is designed to be appropriate to the technical condition of the desired product. If the gold is sufficiently coarse, in association with a +60 mesh gangue, Denver jigs or hydraulic traps provide a compact and thief-proof method for trapping out liberated gold. If below 60 mesh, shaking tables or flotation cells can be used, the latter being especially useful for picking up auriferous sulphides. At this point a comparatively small weight of highly enriched concentrate (Cp and Cs) is removed. In theory, the balance now becomes

$$\text{Dwts. (1)} = \text{Cp Dwts.} + \text{Cs Dwts.} + \text{gold in (6) Tail,}$$

since all the values originally fed into the mill should by point (8) have reported to one or another of the checking points on the right-hand side of the equation.

Here again, error enters because of the hold-up of gold.

In the pump (4) and the classifier beds (5 and 6) as well as in the secondary mill (7) a certain amount of bullion is temporarily lost at the best of times. The factual equation therefore becomes:

$$\text{Dwts. (1) (figure not very reliable)} =$$
$$\text{Dwts. [Cp} + \text{Cs} + \text{Au} \quad 6(-)] + \text{gold held in circuit.}$$

The item Au 6 (−) refers to the tailing from the secondary classifier, which usually leaves the gravity-cum-grinding section and now becomes the head feed to the cyanide section of the plant.

Cp, Cs, and Au (−) can be checked closely by sampling, and Cp + Cs are usually removed from the main plant at this stage and given the rest of their treatment under conditions involving strict surveillance.

Magnet (9).—The gravity devices will probably concentrate part of the iron abraded from the mill machinery and this, if magnetic, can conveniently be removed between the concentrates bin and (10). If non-magnetic iron abounds, as happens when special alloys are used, it must be removed as a special iron concentrate on (10). If the magnetic iron carries gold, the amount removed at this point must be ascertained, accounted for, and if economically significant, recovered.

Wilfley Table (10).—This, or any other concentrator used at this point, removes a certain amount of low-value gangue, some iron and locked gold middlings, and an improved gold concentrate. The table is worked intermittently, batches of concentrate being accumulated from the main plant for treatment during the day shift by specially selected employees. The middling may either be returned to (8) or sent with the concentrate to (11) according to its nature. The tail from (10) is either returned to (8) as shown, or is passed out with Ts to the cyanide section. This applies also to the tail from (11).

Both these tailings carry an appreciable amount of gold. If this were returned to (8) without a preliminary evaluation, the metallurgical balance of the whole operation would be falsified since these tailings would be valued twice over. The amount of gold thus returned must be found by assay and knowledge of the weight of tailing, and must be subtracted from the head value entering to (9) or (10) and added to the new feed (1). The tailing at (10) can be found by difference:

$$\text{Tail Tt} = (\text{Cp} + \text{C3}) - (\text{Magnetics} + \text{Ct}),$$

but the tailing from (11) can only be found by physical methods.

Items 12 *to* 15.—No gold loss, other than that in slags or broken crucibles, should occur in this section. The metallurgical balance can be struck with good precision:

$$\text{Bullion}\,(15) = \text{Ct} - \text{Tails}\,(11).$$

The above general picture shows the strength and weakness of control on this very vulnerable part of the gold mill. Good management, maximum mechanisation, and the employment of a few well-paid shiftsmen rather than a large number of "cheap" labourers are the best safeguard against loss, provided a good basic flow-sheet is being operated according to a strict technical routine. Samples at key points show the assay value, degree of liberation, and solid-liquid ratio.

In many plants the gold is milled not in water but in cyanide solution. In this case the problem of economic and technical control in the grinding section becomes far more complicated, since gold-bearing solutions are withdrawn.

Flotation

Where possible, gold is removed by gravity, amalgamation and/or simple cyanidation. The only flotation in such cases might be that of associated graphite which otherwise could upset amalgamation or reprecipitate values from the pregnant solution, with ensuing loss. Such graphite is also dealt with by ignition (Ashanti) and by surface closure and passivation, using oil. One or two mills use activated charcoal to precipitate the gold contemporaneously with its dissolution by the cyanide, and this auriferous carbon can be recovered by flotation.

The high density of gold (19·2) makes it hard to hold in a flotation froth, though this is by no means impossible. A further difficulty is that even an extremely rich gold ore contains less than 0·0002% of metal, which is quite inadequate for stabilising a froth. For gold occurring in a siliceous gangue flotation would therefore be unsuitable. If, however, part or all of that gold is locked in an auriferous sulphide, sulpharsenide, telluride, etc., a very different picture is presented. Such gold frequently exists as inclusions only a few microns in size, and could not justify the cost of the extremely fine grinding which would be needed to expose it. If all the gold-bearing sulphide can be floated, the exposure of gold in a regrinding treatment is relatively cheap. Alternatively, the float can be sweet-roasted (by the FluoSolids process or in roasting furnaces), the residue being cyanided.

In gold flotation the end-product is so valuable that no risk of loss is taken. Powerful collectors are used, so that every particle tending to float is captured. The bubbles may be stiffened by the use of mineral oil to reinforce the frothing agent, if the natural armouring of mineral particles is insufficient. Mechanical cells are used, and the bubbles are mostly small, providing a stiff mat through which a floated gold particle finds it hard to slip back to the pulp. As a result of so severe a float, the cyanicides which might otherwise run through to the cyanide plant are lifted in the float, which can contain a complex mixture of sulphides. If this is now ground intensively to expose the gold in the auriferous particles, it becomes oxygen-avid, slimy and difficult to thicken or filter. Hence, the cyanidation of a flotation concentrate is more difficult and costly than is that of a straight pulp. Far more intense aeration is required, and the addition of protective alkali must be closely watched, as the sulphides will not only denude the pulp of its oxygen but will release sulphuric acid as a result. Such a pulp can be pre-aerated in water to reduce these troubles, after which it may be run to join the flotation tailings proceeding to cyanidation (Fig. 317).

During pre-aeration ferrous salts in the water are largely oxidised to the ferric state. If an appreciable amount of dissolved (ferrous) iron remains it may be necessary to discard the water used. If sulphides oxidise to sulphites and sulphates and are left in the pulp, they react with cyanide to form thiocyanate. Unoxidised iron may also produce ferrocyanide, again wasting reagent. Soluble sulphides can be precipitated with litharge or lead acetate. The pulp should be aerated at a cyanide strength of between 0·02 % and 0·05 %.

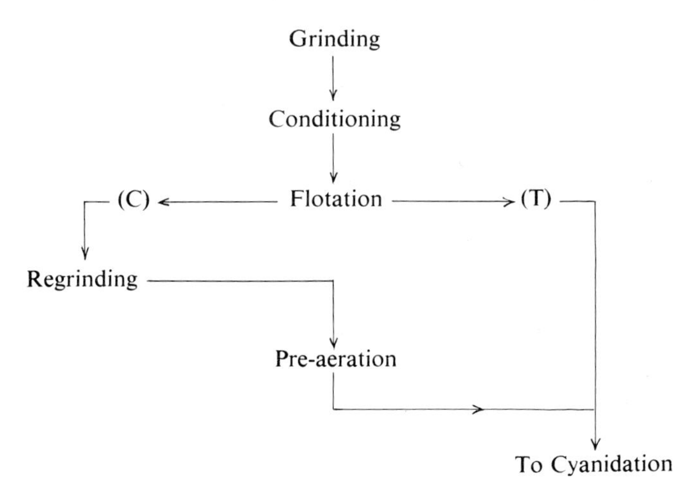

Fig. 317. *Flotation of Auriferous Sulphides*

In a small plant the flotation concentrate might be treated in batches, the solids in the pregnant pulp being settled, decanted, and appropriately rewashed before being run to waste, thus avoiding difficulty in the thickening and filtering of such finely ground material in the main flow-line.

Among the cyanicides are stibnite, realgar, orpiment, and arsenopyrite, which are least maleficent if the pulp is cyanided at a nearly neutral *p*H. Copper minerals will be dissolved, but if the reprecipitation with zinc is staged, a starvation addition will bring down the gold as a rich slime, after which copper can be precipitated with a further addition.

Alternatively, the flotation concentrate may be cyanided without pretreatment to reduce the effects of cyanicides. It must be remembered that a highly enriched fraction of the original ore is being handled and that the tailings assay will be correspondingly high. Where the gold is carried in a valuable metal sulphide, such as that of lead or copper, the concentrate may be shipped to a smelter, but an auriferous pyrite must usually be handled in the plant.

The three broad classes of gold-bearing ore are:

(*a*) Those in which the gold is substantially free, any sulphides having been oxidised away.

(*b*) Those containing part of the gold in a free state and part in pyrite (the commonly occurring type).

(*c*) Those in which the gold is subsidiary, or supplementary, to an economic base metal such as lead, zinc or copper.

Lime, being a depressant for auriferous pyrite as well as for gold, must not be used to modify *p*H, which is controlled as needed by soda-ash. Pine oil-frother can be stiffened with a hardwood creosote or aerofloat 15. If auriferous arsenopyrite is present, it usually responds to activation by copper sulphate (0·5–1.0 lb./ton).

The reagent plan for flotation of gold (and part of what follows applies also to associated silver) depends on which of three broad types of ore is treated. In the first the gold is virtually freed from its gangue by comminution. This type merges into the second, in which part (or even most) of the gold is intimately bound with iron sulphides or occurs as a telluride. In the third type the gold is associated with other metal sulphides of commercial value and is recovered by pyrometallurgical methods from their flotation concentrates.

For ores of the first two types non-selective flotation is used, with hard pulling of a stiff froth. The broad aim is to float all possible auriferous material, even if the resulting concentrate is low in grade. Scavenging is used, but recleaning of any froth (including that from the scavenger cells) is rare. The reasons for this variation from standard practice are *first* that the high density and relative inertness of free gold make it unwise to risk reflotation once it has been successfully lifted; *second* that even a froth assaying about 100 dwt./ton represents a steep upgrading of a typical ore; *third* the small content of gold makes it impossible to overload the froth column with the desired mineral; *fourth* the associated sulphides may have value both as fuel in a roasting treatment of the flotation concentrate and as a source of sulphuric acid in any subsequent chemical treatment (e.g. uranium extraction from cyanide residues). An exception to these considerations would exist in the case of such a combination of ore minerals as auriferous arsenopyrite and non-auriferous stibnite. Here, any reagent plan which would depress the antimony would improve cyanide treatment of the auriferous concentrate. Similarly, treatment to remove cyanide-soluble copper minerals from a bulk-float might be justified, if it reduced cyanide consumption and solution fouling.

Gold, like pyrite, is depressed by lime. If the sulphide minerals are to be activated by copper sulphate in an alkaline circuit (common practice for pyrite and arsenopyrite), sodium carbonate should therefore be used. If flotation is practised on cyanide tailings, the depressing effects of the residual cyanide salts can be countered by a conditioning stage with sulphur dioxide gas, drawn from the roasting plant commonly used to treat the sulphides concentrated in this flotation.

Sulphur dioxide also surface-cleans the particles and sulphatises the hydrated iron in the pulp.

Among the reagents used for the first two types of ore are Aerofloat 208, with a cresylic frother such as Aerofloat 15 or 25. Xanthates (butyl or amyl) are also employed, either alone or with the Aerofloats. Where a little sulphidisation may help, Cyanamid 404 with Aerofloat 242 may be used.[1] Froth may be stiffened by the addition of lubricating oil in the grinding circuit, but a more controllable procedure is to use cresyl ahead of flotation, perhaps with pine oil or one of the proprietary frothers based on the higher alcohols. Pulps are kept thick (30% + solids).

The reagent plan for the third type of ore must conform to the requirement of the associated metal sulphide being floated. If free gold co-exists, the *p*H control should be made with soda-ash, not with lime. The periodical technical literature is rich in descriptions of gold-milling practice.

Working Precautions

The cyanidation of gold (and silver) was considered at the basic level in Chapter 16. Selected flow-sheets in which the process is used, alone or in combination with other methods, are given below.

Mercury, cyanide and prussic acid gas are deadly poisons if absorbed into the system in more than minute quantities. Mercury is most dangerous when present in the atmosphere as the result of leakage during the distillation of amalgam. Good ventilation, meticulous care in sealing the retort, hooding of furnaces, extraction of fume from danger points, and care in handling liquid mercury are the best safeguards. Men working with mercury should wash and change their clothing before eating, and food should not be allowed in the working places. This applies also to cigarettes or any other thing which might serve as a vehicle for the intake of condensed poisonous fume. In addition, periodical medical examination is desirable.

Soluble lead salts—nitrate and acetate—are normally used in the precipitation plant. These are both quick and cumulative poisons. They can be picked up on the skin or from clothing, if lead solutions are handled carelessly. They are unlikely to be inhaled, however, as the lead is not atomised during its use. Cleanliness and periodic medical examination are desirable, as with mercury.

Cyanide can be dangerous in several ways. "Cyanide rash" affects some people seriously, and such individuals should not be employed where cyanide is used. For normal workers the chief danger with the very dilute solutions used is chafing of the flesh by wet clothing. Where this carelessness has caused skin irritation, the remedy is a 1% solution of lime chloride and boracic acid, allowed to dry on the affected part. If accidental contact with cyanide has been made, the affected part should immediately be washed and moistened with either a $2\frac{1}{2}$% solution of Epsom salts or of $2\frac{1}{2}$% acetic acid.

Prussic acid gas is deadly in its effects, and takes less than a minute to kill, if breathed in any quantity. Here, even more emphatically than with the other poison-hazards in the gold mill, prevention is better than cure. Caution in working round piping which may have accumulated a pocket of gas, liberal ventilation, and extraction of fume are important. Swift remedial action is vital if gas has been inhaled. The symptoms are throat irritation, breathing trouble, watering eyes, heaviness, headache, dizziness, an irregular pulse, pallor, and lastly unconsciousness leading to death. These develop in this succession in a matter of minutes.

If breathing has stopped, artificial respiration should be started and continued till the doctor's arrival. Oxygen may be administered, and a capsule of amyl nitrite may be crushed and held close to the patient's nose while

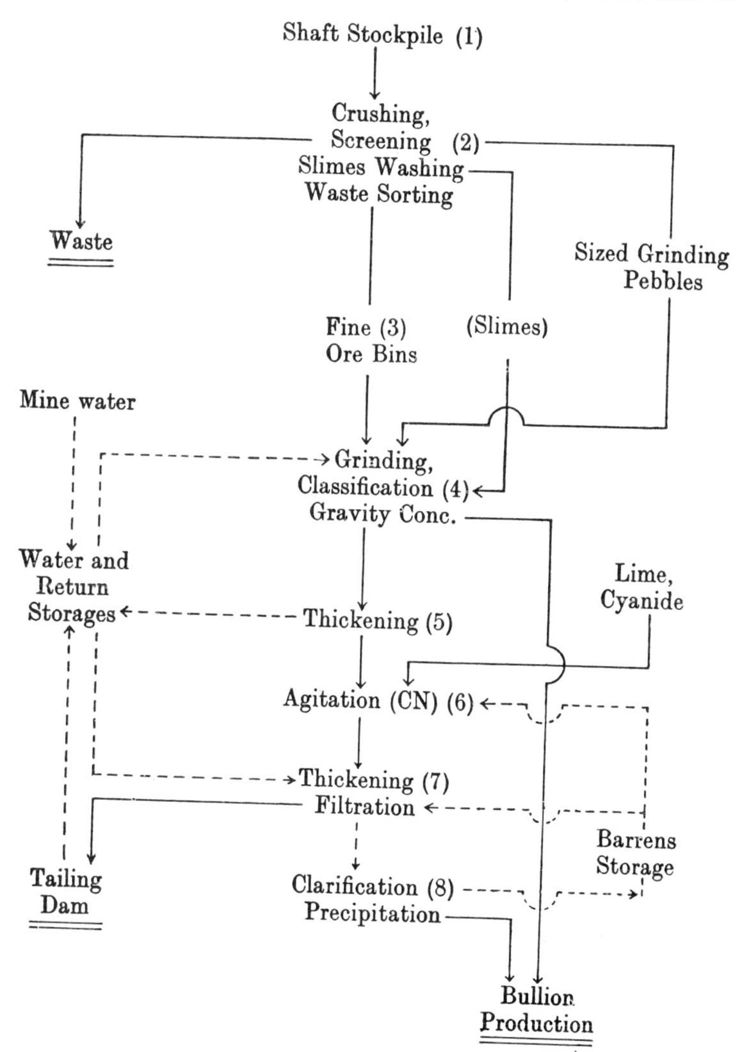

Fig. 318. *General Scheme, Rand Flow-sheet*

using artificial respiration (Holger Nielsen's or Schafer's method). Nothing may be given by mouth to an unconscious person. Clothing contaminated by cyanide should be removed. Keep the patient warm and quiet. Accidents can arise from careless handling of dry cyanide. It should be tipped from its container in such a way that no particles can become airborne. Cyanide solutions used in the dissolution of gold ore are so weak that a few drops splashed into the mouth are unlikely to do harm, but warnings which can be understood by everybody must be displayed if there is any possibility of heedless workers taking a drink of cyanide solution.

The acid-treatment of zinc residues and gold slimes is accompanied by the ebullition of hydrogen, arseniuretted hydrogen and cyanogen gas. These must be drawn off and dispersed well away from any place where they might be breathed.

Some Gold-milling Flow-sheets

A condensed general scheme for a modern Rand plant is shown in Fig. 318. Practice varies from mine to mine, the main causes of variation having to do with whether "hard sorting" (up to 30% waste rejection) is used; whether a flotation stage is incorporated; whether uranium extraction follows cyanidation. Ore is drawn from beneath the main stockpile, which can take the place of the conventional ore bin in this frost-free country. Slimes are washed off by wet-screening and by sprays set above inclined conveyor belts and sent direct to the grinding circuit. The clean ore thus exposed is screened to produce mill pebbles, and hand sorted to remove waste. It then receives up to three stages of grinding, mostly with pebbles in tube mills, though rod mills and primary ball mills are used in several plants. Cyclones have largely replaced the older forms of classifier, but rakes, spirals, Crosse classifiers (a hydraulic appliance mechanically stirred), and spitzkasten are in use. Concurrently with grinding, up to 50% of the gold is concentrated on strakes or tables, together with a valuable osmiridium by-product. These concentrates are cleaned up by tabling and use of magnets. They are then amalgamated, retorted and melted into bar.

The grinding effluent (4) is thickened, either continuously or in batches, overflow water joining mine water for general mill service. The thickened pulp is agitated in batch or series-operated Pachuca tanks, after being thinned with reconstituted cyanide solution from (8) and adjusted for protective alkalinity. From agitation the pulp is thickened, filtered, washed with barren solution to remove the last of the "pregs" from the filter cake, and is then repulped with water or discard solutions and pumped to the slime dam.

Interesting details of the operation of the Van Dyk plant are given by Mokkan.[60] These discuss pipe descaling, care of filters, cyanide strength, etc. Another Paper[61] traces the development of a modern reduction plant in which blocky low-grade ore is mined together with the friable Carbon Leader carrying the bulk of the gold and uranium values. This coarser ore is screened off ("mechanical segregation") from the narrow Leader material and is then upgraded by "reef picking" (the reverse of the sorting away of waste rock). In the final flow-sheet ore at *minus* 4″ is washed on $1\frac{1}{2}''$ screens and the oversize is sorted, adherent grit having been thoroughly

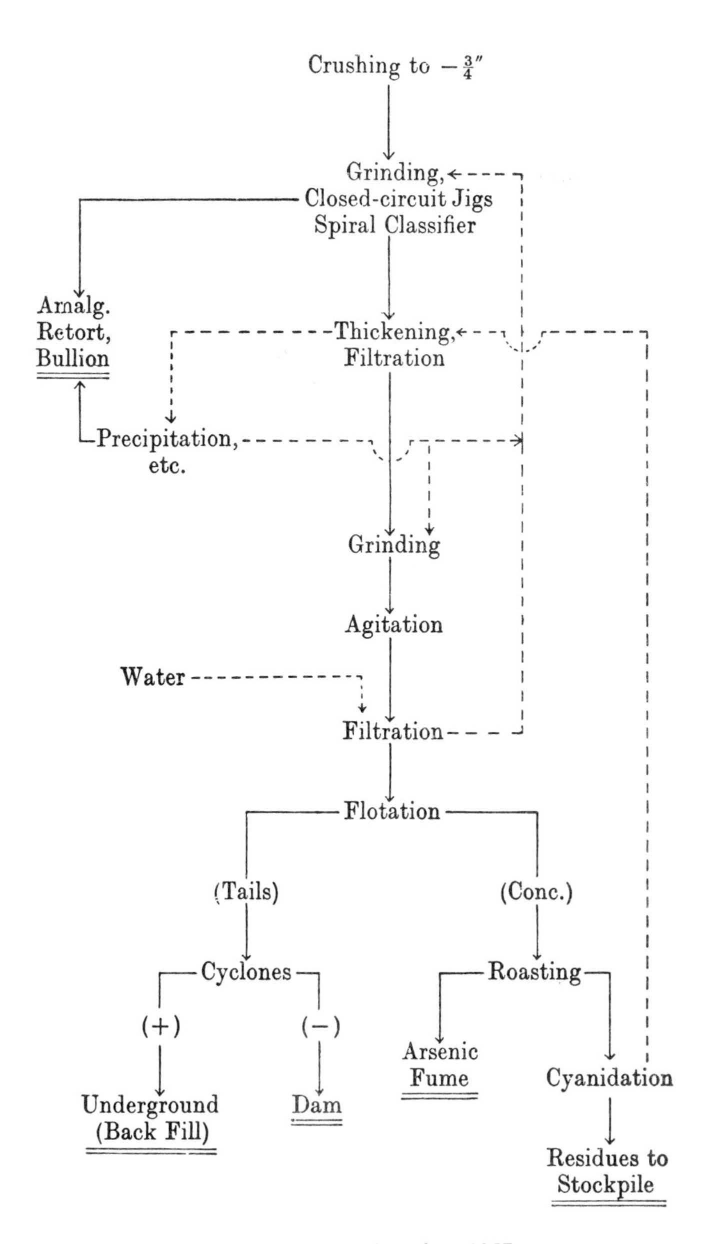

Fig. 319. *Con Flow-sheet* 1957

removed. This Paper describes a number of novel features concerned with automatic control of grinding density, cyclone operation and filtration.

Canadian Practice

The following examples have been condensed from "The Milling of Canadian Ores".[62] They are chosen to illustrate the variations on straight cyanidation which were dictated by differences in the structure of the ores.

Con Mine (Fig. 319) treats an ore consisting mainly of gold, arsenopyrite, pyrite, sphalerite, and stibnite in quartz. Twenty-three and a half per cent of the gold is taken from the jig concentrate, 50½% by cyanidation and 15% after roasting.

Dome stage-crushes to $-\frac{3}{8}''$, and cyanides after removal of a jig concentrate (Fig. 320). An unusual feature is the pre-aeration with lime of a thick grinding pulp in Pachucas, followed by filtration and repulping in cyanide.

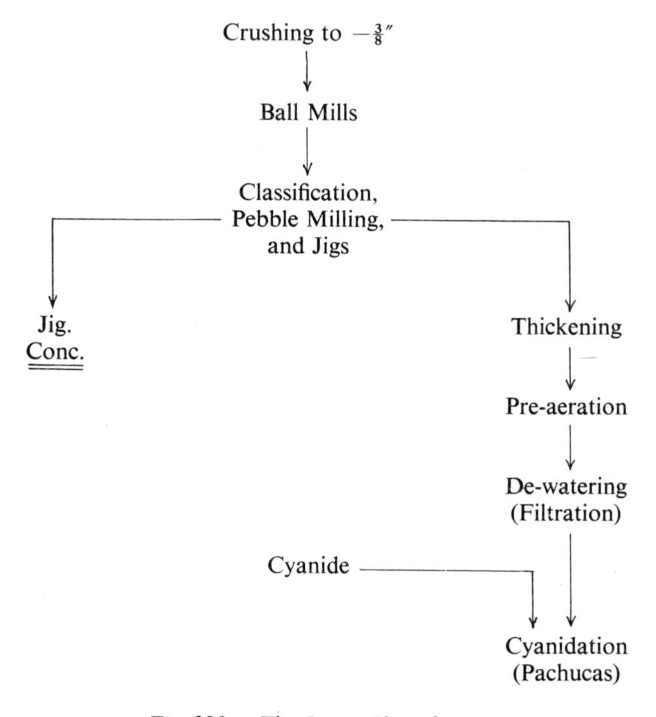

Fig. 320. *The Dome Flow-sheet*

Hollinger starts its 97% recovery with grinding in cyanide to 60% −200 mesh. Here 45% of the gold is extracted. Classifier overflow goes to 90 double-deck shaking tables which remove one sixth of the feed as a pyritic concentrate carrying four fifths of the residual gold. This is ground to 80% −325 mesh, thickened to 55% solids and agitated for 24 to 36 hours in Dorr machines, the tailings then joining the thickened table tailings for 16 hours in Pachucas. Residual solids are split between filtration and C.C.D. (Fig. 321).

Kerr-Addison mines two types of ore. One is a quartzose stock-work in the carbonate zone and yields a 99% gold recovery. The other has auriferous quartz and pyrite in fractured lavas and gives below 95%. In order to avoid trouble in the precipitation section, due to nickel, copper and arsenic, half a ton of foul cyanide is discarded for each ton of ore treated. Nickel must be kept below 0·2 lb./ton. Feed to the roasting section is at 77% solids (Fig. 322).

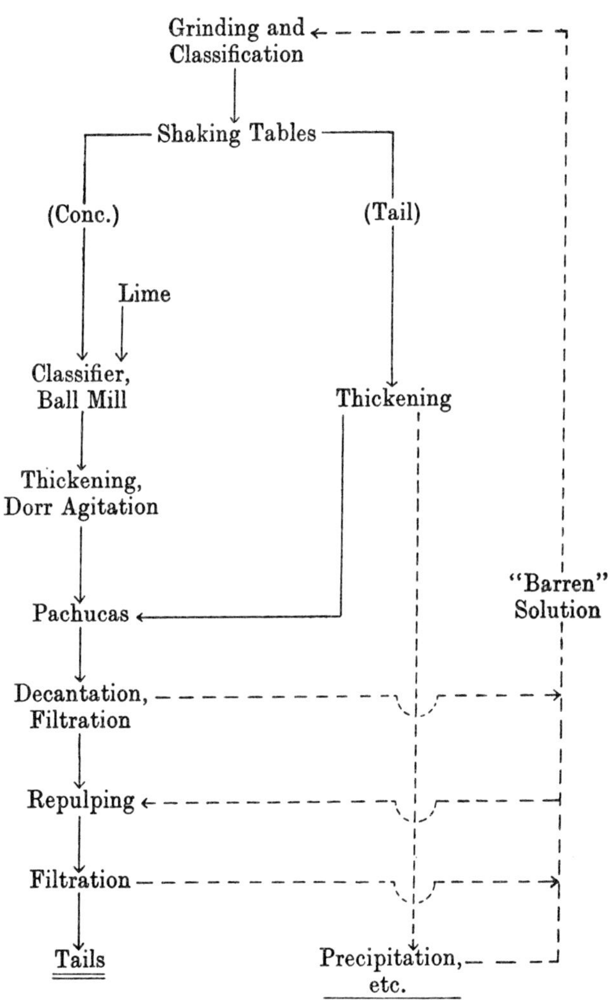

Fig. 321. *The Hollinger Flow-sheet* 1957

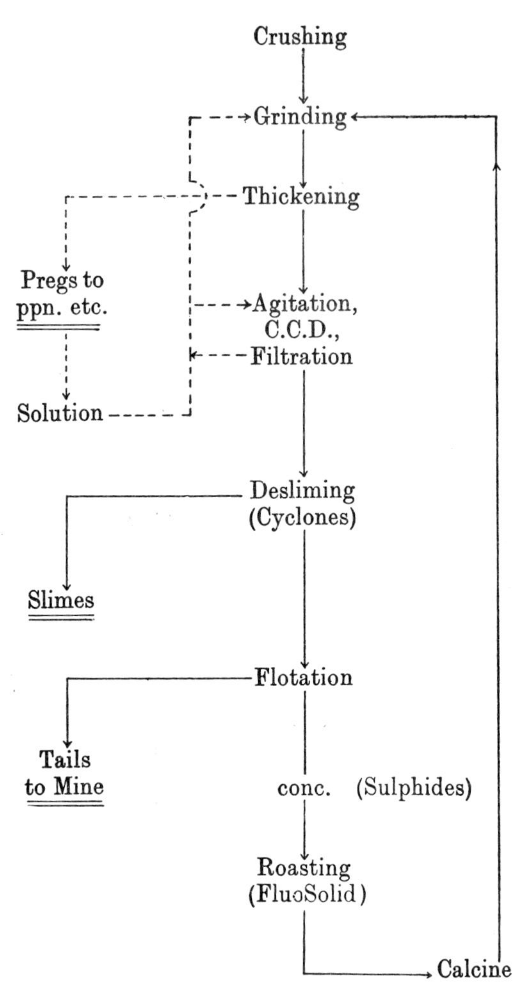

Fig. 322. *Kerr-Addison Flow-sheet* 1958

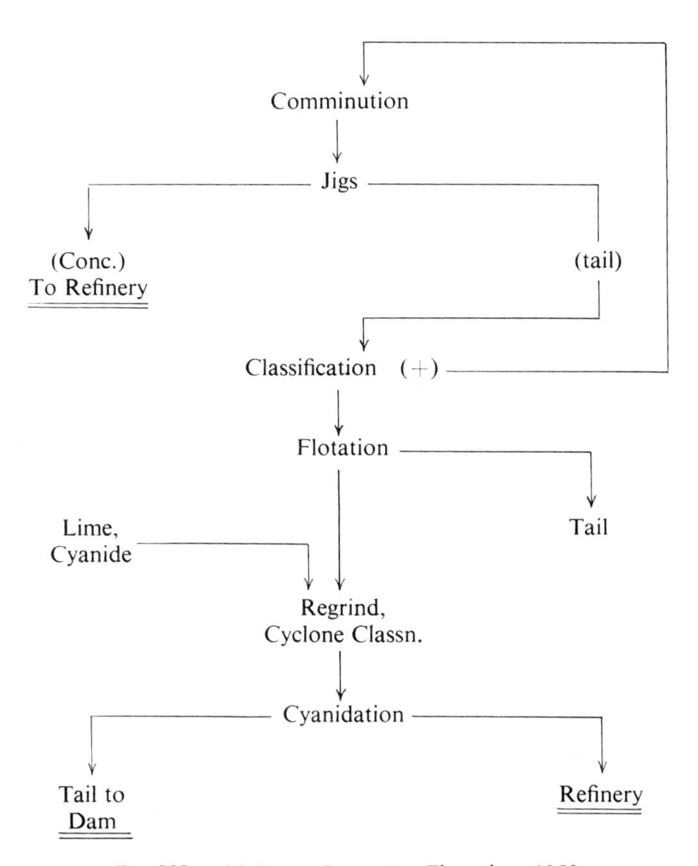

Fig. 323. McIntyre-Porcupine Flow-sheet 1958

McIntyre's gold occurs coarse and fine in quartz and as auriferous pyrite. The flow-sheet (Fig. 323) uses a sequence jig-flotation-cyanidation.

At *Lake Shore* 20% of the gold is associated with tellurium and requires very fine grinding. This is done after flotation of the first cyanide tail (Fig. 324) an operation in which similar tails from two adjacent mines are treated.

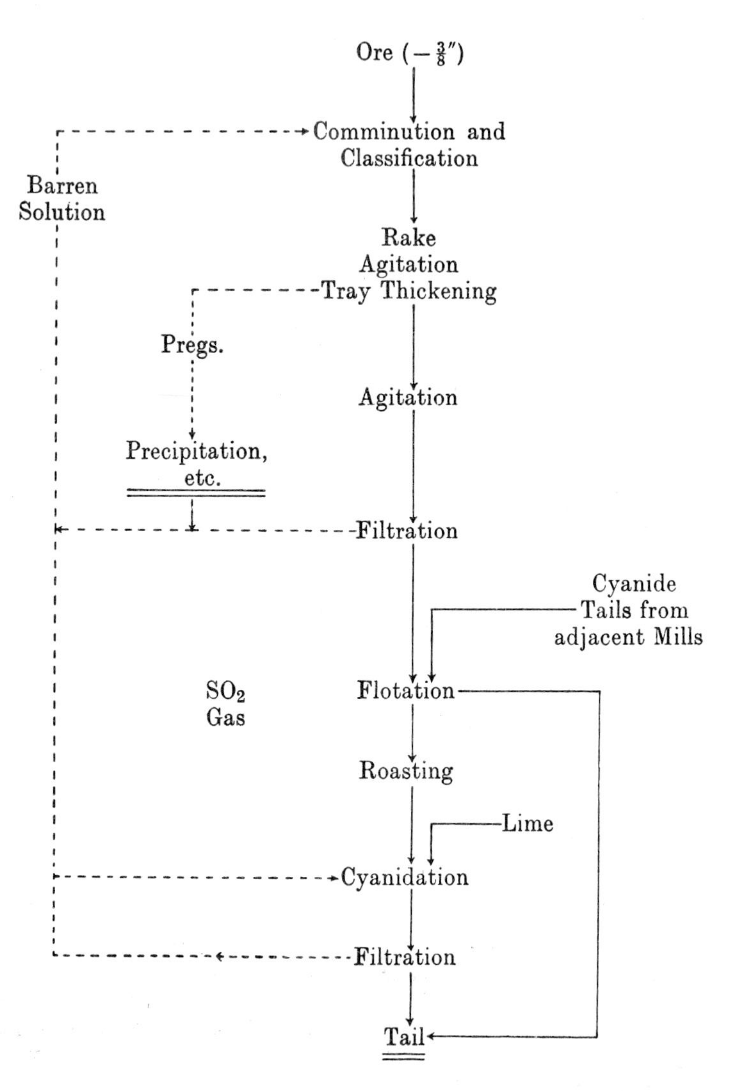

Fig. 324. The Lake Shore Flow-sheet

Falcon Mines, Limited

At Dalney Agitair cells are used to float copper-activated sulphides. This is followed by FluoSolids roasting and cyanidation (Fig. 325).

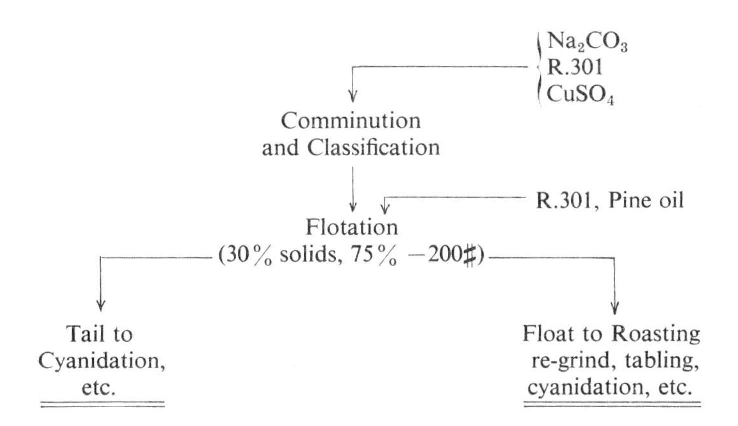

Fig. 325. *The Falcon (Dalney) Flow-sheet*

Graphite

This form of carbon occurs both in coarsely crystalline form and finely divided, with no clear dividing line. The usual treatment starts with a pulp of about *p*H 8, made alkaline with soda-ash, and aided as regards dispersion and selectivity by sodium silicate. The collecting agents for this readily floated mineral are paraffin or diesel oil, perhaps aided by pine tar oil. The frother is pine oil. Pine oil alone may suffice with finely ground graphite. When clean coarse flake can be liberated by primary grinding in a rod mill it is removed by screening. If the product is insufficiently clean, it can be up-graded by some such treatment as is shown in Fig. 326.

In the Kaiserberg mines in Styria two types of graphite schists occur. One yields microcrystalline scales, the other soft and plastic carbon. Associated minerals include mica, sericite, chlorite, quartz, plagioclase, hornblende, tremolite, limonite, and small fractions of other undesired minerals. Hand-sorting and air classification are used to produce a 66% + carbon product. This contains too many abrasives and other impurities for special uses and is pulped, ground finely and floated at *p*H 8 to upgrade it to 90% + carbon, the main impurity being a mica which does not impair its lubricating quality.

Coarse flake commands the best market price, so care is taken to avoid overgrinding. Impact crushing may be followed by screening, with tabling of the screen undersize to produce a coarse flake, tailings and a middling which goes, together with the screen oversize, to rod milling. The mill discharge can then be classified, undersize being floated and oversize tabled.

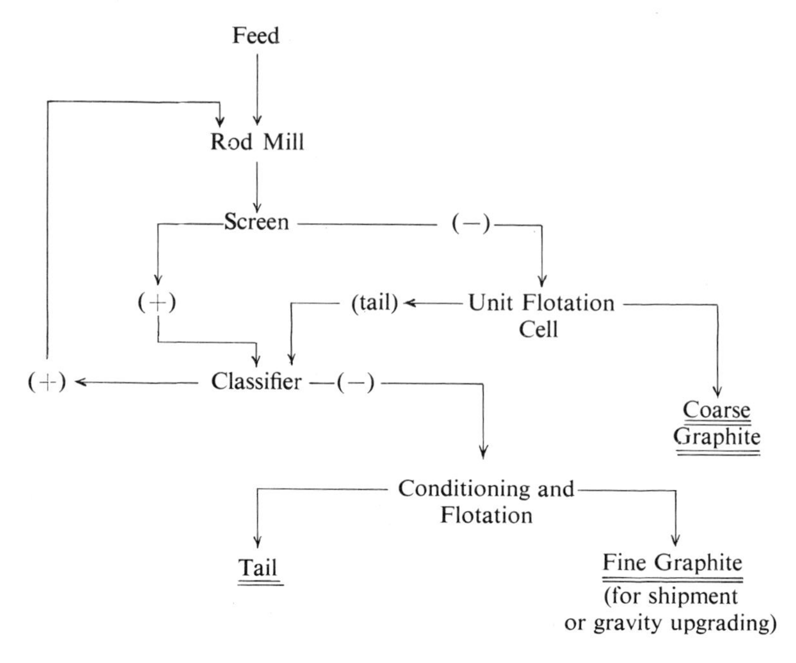

Fig. 326. *Flow-sheet. Recovery of Coarse and Fine Graphite*

It may be found desirable to table the flotation concentrates, since graphite is ductile and tends to smear gangue minerals which then report in the float and can be removed by gravity treatment.

Iron

The main concentrating processes are based on the degree of natural ferro-magnetism in the iron minerals, or on the extent to which this can be developed by reduction roasting. Gravity concentration is also extensively practised and there is large-scale use of the flotation process both for pyritic ores and for those of low ferro-magnetism. Sulphide iron may be floated in connexion with its use in producing sulphuric acid, or for recovery of associated gold. The machines used were described in Chapter 19.

Reduction roasting at research, pilot and operating level was reported in six Papers in the 1963 I.M.P. Conference. Roast-reduction of hematitic quartz was shown as a topochemical process in which reduction proceeds evenly inward from the exposed surface.[63] Concurrently the quartz matrix is micro-cracked, thus reinforcing the thermal stress in the heated mill in which the ore is being autogenously ground at some 600° C., and the iron, now substantially reduced to magnetite, is liberated. Non-magnetic Swedish taconites often contain iron too finely divided for economic liberation by grinding. A method based on grain growth by migration in a mildly reducing atmosphere at from 750° C. to 1,000° C. has been suggested.[64] Solid accretion of migrating iron begins at 500° C., a spongy formation building to form a peripheral layer around each grain. In another study[65] the rate of reduction of *minus* 1 mm. particles in a fluidised bed was found to be rapid between 550° C. and 850° C., but chlorites were not rendered ferro-magnetic at these temperatures. The Lurgi process was described,[66] with its rotary furnace using air or reducing fuel blown in at any desired point along its gently sloping cylindrical shell. It has treated semi-taconites, minettes, siderite and hematite and worked on brown coal, anthracite, town gas, producer and blast furnace gases. In the treatment of complex sideritic veins which carry barite, sulphidic copper, antimony, mercury and silver, magnetic roasting followed by magnetic separation has proved successful.[67]

When iron is concentrated by flotation the product must be pelletised in order to form a suitable blast furnace feed. To produce stable "green" balls for this purpose the flotation product must either be, or be ground down to, 75% *minus* 325 mesh. A suitable pellet is roughly ⅜″ in diameter and contains about 63% iron and 8% silica. This does not apply to concentration for roasting of auriferous pyrite or in sulphuric acid production. The general flotation scheme starts with de-sliming a *minus* 48 mesh feed at a cut point about 20 microns by hydroclassification. The underflow is conditioned at a highish solid content (some 65%) with oleic acid or a tall oil product, or with a petroleum sulphonate or mineral oil refinery residue. After dilution to some 35% solids it is floated, using a frother and, if apatite is to be depressed, sodium fluoride. Sulphuric acid is added in the cleaner section. The low-grade hematite-goethite jaspers of the Marquette Range in Michigan present special problems.[68] They must be ground through 400 mesh to give adequate liberation, and this creates a problem of interfering slimes. The difficulty is mainly concerned with reluctance of the silica slime to float. It has been overcome by selective flocculation of the iron values and their settlement before decantation of the unsettled fraction, ahead of anionic flotation of the rest of the silica. The steps are dispersion by grinding at a *p*H of 10 to 10·5 followed by addition of calcium chloride and starch to flocculate the iron slimes. These are then removed as the underflow from a thickener.

The possibilities of leach treatment for ores not amenable to other methods of upgrading but already containing 35% or more of iron are still in the pilot stage. North American taconites have been upgraded in an hour's hot leach with sodium hydroxide to 65% Fe. Tests reported from France[69] used a 40% to 50% concentration of alkali at 125° C. to 140° C. to give a 30% to 50% upgrading in ½–3 hours. The bulk of the silica, alumina and phosphorus was thus removed.

The following industrial operations illustrate the use of gravity, magnetic concentration and flotation. At Stora Kopperberg's mill in Sweden tests showed that after magnetic removal of magnetite the hematite responded best to concentration with Humphrey's spirals.[70] Ore crushed through 1″ is rod-milled to *minus* 8 mesh, and passed through a wet permanent-magnet installation which removes 70 tons hourly of the 150-ton feed as a 64% concentrate. Thence the pulp is passed to a battery of 80 five-turn spirals which rough out a middling and discard fifty tons of tailing assaying 7½% Fe and 2% P, the phosphorus content of the entering ore being 1%. The middling goes to three-turn spirals for cleaning, its tailing being returned to the rougher spirals. Here 30 tons per hour assaying 63% Fe are recovered. Spirals are worked at a pulp density of 25% to 30% and splitters are set by calibrating gauges. One man operates the whole spiral section in which each machine has a capacity of 1½ tons per hour and stands in an area of one square yard. In Canada's Carol concentrator spirals are used on an ore which is liberated at 14 mesh. Aerofall mills work in closed circuit with 14 mesh screens[71] and the undersize is classified into a coarse product, about a splitting point of 150 mesh. Each split fraction goes to rougher-cleaner-recleaner spirals. An unusual de-sliming concentration is practised with the Moose Mountain ores of Capreol, Ontario, which must be ground substantially through 325 mesh for liberation before producing a 66½% Fe concentrate. After a first-stage magnetic concentration of a rod-mill discharge the concentrate is finely ground and given a second magnetic treatment. The concentrate from this is re-ground, magnetically flocculated and sent to hydro-separators from which the dominantly siliceous non-settling fraction overflows. The operation depends on automated sensing of the separating line between the silica with its density of 2·8 and the zone of flocculated magnetics at S.G. 5·2.[72]

In one Nevada flow-sheet the *minus* ⅛″ magnetics are recovered by dry treatment and the *minus* 48-mesh screened undersizes by wet magnets.[73] Treatment of a quartz-siderite ore assaying 20% to 28% Fe in Hungary by roasting in a reducing atmosphere using natural gas, followed by dry magnetic separation has been used in conjunction with recovery of copper and barite from the ore.[74]

Iron ore flotation was pioneered in Michigan in 1954. At Humboldt where a cherty hematite and magnetite is treated. *Minus* 65 mesh classifier overflow is deslimed and thickened by cyclones, with rejection of some 3½% of the mill feed and 2½% of the iron. The pulp is conditioned at 70% solids with 0·005 lb./ton of Aerosol, 1½ lb. of tall oil and then floated. Concentrates are filtered, stockpiled and drawn to the re-grind section for reduction to at least 75% *minus* 325 mesh, a size essential for formation of a robust "green" ball suitable for pelletising. The need for integration of flotation treatment with pellet production causes the latter to dominate procedure. At Cleveland-Cliffs a rougher concentrate produced by flotation with 1·25 lb./ton of a tall oil fatty acid in Fagergren cells is made, at an average grade of 61·7% Fe.[76] This is re-ground to at least 80% *minus* 325 mesh, to satisfy the requirements for pelletising. It then passes, without further reagent addition, through a series of four conditioning tanks in which it is progressively heated above 98° C. The fatty acids present migrate selectively to the hematite. The effluent is diluted to between 32% and 34% solids and re-floated, iron recovery being nearly 98% and grade 67% Fe.

Lead, Lead-Zinc, etc.

The main dominantly lead-bearing ores are the sulphide (galena, Pbs), partially oxidised sulphide, the sulphate (anglesite, $PbSO_4$), and the carbonate (Cerussite, $PbCO_3$).

Galena usually occurs in association with sphalerite and other sulphides, from which it is differentially floated. Some complex treatments are discussed later. The mineral floats readily with aerofloat or xanthate, in a pulp made alkaline with sodium carbonate. It may be depressed by lime or by a *p*H exceeding 10·4. Potassium dichromate forms a non-reactive coating of lead chromate, and is sometimes used to depress lead from a bulk float. If Aerofloat 25 or 31 is used, little or no pine oil is needed as a frother. Where galena has become tarnished the sulphidising action of R.404 (Cyanamid) may aid collection. With a calcite gangue the adherence of slimed gangue may occur and cause loss unless a dispersant, sodium silicate, is used. If oxidised lead is to be floated, sodium sulphide is commonly used to produce a surface attractive to xanthate collectors. As soluble sulphides are strong depressants for clean lead and silver sulphides, theses minerals must be removed in an earlier flotation operation or they will probably be lost. Since excess sodium sulphide is in any case a depressant, starvation quantities should be added cell by cell in the second flotation treatment. The surfacing effect produced by sulphide addition is transient, and the froth should be removed speedily to avoid reversion of the mineral surface to its oxidic state. The use of copper sulphate to control and stabilise the newly sulphidised mineral particles is favoured by some operators. This is done by two conditioning stages, starting with sulphidising treatment and followed by the use of the copper salt.

In controlling the flotation of lead carbonate, a quick vanning test on the tailings can be made by adding a few drops of sodium sulphide to any whitish heavy mineral, which will turn brown if it consists of insufficiently sulphidised lead.

With most oxidised lead ores it is good practice to take gravity concentration as far as possible and thus to reduce the production of slimed values. Sulphidising agents are used immediately before flotation, with sodium bicarbonate. In one Moroccan mine where Hancock jigs pre-concentrate the flotation feed of a cerussite-limestone ore, considerable improvement was made by cutting down conditioning time and amount of reagents (hydrosulphide and amyl xanthate) and intensifying aeration during flotation.[77] A special impeller was developed, with rotor bars but no stator or deflecting baffles. Since excessive reagent consumption occurs in hot weather conditioning time is cut down at such periods. Aeration in the conditioner is sometimes helpful, the reason not being yet known. Alkaline earths are liable to be harmful since they tend to form coatings of insoluble carbonate on cerussite.[78] By using relatively cheap barium sulphide instead of sodium sulphide, and adding sulphuric acid in the conditioner deep enough to prevent escape of the H_2S evolved one Sardinina operation has improved its metallurgy, the *p*H being controlled between 6·8 and 7·2.

Lithium

Lithium is mainly extracted from spodumene, which has a theoretical lithia content of 8 %, and a practical one of up to 6 %. This pyroxene ($Li.Al.Si_2O_6$) has a density of 3·15 but transforms on heating to 1,050° C. to *beta*-spodumene with a density of 2·4, a state in which it is readily pulverised. Sulphuric ion exchange at above 250° C. is one method of extraction,[85] the acid roasting time being 10 minutes before leaching the resultant lithium sulphate, which is then precipitated as carbonate. In another process the spodumene is sintered for 2 hours with a gypsum-limestone mixture. Quebec Lithium Cpn. reacts β spodumene with about six times its weight of soda-ash at between 130° C. and 300° C. The resulting slurry is cooled and treated with CO_2 under pressure to form soluble lithium bicarbonate. Much of the soda-ash is re-used.[86] On the pilot scale, recovery of a good grade spodumene from Australian pegmatites by flotation with lignin sulphonate and oleic acid has been made.[87]

Manganese

Concentration methods include hand sorting with cobbing (in South Africa), DMS, jigging and tabling for manganese minerals sufficiently liberated at a coarse size for marketing. Where flotation is used the concentrates must reach the specifications required for use in furnaces—48 % Mn, low iron and phosphorus and below 11 % silica *plus* alumina after pelletisation. Three main flotation types of ore are those with a high calcite gangue, those intermediate in calcite and silica, and those with a siliceous gangue. With the first two, the calcite is floated with from 1 to 4 lb./ton of an oleic collector, the pyrolusite or manganite being depressed with $\frac{1}{2}$–3 lb. of dextrin. With the third type, if non-clayey or slimed, the thickened pulp is conditioned at about *p*H 7 with SO_2, tall oil is used as collector and the manganese is then floated after some dilution.

An unusual agglomeration-float of wad has been described[88]. Wad is a complex of manganese oxides which may include psilomelane, hollandite, coronadite, and pyrolusite. This slimes during grinding and part of the flotation problem is concerned with recovery of the slimed manganese. At Three Kids the main gangue minerals are quartz, kaolinite, montmorillonite, calcite, gypsum, and barite. The specific surface of this material can reach 58 m.^{2}/g., and reagent consumption is correspondingly high. The flow-sheet (Fig. 327) requires intense mechanical stirring during conditioning, to break

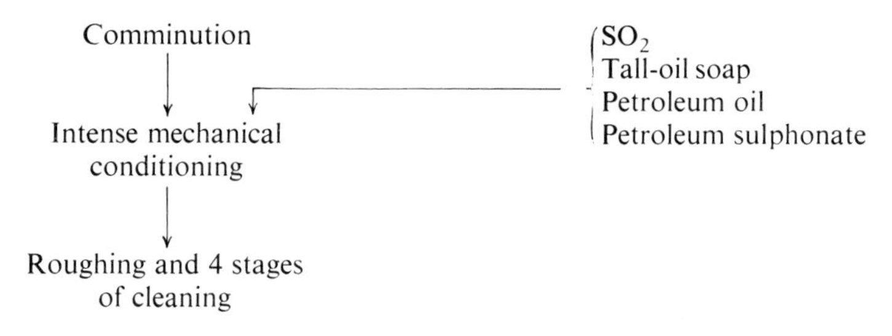

Fig. 327. Three Kids Mine Flow-sheet

up the glomerules formed by the flotation reagents sufficiently for t
reject their non-manganese constituents.

The SO_2 is believed to release Mn ions which sorb to the manganese ....
and provide attractive surfaces for the tall-oil soap in the reagent mixture, which is fed to the pulp as an unstable emulsion.

The intense shearing of this 20% solid pulp kneads the weak heterogeneous flocs into dense glomerules of 100 to 200# size as the non-manganitic fractions are squeezed out. The active components of the soap are fatty-acid and resin-acid sodium soaps. The petroleum oil is a mixture of gas and diesel oil. The wetting agent is a sodium alkyl aryl sulphonate.

Mercury

The main mineral treated is cinnabar, from which the mercury is usually extracted by direct distillation. One Californian producer that has worked a 12 lb./ton deposit intermittently since 1870[89] feeds minus $1\frac{1}{2}''$ ore to rotary furnaces down-sloped $\frac{3}{4}''$ per foot, and turning 1 r.p.m. Heating is by fuel oil, and the mercury-laden gases leaving at between 260° C. and 315° C. are drawn by suction fans to cyclones and condensers. Condensates are collected daily and batch-treated by rabbling with quicklime, when 80% of the mercury drains away. The rest is recovered by batch distillation. Mercury from both operations is next filtered through a bed of quicklime and sold as "prime virgin" mercury, 99·9% or more pure.

Pre-concentration by gravity methods or by flotation may be used with a complex or lowgrade ore before retorting. Slimes and clays must be dispersed prior to effective flotation, but sodium silicate and sodium carbonate have proved unsuitable dispersing agents, as they exercise a depressing effect. An organic chemical based on lignin (called Palcotin), has been successfully used in amounts varying from 0·04 to 0·1 lb./ton. *p*H is not critical, but should be held between *p*H 6 and *p*H 8. The cinnabar readily accepts surface activation by ions of lead or copper. A secondary butyl xanthate (Reagent 301) and Frother 65 are used to complete the work.

Mica

The premium grades of mica are obtained by hand-sorting and trimming, value being proportional to size of sheets. Sizes below 1″ or so, called "punch" and "scrap", can be concentrated by screening and careful comminution. If the compressive crushing force is light, mica is sufficiently flexible to remain unbroken while associated quartz and spar are detached. Jaw crushers, rolls and hammer mills are used, with intermediate screening to separate liberated flake at the largest possible size.

Flotation can be used to remove fluorine-carrying mica from kaolinite, where excessive amounts would constitute a health hazard in the kilning of china ware. A cationic collector agent such as an amine acetate and an alcohol frother are used at a *p*H of up to 11 in a moderately dilute pulp, to float the mica.

Molybdenum

In a number of plants molybdenite is recovered as a by-product bulk-floated with sulphide copper, in which from 0·3% to 2% may occur. In a typical operation in Arizona[90] rougher concentrates are reground through 200 mesh and cleaned. The clean concentrate is thickened and steamed, ferro-cyanide being added to depress the copper. Three stages of flotation are then given, following which the roughed molybdenite froth is reground and recleaned.

The major molybdenite operation is that of Climax, Colorado, where the main selling product is this dominant mineral. Advantage is taken of the strong floatability of MoS_2 in a flow-sheet (Fig. 328) which discards most of

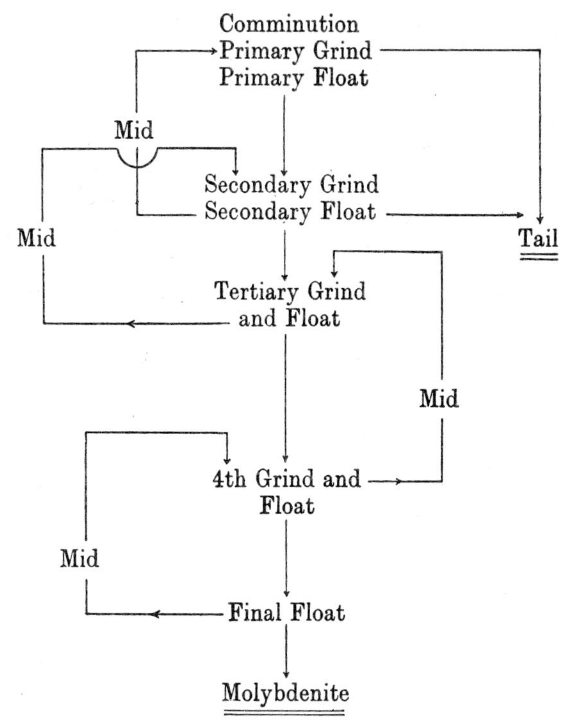

Fig. 328. The Climax Molybdenum Flow-sheet

the gangue at a coarse grind. The ore body is a silicified altered granite, with its molybdenite finely disseminated in stringers and thin veins. Minor quantities of wolframite, monazite, and cassiterite are recovered from the flotation tailings by gravity methods. Because of the flaky nature of molybdenite it is possible to float particles consisting mainly of quartz gangue, using oil only as the collector, aided in its emulsification by a sulphated monoglyceride, and a little pine oil. Cyanide and soda-ash are used in the cleaning stages at *p*H 8·3, to keep down traces of pyrite and chalcopyrite. Primary grinding at 43% + 100♯ is followed by a 25 to 1 concentration in the roughing section, yielding a 9% MoS_2 concentrate. This is thickened and given three further stages of grinding in pebble mills, with staged flotation. Ninety-eight and a half per cent recovery of high-grade molybdenite is maintained. Tailings are sent to a by-product plant where tin and tungsten minerals are recovered by treatment in which gravity, flotation and magnetic separation are used.

More usually molybdenite is bulk-floated with copper sulphides, which latter are then depressed. The method used at Morenci has been described by Papin.[93] Thickened bulk concentrates are dewatered, repulped with fresh water, and brought to *p*H 7·5 with sulphuric acid. The thiophosphate collecting agent is selectively removed from the copper mineral by sodium ferrocyanide, the effects of which are short-lived. Flotation of the molybdenite is performed after brief conditioning, and the rough concentrate thus produced is given further grinding and cleaning, a little sodium cyanide and sulphide being used in this work to aid further additions of ferro-cyanide. In other mills soluble starch has been used as a depressant for molybdenite collected with xanthates, but this did not work with the thiophosphate collector used by Morenci. At Bingham, Utah, steam distillation removes decomposed xanthate selectively from copper, the molybdenite being refloated after this treatment. The general pattern (Fig. 329) is:

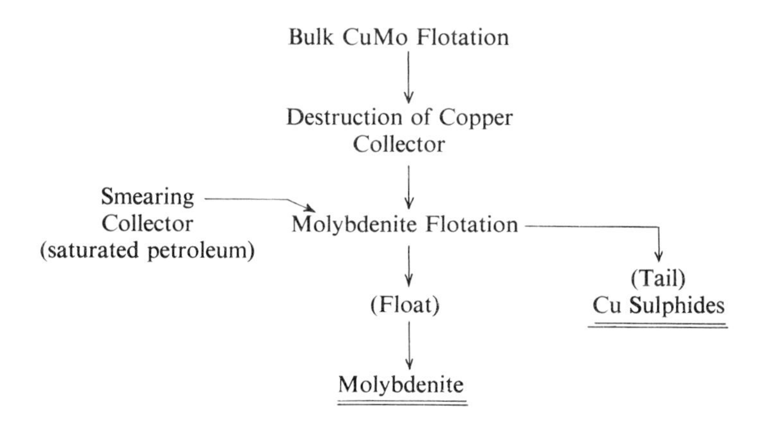

Fig. 329. CuMo Separation

Other methods used include depression of the copper with sodium hypochlorite in an alkaline pulp together with sodium ferrocyanide, fuel oil and an alcohol frother being used to float the molybdenite. Copper minerals can also be depressed by phosphorus, arsenic or antimony salts. The float from such differentiations assays from 3% to 30% MoS_2, and is brought up to selling grade by regrinding, repeated recleaning, roasting, leaching or the use of special reagents.

At Climax, Colorado, a new process has been worked out for treating the oxidised molybdenite ore. After sulphides have been floated high-intensity magnetism is applied to the tailings. The concentrate is added to the slimed oxide values which have been removed by cyclone classification, and acid leaching follows. "Pregs" are treated with pellets of activated charcoal, which is then screened out and stripped of its adsorbed salts with ammonia.[92]

Nickel

The main source of the world's nickel is the copper-nickel sulphide ores typified by those mined at Sudbury, Ontario. The principal sulphide is pentlandite ($NiFeS_2$), which is usually associated with chalcopyrite and iron sulphides. International Nickel (Inco) and Sherritt Gordon follow a

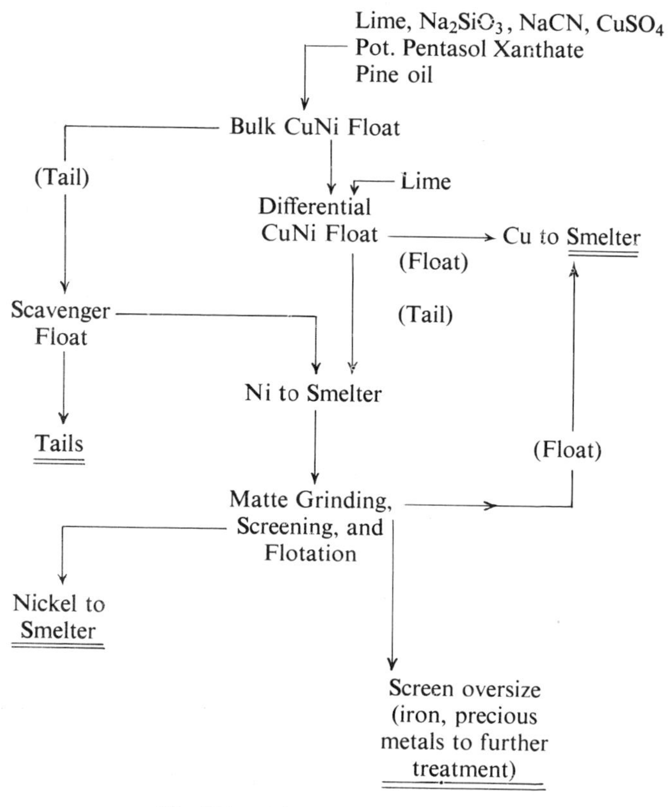

Fig. 330. *The Inco Flow-sheet*

bulk CuNi float by differentiation, while other Canadian producers leave the latter problem to the smelters.

The Inco ore carries pyrrhotite and ferromagnesian silicates. Its pentlandite is intergrown with the pyrrhotite, but the associated chalcopyrite is more readily liberated by grinding. This facilitates differentiation of the bulk float (Fig. 330). Lime controls the *p*H in both bulk and separating floats, a further addition depressing the pentlandite from the copper sulphides in the latter operation. A unique feature of the Inco operation is the use of flotation to separate cuprous sulphide (floated) from nickel sulphide with removal of iron, by grinding the nickel-copper matte produced by smelting the high-nickel concentrate originally depressed from the bulk float. Slow cooling of the matte is essential so that crystals of the sulphides can grow to a suitable size before grinding. At Outokumpu Oy (Finland) a somewhat similar ore is similarly treated, save that lime is not added in the roughing stage and dextrin is added as a depressant for pentlandite when differentiating the bulk float.

The froth in the primary plant at Creighton is judged by fluorescent lighting, daylight being excluded. Winter pulp temperature is 9° C.

The Sherritt Gordon operation,[81] mentioned earlier in this chapter, is complicated by the presence of large amounts of talc which cannot be pre-floated since it would take with it 10% of the nickel and half the copper in the ore. A depressant has been developed which, when used after conditioning with amyl xanthate and a phosphate dispersant, has met the need. Bulk flotation produces a concentrate carrying 70% of the mill-head as free pentlandite and 85% of the copper as chalcopyrite. These are separated in a further flotation stage by depressing the nickel with lime and cyanide. The balance of the mill-head nickel is too intimately associated with pyrrhotite to be liberated economically by grinding. This is upgraded in a second float to 5% Ni and 1% Cu in a circuit made acid with SO_2 (*p*H 6). This concentrate joins the high-grade nickel product, the combined assay being 13·5% Ni and 2% Cu.

Niobium (columbium)

One mill which started production in the Oka (Quebec) district in 1961 removes ferro-magnetic minerals before floating successively the pyrite, apatite, calcite, mica and silicates, leaving pyrochlore as a tailing. Test-work resulting from which a small concentrator is to be built for treating pyrochlore and perovskite to produce a 45% to 50% concentrate of Nb_2O_5 has been described.[95] The process uses mixed amines and di-amines together with a sulphonated wetting agent to produce a low-grade concentrate. Reagents are mixed hot and used as an emulsion in warm water. Ammonium bicarbonate depresses some gangue minerals and aeration is starved, to keep down apatite. In cleaning, a stronger addition of the bicarbonate is made (3%–4%) to depress diopside and mica still further and produce a 10% to 20% grade of concentrate. This is now conditioned with calcium hypochlorite to remove collector, after which the pyrite is floated and the residual pulp is tabled to yield a 50% concentrate of Nb_2O_5.

Phosphorus

An important branch of non-sulphide flotation is concerned with the flotation (froth and agglomeration) of the apatites and phosphatic deposits. The general formula $Ca_3(PO_4)_2$ of true apatite may be modified by intergrowth with carbonate, and the term "phosphate rock" embraces chlorapatite ($Ca_5(PO_4)_3Cl$), fluorapatite ($Ca_5(PO_4)_3F$), and hydroxylapatite ($Ca_5(PO_4)_3(OH)$). The Florida rocks, called collophanes, lie roughly in a thirty-mile circle under a sand cover. They are composed of clay slimes, silica and phosphate pebble in equal proportions. "Pebble" ranges from 1″ down to 400 mesh and varies from well-defined hard particles to soft weathered rock. The clay in this matrix aids pumping of the quarried material to a distant washery. Pebble supplies are low and fine sands are the main source of concentrates, produced by flotation. The deposits are stripped, delivered to sluices feeding a pumping and pipeline system and, at the delivery end, screened. Oversize up to $1\frac{1}{2}''$ is hammer milled to break down pebbles and mud balls. All products then go to screens, hydrosizers and/or log washers. A pebble product (*minus* $\frac{1}{2}''$ *plus* 14 mesh) is recovered by screening and the undersize is de-slimed, *minus* 150 mesh material being rejected. Oversize is classified into *plus* 20 mesh pebble, *minus* 20 *plus* 35 mesh feed for agglomeration tabling and *minus* 35 mesh feed to froth flotation.

Agglomeration feed is again de-slimed and conditioned at 70% to 75% solids in rotary drums with fuel oil, fatty acid and sodium hydroxide. It is then concentrated by tabling, spirals or on specially adapted conveyor belts, silica being discarded. The *minus* 35 mesh feed is conditioned at 65% to 70% solids, diluted to 25% and then floated to produce a 68% + grade of bone phosphate of lime (BPL). Cells must have good resistance to abrasion, such as neoprene linings, good sand relief and copious overflow. To reach a 72% + BPL, the Crago method of reverse flotation is used. The froth product is de-oiled by the use of sulphuric acid and the residual silica is then floated by use of an amine collector. Ideally, the head feed is thickened to 70% solids and agitated with acid, and thoroughly rinsed to remove previous reagents. Fig. 331 gives a general flow-sheet compiled from several sources.

Sun, Snow, and Purcell[98] report that 18-C unsaturated fatty acids are the best collectors for Florida phosphate in the descending order linoleic, oleic, and linolenic acid. A liquid hydrocarbon such as fuel oil is necessary, more being used when less fatty acid is required. The ions of Ca, Mg, Al, or Fe if in the pulp at concentrations exceeding 17 p.p.m. seriously restrict flotation. Best research results were obtained at 20° C. and a *p*H of 8·9.

Hughes[99] in an earlier description of the 30 m. tons/annum industry from which some ten million tons of concentrate are made in Florida, states that after washing and removal of pebble phosphates the treatment head assays *minus* 5% P_2O_5. The −14 + 35# material is agglomerated and the −35 + 50# is floated at *p*H 9 to $9\frac{1}{2}$. Tall oil, caustic alkali, and fuel oil are the reagents, the agglomeration feed being conditioned as a thick pulp in which the phosphates form glomerules with sufficient entrapped air to cause them to ride up and over to the light-fraction discharge on the concentrating appliances.

The standard treatment applied to the fine sands roughs out a phosphate-rich float. This is de-oiled by scrubbing with sulphuric acid, rinsed and floated at *p*H 7·4 with an amine and paraffin, the floating silica being discarded. The Coronet Phosphate Co. treats its coarse sands on 42″ wide belt conveyors. The thick pulp is ploughed toward the centre by bars and baffles as it travels beneath them, and the glomerules which work upward are flushed over the sides by water jets.

Israeli phosphate treatment in the Negev is handicapped by lack of water. A $28\frac{1}{2}$% concentrate of P_2O_5 was made at first by dry methods. These exploit differences in hardness and fracture between the phosphatic oolites, the cementing calcite and the free chert and limestone. Large lumps of the last named are rejected and the balance is crushed through 4″. It is autogenously dry-ground to minus $1\frac{1}{2}$″. Oolites are liberated at about 35 mesh and a fatty acid collector can be used in a de-slimed pulp to float off calcite and leave a 32% phosphate tail.[97]

Potassium

Soluble potassium salts can be extracted by means of "solution mining" where the beds lie between impermeable covers. One such operation, at Kalium in Saskatchewan, draws the resulting mixture of saturated halite (NaCl) and sylvite (KCl) from underground to ponds where solar evaporation produces differential crystallisation. Contamination is minimised by the use of monel piping, iron being undesirable in the product. The physical chemistry of this fractional crystallisation can be studied by means of phase diagrams.[100] At Bonneville the Utah brines and bitterns contain ions of chlorine, sulphate, sodium, potassium, and magnesium. Solar evaporation is aided by dry, hot summers and low rainfall. Ninety per cent of the sodium chloride is crystallised out in a general pond area and the residual bittern is then given further evaporation in a separate pond, where sylvinite (mixed sodium and potassium chlorides) crystallises to yield fertiliser grade potash (96% KCl).

Potash salts can be floated away from those of sodium in a saturated brine. This process is used extensively in the treatment of mined sylvinite ores. After de-sliming to remove clay, which interferes seriously with the use of an amine collector, a cationic float is made. Alternatively the ore may be conditioned with lead chloride to activate the potassium and floated with a 7 to 9 carbon-chain aliphatic acid, such as a mixture of caprylic soap.[101]

Choice between cationic and anionic reagent plans is largely governed by sensitivity to clay dispersed in the feed and by the size of particle which will float. Soaps are not clay-sensitive and do not require careful scrubbing together with cooked starches, natural gums or synthetic polyglycol or acrylamid. Activating lead in soap flotation is added during the wet grinding, up to 4 lb./ton of chloride or sulphate being used. Alcohol, cresylic acid and cooked starch are among the reagents used to control froth volume.

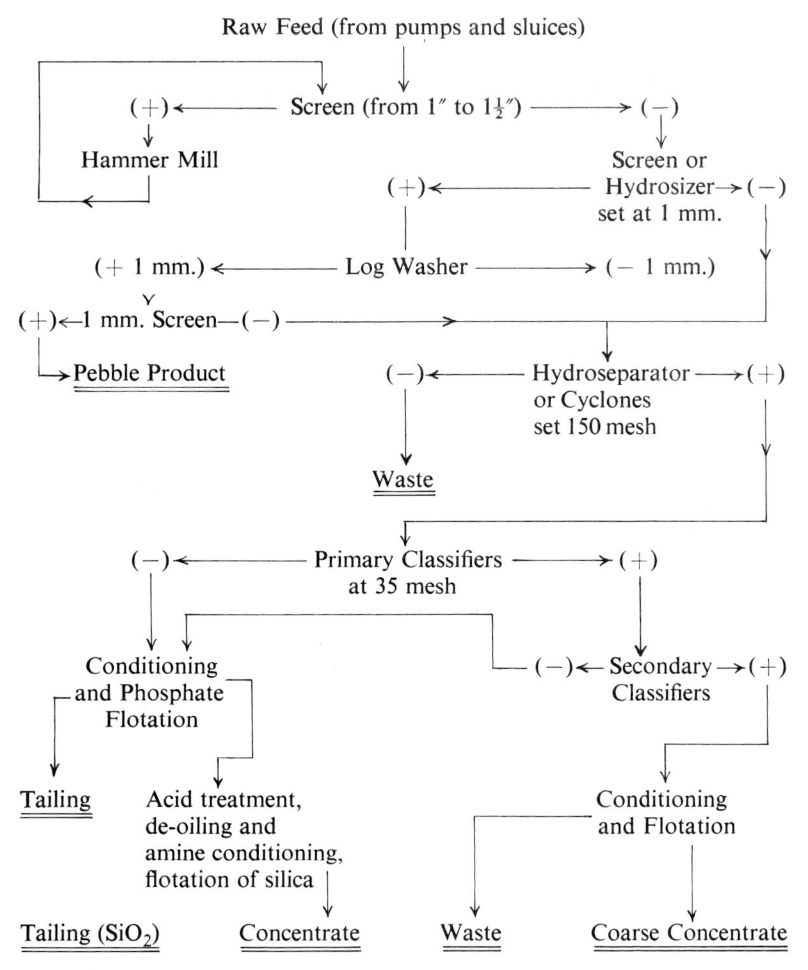

Fig. 331. *Generalised Flow-sheet for processing of Florida pebble phosphates*

Sands and Gravels

Beneficiation of siliceous material has two main objectives—improvement for use in structural engineering and production of high-purity glass sands. Concentration of cassiterite, rutile, etc., from beach and alluvial sands is not discussed in this section. By 1960 28 DMS plants in North America were removing detritus, soft and porous stone from gravels required as concrete aggregate. Specifications vary, partly because some projects entail exposure to severe wintry conditions where absorption of moisture and subsequent freezing and thaw-out might lead to trouble through disintegration. The maximum tolerable percentage of a given type of deleterious stone and/or the overall deleterious content of the aggregate may be specified, together with the size-range of constituent particles. "Deleterious" can include organic material, chert, clay, shale, soft particles or hard but water-absorbing stones. Other undesirable material could be chemically active particles such as ferruginous matter liable to rust and damage the finished structure. Most schists, weathered rock, carbonates and oxides can be dangerous.[102,103]

Silica sands may require simple removal of staining coatings, or flotation of iron and other minerals of which small traces can make them unsuitable for optical glass. Scrubbing with sulphuric acid or sodium hydroxide is used to remove clay and organic coating, and the sand can then be conditioned for floating off of unwanted minerals. One flow-sheet (Fig. 332) shows the use of magnetism.

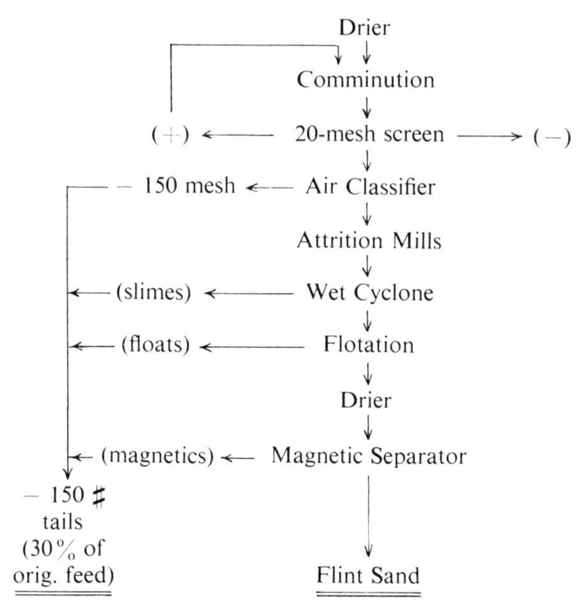

Fig. 332. *Canadian Silica Cpn. Flow-sheet*

Among the minerals which are removed may be iron oxides and stained particles, which respond readily to Cyanamid 400 series reagents, notably 425; garnet, chromium and zirconium minerals, feldspar and mica. Not more than 0·03% of iron can be tolerated in a good optical sand and it may be found profitable to use a warm sulphuric acid leach to dissolve this, when present as Fe_2O_3.

Foundry sands are reclaimed by a thorough scrub with water to remove bonding materials, clays, slimes and fine particles which are classified out and discarded.

Silver

The sulphides include argentite (Ag_2S); polybasite $(Ag_2S \,.\, CuS)_9\,(Sb_2S_3 \,.\, As_2S_3)$; proustite ($3Ag_2S \,.\, As_2S_3$); pyrargyrite ($3Ag_2S \,.\, Sb_2S_3$); and stephanite ($5Ag_2S \,.\, Sb_2S_3$). Flotation characteristics have been studied at length by Leaver, Woolf, and Towne.[104]

Their general finding is that these sulphides, and also tetrahedrite ($4Cu_2S \,.\, Sb_2S_3$), in which part of the copper may be replaced by silver, float well in a natural circuit with ethyl xanthate or Aerofloat 15. Depressants include sodium sulphide, sodium hydrate, lime, and starch. Slimes containing iron oxide or talc interfere with grade and recovery. If free gold is present, Aerofloat 208 may be helpful. Copper sulphate may help in activation if zinc is associated with the silver sulphides. Where selective depression of silver away from base-metal sulphides is needed, this is best performed after bulk flotation, using lime, cyanide, or zinc sulphate. Oxidised silver ores are aided by sulphidising treatment.

Following success in pre-concentration of gold and silver ahead of cyanidation on a small scale, mines in the Pachuca district of Mexico now operate a 1,000 ton flotation plant which makes separate concentrates of lead, zinc and iron, tails being cyanided. Gold and silver contained in these concentrates are paid for by the smelters and cyanidation is improved by removal of cyanicides during flotation. The galena concentrate carries as much refractory gold and silver as possible, together with the copper in the ore. The sphalerite carries bullion values which can be recovered by cyanidation. Iron minerals are also floated, the amenable ones going to cyanidation while those carrying refractory bullion metal are sent to smelting.[105] A pilot plant has shown the feasibility of extraction of copper and silver from a low-grade Peruvian ore by the segregation process.[106]

Sulphur

Although the "inherent floatability" of pure sulphur is arguable, the naturally occurring element is readily floated by creosote or hydrocarbon oils, with sodium silicate as dispersant. Some use has been made of this in stripping disseminated sulphur from water in hot-liquor extraction. In the chief application of flotation to sulphide production (as distinct economically from similar methods used for sulphide removal), the main requirement is a grade of pyrite which will burn strongly; high recovery is a secondary consideration, economy in the use of small quantities of collector, frother, and perhaps acid being of prime importance.

Separative treatment to recover elemental sulphur cemented between the grains of the host rock must take precautions against corrosive attack. Fine crushing in rolls may be followed by pebble milling in a silica-lined mill, using scuffing in open circuit to free the sulphur. Flotation at a *p*H of 2 to 3 in acid-resistant cells follows, perhaps using only an alcohol frother as reagent. Recovery of sulphur from low-grade deposits by simple solvation has been proposed,[107] using dimethyl disulphide as the leaching agent. It might take over where grade was too poor for use of the Frasch process, in which superheated water is pumped down to liquefy and bring to surface the mineral. In Italy kilning is used for this liquefaction, part of the sulphur acting as fuel.

Talc

Talc is mostly processed by dry grinding and air classification. The specification for the paint industry requires more than $98\frac{1}{2}\%$ to be *minus* 325 mesh. Where exceptional purity is needed, flotation is possible, in which a wetting agent (e.g. alkyl sodium sulphonate) can be used.

Tin

Gravity treatment of cassiterite-bearing sands and gravels dominates industrial practice. Where hard-rock mining is used, or a roughed gravity concentrate must be upgraded after further grinding, emphasis is on the avoidance of over-grinding since slimed cassiterite is difficult to recover and tailings losses may be high. A beach-sand flow-sheet was given in an earlier chapter (Fig. 184). Detailed descriptions of current methods in which jigs, tables, spirals and tilting frames are integrated with close screening and classification have been made[108,109].

Among chemical methods which have shown laboratory promise is volatization of tin as its tetrachloride, a process which has had limited use in plant practice. Another proposal is based on the equations

$$Sn + CuCl_2 = Cu + SnCl_2$$

the cassiterite being reduced by carbon monoxide at 850°C before leaching[110].

Tintanium

The minerals mainly concentrated for titanium are rutile (S.G. 4·2, non-magnetic) and ilmenite (S.G. 4·5–5·0, weakly magnetic). Where these occur as beach sands they are separable by mixed H.T. separation and magnetic treatment, as shown in Fig. 266. Ilmenite ores are treated by flotation, possibly with a stage of wet magnetic separation. After conditioning with sulphuric acid, sodium fluoride (to depress associated apatite, and oleic acid or tall oil, a rougher float at pH 6 to 6·5 is made, and cleaned at a lower pH (5–5·5). Papers presented at the 1960 I.M.P.C.[111,112], showed in one treatment the importance of reconditioning the ilmenite rougher float at 70% solids between three successive cleaning operations. Recovery was improved by first desliming the pulp, and tall oil (the collector used with fuel oil) was emulsified before use. The other Paper described agglomeration-flotation of ilmenite without preliminary de-sliming. Upgrading proceeded in a series of twenty conditionings with tall oil and fuel oil, with an alkyl-phenol-polyglycol ether as emulsifier (Etoxol P-19).

Tungsten

The principal ores treated are wolfram (S.G. 7·1 to 7·9) which may be feebly magnetic, and scheelite (S.G. 5·9–6·1). In gravity treatment, care must be taken to concentrate the value at the coarsest possible grind, with avoidance of over-grinding. Jigs, tables, rag frames and spirals are used. Using flotation, scheelite ($CaWO_4$) is the only tungsten ore to have received important commercial treatment. It floats readily at a pH of $10\frac{1}{2}$ in softened water, using a carboxyl collector, sodium silicate as dispersant and tannin as a depressant for associated calcite, fluorite, dolomite, and apatite. The formidable difficulty is in obtaining sufficient separation from the gangue minerals to give an economic grade of concentrate, as the scheelite itself is readily depressed. Spraying of the froth with Aerosol OT has been recommended.

Roasting and/or leaching of tungsten ores and concentrates involves complex treatment to remove impurities which include tin, arsenic, phosphorus, manganese, bismuth, molybdenum, copper, antimony, zinc, lead and iron[113]. A shorter route for scheelite starts with a sulphuric acid leach followed by re-leach of the resulting tungstic acid with ethylene glycol[114].

Uranium

Twenty-four types of uranium deposit have been classified by Everhart[115] in accordance with associated rocks, minerals of value and type of deposit. The more important uranium minerals, as listed by Patching[116] are:

TABLE 44

MORE IMPORTANT URANIUM BEARING MINERALS

Mineral	*Composition*	*S.G.*	*Hardness*
Autunite . .	$CaO.2UO_3.P_2O_5.12H_2O$	3·0–3·2	2–2·5
Brannerite . .	Oxides of Ti, U, Th, etc.	4·5–5·3	4·5
Carnotite . .	$K_2O.2UO_3.V_2O_58H_2O$	4·1	1–2
Davidite . .	Oxides of Fe, Ti, U, etc.	4·46	6
Meta-autunite .	$CaO.2UO_3.P_2O_5.8H_2O$	3.0–3·2	2

TABLE 44—*continued*

Mineral	*Composition*	*S.G.*	*Hardness*
Meta-Torbernite .	$CuO.2UO_3P_2O_58H_2O$	3·68	2
Parsonsite . .	$2PbO.UO_3.P_2O_5.H_2O$	6·23	?
Pyrochlore . .	Na, Ca, Nb_2O_6F	4·2–4·9	5–5·5
Saleeite . .	$MgO.2UO_3.P_2O_5.8H_2O$	W3·3	2–3
Thucholite . .	Hydrocarbon with U, Th, etc.	1·5–2·0	3·5–4
Torbernite . .	$CuO.2UO_3.P_2O_5.12H_2O$	3·2–3·5	2–2·5
Uraninite pitchblende	$xUO_2.yUO_3$	6·5–10·6	5–6
Uranocircite .	$BaO.2UO_3.P_2O_5.8H_2O$	3·5–4·0	2
Uranophane .	$CaO.2UO_3.2SiO_2.6H_2O$	3·8–4·9	2–3
Uranothorite .	$ThO_2.SiO_2.UO_3CaO$, etc.	4·1–4·4	4·5–5

Of these, the most important economically are the oxides pitch-blende and uraninite. Veins of uranium minerals are rarely of mining width, and often are networks. This leads to dilution in mining by waste host rock. The more usual deposits are disseminated among other minerals. Physical pre-treatment aims at reducing the bulk which must be subjected to chemical attack: at removal of interfering ore constituents; and at recovery of associated values which can be separated after appropriate comminution. With the low head value of typical uranium ores (as little as 0·1 %) any roughing step which will remove a substantial fraction without undue loss is useful, since the physical methods employed in mineral dressing are cheaper than leach treatment. Interferants in acid leaching include acid-consuming carbonates (more than about 4%) chlorite, earthy hematite, apatite, and fluorides. Those inimical to carbonate leaching include sulphides (above 0·5% S.). Instead of a discard of gangue, selective sorting has limited use, the notable appliance being the Lapointe picker[117] which exploits the radio-active mineral's ability to signal its position on a miniature conveyor. DMS is used at Radium Hill to separate the heavy sulphides associated with the uranium from an impoverished gangue. The possibilities of froth flotation have received much research attention. Indirect upgrading by removal of gangue and direct flotation of values is an economic process in some plants. One operation, described by Read, Carman, and Gough[118], treats old tailings and current cyanide residues to obtain 40% recovery of U_3O_8 and 90% of the sulphur in a concentrate weighing 6% of the head feed.

The dominant chemical treatments are the acid leach and the carbonate leach. In the former a fairly thick pulp (60% solids) is agitated with sulphuric acid and oxidising agents such as manganese dioxide and sodium chlorate (to maintain the uranium in its reactive state as U_3O_8). The pregnant liquor is then removed and stripped. Where excessive carbonate in the ore precludes this method alkaline leach with sodium carbonate and bicarbonate is used. Oxidation is obtained either by pre-aeration or the use of permanganate, and hot pulps are sometimes needed (70°C or more). Carbonate leaching does not entail the costly materials of construction required to prevent corrosion in an acid leach plant, and treatment of the "pregs" is simpler, but both processes have operating difficulties.

Recovery of uranium from pregnant solutions can be performed by chemical precipitation, static-bed ion exchange, fluid-bed ion exchange or by the use of organic solvents. The first of these methods, based on the use of hydrogen peroxide, fluoride, phosphate, or carbonate in an alkaline liquor, is simple but costly. Ion exchange using anionic resins, is in general use for acid leach liquor. The uranium is seized by the anions on the resins, and periodically removed by nitrates and nitric acid, being finally precipitated with ammonia, NaOH, or magnesia.

Among the expressions used in connection with uranium processing and defined more fully elsewhere by the author[119] are the following:—

(A) Acid treatment process terms.
(S) Solvent extraction process terms.

(A) *Acid cure.* Sulphation to reduce acid consumption during leach.
(S) *Amex process.* Solvent extraction with long-chain amines dissolved in kerosene.
(S) *Aqueous liquor.* Uranium rich; pregnant; filtrate; solution. Sometimes pulp (in R.I.P.). Feed to extraction unit.
(A) *Banks.* Rubber-lined rectangular steel tanks with baffles, in which baskets containing resin are moved up and down through the pregnant solution.
(A) *Beads.* Sized resin spheroids, usually +20 mesh.
(A) *Break through.* Point during loading when uranium ceases to be completely absorbed by the resin column.
(S) *Carrier solvent or diluent.* An inert organic liquid, e.g. kerosene, in which the organic chemical used to extract the uranium from the aqueous liquor is dissolved.
(S) *Dapex process.* Solvent extraction with dialkylphosphoric acid dissolved in kerosene.
(A) *Eluant.* Solution used for elution, the process of dissolving uranium which has been captured by the beads.
(A) *Eluate.* Pregnant solution eluted from loaded resin beads.
(S) *Emulsion breaking rate.* Rate of phase disengagement (organic carrier from aqueous).
(A) *IX.* Ion exchange.
(A) *Loading.* Adsorption of uranium ions from solution by resins during loading cycle.
(A) *Loading factor.* Pounds of U_3O_8 which can be loaded on one cubic foot of new or regenerated resin.
(S) *Phase disengagement.* Separation of solvent (organic) from aqueous solution.
(A) *Poisoning* (1). Loading of resin sites with Mo, Zr, Ti, V, SiO_2, cobalticyanide, polythionate, thereby eliminating them as localities for uranium ion loading.
(S) *Poisoning* (2). Fouling of solvent by Mo, Co, Ti, Th, ferric iron.
(L) *Precipitation.* Removal of uranium compounds from solvent (stripping solution).
(A) *Resins.* Water-insoluble polyelectrolytes with high capacity for absorbing U_3O_8 from leach solution.
(A) *R.I.P.* Resin in pulp. The ion-exchange process applied in an acid-leach slurry from which the coarser particles of sand have been removed.

(A) *Sites.* Exchange positions on surface of bead where ions are captured by resin.

(S) *Solvent extraction.* Selective transfer of metal salts from aqueous solution or pulp to immiscible organic liquid.

Pre-concentration

Of the numerous oxidic uranium minerals, those commercially exploited include autunite, carnotite, and pitchblende or uraninite ($xUO_2.yUO_3$, etc.). All are of variable composition. Thucolite is mixed with hydrocarbons. Tobernite is a hydrated compound with copper phosphate. Uranophane is a compound silicate with calcium. Carnotite is a hydrated compound with vanadate and potassium. The ores treated commercially contain only minute quantities of these uranium-bearing compounds and include quartz, schists, shales, and dolomite. The problem of upgrading prior to leach treatment has engaged much laboratory research, and limited industrial application of its findings has followed. Three lines of attack have been opened up. *First* the uranium mineral might be concentrated. *Second*, associated minerals might be floated, taking with them the uranium. *Third*, gangue minerals (particularly those liable to interfere with subsequent chemical treatment) might be removed, leaving an enriched uranium-bearing tail.

In pilot-scale work Fitzgerald and Kelsall[120] report on a micaceous feldspathic sandstone impregnated with carbonates and carrying uraninite. Levin[121] has described testwork on thucolite in Witwatersrand ores and residues. 20 to 30% of the uranium floated with a frother only. Xanthate brought this figure to between 30 and 70%. With some ores oleic acid was a good collector. Altogether 80% of the mill-head reported in a 20% concentrate, though pre-cyanidation to recover gold reduced flotation efficiency unless counteracted by an acid wash. A valuable review of the research aspects of uranium concentration is given by Patching[116]. Lord and Light[122] describe work on uraninite and three other uranium-bearing minerals which gave high laboratory recovery and upgrading. Read, Carman, and Gough[118] describe the operation of two flotation plants. One treats cyanidation tailings on the Rand to give a concentrate containing 40% of the U_3O_8 in 6% of the feed weight. The other treats the residue from the uranium acid-leach plant. The reagents used are xanthate, Aerofloat 25, and Aerofroth 70.

Butler and Morris[124] have developed a series of organic collectors which give up to 15:1 ratios of concentration from heads assaying 0·1 to 0·5% U_3O_8 under laboratory conditions. In research on low-grade Saskatchewan ores Tinker *et al*[125] have used mercaptans under experimental conditions with promising results.

Laboratory tests on a Canadian ore were made in two stages[126], in which acid-consuming gangue minerals were first floated, and flotation of the uranium minerals followed. Working on ore ground in Faraday Uranium's mill circuit from 91% to 93% of the uranium was recovered in less than half the weight of the head feed. More than a third of the acid consumed in straight leach-extraction was saved. Sodium oleate was used in stage 1, and a proprietary emulsion (Acintol FA-1) in stage 2. Other research on amenability to flotation of various Canadian ores[127] reports encouraging response to the use of fatty acids, petroleum sulphonates and alkyl acid phosphates. Isocetyl acid phosphate was the most promising, at a *p*H of about 1·7. Modifying agents used as gangue depressants included aluminium sulphate, lactic acid and sodium silicate.

A detailed description of the Eldorado practice has been given by Behan[128], together with an account in the same Bulletin by Lillie and Tremblay of the acid-leaching process used. The ore contains pitchblende in complex and shear zones, and this value is either massive or disseminated in veins and stringers. Associated minerals include chalcopyrite, niccolite, pyrite, magnetite, cobalt, bornite, and hematite, with quartz, carbonates, jasper, chlorite, and chert. Pitchblende slimes readily, and the associated minerals are hardly worth the cost of transport from this remote region. The original (1933) treatment relied on sorting, hand-cobbing, and the use of Wilfley tables to recover radium. In 1936 jigging and flotation were added. Today the $+1\frac{1}{2}$ in. material is hand picked to give a high-grade concentrate. Waste is checked by Geiger Muller monitoring and discarded. The finer ore is jigged and barren magnetics are discarded. Further upgrading is produced by regrind, jigging, and flotation to remove copper minerals before leaching the tailing.

The chemistry of acid leaching requires oxidation of the uranium minerals to U_3O_8 as a precondition for reaction with sulphuric acid. Using manganese dioxide to complete the oxidation of uraninite, the reaction is:

$$6H_2SO_4 + 3MnO_2 + 3UO_2 \rightarrow 3UO_2SO_4 + 3MnSO_4 + 6H_2O.$$

Other methods of oxidation are possible, using such reagents as sodium chlorate and iron sulphate (ferric).

$$3H_2SO_4 + NaClO_3 + 3UO_2 \rightarrow 3UO_2SO_4 + NaCl + 3H_2O.$$
$$3Fe_2(SO_4)_3 + 3UO_2 \rightarrow 3UO_2SO_4 + 6FeSO_4.$$

The ferrous sulphate is re-oxidised thus:

$$2FeSO_4 + MnO_2 + 2H_2SO_4 \rightarrow Fe_2(SO_4)_3 + MnSO_4 + 2H_2O.$$

The reaction of a partly oxidised uranium ore with ferric sulphate and sulphuric acid is:

$$UO_2 + 2UO_3 + Fe_2(SO_4)_3 + 2H_2SO_4 \rightarrow 3UO_2SO_4 + 2FeSO_4 + 2H_2O.$$

The equation for the carbonate leach reaction in alkaline solution is:

$$U_3O_8 + \tfrac{1}{2}O_2 + 3CO_3^{--} + 6HCO_3^{-} \rightarrow 3UO_2(CO_3)_3^{4-} + 3H_2O.$$

These equations show how acid leaching works in the absence of excessive carbonate in the gangue. 4% was given above as the economic limit above which neutralisation of sulphuric acid by gangue minerals becomes excessive. At Shiprock, N. Mexico, an "acid-cure" method of pre-treatment was reported in 1955 which raised this limit toward 8%[129]. The ore is pugged with 10% of its weight of water and 20% of sulphuric acid (dry tonnage of feed). It is then left piled for six to eight hours. The uranium, vanadium, most of the lime, and some iron and aluminium go into solution when the material is pulped and agitated while 70% of the residual solids are discarded as a coarse tailing. This is possible because the uranium is interstitial in the grains of sandstone.

Secondary uranium minerals dissolve readily in dilute H_2SO_4, but the tetravalent oxides require oxidising conditions. Where suitable pyritic mineral occurs in the ore in sufficient concentration, it is the source of the sulphate. Otherwise, an oxidising agent such as pyrolusite is necessary. A description of Rand practice has been given by McLean and Prentice[130]. The pregnant liquor from acid leaching carries concentrations of other ions much in excess of the uranium content. Post-war research first worked on straight precipitation methods, but these gave way to the now widely used ion exchange (IX) processes.

Metal sorption on cation-exchange substances (natural zeolites and synthetic resins) has long been used in the removal of Ca and Mg from "hard" water. This form of activated resin is, however, useless because it is non-selective. Uranium recovery is based on anion-exchange resins. The metal sorbs as an anionic complex. When uranyl sulphate solution flows through a bed of strong base anion-exchange resin the uranium is absorbed, while substantially all other metal ions go through. From time to time loading is stopped, and the uranium is displaced by an acidified salt solution. In a pure pregnant sulphuric solution the uranium is present as the uranyl (UO_2^{2+}) ion or an undissociated sulphate (UO_2SO_4). If this is passed over such base IX-resins as De-Acidite FF or Amberlite IRA-400 (chloride form) one UO_2 group and three SO_4 groups are sorbed and four Cl^- ions are displaced from the resin. With a sulphate-form resin the sorption equation is

$$UO_2^{2} + SO_4^{2} + 2R = SO_4 \rightleftharpoons$$
$$R \equiv UO_2(SO_4)_3$$

The sulphate ion does not compete for resin sites, but bisulphate does. Resin can be loaded at *p*H 2 to 7-lb/ft^3, but if the *p*H falls to 0·5 this figure drops to 2-lb/ft^3, sites not occupied by the uranyl in sulphate anion being used by bisulphates. Operation in acid leaching is concerned to give the highest *p*H compatible with effective extraction, together with stable solutions. This requirement is met at a *p*H between 1·8 and 2. If the build-up of ferric sulphate in the circulating leach liquors is excessive a certain amount sorbs to the resin. This ceases if the *p*H is decreased toward 1·4.

At the point where uranium begins to "break through", the resin in the loading column is far from saturated and the uranium is of low purity.

Plants are so designed that two or more columns are in series, and that break through on the second column begins when the first column is saturated. Column 1 is now taken off line, column 2 receives the entering pregnant solution and column 3 replaces column 2 as the save-all. During the last part of the loading to saturation some iron is crowded off the resin.

The loaded resin is next eluted with an acidified mineral salt, the most-used being molar ammonium nitrate containing 0·2 molar nitric acid. The residual ferric ion is the first to be displaced and can be separately discharged to the barren solution storage if higher purity is desired. The uranium is precipitated from the eluate with alkali-ammonia if nitrate is the eluting salt. The cycle of operations is usually fully automatic.

One operating difficulty is the removal of the solids from the acid pulp before IX treatment. Strong flocculants must be used to overcome the dispersing effect of the chemicals, and this, in the case of the finely ground ore such as Rand cyanide tailings, calls for the use of glue or synthetic flocculating agents if filtration is to be fully efficient[131]. A floccule, however, consists largely of water, and this can lead to appreciable loss of pregnant liquor to tail. This dilemma can be avoided in the R.I.P. method, in which extraction is made direct from the pulp after classification to remove sand which might abrade the resin. The anion-exchange beads are placed in an acid-proof wire basket which is jigged slowly up and down in tanks arranged in series ("banks"). At one American mill[132] the pulp, after leaching, is mechanically classified through five Akins machines in series, a counter-current wash being given. Sands are discharged and the overflow, at −300♯ and *p*H 1·8, goes to the R.I.P. system. Its U_3O_8 value is 1 g./l. There are fourteen tanks, each having two baskets of 30♯ Ton-cap stainless steel wire. The load is a 10 in. layer of 20♯ resin and the baskets jig up and down in their bank in balance, at 6·3 cycles per minute. The resin bed expands and contracts during reciprocation. Seven banks are on adsorption and five on elution. One is being washed, first to remove slime and later to remove excess nitrate. The eluate assays 10–12 g./l. and a cycle takes three hours, loading being done counter-current. Pulp is introduced at the most heavily loaded bank and removed at that with the lightest load. Fresh eluant starts at the least loaded bank and leaves at the heaviest. The eluting liquid contains 72 g./l. of ammonium nitrate at *p*H 1·2 produced with sulphuric acid. Pregnant eluate is brought to *p*H 3.5 with lime and then agitated with MgO for four hours at *p*H 6·8 to precipitate the uranium.

A Wyoming mill described by Argall[133] uses tensioned screen material in its R.I.P. process and beads between 16 and 20♯. About 100 lb. of H_2SO_4 is used per ton of ore, with 2 to 5 lb. of MnO_2. Terminal temperature is 25°C and e.m.f. 400 millivolts. R.I.P. feed is at 8% solids and 30%−325♯. The IX feed, at *p*H 1·5, goes through eight banks on loading cycle, the baskets having a 16 in. stroke and 10 cycles/minute. IX tailing carries 0·01 g./l. of U_3O_8. The pregnant eluate is clarified before precipitation with MgO.

The following particulars from Hargrove[134] give a picture of the operation of fixed-bed acid treatment. The mill went to work in mid-1957, and produced its first yellow cake four days later. Construction was acid-resistant wood framing, stainless steel nails and bolts. Rubber covered impellers and pipes, as alternatives to rubber hose, or Saran piping where possible are used. The main uranium mineral is autunite in schistose cross-fractured rock, with some urano-thorite. Grinding releases a 50% solid pulp at −3%+28♯ and +47%−200♯ to Devereaux agitators where it is diluted to 18−20% solids and given 130-lb/ton of H_2SO_4. 96% recovery is obtained, the pulp being heated to 50°C between two stages of agitation at the end of which 75% of the value has been extracted. The pulp is now thickened and the solution goes to stripping. Underflow at 50% solids gets five further stages of agitation, heat and additional acid being used. The discharge is thickened, "pregs" being returned to the first stage of agitation while tails are washed and discharged. Unclarified solution is fed direct to one of four 8 ft × 15 ft columns each equipped with a bottom distributing system of piping covered by coarse sand. The base anion resin resting on this sand occupies 270 ft^3. Three columns are on stream during loading, with a pressure drop of 15 lb between bottom and top and an estimated flow rate of 5 to 6 gal/min/ft^2 of cross-section. A column is considered loaded when its liquor has the same uranium content at both entry and discharge. The second column in line should by then be showing a break-through value of up to 10% of the IX feed. The third column gives a virtually barren effluent.

After loading, the column is taken off stream and given a water-rinse to displace most of the remaining "pregs". This liquor is sent to the next column. Vigorous back-washing is now applied at 200–300 gal/min to loosen the bed and flush out slimes. Any resin carried out by this treatment is trapped on a 100# screen. The column is next eluted with ammonium nitrate in HNO_3, the eluate going to precipitation. The cycle is completed by acid rinse with H_2SO_4 to regenerate the Amberlite 400 anion resin, which has a uranyl nitrate capacity of 4·2 lb/ft^3. The concentration ratio in the IX step is 20:1 from a feed assaying 0·5 to 0·6 g/l. The eluate goes to three pachucas in series, lime slurry being added to the first two and gaseous ammonia to the third. The lime slurry is precipitated as gypsum and takes with it most of the iron as ferric hydrate. The product from the third pachuca is thickened, the underflow, which assays 1–2% uranium, being filtered and returned to the head of the leaching circuit. The filtrate is fed to a series of three pachucas where the *p*H is adjusted to a final 6·8–7·0 with gaseous ammonia. The overflow is thickened, heated to 90°C, re-thickened and filtered. The reagents consumed per ton of feed are 115 lb H_2SO_4, 3·3 lb HNO_3, 6½ lb CaO, 0·28 lb Separan (a Dow flocculant) and 2·8 lb anhydrous ammonia. The power is 35 Kwh/ton. The plant is well "instrumentated" and most controls are air powered.

An alternative to fixed-bed and R.I.P. methods in the stripping of pregnant liquor is the moving-bed[135]. One system has ten columns, six of which are grouped in parallel lines of three for adsorption. Three are used for elution and the tenth back-washes uranium-saturated resin beads. Piping is so arranged that resin in any adsorption column can be transferred to backwash, from backwash to any elution column and from this back to any adsorption column.

For resin recovery the use of flotation has proved interesting in laboratory research[136]. The R.I.P. process cannot tolerate more than some 10% of solids in the pulp, but direct recovery of loaded resin by flotation would not be subject to this limitation, and the problem of resin abrasion by coarser particles might be simpler. Cation-exchange resins are readily floated by amine-type collectors and anion exchange ones by all anionic collectors. Tests showed resin flotation to be clean, rapid and nearly complete while loading capacity and elution were not upset by bead conditioning for flotation. Bead size in the *minus* 16 *plus* 400 mesh range is not critical. Regeneration of cobalt-poisoned resins in which boiling with either thiocyanate or nitrite is

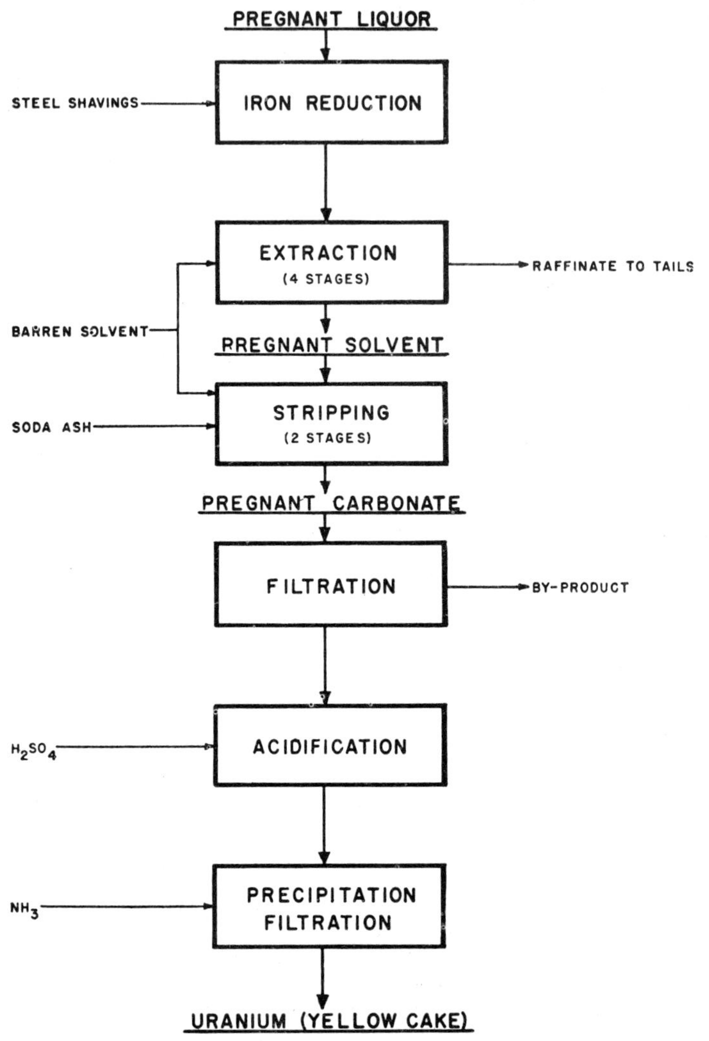

Fig. 333 *The Dapex Process*

used has been described[137] Three Papers on the natural leaching of uranium-bearing Portuguese ores, which for some years have been heap-leached, has been investigated and reported[138]

Solvent extraction as an alternative to IX was described in two papers at the A.I.M.E. New Orleans meeting in February 1957. That by Brown, Crouse, and Coleman described the research on the Amex and Dapex processes. The former uses long chain amines and the latter di-alkyl-phosphoric acid, dissolved in kerosene. The latter has the practical virtue of reagent availability. The second paper by Hazen and Henrickson, describes continuous pilot and plant scale use of the Dapex process. The solvents used are di-2-ethyl hexyl phosphoric acid and tri-butyl hexyl phosphate in a high flash-point kerosene. The flow-sheet (Fig. 333) starts by passing the acid-leach liquor from the thickener overflow through a bed of steel swarf packed 6 ft deep in a 5 ft diameter wooden tank. By keeping the e.m.f. between 275 and 300 millivolts, stripping of this ferrous iron by the organic solvent is avoided. The stripping is performed in four mixers in series, the aqueous phase gravitating down the line while the solvent proceeds counter-current by the aid of air lift. The plant has a flow rate of 120 gal/min from treatment of 350 tons/day. Each mixer-settler is 16 ft in diameter and 6 ft deep. Stripping of the Dapex effluent is in two further mixer-settlers, using a 10% solution of sodium carbonate in water. After destruction of the carbonate with acid, the uranium is precipitated with ammonia or magnesia.

Carbonate leaching in America was first practised at Beaverlodge and has been described by Hannay[139]. The ore is ground to 70%–200# in a solution of sodium carbonate and sulphate, topped up from tailings filtrate to a density of 1·12. Classifier overflow is thickened to 55% solids, heated, oxidised to the hexavalent state and leached (see page 784). Pulp density is 55% to 60% solids and the leach solution contains 30 g/l. Na_2CO_3 and 20 g/l. $NaHCO_3$, the latter maintained by CO_2 from washed flue-gas. Leaching is partly in pressurised vessels and partly in pachucas of special design. The former use 80 lb to 85 lb pressure and 110°C to 115°C. Extraction to 90% < 94% recovery in autoclaves takes 16 hours, and in pachucas, at 77°C, nearly 100 hours. Discharges from both leach circuits are mixed, thickened, and given counter-current wash in tray thickeners, with the flocculating aid of Separan. After final filtration the cake is repulped and deslimed in cyclones, the sands going underground as backfill and the slimes to waste. Leach liquor is clarified, precipitated with NaOH, and filtered. The precipitate is shipped and the barren solution re-carbonated with CO_2 from flue gas before return to circuit.

The use of dodecyl phosphoric acid as the solvent has been described by Black and Koslov.[140] Their paper is abstracted in *The Mining Magazine* for August, 1958, pp. 78–84.

Some Industrial Treatments

Test-work on the uranium deposit discovered in Athabasca in mid-1952 led to the choice of the sulphuric acid—sodium chlorate process for leaching, with recovery by IX followed by precipitation with magnesia. A 60% solids pulp passes through a series of twelve Dorr agitators each of which receives 70 c.f./min of air on the 2,000 ton/day throughput. Sulphuric acid is automatically controlled into the first tank and maintains the *p*H between 1·8 and 3·0. Sodium chlorate is also added here, in a slight excess controlled by measurement of the e.m.f., which is held at 400 millivolts in the first two tanks. Leaching time is about 16 hours.

Bicroft[142], in Canada, started acid leaching in 1956, using conventional fixed-bed columns. The main operating difficulty has been with polythionate poisoning of the resin. The pegmatitic ore yields a low-grade leach liquor at a *p*H of between 1·8 and 2·1, carrying 0·4 g/l. of U_3O_8, ferrous and ferric iron, fluorine, chlorine, thoria, alumina and over 15 p.p.m. of polythionates. Following backwash with warm water the IX columns are eluted with sodium chloride solution acidified with sulphuric acid. Eight bed volumes are sent to precipitation and nine to re-cycling eluate. A "bed volume" is that of the liquor which corresponds to the space occupied in the column by the resin and its voids. The eluate is stripped by precipitation with magnesia, the reaction being $UO_2SO_4 + Mg(OH)_2 = UO_2(OH)_2 + MgSO_4$. The sulphate building up by this reaction is held near 120 g/l. by periodic discharge of neutral eluate. About a quarter of the 0·9 g/l. of fluorine in the pregnant liquor is adsorbed and then eluted with the uranium, from 5% to 10% later precipitating during magnesia reaction while the rest re-cycles with the eluate. Every few days some neutral eluate is withdrawn and agitated with slaked lime at *p*H 9·1 to precipitate this as calcium fluoride, at the same time as sulphate build-up is being controlled.

The IRA-400 resin used at Bicroft shows a plant loading of 3·6 lb/ft^3 of U_3O_8. When fouled by polythionates this falls to 2 lb/ft^3, at which point resin regeneration is necessary. Break through from the first column when the resin is fouled by slime or polythionates starts at some 65% of saturation. The general formula of polythionates is $S_nO_6^{=}$ with n between 2 and 6. They come from any sulphur source, such as sulphide in the ore. One preventative is to maintain the free leaching acidity above 5 g/l., when sulphates instead of polythionates are formed but at Bicroft this would lead to excessive consumption of acid, which is held in a free range between 1 and 3 g/l. Three control variables are of practical use—temperature, retention time and point of addition of sodium chlorate as an oxidising agent. Poisoning is seasonal, being worst in spring and autumn. Resin regeneration is done after these seasonal build-ups of poison by taking a set of four IX columns out of service for cleansing. After elution the column receives a weak NaOH/NaCl wash, the resin being stirred with an air lance to prevent channeling. It is then flushed with the same solution, washed with water, then with normal sulphuric acid and finally with water before being returned to service. The alkali decomposes the polythionates ($2S_5O_6 + 6(OH)^- = 5S_2O_3 + 2H_2O$), the sodium chloride speeding this reaction. Fluoride is removed at the same time ($ThF_4 + 4NaOH = Th(OH)_4 + 4NaF$), the thorium hydroxide dissolving during re-acidification of the resin.

The moving-bed system of IX following acid leach is used by Consolidated Dennison in a 6,000 ton/day plant[143]. Maintenance of anti-acid protection for structures, machines, tanks and piping are an appreciable item of operating cost. The ore minerals include brannerite, uraninite, monazite, sericite, pyrite and rutile. A long leaching period in strong acid at raised temperatures is necessary unless a good addition of chemical oxidants is made. Leach is in three parallel lines, each headed by two pachucas and followed by six mechanical agitators with wood-stave tanks and stainless steel gear. Live steam and strong sulphuric acid are introduced in the pachucas and steam can also be blown in further down-line. Sodium chlorate (2·72 lb/ton milled) is added at the second pachuca and average working temperature along the line is 57°C. The acid strength falls from 85 g/l. to 70 g/l. along the leaching line, and retention time is 50 hours. Agitator maintenance arising from corrosion accounts for a third of the milling cost. The filtered pregnant solution is clarified in leaf filters and carries from 9 to 19 g/l. of U_3O_8. Resin, after loading, is hydraulically transferred from the adsorption columns to the backwash column and thence to elution. The advantages claimed include lower installation cost, higher resin loading, lower consumption of nitrate (4 lb/ton of ore), better concentration with lower cost of precipitation, and better shake-up and slime removal. Against these, attrition of the resin may be somewhat higher and loss of barren solution a little larger. Amberlite IRA-400 is used and loading is of the order of 4·6 lb/c ft at *p*H 2·2. Some 400 tons of eluate, assaying 19 g/l., are produced daily and precipitated in two stages. In the first the *p*H is adjusted with lime slurry to 2, and sulphate is removed as gypsum. Magnesia is then slurried in to bring the *p*H to 3·2 for precipitation of iron and other impurities in a series of three pachucas and a time of five hours. The purified eluate next goes to three mechanical agitators in series, and magnesia slurry raises the *p*H to 5 and at discharge to 6·5. The emerging slurry is thickened for removal of uranium, overflow returning to eluate tankage.

A general description of solvent extraction techniques is given in a trade bulletin[144] and an interesting account of R.I.P. leaching coupled with solvent extraction in another from the same source[145]. The cleaning of process water used in the carbonate leach plant at Homestake has also been described[146].

Zinc

The conventional flotation of sphalerite, usually following that of associated galena, was considered under "Lead". The surface is activated by copper sulphate in a pulp brought to a *p*H of 8–9 with lime, and a straight xanthate float follows. More complicated is the problem of the oxidised zinc ores. These include smithsonite ($ZnCO_3$), hydrozincite ($ZnCO_3Zn(OH)_2$), hemimorphite ($Zn_2SiO_3(OH)_2$), and willemite (ZN_2SiO_4). Research work and practical mill circuits are described by Rey *et al*[147,149]. With a mixed ore of lead and zinc sulphides and oxides, the treatment starts with sulphide flotation. Tails are then conditioned, sulphidised, and given collector conditioning. If amines are used, the slimes must first be removed or neutralised to avoid heavy reagent consumption and contamination of concentrates. They can be neutralised with soda-ash, sodium silicate, and such polyphosphates as calgon. Organic colloids such as starch and carboxymethyl cellulose are also helpful where desliming in cyclones is avoidable. As slimes carry zinc, the question of their handling is economic, and depends partly on the zinc market price. Unlike its action with sulphide minerals, sodium sulphide in reasonable excess has no depressing effect on oxide zinc minerals.

It acts therefore as both a regulating agent and a pH control, working best with amines at between pH $10\frac{1}{2}$ and 11. Rey (*supra*) suggests that the effective collector is a free amine liberated according to the increasing alkalinity of the pulp.

$$R{-}NH_3^+ + OH^- \rightleftharpoons RNH_2 + HOH.$$

Only primary amines have collecting values. They are preferably aliphatic, in the form of soluble acetates or hydrochlorides. Frothers include pine oil, Dowfroth and alcohols. The amine is added with the frother, immediately after the sulphide.

Plant-scale flotation of oxidised zinc ores following sulphidization at 50°C is practised at Ammi[149]. Activation with copper sulphate and flotation with amyl xanthate completes the reagent plan. The method does not suit strongly ferruginous ores. An alternative treatment for the finer fraction of the zinc oxidic minerals uses lauryl amine emulsified with pine oil and a mineral oil with sodium sulphide and a pH of 12.

Some Generalities

This description of typical industrial treatments is necessarily incomplete and has omitted many important ores and elements. Practice improves continually, as new processes are developed and fresh demands are made by the consuming industries served by mineral processing. Some basic principles underlie the planning of most, if not unfortunately all flow-sheets. A summary of the more important is offered in conclusion.

(1) Avoid over-grinding.
(2) Avoid surging and irregular rate of flow at each stage, using automatic aids wisely.
(3) Watch the state of return water and the effect of build-up of detrimental constituents.
(4) Take products, whether values or tails, out of the main circuit coarse and early.
(5) Make good use of roughing followed by cleaning.
(6) Watch the ore and its products for by-product values.
(7) Stockpile any middlings not at present worth treating but of potential value.
(8) Maintain routines, including mill cleanliness and safety practises.
(9) Operate planned maintenance schedules and avoid the need for emergency action.
(10) Break down the cost per ton of each unit operation in terms of labour, material, power and equipment and graph monthly changes.
(11) Make regular analyses of values in each size of tailing, and use the information thus given to improve grinding and treatments in the zones of maximum loss of value.

Mixed ores

The sulphides of lead and zinc, together with lesser amounts of those of other metals, commonly occur in economic quantities and are separately recovered. Where liberation mesh permits, galena can be recovered by jigging and tabling before finer grinding is used to free the species involved at a mesh suitable for differential flotation.

The normal procedure in differential flotation is to activate and float the minerals successively. An alternative commonly practised is to make a bulk float of two or more minerals, and then to differentiate them by depressing one in a secondary flotation. An ore containing payable quantities of copper, lead, zinc, and iron sulphides, together with, say, barytes, might be treated as in Fig. 334.

This example of bulk flotation followed by successive differentiation is taken from pre-war practice at Rammelsberg Mine in the Harz Mountains. Bulk flotation may be a good starting-point when soluble metal salts in the pulp have activated more than one of the sulphides, so that it is easier to lift two together. One sulphide in the mixed concentrate is then deactivated as shown in Fig. 334, where lead is depressed from a bulk copper-lead float. Another case in which bulk flotation might be chosen would be that of an ore in which the valuable sulphides were coarsely aggregated with respect to the gangue but so closely interlocked with one another as to require fine grinding

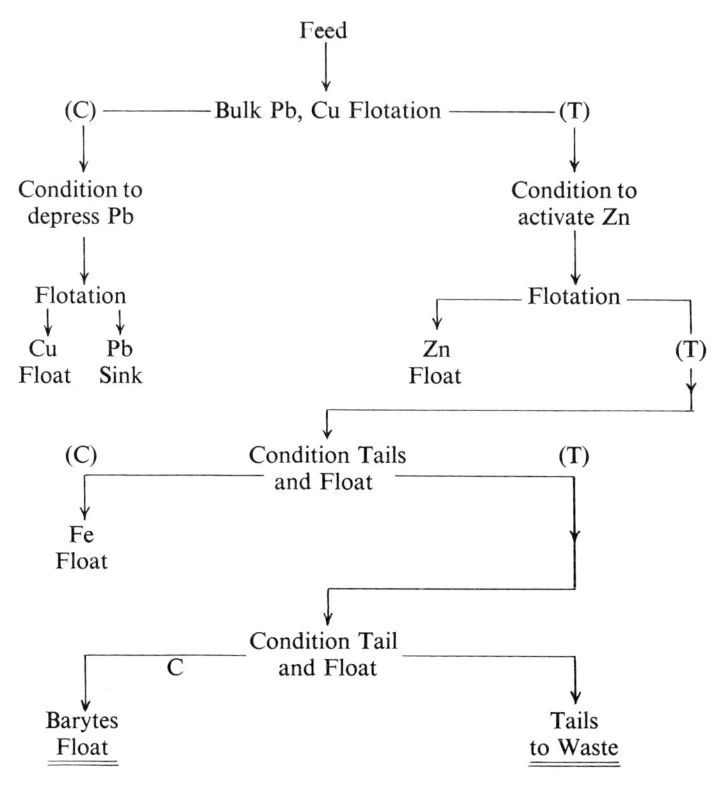

Fig. 334. *The Rammelsberg Flow Sheet*

before separation. Bulk flotation might be used to reject the gangue and thus reduce the tonnage sent to fine grinding. The general principle in differential flotation is to use a collector in starvation quantities, after selectively conditioning the pulp to promote activation of the first mineral to be raised. As a certain amount of activation of the second mineral usually occurs under such conditions, a lively and fragile froth should be maintained. Various frothers derived from alcohol have been developed to aid in obtaining the desired texture.

Lead-Zinc Differentiation

The main factors which influence the behaviour are given by Rey[79] as:

1. Abundance and nature of iron sulphides.
2. Degree of oxidation.
3. Basic or acid nature of the gangue.
4. Presence and nature of copper minerals.

Basic gangue contains sufficient available calcium to neutralise acid and to precipitate as hydroxide any ferrous sulphate derived from oxidation. In aerated water the reaction then follows:

$$2FeSO_4 + 2CaCO_3 + 3H_2O + \tfrac{1}{2}O_2 = 2Fe(OH)_3 + 2CaSO_4 + 2CO_2$$

The sulphates of lead and zinc react similarly, nearly to completion. The gangue tends to control surface contamination of the particles through its influence on the solubility products. If sulphides are oxidised in an acid pulp, they remain bright and highly floatable, but become contaminated by basic salts when oxidised at about *p*H 7. From this Rey proposes a classification into four main ore types, arranged in increasing order of difficulty (as regards differential flotation):

(*A*) Unoxidised ores.
- (1) Galena-sphalerite or galena marmatite.
- (2) The same, plus increased amounts of iron sulphides (pyrite, marcasite, pyrrhotite). This class can be subdivided into ores with low, medium or high content of iron sulphides.

(*B*) Oxidised ores—acid gangue.
- (1) Weakly acid ores.
- (2) Strongly acid ores (galena-sphalerite-anglesite).

(*C*) Oxidised ores—basic gangue (galena-sphalerite-cerussite, possibly smithsonite).
- (1) Without iron sulphides.
- (2) With iron sulphides.

(*D*) Ores containing copper minerals of secondary origin.

A generalised reagent plan for the successive flotation of galena and sphalerite follows these lines. The pulp is conditioned first for activating the galena. *p*H is controlled usually by soda-ash, but sometimes by lime, to a figure between neutral and *p*H 10, and a dispersant such as sodium silicate may also be employed. If there are traces of copper in the ore, from 0·1 lb/ton to 0·5 lb/ton of sodium cyanide may be needed to prevent random activation of the sphalerite during the collector-coating of the galena. To aid in depression of the zinc, a solution of zinc sulphate may also be used. The de-

TABLE 45

UNOXIDISED ORES: USUAL REAGENT FORMULAS AND FLOTATION RESULTS
(*After Rey*)

Proportion of Iron Sulphides	*Alkali (g./ton)*		*Depressant (g./ton)*		*Other Reagents (g./ton)*		*Grade Concentrates*	
None	CaO	0–200	NaCN	0–20	Zn Circuit $CuSO_4$	 250–500	Zn Pb	52/62 0·5/0·1
Pb Circuit	Na_2CO_3	0			Pb Circuit None		Pb An	75/80 0·5/3
Low								
Pb Circuit	Na_2CO_3	0–100	NaCN $ZnSO_4$	0–100 0–200			Pb Zn	60/78 ·7/5
Zn Circuit	CaO	0–750			$CuSO_4$	300–600	Zn Pb	62/60 ·65/·3
Medium								
Pb Circuit	Na_2CO_3 or CaO	0–1000 0–500	NaCN $ZnSO_4$	30–100 0–350			Pb Zn	60/78 2·5/7
Zn Circuit	CaO	500–2500			$CuSO_4$	300–600	Zn Pb	52/60 ·65/·3
High								
Pb Circuit	CaO	200–500 (in mill)	NaCN $ZnSO_4$	150–300 (in mill) 0–500	Na_2SO_3 $NaHSO_3$	or 0–500	Pb Zn	55/70 3/8
Zn Circuit	CaO	500–2500			$CuSO_4$	400–800	Zn Pb	48/57 1/3

* Concentrate grade varies with marmatitic or sphaleric composition.

pressant effect of $ZnSO_4$ is obscure. It either reacts to form zinc cyanide, which concentrates near the surface of the sphalerite and has a strong wetting effect, or the zinc itself makes a common-ion concentration to the lattice, the wetting sulphate making the particle more hydrophilic. When conditioning with these reagents has "complexed" the copper ions, wetted the sphalerite, and cleaned the galena at a *p*H established by test work as most suitable for the specific ore pulp, the collector agent is added sparingly. Ethyl xanthate or an aerofloat is used, the latter being preferred in many cases because it is less powerful and therefore less likely to activate any of the sphalerite. The frothing agent, also used sparingly, may be cresylic acid. The galena float is then taken, rather gently in the rougher section and strongly in the scavenger cells.

The tailings from the galena float are now conditioned with copper sulphate at a *p*H of about 9, reached by adding lime. Lime depresses pyrite, which would otherwise become activated by the copper ions and float. Lime also precipitates most of the copper, leaving only enough of the 1–2 lb/ton normally added, to plate the sphalerite. In sub-arctic conditions steam coils may be used to speed up reaction which at normal mill temperatures takes 10–20 minutes. The reagent addition and *p*H must be so managed that only a trace of copper ion is in solution and that it is renewed as required from the copper hydrate precipitated by the lime. This trace is selectively deposited on the sphalerite, and a collector agent is then added to complete the conditioning. Aerofloats are sometimes used, but ethyl xanthate is the preferred collector. Any good frother such as pine oil is fed in just ahead of aeration.

Rey tabulates in grams per ton the changes in auxiliary reagents which accompany increase in iron sulphide (Table 45). The progressive fall in grade and in crisp separation of the lead from the zinc is accompanied by reagent increases. These are required to correct the effects of soluble salts and to prevent oxidation of iron sulphides to ferrous sulphate, which depresses galena, consumes xanthate, and complexes cyanide. Marmatite (a dark zinc sulphide with ferrous sulphide in solid solution) is less easy to activate and float then sphalerite. Activation may be improved by a short preconditioning with ammonia.

With increasing acidity of ore the lead is partly oxidised toward sulphate, but the zinc remains bright, since its soluble sulphate washes away. Sulphidisation of the oxidised lead is impracticable because of the concentration of ferrous sulphate and acid. Washing, desliming, and mild mechanical attrition may be beneficial.

Ores of Type C (Rey's classification) are hard to differentiate as regards lead and zinc floats. Calcium sulphate is present where difficulties are met. If cerussite is to be floated, it may be floated with the galena using sulphidisation and thus risking trouble later when copper sulphate helps xanthate to entrain gangue into the zinc froth. This might be so serious as to require removal of some water or precipitation of excess sulphide ion with ferrous sulphate. Again, cerussite may be made the final float in the series galena-sphalerite-pyrite-cerussite.

If copper is present in an insoluble form such as chalcopyrite no difficulty should arise, but when the more soluble copper sulphides and oxidised forms occur they ionize. The sphalerite is then surface-activated and preventive cyanide is rapidly consumed, so that staged additions may be desirable. Copper-lead-zinc ores present a variety of problems, particularly when they are required as separate concentrates. Since the copper-surfaced zinc floats with the copper, one method is to make these a bulk float followed by selective separation. Today the more usual practice is to bulk-float the copper with lead and then to separate them by the use either of cyanide, sulphur dioxide or dichromate depression.

At one plant the bulk float is conditioned with SO_2 gas at 35% solids ($1\frac{1}{2}$ to 2 lb/ton) and then for twenty minutes with sodium dichromate (3 to 5 lb/ton) to depress the lead while refloating the copper at a *p*H of 5. When the depressing agent is cyanide, this can be complexed to zinc cyanide if gold would otherwise be leached from the ore and lost. The Cyanamid reagent 675 (a product of zinc oxide, HCN and NH_3) depresses copper. Any of these reagents are consumed at a rate of about $\frac{1}{2}$ lb per unit of copper per ton. McQuiston[80] suggests that choice of the method depends largely on the zinc content and the lead/copper ratio (Table 46).

TABLE 46

After McQuiston

Mine	*Major copper Mineralisation*	*Method of Separation*	*Approximate Lead: Copper Mineral Ratio*
Buchans	Chalcopyrite	SO_2	2:1
Idarado	Chalcopyrite	Cyanide	1:1
San Francisco	Chalcopyrite	SO_2	5:1
Casapalca	Diversified	SO_2	4:1
Mahr	Chalcopyrite	Cyanide	2:1
Tsumeb	Tennantite	Cyanide	2:1

Some mills have difficulty owing to the presence of a slow-floating galena which, having failed to come out in the lead section, floats with the zinc. This adulteration of concentrate may be penalised by the purchasing smelter. In one mill the zinc float is conditioned with cyanide to remove the copper surface from the sphalerite, and then refloated to produce a lead froth and a zinc tailing. Another method is to depress the lead with potassium dichromate in a cleaning stage or in a special refloat circuit. Usually chromate, or dichromate, is used only to depress lead from a bulk copper-lead float. Another type of differential separation of a bulk float is achieved by the destruction of collector agent on one of the sulphides. One commercial method is to steam-heat a flotation concentrate of mixed molybdenite and copper sulphides, when the xanthate can be selectively removed from the copper minerals. The mixed concentrate, after this distillation heating, is refloated, the Mo rising and the Cu sinking. Another method, practised in the Congo, uses an oxidising agent to decondition one of the sulphides in a mixed float, and then refloats and separates them.

Most sulphides can be differentially floated. The separation of copper from iron is difficult with copper ores other than chalcocite, since the steps taken to depress iron have some effect on the iron in the copper mineral.

Tight pH control, and exploitation of the depressing effect of lime upon pyritic minerals, usually solves the problem. Graphite can be floated with diesel oil or paraffin, and may sometimes be taken as a scum from the surface of classifiers if the amount, though a nuisance, is insignificant. Differentiation of copper from zinc is often difficult, owing to the tendency of the latter to acquire a copper surface. The use of cyanide *plus* soda-ash is indicated, care being taken not to exceed the pH at which the copper can sorb collector.

The following examples are taken from modern commercial practice.

Bluebell (*Consolidated Mining and Smelting Co. of Canada, Riondel, B.C.*)

This mine, discovered in 1844, has been working since 1890. The present flotation plant, which treats 700 tons daily, opened up in 1952. The ore is a heavy sulphide replacement (lead and zinc) in sedimentary limestone, with schists and pegmatite. Galena and marmatite are coarsely crystalline, with segregated "chimneys" of pyrrhotite. Minor pyrite, chalcopyrite and arsenopyrite occur. The gangue is limestone, with some coarse quartz, carbonates, and FeMn silicate. Oxidation is extensive.

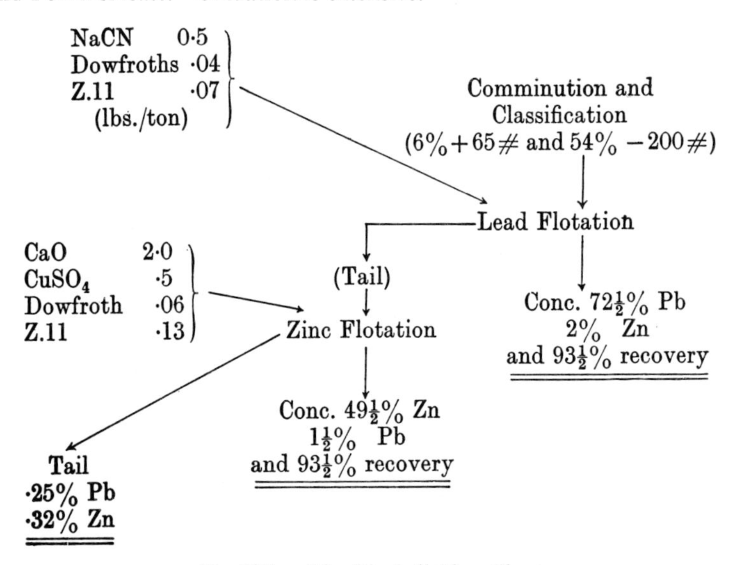

Fig. 335. *The Bluebell Flow Sheet*

This flow-sheet (Fig. 335) has been simplified in block form from one given by Walton[81]. It illustrates the typical sequence in a simple separation, with low reagent consumption and high specific recoveries.

Sullivan (*Consolidated Mining & Smelting Co. of Canada*)

The operation of this, the parent Company of Bluebell, is described by Brock[81]. The mill works a 5-day week at 11,000 tons daily, in a highly developed operation on an intimately associated galena-marmatite-pyrrhotite-pyrite ore with useful by-product constituents. This fine-grained complex calls for fine grinding in controlled stages, and a special "de-zincing" operation in which the lead recleaner concentrate is treated at 40°C with sodium dichromate before removing the bulk of its residual zinc as a float. A block flow-sheet (Fig. 336) shows the increased complexity dictated by the nature of the ore.

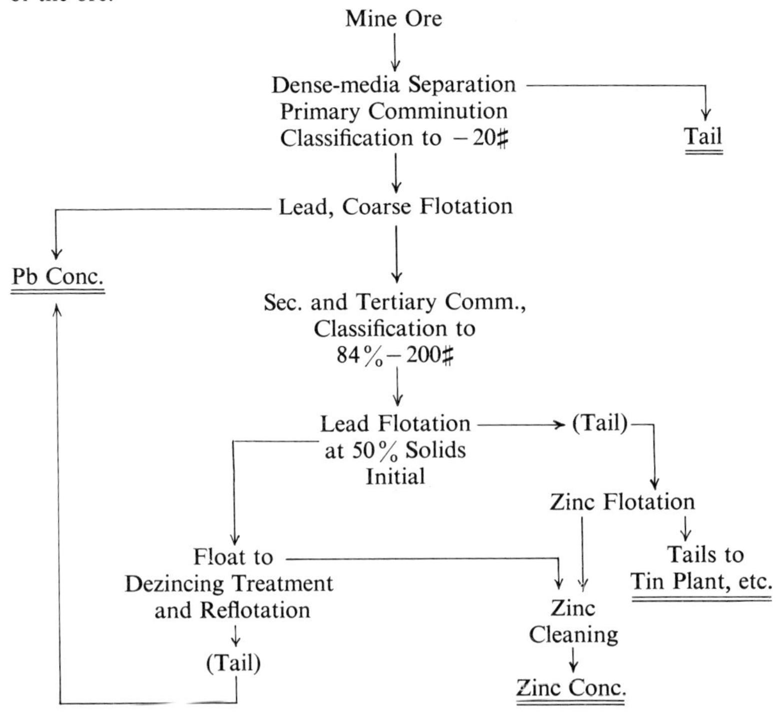

Fig. 336. *Sullivan B.C. Flow Sheet*
(*Note all flotation middlings are returned to secondary comminution.*)

The final lead concentrate assays 65% Pb and 3% Zn. The final zinc averages 45% Zn and 9% Pb. The reagent plan is summarised in Table 48.

TABLE 47

SULLIVAN FLOTATION REAGENTS

lb./ton (original feed)	*Lead Flotation*	*Zinc Flotation*	*Desliming*	*Total*
Cyanide	·063	—	—	·063
Lime	·77	·39	·39	1·55
Z.11 Xanthate	·14	·07	·01	·22
Dowfroth	·006	·008	—	·014
Dichromate	—	—	·063	·063
Cresylic Acid	·008	—	—	·008
Copper Sulphate	—	·58	·29	·87

Sullivan has developed automation and rapid assay methods of control over the years, and valuable contributions to milling literature have been published.

At the H.B. mine (C.M. & S. Co., Salmo, B.C.) talc is removed by flotation with 0·04 lb/ton of methyl isobutyl carbinol ahead of the main flotation[81] in an otherwise conventional treatment. More than one-fifth of the PbZn tailing loss occurs here, despite cyanide depressant.

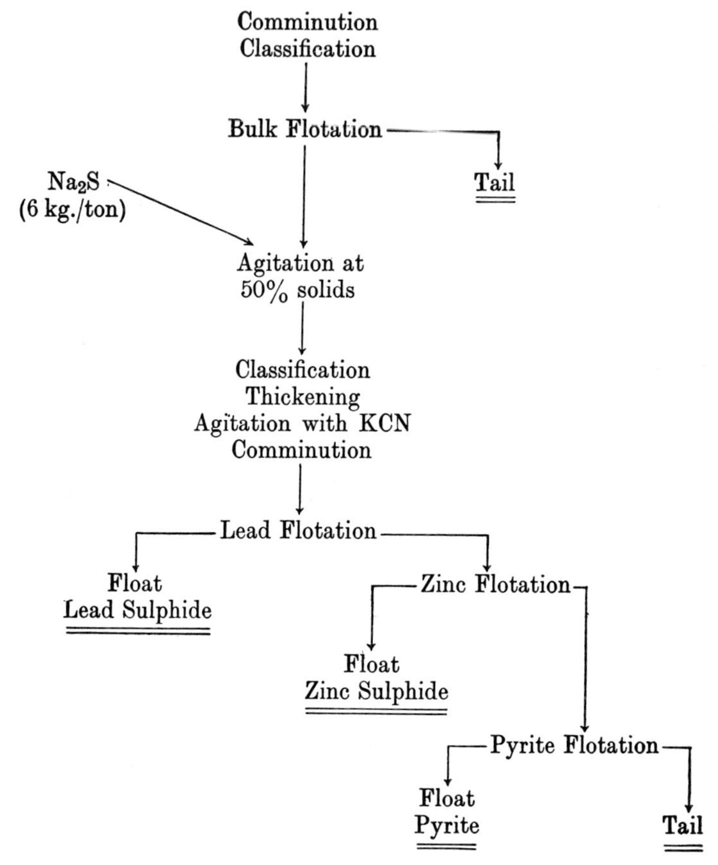

Fig. 337. *The Leninogorsk Flow Sheet* (Pb, Zn, Fe)

Ammeberg, Sweden

Here the ore sulphides are sphalerite, galena and pyrrhotite with quartz, felspars, mica, calcite, garnet, pyroxene, and hornblende. Crystallisation varies, but a liberation limit of 160μ is worked to, liberation of sphalerite from galena being still finer, at 50μ. The mill receives, in addition to its local ore, the sink product from the dense-media plant at the associated mine at

Zinkgruvan. Reagents used include soda-ash, cyanide, zinc sulphate, ethyl xanthate, sodium silicate, copper sulphate, and pine oil[82]. The treatment is on standard lines.

Leninogorsk, U.S.S.R.

A completely different method is used here. It is based on the use of soluble sulphide depression. Treatment starts with bulk flotation of the sulphides (Fig. 337).

This 7000 tons/day operation at 45% −200# replaces a selective method which required a 60% −200# grind. Bulk sulphide concentrates from the recleaning stage are delivered at 40% to 50% solids (to economise on Na_2S) into three conditioners in series, giving a total contact time of 5 to 10 minutes. Here vigorous agitation breaks down residual froth and allows the sodium sulphide to replace the collector agent on the minerals. Most of the reagents are next removed in a wash-operation in two classifiers and a thickener (Fig. 338), at a total use of fresh water of 1·2 m^3/ton of concentrate.

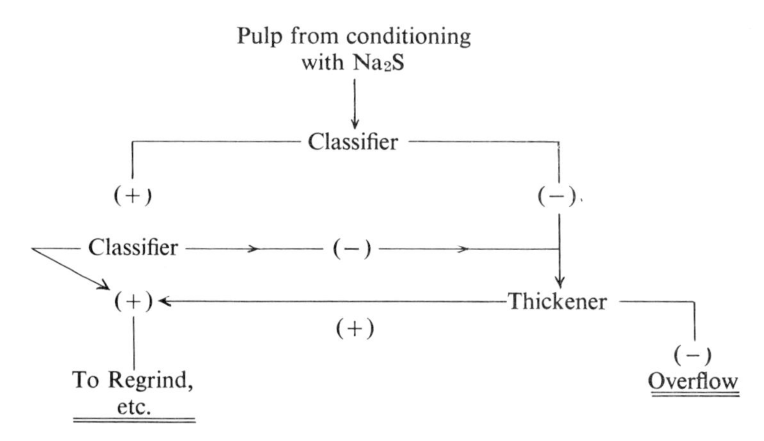

Fig. 338. *Details of Wash Treatment*

Desorption leaves the mineral surface sulphidised and responsive to differential depression. The remaining sodium sulphide in the liquid is removed by adding zinc sulphate

$$Na_2S + ZnSO_4 = ZnS + Na_2SO_4.$$

As the initial tonnage has now been drastically reduced, intensive regrinding and reagent treatment are correspondingly cheapened and simplified. At Leninogorsk the tailings losses of both lead and zinc have been substantially decreased, and concentrate grade improved, while reagent and treatment costs have gone down.

Tsumeb, S.W. Africa

A complex ore is treated in this plant. Work started with the milling of 20-year-old dumps of heavily oxidised Cu-Pb-Zn minerals on the evidence of batch testwork. Much was learned as the mine developed and the flow-sheet was adapted to deal with newly exposed problems. The deposit contains forty-one valuable minerals of copper, lead, zinc, vanadium, and germanium. In the upper levels the copper is oxidised. In the middle zone it becomes chalcocitic and at depth changes to tennantite. In a Paper describing this complex but outstanding operation by Ratledge, Ong, and Boyce[83], stress is laid on the importance of adding the xanthate collector in the sulphide section to the grinding mills rather than in a later conditioning stage. Tests confirmed that with this ore sulphides which have been finely ground do not respond so well as comparatively granular material. Right through the ore body the zinc mineral has been activated by soluble metal salts, and strong depressing action is necessary to inhibit its premature flotation.

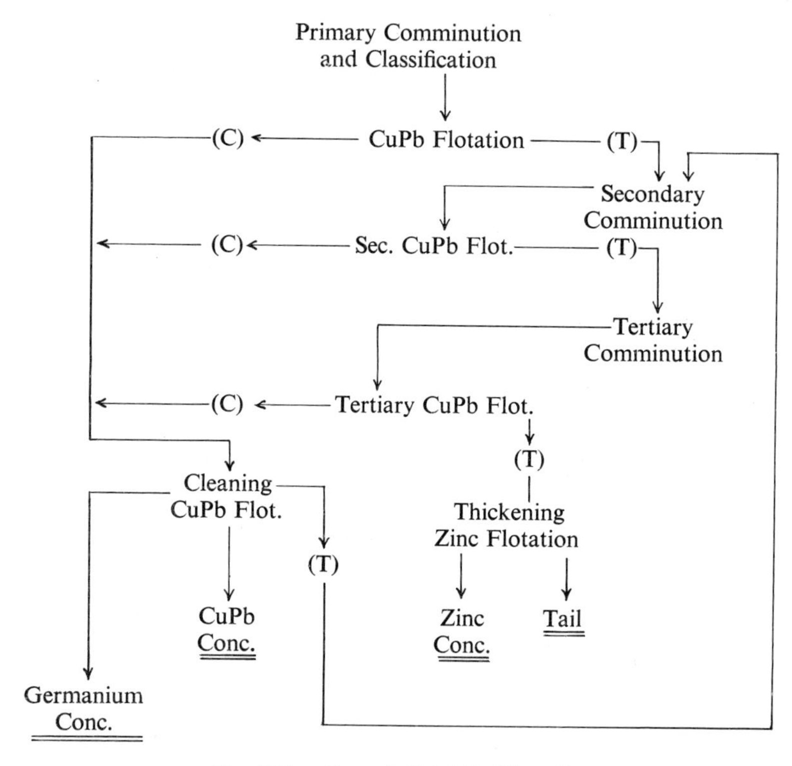

Fig. 339. *Tsumeb Sulphide Flow Sheet*

The sulphide flow-sheet (Fig. 339) was developed after varying the experimental grind. It uses three grind-and-float stages, the third milling stage closing with the secondary flotation section. The CuPb concentrate contains 0·08 % of germanium, mainly as germanite, but with some magnetic renierite. The latter can be checked by placing a sample on a glass slide and moving it over a magnet. The sphalerite is fluorescent and plant control makes use of this and of pilot tables which monitor the flotation tails. In the oxide ore flotation circuits lead sulphide, oxidised lead minerals, slimed lead concentrates, copper sulphide, copper non-sulphides, germanium and zinc products are recovered separately. Preliminary grinding is to 50% *minus* 325 mesh and a bulk sulphide roughing float follows[84]. Magnetic germanium (renierite) is recovered from this float during passage through a series of Franz ferromagnetic filters, but recovery is low as the germanite also contained does not respond to magnetic arrest.

Reagents used in lb/ton of sulphide ore treated in July 1954 were frother (3 cresylic acid to 1 Powell frother) 0·11; collector (50/50 sodium ethyl xanthate and sodium isopropyl xanthate) 0·26; sodium cyanide 0·74; zinc sulphate 1·75; soda-ash 0·76; lime 2·25; and copper sulphate 0·47. The 1954 CuPb concentrate grade was 11·35% Cu and 50·88% Pb, with 86% and $95\frac{1}{4}$% recovery. The zinc concentrate assayed 57·16% Zn, 1·13% Cu, and 2·47% Pb, with a zinc recovery of 62·35%. Germanium is floated from the CuPb concentrate after ten minutes, conditioning at *p*H 5·2 with 1·46 lb/ton of starch and 6 lb (in three stages) of sulphurous acid. A quick roughing float follows, and this product is cleaned after alkalising with 1·46 lb/ton of lime to *p*H $10 < 10\frac{1}{2}$.

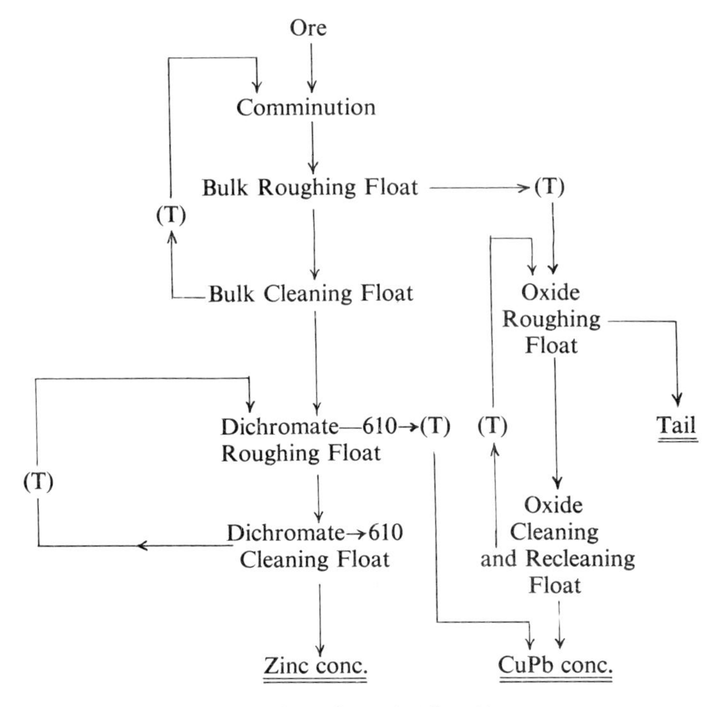

Fig. 340 *Tsumeb High-oxide Flow Sheet*

TABLE 48

BOLIDEN REAGENT PLAN (g./ton)

	Renstrom				Langsele			Boliden		
	CuPb Bulk	*CuPb Sepn.*	*Zn*	*Pyrite*	*Cu*	*Zn*	*Pyrite*	*Cu*	*Pyrite*	*As-Pyrite*
Lime			2800		1300	2900		3300		
Na_2CO_3	250									
H_2SO_4				3500			3500		3200	4500
Sod. diethyl-dibutyl-dithiophosphate	32									
Mercaptobenzthiozol	25									
$Na_2Cr_2O_7$		315								
Ethyl xanthate			440	550		240	600		570	570
Copper sulphate			300			280				
Pine oil	30		35	35	25	40	80	30	30	50
Amyl xanthate					65			95		
*p*H	9·2		12·0	6·7	10·6	12·0	7·5	11·5	8·7	5·7

The reagents used in the "high-oxide" section (Fig. 340) to July 1954 were frother 0·06 lb/ton; Z.4+343 0·46; Z.6 0·53; R.404 0·31; soda-ash 6·29; sulphidizer 11·55; sodium silicate 0·83. The dichromate −610 circuit was temporarily out of action, but 1953 consumption had been 3·61 dichromate and 0·17 R.610.

This flow-sheet utilises the fact that the sphalerite is already highly activated and good recovery of the sulphidic values is possible in a bulk float. Copper and lead are then depressed from this with a combination of R.610 and sodium dichromate, which allows the zinc to float from the depressed CuPb minerals. The bulk float tailings are conditioned with sodium sulphide and then floated. Fresh water is important to success, as cyanides and zinc sulphate, which are present in the mill return water, have deleterious effects.

Boliden, Sweden

This central mill, built in 1951, treats three types of ore. Those from Renström and Langsele yield copper, lead, zinc, and iron sulphides while the Boliden ore body is milled for its copper, pyrite, and arsenopyrite. Part of the auriferous arsenopyrite from Boliden is smelted without pre-treatment in the mill. Renström ore is floated in three stages. First comes bulk $CuFeS_2$ and PbS with precious metals, followed by galena depression with sodium dichromate. Next is flotation of Zn(Fe)S. Finally the FeS_2 is floated for its sulphur (49% in the end product). The flow-sheet for the Langsele ore yields copper, zinc, and pyrite in three stages of flotation. That for Boliden produces successively copper, pyrite, and arsenopyrite. The reagent plan is shown in Table 48.

Sherritt Gordon Mines, Ltd., Canada

At Lynn Lake, 150 miles north of Flin Flon, a nickel-copper-cobalt ore is treated. The values lie in massive sulphides, disseminated sulphides, and stringers. The troublesome gangue minerals are talc and actinolite. Talc is depressed before floating the sulphides by the use of tetra-sodium pyrophosphate (0·4 lb/ton) with the amyl xanthate used in flotation (0·05) for a five-minute conditioning period. A locally patented reagent called Guartec is added (0·3) immediately ahead of flotation. The first flotation product is a high-grade nickel copper concentrate containing 70% of the nickel, mainly as free pentlandite, and 85% of the copper as chalcopyrite. These are differentiated by depression of the nickel with lime and cyanide. The nickel unfloated in the first bulk operation is recovered as a low-grade concentrate after further flotation in a circuit made acid with SO_2. It is too intimately associated with pyrrhotite for separation by grinding.

Differentiation of Non-Sulphides

A research Paper by Ejgeles[154] make the points that selectivity of non-sulphides is achieved by depression rather than activation. The pulp "climate" is important in this respect. Research has shown dextrin and sodium silicate to be of value in the separation of fluorite from calcite; dichromate and dextrin for fluorite from barite; starch and lignin sulphonate for diaspore from pyrophyllite. All this work was done with softened water.

References

1. (1959). *Eng. and Min. J.*, May.
2. Scott, T. R. (1963). *Min. and Chem. Eng. Review*, May.
3. Denver Equip. Co. Bull., M.4–B.68.
4. Finn, W. K. (1963). *World Mining*, Jan.
5. Gartshore, J. L. (1962). *Bull. Can. I.M.M.*, June.
6. (1963). *World Mining*, Feb.
7. Wiser, J. P. (1962). *Can. I.M.M. Bull.*, June.
8. Wilson, J. N. (1960). *I.M.P.C.* (*London*), I.M.M. London.
9. Thom, C. (1962). *Froth Flotation*, 50th Anniv. Vol., A.I.M.E.
10. Clement, M., and Klossel, E. (1963). *I.M.P.C.* (*Cannes*), Pergamon.
11. Everest, D. A., and Napier, E. (1963). *I.M.P.C.* (*Cannes*), Pergamon.
12. Moir, D. N., Collins, D. N., Curwen, H. C., and Manser, R. M. (1963). I.M.P.C. (Cannes), Pergamon.
13. Browning, J. S. (1961). *Min. Eng.* (*A.I.M.M.E.*), July.
14. Misra, R. N., and Bhatnagar, P. P. (1963). *Jnl. Nat. Met. Lab. India.*, Nov.
15. Phillips, W. M. (1963). *Trans. S.M.E.* (*A.I.M.M.E.*), June.
16. Greene, E. W., and Duke, J. B. (1962). *Trans. S.M.E.* (*A.I.M.M.E.*), Dec.
17. Jones, W. I. (1956). *Royal Inst. Chem.*, 3,
18. Geer, M. R., and Yancey, H. F. (1962). U.S. Dept. Int. Bur. of Mines. I.C. 8093.
19. Gieseke, E. W. (1962). *Trans. Amer. S.M.E.*, Dec.
20. Mason, G. A., and Armson, L. (1963). *Steel and Coal*, Feb.
21. Russell, J. A., and Laffin, J. J. (1961). *Can. Min. and Met. Bull.*, Oct.
22. (1963). *Horizon*, Oct.
23. Formanek, V., and Lauvernier, J. (1963). *I.M.P.C.* (*Cannes*), Pergamon.
24. Mitchell, J. S. (1956). *Mining Eng.*, Nov.
25. (1958). *U.S. Bur. of Mines IC* 5388, Feb.
26. Weed, R. C. (1956). *Trans. A.I.M.M.E.*, July.
27. Argall, Jn., G. O. (1964). *World Mining*, Jan.
28. Denver Trefoil. (1962). Spring–Summer.
29. (1961). *Eng. and Min. J.*, June.
30. (1963). *Chamber of Mines Jnl.*, Rhodesia, Dec.
31. Jacobi, J. S. (1963). *Min. Eng.* (*A.I.M.M.E.*), Sept.
32. Markel, W. (1960). *World Mining*, May.
33. Knobler, R. R., and Joseph, W. (1962). *Min. Eng.* (*S.M.E.*), Jan.
34. Bean, J. J. (1960). *Trans. A.I.M.E.*, Dec.
35. Bean, J. J. (1961). *Qly. Colo. School of Mines*, Vol. 56.
36. (1963). *Bull. Can. I.M.M.*, Aug.
37. Last, Stevens and Eaton. (1957). *Trans. A.I.M.E.*, Feb.
38. Ames. (1957). *6th Commonwealth Min. and Met. Congress.*
39. Anon. (1957). *Swedish Min. Dressing Plants, I.M.D.C., Stockholm*, Almqvist & Wiksell.
40. British Patent Spec. 602028.
41. Pryor, E. J., and Lowe, G. M. *Trans. I.M.M.* (*London*), 65, 67.
42. Talbot, H. L. (1946/7). *Trans. I.M.M.* (*London*), 56.
43. Coulter, R. F. (1962). *Can. Min. and Met. Bull.*, June.
44. Hernlund, R. W. (1961). 50*th Anniv. Froth Flotation* (*Colo. Sch. of Mines*), Vol. I.
45. Crabtree, E. H. (1963). *Eng. and Min. J.*, Dec.
46. Deco Trefoil. (1963). May–July.
47. (1960). *Eng. and Min. J.*, Nov.
48. Sutulov, A. (1962). *World Mining*, Aug.
49. Freeman, Rampacek and Evans. (1961). *A.I.M.E.*, Oct.
50. Adamson, R. J. (1959). *Jnl. S. Af. I.M.M.*, Aug.
51. Colvin, E., and Simpson, H. S. (1960). *Jnl. S. Af. I.M.M.*, May.
52. Linari-Linholm, A. A. (1961). *Jnl. S. Af. I.M.M.*, Feb.
53. Nesbitt, A. C., and Weavind, R. G. (1960). *I.M.P.C.* (*London*), I.M.M. London.

References—continued

54. Steygers, L. (1960). *World Mining,* Jan.
55. Mackenzie, D. C. *Mining Magazine,* 56.
56. Rickard, T. A. Trans. Can. I.M.M. 39.
57. Lydon, P. A. (1959). *Mineral Information Service,* California, Nov.
58. Woodcock, J. T. (1962). *C.S.I.R.O.* (*Australia*). "Ore Dressing Inves. Report 628", March.
59. Duschak, L. H., and Schutte, C. N. Bull. U.S. Bur. Min. 222.
60. Mokkan, A. H. (1958). *Jnl. S. Af. I.M.M.,* Feb.
61. Cross, H. E., and Bragman, C. F. (1962). *Jnl. S. Af. I.M.M.,* Oct.
62. (1957). *The Milling of Canadian Ores.* 6th Commonwealth Min. and Met. Congress.
63. Lean, J. B., and Warner, H. K. (1963). *I.M.P.C.* (*Cannes*), Pergamon.
64. Dahlem, D. H., and Sollenberger, C. L. (1963). *I.M.P.C.* (*Cannes*), Pergamon. Press.
65. Boucraut, M., Koskas, R., and Michard, J. (1963), *I.M.P.C.* (*Cannes*), Pergamon Press.
66. Meyer, K. (1963). *I.M.P.C.* (*Cannes*), Pergamon.
67. Hencl, V. (1963). *I.M.P.C.* (*Cannes*), Pergamon.
68. Frommer, D. W. (1964). *Min. Eng.* (*S.M.E.*), April.
69. Herzog, M. E., and Backer, M. L. (1963). *I.M.P.C.* (*Cannes*), Pergamon.
70. Syoberg, S. E. (1961). *Mining World,* Jan.
71. Giambra, S. P. (1963). *Can. Min. and Met. Bull.,* May.
72. (1963). *Bull. C.I.M.M.,* Aug.
73. (1963). *Eng. and Min. Jnl.,* Dec.
74. Gagyi-Palfy, A., Palfy, G., and Halasz, H. (1963). *I.M.P.C.* (*Cannes*), Pergamon.
75. Barkhahn, R. W. (1961). *Mining World,* Nov.
76. Argall, G. O. (1957). *World Mining,* March.
77. Rey, M., de Merre, P., Mancuso, R., and Formanek, V. (1961). 50*th Anniv. Froth Flotation.* Colo. Sch. of Mines Qly., July.
78. Rey, Chataignon, and Formanek. Am. Inst. Min. and Met. Eng., 187.
79. Rey, M. (1957). *I.M.P.C.* (*Stockholm*), Almqvist & Wiksell.
80. McQuiston, F. W. (1957). *I.M.P.C.* (*Stockholm*), Almqvist & Wiksell.
81. (1957). *The Milling of Canadian Ores. 6th Commonwealth Min. and Met. Congress.*
82. Anon. (1957). *Swedish Mineral Dressing Mills,* Almqvist & Wiksell.
83. Ratledge, J. P., Ong, J. N., and Boyce, J. H. (1955). *Trans. A.I.M.M.E.,* 202.
84. Beall, J. V. (1962). *Trans. A.I.M.M.E.,* Dec.
85. Bear, I. J. (1958). *Chem. Eng. and Min. Review,* Feb.
86. Dresher, W. H. (1963). *S.M.E.,* Feb.
87. (1962). *C.S.I.R.O. Report* 717, June.
88. Yates, E. H. (1957). *Min. Eng. A.I.M.E.,* Dec.
89. (1961). Abbott Mercury Mill, Denver Bull. M.4–B.108.
90. Curtis, C. H. (1961). *Min. Eng. A.I.M.E.,* Nov.
91. Hernlund, R. W. (1961). 50th Anniv. Froth Flotation. Colo. School of Mines Qly., July.
92. (1964). *World Mining*, May.
93. Papin, J. E. *Trans. A.I.M.E.,* 202.
94. Gaudin, A. M. (1957). *Flotation*, McGraw-Hill.
95. Noblitt, H. L. (1963). *I.M.P.C.* (*Cannes*), Pergamon.
96. Houston, W. M., and La Venue, W. A. (1962). *Min. Eng. S.M.E.,* Nov.
97. Hoffman, I., and Mariacher, B. C. (1961). *Min. Eng. S.M. Eng.,* May.
98. Sun, S. C., Snow, R. E., and Purcell, V. I. (1957). *Trans. A.I.M.E.,* 208.
99. Hughes, C. V. O. (1956). *Min. Eng.,* Jan.
100. Hadzeriga, P. (1964). *Trans. S.M.E.,* June.
101. Clark, H. N., and Reiter, J. S. (1961). 50th Anniv. Froth Flotation. Colo. Sch. of Mines Qly., Vol. 56, July.
102. Jenkinson, D. W. (1960). *A.I.M.M.E.,* July.

References—continued

103. Hanes, F. E., and Wyman, R. A. (1962). *Can. I.M.M.*, July.
104. (1939/42). U.S. Bur. of Mines Inv. 3436 and 3661.
105. Bryan, R. B. 50th Anniv. Froth Flotation. Colo. Sch. of Mines Qly., Vol. 50.
106. Pollandt, E., and Pease, M.E. (1959/60). *Trans. I.M.M.* (*London*), 69.
107. Huillet, F. D., and Lewis, C. J. (1961). *World Mining*, July.
108. Dalton-Brown, H. (1960/61). *Trans. I.M.M.* (*London*), 70.
109. Chaston, I. R. M. (1961/2). *Trans. I.M.M.* (*London*), 71.
110. Burdon, R. G. (1964). *Mining Magazine*, May.
111. Eidsmo, O., and Mellgren, O. (1960). *I.M.P.C.* (*London*), I.M.M. London.
112. Runolinna, U., Rinne, R., and Kurronen, S. (1960). *I.M.P.C.* (*London*), I.M.M. London.
113. Li, K. C. (1961). *S.M.E.* (*A.I.M.E.*), Preprint 61.B7.
114. Forward, F. A., and Vizsolyi, A. (1963). *I.M.P.C.* (*Cannes*), Pergamon.
115. Everhart, D. L. (1954). *Eng. Min. J.*, 155.
116. Patching, S. W. F. (1956). *Symp. I.M.M.* (*London*), Paper 7.
117. Kaufran, L. A. *Trans. Can. I.M.M.*, 53.
118. Read, Carman and Gough. (1957). *Jnl. S. Af. I.M.M.*, Feb.
119. Pryor, E. J. (1963). *Dictionary of Mineral Technology*, Mining Publications.
120. Fitzgerald and Kelsall. (1956). "Extraction and Refining of the Rarer Minerals". Symposium, I.M.M. (London), March.
121. Levin, J. (1957). *I.M.P.C.* (*Stockholm*), Paper VI.
122. Lord and Light. (1956). *Can. Min. and Met. Bull.*, Feb.
123. Arden, T. V. (1957). "Symposium. Extraction and Refining of Rare Minerals", *I.M.M.*
124. Butler, J. N., and Morris, R. J. (1956). *Trans. A.I.M.M.E.*, Oct.
125. Tinker, et al. (1957). *Precambrian*, 30, Aug.
126. Honeywell, W. R., and Harrison, V. F. (1963). *Can. Min. and Met. Bull.*, Aug.
127. Somnay, J. Y., and Light, D. E. (1963). *Amer. S.M.E.*, March.
128. Behan, E.L. (1956). *Can. Min. and Met. Bull.*, April.
129. (1955). *Mining World*, July.
130. McLean and Prentice. (1955). *Geneva Conf. on Peaceful Use of Atomic Energy*.
131. McCarty, M. F., and Olsen, R. S. (1956). *Eng. Min. J.*., Nov.
132. (1957). *World Mining*, Jan.
133. Argall, G. O. (1957). *Mining World*, Dec.
134. Hargrove, D. (1958). *Eng. Min. J.*, March.
135. Maltby, P. D. R. (1959/60). *Trans. I.M.M.* (*London*), 69.
136. Bhappu, R. B. 50th Anniv. Froth Flotation. Colo. Sch. of Mines Qly., Vol. 56.
137. Nugent, Hoogervorst and Wheeler. (1960). *S. Af. I.M.M.*, Feb.
138. Miller, R. P., Napier, E., Wells, R. A., Audsley, A., and Daborn, G. R. (1963). *Bull. I.M.M.* (*London*), 72.
139. Hannay, R. L. (1956). *Can. Min. Jnl.*, June.
140. Black and Koslov. (1958). *Inst. Chem. Eng., Chicago*.
141. (1963). *Can. Min. Jnl.*, July.
142. Edwards, I. C., and Smith, D. C. (1962). *Bull. Can. I.M.M.*, Oct.
143. (1962). *Trans. Can. I.M.M.*, March.
144. Hazen, W. C. (1963). *Denver Bull. N. T4–B*.32, Oct.
145. Seeton, F. A. (1963). *Denver Bull. No. M4. B*.120, Nov.-Dec.
146. Garman, C. N. (1962). *Trans. A.I.M.E.*, 223, June.
147. Rey *et al.* (1954). *Trans. A.I.M.M.E.*, 199, April.
148. Rey, M., and Formanek, V. (1960). *I.M.P.C.* (*London*), Paper 18.
149. Billi, M., and Quai, V. (1963). *I.M.P.C.* (*Cannes*), Paper 43 J.
150. Lewis, J. L. (1961). 50th Anniv. of Froth Flotation. Colo. School of Mines.
151. Bearce, W. E. (1961). *Ibid.*
152. Wyman, R. A. (1958). Trans. A.I.M.M.E., 211.
153. American Cyanamid Co.
154. Ejgeles, M. A. (1957). *I.M.P.C.* (*Stockholm*), Almqvist & Wiksell.

APPENDIX A

GLOSSARY

Activator.—Substance added to increase the activity of a compound or to speed up chemical reaction, e.g. by heightening the attraction to a specific mineral surface of desired collector ions in flotation.

Agglomeration.—Process using development of surface forces employed in froth flotation, but applied to deslimed sands. Aerophilic particles form glomerules with sufficient imprisoned air to cause them to act as low gravity systems removable by gravity treatment.

Air Lance.—Length of piping down which compressed air is blown, to stir settled sands or to free choked passages.

All-sliming.—Term used for cyanidation by agitation, no coarse sands being separated for leaching.

Amalgamation.—Use of mercury to catch native gold by sorption, forming a liquid or plastic *amalgam* from which mercury is later removed by distillation. *Barrel A, A* in small ball mill. *Plate A* metal plates smeared with Hg trap gold as pulp flows over them.

Amenability.—Characteristic reaction of minerals to basic methods of mineral processing, studied in preliminary testwork on unknown ores.

Angle of Nip.—Angle included between two approaching faces at or below which a particle is seized. Approximately 23° for most minerals.

Anion.—In electrolyte, ion carrying negative charge.

Arrested Crushing.—Comminution in which the product can fall clear of the crushing zone when broken below the "set" (*q.v.*) of the machine, without undergoing further reduction.

Assay Grade.—The percentage of an element or compound in a representative sample, as found by analytical test (assay).

Assay-ton.—For a *long* ton (2240 lb. av.) 32·667 gm., and for a *short* ton (2000 lb. av.) 29·1667 gm. The number of mg. of bullion obtained from one assay-ton equals the number of ounces (troy) of bullion per ton of ore.

Autogenous.—(1) In DMS process, fluid media partly composed of a mineral species selected from ore being treated.
(2) Selectively sized lumps of ore used as grinding media.

Automation.—Process control employing servo-mechanisms or monitoring devices to supplement or replace human control.

Bank (*of cells*).—A row of flotation cells in line.

Banket.—Metamorphosed quartz-pebble conglomerate carrying gold and uranium mined on the Rand.

Barren Solution.—A solution in hydro-metallurgical treatment from which all possible valuable constituents have been removed. See also *Cyanide.*

Batch-treatment.—Treatment of a parcel of material in isolation, as distinct from the treatment of a continuous stream of ore.

Bedding (*Jig*).—The layer of heavy and oversized material placed above the screen in jigging—also called "ragging".

Beilby Layer.—Flow layer resulting from incipient fusion during polishing of mineral surface, and therefore not characteristic of true lattice structure.

Belt Conveyor.—An endless belt on which ore is placed and moved from point to point in the transport system.
Beneficiation.—Treatment of crude ore in order to improve its quality for some required purpose.
Blocking (*of Crusher*).—Obstruction of crushing zone by clayey material or rock which refuses to break down and pass to discharge.
Break.—Optimum mesh-of-grind (*q.v.*).
Bubble.—The dictionary definition of "bubble" is modified in this book to define two physical states:
(*a*) *N-Bubble*. A bubble immersed in water or pulp, with surface tension at or approaching that of water.
(*b*) *F-Bubble* (usually written "bubble" in text). An emerged free bubble of surface tension substantially below that of water and having an area twice that of an N-bubble of equal diameter.
Bubble Pipe.—Tube inserted in pulp at regulated depth, through which compressed air is gently bubbled. The air-pressure indicates the pulp density and provides a means of control.
Bulk Flotation.—The intentional raising as a mineralised froth of more than one mineral in one operation.

Capacitor (*Capacitator*).—An adjustable electric appliance used in circuit with a motor to adjust the power factor.
Cascading (in grinding).—Motion of crushing bodies in a ball mill in which they slide or roll down the surface of the crop load (*q.v.*) after reaching the highest point attainable.
Cataracting.—Motion of crushing bodies in a ball mill in which some fall freely after breaking away from the top of the crop load (*q.v.*) and fall with impact to the toe of the load.
Centrifuging.—Application of centrifugal force to mineral treatment.
Chemical Extraction.—Term taking the place of hydro-metallurgy. Embraces leaching (acid, alkaline, and pressure), ion exchange, solvation-precipitation, calcination.
Choked Crushing.—Comminution in which the discharge arrangements can restrict departure of ore even when it has been broken to the release size of the machine.
Circulating Load.—Retention in *closed circuit* of part of the ore or pulp flowing through the mill, for re-treatment.
Classification.—Sorting of particles in accordance with their rate of fall through a fluid.
Climate (flotation).—Of pulp undergoing froth flotation, the prevailing balance of chemical energy reached by the reacting electrical, physical, and chemical forces.
Close Sizing.—Sizing (*q.v.*) with screens fairly close in size of aperture (mesh). On a Tyler series normal mesh aperture ($\sqrt{2}$ series) is 208, 147, 104, 74 . . . , etc. μ, but in close sizing one or more intermediate apertures might be used.
Closed Circuit.—In mineral dressing, a system in which ore passes from comminution to a sorting device which returns oversize (*q.v.*) for further treatment and releases undersize (*q.v.*) from the closed circuit.
Close-ranged.—Screened or classified between close maximum and minimum limits of size or settlement.
Collector.—Promoter. Heteropolar compound containing a HC group and

an ionised group, chosen for ability to adsorb selectively in froth flotation process and render adsorbing surface relatively hydrophobic.

Colloid Mill.—Grinding appliance such as two discs set close and rotating rapidly in opposite directions, so as to shear or emulsify material passed between them.

Comminution.—Crushing and/or grinding of ore by impact and abrasion. For convenience, the word "crushing" is used for dry methods and "grinding" for wet ones in the text.

Concentrate.—The valuable mineral separated from ore undergoing a specific treatment.

Concentration Criterion.—Relation between density of heavy (S_h) and light (S_l) mineral/s in crushed ore, in a supporting liquid. For water

$$C = \frac{(S_h - 1)}{(S_l - 1)}.$$

Conditioning.—Stage of froth-flotation process in which the surfaces of the mineral species present in a pulp are treated with appropriate chemicals to influence their reaction when the pulp is aerated.

Contact Angle "θ".—The angle across the water phase of an air-water-mineral system, used to measure effect of surface conditioning.

Counter-current.—In mineral dressing, a system in which a stream of ore pulp is progressively impoverished as it proceeds in one direction (usually through a series of machines) while the desired product it contains is progressively enriched as it moves in the opposite direction.

Country Rock.—The valueless rock surrounding a lode.

Coursing Bubble.—One rising freely through cell during froth flotation.

Critical Speed.—The theoretical speed at which a ball is held to the inner surface of the smooth ball mill liners by centrifugal force $\left[n = \frac{76 \cdot 6}{\sqrt{d}}\right]$ where n = r.p.m., and d the mill diameter in feet.

Crop Load.—The mixture of crushing bodies, ore particles, and water being tumbled in the ball mill.

Crystallography.—Science of the shapes, colour, chemical, electrical, physical, and optical properties of crystals.

Customs Mill.—A plant receiving ore for treatment from more than one mine.

Cyanicide.—Any substance present in a pulp which attacks or destroys the cyanide salt being used to dissolve precious metals.

Cyanide.—Usually refers to cyanide solution in circulation in a mill treating gold or silver ores. The "stock" or "solution" is of two main types, "barren" from which all possible value has been extracted, and "pregs" or "pregnant", which is charged with gold or silver and awaits their removal.

Cyclone.—Hydro-cyclone. A classifying (or concentrating) separator into which pulp is fed, so as to take a circular path. Coarser and heavier fractions of solids report at apex of long cone while finer particles overflow from central vortex.

Dense-Media Separation (DMS).—Heavy Media Separation, or sink-float. Separation of sinking heavy from light floating mineral particles in fluid of intermediate density.

Depressing Agent.—One used in flotation process to render selected mineral/s more hydrophilic or to prevent collector action, thus helping to keep them from reporting with the mineralised froth.

Differential Flotation.—Concentration as separate products of a series of minerals removed from an ore by froth flotation.

Dipole.—Co-ordinated valence link between two atoms, or electrical symmetry of a molecule.

Discontinuity Lattice.—See *Lattice.*

Dispersing Agent.—A chemical used to promote the removal of slime from floccules and particle surfaces.

Disseminated Ore.—Ore in which the valuable mineral is fairly evenly distributed through the gangue, as crystals or aggregates of regular size.

Dump.—Deposit of waste material discarded in connexion with ore treatment; stockpile of ore or partly processed material.

Dusting Loss (laboratory sampling).—The loss of part of a sample undergoing test, through leakage of particles into the atmosphere.

Electric Eye, Ear.—Former is photo-electric cell arranged in connexion with monitoring flow, turbidity, height of material in ore bin, etc.; latter is microphonic signal using noise level to check loading of ball mill.

Electrostatic "Bunching".—Flocculation of particles during dry screening due to binding electrical forces at their surfaces.

Electrostatic Separation.—Treatment based on ability of particles to retain or dissipate electric charge after passing through a high-voltage field and while falling between a grounded surface and deflecting electrode.

Elutriation.—Laboratory *classification* (*q.v.*), in which sands are sorted in rising columns of fluid, under precise conditions of control.

Filtration.—Commercially, separation of relatively clear filtrate from pulp, with arrest of solids on suitable membrane, usually moved continuously so as to discharge a ribbon of mineral.

Fines.—The particles of ore below a given size or mesh (undersize), as contrasted with the coarser particles.

Flocculation.—Coalescence of minute particles into floccules (often consisting mainly of water) to accelerate settlement as part of dewatering or thickening of a pulp. Appropriate chemicals are used to promote adhesion.

Flotation, Bulk, and Differential.—Production of a mineralised froth into which the desired ore constituents have been bound firmly enough to ensure the removal of this froth from the original pulp. If this froth carries more than one mineral as a designated main constituent, it is a "bulk float". If it is selective to one constituent of the ore, where more than one will be floated, it is a "differential" float.

Flow-sheet (*Flow-line*).—A diagram showing the sequence of operations, the products made and their respective routes, and any required details of machines, rates, and optimum performance.

Free-milling Gold.—Gold with so clean a surface that it readily amalgamates with mercury after liberation by comminution.

"Free Settling".—As opposed to "hindered settling" in classification, free fall of particle through fluid media.

Frother, Frothing Agent.—Chemical added to pulp before flotation, to promote transient froth in cell.

Gangue.—The fraction of ore rejected as tailing in a separating process. It is usually the valueless portion, but may have some secondary commercial use.

Gape.—The maximum working cross-section of a crushing machine, as regards reception for size reduction of a piece of ore.

Gauss.—Intensity of magnetic field produced by one CGS unit of current (10 amps.) at centre of circle of 1 cm. radius when flowing through arc of 1 cm.

Geiger Muller Counter.—Valve in which a 1000-volt current is discharged by ionising effect. When a charged particle causes this discharge the effect is registered as a clicking noise, or made to activate a counting device.

Go-Devil.—Tool attached to drain-rods, or ball with cutting surface, used to descale pipes.

Gravity Separation.—Exploitation of differences between the densities of particles.

Grindability.—The effect produced on representative pieces of ore by applying standard methods of comminution, assessed comparatively in terms of size reduction and power used.

Hardness.—Of minerals, measurement on Moh's scale according to whether the specimen under test scratches or is scratched by Talc (1), Gypsum (2), Calcite (3), Fluorite (4), Apatite (5), Orthoclase (6), Quartz (7), Topaz (8), Corundum (9), Diamond (10).

"*High-grading*".—Theft of valuable pieces of ore.

High Tension Separation.—Electrostatic separation by use of a D.C. field of from 18,000 to 80,000 volts to charge small particles of ore before separating them on the basis of their retentive quality when in contact with a conductor.

Hindered Settling.—Settlement of particles through a crowded zone, usually in a hydraulic column through which their fall is opposed by rising water.

Impact Grinding.—Shattering of particles by direct fall upon them of crushing bodies.

Interphase.—The zone between two or more phases, or faces.

Iso-electric Point.—Zero potential or point of electrical neutrality; the pH at which particles in aqueous suspension are neutral, and best able to flocculate.

IX.—Ion exchange.

Jigging.—Up-and-down motion of a mass of particles in water by means of pulsion.

Lattice.—The pattern formed by the orderly distribution of atoms in a crystal. *Discontinuity* lattices, whether internal or external, are boundaries of crystals.

Leaching.—Extraction of element or compound from ore by chemical attack causing its dissolution.

Liberation.—Freeing by comminution of particles of a specific mineral from their interlock with other constituents of the ore.

Liberation Mesh.—That particle size at which a specified mineral should in theory become detached from other minerals in the ore during comminution.

Lode.—An ore body occurring as a steep vein or layer filling a rock fissure.

Long-ranged.—Of crushed or ground ore, size distribution covering a wide range of meshes.

Lyophilic.—Having the property of attracting liquids.

Magnetic Separation.—Use of permanent or electro-magnets to remove relatively strong ferro-magnetic particles from para- and dia-magnetic ores.

Mesh.—The aperture framed by the wires of a sieve opening. With square-woven sieves the number of wires per linear inch defines the mesh—*e.g.*, 30 mesh means that there are thirty wires per inch-run.

When crushed ore is fed on to two sieves, the lower being the finer, the coarse sieve defines the *limiting mesh* and the fine one the *retaining mesh* of the sand held between the screens. It can be defined as "$-12+120\#$", the $+12$ for this particular combination being called *oversize* and the $-120\#$ *undersize.*

The terms "long-ranged" or "short-ranged" are sometimes used to indicate roughly the cut-off limits of the sand sizes in an aggregate of particles.

Mesh-of-grind.—Written m.o.g. It states optimum particle size in grinding, in terms of the percentage of the material passing (or alternatively retained on) a given screen-mesh.

Metallurgical Balance Sheet.—Statistical presentation of an operation in terms of the general equation:

Units of value in feed = Units in concentrate + units in tailings + units retained in circuit.

An equation is needed for each valuable mineral undergoing extraction.

Micelle.—(Part). A colloidal aggregate of molecules present in aqueous solutions of many soaps and dyestuffs.

Middlings.—In two-component ore, particles incompletely liberated by comminution into concentrate or gangue. In complex ores, in addition to incomplete liberation, there may be multiphased particles of middling, or intermediate species which react too feebly to treatment to report as concentrate or tailing.

Mill (Milling).—Also concentrator, plant, reduction works. The place where run-of-mine ore is received and treated, and from which end-products are dispatched.

Mill Head.—Ore accepted for treatment in concentrator, after any preliminary rejection such as waste removal.

Mineral Dressing.—Treatment of natural ores or partly processed products derived from such ores in order to segregate or up-grade some or all of their valuable constituents, and/or to remove those not desired by the industrial user. Mineral-dressing processes are applied to industrial waste to retrieve useful by-products.

Mineral Species.—Native element or inorganic substance corresponding generally with ideal mineral having a precise formula and known lattice structure.

Muck-shifting.—Term used for extensive earth-moving operations.

Near-mesh.—Near-sized; grains close in cross-section to a specified screening mesh, which tend to blind apertures and slow down sizing.

Nine-Point Sample.—At final stage of sample preparation, manipulation of small quantity of finely ground mineral. Material is rolled on glazed cloth or paper, flattened into a disc, and divided into eight equal segments. Final sample is drawn equally from each segment and from centre.

Nip, Angle of.—The angle of nip, or wedge angle, is that below which a piece of ore can be seized between two crushing surfaces without slipping.

Opencast.—A working place open to the sky, and from which ore is extracted.
Open Circuit.—A flow-line in which the solid particles pass from one appliance to the next without being screened, classified, or otherwise checked for quality, no fraction being returned for retreatment.
Optimisation.—Co-ordination of various processing factors, controls and specifications to provide best overall conditions for technical and/or economic operation.
Optimum.—The use of this word in mineral dressing is associated with the idea of quality control. As an example, a pulp might be required to be delivered from one section to the next at an optimum *p*H, mesh size, and percentage of solid *plus* or *minus* 10% to ensure smooth working of the receiving section. The + or − tolerance must be acceptable to the treatment which follows.
Ore.—A naturally occurring complex of minerals from which any fraction of commercial value can be extracted and used. "Raw" and "run-of-mine" ore is ore as it comes from the working place.
Overburden.—Valueless earth or rock overlying the ore body.
Over-grinding.—Reduction of size of particles of a given mineral below that aimed at in comminution.
Oversize.—See *Mesh.*

Packing.—Blocking, (*q.v.*).
Panning.—Use of plaque, vanning shovel, pan to fractionate finely ground mineral into light and heavy portions for visual assessment.
Parting Density.—That maintained in the bath in DMS.
Peak Load.—The maximum permissible draft of power from the source of supply.
Penalty.—Reduction in the price per unit paid for concentrate when it fails to reach an agreed specification as regards assay grade, purity, etc.
Permeability Test.—Measurement of the resistance to fluid flow offered by a known volume of particles, as a measure of their "packing" or voids.
Pilot Plant.—A small-scale mill in which representative tonnages of ore can be tested under conditions which foreshadow (or imitate) those of the full-scale operation proposed for a given ore.
Placer.—An alluvial deposit of ore, usually as mineral-bearing gravel or sand.
Plant.—See *Mill.*
Potential-determining Ions.—Those which leave the surface of a solid immersed in aqueous liquid before equilibrium (saturation-point) has been reached, while an electrical double layer is building up and *zeta-potential* develops.
Pregnant Solution.—A value-bearing solution in a hydro-metallurgical operation. And see *Cyanide.*
Pseudo-viscosity.—The viscous effect of friction between the surfaces of particles in a pulp, as distinct from the true viscosity of its liquid component.
Pulp.—A mixture of ground ore and water capable of flowing through suitably graded channels as a fluid. Its dilution or *consistency* is specified either as solid-liquid ratio (by weight) or as a percentage of solids (by weight).

Ragging.—See *Bedding.*

Rank.—Of coal, tabulation of type which shows some degree of metamorphism (from peat to anthracite).
Reef.—A lode (*q.v.*), typically flattish.
Rejects.—Ore minerals removed and discarded at any stage of treatment.
Release Mesh.—Specified mesh-of-grind (m.o.g.) for best conditions for treatment to recover a specific mineral from the ore.
Retaining Mesh.—See *Mesh.*
Reynold's Number.—Dimensionless expression of relation between inertia and viscosity.

$$R_e = \frac{v\delta\rho}{\eta}$$

v is velocity (cm./sec.);
δ is diameter of spherical particle (cm.);
η is absolute viscosity of fluid (gm./cm./sec.);
ρ is density of fluid.

Riffle.—An obstacle to the smooth flow of a pulp along a channel, designed to arrest heavy particles.
Ring Size.—A term for particle size where the piece of ore is too large for screening. It refers to the diameter of the gauge or ring which can be slipped over it.
Roasting.—Heat treatment to produce structural change, normally oxidation of sulphides. If complete "sweet roasting".
Roughing.—The removal of a rough concentrate at the earliest stage of treatment of the ground ore. *Scavenging* is the removal of the last recoverable fraction of value before discarding the final tailing from the treatment section. *Cleaning* is the retreatment of the *rough concentrate* to improve its quality.
Run-of-mill Ore.—Mill-head ore; that accepted for treatment.

Sampling.—Separation of a representative fraction of ore, pulp, or any product for testing or checking purposes.
Scalping.—A milling term for the removing of a mineral during closed-circuit grinding of the ore.
Scavenging.—See *Roughing.*
Scintillometer.—Gamma ray detector.
Screening, or Sizing.—Use of one or more screens (sieves) to separate particles of ore into defined sizes.
Sedimentation.—Method of classification by exploitation of free-falling rates of minute (subsieve) particles. See "Beaker Decantation" in text.
Separation.—See *Concentrate.*
Set.—The discharge opening of a crushing machine. It regulates the size of the largest escaping particle.
Shaking Tables.—Flattish tables oscillated horizontally during separation of minerals fed on to them.
Shape Factor.—Property of a particle which determines the relation between its mass and surface area, and hence its response to frictional restraint.
Size Range.—That between upper (limiting) and lower (retaining) mesh sizes with reference to screened or classified material.
Sizing.—See *Screening.*
Slimes.—"Primary" slimes are extremely fine particles derived from ore, associated rock, clay, or altered rock. They are usually found in old

dumps, and in ore deposits which have been exposed to climatic action and include clay, alumina, hydrated iron, near-colloidal common earths, and weathered feldspars.

"Secondary" slimes are very finely ground minerals from the true ore.

Sluicing.—Separation of minerals in a flowing stream of water.

Solution.—See *Cyanide.*

Solvation.—A combination between solute and solvent (Hackh).

Sorption.—Reaction at or immediately adjacent to, the surface of a solid. *AdS,* monomolecular; *AbS,* deeper physical penetration; *ChemiS,* sorption with chemical action and change.

Sorting.—Removal by hand (hand picking) of selected pieces of rock. Term also applied to classification (*q.v.*) of finely ground pulps.

Specific Population.—Number of particles in unit volume of pulp.

Specific Surface.—Total area of particles in unit-weight of ground material, usually reported in $cm.^2/gm.$

Specimens.—In mineral dressing, unusually rich pieces of ore or characteristic constituents thereof in coarsely crystalline form—not representative samples.

Spills, Spillage.—Ore, pulp, circulating liquor inadvertently discharged from flow-line and requiring appropriate means of recovery or removal.

Stage.—Stage-grinding is successive comminution. Stage-concentration is stage-grinding repeated on the concentrate produced by treatment between grinding stages. Stage-addition in flotation refers to deliberate use of insufficient reagent in the early part of the treatment, in order to increase selectivity of conditioning, followed by further addition at a later point in the process.

Starvation.—As used in flotation, the deliberate inadequate addition of a reagent in order to restrict its effect. In comminution, avoidance of crowding in the machine by restricting rate of feed.

Stope.—Underground working place formed as ore is extracted.

Stope Assay Plans.—Plan showing assay value of exposures of ore in a stope, together with any other data desired.

Subsieve Sizing.—Size analysis of particles too small for efficient grading by use of screens, usually *minus* 200 mesh.

Substrate.—The true lattice of a crystal, as distinct from its discontinuity lattice, or surface.

Superpanner.—Mechanism invented by Professor H. T. Haultain which simulates rocking, bumping, and sluicing action used in panning, and gives precise information as to possibility of gravity treatment of sands. Used in rapid assays and as a research aid.

Surge (tanks, bins).—Receptacles capable of receiving and redispensing small tonnages, thus steadying any fluctuations in a flow-line.

Sweet Roasting.—Removal of sulphur from metal sulphides by ignition.

Tailing.—See *Gangue.*

Teeter.—The dance of a bed of particles in a column of rising water, or teeter column.

Theta (θ).—The symbol for contact-angle in flotation testing.

Thickener.—Large circular tank in which solids settle slowly and form a slurry which is continuously removed from below while fairly clear water overflows.

Thixotropy.—Property of certain mineral suspensions in water (*e.g.*, bentonite) of remaining fluid while agitated, but gelling when quiescent.

Throughput.—Quantity of material passing through a specified treatment in a given time.

Throw.—Lateral displacement of screen, shaking table or crushing surface in motion.

Titer (of fatty acids).—The solidifying range of a liquid fatty acid.

Toggle Action.—Application of crushing force so that the distance moved diminishes without change of input strength, between "gape" and "set". Thus greatest speed of movement of the approaching faces is applied with weakest thrust and *vice versa.*

Tramp Iron.—Pieces of iron, steel, etc., which have been gathered with the ore and which would damage crushing machinery if allowed to enter. *Tramp oversize* is ore which should not have been released from the previous section at the delivery size.

Trash Screen.—Protective screen for removing detritus from the pulp stream.

Turbidimetry.—Measurement of obstruction of light offered by solids in suspension as a guide to their area, settling characteristics and shape.

Tyler.—A sieving system. See Chapter VIII.

Undergrinding.—Insufficient reduction of particle size to produce effective liberation of value.

Undersize.—See *Mesh.*

Unit Cell.—Of crystals, a single complete geometric unit. Of flotation, a single appliance.

Unlocking.—See *Liberation.*

Values.—Compounds separated from run-of-mill ore and worked up to a marketable state; usually expressed as assay grade.

Voids.—Porosity; unfilled space in unit volume of granular material compacted under stated conditions, and expressed as percentage or ratio of solid to void.

Waste Rock.—Barren or submarginal rock or ore which has been mined but is not of sufficient value to warrant treatment and is therefore removed ahead of the milling processes.

Wetting Agent.—A chemical promoting adhesion of a liquid (usually water) to a solid surface.

Zeta-potential Layer.—The zone of shear surrounding a particle immersed in an electrolyte.

APPENDIX B

Extracts from *Recent Developments in Mineral Dressing:* Proceedings of a Symposium arranged by the Institution of Mining and Metallurgy and held in London on 23–25 September, 1952. Published by the I.M.M., 1953.

Paper: "Purpose in Fine Sizing and Comparison of Methods", by E. J. Pryor, H. N. Blyth, and A. Eldridge.

BEAKER DECANTATION

Method of Procedure

No special apparatus is required. The equipment consists of two 2-litre beakers, five 1-litre beakers, a glass rod covered with rubber tubing, a stop-watch and, if available, a mechanical stirrer.

(1) Using the mechanical stirrer, 100 g. of the material to be sized is dispersed in a litre of water in one of the 2-litre beakers for a suitable period, with the aid of a deflocculant. The stirrer is removed when dispersion appears complete.

(2) Water is added until the surface of the suspension is about 5 cm. below the lip of the beaker.

(3) The height from the surface of the pulp to the level of the settled sand is measured (h_1 cm.) and the time for the smallest particle (d microns in diameter) required in the coarsest fraction to settle this distance is calculated according to Stokes's law (x cm./sec., giving a settling time of $\frac{h_1}{x}$ sec.). The pulp is stirred until all particles are in suspension. The maximum possible vertical motion is imparted to the pulp with the minimum of circular motion in a horizontal plane. When the suspension is uniform, the outside of the beaker is sharply tapped and the suspended particles allowed to settle for the calculated time multipled by 1·05 (i.e. $\frac{h_1}{x} \times 1{\cdot}05$ sec.). This increase in settling time over that calculated ensures that all particles at and above the required size have settled after allowing for hindered settling. The beaker is vigorously tapped with the rod for 15 sec. before the end of the settling time. This tapping forms the settled material into a compact cake and ensures that particles built up as a thin layer in the curve between the sides and bottom of the beaker join the main cake.

(4) At the end of $\frac{h_1}{x} \times 1{\cdot}05$ sec, the supernatant pulp is poured quickly into the second 2-litre beaker (B). The settled particles remain as a compact cake on the bottom of beaker A. This cake now contains all the particles of size d and above, but the fraction is still contaminated with residual "fines".

(5) Five 1-litre beakers (C, D, E, F, and G) containing water and dispersant are prepared and the solution from the first of these beakers, C is poured on to the cake in A. The depth from the liquid surface to the sand surface is

measured (h_2 cm.) and the time for a particle of size a to settle this distance is calculated $\frac{h_2}{x}$ sec. . The pulp is brought into uniform suspension by stirring as before, the beaker tapped, and the particles allowed to settle for $\frac{h_2}{x}$ sec., the correction for hindered settling not being required in this thinner pulp; 15 sec. before this time expires the beaker is tapped vigorously. At $\frac{h_2}{x}$ sec. the supernatant pulp containing only those particles finer than d is poured back into the beaker C.

The settled material in A now contains the required fraction with a reduced proportion of that finer material which has settled in the first stage.

(6) Repeating the process carried out with beaker C by using the other four beakers D, E, F, and G, the material less than d contained in the cake is successively reduced.

If, on adding more wash water and stirring, all the particles settle within the calculated time, then theoretically no particles less than d will be found in the cake, and conversely, if, when the final wash water is added, particles are still in suspension at the end of the calculated time, then some particles finer than d will be found in the settled material. Observation indicates that five washes are sufficient.

(7) After the final wash from the fifth beaker (G), the cake is removed from beaker A, dried, and weighed.

(8) Beaker B now contains most of the particles smaller in size than d, the remainder being in the wash solutions. The height from pulp surface to that of the settled sand is measured and the time for a particle $\frac{d}{\sqrt{2}}$ to settle this distance h_3 is calculated from Stokes's law y cm./sec., giving $\frac{h_3}{y}$ sec.

The pulp is thoroughly mixed, tapped, and allowed to settle for $\frac{h_3}{y} \times 1{\cdot}05$ sec. For 15 sec. before this time has elapsed the beaker is tapped vigorously and at the end of the time the supernatant pulp is poured into the now empty beaker A.

(9) Beaker C, which contains only particles finer than d, is used as the first wash for this second cake and the beakers D, E, F, and G are used for the successive washes. Thus, all particles less than d but greater in size than $\frac{d}{\sqrt{2}}$ are transferred from the wash beakers into the $-d$, $+\frac{d}{\sqrt{2}}$ cake contained in beaker B.

(10) The process is repeated for particles of successive $\sqrt{2}$ sizes until a sufficient number of sized fractions have been prepared.

Using 100-g. samples, it was found that a compact cake formed on the bottom of the beaker which remained firmly in position when the supernatant pulp was decanted. Before starting, the pulp was thoroughly dispersed with a mechanical stirrer for 10 min. Fractionation was carried out in tap water at 15° C. containing 0·1 per cent Aerosol A.Y. The results obtained from duplicate head samples are set out in the Table together with the nominal micron sizes calculated from Stokes's law. The total time for one complete fractionation was $5\frac{1}{2}$ hours. Losses were due to the difficulty in collecting all the fines and were therefore added to the *minus* 9-micron fraction.

Cut Number	*Size in Microns (calc.)*	Test 1		Test 2	
		Wt. per cent.	*Cum. per cent. Fines* × ·52	*Wt. per cent.*	*Cum. per cent. Fines* × ·52
1	+59	32·75	35·1	31·20	35·8
2	−59+45	15·55	26·9	16·70	27·1
3	−45+32	14·75	19·2	15·00	19·3
4	−32+21	11·20	13·4	11·85	13·1
5	−21+15	7·60	9·4	7·55	9·2
6	−15+ 9	5·85	6·4	5·25	6·5
	− 9	12·30		12·45	
		100·00		100·00	
Loss		·50		1·00	

Although a total of five washes was given, it was noticed that when treating material finer than 32 microns the final wash still contained particles in suspension.

APPENDIX C

ABBREVIATIONS

Å., A.U. — Angstrom; Angstrom Unit
A.C. — Alternating current.
A.I.M.E.; A.I.M.M.E. — American Institute of Mining, Metallurgical and Petroleum Engineers.
A.S.T.M. — American Society for Testing Materials.
b.p. — Boiling point.
C — Concentrate.
°C. — Degrees, centigrade.
c.g.s. units — Length in cm., mass in grammes, time in seconds.
cm. — Centimetre
cos. — Cosine
d. — Diameter.
dwt. — Pennyweight.
DMS — Dense media separation; heavy media separation; sink-float.
E. — Energy.
e.m.f. — Electromotive force.
f.p. — Freezing point.
G. — Gauss.
g. — Gramme.
g. — Gravitational acceleration.
H. — Magnetic field strength (Henrys).
H.T.S. — High tension separation; electrostatic separation.
IX. — Ion exchange.
K. — Constant.
K.E. — Kinetic energy.
kg. — Kilogramme.
kW.; kWh. — Kilowatt; kilowatt hour.
l. — Length; litre.
M. — Middlings.
m. — Metre; (m^2 = square metre).
M.B.S. — Metallurgical balance sheet.
mega. — One million.
ml. — Millilitre.
m.p. — Melting point.
mμ. — Millimicron.
m.o.g. — Mesh of grind.
*p*H. — Hydrogen-ion exponent.
r. — Radius.
R_n. — Reynolds Number.
S.G. — Specific gravity.
S.M.E. — Society of Mining Engineers (A.I.M.M.E.).
T. — Tailings.
V. — Volt.
v. — Velocity; volume.

γ — gamma; surface tension.
Δ, δ — delta; density, specific gravity.
ζ — zeta; zeta potential.
θ — theta; contact angle.
μ — mu; micron; coefficient of friction; viscosity.
π — pi; ratio of circumference to diameter of circle.
♯ — Mesh size (Tyler).
% — Per centum.
/ — Per; divided by.
$>$ — Greater than.
$<$ — Less than.
$\pm$ — Plus or minus.
$\propto$ — Varies as.
:: — Is proportional to.

A

B

G

H

I

S